LONDON MATHEMATICAL SOCIETY LECTURE NOTE SERIES

The books in the series listed below are available from booksellers, or, in case of difficulty, from Cambridge University Press.

34 Representation theory of Lie groups, M.F. ATIYAH *et al*
36 Homological group theory, C.T.C. WALL (ed)
39 Affine sets and affine groups, D.G. NORTHCOTT
46 p-adic analysis: a short course on recent work, N. KOBLITZ
49 Finite geometries and designs, P. CAMERON, J.W.P. HIRSCHFELD & D.R. HUGHES (eds)
50 Commutator calculus and groups of homotopy classes, H.J. BAUES
57 Techniques of geometric topology, R.A. FENN
59 Applicable differential geometry, M. CRAMPIN & F.A.E. PIRANI
66 Several complex variables and complex manifolds II, M.J. FIELD
69 Representation theory, I.M. GELFAND *et al*
74 Symmetric designs: an algebraic approach, E.S. LANDER
76 Spectral theory of linear differential operators and comparison algebras, H.O. CORDES
77 Isolated singular points on complete intersections, E.J.N. LOOIJENGA
79 Probability, statistics and analysis, J.F.C. KINGMAN & G.E.H. REUTER (eds)
80 Introduction to the representation theory of compact and locally compact groups, A. ROBERT
81 Skew fields, P.K. DRAXL
82 Surveys in combinatorics, E.K. LLOYD (ed)
83 Homogeneous structures on Riemannian manifolds, F. TRICERRI & L. VANHECKE
86 Topological topics, I.M. JAMES (ed)
87 Surveys in set theory, A.R.D. MATHIAS (ed)
88 FPF ring theory, C. FAITH & S. PAGE
89 An F-space sampler, N.J. KALTON, N.T. PECK & J.W. ROBERTS
90 Polytopes and symmetry, S.A. ROBERTSON
91 Classgroups of group rings, M.J. TAYLOR
92 Representation of rings over skew fields, A.H. SCHOFIELD
93 Aspects of topology, I.M. JAMES & E.H. KRONHEIMER (eds)
94 Representations of general linear groups, G.D. JAMES
95 Low-dimensional topology 1982, R.A. FENN (ed)
96 Diophantine equations over function fields, R.C. MASON
97 Varieties of constructive mathematics, D.S. BRIDGES & F. RICHMAN
98 Localization in Noetherian rings, A.V. JATEGAONKAR
99 Methods of differential geometry in algebraic topology, M. KAROUBI & C. LERUSTE
100 Stopping time techniques for analysts and probabilists, L. EGGHE
101 Groups and geometry, ROGER C. LYNDON
103 Surveys in combinatorics 1985, I. ANDERSON (ed)
104 Elliptic structures on 3-manifolds, C.B. THOMAS
105 A local spectral theory for closed operators, I. ERDELYI & WANG SHENGWANG
106 Syzygies, E.G. EVANS & P. GRIFFITH
107 Compactification of Siegel moduli schemes, C-L. CHAI
108 Some topics in graph theory, H.P. YAP
109 Diophantine analysis, J. LOXTON & A. VAN DER POORTEN (eds)
110 An introduction to surreal numbers, H. GONSHOR
111 Analytical and geometric aspects of hyperbolic space, D.B.A. EPSTEIN (ed)
113 Lectures on the asymptotic theory of ideals, D. REES
114 Lectures on Bochner-Riesz means, K.M. DAVIS & Y-C. CHANG
115 An introduction to independence for analysts, H.G. DALES & W.H. WOODIN
116 Representations of algebras, P.J. WEBB (ed)
117 Homotopy theory, E. REES & J.D.S. JONES (eds)
118 Skew linear groups, M. SHIRVANI & B. WEHRFRITZ
119 Triangulated categories in the representation theory of finite-dimensional algebras, D. HAPPEL
121 Proceedings of *Groups - St Andrews 1985*, E. ROBERTSON & C. CAMPBELL (eds)
122 Non-classical continuum mechanics, R.J. KNOPS & A.A. LACEY (eds)
124 Lie groupoids and Lie algebroids in differential geometry, K. MACKENZIE
125 Commutator theory for congruence modular varieties, R. FREESE & R. MCKENZIE
126 Van der Corput's method of exponential sums, S.W. GRAHAM & G. KOLESNIK
127 New directions in dynamical systems, T.J. BEDFORD & J.W. SWIFT (eds)

128 Descriptive set theory and the structure of sets of uniqueness, A.S. KECHRIS & A. LOUVEAU
129 The subgroup structure of the finite classical groups, P.B. KLEIDMAN & M.W.LIEBECK
130 Model theory and modules, M. PREST
131 Algebraic, extremal & metric combinatorics, M-M. DEZA, P. FRANKL & I.G. ROSENBERG (eds)
132 Whitehead groups of finite groups, ROBERT OLIVER
133 Linear algebraic monoids, MOHAN S. PUTCHA
134 Number theory and dynamical systems, M. DODSON & J. VICKERS (eds)
135 Operator algebras and applications, 1, D. EVANS & M. TAKESAKI (eds)
136 Operator algebras and applications, 2, D. EVANS & M. TAKESAKI (eds)
137 Analysis at Urbana, I, E. BERKSON, T. PECK, & J. UHL (eds)
138 Analysis at Urbana, II, E. BERKSON, T. PECK, & J. UHL (eds)
139 Advances in homotopy theory, S. SALAMON, B. STEER & W. SUTHERLAND (eds)
140 Geometric aspects of Banach spaces, E.M. PEINADOR and A. RODES (eds)
141 Surveys in combinatorics 1989, J. SIEMONS (ed)
142 The geometry of jet bundles, D.J. SAUNDERS
143 The ergodic theory of discrete groups, PETER J. NICHOLLS
144 Introduction to uniform spaces, I.M. JAMES
145 Homological questions in local algebra, JAN R. STROOKER
146 Cohen-Macaulay modules over Cohen-Macaulay rings, Y. YOSHINO
147 Continuous and discrete modules, S.H. MOHAMED & B.J. MÜLLER
148 Helices and vector bundles, A.N. RUDAKOV et al
149 Solitons, nonlinear evolution equations and inverse scattering, M.J. ABLOWITZ & P.A. CLARKSON
150 Geometry of low-dimensional manifolds 1, S. DONALDSON & C.B. THOMAS (eds)
151 Geometry of low-dimensional manifolds 2, S. DONALDSON & C.B. THOMAS (eds)
152 Oligomorphic permutation groups, P. CAMERON
153 L-functions and arithmetic, J. COATES & M.J. TAYLOR (eds)
154 Number theory and cryptography, J. LOXTON (ed)
155 Classification theories of polarized varieties, TAKAO FUJITA
156 Twistors in mathematics and physics, T.N. BAILEY & R.J. BASTON (eds)
157 Analytic pro-*p* groups, J.D. DIXON, M.P.F. DU SAUTOY, A. MANN & D. SEGAL
158 Geometry of Banach spaces, P.F.X. MÜLLER & W. SCHACHERMAYER (eds)
159 Groups St Andrews 1989 volume 1, C.M. CAMPBELL & E.F. ROBERTSON (eds)
160 Groups St Andrews 1989 volume 2, C.M. CAMPBELL & E.F. ROBERTSON (eds)
161 Lectures on block theory, BURKHARD KÜLSHAMMER
162 Harmonic analysis and representation theory for groups acting on homogeneous trees, A. FIGA-TALAMANCA & C. NEBBIA
163 Topics in varieties of group representations, S.M. VOVSI
164 Quasi-symmetric designs, M.S. SHRIKANDE & S.S. SANE
165 Groups, combinatorics & geometry, M.W. LIEBECK & J. SAXL (eds)
166 Surveys in combinatorics, 1991, A.D. KEEDWELL (ed)
167 Stochastic analysis, M.T. BARLOW & N.H. BINGHAM (eds)
168 Representations of algebras, H. TACHIKAWA & S. BRENNER (eds)
169 Boolean function complexity, M.S. PATERSON (ed)
170 Manifolds with singularities and the Adams-Novikov spectral sequence, B. BOTVINNIK
172 Algebraic varieties, GEORGE R. KEMPF
173 Discrete groups and geometry, W.J. HARVEY & C. MACLACHLAN (eds)
174 Lectures on mechanics, J.E. MARSDEN
175 Adams memorial symposium on algebraic topology 1, N. RAY & G. WALKER (eds)
176 Adams memorial symposium on algebraic topology 2, N. RAY & G. WALKER (eds)
177 Applications of categories in computer science, M.P. FOURMAN, P.T. JOHNSTONE, & A.M. PITTS (eds)
178 Lower K- and L-theory, A. RANICKI
179 Complex projective geometry, G. ELLINGSRUD, C. PESKINE, G. SACCHIERO & S.A. STRØMME (eds)
180 Lectures on ergodic theory and Pesin theory on compact manifolds, M. POLLICOTT
181 Geometric group theory I, G.A. NIBLO & M.A. ROLLER (eds)
182 Geometric group theory II, G.A. NIBLO & M.A. ROLLER (eds)
183 Shintani zeta functions, A. YUKIE
184 Arithmetical functions, W. SCHWARZ & J. SPILKER
185 Representations of solvable groups, O. MANZ & T.R. WOLF
186 Complexity: knots, colourings and counting, D.J.A. WELSH
187 Surveys in combinatorics, 1993, K. WALKER (ed)
190 Polynomial invariants of finite groups, D.J. BENSON

London Mathematical Society Lecture Note Series. 183

Shintani Zeta Functions

Akihiko Yukie
Oklahoma State University

CAMBRIDGE UNIVERSITY PRESS
Cambridge, New York, Melbourne, Madrid, Cape Town, Singapore, São Paulo

Cambridge University Press
The Edinburgh Building, Cambridge CB2 8RU, UK

Published in the United States of America by Cambridge University Press, New York

www.cambridge.org
Information on this title: www.cambridge.org/9780521448048

First published 1993

A catalogue record for this publication is available from the British Library

ISBN 978-0-521-44804-8 paperback

Transferred to digital printing 2008

To my parents Kenzo and Fumiko Yukie

Table of contents

Preface

The content of this book is taken from my manuscripts 'On the global theory of Shintani zeta functions I–V' which were originally intended for publication in ordinary journals. However, because of its length and the lack of a book on prehomogeneous vector spaces, it has been suggested to publish them together in book form.

It has been more than 25 years since the theory of prehomogeneous vector spaces began. Much work has been done on both the global theory and the local theory of zeta functions. However, we concentrate on the global theory in this book. I feel that another book should be written on the local theory of zeta functions in the future.

The purpose of this book is to introduce an approach based on geometric invariant theory to the global theory of zeta functions for prehomogeneous vector spaces.

This book consists of four parts. In Part I, we introduce a general formulation based on geometric invariant theory to the global theory of zeta functions for prehomogeneous vector spaces. In Part II, we apply the methods in Part I and determine the principal part of the zeta function for Siegel's case, i.e. the space of quadratic forms. In Part III, we handle relatively easy cases which are required to handle the case in Part IV. In Part IV, we use the results in Parts I–III to determine the principal part of the zeta function for the space of pairs of ternary quadratic forms.

We expanded the introduction of the original manuscripts to help non-experts to have a general idea of the subject. What we try to discuss in the introduction is the history of the subject, and what is required to prove the existence of densities of arithmetic objects we are looking for. Even though the theory of prehomogeneous vector spaces involves many topics, we concentrate on two aspects of the theory, i.e. the global theory and the local theory, in the introduction.

Parts I–III of this book correspond to Parts I–III of the above manuscripts, and Part IV of this book corresponds to Parts IV and V of the above manuscripts. Since the manuscripts were originally intended for publication in ordinary journals, certain changes were made to make this book more comprehensible and self-contained.

However, it is impossible to make this book completely self-contained, and we have to require a reasonable background in adelic language, basic group theory, and geometric invariant theory. For this, we assume that the reader is familiar with the following four books and two papers

[1] A. Borel, Some finiteness properties of adele groups over number fields,

[2] A. Borel, *Linear algebraic groups*,

[28] G. Kempf, Instability in invariant theory,

[35] F. Kirwan, *Cohomology of quotients in symplectic and algebraic geometry*,

[46] D. Mumford and J. Fogarty, *Geometric invariant theory*,

[79] A. Weil, *Basic number theory*.

Weil's book [79] is a standard place to learn basic materials on adelic language. Since we do not depend on class field theory, it is enough for the reader to be familiar with the first several chapters of Weil's book. Borel's paper [1] is a place to learn properties of Siegel domains. We need two facts in geometric invariant theory. One is the Hilbert–Mumford criterion of stability, and the other is the rationality of the equivariant Morse stratification. Mumford–Fogarty [46] and Kirwan [35] are the standard books to learn geometric invariant theory and equivariant Morse theory. The rationality of the equivariant Morse stratification was proved by G. Kempf in his paper [28]. However, even though the proofs of the above two facts are technically involved, the statements of these facts are fairly comprehensible and do not require a special background to understand. Therefore, if the reader is unfamiliar with these subjects, I recommend the reader not to worry about the proofs of the statements in this book which we quote from geometric invariant theory and look at the above documents later if necessary.

We have three original results in this book. One is a generalization of 'Shintani's lemma' to $\mathrm{GL}(n)$ concerning estimates of the smoothed Eisenstein series. Shintani proved this lemma for $\mathrm{GL}(2)$ in [64]. The statement of the result is Theorem (3.4.31). The second result is the determination of the principal part of the zeta function for the space of quadratic forms. The statement of the result is Theorem (4.0.1). Shintani himself studied this case and determined the poles of the associated Dirichlet series for quadratic forms which are positive definite in [65]. The last and the main result of this book is the determination of the principal part of the zeta function for the space of pairs of ternary quadratic forms. The statement of the result is Theorem (13.2.2). We discuss the relevance of these results in the introduction.

D. Wright contributed to this book in many places. He suggested the use of 'Wright's principle' in §3.7 after he read the first manuscript of my paper [86]. Also §0.5 is largely from his note. He also found the reference concerning Omar Khayyam when we wrote our paper [84], and helped me to find some references in this book. I would like to give a hearty thanks to him. As I mentioned above, this book is based on geometric invariant theory. For this, I owe a great deal to D. Mumford for teaching me geometric invariant theory and equivariant Morse theory. I was staying at Institute for Advanced Study during the academic year 1989–1990, and at Sonderforschungsbereich 170 Göttingen during the academic year 1990–1991 while I was writing the manuscript of this book. I would like to thank them for their support of this project. This work was partially supported by NSF Grants DMS-8803085, DMS-9101091.

Akihiko Yukie
February 1992, Stillwater, Oklahoma, USA

Notation

For a finite set A, the cardinality of A is denoted by $\#A$. If f, g are functions on a set X and $|f(x)| \le Cg(x)$ for some constant C independent of $x \in X$, we denote $f(x) \ll g(x)$. If $x, y \in \mathbb{R}$, we also use the classical notation $x \ll y$ if y is a much larger number than x. Since we use this classical notation only for numbers, and not for functions, we hope the meaning of this notation will be clear from the context.

Suppose that G is a locally compact group and Γ is a discrete subgroup of G contained in the maximal unimodular subgroup of G. For any left invariant measure dg on G, we choose a left invariant measure dg (we use the same notation, but the meaning will be clear from the context) on $X = G/\Gamma$ so that

$$\int_G f(g)dg = \int_X \sum_{\gamma\in\Gamma} f(g\gamma)dg.$$

We denote the fields of rational, real, and complex numbers by $\mathbb{Q}, \mathbb{R}, \mathbb{C}$ respectively. We denote the ring of rational integers by $\mathbb{Z}$. The set of positive real numbers is denoted by $\mathbb{R}_+$. For any ring R, $R^\times$ is the set of invertible elements of R. Let k be a number field, and o_k its integer ring. Let $\mathfrak{M}, \mathfrak{M}_\infty, \mathfrak{M}_\mathbb{R}, \mathfrak{M}_\mathbb{C}, \mathfrak{M}_f$ be the set of all the places, all the infinite, real, imaginary, finite places of k respectively. Let $\mathbb{A}_f$ (resp. $\mathbb{A}_f^\times$) be the restricted product of the k_v's (resp. $k_v^\times$'s) over $v \in \mathfrak{M}_f$. Let k_∞ (resp. $k_\infty^\times$) be the product of the k_v's (resp. $k_v^\times$'s) over $v \in \mathfrak{M}_\infty$. Then $\mathbb{A} = k_\infty \times \mathbb{A}_f$, $\mathbb{A}^\times = k_\infty^\times \times \mathbb{A}_f^\times$. If $x \in \mathbb{A}$ or $\mathbb{A}^\times$, we denote the finite (resp. infinite) part of x by x_f (resp. x_∞). If V is a vector space over k, we define $V_\mathbb{A}, V_\infty, V_f$ similarly. Let $\mathscr{S}(V_\mathbb{A}), \mathscr{S}(V_\infty), \mathscr{S}(V_f)$ be the spaces of Schwartz–Bruhat functions.

For any place v, k_v is the completion at v. If $v \in \mathfrak{M}_f$, $o_v \subset k_v$ is, by definition, the integer ring of k_v. Let $|\;|$ be the adelic absolute value. The absolute value of k_v is denoted by $|\;|_v$. For $x \in \mathbb{A}^\times$, we denote the product of the $|x|_v$'s over all $v \in \mathfrak{M}_f$ (resp. $v \in \mathfrak{M}_\infty$) by $|x|_f$ (resp. $|x|_\infty$). For $v \in \mathfrak{M}_f$, let π_v be the prime element, and $|\pi_v|_v = q_v^{-1}$. Note that if v is imaginary and $|x|$ is the usual absolute value, $|x|_v = |x|^2$.

Let r_1, r_2 be the numbers of real and imaginary places respectively. Let h, R, and e be the class number, regulator, and the number of roots of unity of k respectively. Let Δ_k be the discriminant of k. Let $\mathfrak{C}_k = 2^{r_1}(2\pi)^{r_2}hRe^{-1}$. We choose a Haar measure dx on $\mathbb{A}$ so that $\int_{\mathbb{A}/k} dx = 1$. For any finite place v, we choose a Haar measure dx_v on k_v so that $\int_{o_v} dx_v = 1$. We use the ordinary Lebesgue measure dx_v for v real, and $dx_v \wedge d\bar{x}_v$ for v imaginary. Then $dx = |\Delta_k|^{-\frac{1}{2}} \prod_v dx_v$ (see [79, p. 91]).

For $\lambda \in \mathbb{R}_+$, let $\underline{\lambda}$ be the idele whose component at v is $\lambda^{\frac{1}{[k:\mathbb{Q}]}}$ if $v \in \mathfrak{M}_\infty$ and 1 if $v \in \mathfrak{M}_f$. Clearly, $|\underline{\lambda}| = \lambda$. We identify $\mathbb{R}_+$ with a subgroup of $\mathbb{A}^\times$ by the map $\lambda \to \underline{\lambda}$. Let $\mathbb{A}^1 = \{x \in \mathbb{A}^\times \mid |x| = 1\}$. Then $\mathbb{A}^\times \cong \mathbb{A}^1/k^\times \times \mathbb{R}_+$, and $\mathbb{A}^1/k^\times$ is compact. We choose a Haar measure $d^\times t^1$ on $\mathbb{A}^1$ so that $\int_{\mathbb{A}^1/k^\times} d^\times t^1 = 1$. Using this measure, we choose a Haar measure $d^\times t$ on $\mathbb{A}^\times$ so that

$$\int_{\mathbb{A}^\times} f(t)d^\times t = \int_0^\infty \int_{\mathbb{A}^1} f(\underline{\lambda}t^1)d^\times \lambda d^\times t^1,$$

where $d^\times\lambda = \lambda^{-1}d\lambda$. For any finite place v, we choose a Haar measure $d^\times t_v$ on $k_v^\times$ so that $\int_{o_v^\times} d^\times t_v = 1$. Let $d^\times t_v(x) = |x|_v^{-1}dx$ if v is real, and $d^\times t_v(x) = |x|_v^{-1}dx \wedge d\bar{x}$ if v is imaginary. Then $d^\times t = \mathfrak{C}_k^{-1} \prod_v d^\times t_v$ (see [79, p. 95]).

Let $<> = \prod_v <>_v$ be a character of $\mathbb{A}/k$. Let v be a finite place. Suppose that $<>_v$ is trivial on $\pi_v^{-c_v}o_v$ and non-trivial on $\pi_v^{-c_v-1}o_v$. Then we define $a_v = \pi_v^{c_v}$. Let $\mathrm{e}(x) = e^{2\pi\sqrt{-1}x}$. If v is a real place, there exists $a_v \in k_v^\times$ such that $< x >_v = \mathrm{e}(a_v x)$, and if v is an imaginary place, there exists $a_v \in k_v^\times$ such that $< x >_v = \mathrm{e}(a_v x + \overline{a_v x})$. For almost all v, $c_v = 0$. Let $\mathfrak{a} = (a_v)_v \in \mathbb{A}^\times$. Then $|\mathfrak{a}| = |\Delta_k|^{-1}$ (see [79, p. 113]). The idele $\mathfrak{a}$ is called the difference idele of k.

Let $\zeta_k(s)$ be the Dedekind zeta function. As in [79], we define

$$Z_k(s) = |\Delta_k|^{\frac{s}{2}} \left(\pi^{-\frac{s}{2}}\Gamma(\frac{s}{2})\right)^{r_1} \left((2\pi)^{-s}\Gamma(s)\right)^{r_2} \zeta_k(s).$$

We define $\mathfrak{R}_k = \mathrm{Res}_{s=1} Z_k(s)$.

For a character ω of $\mathbb{A}^\times/k^\times$, we define $\delta(\omega) = 1$ if ω is trivial, and $\delta(\omega) = 0$ otherwise.

Introduction

§0.1 **What is a prehomogeneous vector space?**

One contribution of Gauss to number theory in the early nineteenth century was the discovery of the correspondence between equivalence classes of integral binary quadratic forms and ideal classes of quadratic fields. This correspondence can be described as follows.

Let $f(v) = f(v_1, v_2) = x_0v_1^2 + x_1v_1v_2 + x_2v_2^2$ be a binary quadratic form such that x_0, x_1, x_2 are rational integers. We define an action of the group $\{\pm 1\} \times \mathrm{GL}(2, \mathbb{Z})$ on the set of integral binary quadratic forms so that if $g = (t, g_1)$ where $t = \pm 1, g_1 \in \mathrm{GL}(2, \mathbb{Z})$, $gf(v) = tf(vg_1)$. We consider equivalence classes of integral binary quadratic forms with respect to this action. It is easy to see that the discriminant $x_1^2 - 4x_0x_2$ is invariant under such an action. On the other hand, let m be a square free integer, and consider a non-zero ideal $\mathfrak{a}$ of the ring of algebraic integers in the field $k = \mathbb{Q}(\sqrt{m})$. The discriminant Δ_k of k is m if $m \equiv 1 \bmod 4$ and $4m$ if $m \equiv 2$ or $3 \bmod 4$. As a module over $\mathbb{Z}$, $\mathfrak{a}$ is generated by two elements, say α, β, because $\mathfrak{a}$ is a torsion free rank two module over $\mathbb{Z}$. Consider the binary quadratic form $f_{\mathfrak{a}}(v) = N(\mathfrak{a})^{-1}N(\alpha v_1 + \beta v_2)$, where $N(\mathfrak{a}), N(\alpha v_1 + \beta v_2)$ are the norms. It is easy to see that $f_{\mathfrak{a}}$ depends only on the ideal class of $\mathfrak{a}$. Moreover, it turns out that ideal classes of k correspond bijectively to equivalence classes of primitive integral binary quadratic forms with discriminant Δ_k by the map $\mathfrak{a} \to f_{\mathfrak{a}}$.

Gauss established this correspondence in [16], and the reader can see a modern proof in Theorem 4 [3, p. 142]. Here, we consider a natural question: why do we consider such a correspondence? One conceptual reason is that it gives us a parametrization of ideal classes of quadratic fields in terms of a group action on a vector space. We can use this parametrization to actually compute the class numbers of quadratic fields. But what we are interested in in this book is a more analytic question. In order to illustrate our purpose, let us describe the conjecture of Gauss.

Let h_d be the number of $\mathrm{SL}(2, \mathbb{Z})$-equivalence classes of primitive integral binary quadratic forms which are either positive definite or indefinite. Then Gauss conjectured the asymptotic property of the average of h_d. However, an integral form in the sense of Gauss is a form $x_0v_1^2 + 2x_1v_1v_2 + x_2v_2^2$ such that x_0, x_1, x_2 are integers. Here we consider $x_0v_1^2 + x_1v_1v_2 + x_2v_2^2$ such that x_0, x_1, x_2 are integers. With this understanding, we have the following asymptotic formula

$$(0.1.1) \qquad \sum_{0<-d<x} h_d \sim \frac{\pi}{18\zeta(3)} x^{\frac{3}{2}},$$

$$\sum_{0<d<x} h_d \log \epsilon_d \sim \frac{\pi^2}{18\zeta(3)} x^{\frac{3}{2}}.$$

where $\epsilon_d = \frac{1}{2}(t + u\sqrt{d})$ and (t, u) is the smallest positive integral solution of the equation $t^2 - du^2 = 1$.

This conjecture was first proved by Lipschitz [42] for $d < 0$, and by Siegel [69] for $d > 0$, and much work has been done on the error term estimate also (see [65, pp. 44,45] for example). However, we are allowing all integers d here, and if $d = m^2d'$ and d' is a square free integer, $h_d, h_{d'}$ are related by a simple relation. Therefore, we are counting essentially the same object infinitely many times in (0.1.1). If k is a quadratic field over $\mathbb{Q}$, let h_k, R_k be the class number and the regulator respectively. If d is a square free integer, h_d is the number of ideal classes with respect to multiplication by elements with positive norms. Therefore, this h_d is slightly different from h_k of a number field of discriminant d even though they are closely related.

The problem of counting $h_k R_k$ of quadratic fields k was first settled by Goldfeld–Hoffstein [17] and was slightly generalized by Datskovsky [9] from our viewpoint recently. Here, we quote Datskovsky's statement for the simplest case.

$$(0.1.2) \qquad \sum_{\substack{0<-\Delta_k<x \\ [k:\mathbb{Q}]=2}} h_k \sim \frac{\zeta(2)}{3\pi} \prod_p (1-p^{-2}-p^{-3}+p^{-4}) x^{\frac{3}{2}},$$

$$\sum_{\substack{0<\Delta_k<x \\ [k:\mathbb{Q}]=2}} h_k R_k \sim \frac{\zeta(2)}{3} \prod_p (1-p^{-2}-p^{-3}+p^{-4}) x^{\frac{3}{2}},$$

where k runs through quadratic fields and Δ_k is the discriminant. Note that $\frac{\zeta(2)}{3} = \frac{\pi^2}{18}$. Therefore, (0.1.1) and (0.1.2) are very similar except for the difference between $\frac{1}{\zeta(3)}$ and $\prod_p (1-p^{-2}-p^{-3}+p^{-4})$.

Statements like (0.1.2) are the kind of density theorems we are looking for. We discuss the difference between (0.1.1) and (0.1.2) later in the introduction, and we go back to the space of binary quadratic forms again. The main ingredients of the above correspondence were the group $G = \mathrm{GL}(2)$ acting on the vector space V of binary quadratic forms, and the polynomial $\Delta(x) = x_1^2 - 4x_0x_2$ $(x = (x_0, x_1, x_2))$ which satisfies the property $\Delta(gx) = \det g \Delta(x)$ for $g \in \mathrm{GL}(2), x \in V$. Moreover, if we consider this vector space over an algebraically closed field, the generic point is a single G-orbit. The fundamental reason why one can prove results like (0.1.1), (0.1.2) is that we can use the Fourier analysis on the vector space V. Also when we consider the averages as (0.1.1) or (0.1.2), we can use the value of $\Delta(x)$ to average over. The fact that the generic point is a single orbit assures us that there is essentially one choice of such a polynomial.

Sato and Shintani introduced the notion of prehomogeneous vector spaces in [60] and generalized the situation as the above example. We now state the definition of prehomogeneous vector spaces from our viewpoint.

Let k be an arbitrary field. Let G be a connected reductive group, V a representation of G, and χ_V an indivisible non-trivial rational character of G, all defined over k.

Definition (0.1.3) *The triple (G, V, χ_V) is called a prehomogeneous vector space if the following two conditions are satisfied.*
(1) There exists a Zariski open orbit.
(2) There exists a polynomial $\Delta \in k[V]$ such that $\Delta(gx) = \chi'(g)\Delta(x)$ where χ' is a rational character and $\chi' = \chi_V^a$ for some positive integer a.

Note that if Δ_1, Δ_2 are two polynomials as in the above definition, there exist positive integers a, b such that $\frac{\Delta_1^a}{\Delta_2^b}$ is a G-invariant rational function. Since there exists an open orbit, this implies that $\frac{\Delta_1^a}{\Delta_2^b}$ is a constant function. Therefore, for any

k-algebra R, the set $V_R^{ss} = \{x \in V_R \mid \Delta(x) \neq 0\}$ does not depend on the choice of Δ, and we call it the set of semi-stable points. A polynomial Δ which satisfies the property (2) is called a relative invariant polynomial. If V is an irreducible representation, the center of the image $G \to \mathrm{GL}(V)$ has split rank one by Schur's lemma. Therefore the choice of χ_V is unique. So we call (G, V) a prehomogeneous vector space also.

For the space of binary quadratic forms, let $G = \mathrm{GL}(1) \times \mathrm{GL}(2)$, and $\chi_V(t, g) = t \det g$ for $g = (t, g)$. Then (G, V, χ_V) is a prehomogeneous vector space in the above sense.

Before we start the discussion on prehomogeneous vector spaces, let us make one more historical remark.

There is no doubt that Gauss was the first mathematician who recognized the group theoretic approach to number theory. But one particular prehomogeneous vector space had appeared already in the eleventh century.

The solution of cubic and quartic equations by radicals has been known for a long time. But before the solution was found, there was a poet-mathematician Omar Khayyam in medieval Persia who worked on this problem. He did not think it was possible to solve cubic equations by radicals, and instead he tried to express the solutions to cubic equations geometrically. For example, the solution of the equation $x^3 = N$ can be realized as the intersection of two parabolas $y = x^2, y^2 = Nx$. After the solution by radicals was found, his work has long been forgotten. However, it is surprisingly related to the theory of prehomogeneous vector spaces.

What makes the space of binary quadratic forms so interesting is that we can associate a quadratic field to a generic point of the vector space. More precisely, if G is $\mathrm{GL}(1) \times \mathrm{GL}(2)$, V is the space of binary quadratic forms and $\chi_V(t, g) = t \det g$, there is a map from $G_{\mathbb{Q}} \setminus V_{\mathbb{Q}}^{ss}$ to the set of isomorphism classes of fields of degree less than or equal to 2 over $\mathbb{Q}$. This map is clearly surjective, and this surjectivity is the reason why we count the class number of all the quadratic fields in (0.1.2).

Now, let us consider the group $G = \mathrm{GL}(3) \times \mathrm{GL}(2)$, and the vector space V of pairs of ternary quadratic forms. If we define $\chi_V(g_1, g_2) = (\det g_1)^4 (\det g_2)^3$, the triple (G, V, χ_V) is a prehomogeneous vector space (see [59] or Chapter 8). If $x = (Q_1, Q_2) \in V_k$ and Q_1, Q_2 are ternary quadratic forms, we can consider the set $\mathrm{Zero}(x)=\{v \in \mathbb{P}^2 \mid Q_1(v) = Q_2(v) = 0\}$. We call $\mathrm{Zero}(x)$ the zero set of x. We will show in Chapter 8 that (G, V, χ_V) is a prehomogeneous vector space and V^{ss} consists of points whose zero sets are four distinct points in $\mathbb{P}^2$. It is easy to see that the field generated by the residue fields of points in $\mathrm{Zero}(x)$ is a splitting field of a quartic equation. Now the question is if we can get all such fields from pairs of ternary quadratic forms. This is easy because any quartic equation $x^4 + a_1 x^3 + a_2 x^2 + a_3 x + a_4 = 0$ can be written as an intersection of two conics as follows.

$$y = x^2, y^2 + a_1 xy + a_2 x^2 + a_3 x + a_4 = 0.$$

But this is what Omar Khayyam did about 900 years ago, and he essentially proved the surjectivity of the map from $G_{\mathbb{Q}} \setminus V_{\mathbb{Q}}^{ss}$ to the isomorphism classes of splitting fields of quartic equations. For the works of Omar Khayyam, the reader should see [74]. The analytic theory of this prehomogeneous vector space is the main topic of this book, and we handle the global zeta function for this case in Part IV.

§0.2 The classification

In this section, we discuss the classification of irreducible prehomogeneous vector spaces over an algebraically closed field. Throughout this section, k is an algebraically closed field of characteristic zero.

First, we show that from a given prehomogeneous vector space, we can make infinitely many prehomogeneous vector spaces which are essentially the same as the original prehomogeneous vector space.

Let G be a reductive group, and V a representation of G. Suppose that the dimension of V is n. For an integer $0 < m < n$, we consider two representations $(G \times \mathrm{GL}(m), V \otimes k^m)$, $(G \times \mathrm{GL}(n-m), V \otimes k^{n-m})$. Then generic $\mathrm{GL}(m)_k$-orbits of $V_k \otimes k^m$ correspond bijectively with generic $\mathrm{GL}(n-m)_k$-orbits of $V_k \otimes k^{n-m}$ because the Grassmann variety of m planes in V can be identified with the Grassmann variety of $n-m$ planes in V. Therefore, $(G \times \mathrm{GL}(m), V \otimes k^m)$ is a prehomogeneous vector space if and only if $(G \times \mathrm{GL}(n-m), V \otimes k^{n-m})$ is a prehomogeneous vector space, and the sets of generic orbits coincide. If two prehomogeneous vector spaces are related in this way, we identify two such representations and consider the equivalence relation determined by this identification. If the dimension of (G, V) is the smallest among prehomogeneous vector spaces which are equivalent to (G, V), we say that (G, V) is reduced. Also we identify two prehomogeneous vector spaces $(G, V), (G', V)$ if the images of G, G' in $\mathrm{GL}(V)$ are the same.

Sato and Kimura proved in [59] that the following is the list of all the irreducible reduced prehomogeneous vector spaces.

(1) $G = \mathrm{GL}(n) \times H$, $V = M(n,n)_k$ where $H \subset \mathrm{GL}(n)$ is any reductive subgroup such that the k^n is an irreducible representation of H.

(2) $G = \mathrm{GL}(1) \times \mathrm{GL}(n)$, $V = \mathrm{Sym}^2 k^n$.

(3) $G = \mathrm{GL}(1) \times \mathrm{GL}(2n)$, $V = \wedge^2 k^{2n}$.

(4) $G = \mathrm{GL}(1) \times \mathrm{GL}(2)$, $V = \mathrm{Sym}^3 k^2$.

(5), (6), (7) $G = \mathrm{GL}(1) \times \mathrm{GL}(n)$, $V = \wedge^3 k^n$ where $n = 6, 7, 8$.

(8) $G = \mathrm{GL}(3) \times \mathrm{GL}(2)$, $V = \mathrm{Sym}^2 k^3 \otimes k^2$.

(9) $G = \mathrm{GL}(6) \times \mathrm{GL}(2)$, $V = \wedge^2 k^6 \otimes k^2$.

(10), (11) $G = \mathrm{GL}(n) \times \mathrm{GL}(5)$, $V = k^n \otimes \wedge^2 k^5$ where $n = 3, 4$.

(12) $G = \mathrm{GL}(3) \times \mathrm{GL}(3) \times \mathrm{GL}(2)$, $V = k^3 \otimes k^3 \otimes k^2$.

(13) $G = \mathrm{GSp}(2n) \times \mathrm{GL}(2m)$, $V = k^{2n} \otimes k^{2m}$ where $n \geq 2m \geq 2$.

(14) $G = \mathrm{GL}(1) \times \mathrm{GSp}(6)$, V is a 14 dimensional representation of G.

(15) $G = \mathrm{GO}(n) \times \mathrm{GL}(m)$, $V = k^n \otimes k^m$ where $n \geq 3$, $\frac{n}{2} \geq m \geq 1$.

(16), (17), (18) $G = \mathrm{GSpin}(7) \times \mathrm{GL}(n)$, $V = \mathrm{spin}_7 \otimes k^n$ where $n = 1, 2, 3$ and spin_7 is the spin representation.

(19), (22) $G = \mathrm{GSpin}(n)$, $V = \mathrm{spin}_n$ where $n = 9, 11$.

(20), (21) $G = \mathrm{GSpin}(10) \times \mathrm{GL}(n)$, $V = \mathrm{halfspin}_{10} \otimes k^n$ where $\mathrm{halfspin}_{10}$ is the halfspin representation and $n = 2, 3$.

(23), (24) $G = \mathrm{GL}(1) \times \mathrm{GSpin}(n)$, $V = \mathrm{halfspin}_n$ where $n = 12, 14$.

(25), (26) $G = G_2 \times \mathrm{GL}(n)$, $V = k^7 \otimes k^n$ where k^7 is a representation of G_2 and $n = 1, 2$.

(27), (28) $G = E_6 \times \mathrm{GL}(n)$, $V = k^{27} \otimes k^n$ where k^{27} is a representation of E_6 and $n = 1, 2$.

(29) $G = \mathrm{GL}(1) \times E_7$, V is a 56 dimensional representation of E_7

(30) $G = \mathrm{GSp}(2n) \times \mathrm{GO}(3)$, $V = k^{2n} \otimes k^3$.

The cases (1)–(29) are what we call regular prehomogeneous vector spaces. For the definition of the regularity, the reader should see [59]. Even though it does not make any difference over an algebraically closed field, we have included the GL(1) factor in (2)–(5) etc. and used groups like $\mathrm{GSp}(2n), \mathrm{GO}(n)$ instead of $\mathrm{Sp}(2n), \mathrm{O}(n)$ etc., because it is more natural number theoretically. Most of these representations are what we call prehomogeneous vector spaces of parabolic type classified by Rubenthaler in his thesis [52]. This is the kind of prehomogeneous vector spaces which one can construct from parabolic subgroups of reductive groups as follows.

Let G be a reductive group, and $P = MU$ a standard parabolic subgroup where M is the Levi component and U is the unipotent radical. The reductive part M acts on U by conjugation, and therefore on $V = U/[U, U]$ also. Since V can be considered as a vector space, we have a representation of a reductive group M. Vinberg [75] proved that there is a Zariski open orbit. Therefore, if there exists a relatively invariant polynomial, (M, V) is a prehomogeneous vector space by choosing a relative invariant polynomial and is called a prehomogeneous vector space of parabolic type.

For example, if we consider the Siegel parabolic subgroup P of $\mathrm{GSp}(2n)$, $M = \mathrm{GL}(1) \times \mathrm{GL}(n)$ and V is the space of quadratic forms in n variables. If G is a type C_n group etc., we say that (M, V) is of type C_n etc. Then (2) is C_n type, (3) is D_{2n} type, (4) is G_2 type, (5), (6), (7) are of E_6, E_7, E_8 types, (8) is F_4 type, (9) is E_7 type, (10), (11) are E_7, E_8 types, (12) is E_6 type, (13) is C_{n+m} type, (14) is F_4 type, (15) is B, D type, (16) is F_4 type, (20), (23), (25), (27) are E_7 type, (21), (24), (26), (28) are E_8 type (29) is E_8 type. (1) is not always of parabolic type. (17), (18) (19), (22), (25), (26) are not in Table 1 [52, pp. 35–38].

For the details on prehomogeneous vector spaces of parabolic type, the reader should see [52].

§0.3 The global zeta function

In this section, we discuss the meromorphic continuation and the functional equation of the zeta function, restricting ourselves to irreducible prehomogeneous vector spaces (G, V, χ_V) for simplicity. The reader should see §3.1 for the general definition of the zeta function. For the rest of this section, k is a number field.

For simplicity, we assume that there exists a one dimensional split torus $T_0 \cong \mathrm{GL}(1)$ in the center of G acting on V by the ordinal multiplication by $t^{e_0} \in \mathrm{GL}(1)$ and $\chi_V(t) = t^e$ for $t \in T_0$ where $e_0, e > 0$ are positive integers. Let Δ be a relative invariant polynomial, and d the degree of Δ. Then $|\Delta(gx)| = |\chi_V(g)|^{\frac{e_0 d}{e}} |\Delta(x)|$. Let N be the dimension of V.

We assume that the representation $G \to \mathrm{GL}(V)$ is faithful. Therefore, in terms of the list in §0.2, we are considering $(G/\widetilde{T}, V)$ where $\widetilde{T}$ is the kernel of the homomorphism $G \to \mathrm{GL}(V)$. We fix a Haar measure dg on $G_{\mathbb{A}}$. Moreover, we assume that dg is of the form $dg = \prod_v dg_v$ where dg_v is a Haar measure on G_{k_v} for $v \in \mathfrak{M}$. Let $L \subset V_k^{\mathrm{ss}}$ be a G_k-invariant subset. For $\Phi \in \mathscr{S}(V_{\mathbb{A}})$ and a complex variable s, we define

$$Z_L(\Phi, s) = \int_{G_{\mathbb{A}}/G_k} |\chi_V(g)|^{\frac{e_0 s}{e}} \sum_{x \in L} \Phi(gx)\, dg, \tag{0.3.1}$$

$$Z_{L+}(\Phi,s)=\int_{\substack{G_{\mathbb{A}}/G_k\\ |\chi_V(g)|\geq 1}}|\chi_V(g)|^{\frac{e_0 s}{e}}\sum_{x\in L}\Phi(gx)dg,$$

if these integrals are well defined.

We say that (G,V,χ_V) is of complete type if $Z_{V_k^{ss}}(\Phi,s)$ converges absolutely for $\mathrm{Re}(s)\gg 0$ and $Z_{V_k^{ss}+}(\Phi,s)$ is an entire function for all $\Phi\in\mathscr{S}(V_{\mathbb{A}})$. It is very likely that the first condition implies the second condition. We say that (G,V,χ_V) is of incomplete type if it is not of complete type. If the stabilizer G_x contains a split torus in its center for some $x\in V_k^{ss}$, (G,V,χ_V) is of incomplete type. This applies to the cases (2) $n=2$, (12), (15) $m=1,2$, (17) in §2. If G_x does not contain a split torus in its center for any $x\in V_k^{ss}$, it is very likely that it is of complete type even though this has yet to be proved. Some examples are known. Siegel showed that the case (2) in §0.2 for $n\geq 3$ is of complete type. In general, if $\dim G=\dim V$, it is of complete type, as we show in this book. This applies to the cases (4), (8), (11) in §0.2 (the case (4) is due to Davenport and Shintani). If (G,V,χ_V) is an irreducible prehomogeneous vector space of complete type, we choose V_k^{ss} as L in the definition of the zeta function and use the notation $Z(\Phi,s)$, $Z_+(\Phi,s)$.

We fix a measure $dx=\prod_v dx_v$ on $V_{\mathbb{A}}$. Then

$$d(gx)=|\chi_V(g)|^{\frac{e_0 N}{e}}dx,\ |\Delta(gx)|=|\chi_V(g)|^{\frac{e_0 d}{e}}|\Delta(x)|.$$

First we show that $Z_L(\Phi,s)$ decomposes into a summation over rational orbits. Suppose $\Phi=\otimes_v\Phi_v\in\mathscr{S}(V_{\mathbb{A}})$. Let $x\in V_k^{ss}$. Let G_x be the stabilizer of x, and G_x^0 its connected component of 1. Let $o(x)=[G_{xk}:G_{xk}^0]$. Then by an obvious consideration,

$$Z_L(\Phi,s)=\sum_{x\in G_k\backslash L}\frac{1}{o(x)}\int_{G_{\mathbb{A}}/G_{xk}^0}|\chi_V(g)|^{\frac{e_0 s}{e}}\Phi(gx)dg.$$

Let $x\in L$, and $v\in\mathfrak{M}$. We choose a left G_{k_v}-invariant measure $dg'_{x,v}$ on $G_{k_v}/G_{xk_v}^0$ and a Haar measure $dg'_{x,v}$ on $G_{xk_v}^0$ so that $dg_v=dg'_{x,v}dg''_{x,v}$ for all v and $dg'_x=\prod_v dg'_{x,v}$, $dg''_x=\prod_v dg''_{x,v}$ are well defined. Then there exists a constant $b_{x,v}>0$ such that

$$\int_{G_{k_v}/G_{xk_v}^0}\Psi(g'_{x,v}x)dg'_{x,v}=b_{x,v}\int_{G_{k_v}x}\Psi(y_v)|\Delta(y_v)|_v^{-\frac{N}{d}}dy_v$$

for any measurable function Ψ on $G_{k_v}x$. We discuss the choice of the measures $dg'_{x,v}$, $dg''_{x,v}$ for some cases in §0.5.

Let $\mu(x)$ be the volume of $G_{x\mathbb{A}}^0/G_{xk}^0$ with respect to the measure dg''_x. We define

$$X_{x,v}(\Phi,s)=\int_{G_{k_v}x}\Phi_v(y_v)|\Delta(y_v)|_v^{\frac{s-N}{d}}dy_v,$$

$$Z_{x,v}(\Phi,s)=\int_{G_{k_v}/G_{xk_v}^0}|\chi_V(g'_{x,v})|^{\frac{e_0 s}{e}}\Phi_v(g'_{x,v}x)dg'_{x,v}.$$

Note that these distributions depend only on the orbit $G_{k_v}x$ and Φ_v. These distributions are related as follows.

$$
\begin{aligned}
Z_{x,v}(\Phi, s) &= |\Delta(x)|^{-\frac{s}{d}} \int_{G_{k_v}/G^0_{x k_v}} |\Delta(x)|^{\frac{s}{d}} |\chi_V(g'_{x,v})|^{\frac{e_0 s}{e}} \Phi_v(g'_{x,v}x) dg'_{x,v} \\
&= |\Delta(x)|^{-\frac{s}{d}} \int_{G_{k_v}/G^0_{x k_v}} |\Delta(g'_{x,v}x)|^{\frac{s}{d}} \Phi_v(g'_{x,v}x) dg'_{x,v} \\
&= b_{x,v} |\Delta(x)|_{v_0}^{-\frac{s}{d}} X_{x,v}(\Phi, s).
\end{aligned}
$$

We choose an infinite place v_0. Then the above decomposition becomes

$$
\begin{aligned}
(0.3.2)\quad Z_L(\Phi, s) &= \sum_{x \in G_k \backslash L} \frac{\mu(x)}{o(x)} \prod_{v \in \mathfrak{M}} Z_{x,v}(\Phi, s) \\
&= \sum_{x \in G_k \backslash L} \frac{\mu(x)}{o(x)|\Delta(x)|_{v_0}^{\frac{s}{d}}} b_{x,v_0} X_{x,v_0}(\Phi, s) \prod_{v \in \mathfrak{M} \backslash \{v_0\}} Z_{x,v}(\Phi, s).
\end{aligned}
$$

Let $V_1, \cdots, V_l$ be $G_{k_{v_0}}$-orbits of $V^{ss}_{k_{v_0}}$. Note that even though the group G is assumed to be connected as an algebraic group, $G_{k_{v_0}}$ may not be connected in general, for example GL(1). The distribution $X_{x,v_0}(\Phi, s)$ only depends on the sets $V_1, \cdots, V_l$, and we denote it by $X_1(\Phi, s), \cdots, X_l(\Phi, s)$. We define

$$
\begin{aligned}
(0.3.3)\qquad X(\Phi, s) &= (X_1(\Phi, s), \cdots, X_l(\Phi, s)), \\
\xi_{L,i}(\Phi, s) &= \sum_{x \in G_k \backslash V_i \cap L} \frac{\mu(x) b_{x,v_0}}{o(x)|\Delta(x)|_{v_0}^{\frac{s}{d}}} \prod_{v \in \mathfrak{M} \backslash \{v_0\}} Z_{x,v}(\Phi, s), \\
\xi_L(\Phi, s) &= \begin{pmatrix} \xi_{L,1}(\Phi, s) \\ \vdots \\ \xi_{L,l}(\Phi, s) \end{pmatrix}.
\end{aligned}
$$

Then

$$
Z_L(\Phi, s) = X(\Phi, s)\xi_L(\Phi, s).
$$

If $L = V^{ss}_k$, we drop L and use the notation $\xi(\Phi, s)$ etc.

Let V^* be the dual space of V and (x, y) the natural pairing of $x \in V, y \in V^*$. For $g \in G$, let $g^* \in \mathrm{GL}(V^*)$ be the element such that $(gx, y) = (x, g^*y)$ for all $x \in V, y \in V^*$. Then $g \to (g^*)^{-1}$ is a representation of G on V^* and is called the contragredient representation of (G, V). If (G, V) is a prehomogeneous vector space, (G, V^*) is also a prehomogeneous vector space, and there exists a relative invariant polynomial Δ^* of the same degree d. Moreover $\chi_{V^*}(g) = \chi_V(g)^{-1}$. Also $V^*_{v_0}$ has the same number of $G_{k_{v_0}}$-orbits $V^*_1, \cdots, V^*_l$.

If $L^* \subset V^*_k$ is a G_k-invariant subset, we define functions $X^*_i(\Phi^*, s), \xi^*_{L^*,i}(\Phi^*, s)$ for $i = 1, \cdots, l$ and $X^*(\Phi^*, s), \xi^*_{L^*}(\Phi^*, s)$ similarly as in (0.3.3) for V^*.

Let $\Delta^*(\partial), \Delta(\partial)$ be the differential operators with constant coefficients on V, V^* which correspond to Δ^*, Δ. The local theory at infinite places is based on the following fact.

Lemma (0.3.4) *There exists a polynomial $b(s)$ (resp. $b^*(s)$) such that*

$$
\Delta^*(\partial)\Delta(x)^s = b(s)\Delta(x)^{s-1} \ (\textit{resp. } \Delta(\partial)\Delta^*(x)^s = b^*(s)\Delta^*(x)^{s-1})
$$

for any natural number s.

This lemma follows from the fact that $\Delta(\partial), \Delta^*(\partial)$ are invariant differential operators. These polynomials $b(s), b^*(s)$ are called the b-functions for Δ, Δ^* respectively. For these facts, the reader should see [60], [64].

Once we know (0.3.4), the meromorphic continuation of $X_i(\Phi, s)$ follows by the relations

$$\begin{aligned} X_i(\Delta^*(\partial)\Phi, s) &= \pm b(\tfrac{s-N}{d})X_i(\Phi, s-d) \qquad v_0 \in \mathfrak{M}_{\mathbb{R}}, \\ X(\Delta^*(\partial)\overline{\Delta^*(\partial)}\Phi, s) &= \pm b(\tfrac{s-N}{d})b(\tfrac{\overline{s-N}}{d})X(\Phi, s-d) \quad v_0 \in \mathfrak{M}_{\mathbb{C}}. \end{aligned}$$

For $\Phi \in \mathscr{S}(V_{\mathbb{A}})$, we define its Fourier transform $\widehat{\Phi} \in \mathscr{S}(V_{\mathbb{A}}^*)$ by

$$\widehat{\Phi}(y) = \int_{V_k} \Phi(x) < (x, y) > dx.$$

For $\Phi^* \in \mathscr{S}(V_{\mathbb{A}}^*)$, we define its Fourier transform $\widehat{\Phi}^* \in \mathscr{S}(V_{\mathbb{A}})$ similarly. If Φ has a product form $\Phi = \otimes_v \Phi_v$, $\widehat{\Phi}$ has the product form $\widehat{\Phi} = \otimes_v \widehat{\Phi}_v$ where $\widehat{\Phi}_v$ is the Fourier transform of Φ_v with respect to $< (x, y) >_v$.

The following theorem was proved by Sato and Shintani in [60].

Theorem (0.3.5) (Sato–Shintani) *The distributions* $X_i(\Phi, s), X_i^*(\Phi^*, s)$ *can be continued meromorphically to the entire* $s \in \mathbb{C}$ *for all* i, *and satisfy a functional equation*

$$X^*(\widehat{\Phi}, N-s) = X(\Phi, s)C(s),$$

where $C(s) = (c_{ij}(s))_{1 \leq i,j \leq k}$ *is a function which is meromorphic everywhere.*

We do not depend on (0.3.5) in this book, so we explain only briefly why (0.3.5) should be true.

If $g \in G_{k_{v_0}}$ and $\Phi_1(x) = \Phi(gx)$,

$$X_i(\Phi_1, s) = |\chi_V(g)|^{-\frac{e_0 s}{e}} X_i(\Phi, s), \; X_i^*(\widehat{\Phi}_1, N-s) = |\chi_V(g)|^{-\frac{e_0 s}{e}} X_i^*(\widehat{\Phi}, N-s)$$

for all i.

Choose $x_i \in V_i$ for $i = 1, \cdots, l$. Then $V_i \cong G_{k_{v_0}}/G_{x_i, k_{v_0}}$. So if $\Phi \in \mathscr{S}(V_{k_{v_0}})$ is compactly supported, (0.3.5) follows from the uniqueness of left $G_{k_{v_0}}$-invariant measures on orbits $V_1, \cdots, V_l$. For the proof of (0.3.5) for general Φ_{v_0}, the reader should see [60].

A similar statement to the above theorem for reducible prehomogeneous vector spaces was proved by F. Sato. As we will see below, the functional equation of the global zeta function follows from (0.3.5) for complete types. Therefore, one can deduce a similar global functional equation under the assumption that the zeta function is defined for $V_k^{\rm ss}$. For this, the reader should see [56]–[58]. For finite places, a similar statement to the above theorem was proved by Igusa in [22].

Suppose that $(G, V), (G, V^*)$ are of complete type. We denote the zeta function for V^* by $Z^*(\Phi^*, s), Z_+^*(\Phi^*, s)$. Let $G_{\mathbb{A}}^0 = \{g \in G_{\mathbb{A}} \mid |\chi_V(g)| = 1\}$. For $\Phi \in \mathscr{S}(V_{\mathbb{A}})$ and $\lambda \in \mathbb{R}_+$, we define $\Phi_\lambda \in \mathscr{S}(V_{\mathbb{A}})$ by the formula $\Phi_\lambda(x) = \Phi(\underline{\lambda} x)$. We choose a Haar measure dg^0 on $G_{\mathbb{A}}^0$ so that

$$Z(\Phi, s) = \int_0^\infty \int_{G_{\mathbb{A}}^0/G_k} \lambda^s \sum_{x \in V_k^{\rm ss}} \Phi(\underline{\lambda} g^0 x) d^\times \lambda dg^0.$$

For any $\Phi \in \mathscr{S}(V_{\mathbb{A}})$, we define

$$J(\Phi, g) = \left(|\chi(g)|^{-\frac{N}{e}} \sum_{x \in V_k^* \setminus V_k^{*ss}} \widehat{\Phi}((g^*)^{-1}x) - \sum_{x \in V_k \setminus V_k^{ss}} \Phi(gx) \right),$$

$$I^0(\Phi) = \int_{G^0_{\mathbb{A}}/G_k} J(\Phi, g^0) dg^0,$$

$$I(\Phi, s) = \int_0^1 \lambda^s I^0(\Phi_\lambda) d^\times \lambda.$$

Then by the Poisson summation formula,

$$Z(\Phi, s) = Z_+(\Phi, s) + Z_+^*(\widehat{\Phi}, N - s) + I(\Phi, s). \tag{0.3.6}$$

Sato and Shintani proved the following theorem under a strong assumption in [60]. Shintani later used a slightly different argument and proved that the meromorphic continuation and the functional equation of the global zeta function follow from Theorem (0.3.5). The reader can see his argument in [65] where he only considers the space of quadratic forms, but his argument works for the general case as follows.

Theorem (0.3.7) *Suppose that $(G, V), (G, V^*)$ are of complete type. Then the zeta functions $Z(\Phi, s), Z^*(\Phi^*, s)$ can be continued meromorphically everywhere for all $\Phi \in \mathscr{S}(V_{\mathbb{A}}), \Phi^* \in \mathscr{S}(V_{\mathbb{A}}^*)$ and satisfy a functional equation*

$$Z(\Phi, s) = Z^*(\widehat{\Phi}, N - s).$$

Proof. Consider $v_0 \in \mathfrak{M}_\infty$ in (0.3.2). Suppose $v_0 \in \mathfrak{M}_{\mathbb{R}}$. We fix $1 \leq i \leq l$ and prove that $\xi_i(\Phi, s)$ can be continued meromorphically everywhere. We choose $\Phi = \otimes_v \Phi_v$ so that $\Phi_{v_0}(x) \in C_0^\infty(V_i)$. Let $\Phi_1 = \otimes_v \Phi_{1v} \in \mathscr{S}(V_{\mathbb{A}})$ be the function such that $\Phi_{1v_0}(x) = \Delta^*(\partial)\Phi_{v_0}(x)$, and $\Phi_{1v} = \Phi_v$ for all $v \in \mathfrak{M} \setminus \{v_0\}$. Then $\widehat{\Phi}_{1v_0}(x) = (2\pi\sqrt{-1})^d \Delta^*(x)\Phi_{v_0}(x)$, so $\Phi_{1v_0}(x) = 0, \widehat{\Phi}_{1v_0}(y) = 0$ for $x \in V_{k_{v_0}} \setminus V_{k_{v_0}}^{ss}, y \in V_{k_{v_0}}^* \setminus V_{k_{v_0}}^{*ss}$. Then by (0.3.6), $Z(\Phi_1, s) = Z_+(\Phi_1, s) + Z_+^*(\widehat{\Phi}_1, N - s)$. Similarly, $Z^*(\widehat{\Phi}_1, s) = Z_+^*(\widehat{\Phi}_1, s) + Z_+(\Phi_1, N - s)$. Therefore, $Z(\Phi_1, s), Z^*(\widehat{\Phi}_1, s)$ are entire functions and $Z(\Phi_1, s) = Z^*(\widehat{\Phi}_1, N - s)$. Since $X_i(\Phi_1, s) = \pm b(\frac{s-N}{d}) X_i(\Phi, s - d)$, we can choose Φ_{v_0} so that $X_i(\Phi_1, s) \neq 0$. By the choice of Φ_{v_0}, $X_j(\Phi_1, s) = 0$ for $j \neq i$. Therefore, $X_i(\Phi_1, s)\xi_i(\Phi, s)$ is an entire function. This proves the meromorphic continuation of $\xi_i(\Phi, s)$. The meromorphic continuation of $\xi_i^*(\Phi^*, s)$ is similar.

By (0.3.5) and what we have just shown,

$$X^*(\widehat{\Phi}_1, N - s)\xi^*(\widehat{\Phi}, N - s) = X(\Phi_1, s)\xi(\Phi, s) = X(\Phi_1, s)C(s)\xi^*(\widehat{\Phi}, N - s).$$

Since $X_j(\Phi_1, s) = 0$ for $j \neq i$,

$$\xi_i(\Phi, s) = \sum_{j=1}^{l} C_{ij}(s)\xi_j^*(\widehat{\Phi}, N - s). \tag{0.3.8}$$

Therefore,

$$\xi(\Phi, s) = C(s)\xi^*(\widehat{\Phi}, N - s).$$

Since $\xi(\Phi, s)$ and $\xi^*(\widehat{\Phi}, N - s)$ do not depend on the v_0 part, the above equation is true for any $\Phi \in \mathscr{S}(V_{\mathbb{A}})$.

Now we consider an arbitrary Φ. The meromorphic continuation of

$$Z(\Phi, s), Z^*(\Phi^*, s)$$

follows from that of $X(\Phi, s)$, $\xi(\Phi, s)$ etc. Using the relation (0.3.8),

$$\begin{aligned} Z(\Phi, s) = X(\Phi, s)\xi(\Phi, s) &= X(\Phi, s)C(s)\xi^*(\widehat{\Phi}, N - s) \\ &= X^*(\widehat{\Phi}, N - s)\xi^*(\widehat{\Phi}, N - s) \\ &= Z^*(\widehat{\Phi}, N - s). \end{aligned}$$

Thus the functional equation for general Φ follows.

If $v_0 \in \mathfrak{M}_{\mathbb{C}}$, the proof is similar, except that we consider either $\Phi_{1v_0}(x) = \Delta^*(\partial)\Phi_{v_0}(x)$ or $\overline{\Delta^*(\partial)}\Phi_{v_0}(x)$. Then

$$\widehat{\Phi_{1v_0}}(x) = (2\pi\sqrt{-1})^d\Delta^*(x)\Phi_{1v_0}(x) \text{ or } (2\pi\sqrt{-1})^d\overline{\Delta^*(x)}\Phi_{1v_0}(x).$$

It is easy to see that

$$X(\Phi_1, s) = \begin{cases} \pm b(\frac{s-N}{d}) \int_{V^{ss}_{v_0}} \Phi(y_{v_0})\Delta(y_{v_0})^{\frac{s-d-N}{d}}\overline{\Delta(y_{v_0})}^{\frac{s-N}{d}} dy_{v_0} \text{ or} \\ \pm \overline{b(\frac{\bar{s}-N}{d})} \int_{V^{ss}_{v_0}} \Phi(y_{v_0})\Delta(y_{v_0})^{\frac{s-N}{d}}\overline{\Delta(y_{v_0})}^{\frac{s-d-N}{d}} dy_{v_0}. \end{cases}$$

Therefore, $X(\Phi_1, s)$ is an entire function, and we can use the same argument.

Q.E.D.

Let Φ_1 be as in the proof of the above theorem. Let $v_0 \in \mathfrak{M}_{\mathbb{R}}$.

By the above consideration,

$$\xi_i(\Phi, s) = \frac{Z(\Phi_1, s)}{X_i(\Phi_1, s)} = \pm\frac{Z(\Phi_1, s)}{b(\frac{s-N}{d})X_i(\Phi, s-d)}$$

and $Z(\Phi_1, s)$ is an entire function. Therefore, if $s = s_0$ is a pole of $\xi_i(\Phi, s)$, it has to be a zero of $b(\frac{s-N}{d})X_i(\Phi, s-d)$. However, since Φ_{v_0} is compactly supported, $X_i(\Phi, s-d)$ is an entire function, and we can choose Φ_{v_0} so that $X_i(\Phi, s_0 - d) \neq 0$. Therefore, the order of the pole $s = s_0$ of $\xi_i(\Phi, s)$ does not exceed the multiplicity of $s - s_0$ in the polynomial $b(\frac{s-N}{d})$. If $v_0 \in \mathfrak{M}_{\mathbb{C}}$, the order of the pole $s = s_0$ does not exceed the order of multiplicity of $s - s_0$ in the polynomials $b(\frac{s-N}{d})$ and $\overline{b(\frac{\bar{s}-N}{d})}$ by a similar consideration.

Also by the above consideration, if $s = s_0$ is an order l pole of $\xi_i(\Phi, s)$, it is at most an order l pole of $Z(\Phi, s)$ by choosing some Φ_{v_0}.

Kimura and Ozeki computed b-functions for irreducible reduced regular prehomogeneous vector spaces (see [31] for example). Unfortunately, the orders of the poles and the multiplicities of the roots of $b(s)$ do not coincide, for example in the case of pairs of ternary quadratic forms. Therefore, the theory of b-functions does not predict the correct pole structures of zeta functions in general.

Let $f(s)$ be a meromorphic function of s. Suppose

$$f(s) = \frac{a_{-n}}{(s-s_0)^n} + \cdots + \frac{a_{-1}}{s-s_0} + g(s),$$

where $g(s)$ is holomorphic around s_0. We call

$$\frac{a_{-n}}{(s-s_0)^n} + \cdots + \frac{a_{-1}}{s-s_0}$$

the principal part of f around s_0.

If $(G,V),(G,V^*)$ are of complete type, the set of poles of $Z(\Phi,s)$ is contained in the set of roots of the b-function. Therefore, $Z(\Phi,s)$ has a finite number of finite order poles. An easy consideration as in the following proposition shows that we can expect a principal part formula if $(G,V),(G,V^*)$ are of complete type.

Proposition (0.3.9) *Suppose that $(G,V),(G,V^*)$ are of complete type. Let $s = s_1,\cdots,s_n$ be the poles of $Z(\Phi,s)$, and $f_i(\Phi,s)$ the principal part of $Z(\Phi,s)$ around s_i. Then*

$$Z(\Phi,s) = Z_+(\Phi,s) + Z_+^*(\widehat{\Phi},N-s) + \sum_{i=1}^n f_i(\Phi,s).$$

Proof. We define

$$\widetilde{I}(\Phi,s) = I(\Phi,s) - \sum_{i=1}^n f_i(\Phi,s).$$

We also define $\widetilde{I}^*(\Phi^*,s)$ etc. for V^* similarly. Then it is easy to see that $\widetilde{I}(\Phi,s) = \widetilde{I}^*(\widehat{\Phi},N-s)$. By assumption, $|\chi_V(g)|^{\frac{e_0\mathrm{Re}(s)}{e}}|J(\Phi,g)|$ is an integrable function if $\mathrm{Re}(s) \gg 0$. Moreover if $|\chi_V(g)| \leq 1$, this function is decreasing with respect to $\mathrm{Re}(s)$ and

$$\lim_{\mathrm{Re}(s)\to\infty} |\chi_V(g)|^{\frac{e_0\mathrm{Re}(s)}{e}}|J(\Phi,g)| = 0$$

if $|\chi_V(g)| < 1$. Therefore, by the bounded convergence theorem, $I(\Phi,s)$ is bounded if $\mathrm{Re}(s) \gg 0$ and

$$\lim_{\mathrm{Re}(s)\to\infty} I(\Phi,s) = 0.$$

The function $\sum_{i=1}^n f_i(\Phi,s)$ clearly satisfies similar properties. Therefore, $\widetilde{I}(\Phi,s)$ is bounded if $\mathrm{Re}(s) \gg 0$ and

$$\lim_{\mathrm{Re}(s)\to\infty} \widetilde{I}(\Phi,s) = 0.$$

The same is true for $\widetilde{I}^*(\Phi^*,s)$ also for any $\Phi^* \in \mathscr{S}(V_{\mathbb{A}}^*)$. Since

$$\widetilde{I}(\Phi,s) = \widetilde{I}^*(\widehat{\Phi},N-s),$$

$I(\Phi,s)$ is bounded if $\mathrm{Re}(s) \ll 0$ and

$$\lim_{\mathrm{Re}(s)\to-\infty} I(\Phi,s) = 0$$

also.

Therefore, $\widetilde{I}(\Phi, s)$ is an entire function, and is bounded if $\mathrm{Re}(s) \gg 0$ or $\mathrm{Re}(s) \ll 0$. So, by Lindelöf's theorem, $\widetilde{I}(\Phi, s)$ is bounded everywhere. This implies that $\widetilde{I}(\Phi, s)$ is a constant function by Liouville's theorem. Since

$$\lim_{\mathrm{Re}(s)\to\infty} \widetilde{I}(\Phi, s) = 0,$$

$\widetilde{I}(\Phi, s) = 0$.

Q.E.D.

We call a formula of the form (0.3.9) a principal part formula.

The statement of the above proposition is very likely to be false for prehomogeneous vector spaces of incomplete type. For example, in the case of the space of binary quadratic forms, the zeta function is defined for the set of forms without rational factors and it has infinitely many poles from the zeros of the Riemann zeta function. We will discuss this case in §4.2.

Suppose that there is an involution $g \to g^\iota$ of G and a bilinear form $[\ ,\ ]$ defined over k such that $[gx, g^\iota y] = [x, y]$. We use this bilinear form to identify V with V^*. Then ι composed with the homomorphism $G \to \mathrm{GL}(V^*)$ gives us the contragredient representation of V. Therefore, we can consider $\widehat{\Phi} \in \mathscr{S}(V_{\mathbb{A}})$ under this condition. So if (G, V) is of complete type, (G, V^*) is of complete type also, and we can consider $Z^*(\widehat{\Phi}, s) = Z(\widehat{\Phi}, s)$. For groups over $\mathbb{R}$, the existence of such an involution is known (see [64] for example).

If G is a product of $\mathrm{GL}(n)$'s like $G = \mathrm{GL}(n_1) \times \cdots \times \mathrm{GL}(n_f)$, there is a canonical involution $g = (g_1, \cdots, g_f) \to g^\iota = ({}^t g_1^{-1}, \cdots, {}^t g_f^{-1})$. Then the weights of the representation $(g, x) \to g^\iota x$ and the weights of V^* are the same. Therefore, these representations are equivalent. This implies the existence of a bilinear form as above.

For the rest of this section, we assume that G is a split group over $\mathbb{Q}$ for simplicity. Suppose $\Phi = \otimes \Phi_v$ and Φ_v is the characteristic function of V_{o_v} for all $v \in \mathfrak{M}_f$. Then Φ is clearly invariant under the action of $K_f = \prod_{v \in \mathfrak{M}_f} G_{o_v}$. Since $K_f \setminus G_{\mathbb{A}}/G_{\mathbb{Q}} \cong G_{\mathbb{R}}/G_{\mathbb{Z}}$,

$$Z_L(\Phi, s) = \int_{G_{\mathbb{R}}/G_{\mathbb{Z}}} |\chi_V(g_\infty)|_\infty^{\frac{e_0 s}{e}} \sum_{x \in L} \Phi(g_\infty x) dg_\infty.$$

But since $g_\infty \in G_{\mathbb{R}}$, $\Phi(g_\infty x) = 0$ unless $x \in V_{\mathbb{Z}}$. Therefore,

$$Z_L(\Phi, s) = \int_{G_{\mathbb{R}}/G_{\mathbb{Z}}} |\chi_V(g_\infty)|_\infty^{\frac{e_0 s}{e}} \sum_{x \in V_{\mathbb{Z}} \cap L} \Phi_\infty(g_\infty x) dg_\infty.$$

This is the classical definition of the zeta function as was considered by Shintani in [64], [65]. We can apply a similar argument and this time, we can write $Z(\Phi, s)$ in terms of Dirichlet series. Let $X_1(\Phi, s), \cdots, X_l(\Phi, s)$ be as before. Let $\mu_\infty(x)$ be the volume of $G^0_{x,\mathbb{R}}/G^0_{x,\mathbb{Z}}$ with respect to the measure $dg''_{x,\infty}$ and $o_\infty(x) = [G_{\mathbb{Z}}; G^0_{\mathbb{Z}}]$. Then

$$(0.3.10) \qquad Z_L(\Phi, s) = \sum_{i=1}^{l} X_i(\Phi, s) \sum_{x \in G_{\mathbb{Z}} \setminus V_i \cap V_{\mathbb{Z}} \cap L} \frac{\mu_\infty(x) b_{x,\infty}}{o_\infty(x) |\Delta(x)|_\infty^{\frac{s}{d}}}.$$

This implies that

$$\xi_i(\Phi, s) = \sum_{x \in G_{\mathbb{Z}} \backslash V_i \cap V_{\mathbb{Z}} \cap L} \frac{\mu_\infty(x) b_{x,\infty}}{o_\infty(x) |\Delta(x)|_\infty^{\frac{s}{d}}}$$

for all i.

Many statements in analytic number theory are based on the following theorem.

Theorem (0.3.11) (Tauberian Theorem) *Let $f(s) = \sum_n \frac{a_n}{n^s}$ be a Dirichlet series such that $a_n \geq 0$ for all n. Suppose that $f(s)$ has an analytic continuation to $Re(s) \geq a$ and is holomorphic everywhere except for a pole at $s = 1$. Moreover, we assume that $f(s) = g(s)(s-1)^a$ for some positive integer a where $g(s)$ is a function which is holomorphic around $s = 1$. Then*

$$\sum_{n=1}^{X} a_n \sim \frac{g(1)}{\Gamma(a)} X (\log X)^{a-1}.$$

This is the plain Tauberian theorem. For the proof of this theorem, the reader should see Theorem I in [46, p. 464]. The version of Sato and Shintani in [60] was specially designed for prehomogeneous vector spaces and is much stronger. One can use Sato–Shintani's version to actually give some error term estimates in some cases. For this, the reader should see [65].

Let us first consider the associated Dirichlet series, and discuss what we can deduce for $G_{\mathbb{Z}}$-equivalence classes of generic integral orbits.

If $(G, V), (G, V^*)$ are of complete type, the poles of $\xi_i(\Phi, s)$'s are contained in the set of poles of $Z(\Phi, s)$. Therefore if $Z(\Phi, s)$ satisfies the condition of (0.3.11), we obtain the existence of constants C, a, b such that

$$\sum_{\substack{x \in G_{\mathbb{Z}} \backslash V_i \cap V_{\mathbb{Z}} \cap L \\ |\Delta(x)|_\infty \leq X}} \frac{\mu_\infty(x) b_{x,\infty}}{o_\infty(x)} \sim C X^a (\log X)^b.$$

§0.4 The orbit space $G_k \backslash V_k^{\mathrm{ss}}$

In the last section, we saw two decompositions of the zeta function into summations over rational and integral orbits. The main purpose of the analytic theory of prehomogeneous vector spaces is to use these decompositions and the Tauberian theorem and study the distribution of arithmetic objects represented by these orbit spaces. Therefore, it is natural to compare the difference between these two decompositions, and consider the meanings of these orbit spaces.

Let k be any field of characteristic zero. Let (G, V, χ) be a prehomogeneous vector space over k. We choose a point $w \in V_k^{\mathrm{ss}}$. This is possible because k is an infinite field and $V^{\mathrm{ss}} \subset V$ is a Zariski open set. Let $x \in V_k^{\mathrm{ss}}$. Then x is in the G-orbit of w over the algebraic closure $\bar{k}$. We choose $g_x \in G_{\bar{k}}$ such that $gw = x$. For $\sigma \in \mathrm{Gal}(\bar{k}/k)$, we define $\tau_{x,\sigma} = g_x^{-1} g_x^{\sigma}$. Then $c_x = \{\tau_{x,\sigma}\}$ defines an element of $H^1(\bar{k}/k, G_w)$ (the first Galois cohomology set). Let $\rho : H^1(\bar{k}/k, G_x) \to H^1(\bar{k}/k, G)$ be the natural map. The following proposition was proved by Igusa in [24].

Proposition (0.4.1) (Igusa) *By the map $x \to c_x$, $G_k \backslash V_k^{\mathrm{ss}}$ corresponds bijectively with $\rho^{-1}(1)$.*

Proof. Let $x, y \in V_k^{ss}$. Suppose $c_x = c_y$. Then there exists an element $h \in G_{w\bar{k}}$ such that $g_x^{-1}g_x^\sigma = h^{-1}g_y^{-1}g_y^\sigma h^\sigma$. This implies that the element $h' = g_y h g_x^{-1}$ is invariant under the action of the Galois group. Therefore, $h \in G_k$ and $hx = y$. Following the argument backward, the converse is true also. This shows that the map $G_k \setminus V_k^{ss} \to \rho^{-1}(1)$ is well defined and is injective. Let $\{\tau_\sigma\} \in \rho^{-1}(1)$. This means that there exists $g \in G_{\bar{k}}$ such that $\tau_\sigma = g^{-1}g^\sigma$ for all $\sigma \in \mathrm{Gal}(\bar{k}/k)$. Then $(gw)^\sigma = g^\sigma w = g\tau_\sigma w = gw$ for all σ. Therefore $gw \in V_k$. This proves the surjectivity.

Q.E.D.

We consider some examples. If G is a product of $\mathrm{GL}(n)$'s, $H^1(\bar{k}/k, G) = 0$ (see [38, p.16] for example). Therefore, if $w \in V_k^{ss}$, the quotient space $G_k \setminus V_k^{ss}$ corresponds bijectively with $H^1(\bar{k}/k, G_w)$. We consider the cases (2) $n = 2$, (4), (8), (9), (11), (12), (15) $n = 4, 6, m = 2$ of the classification. For these cases, D. Wright and the author proved in [84] that there exists a point $w \in V_k^{ss}$ such that the connected component G_w^0 of 1 is a product of $\mathrm{GL}(n)$'s and the quotient G_w/G_w^0 is isomorphic to $\mathfrak{S}_i$ for some i where $\mathfrak{S}_i$ is the permutation group with the trivial Galois group action. The number i is 2 for the cases (2) $n = 2$, (15) $n = 4, 6$, 3 for the cases (4), (9), (12), 4 for the case (8), and 5 for the case (11). Since the Galois group action on $\mathfrak{S}_i$ is trivial, $H^1(\bar{k}/k, \mathfrak{S}_i)$ is the set of conjugacy classes of homomorphisms from $\mathrm{Gal}(\bar{k}/k)$ to $\mathfrak{S}_i$. If $x \in V_k^{ss}$, let p_x be the homomorphism which corresponds to x. The kernel of p_x is a closed subgroup of finite index, and therefore determines a Galois extension of k, which we denote by $k(x)$. We also proved in [84] that $k(x)$'s are splitting fields of degree i equations without multiple roots and all such fields arise in this way. We already discussed the construction of $k(x)$ for the case in §0.1. As is illustrated by this example, the construction of $k(x)$ for other cases was geometric also. This consideration shows the following theorem.

Theorem (0.4.2) *Consider the cases* (2) $n = 2$, (4), (8), (9), (11), (12), (15) $n = 4, 6, m = 2$ *of the classification. Let* $x, y \in V_k^{ss}$. *Then* $G_k x = G_k y$ *if and only if there exists a permutation* r *such that* $p_x = r p_y r^{-1}$.

For the details, the reader should see our paper [84].

Next, we consider the case (15) in general (i.e. not necessarily split). We fix a quadratic form Q on $W = k^n$ and consider the group $\mathrm{GO}(Q)$ as $\mathrm{GO}(n)$ in (15). We identify V with the space of linear forms in m variables $y = (y_1, \cdots, y_m)$ with coefficients in the space W. To $x = y_1 w_1 + \cdots + y_m w_m$ where $w_1, \cdots, w_m \in W$, we associate a quadratic form in m variables $Q(y_1 w_1 + \cdots + y_m w_m)$. This means that we are considering quadratic forms which can be expressed in terms of a fixed quadratic form Q. So this is the situation of the Siegel–Weil formula. However, the zeta function theory may give us a different statement from the Siegel–Weil formula.

In [24], Igusa considered some exceptional cases and proved that points in $G_k \backslash V_k^{ss}$ correspond with division algebras. Igusa also proved in [24] that the orbit space $G_k \setminus V_k^{ss}$ consists of one point for the cases (1), (3), (13), (15) n even and $m = 1$, (16), (19), (20), (27). If k is a number field, (6) falls into this category also. We call these cases single orbit cases. As far as rational orbits are considered, these are the cases where the counting problem is trivial.

The reader can probably see from these examples that the space $G_k \backslash V_k^{ss}$ parametrizes interesting arithmetic objects. But what are we counting? In the case of

$G_{\mathbb{Z}}$-orbits, we were counting the quantity $\frac{\mu_\infty(x)}{o_\infty(x)}$. If we make an analogy when k is a number field, we may be counting the quantity $\frac{\mu(x)}{o(x)}$. In the case $G = \mathrm{GL}(1) \times \mathrm{GL}(2)$, $V = \mathrm{Sym}^2 k^2$, $\mu(x)$ gives us the class number times the regulator by the natural choice of the measures dg'_x, dg''_x as was proved by Datskovsky in [9]. If we consider the group $G = \mathrm{GL}(1) \times \mathrm{SL}(2)$ and the space of binary quadratic forms, we obtain the statement (0.1.1) by the consideration of $G_{\mathbb{Z}}$-orbits. Therefore, the difference between rational orbits and integral orbits is roughly speaking the difference between (0.1.1) and (0.1.2). So by considering $G_{\mathbb{Z}} \setminus V_{\mathbb{Z}}^{\mathrm{ss}}$, we may be counting essentially the same object infinitely many times.

For the above reason, we are more interested in rational orbits. However, one big problem arises now. In the decomposition (0.3.2), the coefficient of $\frac{\mu(x)}{o(x)}$ is not a constant, and the quantity to average over is not so clear. Therefore, we cannot directly apply the Tauberian theorem. If the rightmost pole is a simple pole, Datskovsky and Wright [11] and Datskovsky [9] formulated a process called the 'filtering process' to deduce density theorems for G_k-orbits. We discuss this process in the next section.

§0.5 The filtering process and the local theory: a note by D. Wright

The content of this section is largely from D. Wright's note, with a slight modification according to Datskovsky's formulation in [9]. In this section and §0.6, we assume that k is a number field.

We have a function $Z(\Phi, s)$ called the zeta function. It has some group theoretic properties. However, the filtering process depends only on the formal properties of $Z(\Phi, s)$. So we formulate the filtering process in an abstract way. We consider an arbitrary tempered distribution $Z(\Phi, s)$, i.e. a continuous linear functional on $\mathscr{S}(V_{\mathbb{A}})$. We have to make certain assumptions.

The first assumption is the following.

Assumption (0.5.1)

(1) *The function* $Z(\Phi, s)$ *has an analytic continuation to* $\mathrm{Re}(s) \geq \kappa > 0$ *and is holomorphic everywhere except for a simple pole at* $s = \kappa$ *with the residue* $R(\Phi)$.

(2) *The function* $Z(\Phi, s)$ *has the following decomposition*

$$Z(\Phi, s) = \sum_{x \in I} \frac{c(x)}{|\Delta_x|^s} L_x(\Phi, s),$$

where I *is some index set,* $c(x) \geq 0, \Delta_x \in \mathbb{Z}$ *are constants depending on* x. *This* Δ_x *should not be confused with* $\Delta(x)$.

(3) *If* $\Phi = \otimes_v \Phi_v \in \mathscr{S}(V_{\mathbb{A}})$, *the function* $L_x(\Phi, s)$ *has a product form*

$$L_x(\Phi, s) = \prod_v L_{x,v}(\Phi_v, s).$$

(4) *If* $v \in \mathfrak{M}_f$ *and* Φ_v *is the characteristic function (which we denote by* Φ_{o_v}*) of* V_{o_v}, $L_{x,v}(\Phi_v, s)$ *has a form*

$$L_{x,v}(\Phi_v, s) = 1 + \sum_{m \geq 1} a_{m,v} q_v^{-ms},$$

where $a_{m,v} \geq 0$ *for all* m.

We define $L_{x,v}(s) = L_{x,v}(\Phi_{o_v}, s)$. For any finite set $\mathfrak{M}_\infty \subset S \subset \mathfrak{M}$, we define

$$\Phi_S = \prod_{v \in S} \Phi_v,$$
$$L_{x,S}(\Phi, s) = \prod_{v \notin S} L_{x,v}(\Phi_v, s),$$
$$\xi_{x,S}(\Phi, s) = \prod_{v \in S} L_{x,v}(\Phi_v, s),$$
$$L_{x,S}(s) = \prod_{v \notin S} L_{x,v}(s).$$

By the assumption (0.5.1) (4), we may write

$$L_{x,S}(\Phi, s) = 1 + \sum_{m \geq 2} \frac{a_{x,S}(m)}{m^s}.$$

Obviously, $a_{x,S}(m) \geq 0$ for all x, S, m.

Let A_v be the index set of all the possible types of Euler factors $L_{x,v}(\Phi_v, s)$, i.e. I has a decomposition $I = \coprod_{\alpha_v \in A_v} I_{\alpha_v}$ indexed by A_v such that if $x, y \in \alpha_v$, $L_{x,v}(\Phi_v, s) = L_{y,v}(\Phi_v, s)$. For $\alpha_v \in A_v$, let $L_{\alpha_v}(\Phi_v, s)$ be the corresponding Euler factor. Let $L_{\alpha_v}(s) = L_{\alpha_v}(\Phi_{o_v}, s)$. Let $A_S = \prod_{v \in S} A_v$. For $\alpha = (\alpha_v) \in A_S$, we define

$$\zeta_\alpha(\Phi_S, s) = \prod_{v \in S} L_{\alpha_v}(\Phi_v, s).$$

For $x \in I$, we say that $x \in \alpha \in A_S$ if $x \in I_{\alpha_v}$ for all $v \in S$.

Suppose that $\Phi = \otimes_v \Phi_v$ and $\Phi_v = \Phi_{o_v}$ for $v \notin S$. Then

$$Z(\Phi, s) = \sum_{\alpha \in A_S} \zeta_\alpha(\Phi_S, s) \sum_{x \in \alpha} \frac{c(x)}{|\Delta_x|^s} L_{x,S}(s).$$

We define

$$\xi_{\alpha,S}(s) = \sum_{x \in \alpha} \frac{c(x)}{|\Delta_x|^s} L_{x,S}(s).$$

Assumption (0.5.2)

(1) *For any* S *and* $\alpha \in A_S$, Φ_S *may be chosen so that* $\zeta_\alpha(\Phi_S, s) \neq 0$ *while* $\zeta_\beta(\Phi_S, s) = 0$ *for all* $\beta \neq \alpha$.

(2) *There is a constant* $r_\alpha > 0$ *for each* α *such that*

$$R(\Phi) = \sum_{\alpha \in A_S} r_\alpha \zeta_\alpha(\Phi_S, \kappa).$$

These two assumptions imply that $\mathrm{Res}_{s=\kappa}\, \xi_{\alpha,S}(s) = r_\alpha$.

Let $S \subset T \subset \mathfrak{M}$ be another finite set. For $\beta = (\beta_v) \in A_T$, we define $\beta|_S = (\beta_v)_{v\in S} \in A_S$. For $\alpha \in A_S$, we define

$$\begin{aligned}\xi_{\alpha,T}(s) &= \sum_{x\in\alpha} \frac{c(x)}{|\Delta_x|^s} L_{x,T}(s)\\ &= \sum_{\beta\in A_T,\beta|_S=\alpha}\ \sum_{x\in\beta} \frac{c(x)}{|\Delta_x|^s} L_{x,T}(s)\\ &= \sum_{\beta\in A_T,\beta|_S=\alpha} \xi_{\beta,T}(s).\end{aligned}$$

We need another condition of uniformity.

Assumption (0.5.3) *There is a Dirichlet series $1 + \sum_{m\geq 2} \frac{a_T(m)}{m^s}$ which is holomorphic in $\mathrm{Re}(s) \geq \kappa$ such that $|a_{x,T}(m)| \leq a_T(m)$ for all $x \in I$. Also, for any $N > 0$, we may choose T sufficiently large so that $a_T(m) = 0$ for $1 < m \leq N$. Finally, $a_T(m) \leq a_S(m)$ if $S \subset T$.*

Under these assumptions, the following proposition was proved in [9], [11].

Proposition (0.5.4)

$$\lim_{X\to\infty} \frac{1}{X^\kappa} \sum_{|\Delta_x|\leq X} c(x) = \frac{1}{\kappa} \lim_{T\to\mathfrak{M}} \left[\sum_{\beta\in A_T,\beta|_S=\alpha} r_\beta \right]$$

if the right hand side converges.

Proof. By the Tauberian theorem,

$$\frac{1}{X^\kappa} \sum_{\substack{x\in\alpha\\ |\Delta_x|m\leq X}} c(x)a_{x,T}(m) \sim \frac{1}{\kappa} \sum_{\beta\in A_T,\beta|_S=\alpha} r_\beta. \tag{0.5.5}$$

An easy consideration shows that

$$\sum_{\substack{x\in\alpha\\ |\Delta_x|m\leq X}} c(x)a_{x,T}(m) = \sum_{\substack{x\in\alpha\\ |\Delta_x|\leq X}} c(x) + \sum_{\substack{x\in\alpha\\ 1<m\leq \frac{X}{|\Delta_x|}}} c(x)a_{x,T}(m). \tag{0.5.6}$$

By the assumption (0.5.3),

$$\begin{aligned}\left| \sum_{\substack{x\in\alpha\\ 1<m\leq \frac{X}{|\Delta_x|}}} c(x)a_{x,T}(m) \right| &\leq \sum_{\substack{x\in\alpha\\ 1<m\leq \frac{X}{|\Delta_x|}}} c(x)a_T(m)\\ &= \sum_{1<m\leq X} a_T(m) \sum_{\substack{x\in\alpha\\ |\Delta_x|\leq \frac{X}{m}}} c(x).\end{aligned}$$

Since $a_{x,T}(m) \geq 0$, by (0.5.5), (0.5.6), there exists a constant $C > 0$ such that

$$\sum_{|\Delta_x|\leq X} c(x) \leq CX^\kappa$$

for all $X > 0$. So applying this to $\frac{X}{m}$,

$$\sum_{1<m\leq X} a_T(m) \sum_{|\Delta_x|\leq \frac{X}{m}} c(x) \leq CX^{\kappa} \sum_{1<m\leq X} \frac{a_T(m)}{m^{\kappa}}$$
$$\leq CX^{\kappa}(L_T(\kappa)-1),$$

where $L_T(s) = 1 + \sum_{m=2}^{\infty} \frac{a_T(m)}{m^s}$.

Considering the limit $T \to \mathfrak{M}$, we get the statement of the proposition.

Q.E.D.

From this consideration, it is clear that in order to apply this process, we have to know the residue $R(\Phi)$ because otherwise we have no control of the value r_α. This means that without knowing $R(\Phi)$, we do not even know the existence of a statement of the form (0.1.2). Therefore, the computation of the principal part of the global zeta function is not just for determining the constant in the asymptotic formula for $G_{\mathbb{Z}}$-orbits for which we basically know the existence.

As far as the local theory is concerned, we have to answer the following questions.

(1) Estimate $Z_{x,v}(\Phi, s)$ on the domain $\{s \mid \mathrm{Re}(s) > \sigma\}$ for some $\sigma < \kappa$ and study its q-expansion.

(2) Find the value $b_{x,v}$ for sufficiently many x, v.

(3) Find the value r_α for sufficiently many α.

Therefore, we do not necessarily have to know the poles of $Z_{x,v}(\Phi, s)$, because it is very likely that $Z_{x,v}(\Phi, s)$ is holomorphic at the rightmost pole of the global zeta function. However, we may have to explicitly compute it in order to deduce the assumptions (0.5.1), (0.5.3).

Except for the single orbit cases, this process has been carried out for only two cases. One is the space of binary cubic forms where the statement for G_k-orbits is a slight generalization of the theorem of Davenport and Heilbronn:

$$\sum_{\substack{|\Delta_k|<x \\ [k:\mathbb{Q}]=3}} 1 \sim \frac{x}{\zeta(3)},$$

where k runs through all the cubic fields and Δ_k is the discriminant.

The other case is the space of binary quadratic forms where the statement for G_k-orbit is the theorem of Goldfeld–Hoffstein–Datskovsky (0.1.2). Except for these two cases and the single orbit cases, we do not even know if an asymptotic formula like (0.1.2) exists or not. These cases are good examples to illustrate the filtering process, so we review these two cases. The representations we consider are $(\mathrm{GL}(1) \times \mathrm{GL}(2)/\widetilde{T}, V)$ where T is the kernel of the homomorphism $\mathrm{GL}(1) \times \mathrm{GL}(2) \to \mathrm{GL}(V)$ and V is the space of binary cubic or quadratic forms.

We consider $L = V_k^{\mathrm{ss}}$ in the case of binary cubic forms, and the set of forms without rational factors in the case of binary quadratic forms. Let $v \in \mathfrak{M}$. We choose the index set A_v as the set of G_{k_v}-orbits in $V_{k_v}^{\mathrm{ss}}$. For each $\alpha_v \in A_v$, we choose a 'good' representative in $V_{k_v}^{\mathrm{ss}}$ and use the same notation α_v. This element α_v corresponds to a field $k_v(\alpha_v)$ of degree ≤ 3 or ≤ 2 over k_v and has the property that $\Delta(\alpha_v)$ is the local discriminant of $k_v(\alpha_v)/k_v$.

In the case of binary cubic forms, G_x^0 is trivial. In the case of binary quadratic forms, we choose a measure dg''_{α_v} on $G_{\alpha_v}^0$ so that the volume of $G_{\alpha_v \mathcal{O}_v}^0$ is 1 if $v \in \mathfrak{M}_f$.

If $v \in \mathfrak{M}_\infty$, we choose the 'canonical measure' on $G^0_{\alpha_v}$, which is isomorphic to either $\mathbb{R}^\times$, a circle, or $\mathbb{C}^\times$. This formulation is due to Datskovsky.

If $x \in L$ is in the orbit of α_v, we choose an element $h_x \in G_{k_v}$ so that $x = h_x \alpha_v$. Clearly, $G^0_{x k_v} = h_x G^0_{\alpha_v k_v} h_x^{-1}$. We define $dg''_{x,v}$ to be the measure induced from $G^0_{\alpha_v k_v}$. Then $b_{x,v} = b_{\alpha_v,v}$. (We defined $b_{x,v}$ for $x \in L$, but we can extend the definition to elements of $V^{\text{ss}}_{k_v}$.) Therefore,

$$Z_{x,v}(\Phi, s) = \frac{b_{x,v}}{|\Delta(x)|_v^{\frac{s}{3}}} X_{x,v}(\Phi, s) = \frac{b_{x,v}}{|\Delta(x)|_v^{\frac{s}{3}}} X_{\alpha_v,v}(\Phi, s) = \frac{|\Delta_{\alpha_v}|_v^{\frac{s}{3}}}{|\Delta(x)|_v^{\frac{s}{3}}} Z_{\alpha_v,v}(\Phi, s)$$

in the case of binary quadratic forms, and

$$Z_{x,v}(\Phi, s) = \frac{|\Delta_{\alpha_v}|_v^{\frac{s}{4}}}{|\Delta(x)|_v^{\frac{s}{4}}} Z_{\alpha_v,v}(\Phi, s)$$

in the case of binary cubic forms by a similar consideration.

Let $k(x)$ be the field generated by a root of the form, and Δ_x the discriminant of the field $k(x)$. Since $|\Delta(x)| = 1$,

$$\prod_v Z_{x,v}(\Phi, s) = \begin{cases} |\Delta_x|^{-\frac{s}{4}} \prod_v Z_{\alpha_v,v}(\Phi, s), \text{ or} \\ |\Delta_x|^{-\frac{s}{3}} \prod_v Z_{\alpha_v,v}(\Phi, s), \end{cases}$$

where $k(x)$ is the field which corresponds to x and $G_{k_v}\alpha_v = G_{k_v} x$ for all v. We choose $Z_{\alpha_v,v}(\Phi, s)$ as $L_{x,v}(\Phi_v, s)$, and $\frac{\mu(x)}{o(x)}$ as $c(x)$. Then all the assumptions are satisfied.

The computation of r_α reduces to the computation of $(1 - q_v^{-1})(1 - q_v^{-2})|\Delta_{\alpha_v}|_v$ in the case of binary cubic forms and $\text{vol}(K_v \alpha_v)$ in the case of binary quadratic forms. In [11], Datskovsky and Wright proved that if $v \in \mathfrak{M}_f$,

$$(1 - q_v^{-1}) \sum_{\alpha_v} \frac{|\Delta_{\alpha_v}|_v}{o(\alpha_v)} = 1 - q_v^{-3},$$

where $o(\alpha_v)$ is the order of the stabilizer in G_{k_v} in the case of binary cubic forms, and in [9], Datskovsky proved that if $v \in \mathfrak{M}_f$,

$$\sum_{\alpha_v} \text{vol}(K_v \alpha_v) = 1 - q_v^{-2} - q_v^{-3} + q_v^{-4}$$

in the case of binary quadratic forms. This explains the Euler factor in (0.1.2) or in the theorem of Davenport and Heilbronn. In both cases, we have $\zeta_k(2)$ from the residue of the zeta function. In the case of binary cubic forms, it cancels out with $(1 - q_v^{-2})$ in the above consideration. For the details, the reader should see [9]–[11].

As far as the principal part of the zeta function is concerned, Shintani handled two cases: (a) the space of binary cubic forms; (b) the space of binary quadratic forms and the space of positive definite quadratic forms in $n \geq 3$ variables. In this book, we prove principal part formulas for two more cases: (c) the space of quadratic forms in $n \geq 3$ variables (not necessarily positive definite); (d) the space of pairs of ternary quadratic forms. Igusa studied the Euler factors of the single orbit cases and the zeta function is basically a product of Dedekind zeta functions. In case (c), the rightmost pole of the zeta function is simple. So if one can work out the local theory, it is possible to deduce a density theorem for G_k-orbits. However, this may follow easily from Siegel's Mass formula without using zeta function theory. In case (d), the rightmost pole is of order 2, so the filtering process has to be improved in the future.

Besides the single orbit cases, the above cases are all the split irreducible reduced prehomogeneous vector spaces for which the principal part of the global zeta function is known. Since the cases (12), (15) $m = 1, 2$, (17) are prehomogeneous vector spaces of incomplete type, the meromorphic continuation of the zeta function (or the definition of the zeta function) is unknown and it does not follow from (0.3.7). The case (2) $n = 2$ is of incomplete type, but was handled by Shintani in [65].

In many cases, integrals of the form

$$\int_{V_{o_v}} |\Delta(y_v)|_v^s dy_v$$

were computed. For this, the reader should see a survey by Igusa [26] or Kimura–Sato–Zhu [34]. However, in order to apply the filtering process, one has to consider integrals of the form

$$\int_{G_{k_v} x} \Phi_v(y_v)|\Delta(y_v)|_v^s dy_v,$$

where $x \in V_{k_v}^{ss}$ and Φ_v is a Schwartz–Bruhat function. This kind of integrals are not known for many cases. Also we have yet to answer to questions (2), (3).

§0.6 The outline of the general procedure

Before we start the discussion on global zeta functions, it may help the reader to give an outline of the general procedure in this book.

Let (G, V, χ_V) be an irreducible prehomogeneous vector space of complete type. Roughly speaking, we will try to write the principal part of $Z(\Phi, s)$ in terms of special values of similar integrals for different prehomogeneous vector spaces. One can immediately notice that special values of zeta functions are homogeneous in the sense that $Z(\Phi_\lambda, s) = \lambda^{-s} Z(\Phi, s)$. Therefore, let us consider a slightly more abstract but simpler situation.

Let $f_i(\Phi, s)$ be a meromorphic function on the entire plane for $i = 1, \cdots, 4$ such that $f_i(\Phi_\lambda, s) = \lambda^{a_i s + b_i} f_i(\Phi, s)$ where $a_i, b_i \in \mathbb{C}$. Moreover, we assume the existence of an entire function $g_i(\Phi, s)$ such that

$$f_i(\Phi, s) = g_i(\Phi, s) + \int_0^1 \lambda^s f_{i+1}(\Phi_\lambda, c_i) d^\times \lambda \tag{0.6.1}$$

for $i = 1, 2, 3$ where $c_i \in \mathbb{C}$. Then by the homogeneity of $f_i(\Phi, s)$, we get the principal part formula

$$f_i(\Phi, s) = g_i(\Phi, s) + \frac{f_{i+1}(\Phi, c_i)}{s + a_{i+1} c_i + b_{i+1}} \tag{0.6.2}$$

provided that c_i is not a pole of $f_{i+1}(\Phi, s)$. If these formulas are true, by substituting successively, we get

$$(0.6.3)\quad \begin{aligned} f_2(\Phi, c_1) &= g_2(\Phi, c_1) + \frac{f_3(\Phi, c_2)}{c_1 + a_3 c_2 + b_3} \\ &= g_2(\Phi, c_1) + \frac{g_3(\Phi, c_2)}{c_1 + a_3 c_2 + b_3} + \frac{f_4(\Phi, c_3)}{(c_1 + a_3 c_2 + b_3)(c_2 + a_4 c_3 + b_4)}. \end{aligned}$$

Life would be easier if we could prove (0.6.1) for $i = 1$ directly without using other cases. But unfortunately, what we can prove is a formula like (0.6.1) except $f_2(\Phi, c_1)$ is replaced by the right hand side of (0.6.3). Then only after proving that the right hand side of (0.6.3) is $f_2(\Phi, c_1)$ can we determine the principal part of $f_1(\Phi, s)$. For the sake of the filtering process, we only need the rightmost pole. However, this argument shows that we pretty much have to determine all the poles simultaneously in order to determine one of the poles.

Let us be slightly more explicit. Our approach is based on the use of the Poisson summation formula as is often the case in number theory. Consider (0.3.6). Here, the issue is the term $I^0(\Phi)$. Now we face the following two questions.

(1) Can we find a nice stratification of the singular set $V_k \setminus (V_k^{ss} \cup \{0\})$?

(2) If the answer to the question (1) is yes, can we separate the integral $I^0(\Phi)$ according to this stratification?

Kimura and Ozeki determined the orbit decompositions of prehomogeneous vector spaces over an arbitrary algebraically closed field of characteristic zero by representation theoretic methods (see [31]–[33] for example). However, they did not consider the rationality question or the inductive structure of orbits. Question (1) can now be answered by a branch of geometric invariant theory called equivariant Morse theory.

In the early 1980s, Kempf [28], Kirwan [35], and Ness [48] established the notion of equivariant Morse stratification for algebraic situations. It was mainly intended for non-prehomogeneous representations, because their interest was moduli. However, if we restrict ourselves to prehomogeneous vector spaces, it gives us a systematic way of handling orbit decompositions of prehomogeneous vector spaces. For example, it is possible to write a computer program to find all the necessary data. Moreover, the rationality of the inductive structure of rational orbits is guaranteed by Kempf's theorem [28]. This is one of the reasons why we chose the formulation of geometric invariant theory in this book. We discuss invariant theoretic background in §3.2.

Consider question (2). Suppose that we have a 'nice' stratification

$$V_k \setminus \{0\} = V_k^{ss} \coprod \coprod_{i=1}^{l} S_{ik}.$$

The main analytic difficulty is that the theta series $\sum_{x \in S_{ik}} \Phi(gx)$ is not integrable on $G_{\mathbb{A}}^1 = \{g \in G_{\mathbb{A}} \mid |\chi_V(g)| = 1\}$ in general. For $\mathrm{GL}(n)$ let $\mathrm{GL}(n)_{\mathbb{A}}^0 = \{g \in \mathrm{GL}(n)_{\mathbb{A}} \mid |\det g| = 1\}$.

Let $G = \mathrm{GL}(n)$. In order to handle this difficulty for the space of binary cubic forms, Shintani [64] constructed a function $\mathscr{E}(g, w)$ on $G_{\mathbb{A}}^0 \times \mathbb{C}$ with the following properties for the case $n = 2$.

(1) If $f(g^0)$ is an integrable function on $G^0_{\mathbb{A}}/G_k$, $\int_{G^0_{\mathbb{A}}/G_k} f(g^0)\mathscr{E}(g^0,w)dg^0$ becomes a meromorphic function of w and the residue at a certain pole gives the original integral $\int_{G^0_{\mathbb{A}}/G_k} f(g^0)dg^0$.

(2) If $f(g^0)$ is a slowly increasing function, $\int_{G^0_{\mathbb{A}}/G_k} f(g^0)\mathscr{E}(g^0,w)dg^0$ converges absolutely for $\mathrm{Re}(w)$ sufficiently large.

By the property (1), by considering this kind of integrals, we can recover the original integral. By the property (2), if we have a finite number of slowly increasing functions $f_1(g),\cdots,f_l(g)$ on $G_{\mathbb{A}}/G_k$ such that the sum $f(g)=f_1(g)+\cdots+f_l(g)$ is integrable,

$$\int_{G^0_{\mathbb{A}}/G_k} f(g^0)\mathscr{E}(g^0,w)dg^0 = \sum_{i=1}^{l}\int_{G^0_{\mathbb{A}}/G_k} f_i(g^0)\mathscr{E}(g^0,w)dg^0.$$

Using this formula, we can associate a certain distribution to each stratum S_i of $V_k \setminus (V_k^{\mathrm{ss}} \cup \{0\})$.

We prove in Part I of this book that the smoothed version of Eisenstein series satisfies the above properties when the group is a product of $\mathrm{GL}(n)$'s. For this reason, we have to restrict ourselves to such groups. However, there are many interesting cases where the group is a product of $\mathrm{GL}(n)$'s, and we hope this is not such a heavy restriction.

In a way, we are going to establish cancellations of divergent integrals indirectly in terms of the smoothed Eisenstein series. We will see this feature in Chapters 4, 10, 12, 13.

Now we start our discussion on global zeta functions, and we concentrate on the global theory for the rest of this book.

Part I The general theory

Chapter 1 Preliminaries

§1.1 An invariant measure on $\mathrm{GL}(n)$

In this section, we choose an invariant measure on $G = \mathrm{GL}(n)$. For the rest of this book, we assume that k is a number field unless otherwise stated.

For any group G over k, let $X^*(G), X_*(G)$ be the groups of rational characters and of one parameter subgroups (which we abbreviate to '1PS' from now on) respectively. For any split torus $T \cong \mathrm{GL}(1)^h$, T_+ is the subset of $T_{\mathbb{A}}$ which corresponds to $\mathbb{R}_+^h$ by the above identification.

Let $G = \mathrm{GL}(n)$ for the rest of this section. Let $T \subset G$ be the set of diagonal matrices, and $N \subset G$ the set of lower triangular matrices whose diagonal entries are 1. Let N^- be the set of upper triangular matrices with diagonal entries 1. Then $B = TN$ is a Borel subgroup of G. We use the notation

$$(1.1.1) \qquad t = a_n(t_1, \cdots, t_n) = \begin{pmatrix} t_1 & & \\ & \ddots & \\ & & t_n \end{pmatrix}, \ t_1, \cdots, t_n \in \mathbb{A}^\times,$$

$$n_n(u) = \begin{pmatrix} 1 & & 0 \\ & \ddots & \\ u & & 1 \end{pmatrix}, \ u = (u_{ij})_{i>j} \in \mathbb{A}^{\frac{n(n-1)}{2}},$$

for elements in $T_{\mathbb{A}}, N_{\mathbb{A}}$ respectively. We also use the notation $a(t_1, \cdots, t_n), n(u)$ when there is no confusion. Clearly, $T_+ = \{t = a_n(\underline{t}_1, \cdots, \underline{t}_n) \mid t_1, \cdots, t_n > 0\}$ for the above T.

Let $K = \prod_{v \in \mathfrak{M}} K_v$, where $K_v = \mathrm{O}(n)$ if $v \in \mathfrak{M}_{\mathbb{R}}$, $K_v = \mathrm{U}(n)$ if $v \in \mathfrak{M}_{\mathbb{C}}$, and $K_v = \mathrm{GL}(n, o_v)$ if $v \in \mathfrak{M}_f$. The group $G_{\mathbb{A}}$ has the Iwasawa decomposition $G_{\mathbb{A}} = KT_{\mathbb{A}}N_{\mathbb{A}}$. So any element $g \in G_{\mathbb{A}}$ can be written as $g = k(g)t(g)n(u(g))$, where $k(g) \in K$, $t(g) = a_n(t_1(g), \cdots, t_n(g)) \in (\mathbb{A}^\times)^n$, and $u(g) \in \mathbb{A}^{\frac{n(n-1)}{2}}$. Let $\mathfrak{t} = X_*(T) \otimes \mathbb{R}, \mathfrak{t}^* = X^*(T) \otimes \mathbb{R}$, and $\mathfrak{t}^*_{\mathbb{C}} = \mathfrak{t}^* \otimes \mathbb{C}$ etc. For $s \in \mathfrak{t}^*_{\mathbb{C}}$, we define t^s in the usual manner. We identify $\mathfrak{t}^*_{\mathbb{C}}$ with $\mathbb{C}^n$ so that $t^z = |t_1|^{z_1} \cdots |t_n|^{z_n}$ for $z = (z_1, \cdots, z_n) \in \mathbb{C}^n$, $t = a_n(t_1, \cdots, t_n) \in T_{\mathbb{A}}$. Also let $t_v^z = |t_1|_v^{z_1} \cdots |t_n|_v^{z_n}$ for a place $v \in \mathfrak{M}$. The Weyl group of G is isomorphic to the group of permutations of n numbers $\{1, \cdots, n\}$. For two permutations τ_1, τ_2 of n numbers $\{1, \cdots, n\}$, we define the product $\tau_1\tau_2$ by $\tau_1\tau_2(i) = \tau_2(\tau_1(i))$. For a permutation τ, we consider a permutation matrix whose $(i, \tau(i))$-entry is 1 for all i, and denote it by τ also. For a permutation τ and z as above, we define $\tau z \in \mathbb{C}^n$ by $\tau z_i = z_{\tau(i)}$.

Let ρ be half the sum of the weights of N with respect to conjugations by elements of T. Let $du = \prod_{i>j} du_{ij}$. This is an invariant measure on $N_{\mathbb{A}}$. We choose an invariant measure dk on K so that $\int_K dk = 1$. Let $d^\times t = d^\times t_1 \cdots d^\times t_n$ ($t = a_n(t_1, \cdots, t_n)$). Let $db = t^{-2\rho} d^\times t du$ ($t^{-2\rho} = \prod_{i<j} |t_i^{-1} t_j|$). We choose an invariant measure on $G_{\mathbb{A}}$ by $dg = t^{-2\rho} dk d^\times t du$. Let $G^0_{\mathbb{A}} = \{g \in G_{\mathbb{A}} \mid |\det g| = 1\}$, and $c_n(\lambda) = a_n(\underline{\lambda}, \cdots, \underline{\lambda})$. Let dg^0 be the invariant measure on $G^0_{\mathbb{A}}$ such that for any

measurable function $f(g)$ on $G_{\mathbb{A}}$,

$$\int_{G_{\mathbb{A}}} f(g)dg = n\int_0^\infty \int_{G^0_{\mathbb{A}}} f(c_n(\lambda)g^0)d^\times\lambda dg^0,$$

where $d^\times\lambda = \lambda^{-1}d\lambda$. We define invariant measures on $G_{k_v}, K_v, B_{k_v}, N_{k_v}, T_{k_v}$ similarly, and denote them by $dg_v, dk_v, db_v, du_v, d^\times t_v$ respectively. Then $du = |\Delta_k|^{-\frac{n(n-1)}{4}}\prod_v du_v$, and $d^\times t = \mathfrak{C}_k^{-n}\prod_v d^\times t_v$.

Let

$$(1.1.2) \qquad \phi(s) = \frac{Z_k(s)}{Z_k(s+1)},\ \phi_n(s) = \prod_{i=1}^n \phi(s+i-1) \text{ for } n \geq 2.$$

We define $\varrho = \mathrm{Res}_{s=1}\,\phi(s)$. Let

$$(1.1.3) \qquad \mathfrak{V}_n = \frac{Z_k(2)\cdots Z_k(n)}{\mathfrak{R}_k^{n-1}}.$$

The constant $\mathfrak{V}_n$ is the volume of $G^0_{\mathbb{A}}/G_k$ with respect to the measure dg^0. We define $\mathfrak{V}_1 = 1$ for convenience.

Let $\Omega \subset B_{\mathbb{A}}$ be a compact set. For a constant $\eta > 0$, we define

$$T_\eta = \{a_n(t_1,\cdots,t_n) \in T_{\mathbb{A}} \mid |t_i t_{i+1}^{-1}| \geq \eta\},$$

and $T_{\eta+} = T_+ \cap T_\eta$. Let $T^0_\eta = G^0_{\mathbb{A}} \cap T_\eta$, and $T^0_{\eta+} = T_+ \cap T^0_\eta$. Then for suitable Ω, η, $\mathfrak{S} = KT_{\eta+}\Omega$ surjects to $G_{\mathbb{A}}/G_k$. Let $\mathfrak{S}^0 = \mathfrak{S} \cap G^0_{\mathbb{A}}$. Then $\mathfrak{S}^0$ also surjects to $G^0_{\mathbb{A}}/G_k$. We call $\mathfrak{S}$ (resp. $\mathfrak{S}^0$) a Siegel domain for $G_{\mathbb{A}}/G_k$ (resp. $G^0_{\mathbb{A}}/G_k$). There also exists a compact set $\widehat{\Omega} \subset G_{\mathbb{A}}$ such that $\mathfrak{S} \subset \widehat{\Omega}T_{\eta+}, \mathfrak{S}^0 \subset \widehat{\Omega}T^0_{\eta+}$. Note that since Ω is a compact set, $|t_i(g)t_{i+1}(g)^{-1}|$ is bounded below for all i for $g \in \mathfrak{S}$.

§1.2 Some adelic analysis

For later purposes, we prove some elementary estimates of various theta series.

Let m be a positive integer. For $x = (x_1,\cdots,x_m) \in (k_\infty)^m$, we define

$$\|x\|_m = \sum_i \left(\sum_{v\in\mathfrak{M}_{\mathbb{R}}} |x_i|_v^{[k:\mathbb{Q}]} + \sum_{v\in\mathfrak{M}_{\mathbb{C}}} |x_i|_v^{\frac{[k:\mathbb{Q}]}{2}} \right).$$

The following lemma follows from the integral test.

Lemma (1.2.1) *Let m be a positive integer, $N > m$, and $\nu \in \mathbb{R}_+$. Let $L \subset k_\infty$ be a lattice. Then there exists a constant $C(N,m,L)$ such that*

$$(1) \qquad \sum_{x\in L^m} (1+\nu\|x\|_m)^{-N} \leq C(N,m,L)\sup(1,\nu^{-m}),$$

$$(2) \qquad \sum_{x\in L^m\setminus\{0\}} (1+\nu\|x\|_m)^{-N} \leq C(N,m,L)\inf(\nu^{-m},\nu^{-N}).$$

If $m \leq N' \leq N$, then $\inf(\nu^{-m}, \nu^{-N}) \leq \nu^{-N'}$. Therefore, if a function $f(\nu)$ is bounded by a constant multiple of functions of the form (1) for all $N > m$, then $f(\nu) \ll \nu^{-N}$ for all $N \geq m$.

When we consider estimates on Siegel domains, a typical situation is that we have an action of elements which are products of diagonal elements and elements from a fixed compact set. Weil proved many useful statements in [77] to prove certain estimates in this kind of situation. The following lemma is Lemma 4 in [77, p. 193].

Lemma (1.2.2) *For any sequence $(a_l)_{l\in\mathbb{N}}$ of real numbers $a_l > 0$, there exists a Schwartz–Bruhat function $\Phi \in \mathscr{S}(\mathbb{R})$ such that*

$$\inf_{l\in\mathbb{N}}(a_l(1+|x|)^{-l}) \leq \Phi(x)$$

for all $x \in \mathbb{R}$.

Proof. Let $f(x) = \inf_{l\in\mathbb{N}}(a_l(1+|x|)^{-l})$. We choose a function $0 \leq g \in C^\infty(\mathbb{R})$ so that it is supported on $[-1,1]$ and $\int_{\mathbb{R}} g(x)dx = 1$. Let h be the convolution $h = f * g$. Then

$$f(x-1) \geq h(x) \geq f(x+1) \text{ for } x \geq 1,$$
$$f(x-1) \leq h(x) \leq f(x+1) \text{ for } x \leq -1.$$

Let $D = \frac{d}{dx}$. We have $D^p h = f * D^p g$ for all $p \geq 0$, so $|x^n D^p h(x)|$ is bounded for all $n, p \geq 0$. Therefore, $h \in \mathscr{S}(\mathbb{R})$. Choose $h_0 \in \mathscr{S}(\mathbb{R})$ so that $h_0(x) \geq 0$ for all x and $h_0(x) \geq a_0$ on $[-2,2]$. Then

$$f(x) \leq h(x-1) + h(x+1) + h_0(x).$$

Q.E.D.

The following lemma is a special case of Lemma 5 in [77, p. 194].

Proposition (1.2.3) *Let $C \subset \mathrm{GL}(V_{\mathbb{A}})$ be a compact set, and $\Phi \in \mathscr{S}(V_{\mathbb{A}})$. Then there exists $\Psi \geq 0$ such that $|\Phi(gx)| \leq \Psi(x)$ for all $g \in C, x \in V_{\mathbb{A}}$.*

Proof. We can assume that Φ has a product form $\Phi = \otimes_v \Phi_v$. Let $\Phi_f = \otimes_{v\in\mathfrak{M}_f}\Phi_v$. Since C is a compact set, there exists a compact open subgroup $L \subset V_f$ such that $\Phi_f(gx) = 0$ for $g \in C, x_f \in V_f \setminus L$. Therefore, there exists $\Psi_f \in \mathscr{S}(V_f)$ such that $|\Phi_f(gx)| \leq \Psi_f(x)$ for $g \in C, x_f \in V_f \setminus L$.

We consider infinite places. We identify $V_\infty \cong \mathbb{R}^n$ for some n, and write $x_\infty = (x_1, \cdots, x_n)$. Let $r(x_\infty) = \sum_{i=1}^n x_i^2$. We choose a_l in (1.2.2) as follows.

$$a_l = \sup_{x\in V_{\mathbb{A}}, g\in C}(1+r(x_\infty))^l|\Phi_\infty(g_\infty x_\infty)|.$$

Since C is compact, $\{g^{-1} \mid g \in C\}$ is also a compact set. Therefore, $a_l < \infty$ for all l. Then by (1.2.2), there exists a Schwartz–Bruhat function $\widetilde{\Psi}_\infty \in \mathscr{S}(\mathbb{R})$ such that

$$\inf_{l\in\mathbb{N}} a_l(1+r(x_\infty))^{-l} \leq \widetilde{\Psi}_\infty(r(x_\infty)).$$

It is easy to see that

$$|\Phi_\infty(g_\infty x_\infty)| \leq \widetilde{\Psi}_\infty(r(x_\infty))$$

for $x \in V_{\mathbb{A}}, g \in C$.

Let $\Psi_\infty(x) = \widetilde{\Psi}_\infty(r(x_\infty))$. Then $\Psi = \Psi_\infty \Psi_f$ satisfies the condition of our lemma.

Q.E.D.

In particular, for any $\Phi \in \mathscr{S}(V_{\mathbb{A}})$, there exists $\Psi \geq 0$ such that $|\Phi(x)| \leq \Psi(x)$ for all $x \in V_{\mathbb{A}}$.

Now we can easily prove the following two lemmas using (1.2.2), (1.2.3).

Lemma (1.2.4) *For any $\Phi \in \mathscr{S}(\mathbb{R})$ and a positive integer n, there exists $0 \leq \Psi \in \mathscr{S}(\mathbb{R})$ such that $\Psi(-x) = \Psi(x)$, $|\Phi(x)|^{\frac{1}{n}} \leq \Psi(x)$, and Ψ is decreasing on $[0, \infty)$.*

Proof. We can assume that $\Phi(-x) = \Phi(x)$. In (1.2.2), we choose $a_l = \sup_{x\in\mathbb{R}}(1 + |x|)^l |\Phi(x)|^{\frac{1}{n}}$. Then there exists $\Psi_1 \in \mathscr{S}(\mathbb{R})$ such that

$$\inf_{l\in\mathbb{N}}((1+|x|)^{-l} \sup_{y\in\mathbb{R}}(1+|y|)^l |\Phi(y)|^{\frac{1}{n}}) \leq \Psi_1(x)$$

for all x. But

$$\Phi(x)|^{\frac{1}{n}} \leq \inf_{l\in\mathbb{N}}((1+|x|)^{-l} \sup_{x\in\mathbb{R}}(1+|y|)^l |\Phi(y)|^{\frac{1}{n}})$$

for all $x \in \mathbb{R}$.

Since $\Psi_1 \in \mathscr{S}(\mathbb{R})$,

$$\Psi_1(x) = -\int_x^\infty \frac{d}{dy}\Psi_1(y)dy.$$

We choose $\Psi_2 \in \mathscr{S}(\mathbb{R})$ so that $|\frac{d}{dx}\Psi_1(x)| \leq \Psi_2(x)$ for all x. We define

$$\Psi_3(x) = \int_x^\infty \frac{d}{dy}\Psi_2(y)dy$$

for $x \geq 1$. Then $\Psi_3(x) \geq \Psi_1(x)$ for $x \geq 1$ and Ψ_3 is decreasing on $[1, \infty)$. Now we can extend Ψ_3 to an even Schwartz–Bruhat function on $\mathbb{R}$ decreasing on $[0, \infty)$. Then $\Psi_3(x) \geq |\Phi(x)|^{\frac{1}{n}}$ for $x \geq 1$ and $x \leq -1$. Since $|\Phi(x)|^{\frac{1}{n}}$ is bounded on $[-1, 1]$, we can choose Ψ_3 so that $\Psi_3(x) \geq |\Phi(x)|^{\frac{1}{n}}$ on $[-1, 1]$ also (for example, add a constant multiple of the function e^{-x^2}).

Q.E.D.

Lemma (1.2.5) *Suppose $\Phi \in \mathscr{S}(\mathbb{A}^n)$. Then there exist Schwartz–Bruhat functions $\Phi_1, \cdots,$
$\Phi_n \geq 0$ such that $|\Phi(x_1, \cdots, x_n)| \leq \Phi_1(x_1) \cdots \Phi_n(x_n)$.*

Proof. We can assume that Φ has a product form $\Phi = \otimes_v \Phi_v$. It is easy to see that the finite part Φ_f of Φ is bounded by a product of Schwartz–Bruhat functions on $\mathbb{A}_f$. So we consider infinite places. Since the argument is similar, we only consider real places.

Since $\mathrm{O}(n, \mathbb{R})$ is a compact group, there exists $\Psi_1 \in \mathscr{S}(\mathbb{R})$ such that $\Psi_1 \geq 0$ and

$$|\Phi(x_1, \cdots, x_n)| \leq \Psi_1(\sqrt{x_1^2 + \cdots + x_n^2})$$

for all $x = (x_1, \cdots, x_n)$.

We choose $\Psi_2 \in \mathscr{S}(\mathbb{R})$ so that $\Psi_2 \geq 0$, Ψ_2 is decreasing on $[0,\infty)$, and $\Psi_1(r) \leq \Psi_2(r)^n$ for all $r \in \mathbb{R}$. Then

$$\Psi_1(\sqrt{x_1^2+\cdots+x_n^2}) \leq \Psi_2(\sqrt{x_1^2+\cdots+x_n^2})^n \leq \Psi_2(x_1)\cdots\Psi_2(x_n).$$

Q.E.D.

In the following three lemmas, we consider the following situation. Let V_{ij_i} be a vector space of dimension $m_{ij_i} \geq 0$ for $i=1,2,\ j_i=1,\cdots,k_i$. Let $V=\oplus_{j_1=1}^{k_1} V_{1j_1} \oplus \oplus_{j_2=1}^{k_2} V_{2j_2}$. We write an element of V in the form (y,z), where $y=(y_1,\cdots,y_{k_1}) \in \oplus_{j_1=1}^{k_1} V_{1j_1}$ etc. Let $\nu_{i1},\cdots,\nu_{ik_i} \in \mathbb{R}_+$ for $i=1,2$. Let $\nu_i = (\underline{\nu}_{i1},\cdots,\underline{\nu}_{ik_i})$ for $i=1,2$. We define $\nu_1 y = (\underline{\nu}_{11}y_1,\cdots,\underline{\nu}_{1k_1}y_{k_1})$ for $y \in \oplus_{j_1=1}^{k_1} V_{1j_1\mathbb{A}}$. We define $\nu_2 z$ similarly. Let $\nu=(\nu_1,\nu_2)$. We define $\nu(y,z)=(\nu_1 y,\nu_2 z)$. We identify $V_{ij_i} \cong k^{m_{ij_i}}$, and write $y_1=(y_{11},\cdots,y_{1m_{11}})$ etc.

Lemma (1.2.6) *Let Φ be a Schwartz–Bruhat function on $V_{\mathbb{A}}$. Then for any $N_{ij} \geq m_{ij}$,*

$$\sum_{\substack{1\leq j_1\leq k_1\\ 1\leq j_2\leq k_2}} \sum_{\substack{y_{j_1}\in V_{1j_1k}\setminus\{0\}\\ z_{j_2}\in V_{2j_2k}}} |\Phi(\nu_1 y,\nu_2 z)| \ll \prod_{1\leq j_1\leq k_1} \nu_{1j_1}^{-N_{1j_1}} \prod_{1\leq j_2\leq k_2} \sup(1,\nu_{2j_2}^{-m_{2j_2}}).$$

Proof. There exist a lattice $L \subset k_\infty$, and $0 \leq \Phi_\infty \in \mathscr{S}(V_\infty)$ such that

$$\sum_{\substack{1\leq j_1\leq k_1\\ 1\leq j_2\leq k_2}} \sum_{\substack{y_{j_1}\in V_{1j_1k}\setminus\{0\}\\ z_{j_2}\in V_{2j_2k}}} |\Phi(\nu_1 y,\nu_2 z)| \ll \sum_{\substack{1\leq j_1\leq k_1\\ 1\leq j_2\leq k_2}} \sum_{\substack{y_{j_1}\in V_{1j_1L}\setminus\{0\}\\ z_{j_2}\in V_{2j_2L}}} \Phi_\infty(\nu_1 y,\nu_2 z),$$

where V_{1j_1L}, V_{2j_2L} are the subsets of points whose coordinates belong to L.

Let M be a sufficiently large number. Since Φ_∞ is rapidly decreasing,

$$\begin{aligned}&\sum_{\substack{1\leq j_1\leq k_1\\ 1\leq j_2\leq k_2}} \sum_{\substack{y_{j_1}\in V_{1j_1L}\setminus\{0\}\\ z_{j_2}\in V_{2j_2L}}} \Phi_\infty(\nu_1 y,\nu_2 z)\\ &\ll \sum_{\substack{1\leq j_1\leq k_1\\ 1\leq j_2\leq k_2}} \sum_{\substack{y_{j_1}\in V_{1j_1L}\setminus\{0\}\\ z_{j_2}\in V_{2j_2L}}} \prod_{1\leq j_1\leq k_1} (1+\nu_{1j_1}\|y_{j_1}\|_{m_{1j_1}})^{-M}\\ &\qquad\times \prod_{1\leq j_2\leq k_2} (1+\nu_{2j_2}\|z_{j_2}\|_{m_{2j_2}})^{-M}.\end{aligned}$$

Then (1.2.6) is clear from (1.2.1) and the remark after that.

Q.E.D.

Lemma (1.2.7) *Let $\Phi \in \mathscr{S}(V_{\mathbb{A}})$. Let $p_{j_2}=p_{j_2}(\nu,y,z) \in \mathbb{A}$ be a function which does not depend on $z_{j_2},\cdots,z_{k_2}$ for $j_2=1,\cdots,k_2$. Then there exists a Schwartz–Bruhat function $\Phi_1 \geq 0$ such that*

$$\begin{aligned}&\sum_{\substack{1\leq j_1\leq k_1\\ 1\leq j_2\leq k_2}} \sum_{\substack{y_{j_1}\in V_{1j_1k}\setminus\{0\}\\ z_{j_2}\in V_{2j_2k}}} |\Phi(\nu_1 y,\underline{\nu}_{21}(z_1+p_1),\cdots,\underline{\nu}_{2k_2}(z_{k_2}+p_{k_2}))|\\ &\ll \sum_{\substack{1\leq j_1\leq k_1\\ 1\leq j_2\leq k_2}} \sum_{\substack{y_{j_1}\in V_{1j_1k}\setminus\{0\}\\ z_{j_2}\in V_{2j_2k}}} \Phi_1(\nu_1 y,\nu_2 z).\end{aligned}$$

Proof. We may assume that $\Phi \geq 0$. We fix a non-degenerate bilinear form $[\ ,\ ]_{j_2}$ on V_{2j_2} defined over k for $j_2 = 1, \cdots, k_2$. By the Poisson summation formula,

$$\sum_{z_{k_2} \in V_{2k_2 k}} \Phi(\nu_1 y, \underline{\nu}_{21}(z_1 + p_1), \cdots, \underline{\nu}_{2k_2}(z_{k_2} + p_{k_2}))$$

is equal to $\nu_{2k_2}^{-m_{2k_2}}$ times the following sum

$$\sum_{z_{k_2} \in V_{2k_2 k}} \Psi(\nu_1 y, \underline{\nu}_{21}(z_1 + p_1), \cdots, \underline{\nu}_{2k_2-1}(z_{k_2-1} + p_{k_2-1}), \underline{\nu}_{2k_2}^{-1} z_{k_2}) < -z_{k_2} p_{k_2} >,$$

where $\Psi(y, z)$ is the Fourier transform of Φ with respect to z_{k_2} and $[\ ,\ ]_{k_2}$. We can find a Schwartz–Bruhat function $\Psi_1 \geq 0$ so that $\Psi_1 \geq |\Psi|$. Then the above sum is bounded by a constant multiple of

$$\nu_{2k_2}^{-m_{2k_2}} \sum_{z_{k_2} \in V_{2k_2 k}} \Psi_1(\nu_1 y, \underline{\nu}_{21}(z_1 + p_1), \cdots, \underline{\nu}_{2k_2-1}(z_{k_2-1} + p_{k_2-1}), \underline{\nu}_{2k_2}^{-1} z_{k_2}).$$

By applying the Poisson summation formula successively, there exists a Schwartz–Bruhat function $\Psi_2 \geq 0$ such that

$$\begin{aligned} &\sum_{\substack{1 \leq j_1 \leq k_1 \\ 1 \leq j_2 \leq k_2}} \sum_{\substack{y_{j_1} \in V_{1j_1 k} \setminus \{0\} \\ z_{j_2} \in V_{2j_2 k}}} |\Phi(\nu_1 y, \underline{\nu}_{21}(z_1 + p_1), \cdots, \underline{\nu}_{2k_2}(z_{k_2} + p_{k_2}))| \\ &\ll \nu_{21}^{-m_{21}} \cdots \nu_{2k_2}^{-m_{2k_2}} \sum_{\substack{1 \leq j_1 \leq k_1 \\ 1 \leq j_2 \leq k_2}} \sum_{\substack{y_{j_1} \in V_{1j_1 k} \setminus \{0\} \\ z_{j_2} \in V_{2j_2 k}}} \Psi_2(\nu_1 y, \nu_2^{-1} z). \end{aligned}$$

Let Ψ_3 be the Fourier transform of Ψ_2 with respect to $[\ ,\ ]_1, \cdots, [\ ,\ ]_{k_2}$. Then if $\Phi_1 \geq |\Psi_3|$, Φ_1 satisfies the statement of (1.2.7).

Q.E.D.

The following lemma follows from (1.2.6) and (1.2.7).

Lemma (1.2.8) *Let* $\Phi, y, z, \nu = (\nu_1, \nu_2), p_{j_2}$, $j_2 = 1, \cdots, k_2$ *be as in* (1.2.4). *Then for any* $N_{1j_i} \geq m_{1j_i}$, $j_1 = 1, \cdots, k_1$,

$$\begin{aligned} &\sum_{\substack{1 \leq j_1 \leq k_1 \\ 1 \leq j_2 \leq k_2}} \sum_{\substack{y_{j_1} \in V_{1j_1 k} \setminus \{0\} \\ z_{j_2} \in V_{2j_2 k}}} |\Phi(\nu_1 y, \underline{\nu}_{21}(z_1 + p_1), \cdots, \underline{\nu}_{2k_2}(z_{k_2} + p_{k_2}))| \\ &\ll \prod_{1 \leq j_1 \leq k_1} \nu_{1j_1}^{-N_{1j_1}} \prod_{1 \leq j_2 \leq k_2} \sup(1, \nu_{2j_2}^{-m_{2j_2}}). \end{aligned}$$

Chapter 2 Eisenstein series on GL(n)

This chapter and the next chapter are the technical heart of this book. In this chapter, we estimate the Eisenstein series on GL(n) for the Borel subgroup B. This reduces to the estimate of Whittaker functions on GL(n). The estimate of the Whittaker functions consists of two problems. One is at finite places, and the other is at infinite places. We prove the explicit formula for the p-adic Whittaker functions in §2.3 following Shintani [66]. Even though some estimate of Whittaker functions are known, the uniformity of the estimate is a big issue in this book, because we intend to consider contour integrals. Therefore, it is necessary to prove an estimate which is of polynomial growth with respect to the parameter of the Lie algebra. For this purpose, we generalize Shintani's approach in [64], which is the naive use of integration by parts. As far as GL(3) is concerned, Bump's estimate [5] is the best possible estimate. Our estimate, even though it is enough for our purposes for the moment, may not be the optimum estimate, and we may have to prove a better estimate in the future to handle more prehomogeneous vector spaces.

§2.1 The Fourier expansion of automorphic forms on GL(n)

In this section, we review the notion of the Fourier expansion of automorphic functions on GL(n) following Piatecki–Shapiro in [51].

We assume that $G = \mathrm{GL}(n)$ in this section. Let $K \subset G_{\mathbb{A}}$ be the maximal compact subgroup which we defined in §1.1. We use the notation

$$\begin{aligned}\psi_\alpha(n(u)) &= <\alpha_1 u_{21}> \cdots <\alpha_{n-1}u_{nn-1}> \ \text{for}\ \alpha \in (\mathbb{A}^\times)^{n-1}, n(u) \in N_{\mathbb{A}},\\ \psi_{\alpha,v}(n(u)) &= <\alpha_1 u_{21}>_v \cdots <\alpha_{n-1}u_{nn-1}>_v \ \text{for}\ \alpha \in (k_v^\times)^{n-1}, n(u) \in N_{k_v}.\end{aligned}$$

If $\alpha \in k^{n-1}$, ψ_α is a character of $N_{\mathbb{A}}/N_k$.

Let f be an automorphic form on $G_{\mathbb{A}}/G_k$, and $\alpha = (\alpha_1, \cdots, \alpha_{n-1}) \in k^{n-1}$.

Definition (2.1.1)

$$f_\alpha(g) = f_{\alpha_1,\cdots,\alpha_{n-1}}(g) = \int_{N_{\mathbb{A}}/N_k} f(gn(u))\psi_\alpha(n(u))du.$$

Definition (2.1.2) *For $1 \le i \le n-1$, we define*

$$\begin{aligned}\Gamma_i &= \left\{ \begin{pmatrix} A & 0 \\ 0 & I_{n-i}\end{pmatrix} \middle| A \in \mathrm{GL}(i)_k \right\},\\ \Gamma_i^\infty &= \left\{ \begin{pmatrix} A & 0 & 0 \\ c & d & 0 \\ 0 & 0 & I_{n-i}\end{pmatrix} \middle| A \in \mathrm{GL}(i-1)_k,\ c \in k^{i-1},\ d \in k^\times \right\},\end{aligned}$$

where I_{n-i} is the unit matrix of dimension $n-i$.

One difficulty of harmonic analysis on GL(n) is that the unipotent radical N of the Borel subgroup is not abelian. So we cannot directly use the the Fourier expansion of functions on affine spaces over $\mathbb{A}$. However, in the case of GL(n), it turns out that one can reduce the consideration to integrations with respect to a character of $N_{\mathbb{A}}/N_k$ as the following proposition shows.

Proposition (2.1.3) *Let $f(g)$ be a continuous function on $G_{\mathbb{A}}/G_k$. Then*

$$f(g) = \sum_{\alpha \in k^{n-1}} \sum_{\substack{\gamma_i \in \Gamma_i/\Gamma_i^\infty \text{ if } \alpha_i \neq 0 \\ \gamma_i = 1 \text{ if } \alpha_i = 0}} f_\alpha(g\gamma_{n-1}\cdots\gamma_1)$$

if the right hand side converges absolutely and locally uniformly.

Proof. For $u_n = (u_{n1}, \cdots, u_{nn-1}) \in \mathbb{A}^{n-1}$, we define

$$n^{(1)}(u_n) = \begin{pmatrix} 1 & & & \\ & 1 & & \\ & & \ddots & \\ u_{n1} & \cdots & u_{nn-1} & 1 \end{pmatrix}.$$

Let $N_1 \subset N$ be the subgroup which consists of elements of the form $n^{(1)}(u_n)$. We define a measure du_n on $N_{1\mathbb{A}}$ in the usual manner. For $\alpha^{(1)} = (\alpha_{n1}, \cdots, \alpha_{nn-1}) \in k^{n-1}$, we define

$$f^1_{\alpha^{(1)}}(g) = \int_{N_{1\mathbb{A}}/N_{1k}} f(gn^{(1)}(u_n)) < \alpha_{n1}u_{n1} + \cdots + \alpha_{nn-1}u_{nn-1} > du_n.$$

Since $N_1 \subset N$ is an abelian subgroup, f has the following Fourier expansion with respect to N_1:

$$f(g) = \sum_{\alpha^{(1)} \in k^{n-1}} f^1_{\alpha^{(1)}}(g).$$

If we just assume the continuity of f, this equality is true for almost all g, and the convergence may not be uniform. However, each Fourier coefficient $f^1_{\alpha^{(1)}}(g)$ is a continuous function again.

Let $f_{1,\alpha_{nn-1}}(g) = f^1_{(0,\cdots,0,\alpha_{nn-1})}(g)$. Consider the matrix $\gamma = \begin{pmatrix} A & 0 \\ 0 & 1 \end{pmatrix}$ where $A \in \mathrm{GL}(n-1)_k$. Then

$$\begin{aligned} f_{1,\alpha_{nn-1}}(g\gamma) &= \int_{N_{1\mathbb{A}}/N_{1k}} f(g\gamma n^{(1)}(u_n)) < \alpha_{nn-1}u_{nn-1} > du_n \\ &= \int_{N_{1\mathbb{A}}/N_{1k}} f(g\gamma n^{(1)}(u_n)\gamma^{-1}) < \alpha_{nn-1}u_{nn-1} > du_n, \end{aligned}$$

because f is a function on $G_{\mathbb{A}}/G_k$. An easy computation shows that $\gamma n^{(1)}(u_n)\gamma^{-1} = n^{(1)}(u'_n)$ where $u'_n = u_n A^{-1}$. Therefore, $u_{nn-1} = a_{1n-1}u'_{n1} + \cdots + a_{n-1n-1}u'_{nn-1}$ where ${}^t(a_{1n-1}, \cdots, a_{n-1n-1})$ is the $(n-1)$-st column of A. This implies that $f_{1,\alpha_{nn-1}}(g\gamma) = f^1_{(a_{1n-1}\alpha_{nn-1}, \cdots, a_{n-1n-1}\alpha_{nn-1})}(g)$. Hence,

$$f(g) = \sum_{\substack{\gamma_{n-1} \in \Gamma_{n-1}/\Gamma^\infty_{n-1} \\ \alpha_{nn-1} \in k^\times}} f_{1,\alpha_{nn-1}}(g\gamma_{n-1}) + f_{1,0}(g).$$

For $u_{n-1} = (u_{n-11}, \cdots, u_{n-1n-2}) \in \mathbb{A}^{n-2}$, we define

$$n^{(2)}(u_{n-1}) = \begin{pmatrix} 1 & & & & \\ & \ddots & & & \\ & & 1 & & \\ u_{n-11} & \cdots & u_{n-1n-2} & 1 & \\ & & & & 1 \end{pmatrix}.$$

Let $N^2 \subset N$ be the subgroup which consists of elements of the form $n^{(2)}(u_{n-1})$. We define a measure du_{n-1} on $N_{2\mathbb{A}}$ in the usual manner. Let

$$f^2_{\alpha_{nn-1},(\alpha_{n-11},\cdots,\alpha_{n-1n-2})}(g) = \int_{N_{2\mathbb{A}}/N_{2k}} f_{1,\alpha_{nn-1}}(gn^{(2)}(u_{n-1})) \times < \alpha_{n-11}u_{n-11} + \cdots + \alpha_{n-1n-2}u_{n-1n-2} > du_{n-1}.$$

Since $N_{2k} \subset \Gamma^{\infty}_{n-1}$, $f_{1,\alpha_{nn-1}}(gu_{n-1}) = f_{1,\alpha_{nn-1}}(g)$ for $u_{n-1} \in k^{n-2}$. So the Fourier expansion of $f_{1,\alpha_{nn-1}}(g)$ with respect to N_2 is as follows.

$$f_{1,\alpha_{nn-1}}(g) = \sum_{\alpha_{n-11},\cdots,\alpha_{n-1n-2} \in k} f^2_{\alpha_{nn-1},(\alpha_{n-11},\cdots,\alpha_{n-1n-2})}(g).$$

As in the previous step, we have

$$f_{1,\alpha_{nn-1}}(g) = \sum_{\substack{\gamma_{n-2} \in \Gamma_{n-2}/\Gamma^{\infty}_{n-2} \\ \alpha_{n-1n-2} \in k^{\times}}} f_{2,(\alpha_{n-1n-2},\alpha_{nn-1})}(g\gamma_{n-2}) + f_{2,(0,\alpha_{nn-1})}(g),$$

where

$$f_{2,(\alpha_{n-1n-2},\alpha_{nn-1})}(g) = f^2_{\alpha_{nn-1},(0,\cdots,0,\alpha_{n-1n-2})}(g).$$

The equality (2.1.3) for almost all g follows by continuing this process. Since we are assuming that the right hand side of (2.1.3) converges absolutely and locally uniformly, both sides of (2.1.3) are continuous functions. Therefore, the equality (2.1.3) is true for all g.

Q.E.D.

We are going to apply this proposition to Eisenstein series on GL(n) in §§2.3–2.4. It turns out that the assumption in (2.1.3) is satisfied for Eisenstein series on GL(n). Therefore, we do not have a problem with the convergence of (2.1.3) for our purposes.

Next, we consider the representative $\gamma_{n-1} \cdots \gamma_1$ more explicitly.

Definition (2.1.4) *For a permutation τ, we define*

$$N_\tau = N \cap \tau N \tau^{-1}, \; N^-_\tau = N \cap \tau N^- \tau^{-1}.$$

Note that $N_\tau \cap N^-_\tau = \{1\}$.

Definition (2.1.5) *We call $l(\tau) = \dim N/N_\tau$ the length of τ.*

For a permutation τ, let

$$I_\tau = \{(i,j) \mid 1 \le j < i \le n,\ \tau(i) < \tau(j)\}.$$

For a pair of numbers (i,j), we define $\tau(i,j) = (\tau(i), \tau(j))$. An element $n(u) \in N$ belongs to N_τ (resp. N_τ^-) if $u_{ij} = 0$ for $(i,j) \in I_\tau$ (resp. $(i,j) \notin I_\tau$). Therefore, $l(\tau) = \# I_\tau$.

We use induction with respect to $l(\tau)$ very often in Chapters 2 and 3. For that purpose, we have to know how $l(\tau)$ behaves when τ is a product of two elements.

Lemma (2.1.6) *Suppose* $\tau = \tau_1\tau_2$. *Then*
(1) $l(\tau) \le l(\tau_1) + l(\tau_2)$ *if* $\tau = \tau_1\tau_2$,
(2) $l(\tau) = l(\tau_1) + l(\tau_2)$ *if and only if* $I_\tau = I_{\tau_1} \coprod \tau_1^{-1} I_{\tau_2}$.

Proof. Suppose $\tau = \tau_1\tau_2$. Then

$$\begin{aligned} l(\tau) &\le \dim N/N \cap \tau_1 N \tau_1^{-1} \cap \tau N \tau^{-1} \\ &= l(\tau_1) + \dim N_{\tau_1}/N_{\tau_1} \cap \tau N_{\tau_2} \tau^{-1}. \end{aligned}$$

There is an imbedding of $N_{\tau_1}/N_{\tau_1} \cap \tau N_{\tau_2}\tau^{-1}$ into $\tau_1 N \tau_1^{-1}/\tau_1 N_{\tau_2} \tau_1^{-1}$. This proves that $l(\tau) \le l(\tau_1) + l(\tau_2)$ in general, and $l(\tau) = l(\tau_1) + l(\tau_2)$ if and only if $N_\tau \subset N_{\tau_1}$ and N_{τ_1}/N_τ is isomorphic to $\tau_1 N \tau_1^{-1}/\tau_1 N_{\tau_2}\tau_1^{-1}$. The first condition is equivalent to the condition $I_{\tau_1} \subset I_\tau$. If this is true, the second condition is equivalent to the condition $I_\tau = I_{\tau_1} \cup \tau_1^{-1} I_{\tau_2}$. But since $l(\tau) \le l(\tau_1) + l(\tau_2)$, if this happens, the union has to be a disjoint union.

Q.E.D.

Lemma (2.1.7)
(1) *For all* τ, $N = N_\tau^- N_\tau$.
(2) *Suppose that* $\tau = \tau_1\tau_2$ *and* $l(\tau) = l(\tau_1) + l(\tau_2)$. *Then* $N_\tau^- = N_{\tau_1}^-(\tau_1 N_{\tau_2}^- \tau_1^{-1})$.

Proof. If $l(\tau) = 1$, (1) is clear. Suppose that $\tau = \tau_1\tau_2$, $\tau_1, \tau_2 \ne 1$, and $l(\tau) = l(\tau_1) + l(\tau_2)$. Let $n \in N$. By induction on $l(\tau)$, we can assume that $n = n_1 n_2$, $n_2 = \tau_1 n_3 \tau_1^{-1}$, and $n_3 = n_4 n_5$ where $n_1 \in N_{\tau_1}^-$, $n_2 \in N_{\tau_1}$, $n_4 \in N_{\tau_2}^-$, and $n_5 \in N_{\tau_2}$. Since $I_\tau = I_{\tau_1} \coprod \tau_1^{-1} I_{\tau_2}$, $N_{\tau_1}^-, \tau_1 N_{\tau_2}^- \tau_1^{-1} \subset N_\tau^-$. Therefore, $n_1(\tau_1 n_4 \tau_1^{-1}) \in N_\tau^-$. This implies that $\tau_1 n_5 \tau_1^{-1} \in N$. Since $n_5 \in N_{\tau_2} \subset \tau_2 N \tau_2^{-1}$, $\tau_1 n_5 \tau_1^{-1} \in \tau_1\tau_2 N \tau_2^{-1}\tau_1^{-1} = \tau N \tau^{-1}$. So $\tau_1 n_5 \tau_1^{-1} \in N_\tau$. If $n \in N_\tau^-$, $\tau_1 n_5 \tau_1^{-1} \in N_\tau^- \cap N_\tau = 1$. Therefore, $N_\tau^- = N_{\tau_1}^-(\tau_1 N_{\tau_2}^- \tau_1^{-1})$. This proves the lemma.

Q.E.D.

Remark (2.1.8) If $\tau = \tau_1\tau_2\tau_3$ and $l(\tau) = l(\tau_1) + l(\tau_2) + l(\tau_3)$, then

$$l(\tau_1\tau_2) = l(\tau_1) + l(\tau_2),\ l(\tau_2\tau_3) = l(\tau_2) + l(\tau_3),$$

using (1) of the above lemma.

In (2.1.9), (2.1.10), we consider a subset $I = \{i_1, \cdots, i_h\} \subset \{1, \cdots, n-1\}$, where $1 \le i_1 < \cdots < i_h \le n-1$.

Proposition (2.1.9) *Let* $f(g)$ *be as in* (2.1.3). *Then* $f(g)$ *has the following expansion*

$$f(g) = \sum_I \sum_{\substack{\alpha \in k^{n-1} \\ \alpha_i \ne 0\ i \in I \\ \alpha = 0\ i \notin I}} \sum_\tau \sum_{\gamma \in N_{\tau k}^-} f_\alpha(g\gamma\tau),$$

where τ runs through all the permutations satisfying $\tau^{-1}(i) < \tau^{-1}(j)$ if $i < j$, $j \notin I$, *provided the right hand side converges absolutely and locally uniformly.*

Lemma (2.1.10)
(1) *Suppose that τ_j^{-1} fixes $i_j+1,\cdots,n$, and does not change the order of $m_1 < m_2$ for $m_2 \neq i_j$. Then $\tau = \tau_h \cdots \tau_1$ satisfies the condition in* (2.1.9), *and* $l(\tau) = l(\tau_h) + \cdots + l(\tau_1)$.
(2) *Conversely, any τ satisfying the condition in* (2.1.9) *can be written uniquely in the above form.*

Proof. We use induction on $\#I$.

Let $I' = \{i_1, \cdots, i_{h-1}\}$, and $\tau' = \tau_{h-1} \cdots \tau_1$. Suppose that $(i,j) \in I_{\tau_h}$. Then $j = \tau_h^{-1}(i_h)$. So $\tau(i,j) = \tau'(i-1, i_h)$. Since $i - 1 < i_h$, $\tau'(i-1) < \tau'(i_h) = i_h$. Therefore, $I_{\tau_h} \subset I_\tau$.

Suppose that $(i,j) \in \tau_h^{-1} I_{\tau'}$. Then $(\tau_h(i), \tau_h(j)) \in I_{\tau'}$. So $\tau'(\tau_h(i)) < \tau'(\tau_h(j)) \leq i_{h-1}$. Therefore, $I_{\tau_h} \cap \tau_h^{-1} I_{\tau'} = \emptyset$, and $\tau(i) < \tau(j)$. If $i < j$, since $\tau_h(i) > \tau_h(j)$, $(j,i) \in I_{\tau_h}$. This implies that $i = \tau_h^{-1}(i_h)$. Therefore, $(i_h, \tau_h(j)) \in I_{\tau'}$. This is a contradiction. Hence, $\tau_h^{-1} I_{\tau'} \subset I_\tau$, and $l(\tau) = l(\tau_h) + l(\tau')$.

Suppose that $i < j, \tau^{-1}(i) > \tau^{-1}(j)$. Then $(\tau^{-1}(i), \tau^{-1}(j)) \in I_\tau$. This means that if $(\tau^{-1}(i), \tau^{-1}(j)) \in I_{\tau_h}$, $\tau^{-1}(j) = \tau_h^{-1}(\tau'^{-1}(j)) = \tau_h^{-1}(i_h)$. This implies that $j = i_h$. If $(\tau^{-1}(i), \tau^{-1}(j)) \in \tau_h^{-1} I_{\tau'}$, $(\tau'^{-1}(i), \tau'^{-1}(j)) \in I_{\tau'}$. Therefore, by induction, $j \in I'$. This proves (1).

Let τ_h be a permutation such that (i) $\tau_h^{-1}(i) = i$ for $i > i_h$, (ii) if $i < j < i_h$, $\tau_h^{-1}(i) < \tau_h^{-1}(i)$, and (iii) $\tau_h^{-1}(i_h) = \tau^{-1}(i_h)$. Let $\tau' = \tau_h^{-1}\tau$. Then $\tau'(i_h) = i_h$. Suppose that $i_h > i > i_{h-1}$ and $i > j$. Then $\tau^{-1}(i) > \tau^{-1}(j)$. If $\tau_h(\tau^{-1}(i)) < \tau_h(\tau^{-1}(j))$, $\tau'^{-1}(j) = \tau_h(\tau^{-1}(j)) = i_h$, so $j = i_h$. This is a contradiction. Therefore, $\tau'^{-1}(i) \geq i$. This implies that $\tau'(i) = i$ for $i > i_{h-1}$. By induction, this proves (2).

Q.E.D.

Proof of (2.1.9). Take $\tau = \tau_h \cdots \tau_1$ as in (1) of (2.1.11). Let $\overline{N}_m = N_k \cap \Gamma_m$. $\overline{N}_m$ is a subgroup of Γ_m^∞.

Consider the natural map

$$\overline{N}_{i_j} \tau_j \overline{N}_{i_j} / \overline{N}_{i_j} \to \overline{N}_{i_j} \tau \Gamma_{i_j}^\infty / \Gamma_{i_j}^\infty.$$

We show that this map is bijective. Suppose that $n(u) \in \overline{N}_{i_j}$, and $n(u)\tau_j\Gamma_{i_j}^\infty = \tau_j\Gamma_{i_j}^\infty$. Then $n(u) \in \overline{N}_{i_j} \cap \tau_j \Gamma_{i_j}^\infty \tau_j^{-1}$. This means that $u_{\tau_j^{-1}(i)\tau_j^{-1}(i_j)} = 0$ for $i < i_j$. If $i < i' < i_j$, $\tau_j^{-1}(i) < \tau_j^{-1}(i')$ because of the condition on τ_j. So $u_{\tau_j^{-1}(i)\tau_j^{-1}(i')} = 0$. This means that $n(u) \in \overline{N}_{i_j} \cap \tau_j \overline{N}_{i_j} \tau_j^{-1}$. Therefore, $\overline{N}_{i_j} \cap \tau_j \Gamma_{i_j}^\infty \tau_j^{-1} = \overline{N}_{i_j} \cap \tau_j \overline{N}_{i_j} \tau_j^{-1}$.

The set $\overline{N}_{i_j} / \overline{N}_{i_j} \cap \tau_j \overline{N}_{i_j} \tau_j^{-1}$ corresponds bijectively with the set $N_{\tau_j k}^-$. Since $\overline{N}_{i_j} \setminus \Gamma_{i_j} / \Gamma_{i_j}^\infty$ is represented by the collection of such τ_j's we get (2.1.9) by (2.1.8), (2.1.10).

Q.E.D.

§2.2 The constant terms of Eisenstein series on GL(n)

In order to make this book accessible to non-experts, we include a discussion on the constant terms of Eisenstein series on $G = \mathrm{GL}(n)$ in this section. We start with the definition.

We fix a place $v \in \mathfrak{M}$. For $u \in k_v$, we consider the Iwasawa decomposition of the element ${}^t n_2(u)$.

Definition (2.2.1) *For $u \in k_v$, we define*

$$c_v(u) = \begin{cases} (1+|u|_v^2)^{-\frac{1}{2}}, \; s_v(u) = uc_v(u) & v \in \mathfrak{M}_{\mathbb{R}}, \\ c_v(u) = (1+|u|_v)^{-\frac{1}{2}}, \; s_v(u) = \bar{u}c_v(u) & v \in \mathfrak{M}_{\mathbb{C}}, \\ c_v(u) = \begin{Bmatrix} 1 & u \in o_v \\ u^{-1} & u \notin o_v \end{Bmatrix}, \; s_v(u) = \begin{Bmatrix} 0 & u \in o_v \\ 1 & u \notin o_v \end{Bmatrix} & v \in \mathfrak{M}_f. \end{cases}$$

Let $h_v(u)$ be the matrix

$$\begin{pmatrix} c_v(u) & s_v(u) \\ -s_v(u) & c_v(u) \end{pmatrix}, \begin{pmatrix} c_v(u) & \overline{s_v(u)} \\ -s_v(u) & c_v(u) \end{pmatrix}, \begin{pmatrix} 0 & s_v(u) \\ -s_v(u) & c_v(u) \end{pmatrix}$$

for $v \in \mathfrak{M}_{\mathbb{R}}, \mathfrak{M}_{\mathbb{C}}, \mathfrak{M}_f$ respectively. Then it is easy to verify the following equality

$${}^t n_2(u) = h_v(u) \begin{pmatrix} c_v(u) & 0 \\ 0 & c_v(u)^{-1} \end{pmatrix} \begin{pmatrix} 1 & 0 \\ c_v(u)s_v(u) & 1 \end{pmatrix}.$$

Definition (2.2.2) *We define*
(1) $\alpha_v(u_v) = |c_v(u_v)|_v$ *for* $u_v \in k_v$,
(2) $\alpha_\infty(u_\infty) = \prod_{v \in \mathfrak{M}_\infty} \alpha_v(u_v)$ *for* $u_\infty = (u_v) \in k_\infty$,
(3) $\alpha_f(u_f) = \prod_{v \in \mathfrak{M}_f} \alpha_v(u_v)$ *for* $u_f = (u_v) \in \mathbb{A}_f$,
(4) $\alpha(u) = \prod_v \alpha_v(u_v)$ *for* $u = (u_v)_v \in \mathbb{A}$.

If $u = (u_v)_v \in \mathbb{A}$, we use the notation $\alpha_\infty(u), \alpha_f(u)$ for $\alpha_\infty(u_\infty), \alpha_f(u_f)$ also.

The function $\alpha_v(u_v)$ has the following explicit description

$$\alpha_v(u_v) = \begin{cases} (1+|u_v|_v^2)^{-\frac{1}{2}} & v \in \mathfrak{M}_{\mathbb{R}}, \\ (1+|u_v|_v)^{-1} & v \in \mathfrak{M}_{\mathbb{C}}, \\ \sup(1,|u_v|_v)^{-1} & v \in \mathfrak{M}_f. \end{cases}$$

We now define the Eisenstein series for the Borel subgroup of G. For $g \in G_{\mathbb{A}}$, let $g = k(g)t(g)n(u(g))$ be the Iwasawa decomposition as before. Consider the space $\{z = (z_1, \cdots, z_n) \in \mathbb{C}^n \mid z_1 + \cdots + z_n = 0\}$. We can consider ρ (half the sum of the positive weights) as an element of this space.

Definition (2.2.3) *For $g^0 \in G^0_{\mathbb{A}}$ and z as above, we define*

$$E_B(g^0, z) = \sum_{\gamma \in G_k/B_k} t(g^0\gamma)^{z+\rho}.$$

Our first task is to prove the convergence of $E_B(g^0, z)$. We do it by reducing the problem to the case $n = 2$. First, observe that $N_k \setminus G_k/B_k$ is represented by Weyl group elements. If τ is a permutation, an element $n \in N_k$ fixes the coset τB_k if and only if $n \in N_{\tau k}$. Therefore, the Eisenstein series has the following decomposition

$$E_B(g^0, z) = \sum_{\tau} \sum_{\gamma \in N_k / N_{\tau k}} t(g^0 \gamma \tau)^{z+\rho}. \tag{2.2.4}$$

We are going to estimate each sum in (2.2.4). For this purpose, we need a lemma.

Lemma (2.2.5) *Let $\epsilon > 0$ be a constant. Then there exists a continuous function $c(\sigma)$ of $\sigma \in \mathbb{R}$ such that for any $t \in \mathbb{A}^\times, u \in \mathbb{A}$,*

$$\sum_{\gamma \in k} \alpha(t(u+\gamma))^{\sigma+1} \leq C(\sigma) \sup(1, |t|^{-1})$$

for $\sigma \geq 1 + \epsilon$.

Proof. We choose compact subsets $U_1 \subset \mathbb{A}^1, U_2 \subset \mathbb{A}$ so that U_1, U_2 surjects to $\mathbb{A}^1/k^\times, \mathbb{A}/k$ respectively. Since we are trying to prove an estimate which is uniform with respect to u, we can assume that $t = \underline{\lambda} t'$ and $t' \in U_1, u \in U_2$. We write $\gamma = ab^{-1}$ where $a \in o_k, b \in o_k \setminus \{0\}$. The inequality $\alpha_f(\underline{\lambda} t'(u+\gamma)) \leq 1$ is always satisfied. Therefore,

$$\alpha(\underline{\lambda} t'(u+\gamma))^{\sigma+1} \leq \alpha_\infty(\underline{\lambda} t'(u+\gamma))^{\sigma+1},$$

and this is bounded by a constant multiple of

$$|b|_f^{\sigma+1} \prod_{v \in \mathfrak{M}_\mathbb{R}} (|b|_v^2 + |\underline{\lambda} t'(u+a)|_v^2)^{-\frac{\sigma+1}{2}} \prod_{v \in \mathfrak{M}_\mathbb{C}} (|b|_v + |\underline{\lambda} t'(u+a)|_v)^{-(\sigma+1)}$$
$$\leq \zeta_k(\sigma+1) \prod_{v \in \mathfrak{M}_\mathbb{R}} (|b|_v^2 + |\underline{\lambda} t'(u+a)|_v^2)^{-\frac{\sigma+1}{2}} \prod_{v \in \mathfrak{M}_\mathbb{C}} (|b|_v + |\underline{\lambda} t'(u+a)|_v)^{-(\sigma+1)}.$$

Since U_1 is compact, there exist constants $0 < C_1 < C_2$ such that $C_1 \leq |t'|_v \leq C_2$ for $t' \in U_1$. Also $|t'|_v = 1$ for almost all v and for all $t' \in U_1$. Therefore, the above function is bounded by a constant multiple of

$$\zeta_k(\sigma+1) \prod_{v \in \mathfrak{M}_\mathbb{R}} (|b|_v^2 + |\underline{\lambda}(u+a)|_v^2)^{-\frac{\sigma+1}{2}} \prod_{v \in \mathfrak{M}_\mathbb{C}} (|b|_v + |\underline{\lambda}(u+a)|_v)^{-(\sigma+1)}.$$

Now (2.2.5) follows by the integral test.

Q.E.D.

Let $t \in T_\mathbb{A}$, $\gamma \in N_{\tau k}^-$, and $n(u) \in N_\mathbb{A}$.

Lemma (2.2.6) *There exists a constant $M > 0$ such that the sum*

$$\sum_{\gamma \in N_{\tau k}^-} t(tn(u)\gamma t^{-1}\tau)^z$$

converges absolutely and locally uniformly with respect to t, z, uniformly with respect to u, if for any pair $(i,j) \in I_\tau$, $\mathrm{Re}(z_{\tau(i)} - z_{\tau(j)}) \geq M$. Moreover this sum is bounded by a finite sum of functions of the form

$$c(z) \prod_{(i,j) \in I_\tau} |t_i t_j^{-1}|^{c_{ij}},$$

where c_{ij} is a constant independent of u, t, z, and $c(z)$ is a function of polynomial growth. (So $n(u)$ can depend on t.)

Proof. We can assume that z is real. We use induction on $l(\tau)$. Let $\tau = \tau_1\tau_2$, $l(\tau) = l(\tau_1) + l(\tau_2)$, and $l(\tau_1) = 1$. We can write $\gamma = \gamma_1\tau_1\gamma_2\tau_1^{-1}$, where $\gamma_1 \in N^-_{\tau_1 k}, \gamma_2 \in N^-_{\tau_2 k}$. Let $tn(u)\gamma_1 t_1^{-1}\tau_1 = kan_1$, where

$$k \in K,\ a = a(t, u, \gamma_1) \in T_{\mathbb{A}},\ n_1 = n_1(t, u, \gamma_1) \in N_{\mathbb{A}}.$$

For $t \in T_{\mathbb{A}}$, let $t^\tau = \tau^{-1}t\tau$. Then

$$\begin{aligned} tn(u)\gamma_1\tau_1\gamma_2\tau_1^{-1}t^{-1}\tau &= tn(u)\gamma_1 t^{-1}\tau_1 t^{\tau_1}\gamma_2\tau_1^{-1}t^{-1}\tau \\ &= kan_1 t^{\tau_1}\gamma_2\tau_2(t^{-1})^\tau = kat^{\tau_1}n_2\gamma_2\tau_2(t^{-1})^\tau, \end{aligned}$$

for some $n_2 = n_2(t, u, \gamma_1) \in N_{\mathbb{A}}$. This is equal to

$$kat^{\tau_1}n_2\gamma_2 a^{-1}(t^{-1})^{\tau_1}\tau_2 a^{\tau_2}(t^{\tau_1})^{\tau_2}(t^{-1})^\tau = kat^{\tau_1}n_2\gamma_2 a^{-1}(t^{-1})^{\tau_1}\tau_2 a^{\tau_2}.$$

So

$$t(tn(u)\gamma t^{-1}\tau)^z = t(at^{\tau_1}n_2\gamma_2 a^{-1}(t^{-1})^{\tau_1}\tau_2)^z a^{\tau_2 z}.$$

Suppose that for any $(i,j) \in I_\tau$, $z_{\tau(i)} - z_{\tau(j)} \gg 0$. Since $I_\tau = I_{\tau_1} \coprod \tau_1^{-1}I_{\tau_2}$, if $(i,j) \in I_{\tau_2}$, then $\tau_1^{-1}(i,j) \in I_\tau$. Since $\tau(\tau_1^{-1}(i)) = \tau_2(i)$, the condition on z is satisfied for τ_2. So by induction,

$$\sum_{\gamma_2 \in N^-_{\tau_2 k}} t(at^{\tau_1}n_2\gamma_2 a^{-1}(t^{-1})^{\tau_1}\tau_2)^z a^{\tau_2 z}$$

is bounded by a finite sum of functions of the form

$$c_1(z)a^{\tau_2 z}\prod_{(i,j)\in I_{\tau_2}} |a_i a_j^{-1} t_{\tau_1^{-1}(i)} t^{-1}_{\tau_1^{-1}(j)}|^{c'_{ij}},$$

where $c_1(z)$ is a function of polynomial growth and c'_{ij} is a constant for all $(i,j) \in I_{\tau_2}$. If $(i,j) \in I_{\tau_2}$, $(\tau_1^{-1}(i), \tau_1^{-1}(j)) \in I_\tau$. So we can ignore

$$\prod_{\substack{i>j \\ \tau_2(i)<\tau_2(j)}} |t_{\tau_1^{-1}(i)} t^{-1}_{\tau_1^{-1}(j)}|^{c'_{ij}}.$$

By definition, $(\tau_2 z)_{\tau(i)} = z_{\tau_1(i)}$. So the condition on $\tau_2 z$ is satisfied for τ_1. Note that c'_{ij} does not depend on z. So by induction,

$$\sum_{\gamma_1 \in N^-_{\tau_1 k}} a^{\tau_2 z} \prod_{\substack{i>j \\ \tau_2(i)<\tau_2(j)}} |a_i a_j^{-1}|^{c'_{ij}}$$

is bounded by a finite sum of functions of the form

$$c_2(z) \prod_{\substack{i>j \\ \tau_1(i)<\tau_1(j)}} |t_i t_j^{-1}|^{c''_{ij}},$$

where $c_2(z)$ is a function of polynomial growth. Since $I_{\tau_1} \subset I_\tau$, we have reduced the problem to the case $l(\tau) = 1$.

If $l(\tau) = 1$, N_τ is a normal subgroup of N. So $n(u)\gamma = n(u_1)\gamma n(u_2(\gamma))$, where $n(u_1) \in N_{\tau\mathbb{A}}^-$, $n(u_2(\gamma)) \in N_{\tau\mathbb{A}}$. Therefore,

$$t(tn(u)\gamma t^{-1}\tau) = t(tn(u_1)\gamma t^{-1}\tau),$$

and we have reduced the problem to the case $n = 2$. But the case $n = 2$ is (2.2.5).

Q.E.D.

Now the convergence of the Eisenstein series is an easy consequence of (2.2.6).

Lemma (2.2.7) *The sum $E_B(g^0, z)$ converges absolutely and locally uniformly if* $\mathrm{Re}(z_i - z_{i+1}) \gg 0$ *for all i.*

Proof. We modify each term of (2.2.4) as follows.

$$\begin{aligned}
\sum_{\gamma \in N_k/N_{\tau k}} t(g^0\gamma\tau)^{z+\rho} &= \sum_{\gamma \in N_k/N_{\tau k}} t(t(g^0)n(u(g^0))\gamma\tau)^{z+\rho} \\
&= \sum_{\gamma \in N_k/N_{\tau k}} t(t(g^0)n(u(g^0))\gamma t(g^0)^{-1}t(g^0)\tau)^{z+\rho} \\
&= \sum_{\gamma \in N_k/N_{\tau k}} t(g^0)^{\tau z + \tau\rho} t(t(g^0)n(u(g^0))\gamma t(g^0)^{-1}\tau)^{z+\rho}.
\end{aligned}$$

Then the convergence of this sum follows from (2.2.6).

Q.E.D.

It turns out that $E_B(g^0, z)$ is holomorphic if $\mathrm{Re}(z_i - z_{i+1}) > 1$ for all i. But since we do not need such a convergence, we do not try to prove the convergence $E_B(g, z)$ on the above domain.

The Fourier expansion of the Eisenstein series $E_B(g^0, z)$ and subsequent estimates are the issue of this chapter. For the rest of this section, we consider the constant terms of $E_B(g, z)$.

Definition (2.2.8) *For $z \in \mathbb{C}$, we define*

$$M_v(z) = \int_{k_v} |c_v(u_v)|_v^{z+1} du_v.$$

It is easy to verify that this integral converges absolutely for $\mathrm{Re}(z) > 0$. We evaluate this integral explicitly here.

Proposition (2.2.9)

(1) *If* $v \in \mathfrak{M}_{\mathbb{R}}$, $M_v(z) = \dfrac{\pi^{-\frac{z}{2}}\Gamma(\frac{z}{2})}{\pi^{-\frac{z+1}{2}}\Gamma(\frac{z+1}{2})}$.

(2) *If* $v \in \mathfrak{M}_{\mathbb{C}}$, $M_v(z) = \dfrac{(2\pi)^z\Gamma(z)}{(2\pi)^{z+1}\Gamma(z+1)}$.

(3) *If* $v \in \mathfrak{M}_f$, $M_v(z) = \dfrac{1 - q_v^{-(z+1)}}{1 - q_v^{-z}}$.

Proof. Let $v \in \mathfrak{M}_{\mathbb{R}}$. Then

$$\begin{aligned}\Gamma\left(\frac{z+1}{2}\right) M(z) &= \int_{\mathbb{R}_+}\int_{\mathbb{R}} e^{-t}t^{\frac{z+1}{2}}(1+u)^{-\frac{z+1}{2}} d^\times t du \\ &= \int_{\mathbb{R}_+}\int_{\mathbb{R}} e^{-t(1+u^2)}t^{\frac{z+1}{2}} d^\times t du = \sqrt{\pi}\int_{\mathbb{R}_+} e^{-t}t^{\frac{z}{2}} d^\times t \\ &= \sqrt{\pi}\Gamma\left(\frac{z}{2}\right).\end{aligned}$$

This proves (1).

The case $v \in \mathfrak{M}_{\mathbb{C}}$ is similar to the real case and is left to the reader. Again, notice that $|x|_v$ is the square of the usual absolute value.

Let $v \in \mathfrak{M}_f$. Then

$$\begin{aligned}M(z) &= \int_{\mathcal{O}_v} du_v + \sum_{i=1}^{\infty}\int_{\pi_v^{-i}\mathcal{O}_v^\times} q_v^{-i(z+1)} du_v \\ &= 1 + \sum_{i=1}^{\infty}(1-q_v^{-1})q_v^{-iz} = 1 + \frac{(1-q_v^{-1})q_v^{-z}}{1-q_v^{-z}} \\ &= \frac{1-q_v^{-(z+1)}}{1-q_v^{-z}}.\end{aligned}$$

Q.E.D.

We define

$$M(z) = \int_{\mathbb{A}} \alpha(u)^{z+1} du. \tag{2.2.10}$$

Then by the choice of the measure on $\mathbb{A}$, $M(z) = |\Delta_k|^{-\frac{1}{2}}\prod_v M_v(z)$ where Δ_k is the discriminant of k. Therefore, $M(z) = \frac{Z_k(z)}{Z_k(z+1)} = \phi(z)$ (see (1.1.2)). Since we know that the functions

$$\frac{\pi^{-\frac{z}{2}}\Gamma(\frac{z}{2})}{\pi^{-\frac{z+1}{2}}\Gamma(\frac{z+1}{2})},\ \frac{(2\pi)^z\Gamma(z)}{(2\pi)^{z+1}\Gamma(z+1)},\ \zeta_k(z),\ \frac{1}{\zeta_k(z+1)}$$

are at most of polynomial growth on any vertical strip contained in $\{z \mid \mathrm{Re}(z) > 0\}$ when $\mathrm{Im}(z) \gg 0$, the function $M(z)$ is at most of polynomial growth on such a vertical strip when $\mathrm{Im}(z) \gg 0$.

We fix a Haar measure du on $N_{\tau\mathbb{A}}^-$ so that the volume of $N_{\tau\mathbb{A}}^-/N_{\tau k}^-$ is 1. For a permutation τ, we define

$$M_\tau(z) = \int_{N_{\tau\mathbb{A}}^-} t(n(u)\tau)^{z+\rho} du. \tag{2.2.11}$$

Lemma (2.2.12) *If $t \in T_{\mathbb{A}}^0$ and τ is a Weyl group element,*

$$\int_{N_{\tau\mathbb{A}}^-} t(tn(u)\tau)^{z+\rho} du = t^{\tau z+\rho} M_\tau(z).$$

Proof. Let $t = a_n(t_1, \cdots, t_n)$. Then

$$\begin{aligned}
\int_{N^-_{\tau\mathbb{A}}} t(tn(u)\tau)^{z+\rho} du &= \int_{N^-_{\tau\mathbb{A}}} t(tn(u)t^{-1}\tau\tau^{-1}t\tau)^{z+\rho} du \\
&= t^{\tau z + \tau\rho} \int_{N^-_{\tau\mathbb{A}}} t(tn(u)t^{-1}\tau)^{z+\rho} du \\
&= t^{\tau z + \tau\rho} \left(\prod_{\substack{i>j \\ \tau(i)<\tau(j)}} |t_i t_j^{-1}|^{-1} \right) \int_{N^-_{\tau\mathbb{A}}} t(n(u)\tau)^{z+\rho} du.
\end{aligned}$$

We simplify the coefficient of the integral as follows.

$$\begin{aligned}
t^{\tau\rho} \prod_{\substack{i>j \\ \tau(i)<\tau(j)}} |t_i t_j^{-1}|^{-1} &= \left(\prod_{i<j} |t_{\tau^{-1}(i)} t^{-1}_{\tau^{-1}(j)}| \right)^{\frac{1}{2}} \prod_{\substack{i>j \\ \tau(i)<\tau(j)}} |t_i t_j^{-1}|^{-1} \\
&= \left(\prod_{\tau(i)<\tau(j)} |t_i t_j^{-1}| \right)^{\frac{1}{2}} \prod_{\substack{i>j \\ \tau(i)<\tau(j)}} |t_i t_j^{-1}|^{-1} \\
&= \left(\prod_{\substack{i<j \\ \tau(i)<\tau(j)}} |t_i t_j^{-1}| \right)^{\frac{1}{2}} \left(\prod_{\substack{i>j \\ \tau(i)<\tau(j)}} |t_i t_j^{-1}| \right)^{-\frac{1}{2}}.
\end{aligned}$$

By exchanging i, j in the second factor, this is equal to t^ρ. This proves the lemma. Q.E.D.

Proposition (2.2.13)

$$M_\tau(z) = \prod_{\substack{i>j \\ \tau(i)<\tau(j)}} M(z_{\tau(i)} - z_{\tau(j)}).$$

Proof. By induction on the length $l(\tau)$. We have already considered the case $l(\tau) = 1$. Suppose that $\tau = \tau_1\tau_2$, $\tau_1, \tau_2 \neq 1$, and $l(\tau) = l(\tau_1) + l(\tau_2)$. This implies that $N^-_{\tau\mathbb{A}} = N^-_{\tau_1\mathbb{A}}(\tau_1 N^-_{\tau_1\mathbb{A}} \tau_1^{-1})$. Therefore,

$$\begin{aligned}
M_\tau(z) &= \int_{N^-_{\tau_1\mathbb{A}} \times N^-_{\tau_1\mathbb{A}}} t(n(u_1)(\tau_1 n(u_2)\tau_1^{-1})\tau)^{z+\rho} du_1 du_2 \\
&= \int_{N^-_{\tau_1\mathbb{A}} \times N^-_{\tau_1\mathbb{A}}} t(n(u_1)\tau_1 n(u_2)\tau_2)^{z+\rho} du_1 du_2 \\
&= \int_{N^-_{\tau_1\mathbb{A}} \times N^-_{\tau_1\mathbb{A}}} t(t(n(u_1)\tau_1)n(u_2)\tau_2)^{z+\rho} du_1 du_2 \\
&= \int_{N^-_{\tau_1\mathbb{A}}} t(n(u_1)\tau_1)^{\tau_2 z+\rho} du_1 \int_{N^-_{\tau_2\mathbb{A}}} t(n(u_2)\tau_2)^{z+\rho} du_2 \\
&= M_{\tau_1}(\tau_2 z) M_{\tau_2}(z).
\end{aligned}$$

Q.E.D.

By (2.2.4) and (2.2.13),

$$\begin{aligned}\int_{N_{\mathbb{A}}/N_k} E_B(g^0 n(u), z)du &= \sum_\tau \int_{N_{\mathbb{A}}/N_{\tau k}} t(g^0 n(u)\tau)^{z+\rho} du \\ &= \sum_\tau \int_{N_{\mathbb{A}}/N_{\tau \mathbb{A}}} t(t(g^0)n(u)\tau)^{z+\rho} du \\ &= \sum_\tau \int_{N^-_{\tau\mathbb{A}}} t(t(g^0)n(u)\tau)^{z+\rho} du \\ &= \sum_\tau M_\tau(z) t(g^0)^{\tau z+\rho}.\end{aligned}$$

Therefore, we have the following proposition.

Proposition (2.2.14)

$$\int_{N_{\mathbb{A}}/N_k} E_B(g^0 n(u), z)du = \sum_\tau M_\tau(z) t(g^0)^{\tau z+\rho}.$$

§2.3 The Whittaker functions

We estimate the Whittaker functions on $G = \mathrm{GL}(n)$ in this section. They appear as the Fourier coefficients of Eisenstein series, which we will consider in §2.4.

Part of our computation is similar to Stade [71]. However, our computation is based on a naive use of integration by parts. Let τ_G be the longest element among the permutations of $\{1, \cdots, n\}$, i.e. $\tau_G = (1,n)(2,n-1)\cdots$. Let $K = \prod_v K_v$ be the maximal compact subgroup of $G_{\mathbb{A}}$ which we defined in §1.1. Consider $z = (z_1, \cdots, z_n) \in \mathbb{C}^n$ such that $z_1 + \cdots + z_n = 0$.

Definition (2.3.1) *Let v be a place of k. We define*

(1) $\quad W_{n,v}(\alpha_v, z) = \displaystyle\int_{N^-_{\tau_G k_v}} t(u\tau_G)_v^{z+\rho} \psi_{v,\alpha_v}(n_n(u_v)) du_v$ *for* $\alpha \in (k_v^\times)^{n-1}$,

(2) $\quad W_{n,\infty}(\alpha_\infty, z) = \displaystyle\prod_{v\in\mathfrak{M}_\infty} W_{n,v}(\alpha_v, z)$ *for* $\alpha_\infty = (\alpha_v)_{v\in\mathfrak{M}_\infty} \in (\mathbb{A}_\infty^\times)^{n-1}$,

(3) $\quad W_{n,f}(\alpha_f, z) = \displaystyle\prod_{v\in\mathfrak{M}_f} W_{n,v}(\alpha_v, z)$ *for* $\alpha_f = (\alpha_v)_{v\in\mathfrak{M}_f} \in (\mathbb{A}_f^\times)^{n-1}$,

(4) $\quad W_n(\alpha, z) = \displaystyle\prod_{v\in\mathfrak{M}} W_{n,v}(\alpha_v, z)$ *for* $\alpha = (\alpha_v)_v \in (\mathbb{A}^\times)^{n-1}$.

If $\alpha = (\alpha_v)_v \in \mathbb{A}^\times$, we also use the notation $W_{n,\infty}(\alpha, z)$ etc. We fix $v \in \mathfrak{M}$. For the rest of this section, we drop the index v from $c_v(u), s_v(u)$ etc. if there is no confusion.

We first prove an inductive formula for $W_{n,v}(\alpha, z)$. Let $\widetilde{z}_i = z_i + \frac{z_n}{n-1}$ for $i = 1, \cdots, n$ and $\widetilde{z} = (\widetilde{z}_1, \cdots, \widetilde{z}_{n-1})$. Clearly, $\widetilde{z}_i - \widetilde{z}_j = z_i - z_j$ for all $i, j \le n$, and $\widetilde{z}_1 + \cdots + \widetilde{z}_{n-1} = 0$.

Proposition (2.3.2) *Let $v \in \mathfrak{M}_\infty$. Then*

$$W_{n,v}(\alpha, z) = \int_{k_v^{n-1}} |c(u_{21})|_v^{z_{n-1}-z_n+1} \cdots |c(u_{n1})|_v^{z_1-z_n+1}$$
$$\times W_{n-1,v}(\alpha_2 c(u_{21})c(u_{31})^{-1}, \cdots, \alpha_{n-1}c(u_{n-11})c(u_{n1})^{-1}, \widetilde{z})$$
$$\times < \alpha_1 u_{21} + \alpha_2 s(u_{21})u_{31} + \cdots + \alpha_{n-1}s(u_{n-11})u_{n1} >_v du_{21} \cdots du_{n1}.$$

Proof. Let

$$\widetilde{\tau}_1 = (1,2), \widetilde{\tau}_2 = (2,3), \cdots, \widetilde{\tau}_{n-1} = (n-1,n),\ \nu_{n-1} = \tau_{\mathrm{GL}(n-1)}.$$

Then

$$\tau_G = \widetilde{\tau}_1 \cdots \widetilde{\tau}_{n-1}\nu_{n-1}, \quad l(\tau_G) = l(\widetilde{\tau}_1) + \cdots + l(\widetilde{\tau}_{n-1}) + l(\nu_{n-1}).$$

Let $A_{i1}(u), A_{i2}(u), A_{i3}(u), A_{i4}(u)$ be the following matrices

$$\begin{pmatrix} I_{i-2} & & & \\ & 1 & & \\ & u & 1 & \\ & & & I_{n-i} \end{pmatrix}, \begin{pmatrix} I_{i-2} & & & \\ & c(u) & & \\ & s(u) & c(u) & \\ & & & I_{n-i} \end{pmatrix},$$

$$\begin{pmatrix} I_{i-2} & & & \\ & c(u) & & \\ & & c(u)^{-1} & \\ & & & I_{n-i} \end{pmatrix}, \begin{pmatrix} I_{i-2} & & & \\ & 1 & & \\ & c(u)s(u) & 1 & \\ & & & I_{n-i} \end{pmatrix}$$

respectively. Let $\mu_i = \widetilde{\tau}_i \cdots \widetilde{\tau}_{n-1}\nu_{n-1}$. Also let

$$B_i(u) = \begin{pmatrix} 1 & & & & & & & & \\ u_{32} & 1 & & & & & & & \\ \vdots & \vdots & \ddots & & & & & & \\ u_{i-12} & \cdots & u_{i-1i-2} & 1 & & & & & \\ 0 & \cdots & \cdots & 0 & 1 & & & & \\ u_{i2} & \cdots & \cdots & u_{ii-1} & u_{i1} & 1 & & & \\ \vdots & \vdots & \vdots & \vdots & \vdots & u_{i+1i} & 1 & & \\ \vdots & \vdots & \vdots & \vdots & \vdots & \vdots & \vdots & \ddots & \\ u_{n2} & \cdots & \cdots & u_{ni-1} & u_{n1} & u_{ni} & \cdots & u_{nn-1} & 1 \end{pmatrix}.$$

Then

$$B_i(u)\mu_{i-1}$$
$$= A_{i1}(u_{i1}) \begin{pmatrix} 1 & & & & & & & & \\ u_{32} & 1 & & & & & & & \\ \vdots & \vdots & \ddots & & & & & & \\ u_{i-12} & \cdots & u_{i-1i-2} & 1 & & & & & \\ 0 & \cdots & \cdots & 0 & 1 & & & & \\ u_{i2} & \cdots & \cdots & u_{ii-1} & 0 & 1 & & & \\ \vdots & \vdots & \vdots & \vdots & \vdots & u_{i+1i} & 1 & & \\ \vdots & \vdots & \vdots & \vdots & \vdots & \vdots & \vdots & \ddots & \\ u_{n2} & \cdots & \cdots & u_{ni-1} & u_{n1} & u_{ni} & \cdots & u_{nn-1} & 1 \end{pmatrix} \mu_{i-1}$$
$$= A_{i1}(u_{i1})\widetilde{\tau}_{i-1}B_{i+1}(u)\mu_i = hA_{i2}(u_{i1})B_{i+1}(u)\mu_i = hA_{i3}(u_{i1})A_{i4}(u_{i1})B_{i+1}(u)\mu_i$$
$$= hA_{i3}(u_{i1})B_{i+1}(\widetilde{u})\mu_i n(u'),$$

where $h \in K_v, n(u') \in N_{\mathbb{A}}$, and $n(\widetilde{u})$ is obtained by replacing u_{ji} by

$$u_{ji} - c(u_{i1})s(u_{i1})u_{j1}$$

for $j = i+1, \cdots, n$.

Let

$$\begin{aligned}&\psi_i(\alpha, u)\\ &=< \alpha_2 c(u_{21})c(u_{31})^{-1}u_{32} + \cdots + \alpha_{i-2}c(u_{i-21})c(u_{i-11})^{-1}u_{i-1i-2}\\ &\quad + \alpha_{i-1}c(u_{i-11})u_{ii-1} + \alpha_1 u_{21} + \alpha_2 s(u_{21}u_{31}) + \cdots + \alpha_{i-1}s(u_{i-11})u_{i1} >_v .\end{aligned}$$

Then by the above relation,

$$\begin{aligned}&\int_{k_v^{n-1}} |c(u_{21})|_v^{z_{n-1}-z_n+1} \cdots |c(u_{i-11})|_v^{z_{n-i+2}-z_n+1} t(B_i(u)\mu_{i-1})^{z+\rho}\\ &\qquad \times \psi_i(\alpha, u)du_{21}\cdots du_{n1}\\ &= \int_{k_v^{n-1}} |c(u_{21})|_v^{z_{n-1}-z_n+1} \cdots |c(u_{i1})|_v^{z_{n-i+1}-z_n+1} t(B_{i+1}(u)\mu_i)^{z+\rho}\\ &\qquad \times \psi_{i+1}(\alpha, u)du_{21}\cdots du_{n1}\end{aligned}$$

for all i. Now (2.3.2) follows by induction.

Q.E.D.

We first consider finite places. The explicit formula for p-adic Whittaker functions was first proved for GL(n) by Shintani in [66], and was generalized to arbitrary quasi-split groups by Casselman–Shalika in [7] . We only consider GL(n), for which we use an explicit computation to compute the p-adic Whittaker functions.

Let $v \in \mathfrak{M}_f$. Let $\mathfrak{a} = (a_v)_v$ be the difference idele. Then the character $x \to < a_v x >_v$ is a character of index zero on k_v, so we assume that $< >_v$ is of index zero to start with.

Lemma (2.3.3)
(1) $\int_{\pi_v^i o_v} < u >_v du = 0$ *if* $i < 0$.
(2) $\int_{\pi_v^i o_v^\times} < u >_v du = 0$ *if* $i < -1$.

Proof. Since the character $< >_v$ is non-trivial on the compact group $\pi_v^i o_v$, (1) follows. Since $\pi_v^i o_v^\times = \pi_v^i o_v \setminus \pi_v^{i+1} o_v$, (2) follows from (1).

Q.E.D.

We first consider the case $n = 2$.

Proposition (2.3.4) *Let* $v \in \mathfrak{M}_f$, *and* $\alpha \in k_v^\times$. *Then*
(1) *if* $\alpha \notin o_v$, $W_{2,v}(\alpha, z) = 0$,
(2) *if* $\alpha \in \pi_v^i o_v^\times$ *for* $i \geq 0$, $W_{2,v}(\alpha, z) = \dfrac{1 - q_v^{-(z+1)}}{1 - q_v^{-z}}(1 - q_v^{-(i+1)z})$.

Proof. Since $|c(u)|_v$ depends only on $|u|_v$, we assume that $\alpha = \pi_v^i$ for $i \in \mathbb{Z}$. Suppose $i < 0$. Then

$$(2.3.5) \qquad W_{n,v}(\alpha, z) = \int_{o_v} < \pi_v^i u >_v du + \sum_{l=1}^{\infty} \int_{\pi_v^{-l} o_v^\times} |c(u)|_v^{z+1} < \pi_v^i u >_v du.$$

Both terms are zero by (2.3.3). So (1) follows.

Suppose $i \geq 0$. The equation (2.3.5) is still valid, and the first term is 1 this time. If $l \geq i+2$,

$$\int_{\pi_v^{-l} o_v^\times} |c(u)|_v^{z+1} < \pi_v^i u >_v du = 0$$

for a similar reason as above. Therefore,

$$W_{n,v}(\alpha, z) = 1 + \sum_{l=1}^{i} (1 - q_v^{-1}) q_v^{-lz} + q_v^{-(i+1)(z+1)} \int_{\pi_v^{-(i+1)} o_v^\times} < \pi_v^i u >_v du.$$

By (2.3.3),

$$\int_{\pi_v^{-(i+1)} o_v} < \pi_v^i u >_v du = 0.$$

This implies that the last term is equal to

$$-q_v^{-(i+1)(z+1)} \mathrm{vol}(\pi_v^{-i} o_v) = -q_v^{-(i+1)z-1}.$$

So

$$W_{n,v}(\alpha, z) = 1 + \sum_{l=1}^{i} (1 - q_v^{-1}) q_v^{-lz} - q_v^{-(i+1)z-1}.$$

Now it is easy to verify that this is the right hand side of (2).

Q.E.D.

We continue to assume that $< >_v$ is of index zero. We choose $t = a_n(t_1, \cdots, t_n) \in T_{k_v}$ so that

$$\alpha_1 = t_1 t_2^{-1}, \cdots, \alpha_{n-1} = t_{n-1} t_n^{-1}. \tag{2.3.6}$$

Let $C_n(t, z)$ be the $n \times n$ matrix whose (i, j)-entry is $|\pi_v^{n-j} t_j|_v^{z_{n-i+1}}$, i.e.

$$C_n(t, z) = \begin{pmatrix} |\pi_v^{n-1} t_1|_v^{z_n} & |\pi_v^{n-2} t_2|_v^{z_n} & \cdots & |t_n|_v^{z_n} \\ \vdots & \vdots & \vdots & \vdots \\ |\pi_v^{n-1} t_1|_v^{z_1} & |\pi_v^{n-2} t_2|_v^{z_1} & \cdots & |t_n|_v^{z_1} \end{pmatrix}.$$

We define

$$\lambda_{n,v}(t, z) = t^{-\tau_G z} q_v^{-\frac{1}{2} \sum_{i<j} (z_i - z_j)} \det C_n(t, z) \prod_{i<j} \frac{1 - q_v^{-(z_i - z_j + 1)}}{1 - q_v^{-(z_i - z_j)}}. \tag{2.3.7}$$

Note that this function depends only on α. Also $\lambda_{n,v}(t, z) = 0$ if $\alpha_i \in \pi_v^{-1} o_v^\times$ for some i.

The following theorem is due to Shintani [66], and Casselman–Shalika [7].

Theorem (2.3.8)
(1) *If* $\alpha_i \notin o_v$ *for some* i, $W_{n,v}(\alpha, z) = 0$.
(2) *If* $\alpha_i \in o_v$ *for all* i, $W_{n,v}(\alpha, z) = \lambda_{n,v}(\alpha, z)$.

Proof. We use induction on n. We prove (1) first.

Lemma (2.3.9) *$W_{n,v}(\alpha, z) = 0$ if $\alpha_i \notin o_v$ for some i.*

Proof. By induction on n. Suppose $\alpha_i \notin o_v$. If $u_{i1} \in o_v$, $\alpha_i c(u_{i1})c(u_{i+11})^{-1} \notin o_v$. Therefore, by induction, (2.3.2) is equal to the integral over the set $\{(u_{21}, \cdots, u_{n1}) \in k_v^{n-1} \mid u_{i1} \notin o_v\}$. This implies that $s(u_{i1}) = 1$. By definition,

$$\begin{aligned}
&\int_{k_v} |c(u_{i+11})|_v^{z_{n-i}-z_n+1} < \alpha_i u_{i+1} + \alpha_{i+1}s(u_{i+11})u_{i+21} > \\
&\qquad \times W_{n-1,v}(\alpha_2 c(u_{21})c(u_{31})^{-1}, \cdots, \alpha_{n-1}c(u_{n-11})c(u_{n1})^{-1}, \widetilde{z})du_{i+11} \\
&= \int_{o_v} < \alpha_i u >_v du \times W_{n-1,v}(\cdots, \alpha_i c(u_{i1}), \alpha_{i+1}c(u_{i+11})^{-1}, \cdots, \widetilde{z}) \\
&\quad + \sum_{j=1}^{\infty} q_v^{-j(z_{n-i}-z_n+1)} \int_{o_v^\times} < \alpha_i \pi_v^{-j} u >_v du < \alpha_{i+1}u_{i+21} >_v \\
&\qquad \times W_{n-1,v}(\cdots, \alpha_i c(u_{i1})\pi_v^{-j}, \pi_v^j \alpha_{i+1}c(u_{i+21})^{-1}, \cdots, \widetilde{z}).
\end{aligned}$$

Since $\alpha_i \notin o_v$, all these terms are zero by (2.3.3).

Q.E.D.

Lemma (2.3.10) *Let $\alpha_i \in o_v$ for all i. Then in (2.3.2), we can replace $s(u_{i1})$ by 1 for all i without changing the value of the integral.*

Proof. By induction on i. Suppose that we can replace $s(u_{j1})$ by 1 for all $j < i$ without changing the value of (2.3.2). Then

$$\begin{aligned}
W_{n,v}(\alpha, z) = &\int_{k_v^{n-1}} |c(u_{21})|_v^{z_{n-1}-z_n+1} \cdots |c(u_{n1})|_v^{z_1-z_n+1} \\
&\times W_{n-1,v}(\alpha_2 c(u_{21})c(u_{31})^{-1}, \cdots, \alpha_{n-1}c(u_{n-11})c(u_{n1})^{-1}, \widetilde{z}) \\
&\times < \alpha_1 u_{21} + \cdots + \alpha_{i-1}u_{i1} + \alpha_i s(u_{i1})u_{i+11} + \cdots >_v du_{21} \cdots du_{n1}.
\end{aligned}$$

If $u_{i1} \notin o_v$, $s(u_{i1}) = 1$. So we compare the integral over the set $\{(u_{21}, \cdots, u_{n1}) \mid u_{i1} \in o_v\}$. If $\alpha_i u_{i+11} \in o_v$, $< \alpha_i u_{i+11} >_v = < \alpha_i s(u_{i1})u_{i+11} >_v = 1$. If $\alpha_i u_{i+11} \notin o_v$, $\alpha_i c(u_{i1})c(u_{i+11})^{-1} \notin o_v$, because $|c(u_{i1})|_v = 1$. Therefore,

$$W_{n-1,v}(\alpha_2 c(u_{21})c(u_{31})^{-1}, \cdots, \alpha_{n-1}c(u_{n-11})c(u_{n1})^{-1}, \widetilde{z}) = 0.$$

Thus, we can replace $s(u_{i1})$ by 1.

Q.E.D.

We continue the proof of (2.3.8). Let

$$\widetilde{t}(u) = (\widetilde{t}_1(u), \cdots, \widetilde{t}_{n-1}(u)) = (t_2 c(u_{21}), \cdots, t_n c(u_{n1})).$$

Then

$$\widetilde{t}_1(u)\widetilde{t}_2(u)^{-1} = \alpha_2 c(u_{21})c(u_{31})^{-1}, \cdots, \widetilde{t}_{n-2}(u)\widetilde{t}_{n-1}(u)^{-1} = \alpha_{n-1}c(u_{n-11})c(u_{n1})^{-1}.$$

Lemma (2.3.11) *Let $\alpha_i \in o_v$ for all i. Then*

$$\begin{aligned}
W_{n,v}(\alpha, z) = &\int_{k_v^{n-1}} |c(u_{21})|_v^{z_{n-1}-z_n+1} \cdots |c(u_{n1})|_v^{z_1-z_n+1} \lambda_{n-1,v}(\widetilde{t}(u), \widetilde{z}) \\
&\times < \alpha_1 u_{21} + \alpha_2 u_{31} + \cdots + \alpha_{n-1}u_{n1} >_v du_{21} \cdots du_{n1}.
\end{aligned}$$

Proof. We can assume that $\alpha_1 = \pi_v^{h_1}, \cdots, \alpha_{n-1} = \pi_v^{h_{n-1}}$. Let $D_{j_1,\cdots,j_{n-1}} = \pi_v^{j_1} o_v^\times \times \cdots \times \pi_v^{j_{n-1}} o_v^\times$. We compare integrals over the set $D_{j_1,\cdots,j_{n-1}}$.

If $\alpha_i c(u_{i1})c(u_{i+11})^{-1} \in o_v$ for $i = 2, \cdots, n-1$,

$$W_{n-1,v}(\alpha_2 c(u_{21})c(u_{31})^{-1}, \cdots, \alpha_{n-1}c(u_{n-11})c(u_{n1})^{-1}, \widetilde{z}) = \lambda_{n-1,v}(\widetilde{t}(u), \widetilde{z})$$

on $D_{j_1,\cdots,j_{n-1}}$. Suppose that $\alpha_i c(u_{i1})c(u_{i+11})^{-1} \in \pi_v^l$ where $l < 0$ for some i. Then

$$W_{n-1,v}(\alpha_2 c(u_{21})c(u_{31})^{-1}, \cdots, \alpha_{n-1}c(u_{n-11})c(u_{n1})^{-1}, \widetilde{z}) = 0.$$

Since $c(u_{i1})$ is always integral, $h_1 + j_{i+1} \leq l$. If $l < -1$,

$$\int_{D_{j_1,\cdots,j_{n-1}}} |c(u_{21})|_v^{z_{n-1}-z_n+1} \cdots |c(u_{n1})|_v^{z_1-z_n+1} \lambda_{n-1,v}(\widetilde{t}(u), \widetilde{z})$$
$$\times < \alpha_1 u_{21} + \alpha_2 u_{31} + \cdots + \alpha_{n-1}u_{n1} >_v du_{21} \cdots du_{n1} = 0$$

by applying (2.3.3) to u_{i+11}. If $l = -1$, and $u_{i1} \notin o_v$, $h_1 + j_{i+1} < -1$, so again, the above integral is zero. If $l = -1$, and $u_{i1} \in o_v$, $h_1 + j_{i+1} = -1$, and $\lambda_{n-1,v}(\widetilde{t}(u), \widetilde{z}) = 0$ by the remark after (2.3.7).

Q.E.D.

Now we can complete the proof of (2.3.8).

By definition,

$$C_{n-1}(\widetilde{t}(u), \widetilde{z}) = (|\pi_v^{n-j-1} t_{j+1} c(u_{j+11})|_v^{\widetilde{z}_{n-i}})_{i,j},$$

i.e.

$$C_{n-1}(\widetilde{t}(u), \widetilde{z}) = \begin{pmatrix} |\pi_v^{n-2} t_2 c(u_{21})|_v^{\widetilde{z}_{n-1}} & \cdots & |t_n c(u_{n1})|_v^{\widetilde{z}_{n-1}} \\ \vdots & \vdots & \vdots \\ |\pi_v^{n-2} t_2 c(u_{21})|_v^{\widetilde{z}_1} & \cdots & |t_n c(u_{n1})|_v^{\widetilde{z}_1} \end{pmatrix}.$$

Let $\nu_{n-1} = \tau_{\mathrm{GL}(n-1)}$. An easy computation shows that

$$|c(u_{21})|_v^{z_{n-1}-z_n+1} \cdots |c(u_{n1})|_v^{z_1-z_n+1} \widetilde{t}(u)^{-\nu_{n-1}\widetilde{z}}$$
$$= \prod_{j=1}^{n-1} t_{j+1}^{-\widetilde{z}_{n-j}} \prod_{j=1}^{n-1} |c(u_{j+11})|_v^{-\widetilde{z}_{n-j}+z_{n-j}-z_n+1}.$$

Therefore,

$$\prod_{j=1}^{n-1} t_{j+1}^{-\widetilde{z}_{n-j}} \prod_{j=1}^{n-1} |c(u_{j+11})|_v^{-\widetilde{z}_{n-j}+z_{n-j}-z_n+1} \det C_{n-1}(\widetilde{t}(u), \widetilde{z})$$
$$= \det D^{(1)}(t, u, z) \prod_{j=1}^{n-1} t_{j+1}^{-\widetilde{z}_{n-j}},$$

where $D^{(1)}(t, u, z)$ is the following matrix

$$\begin{pmatrix} |\pi_v^{n-2} t_2|_v^{\widetilde{z}_{n-1}} |c(u_{21})|_v^{z_{n-1}-z_n+1} & \cdots & |t_n|_v^{\widetilde{z}_{n-1}} |c(u_{n1})|_v^{z_{n-1}-z_n+1} \\ \vdots & \vdots & \vdots \\ |\pi_v^{n-2} t_2|_v^{\widetilde{z}_1} |c(u_{21})|_v^{z_1-z_n+1} & \cdots & |t_n|_v^{\widetilde{z}_1} |c(u_{n1})|_v^{z_1-z_n+1} \end{pmatrix}.$$

It is easy to see that

$$\int_{k_v^{n-1}} \det D^{(1)}(t,u,z) < \alpha_1 u_{21} + \alpha_2 u_{31} + \cdots + \alpha_{n-1} u_{n1} >_v du_{21} \cdots du_{n1}$$
$$= \det D^{(2)}(t,z) \prod_{i=1}^{n-1} \frac{1 - q_v^{-(z_i - z_n + 1)}}{1 - q_v^{-(z_i - z_n)}},$$

where $D^{(2)}(t,z)$ is the following matrix

$$\begin{pmatrix} (1 - |\pi_v t_1 t_2^{-1}|_v^{z_{n-1}-z_n}) |\pi_v^{n-2} t_2|_v^{\widetilde{z}_{n-1}} & \cdots & (1 - |\pi_v t_{n-1} t_n^{-1}|_v^{z_{n-1}-z_n}) |t_n|_v^{\widetilde{z}_{n-1}} \\ \vdots & \vdots & \vdots \\ (1 - |\pi_v t_1 t_2^{-1}|_v^{z_1 - z_n}) |\pi_v^{n-2} t_2|_v^{\widetilde{z}_1} & \cdots & (1 - |\pi_v t_{n-1} t_n^{-1}|_v^{z_1 - z_n}) |t_n|_v^{\widetilde{z}_1} \end{pmatrix}.$$

Let v_j be an $(n-1)$-dimensional column vector whose i-entry is $|\pi_v^{n-j} t_j|_v^{\widetilde{z}_{n-i}}$. Let $d_j = |\pi_v t_j t_{j+1}^{-1}|_v^{-\widetilde{z}_n}$. Then,

$$D^{(2)}(t,z) = (v_2 - d_1 v_1, v_3 - d_2 v_2, \cdots, v_n - d_{n-1} v_{n-1}).$$

Therefore, $\det D^{(2)}(t,z)$ is equal to

$$\begin{aligned} &\det(v_2, \cdots, v_n) - d_1 \det(v_1, v_3, \cdots, v_n) \\ &\quad + \cdots + (-1)^{n-1} d_1 \cdots d_{n-1} \det(v_1, \cdots, v_{n-1}) \\ &= \det \begin{pmatrix} 1 & d_1 & d_1 d_2 & \cdots & d_1 \cdots d_{n-1} \\ v_1 & v_2 & v_3 & \cdots & v_n \end{pmatrix}, \end{aligned}$$

and

$$d_1 \cdots d_j = |t_1|^{-\widetilde{z}_n} |\pi_v|_v^{-(n-1)\widetilde{z}_n} |\pi_v^{n-j-1} t_{j+1}|_v^{\widetilde{z}_n}.$$

Note that $(n-1)\widetilde{z}_n = n z_n$.

It is easy to see that

$$t_1^{-\widetilde{z}_n} t_2^{-\widetilde{z}_{n-1}} \cdots t_n^{-\widetilde{z}_1} = t^{-\tau_G z} (t_1 \cdots t_n)^{-\frac{z_n}{n-1}}.$$

Therefore, $W_{n,v}(\alpha, z)$ is equal to

$$t^{-\tau_G z} (t_1 \cdots t_n)^{-\frac{z_n}{n-1}} |\pi_v|_v^{-n z_n} q_v^{-\frac{1}{2} \sum_{i<j\le n-1} (z_i - z_j)} \det D^{(3)}(t,z) \prod_{i<j} \frac{1 - q_v^{-(z_i - z_j + 1)}}{1 - q_v^{-(z_i - z_j)}},$$

where $D^{(3)}(t,z)$ is the following matrix

$$\begin{pmatrix} |\pi_v^{n-1} t_1|_v^{\widetilde{z}_n} & \cdots & |t_n|_v^{\widetilde{z}_n} \\ \vdots & \vdots & \vdots \\ |\pi_v^{n-1} t_1|_v^{\widetilde{z}_1} & \cdots & |t_n|_v^{\widetilde{z}_1} \end{pmatrix}.$$

By an easy computation,

$$\det D^{(3)}(t,z) = (t_1 \cdots t_n)^{\frac{z_n}{n-1}} |\pi_v|_v^{\frac{n}{2} z_n} \det C_n(t,z).$$

Since

$$-\frac{1}{2}\sum_{i<j\leq n-1}(z_i-z_j)+\frac{n}{2}z_n=-\frac{1}{2}\sum_{i<j\leq n}(z_i-z_j),$$

this proves (2.3.8).

Q.E.D.

We have been using the assumption that the character $< >_v$ is of index zero. In general, if $\mathfrak{a}=(a_v)_v$ is the difference idele, $< a_v\cdot >_v$ is a character of index zero. So for a finite place v, we define a function $\sigma_{n,v}(\alpha,z)$ of $\alpha\in(k_v^\times)^{n-1}$ and z as above by the formula

$$W_{n,v}(a_v\alpha,z)=\sigma_{n,v}(\alpha,z)\prod_{i<j}(1-q_v^{-(z_i-z_j+1)}). \tag{2.3.12}$$

Definition (2.3.13) *For $\alpha=(\alpha_v)_v\in\mathbb{A}^\times$, we define*

$$\sigma_n(\alpha,z)=\prod_{v\in\mathfrak{M}_f}\sigma_{n,v}(\alpha_v,z).$$

By definition,

$$W_{n,f}(\mathfrak{a}\alpha,z)=\frac{\sigma_n(\alpha,z)}{\prod_{i<j}\zeta_k(z_i-z_j+1)}.$$

Lemma (2.3.14) *Let $\alpha=(\alpha_1,\cdots,\alpha_{n-1})\in(k_v^\times)^{n-1}$ and $\alpha_j\in\pi_v^{\lambda_j}$ for all j. Then $\sigma_{n,v}(\alpha,z)$ satisfies the following inductive relation*

$$\sigma_{n,v}(\alpha,z)=\sum_{j=1}^{n-1}\sum_{c_j=0}^{\lambda_j}q_v^{-c_1(z_{n-1}-z_n)}\cdots q_v^{-c_{n-1}(z_1-z_n)}$$
$$\times\,\sigma_{n-1,v}(\alpha_2\pi_v^{c_1-c_2},\cdots,\alpha_{n-1}\pi_v^{c_{n-2}-c_{n-1}},\widetilde{z}_1,\cdots,\widetilde{z}_{n-1}).$$

Proof. We choose t as in (2.3.6). Let

$$c=(c_1,\cdots,c_{n-1}),\ \text{and}\ \widetilde{t}(c)=(\pi_v^{c_1}t_2,\cdots,\pi_v^{c_{n-1}}t_n).$$

By (2.3.8), the right hand side is equal to

$$\sum_{j=1}^{n-1}\sum_{c_j=0}^{\lambda_j}q_v^{-c_1(z_{n-1}-z_n)}\cdots q_v^{-c_{n-1}(z_1-z_n)}\frac{\widetilde{t}(c)^{-\nu_{n-1}\widetilde{z}}|\pi_v|_v^{\frac{1}{2}\sum_{i<j\leq n-1}(z_i-z_j)}}{\prod_{i<j\leq n-1}(1-q_v^{-(z_i-z_j)})}$$
$$\times\det C_{n-1}(\widetilde{t}(c),\widetilde{z}),$$

where $\nu_{n-1}=\tau_{\mathrm{GL}(n-1)}$.

Let

(2.3.15)

$$B(t,z)=\sum_{i=1}^{n-1}\sum_{c_j=0}^{\lambda_j}q_v^{-c_1(z_{n-1}-z_n)}\cdots q_v^{-c_{n-1}(z_1-z_n)}\widetilde{t}(c)^{-\nu_{n-1}\widetilde{z}}\det C_{n-1}(\widetilde{t}(c),\widetilde{z}).$$

The (i,j)-entry of $C_{n-1}(\widetilde{t}(c),\widetilde{z})$ is

$$|\pi_v^{n-j-1}\pi_v^{c_j}t_{j+1}|^{\widetilde{z}_{n-i}} = q_v^{-c_j\widetilde{z}_{n-i}}|\pi_v^{n-j-1}t_{j+1}|^{\widetilde{z}_{n-i}}.$$

Also

$$\widetilde{t}(c)^{-\nu_{n-1}\widetilde{z}}q_v^{-c_1(z_{n-1}-z_n)}\cdots q_v^{-c_{n-1}(z_1-z_n)} = t_2^{-\widetilde{z}_{n-1}}\cdots t_n^{-\widetilde{z}_1}q_v^{(c_1+\cdots+c_{n-1})\widetilde{z}_n}.$$

Let

$$D(t,z,c) = \left(q_v^{-c_j(z_{n-i}-z_n)}|\pi_v^{n-j-1}t_{j+1}|_v^{\widetilde{z}_{n-i}}\right)_{1\le i,j\le n-1}.$$

Then

$$B(t,z) = \sum_{j=1}^{n-1}\sum_{c_j=0}^{\lambda_j} t_2^{-\widetilde{z}_{n-1}}\cdots t_n^{-\widetilde{z}_1}\det D(t,z,c).$$

The order of the summation and the determinant can be changed.

Since

$$\sum_{c_j=0}^{\lambda_j} q_v^{-c_j(z_{n-i}-z_n)} = \frac{1-q_v^{-(\lambda_j+1)(z_i-z_n)}}{1-q_v^{-(z_i-z_n)}},$$

$$B(t,z) = t_2^{-\widetilde{z}_{n-1}}\cdots t_n^{-\widetilde{z}_1}\det D^{(2)}(t,z)\prod_{i=1}^{n-1}\frac{1}{1-q_v^{-(z_i-z_n)}},$$

where $D^{(2)}(t,z)$ is as in the proof of (2.3.8). The rest of the proof is similar to the proof of (2.3.8).

Q.E.D.

Definition (2.3.16) *For $r=(r_1,\cdots,r_{n-1})\in\mathbb{R}^{n-1}$, we define $D(n,r)$ to be the following set*

$$\{x=(x_1,\cdots,x_n)\in\mathbb{R}^n \mid x_1+\cdots+x_n=0,\ x_i-x_{i+1}>r_i \text{ for } i=1,\cdots,n-1\}.$$

Lemma (2.3.17) *We fix r. Suppose $\alpha=(\alpha_1,\cdots,\alpha_{n-1})\in(o_v\setminus\{0\})^{n-1}$. Then there exist constants $N_1,\cdots,N_{n-1}<0$ independent of v such that*

$$|\sigma_{n,v}(\alpha,z)|\le|\alpha_1|_v^{N_1}\cdots|\alpha_{n-1}|_v^{N_{n-1}}$$

for $\mathrm{Re}(z)\in D(n,r)$.

Proof. We use induction on n. Let $c_1,\cdots,c_{n-1}$ be as in the proof of (2.3.14).

There exist $M_1,\cdots,M_{n-2}<0$ such that

$$\begin{aligned}&|\sigma_{n-1,v}(\alpha_2\pi_v^{c_1-c_2},\cdots,\alpha_{n-1}\pi_v^{c_{n-2}-c_{n-1}},\widetilde{z}_1,\cdots,\widetilde{z}_{n-1})|\\ &\le|\alpha_2\pi_v^{c_1-c_2}|_v^{M_1}\cdots|\alpha_{n-1}\pi_v^{c_{n-2}-c_{n-1}}|_v^{M_{n-2}}.\end{aligned}$$

But $|\alpha_1|_v\le|\pi_v^{c_1}|_v\le|\pi_v^{c_1-c_2}|_v,\cdots,|\alpha_{n-2}|_v\le|\pi_v^{c_{n-2}}|_v\le|\pi_v^{c_{n-2}-c_{n-1}}|_v$. We define $M_0=\sup_i\{\mathrm{Re}(z_i-z_n),1\}$. Then

$$|\pi_v^{-c_1(z_{n-1}-z_n)}\cdots\pi_v^{-c_{n-1}(z_1-z_n)}|_v\le|\alpha_1\cdots\alpha_{n-1}|_v^{-M_0}.$$

Also the number of terms in (2.3.14) is bounded by $|\alpha_1 \cdots \alpha_{n-1}|_v^{-1}$. Hence,

$$|\sigma_{n,v}(\alpha, z)| \leq |\alpha_1 \cdots \alpha_{n-1}|_v^{-M_0-1} |\alpha_1\alpha_2|_v^{M_1} \cdots |\alpha_{n-2}\alpha_{n-1}|_v^{M_{n-2}}.$$

Q.E.D.

For infinite places, some integral formulas are known, for example [71]. However, we have a problem of uniformity. We illustrate the difficulty of infinite places by the example of GL(2). For details on special functions, the reader should see the classic book of Whittaker–Watson [82] (also see [18, pp. 50, 55])

Let $K_\nu(x)$ be the K-Bessel function, i.e.

$$K_\nu(x) = \frac{1}{2}\int_{\mathbb{R}_+} e^{-\frac{x}{2}(t+\frac{1}{t})} t^\nu d^\times t \tag{2.3.18}$$

for $\nu \in \mathbb{C}, x \in \mathbb{R}_+$.

Let $\mathfrak{a} = (a_v)_v$ be the difference idele, and we assume that $a_v = 1$ for $v \in \mathfrak{M}_\infty$.

Proposition (2.3.19)

(1) $$W_{2,v}(\alpha, z) = \frac{2|\alpha|_v^{\frac{z}{2}}}{\pi^{-\frac{z+1}{2}}\Gamma(\frac{z+1}{2})} K_{\frac{z}{2}}(2\pi|\alpha|_v) \quad \textit{for } v \in \mathfrak{M}_\mathbb{R},$$

(2) $$W_{2,v}(\alpha, z) = \frac{2|\alpha|_v^{\frac{z}{2}}}{(2\pi)^{-(z+1)}\Gamma(z+1)} K_{\frac{z}{2}}(4\pi\sqrt{|\alpha|_v}) \quad \textit{for } v \in \mathfrak{M}_\mathbb{C}.$$

Proof. Since the proof is similar, we only consider real places. Let $v \in \mathfrak{M}_\mathbb{R}$. We use the usual trick of multiplying the Gamma function as follows.

$$\begin{aligned}
\Gamma\left(\frac{z+1}{2}\right) W_{2,v}(\alpha, z) &= \int_{\mathbb{R}_+}\int_{\mathbb{R}} e^{-t} t^{\frac{z+1}{2}} \frac{e^{2\pi\sqrt{-1}\alpha}}{1+u^2} d^\times t du \\
&= \int_{\mathbb{R}_+}\int_{\mathbb{R}} e^{-t(1+u^2)+2\pi\sqrt{-1}\alpha} t^{\frac{z+1}{2}} d^\times t du \\
&= \sqrt{\pi}\int_{\mathbb{R}_+} e^{-t-\frac{\pi^2\alpha^2}{t}} t^{\frac{z}{2}} d^\times t \\
&= \pi^{\frac{z+1}{2}} |\alpha|^{\frac{z}{2}} \int_{\mathbb{R}_+} e^{-\pi|\alpha|(t+\frac{1}{t})} t^{\frac{z}{2}} d^\times t.
\end{aligned}$$

Q.E.D.

This formula looks superficially nice, and we seem to have a good estimate concerning the growth with respect to α. However, we now have a Gamma function in the denominator. Because of Stirling's formula, $\Gamma(\frac{z+1}{2})^{-1}$ is a function of exponential growth with respect to $\mathrm{Im}(z)$. Since we intend to consider contour integrals along vertical lines, this is very inconvenient. So this formula is not directly applicable for our purposes. On the other hand, in the cases $n = 2, 3$, it is possible to compute the Mellin transform of $W_{n,v}(\alpha, z)$ and actually get a polynomial growth estimate (see Bump [5] for example). But in order to generalize this method, one may have to prove an estimate similar to Stirling's formula for general hypergeometric functions, which is an open problem.

The asymptotic expansion of the K-Bessel function is known to be as follows.

$$K_\nu(x) = \left(\frac{\pi}{2x}\right)^{\frac{1}{2}} e^{-x}\left[1+\sum_{r=1}^{\infty}\frac{(4\nu^2-1)(4\nu^2-3^2)\cdots(4\nu^2-(2r-1)^2)}{r!2^{3r}x^r}\right]$$

for $x \gg 0$. However, we are going to prove that $W_{n,v}(\alpha,z)$ has an estimate of polynomial growth with respect to $\mathrm{Im}(z)$. Since

$$K_{\frac{z}{2}}(2\pi|\alpha|_v) = \frac{1}{2}\pi^{\frac{z+1}{2}}|\alpha|_v^{-\frac{z}{2}}\Gamma(\frac{z+1}{2})W_{2,v}(\alpha,z)$$

for $v \in \mathfrak{M}_{\mathbb{R}}$, $K_\nu(x)$ has an estimate of exponential decay with respect to $\mathrm{Im}(\nu)$. Therefore, either the convergence of the above asymptotic expansion is not uniform with respect to $\mathrm{Im}(\nu)$ or the error term is significant as $\mathrm{Im}(\nu)\to\infty$. So for the sake of a uniform estimate, we cannot use the asymptotic expansion either.

Shintani avoided these difficulties in [64], and used the naive integration by parts. This is the idea of our approach to handle Whittaker functions at infinite places. Since our idea is based on the use of integration by parts, it may be interesting to interpret our computation from the viewpoint of differential operators, for which we do not know if it is possible to do so.

In (2.3.20)–(2.3.22), $v \in \mathfrak{M}_\infty$.

In order to estimate $W_{n,v}(\alpha,z)$, it is easier to consider a slightly more general function. When $v \in \mathfrak{M}_{\mathbb{R}}$ (resp. $v \in \mathfrak{M}_{\mathbb{C}}$) and $a_v = 1$, we denote $W_{n,v}(\alpha,z)$ by $W_{n,\mathbb{R}}(\alpha,z)$ (resp. $W_{n,\mathbb{C}}(\alpha,z)$).

Definition (2.3.20) *Let $\alpha = (\alpha_1,\cdots,\alpha_{n-1}) \in (\mathbb{R}_+)^{n-1}$. Let $z, \widetilde{z}$ be as in (2.3.2). For a non-negative integer j, we define*

$$\begin{aligned}W_{n,j,\mathbb{R}}(\alpha,z) = \int_{\mathbb{R}^{n-1}} & u_1^j c(u_1)^{z_{n-1}-z_n+1}\cdots c(u_{n-1})^{z_1-z_n+1}\\ & \times W_{n-1,\mathbb{R}}(\alpha_2 c(u_1)c(u_2)^{-1},\cdots,\alpha_{n-1}c(u_{n-1})c(u_n)^{-1},\widetilde{z}_1,\cdots,\widetilde{z}_{n-1})\\ & \times \mathrm{e}(\alpha_1u_1+\alpha_2 s(u_1)u_2+\cdots+\alpha_{n-1}s(u_{n-2})u_{n-1})du_1\cdots du_{n-1}.\end{aligned}$$

We define $W_{n,j,\mathbb{C}}(\alpha,z)$ in the same way replacing u_1^j in the above formula by $\mathrm{Re}(u_1)^j$, $c(u_i)^{z_{n-1}-z_n+1}$ etc. by $c(u_i)^{2(z_{n-1}-z_n+1)}$ etc., and $e(\alpha_1u_1+\cdots)$ by $e(2\mathrm{Re}(\alpha_1u_1+\cdots))$.

Obviously,

$$\frac{d^j}{d\alpha_1^j}W_{n,\mathbb{R}}(\alpha,z) = (2\pi\sqrt{-1})^jW_{n,j,\mathbb{R}}(\alpha,z),$$
$$\frac{d^j}{d\alpha_1^j}W_{n,\mathbb{C}}(\alpha,z) = (4\pi\sqrt{-1})^jW_{n,j,\mathbb{C}}(\alpha,z).$$

(We consider the derivation considering $\alpha_1 \in \mathbb{R}$.)

Consider functions of the form $g(u) = f(u)c(u)^a$, where $f(u)$ is a polynomial, and a is an integer. We define $\deg g = \deg f - a$. This is well defined. If $\deg g \le 0$, $\sup|g(u)| < \infty$.

Now we estimate $W_{n,j,\mathbb{R}}(\alpha,z)$ by induction on n. For $n=2$, we can obtain a precise result as in Shintani [64] as follows.

Lemma (2.3.21) *Let $l \geq 1$ be an integer. Then $W_{2,\mathbb{R},j}(\alpha_1, z), W_{2,\mathbb{C},j}(\alpha_1, z)$ are entire functions, and for any such l satisfying $\mathrm{Re}(z_1 - z_2) + l > j$ for the real case and $2(\mathrm{Re}(z_1 - z_2) + l) > j$ for the imaginary case, there exists a function $c_{j,l}(z)$ of polynomial growth such that*

$$|W_{2,j,\mathbb{R}}(\alpha_1, z)| \leq c_{j,l}(z)\alpha_1^{-l},$$
$$|W_{2,j,\mathbb{C}}(\alpha_1, z)| \leq c_{j,l}(z)\alpha_1^{-2l}.$$

Proof. In the real case, if $f(u) = u^j c(u)^{z+1}$, the l-th derivative of $f(u)$ is bounded by a function of the form $g(u)c(u)^{\mathrm{Re}(z)+1}$, where $\deg g \leq j - l$. Thus, one just has to use integration by parts l times. The imaginary case is similar.

Q.E.D.

Consider the domain $\{z \mid \mathrm{Re}(z) \in D(2,0)\}$. If $j = 0$, by the remark after (1.2.1), the above estimate is true for any real number $l \geq 1$.

Now we consider general n.

Proposition (2.3.22) *Let r be as above. Then*
(1) $W_{n,j,\mathbb{R}}(\alpha, z), W_{n,j,\mathbb{C}}(\alpha, z)$ are entire functions,
(2) There exist $M_1 = M_1(r), \cdots, M_{n-1} = M_{n-1}(r) \in \mathbb{R}$, independent of j such that if $l_1, \cdots, l_{n-1}$ are positive integers satisfying

$$l_1 - l_2 > M_1 + j,\ l_2 - l_3 > M_2, \cdots, l_{n-2} - l_{n-1} > M_{n-2},\ l_{n-1} > M_{n-1},$$

then there exist estimates of the form

$$|W_{n,j,\mathbb{R}}(\alpha, z)| \leq c_{r,j,l}(z)\alpha_1^{-l_1} \cdots \alpha_{n-1}^{-l_{n-1}},$$
$$|W_{n,j,\mathbb{C}}(\alpha, z)| \leq c_{r,j,l}(z)\alpha_1^{-2l_1} \cdots \alpha_{n-1}^{-2l_{n-1}},$$

where $\mathrm{Re}(z) \in D(n, r)$, and $c_{r,j,l}(z)$ is a function of polynomial growth.

Proof. We use induction on n. We first consider the real case. Consider the function $W_{n+1,j,\mathbb{R}}(\alpha, z)$. Let $r = (r_1, \cdots, r_n) \in \mathbb{R}^n$ and $r' = (r_1, \cdots, r_{n-1})$. Let $z = (z_1, \cdots, z_{n+1}) \in \mathbb{C}^{n+1}$, $z_1 + \cdots z_{n+1} = 0$, and $\mathrm{Re}(z) = x = (x_1, \cdots, x_{n+1}) \in D(n+1, r)$. Let l_1 be a positive integer, and $0 \leq m_1, m_2, m_3 \leq l_1$.

Let

$$\beta(\alpha, u) = (\alpha_2 c(u_1)c(u_2)^{-1}, \cdots, \alpha_n c(u_{n-1})c(u_n)^{-1}).$$

Let $\widetilde{z}_i = z_i + \frac{z_{n+1}}{n}$ and $\widetilde{x}_i = \mathrm{Re}(\widetilde{z}_i)$ for $i = 1, \cdots, n$. Consider a function of the form

$$g(u_1)c(u_1)^m(\alpha_2 c(u_1)c(u_2)^{-1})^{m'} W_{n,m',\mathbb{R}}(\beta(\alpha, u), \widetilde{z}_1, \cdots, \widetilde{z}_n).$$

The derivative of the above function with respect to u_1 is

$$g_1(u_1)c(u_1)^{m+1}(\alpha_2 c(u_1)c(u_2)^{-1})^{m'} W_{n,m',\mathbb{R}}(\beta(\alpha, u), z_1, \cdots, z_n)$$
$$+ g_2(u_1)c(u_1)^{m+1}(\alpha_2 c(u_1)c(u_2)^{-1})^{m'+1} W_{n,m'+1,\mathbb{R}}(\beta(\alpha, u), \widetilde{z}_1, \cdots, \widetilde{z}_n),$$

where

$$g_1(u_1) = g'(u_1)c(u_1)^{-1} - mg(u_1)u_1c(u_1) - m'g(u_1)u_1c(u_1),$$
$$g_2(u_1) = -2\pi\sqrt{-1}g(u_1)u_1c(u_1).$$

Clearly $\deg g_1, \deg g_2 \le \deg g$. By induction, the m_2-th derivative of

$$W_{n,\mathbb{R}}(\beta(\alpha,u),\widetilde{z}_1,\cdots,\widetilde{z}_n)$$

with respect to u_1 is bounded by a finite linear combination of functions of the form

$$c(u_1)^{m_2}(\alpha_2 c(u_1)c(u_2)^{-1})^{m_4}|W_{n,m_4,\mathbb{R}}(\beta(\alpha,u),\widetilde{z}_1,\cdots,\widetilde{z}_n)|,$$

where $0 \le m_4 \le m_2$.

Similarly, the m_3-th derivative of $e(\alpha_2 s(u_1)u_2)$ with respect to u_1 is bounded by a finite linear combination of functions of the form $c(u_1)^{m_3}(\alpha_2 c(u_1)c(u_2)^{-1})^{m_5}$ where $0 \le m_5 \le m_3$.

We apply integration by parts with respect to $e(\alpha_1 u_1)$ l_1 times to $W_{n+1,j,\mathbb{R}}(\alpha,z)$. Then, by the above remark, $W_{n+1,j,\mathbb{R}}(\alpha,z)$ is bounded by a finite linear combination of functions of the form

$$\begin{aligned}\alpha_1^{-l_1}\int_{\mathbb{R}^n} & c(u_1)^{x_n-x_{n+1}+1-j+l_1}c(u_2)^{x_{n-1}-x_{n+1}+1}\cdots c(u_n)^{x_1-x_{n+1}+1}\\ &\times(\alpha_2 c(u_1)c(u_2)^{-1})^{m_4+m_5}|W_{n,m_4,\mathbb{R}}(\beta(\alpha,u),\widetilde{x}_1,\cdots,\widetilde{x}_n)|du_1\cdots du_n\end{aligned}$$

where $0 \le m_4+m_5 \le l_1$.

Since $\mathrm{Re}(\widetilde{z}_1,\cdots,\widetilde{z}_n) \in D(n,r')$, by induction, we can choose $\widetilde{M}_2,\cdots,\widetilde{M}_n$ independent of m_4 (and hence independent of l_1, j) so that if $\widetilde{l}_2, l_3,\cdots,l_n$ are positive integers satisfying the condition

$$\widetilde{l}_2 - l_3 > \widetilde{M}_2 + m_4,\ l_3 - l_4 > \widetilde{M}_3,\cdots,l_{n-1}-l_n > \widetilde{M}_{n-1},\ l_n > \widetilde{M}_n,$$

then there exists an estimate of the form

$$\begin{aligned}&|W_{n,m_4,\mathbb{R}}(\beta(\alpha,u),\widetilde{x}_1,\cdots,\widetilde{x}_n)|\\ &\le c_{r',j,l}(z)(\alpha_2 c(u_1)c(u_2)^{-1})^{-\widetilde{l}_2}\cdots(\alpha_n c(u_{n-1})c(u_n)^{-1})^{-l_n},\end{aligned}$$

where $c_{r',j,l}(z)$ is a function of polynomial growth (since m_4 is an integer between 0 and l_1 and $c_{r',j,l}(z)$ depends on l_1, we do not have to write the dependency on m_4). Then $|W_{n+1,j,\mathbb{R}}(\alpha,z)|$ is bounded by a finite linear combination of functions of the form

$$\begin{aligned}& c'_{r',j,l}(z)\alpha_1^{-l_1}\alpha_2^{-(\widetilde{l}_2-m_4-m_5)}\alpha_3^{-l_3}\cdots\alpha_n^{-l_n}\\ &\times\int_{\mathbb{R}^n} c(u_1)^{x_n-x_{n+1}+1-j+l_1-\widetilde{l}_2+m_4+m_5}c(u_2)^{x_{n-1}-x_{n+1}+1+\widetilde{l}_2-m_4-m_5-l_3}\\ &\qquad\times c(u_3)^{x_{n-2}-x_{n+1}+1+l_3-l_4}\cdots c(u_n)^{x_1-x_{n+1}+1+l_n}du_1\cdots du_n,\end{aligned}$$

where $c'_{r',j,l}(z)$ is a function of polynomial growth.

By replacing $\widetilde{M}_3,\cdots,\widetilde{M}_n$ if necessary, we can assume that the integral over $u_3,\cdots,u_n$ converges absolutely for $x \in D(n,r)$. The only condition that $\widetilde{l}_2$ has to satisfy is that $\widetilde{l}_2 - l_3 > \widetilde{M}_2 + m_4$. So $\widetilde{l}_2 - l_3 - m_4 - m_5$ can be any integer greater than $\widetilde{M}_2$.

Let $l_2 = \widetilde{l}_2 - m_4 - m_5$. Then if $l_1 - l_2 - j$, $l_2 - l_3$ are sufficiently large, the integral over u_1, u_2 converges absolutely for $x \in D(n,r)$. Therefore, there exist $M_1, \cdots, M_n$, such that if $l_1 - l_2 > M_1 + j$, $l_2 - l_3 > M_2, \cdots, l_{n-1} - l_n > M_{n-1}$, $l_n > M_n$,

$$|W_{n+1,j,\mathbb{R}}(\alpha, z)| \leq c_{r,j,l}(z)\alpha_1^{-l_1} \cdots \alpha_n^{-l_n},$$

where $c_{r,j,l}(z)$ is a function of polynomial growth. This proves (2.3.22) for $n+1$.

Since α is real, $\mathrm{Re}(\alpha_1 u_1) = \alpha_1 \mathrm{Re}(u_1)$ etc. So, in order to estimate $W_{n,j,\mathbb{C}}(\alpha, z)$, we apply integration by parts to $\mathrm{Re}(u_1)$, and the rest of the argument is similar to the real case.

Q.E.D.

By the remark after (1.2.1), $l_1, \cdots, l_{n-1}$ can be arbitrary real numbers satisfying $l_1 \gg \cdots \gg l_{n-1} \gg 0$.

Lemma (2.3.23) *Let $v \in \mathfrak{M}$. Then*

$$W_{n,v}(\alpha, z) = W_{n,v}((-\alpha_{n-1}, \cdots, -\alpha_1), -\tau_G z).$$

Proof. If the diagonal part of the Iwasawa decomposition of $n(u)\tau_G$ for $u \in k_v^{\frac{n(n+1)}{2}}$ is $a(u)$, the diagonal part of the Iwasawa decomposition of $\tau_G {}^t n(u)^{-1}$ is $(a(u)^{-1})^{\tau_G}$. So

$$t(\tau_G {}^t n(u)^{-1})^{z+\rho} = (a(u)^{-1})^{\tau_G(z+\rho)} = t(n(u)\tau_G)^{-\tau_G(z+\rho)} = t(n(u)\tau_G)^{-\tau_G z+\rho}.$$

Let $\widetilde{u} \in k_v^{\frac{n(n+1)}{2}}$ be the element such that $n(\widetilde{u}) = \tau_G {}^t n(u)^{-1}\tau_G$. Then

$$\begin{aligned} W_{n,v}(\alpha, z) &= \int_{N_{k_v}} t(n(\widetilde{u})\tau_G)^{-\tau_G z+\rho}\psi_\alpha(n(u))du \\ &= \int_{N_{k_v}} t(n(\widetilde{u})\tau_G)^{-\tau_G z+\rho}\psi_{(-\alpha_{n-1},\cdots,-\alpha_1)}(n(\widetilde{u}))d\widetilde{u} \\ &= W_{n,v}((-\alpha_{n-1}, \cdots, -\alpha_1), -\tau_G z). \end{aligned}$$

This proves the lemma.

Q.E.D.

Since $-\tau_G z = (-z_n, \cdots, -z_1)$, the condition $\mathrm{Re}(z) \in D(n,r)$ implies the condition $\mathrm{Re}(-\tau_G z) \in D(n, r_{n-1}, \cdots, r_1)$.

Corollary (2.3.24)
(1) *If $0 \ll l_1 \ll l_2 \ll \cdots \ll l_{n-1}$ or $l_1 \gg l_2 \gg \cdots \gg l_{n-1} \gg 0$, there exists an estimate of the form*

$$|W_{n,\infty}(\alpha, z)| \leq c_{r,l}(z)|\alpha_1|_\infty^{-l_1} \cdots |\alpha_{n-1}|_\infty^{-l_{n-1}},$$

for $\mathrm{Re}(z) \in D(n,r)$, where $c_{r,l}(z)$ is a function of polynomial growth ($l_1, \cdots, l_{n-1}$ depend on r).
(2) *Suppose $n = 2$. Then, for any positive integer l satisfying $l + r > 0$, the estimate in (1) is true for $\mathrm{Re}(z) \in D(2,r)$.*

Proof. Since $W_{n,v}(\alpha, z) = W_{n,\mathbb{R}}(a_v\alpha, z)$ or $W_{n,\mathbb{C}}(a_v\alpha, z)$, we can assume that $a_v = 1$. Let $t \in T_{k_v}$ such that $t_it_{i+1}^{-1} = \alpha_i$ for $i = 1, \cdots, n-1$. Since

$$W_{n,v}(\alpha, z) = t^{-\tau_G-\rho}\int_{N_{k_v}} t(tn(u)\tau_G)^{z+\rho}\psi_{v,(1,\cdots,1)}(n(u))du$$

and the right hand side depends only on $|t_1|_v, \cdots, |t_n|_v$, we may assume that $\alpha_1, \cdots, \alpha_{n-1} \in \mathbb{R}_+$. Therefore, we can use (2.3.21)–(2.3.23).

Q.E.D.

Now, we can obtain an estimate of the global Whittaker function as follows. For $\alpha = (\alpha_1, \cdots, \alpha_{n-1}) \in (\mathbb{A}^\times)^{n-1}$, and $t = a_n(t_1, \cdots, t_n) \in T_\mathbb{A}$, we define

$$\delta(\alpha, t) = (\alpha_1t_1t_2^{-1}, \cdots, \alpha_{n-1}t_{n-1}t_n^{-1}) \in (\mathbb{A}^\times)^{n-1}. \tag{2.3.25}$$

Proposition (2.3.26) *Let* $r = (r_1, \cdots, r_{n-1}) \in \mathbb{R}_+^{n-1}$.
(1) *If* $l_1, \cdots, l_{n-1}$ *are real numbers satisfying* $0 \ll l_1 \ll l_2 \ll \cdots \ll l_{n-1}$ *or* $l_1 \gg l_2 \gg \cdots \gg l_{n-1} \gg 0$, *then there exists an estimate of the form*

$$\sum_{i=1}^{n-1}\sum_{\alpha_i\in k^\times} |W_n(\delta(\alpha, t), z)| \le c_{r,l}(z)|t_1t_2^{-1}|^{-l_1}\cdots|t_{n-1}t_n^{-1}|^{-l_{n-1}},$$

for all $t \in T_\mathbb{A}$, $\mathrm{Re}(z) \in D(n, r)$, *and* $c_l(z)$ *is a function of polynomial growth.*
(2) *Suppose* $n = 2$. *Then for any real number* $l > 1$,

$$\sum_{\alpha_1\in k^\times} |W_2(\delta(\alpha, t), z)| \le c_l(z)|t_1t_2^{-1}|^{-l},$$

where $t \in T_\mathbb{A}$, $\mathrm{Re}(z) \in D(2, r)$, *and* $c_l(z)$ *is a function of polynomial growth.*

Proof. By (2.3.22),

$$|W_{n,\infty}(\delta(\alpha, t), z)| \le c_l'(z)|\alpha_1t_1t_2^{-1}|_\infty^{-l_1}\cdots|\alpha_{n-1}t_{n-1}t_n^{-1}|_\infty^{-l_{n-1}},$$

for some function $c_l'(z)$ of polynomial growth. So $\sum_{i=1}^{n-1}\sum_{\alpha_i\in k^\times}|W_n(\delta(\alpha, t), z)|$ is bounded by

$$c_l'(z)\sum_{i=1}^{n-1}\sum_{\substack{\alpha_i\in k^\times \\ \mathfrak{a}\alpha_it_it_{i+1}^{-1}\ \text{integral}}} \frac{|\sigma_n(\mathfrak{a}\delta(\alpha, t), z)|}{|\prod_{i<j}\zeta(z_i - z_j + 1)|}|\alpha_1t_1t_2^{-1}|_\infty^{-l_1}\cdots|\alpha_{n-1}t_{n-1}t_n^{-1}|_\infty^{-l_{n-1}}.$$

By (2.3.17), there exist constants $N_1, \cdots, N_{n-1}$ such that

$$|\sigma_n(\mathfrak{a}\delta(\alpha, t), z)| \le |\mathfrak{a}_f|_f^{N_1+\cdots+N_{n-1}}|\alpha_1t_1t_2^{-1}|_f^{N_1}\cdots|\alpha_{n-1}t_{n-1}t_n^{-1}|_f^{N_{n-1}}.$$

Since $|\zeta(z_i - z_j + 1)^{-1}|$ is of logarithmic growth with respect to $\mathrm{Im}(z)$ for $\mathrm{Re}(z) \in D(n, r)$, the above sum is bounded by

$$c_l''(z)\sum_{i=1}^{n-1}\sum_{\substack{\alpha_i\in k^\times \\ \mathfrak{a}\alpha_it_it_{i+1}^{-1}\ \text{integral}}} |\mathfrak{a}\alpha_1t_1t_2^{-1}|_f^{l_1+N_1}\cdots|\mathfrak{a}\alpha_{n-1}t_{n-1}t_n^{-1}|_f^{l_{n-1}+N_{n-1}}$$
$$\times|t_1t_2^{-1}|^{-l_1}\cdots|t_{n-1}t_n^{-1}|^{-l_{n-1}},$$

for some function $c_l''(z)$ of polynomial growth.

If $l_i + N_i > 1$,

$$\sum_{\substack{\alpha_i \in k^\times \\ \mathfrak{a}\alpha_i t_i t_{i+1}^{-1} \text{ integral}}} |\mathfrak{a}\alpha_i t_i t_{i+1}^{-1}|_f^{l_i+N_i} \leq \zeta(l_i + N_i).$$

This proves (1).

For (2), we can be slightly more precise. Let $\mathrm{id}(\alpha t_1 t_2^{-1}) \subset o_k$ be the ideal determined by $\alpha t_1 t_2^{-1}$. Then

$$\begin{aligned}
&\sum_{\alpha \in k} |\sigma_2(\mathfrak{a}\delta(\alpha, t), z)| |\alpha t_1 t_2^{-1}|_\infty^{-l} \\
&\leq \sum_{\substack{\mathfrak{c} \subset o_k \text{ideal} \\ \mathfrak{c} | \text{ideal}(\alpha t_1 t_2^{-1})}} \sum_{\substack{\alpha \in k \\ \mathfrak{a}\alpha t_1 t_2^{-1} \text{ integral}}} N(\mathfrak{c})^{-\mathrm{Re}(z)} |\alpha t_1 t_2^{-1}|_f^l |t_1 t_2^{-1}|^{-1} \\
&\leq |t_1 t_2^{-1}|^{-1} \sum_{c, d \subset o_k \text{ideal}} N(c)^{-(\mathrm{Re}(z)+l)} N(d)^{-l} \\
&\leq \zeta(\mathrm{Re}(z) + l) \zeta(l).
\end{aligned}$$

Q.E.D.

§2.4 The Fourier expansion of Eisenstein series on GL(n)

In this section, we estimate the Fourier coefficients of $E_B(g^0, z)$.

We fix a subset $I = \{i_1, \cdots, i_l\} \subset \{1, \cdots, n-1\}$. Consider $\alpha = (\alpha_1, \cdots, \alpha_{n-1}) \in k^{n-1}$ such that $\alpha_i \neq 0$ if and only if $i \in I$. Let $g^0 = k(g^0)t(g^0)n(u(g^0))$ be the Iwasawa decomposition of g^0 as before. Then

$$\begin{aligned}
E_B(g^0, z)_\alpha &= \psi_\alpha(-n(u(g^0))) \int_{N_\mathbb{A}/N_k} \sum_{\gamma \in G_k/B_k} t(t(g^0)n(u)\gamma)^{z+\rho} \psi_\alpha(n(u)) du \\
&= \psi_\alpha(-n(u(g^0))) \sum_{\tau \in N_k \backslash G_k/B_k} \int_{N_\mathbb{A}/N_{\tau k}} t(t(g^0)n(u)\tau)^{z+\rho} \psi_\alpha(n(u)) du.
\end{aligned}$$

The double coset $N_k \setminus G_k/B_k$ is represented by the set of all the permutation matrices. Let τ be a permutation. If $n(u) \in N_{\tau\mathbb{A}}$, $t(t(g^0)n(u)\tau)^{z+\rho} = t(t(g^0)\tau)^{z+\rho}$. So if there exists $i \in I$ such that $\tau(i) < \tau(i+1)$, the above integral is zero. Therefore, we only consider τ such that $\tau(i+1) < \tau(i)$ for all $i \in I$.

For any such τ,

$$\begin{aligned}
&\int_{N_\mathbb{A}/N_{\tau k}} t(t(g^0)n(u)\tau)^{z+\rho} \psi_\alpha(n(u)) du \\
&= \int_{N_{\tau\mathbb{A}}^-} t(t(g^0)n(u)\tau)^{z+\rho} \psi_\alpha(n(u)) du \\
&= t(g^0)^{\tau z+\rho} \int_{N_{\tau\mathbb{A}}^-} t(n(u)\tau)^{z+\rho} \psi_{\delta(\alpha, t(g^0))}(n(u)) du.
\end{aligned}$$

First, we describe integrals of the above form in terms of the Whittaker functions. For that purpose, we prove a combinatorial lemma. Suppose that

$$I = \{j_1, j_1+1, \cdots, j_1+k_1, j_2, j_2+1, \cdots, j_2+k_2, \cdots, j_r, \cdots, j_r+k_r\},$$

where $j_2 > j_1+k_1+1, \cdots, j_r > j_{r-1}+k_{r-1}+1$.

Consider τ such that $\tau(i+1) < \tau(i)$ for all $i \in I$. Let τ_{1m} be the longest element among permutations of $\{j_m, \cdots, j_m+k_m\}$, i.e. τ_{1m} is the product of transpositions (j_m, j_m+k_m), $(j_m+1, j_m+k_m-1), \cdots$. Let $\tau_1 = \prod \tau_{1m}$, and $\tau_2 = \tau_1^{-1}\tau$.

Lemma (2.4.1) $l(\tau) = l(\tau_1) + l(\tau_2)$.

Proof. The permutation τ changes the orders of $\{j_1, j_1+1\}, \cdots, \{j_1+k_1, j_1+k_1+1\}, \{j_2, j_2+1\}, \cdots$. So if $j_m \le i < j \le j_m+k_m+1$, $1 \le m \le r$, $\tau(i) > \tau(j)$. Therefore, $I_{\tau_1} \subset I_\tau$.

Consider $(i,j) \in \tau_1^{-1} I_{\tau_2}$. The pair (i,j) satisfies two conditions $\tau_1(i) > \tau_1(j)$ and $\tau(i) < \tau(j)$. If $(i,j) \in I_{\tau_1}$, $\tau_1(i) < \tau_1(j)$, so this cannot happen. So $I_{\tau_1} \cap \tau_1^{-1} I_{\tau_2} = \emptyset$. If $i < j$, then $(j,i) \in I_{\tau_1}$. So $j_m \le i < j \le j_m+k_m+1$ for some m. But by the assumption on τ, $\tau(j) < \tau(i)$, so this cannot happen either. This means that $i > j$, and $(i,j) \in I_\tau$. Hence, $I_{\tau_1} \coprod \tau_1^{-1} I_{\tau_2} \subset I_\tau$. So $l(\tau) \ge l(\tau_1) + l(\tau_2)$. But by (2.1.6), $l(\tau) \le l(\tau_1) + l(\tau_2)$. This proves the lemma.

Q.E.D.

Definition (2.4.2) *For* $\alpha = (\alpha_1, \cdots, \alpha_{n-1}) \in (\mathbb{A}^\times)^{n-1}$ *and* $z = (z_1, \cdots, z_n) \in \mathbb{C}^n$, *we define*

$$W_\tau(\alpha, z) = \int_{N^-_{\tau\mathbb{A}}} t(n(u)\tau)^{z+\rho} \psi_\alpha(n(u)) du.$$

We do not restrict ourselves to z satisfying $z_1 + \cdots + z_n = 0$ for convenience. However, we can always add a scalar multiple of $(1, \cdots, 1)$ to z and make it satisfy the above condition. Note that this does not change the value of $W_\tau(\alpha, z)$ or $z_i - z_{i+1}$ for $i = 1, \cdots, n-1$.

Consider the decomposition $\tau = \tau_1\tau_2$ in (2.4.1). Then by a standard argument,

$$(2.4.3)\quad W_\tau(\alpha, z) = \prod_{\substack{i>j \\ \tau_2(i)<\tau_2(j)}} \phi(z_{\tau_2(i)} - z_{\tau_2(j)}) \int_{N^-_{\tau_1\mathbb{A}}} t(n(u_1)\tau_1)^{\tau_2 z+\rho} \psi_\alpha(n(u_1)) du_1.$$

In terms of τ,

$$\prod_{\substack{i>j \\ \tau_2(i)<\tau_2(j)}} \phi(z_{\tau_2(i)} - z_{\tau_2(j)}) = \prod_{\substack{\tau_1(i)>\tau_1(j) \\ \tau(i)<\tau(j)}} \phi(z_{\tau(i)} - z_{\tau(j)}).$$

Clearly,

$$(2.4.4)\qquad \int_{N^-_{\tau_1\mathbb{A}}} t(n(u_1)\tau_1)^{\tau_2 z+\rho} \psi_\alpha(n(u_1)) du_1 = \prod_{m=1}^{r} W_{\tau_{1m}}(\alpha, \tau_2 z).$$

Example (2.4.5)

Consider $G = \mathrm{GL}(9)$. Let $\alpha_1, \alpha_2, \alpha_3, \alpha_5, \alpha_6, \alpha_8 \neq 0$. Then in the figure below, u_{ij} in $*$ contributes $\phi(z_{\tau(i)} - z_{\tau(j)})$.

$$\begin{pmatrix} 1 & & & & & & & & \\ & 1 & & & & & & & \\ & & 1 & & & & 0 & & \\ & & & 1 & & & & & \\ & & & & 1 & & & & \\ & & & & & 1 & & & \\ & & * & & & & 1 & & \\ & & & & & & & 1 & \\ & & & & & & & & 1 \end{pmatrix}$$

Because of the choice of our measure, $W_{\tau_G}(\alpha, z) = |\Delta_k|^{-\frac{n(n-1)}{4}} W_n(\alpha, z)$.

Let $I = \{i_1, \cdots, i_h\} \subset \{1, \cdots, n-1\}$, where $i_1 < \cdots < i_h$ as before. Let τ be a permutation which satisfies the condition that $\tau(i_j + 1) < \tau(i_j)$ for $j = 1, \cdots, h$. Let ν be a permutation which satisfies the condition that if $i < j$, $j \notin I$, then $\nu^{-1}(i) < \nu^{-1}(j)$. The element ν can be written in the form $\nu = \nu_h \cdots \nu_1$, where ν_j fixes $i > i_j$, and ν_j^{-1} preserves the order of $\{1, \cdots, i_j - 1\}$.

Definition (2.4.6) *We define the function $E_{B,I,\tau,\nu}(g^0, z)$ by the following sum*

$$\sum_{\substack{\alpha_i \in k^\times \text{ if } i \in I \\ \alpha_i = 0 \text{ if } i \notin I}} \sum_{\gamma \in N_{\nu k}^-} t(g^0\gamma\nu)^{\tau z + \rho} \psi_\alpha(n(-u(g^0\gamma\nu))) W_\tau(\delta(\alpha, t(g^0\gamma\nu)), z).$$

Then by (2.1.9),

$$E_B(g^0, z) = \sum E_{B,I,\tau,\nu}(g^0, z)$$

if the right hand side converges absolutely and locally uniformly.

We estimate $E_{B,I,\tau,\nu}(g^0, z)$, and prove the required convergence also. Let $\tau = \tau_1\tau_2$ be the decomposition as in (2.4.1). Let

$$E_{B,I,\tau,\nu}(g^0, z) = \prod_{\substack{\tau_1(i) > \tau_1(j) \\ \tau(i) < \tau(j)}} \phi(z_{\tau(i)} - z_{\tau(j)}) \overline{E}_{B,I,\tau,\nu}(g^0, z).$$

By the remark after (2.4.5),

$$W_\tau(g^0, z) = \prod_{\substack{\tau_1(i) > \tau_1(j) \\ \tau(i) < \tau(j)}} \phi(z_{\tau(i)} - z_{\tau(j)}) \prod_{m=1}^{l} W_{\tau_{1m}}(g^0, \tau_2 z).$$

The condition $\mathrm{Re}(z_{\tau_2(i)} - z_{\tau_2(i+1)}) > 0$ for all $i \in I$, is equivalent to the condition $\mathrm{Re}(z_{\tau(i+1)} - z_{\tau(i)}) > 0$ for all $i \in I$.

Definition (2.4.7) *Let $\delta > 0$. We consider $I, \tau = \tau_1\tau_2$ as in (2.4.2). Let $\delta > 0$. We define $D_\tau, D_{\tau,\delta}, D_{I,\tau}, D_{I,\tau,\delta}$ to be the following sets (1)–(4)*

$$(1)\quad \{x = (x_1, \cdots, x_n) \mid x_1 + \cdots + x_n = 0, x_{\tau(i)} - x_{\tau(j)} > 1 \text{ for } (i,j) \in I_\tau\},$$

$$(2)\quad \{x = (x_1, \cdots, x_n) \mid x_1 + \cdots + x_n = 0, x_{\tau(i)} - x_{\tau(j)} > 1 + \delta \text{ for } (i,j) \in I_\tau,$$

$$(3)\quad \left\{ x = (x_1, \cdots, x_n) \,\middle|\, \begin{array}{c} x_1 + \cdots + x_n = 0, x_{\tau(i+1)} - x_{\tau(i)} > 0 \text{ for } i \in I, \\ x_{\tau(i)} - x_{\tau(j)} > 1 \text{ for } (i,j) \in \tau_1^{-1} I_{\tau_2} \end{array} \right\},$$

$$(4)\quad \left\{ x = (x_1, \cdots, x_n) \,\middle|\, \begin{array}{c} x_1 + \cdots + x_n = 0, x_{\tau(i+1)} - x_{\tau(i)} > \delta \text{ for } i \in I, \\ x_{\tau(i)} - x_{\tau(j)} > 1 + \delta \text{ for } (i,j) \in \tau_1^{-1} I_{\tau_2} \end{array} \right\}$$

respectively.

By (2.3.24), if $0 \ll l_1 \ll \cdots \ll l_h$ or $l_1 \gg \cdots \gg l_h \gg 0$,

$$|E_{B,I,\tau,\nu}(g^0, z)| \le c_l(z) \sum_{\gamma \in N_{\nu k}^-} t(g^0\gamma\nu)^{\tau x + \rho} \prod_{j=1}^{h} |t_{i_j}(g^0\gamma\nu) t_{i_j+1}(g^0\gamma\nu)^{-1}|^{-l_j},$$

for $x = \mathrm{Re}(z) \in D_{I,\tau,\delta}$, where $c_l(z)$ is a function of polynomial growth.

Let

$$e_i = (\overbrace{0, \cdots, 0, 1}^{i}, -1, 0, \cdots, 0),\ \bar{e}_{ij} = (\underbrace{\overbrace{0, \cdots, 0, 1}^{i}, 0, \cdots, 0, -1}_{j}, 0, \cdots, 0).$$

We consider $\bar{e}_{ij}$ for $i > j$ also. Clearly, $\bar{e}_{ij} = -\bar{e}_{ji}$. Then the right hand side of the above inequality is

$$c_l(z) \sum_{\gamma \in N_{\nu k}^-} t(t(g^0)\gamma\nu)^{\tau x + \rho - \sum_{j=1}^h l_j e_{i_j}},$$

which is equal to

$$c_l(z) \sum_{\gamma \in N_{\nu k}^-} t(t(g^0)\gamma t(g^0)^{-1}\nu)^{\tau x + \rho - \sum_{j=1}^h l_j e_{i_j}} \cdot t(g^0)^{\nu(\tau x + \rho - \sum_{j=1}^h l_j e_{i_j})}.$$

Proposition (2.4.8) *Let $f = (f_1, \cdots, f_n) = -\sum l_j e_{i_j}$. Then, for any $M > 0$, there exist $0 \ll l_1 \ll \cdots \ll l_h$ satisfying the following two conditions.*
(1) For any $(i,j) \in I_\nu$, $f_{\nu(i)} - f_{\nu(j)} \ge M$.
(2) There exist $A_{ij} \ge 0$ for $i > j$ and $A_{ij} \ge M$ for $(i,j) \in I_\nu$ such that

$$\nu(f) = \sum_{i > j} A_{ij} \bar{e}_{ij}.$$

Proof. We use induction on h. Let $\nu' = \nu_{h-1} \cdots \nu_1$. Let

$$f' = (f_1', \cdots, f_n') = -\sum_{j=1}^{h-1} l_j e_{i_j}.$$

Since $l(\nu) = l(\nu_h) + l(\nu')$, $I_\nu = I_{\nu_h} \coprod \nu_h^{-1} I_{\nu'}$. If $(i,j) \in I_{\nu'}$,

$$\nu(\nu_h^{-1}(i), \nu_h^{-1}(j)) = (\nu'(i), \nu'(j)),\ \nu'(j) \le i_{h-1} < i_h.$$

So adding $-l_h e_{i_h}$ to f' does not change these coordinates. Therefore, by induction, we can assume that $f_{\nu(i)} - f_{\nu(j)} \ge M$ for $(i,j) \in \nu_h^{-1} I_{\nu'}$. If $(i,j) \in I_{\nu_h}$, $\nu_h(j) = i_h$. So we only have to take $l_h \gg l_1, \cdots, l_{h-1}$. This proves (1).

Note that for any τ, $\bar{e}_{\tau^{-1}(i)\tau^{-1}(j)} = \tau \bar{e}_{ij}$. Suppose $\nu_h^{-1}(i_h) = \bar{i}_h$. Then $\nu_h(i) = i$ for $i < \bar{i}_h$, $i > i_h$, $\nu_h(\bar{i}_h) = i_h$, and $\nu_h(i) = i-1$ for $i = \bar{i}_h + 1, \cdots, i_h$. By induction,

$$\nu'(f_1) = \sum_{i>j} B_{ij} \bar{e}_{ij},$$

where $B_{ij} \ge 0$, and if $(i,j) \in I_{\nu'}$, $B_{ij} \ge M$.

$\nu(f) = \nu_h(\nu'(f_1)) - \nu(l_h e_{i_h}) = \nu_h(\nu'(f_1)) - \nu_h(l_h e_{i_h})$, and $\nu_h(\nu'(f_1))$ is equal to

$$\sum_{(i,j)\in I_{\nu'}} B_{ij} \bar{e}_{\nu_h^{-1}(i)\nu_h^{-1}(j)} + \sum_{\substack{(i,j)\notin I_{\nu'} \\ \nu_h^{-1}(i) > \nu_h^{-1}(j)}} B_{ij} \bar{e}_{\nu_h^{-1}(i)\nu_h^{-1}(j)} + \sum_{\substack{(i,j)\notin I_{\nu'} \\ \nu_h^{-1}(i) < \nu_h^{-1}(j)}} B_{ij} \bar{e}_{\nu_h^{-1}(i)\nu_h^{-1}(j)}.$$

Since $\nu_h^{-1} I_{\nu'} \subset I_\nu$, the condition on the coefficients is satisfied for pairs in the first term. We can ignore the second term, because $\nu_h^{-1}(i) > \nu_h^{-1}(j)$. In the third term, $(\nu_h^{-1}(j), \nu_h^{-1}(i)) \in I_{\nu_h}$. So $\nu_h^{-1}(i) = \bar{i}_h$, $\bar{i}_h \le \nu_h^{-1}(j) \le i_h$. But

$$-\nu_h(e_{i_h}) = -(e_{\bar{i}_h} + \cdots + e_{i_h}) = \bar{e}_{\bar{i}_h+1\bar{i}_h} + \cdots + \bar{e}_{i_h+1 i_h}.$$

So if $l_h \gg l_{h-1}$,

$$-l_h \nu_h(e_{i_h}) - \sum_{\substack{(i,j)\notin I_{\nu'} \\ \nu_h^{-1}(i) < \nu_h^{-1}(j)}} B_{ij} \bar{e}_{\nu_h^{-1}(j)\nu_h^{-1}(i)} = C_{\bar{i}_h+1\bar{i}_h} e_{\bar{i}_h+1\bar{i}_h} + \cdots + C_{i_h+1 i_h} \bar{e}_{i_h+1 i_h},$$

where $C_{\bar{i}_h+1\bar{i}_h}, \cdots, C_{i_h+1 i_h} \ge M$. This proves (2).

Q.E.D.

Lemma (2.4.9) *There exists a linear function $A_{ij}(x)$ on $\{x = (x_1, \cdots, x_n) \in \mathbb{C}^n \mid x_1 + \cdots + x_n = 0\}$ for $(i,j) \in I_\nu$ such that*

$$\nu(x) - x = \sum_{(i,j)\in I_\nu} A_{ij}(x) \bar{e}_{ij}.$$

Proof. Let $\nu = \nu_h \cdots \nu_1$ and $\nu' = \nu_{h-1} \cdots \nu_1$. By induction on h, there exists a linear function $A'_{ij}(x)$ for $(i,j) \in I_{\nu'}$ such that

$$\nu'(x) - x = \sum_{(i,j)\in I_{\nu'}} A'_{ij}(x) \bar{e}_{ij}.$$

It is easy to see that

$$\nu(x) - x = \sum_{(i,j)\in I_{\nu'}} A'_{ij}(x) \bar{e}_{\nu_h^{-1}(i)\nu_h^{-1}(j)} + \nu_h(x) - x.$$

Since $I_\nu = I_{\nu_h} \coprod \nu_h^{-1} I_{\nu'}$, we can ignore the first term.

Suppose $\nu_h(\bar{i}_h) = i_h$. Then

$$\begin{aligned}\nu_h(x) - x &= x_{\bar{i}_h} \bar{e}_{\bar{i}_h+1\bar{i}_h} + x_{\bar{i}_h+1} \bar{e}_{\bar{i}_h+2\bar{i}_h+1} + \cdots + x_{i_h} \bar{e}_{\bar{i}_h i_h} \\ &= x_{\bar{i}_h}(\bar{e}_{i_h\bar{i}_h} - \bar{e}_{i_h\bar{i}_h+1}) + \cdots + x_{i_h-1}\bar{e}_{i_h i_h-1} - x_{i_h}\bar{e}_{i_h\bar{i}_h}.\end{aligned}$$

Since $I_{\nu_h} = \{(i_h, \bar{i}_h), \cdots, (i_h, i_h - 1)\}$, the lemma follows.

Q.E.D.

These considerations and (2.2.6) show the following proposition.

Proposition (2.4.10)
(1) *There exist constants c_{mij} for a finite number of m's and $(i,j) \in I_\tau$ such that if $D \subset D_{I,\tau,\delta}$ is a bounded set, then there exists a function $c_{D,l}(z)$ of polynomial growth such that if $0 \ll l_1 \ll \cdots \ll l_h$,*

$$|E_{B,I,\tau,\nu}(g^0, z)| \le c_{D,l}(z) t(g^0)^{\nu(\tau \mathrm{Re}(z) + \rho - \sum l_j e_{i_j})} \sum_m \prod_{(i,j)\in I_\nu} |t_i(g^0) t_j(g^0)^{-1}|^{c_{mij}}$$

for $\mathrm{Re}(z) \in D$
(2) *Let $\delta > 0$. If $\nu = 1$ and $0 \ll l_1 \ll \cdots \ll l_h$ or $l_1 \gg \cdots \gg l_h \gg 0$*

$$|E_{B,I,\tau,\nu}(g^0, z)| \le c_{\delta,l}(z) t(g^0)^{\tau \mathrm{Re}(z) + \rho - \sum l_j e_{i_j}}$$

for $\mathrm{Re}(z) \in D_{I,\tau,\delta}$.

The reason why we have to restrict ourselves to a bounded set D in (1) is that the choice of l in (2.4.8) depends on z. If $\nu = 1$, we do not have to use (2.4.8), and therefore we can get a slightly stronger statement (2).

By (2.4.9) and replacing M in (2.4.8) if necessary, we get the following proposition.

Proposition (2.4.11) *Let $D \subset D_{I,\tau}$ be a bounded set. For any $M > 0$, there exists a function $c_{D,M}(z)$ of polynomial growth such that*

$$|E_{B,I,\tau,\nu}(g^0, z)| \le c_{D,M}(z) t(g^0)^{\tau \mathrm{Re}(z) + \rho} \prod_{i>j} |t_i(g^0) t_j(g^0)^{-1}|^{A_{ij}},$$

for $x \in D$, where $A_{ij} \ge 0$, and if $(i,j) \in I_\nu$, $A_{ij} \ge M$.

In particular, if g^0 is in the Siegel domain, $|t_i(g^0) t_j(g^0)^{-1}|$ is bounded. Therefore, we get the following proposition.

Proposition (2.4.12) *In the same situation as in* (2.4.10), *there exists a function of polynomial growth $c_D(z)$ such that*

$$|E_{B,I,\tau,\nu}(g^0, z)| \le c_D(z) t(g^0)^{\tau \mathrm{Re}(z) + \rho}$$

for $\mathrm{Re}(z) \in D, g^0 \in K T^0_{\eta+} \Omega$.

Note that we only have to choose one such $M > 0$ for (2.4.11), and the choice depends on D, so we only have to write the dependence of $c_D(z)$ on D. However, the function $t(g^0)^{\tau \mathrm{Re}(z) + \rho}$ does not depend on D.

For later purposes, we prove an easy lemma.

Lemma (2.4.13) $E_B(\tau_G{}^t(g^0)^{-1}\tau_G, z) = E_B(g^0, -\tau_G z)$.

Proof. Similarly as in (2.2.15),

$$\begin{aligned} E_B(\tau_G{}^t(g^0)^{-1}\tau_G, z) &= \sum_{\gamma \in G_k/B_k} t(\tau_G{}^t g(g^0)^{-1}\tau_G\gamma)^{z+\rho} \\ &= \sum_{\gamma \in G_k/B_k} (\tau_G t(g^0\tau_G{}^t\gamma^{-1}\tau_G)\tau_G)^{z+\rho} \\ &= \sum_{\gamma' \in G_k/B_k} t(g^0\gamma')^{-\tau_G(z+\rho)} \\ &= E_B(g^0, -\tau_G z), \end{aligned}$$

because $-\tau_G\rho = \rho$. This proves the lemma.

Q.E.D.

Chapter 3 The general program

In this chapter, we define the zeta function and describe our general program to determine the principal part of the zeta function. One goal is to prove Shintani's lemma (3.4.31), (3.4.34) for GL(n). Using Shintani's lemma, we prove that certain distributions can be associated to some Morse stratum.

§3.1 The zeta function

Let (G, V, χ_V) be a prehomogeneous vector space. Let $G' = \mathrm{Ker}\,(\chi_V)$. Since G is connected and χ_V is indivisible, G' is connected also. For practical purposes, we assume that there exists a one dimensional split torus T_0 contained in the center of G such that if $\alpha \in T_0$, it acts on V by multiplication by α and $\chi_V(\alpha) = \alpha^e$ for some $e > 0$. In other words, we assume that the constant e_0 in §0.3 is 1.

We choose a maximal split torus $T \subset G$. Let $T' = T \cap G'$. We define $\mathfrak{t} = X_*(T') \otimes \mathbb{R}$, $\mathfrak{t}^* = X^*(T') \otimes \mathbb{R}$. We can identify $\mathfrak{t}$ with the Lie algebra of T'_+, and $\mathfrak{t}^*$ with the dual space of $\mathfrak{t}$. Let $\mathfrak{t}_{\mathbb{Q}} = X_*(T') \otimes \mathbb{Q}$, $\mathfrak{t}_{\mathbb{C}} = X_*(T') \otimes \mathbb{C}$ etc. Elements in $\mathfrak{t}_{\mathbb{Q}}, \mathfrak{t}^*_{\mathbb{Q}}$ are called rational elements. We choose a minimal parabolic subgroup $T \subset P$.

We define $G^0_{\mathbb{A}} = \{g \in G_{\mathbb{A}} \mid |\chi(g)| = 1 \text{ for all } \chi \in X^*(G)\}$. Let $G^1_{\mathbb{A}} = \{g \in G_{\mathbb{A}} \mid |\chi_V(g)| = 1\}$, $T^1_{\mathbb{A}} = T_{\mathbb{A}} \cap G^1_{\mathbb{A}}$, and $T^0_{\mathbb{A}} = T_{\mathbb{A}} \cap G^0_{\mathbb{A}}$. Let $T^1_+ = T_+ \cap T^1_{\mathbb{A}}, T^0_+ = T_+ \cap T^0_{\mathbb{A}}$. Let ω be a character of $G^1_{\mathbb{A}}/G_k$. A principal quasi-character of $G^1_{\mathbb{A}}/G_k$ is a function of the form $|\xi(g)|^s$ where ξ is a rational character of G and $s \in \mathbb{C}$. If χ is a principal quasi-character of $G^1_{\mathbb{A}}/G_k$, we extend it to $G_{\mathbb{A}}$ so that it is trivial on T_{0+}. Let $\chi = (\chi_1, \cdots, \chi_h)$ be principal quasi-characters. We define $\chi(g) = \prod \chi_i(g)$.

For later purposes, we introduce some notation.

Definition (3.1.1) *Let $L \subset V_k$ be a G_k-invariant subset. For $\Phi \in \mathscr{S}(V_{\mathbb{A}})$, we define $\Theta_L(\Phi, g) = \sum_{x \in L} \Phi(gx)$.*

Definition (3.1.2) *For a G_k-invariant subset $L \subset V^{\mathrm{ss}}_k$ and a Schwartz–Bruhat function $\Phi \in \mathscr{S}(V_{\mathbb{A}})$, we define*

$$(1) \qquad Z_L(\Phi, \omega, \chi, s) = \int_{G_{\mathbb{A}}/G_k} \omega(g)\chi(g)|\chi_V(g)|^{\frac{s}{e}} \Theta_L(\Phi, g)dg,$$

$$(2) \qquad Z_{L+}(\Phi, \omega, \chi, s) = \int_{\substack{G_{\mathbb{A}}/G_k \\ |\chi_V(g)| \geq 1}} \omega(g)\chi(g)|\chi_V(g)|^{\frac{s}{e}} \Theta_L(\Phi, g)dg.$$

if these integrals converge absolutely.

If $X^*(G)$ is generated by one element, we do not have to consider χ, and therefore, we write $Z_L(\Phi, \omega, s)$ etc.

Our first task is to prove the convergence of the integrals in (3.1.2) for some subset $L \subset V^{\mathrm{ss}}_k$.

Let $\alpha_1, \cdots, \alpha_r$ be the set of simple roots of P. Let $T_\eta = \{t \in T_{\mathbb{A}} \mid t^{\alpha_1}, \cdots, t^{\alpha_r} \geq \eta\}$, and $T^1_\eta = T_\eta \cap T^1_{\mathbb{A}}, T^0_\eta = T_\eta \cap T^0_{\mathbb{A}}$. Also let $T_{\eta+} = T_\eta \cap T_+, T^1_{\eta+} = T_\eta \cap T^1_+, T^0_{\eta+} = T_\eta \cap T^0_+$. Let ρ be half the sum of the positive weights. Let K be a special maximal compact subgroup. So $G_{\mathbb{A}} = KP_{\mathbb{A}}$. Let $P^0_{\mathbb{A}} = \{p \in P_{\mathbb{A}} \mid |\chi(p)| = 1 \text{ for } \chi \in X^*(P)\}$. We take a compact subset $\Omega \subset P^0_{\mathbb{A}}$, and consider $\mathfrak{S} = KT_{\eta+}\Omega, \mathfrak{S}^1 = KT^1_{\eta+}\Omega$. Then if η is sufficiently small and Ω is sufficiently large, $\mathfrak{S}, \mathfrak{S}^1$ surject to $G_{\mathbb{A}}/G_k, G^1_{\mathbb{A}}/G_k$

respectively. There also exists a compact set $\widehat{\Omega} \subset G_{\mathbb{A}}$ such that $\mathfrak{S} \subset \widehat{\Omega} T_{\eta+}, \mathfrak{S}^1 \subset \widehat{\Omega} T^1_{\eta+}$. We call $\mathfrak{S}$ (resp. $\mathfrak{S}^1$) a Siegel domain for $G_{\mathbb{A}}/G_k$ (resp. $G^1_{\mathbb{A}}/G_k$).

We choose a coordinate system $x = (x_1, \cdots, x_N)$ whose coordinate vectors are eigenvectors of T. Let $\gamma_i \in \mathfrak{t}^*$ be the weights of x_i. For $x \in V_k \setminus \{0\}$, we define $I_x = \{i \mid x_i \neq 0\}$. Let C_x be the convex hull of the set $\{\gamma_i \mid i \in I_x\}$.

Definition (3.1.3) *A point $x \in V_k \setminus \{0\}$ is called k-stable if for any $g \in G_k$, C_{gx} contains a neighborhood of the origin.*

We denote the set of k-stable points by V_k^{s}. We define $V_{\mathrm{st}k}^{\mathrm{ss}} = V_k^{\mathrm{ss}} \setminus V_k^{\mathrm{s}}$ ('st' stands for 'strictly semi-stable').

Proposition (3.1.4) *Let $L = V_k^{\mathrm{s}}$.*
(1) There exists a constant $C = C(\chi)$ such that the integral (1) in (3.1.2) converges absolutely and locally uniformly if $\mathrm{Re}(s) \geq C$.
(2) The integral (2) in (3.1.2) converges absolutely and locally uniformly for all s.

Proof. Let r be the dimension of T'. We choose an isomorphism $d : \mathbb{R}_+^r \to T_+^1$. Also we choose an isomorphism $c : \mathbb{R}_+ \to T_{0+}$ (T_0 is as in the introduction). Clearly, $T_+ = T_+^1 T_{0+}$.

Definition (3.1.5) *Let i be a positive integer. For $\mu = (\mu_1, \cdots, \mu_i) \in \mathbb{R}_+^i$ and a positive number M, we define $\mathrm{rd}_{i,M}(\mu) = \inf(\mu_1^{\pm M} \cdots \mu_i^{\pm M})$, where we consider all the possible $\pm$.*

Clearly, if $M_1 \geq M_2$, $\mathrm{rd}_{i,M_1}(\mu) \leq \mathrm{rd}_{i,M_2}(\mu)$.

Lemma (3.1.6) *There exists a finite number of positive numbers $c_1, \cdots, c_a$ such that for any $M \geq 1$,*

$$|\Theta_{V_k^{\mathrm{s}}}(\Phi, kc(t_0)d(t^1))| \ll \sum_{i=1}^{a} t_0^{-Mc_i} \mathrm{rd}_{r,M}(t^1)$$

for $k \in \widehat{\Omega}, t_0 \in \mathbb{R}_+$, and $t^1 \in \mathbb{R}_+^r$.

Proof. By (1.2.3), there exists a Schwartz–Bruhat function $0 \leq \Psi \in \mathscr{S}(V_{\mathbb{A}})$ such that $|\Phi(kx)| \leq \Psi(x)$ for all $k \in \widehat{\Omega}, x \in V_{\mathbb{A}}$. Let $t^1 = (t_1, \cdots, t_r) \in \mathbb{R}_+^r$, and $t_0 \in \mathbb{R}_+$. Since

$$|\Theta_{V_k^{\mathrm{s}}}(\Phi, kc(t_0)d(t^1))| \leq \Theta_{V_k^{\mathrm{s}}}(\Psi, c(t_0)d(t^1))$$

for $k \in \widehat{\Omega}$, we estimate $\Theta_{V_k^{\mathrm{s}}}(\Psi, c(t_0)d(t^1))$.

Let $I \subset \{1, \cdots, N\}$ be a subset such that $\{\gamma_i \mid i \in I\}$ contains a neighborhood of the origin. We define

$$\Theta_I(\Psi, c(t_0)d(t^1)) = \sum_{\substack{x_i \in k^\times \text{ for } i \in I \\ x_i = 0 \text{ for } i \notin I}} \Psi(c(t_0)d(t^1)x).$$

By the definition of V_k^{s},

$$\Theta_{V_k^{\mathrm{s}}}(\Psi, c(t_0)d(t^1)) \leq \sum_I \Theta_I(\Psi, c(t_0)d(t^1)).$$

Let $t_1, \cdots, t_r \geq 1$. Since the convex hull of $\{\gamma_i \mid i \in I\}$ contains a neighborhood of the origin, there exist constants $e_i \geq 1$ for $i \in I$ and $f_i \geq 1$ for $i = 1, \cdots, r$ such that

$$d(t^1)^{\sum_{i \in I} e_i \gamma_i} = \prod_{i=1}^{r} t_i^{-f_i}.$$

Then

$$(c(t_0)d(t^1))^{\sum_{i \in I} e_i \gamma_i} = t_0^{-\sum_{i \in I} e_i} \prod_{i=1}^{r} t_i^{-f_i}.$$

By (1.2.6), for any $M \geq 1$,

$$\Theta_I(\Psi, c(t_0)d(t^1)) \ll t_0^{-M \sum_{i \in I} e_i} \prod_{i=1}^{r} t_i^{-M f_i}.$$

The argument is similar for other cases. Therefore, there exist a finite number of positive numbers $c'_1, \cdots, c'_b$ such that for any $M \geq 1$,

$$\Theta_I(\Psi, c(t_0)d(t^1)) \ll \sum_{i=0}^{b} t^{-M c'_i} \mathrm{rd}_{r,M}(t^1).$$

Since there are finitely many possibilities for I, the lemma follows.

Q.E.D.

We continue the proof of (3.1.4).

We consider the same Ψ as in (3.1.6). Let $\sigma = \mathrm{Re}(s)$. Since $\mathfrak{S} \subset \widehat{\Omega} T_{\eta +}$,

$$|Z_{V_k^s}(\Phi, \omega, \chi, s)| \ll \int_{T_{\eta+}} t_0^{\sigma} |\chi(d(t^1))| \Theta_{V_k^s}(\Psi, c(t_0)d(t^1))(t^1)^{-2\rho} d^\times t_0 d^\times t^1,$$

where $d^\times t_0, d^\times t^1$ are Haar measures on T_{0+}, T^1_+ respectively.

Consider $c_1, \cdots, c_a$ in (3.1.5). If $M \gg 0$, $|\chi(d(t^1))|(t^1)^{-2\rho}\mathrm{rd}_{r,M}(t^1)$ is integrable on T^1_+. Note that this M depends on χ. Moreover, if $t_0 \geq 1$, $\sum_{i=1}^{a} t_0^{\sigma - M c_i}$ is integrable if $M c_i > \sigma$ for all i. This proves (3.1.4)(2).

Therefore, for (1), we only have to consider the subset $\{t_0 \leq 1\}$.

We fix M so that $|\chi(d(t^1))|(t^1)^{-2\rho}\mathrm{rd}_{r,M}(t^1)$ is integrable on T^1_+. Then the integral over the set $\{t_0 \leq 1\}$ converges if $\sigma > M c_i$ for all i. This proves (3.1.4)(2).

Q.E.D.

We say that (G, V, χ_V) is of complete type if for any χ, there exists a constant $C(\chi)$ such that $Z_{V_k^{ss}}(\Phi, \omega, \chi, s)$ converges absolutely and locally uniformly for all Φ, ω, and $\mathrm{Re}(s) > C(\chi)$. Otherwise we say that it is of incomplete type. If (G, V, χ_V) is of complete type, we write $Z_V(\Phi, \omega, \chi, s)$ or simply $Z(\Phi, \omega, \chi, s)$ for $Z_{V_k^{ss}}(\Phi, \omega, \chi, s)$ etc.

For the rest of this section, we consider groups of the form $G/\widetilde{T}$, where $G = \mathrm{GL}(n_1) \times \cdots \times \mathrm{GL}(n_f)$ and $\widetilde{T}$ is a (split) torus contained in the center of G. Let $G_i = \mathrm{GL}(n_i)$ and T_i the subgroup of diagonal matrices. Let $T = T_1 \times \cdots \times T_f$. Then T is a maximal torus of G. Let $N_i \subset G_i$ be the subgroup of lower triangular matrices with diagonal entries 1. Let $N = N_1 \times \cdots \times N_f$. Then $B = TN$ is a Borel subgroup of G. Let K be the product of maximal compact subgroups of $\mathrm{GL}(n_i)_{\mathbb{A}}$ which we defined in §1.1. The group $G_{\mathbb{A}}$ has the Iwasawa decomposition $G_{\mathbb{A}} = KT_{\mathbb{A}}N_{\mathbb{A}}$, and we write the Iwasawa decomposition of an element $g \in G_{\mathbb{A}}$ as $g = k(g)t(g)n(u(g))$, where $t(g) = (t_1(g), \cdots, t_f(g)), t_i(g) = a_{n_i}(t_{i1}(g), \cdots, t_{in_i}(g)) \in (\mathbb{A}^\times)^{n_i}$ for all i, and $u(g) = (u_1(g), \cdots, u_f(g)), u_i(g) \in \mathbb{A}^{\frac{n_i(n_i-1)}{2}}$ for all i. Let $\rho_1, \cdots, \rho_f$ be half the sum of the positive weights of G_i and $\rho = (\rho_1, \cdots, \rho_f)$. Clearly, $T_+ = T_{1+} \times \cdots \times T_{f+}$. We define $G^0_{\mathbb{A}} = G^0_{1\mathbb{A}} \times \cdots \times G^0_{f\mathbb{A}}$ and $T^0_{\mathbb{A}} = T_{\mathbb{A}} \cap G^0_{\mathbb{A}}$ etc.

Let $\omega_1, \cdots, \omega_f$ be characters of $\mathbb{A}^\times / k^\times$, and $\omega = (\omega_1, \cdots, \omega_f)$. We define

$$\omega(g^0) = \prod_i \omega_i(\det g_i) \tag{3.1.7}$$

for $g^0 = (g_1, \cdots, g_f) \in G^0_{\mathbb{A}}$. Let $\delta_\#(\omega) = \delta(\omega_1) \cdots \delta(\omega_f)$. Let dg_i be the measure on $\mathrm{GL}(n_i)^0_{\mathbb{A}}$ which we defined in §1.1. Let $dg^0 = \prod_i dg_i$

Let $c_{n_i}(\lambda_i)$ be as in §1.1 for $\lambda_i \in \mathbb{R}_+$. For $\lambda = (\lambda_1, \cdots, \lambda_f) \in \mathbb{R}^f_+$, let $c(\lambda) = (c_{n_1}(\lambda_1), \cdots, c_{n_f}(\lambda_f))$. By c, we identify $\mathbb{R}^f_+$ with a subgroup of $G_{\mathbb{A}}$. Any element of $G_{\mathbb{A}}$ is of the form $c(\lambda)g^0$ where $g^0 \in G^0_{\mathbb{A}}$. We define $T^1_+ = \{t \in T_+ \mid \chi_V(t) = 1\}$. We choose a subgroup $\overline{T}_+ \subset \mathbb{R}^f_+$ so that $\overline{T}_+ \cong T^1_+/\widetilde{T}_+$. Let $g^1 = \bar{t}g^0$ where $\bar{t} \in \overline{T}$, $g^0 \in G^0_{\mathbb{A}}$. Let $G^1_{\mathbb{A}} = \overline{T}G^0_{\mathbb{A}}$, and $\widetilde{G}_{\mathbb{A}} = \mathbb{R}_+ \times G^1_{\mathbb{A}}$. We identify T_{0+} with $\mathbb{R}_+$. Let $d^\times \bar{t}$ be a Haar measure on $\overline{T}$, and $dg^1 = d^\times \bar{t} dg^0$. Let $\widetilde{g} = (\lambda, g^1)$, where $\lambda \in \mathbb{R}_+$. We define $d\widetilde{g} = d^\times \lambda dg^1$. If we choose $d^\times \bar{t}$ suitably, $Z_L(\Phi, \omega, \chi, s)$ is equal to the following integral

$$\int_{\mathbb{R}_+ \times G^1_{\mathbb{A}}/G_k} \lambda^s \omega(g^1)\chi(g^1)\Theta_L(\Phi, \widetilde{g})d\widetilde{g}. \tag{3.1.8}$$

We have a similar expression for $Z_{L+}(\Phi, \omega, \chi, s)$ also.

If $X^*(G/\widetilde{T})$ is generated by one element, $G^1_{\mathbb{A}} = G^0_{\mathbb{A}}$, and $\overline{T}_+$ is trivial. Therefore, in this case, we use the measure on $G^0_{\mathbb{A}}$ which we defined in §1.1. Also, we do not have to consider χ.

We define a bilinear form to define an appropriate Fourier transform. We identify the Weyl group of G_i with permutations of $\{1, \cdots, n_i\}$. Let τ_G be the longest element of the Weyl group of G. Consider the coordinate $x = (x_1, \cdots, x_N)$ as before. As we mentioned in the introduction, the weights of the contragredient representation V^* and the weights of the representation $g \to {}^tg^{-1}$ on V are the same. Therefore, there exists a bilinear form $[\ ,\ ]'_V$ such that $[gx, {}^tg^{-1}y]'_V = [x, y]'_V$ for all $x, y \in V$. We define a bilinear form $[\ ,\ ]_V$ on V by the formula

$$[x, y]_V = [x, \tau_G y]'_V. \tag{3.1.9}$$

We define a Fourier transform $\mathscr{F}_V\Phi$ of $\Phi \in \mathscr{S}(V_{\mathbb{A}})$ using $[\ ,\ ]_V$, i.e.

$$\mathscr{F}_V\Phi(x) = \int_{V_{\mathbb{A}}} \Phi(y) < [x, y]_V > dy. \tag{3.1.10}$$

If there is no confusion, we write $\widehat{\Phi}$ instead of $\mathscr{F}_V\Phi$ also.

We define an involution ι on $\widetilde{G}_{\mathbb{A}}$ by $\widetilde{g}^\iota = (\lambda^{-1}, \tau_G {}^t(g^1)^{-1}\tau_G)$. An easy consideration shows that $[\widetilde{g}x, \widetilde{g}^\iota y]_V = [x, y]_V$. One advantage of using this involution is that the Borel subgroup B maps to itself.

The determinant of $\mathrm{GL}(V)$ defines a rational character of G, which we call $\bar{\kappa}_V$. Let κ_V be the principal quasi-character on $G^1_{\mathbb{A}}$ defined by $\kappa_V(g^1) = |\bar{\kappa}_V(g^1)|^{-1}$. We extend κ_V to $\widetilde{G}_{\mathbb{A}}$ so that it is trivial on T_{0+}. The Fourier transform of the function $\Phi(g^1\cdot)$ is $\kappa_V(g^1)\widehat{\Phi}((g^1)^{\iota}\cdot)$ for $g^1 \in G^1_{\mathbb{A}}$.

Let dk be the measure on K such that the volume of K is 1. For $\Phi \in \mathscr{S}(V_{\mathbb{A}})$ and ω as above, we define

$$M_{V,\omega}\Phi(x) = \int_K \omega(k)\Phi(kx)dk. \tag{3.1.11}$$

The operator $M_{V,\omega}$ satisfies similar properties to those in Lemma 5.1 in [83] as follows.

Lemma (3.1.12)
(1) *For any* $k \in K$, $M_{V,\omega}\Phi(kx) = \omega(k)^{-1}M_{V,\omega}\Phi(x)$.
(2) $M_{V,\omega}M_{V,\omega} = M_{V,\omega}$.
(3) $\mathscr{F}_V M_{V,\omega}\Phi = M_{V,\bar{\omega}}\mathscr{F}_V\Phi$.

Clearly,

$$Z_L(\Phi,\omega,\chi,s) = Z_L(M_{V,\omega}\Phi,\omega,\chi,s).$$

Therefore, in this book, we assume that $\Phi = M_{V,\omega}\Phi$. In particular, if ω is trivial, Φ is K-invariant.

§3.2 The Morse stratification

In this section, we review invariant theory and equivariant Morse theory. We do not restrict ourselves to prehomogeneous representations in this section. We consider an arbitrary perfect field k.

Let G be a split connected reductive group and V a representation of G both defined over k. This G does *not* correspond to the group G in §3.1. Instead, it corresponds to $G' = \mathrm{Ker}(\chi_V)$.

We choose a maximal split torus $T \subset G$. We define $\mathfrak{t}$ etc. as in §3.1. We fix a Weyl group invariant inner product $(\ ,\ )$ on $\mathfrak{t}^*$. We assume that if $z, z' \in \mathfrak{t}^*_{\mathbb{Q}}$, then $(z,z') \in \mathbb{Q}$. Let $\|\ \|$ be the metric defined by this inner product. We choose a split Borel subgroup $T \subset B$ and a Weyl chamber $\mathfrak{t}^*_- \subset \mathfrak{t}^*$ so that the weights of the unipotent radical of B belong to $\mathfrak{t}^*_-$ by the conjugation $b \to tbt^{-1}$. We can identify $\mathfrak{t}, \mathfrak{t}^*$ by this inner product, and therefore, we consider $\|\ \|, \mathfrak{t}_-$ for $\mathfrak{t}$ also.

We recall the definition of stability over $\bar{k}$. Let $\pi : V \setminus \{0\} \to \mathbb{P}(V)$ be the natural projection map. Let $\bar{k}[V]^{G_{\bar{k}}}$ be the ring of polynomials invariant under the action of $G_{\bar{k}}$. Suppose that $f \in \bar{k}[V]^{G_{\bar{k}}}$ is a homogeneous polynomial. We define $\mathbb{P}(V)_f = \{\pi(x) \mid f(x) \neq 0\}$.

Definition (3.2.1) *Let* $y \in \mathbb{P}(V)_{\bar{k}}$. *Then we define*
(1) y *is semi-stable if there exists a homogeneous polynomial* $f \in \bar{k}[V]$ *invariant under* $G_{\bar{k}}$ *such that* $y \in \mathbb{P}(V)_f$,
(2) y *is properly stable if there exists a homogeneous polynomial* $f \in \bar{k}[V]$ *invariant under* $G_{\bar{k}}$ *such that* $y \in \mathbb{P}(V)_f$, *all the orbits in* $\mathbb{P}(V)_f$ *are closed, and the stabilizer of* y *in* $G_{\bar{k}}$ *is finite,*
(3) y *is unstable if it is not semi-stable.*

We use the notation $\mathbb{P}(V)^{\mathrm{ss}}_{\bar{k}}$ (resp. $\mathbb{P}(V)^{s}_{(0)\bar{k}}$) for the set of semi-stable (resp. properly stable) points. These are $\mathrm{Gal}(\bar{k}/k)$-invariant open sets of $\mathbb{P}(V)_{\bar{k}}$. However, we will consider the set $\mathbb{P}(V)^{s}_{(0)\bar{k}}$ in number theoretic situations only when $\mathbb{P}(V)^{\mathrm{ss}}_{\bar{k}} = \mathbb{P}(V)^{s}_{(0)\bar{k}}$.

The fundamental theorem of geometric invariant theory is the following Hilbert–Mumford criterion of stability.

Theorem (3.2.2) (Mumford) *Let $y \in \mathbb{P}(V)_{\bar{k}}$. Then*
(1) y is semi-stable if and only if it is semi-stable for all the non-trivial 1PS's of G,
(2) y is properly stable if and only if it is properly stable for all the non-trivial 1PS's of G.

We can diagonalize the action of T because k is a perfect field. Let $(x_0, \cdots, x_N)$ be a coordinate system of V whose coordinate vectors are eigenvectors of T. Let $\gamma_i \in \mathfrak{t}^*$ be the weight determined by the i-th coordinate.

Consider the case when $G = \mathrm{GL}(1)$. Suppose that the action of $\mathrm{GL}(1)$ on the i-th coordinate is by α^{j_i} for $\alpha \in \mathrm{GL}(1)$ and $j_1 \leq \cdots \leq j_N$. For $x \in V_{\bar{k}} \setminus \{0\}$, we define $I_x \subset \{1, \cdots, N\}$ as in §3.1. If f is a polynomial as in (3.2.1)(1), f is a summation of monomials whose factors do not entirely consist of x_i's such that $j_i > 0$. Therefore, if x is semi-stable, I_x should contain i such that $j_i \leq 0$. Moreover, if x is properly stable, I_x should contain i_1, i_2 such that $j_{i_1} < 0, j_{i_2} > 0$. For, if $j_i \leq 0$ for all $i \in I_x$, I_x must contain i such that $j_i = 0$ considering the action of α^{-1}. Since $y = \lim_{\alpha^{-1} \to 0} \alpha x$ should be in the orbit of x and the stabilizer of y is infinite, this is a contradiction.

If $\pi(x)$ is semi-stable or properly stable, the above property should be true for all the non-trivial 1PS's of T and gx for all $g \in G$. This implies that if $\pi(x)$ is semi-stable (resp. properly stable), the convex hull of $\{\gamma_i \mid i \in I_{gx}\}$ contains the origin (resp. a neighborhood of the origin) for all $g \in G$. The definition of k-stable points in §3.1 was a number theoretic analogy of the above definition. In particular, if (G, V, χ_V) is a prehomogeneous vector space and $\dim G = \dim V$, the subset of $\mathbb{P}(V)$ which consists of positive dimensional stabilizers is a G-invariant subset and is equal to the set of unstable points. Since $\dim \mathrm{Ker}(\chi_V) = \dim \mathbb{P}(V)$, all the semi-stable points are properly stable. This implies that $V^{\mathrm{ss}}_k = V^{\mathrm{s}}_k$. Therefore, (G, V) is of complete type. However, the reader should note that we are only using the trivial direction (only if part) of Theorem (3.2.2). The true value of the Hilbert–Mumford criterion of stability for our situation is the rationality of the Morse stratification. We recall the definition of the equivariant Morse stratification for the rest of this section.

The following definition is due to Kirwan [35].

Definition (3.2.3) *A point $\beta \in \mathfrak{t}^*_{\mathbb{Q}}$ is called a minimal combination of weights if β is the closest point to the origin of the convex hull of a finite subset of $\{\gamma_1, \cdots, \gamma_N\}$.*

We denote the set of minimal combination of weights which lie in $\mathfrak{t}^*_-$ by $\mathfrak{B}$. $\mathfrak{B}$ is the index set of the stratification. Since $\mathfrak{B}$ is a finite set, this means that the stratification is a finite stratification. Let $\beta \in \mathfrak{B}$. We describe the stratum S_β. The reader should note that S_β can be the empty set.

We change the ordering of $\{\gamma_1, \cdots, \gamma_N\} = \{\gamma'_1, \cdots, \gamma'_N\}$ so that

$$(\gamma'_1, \beta) \leq \cdots \leq (\gamma'_N, \beta).$$

We assume that

$$((\gamma'_1,\beta),\cdots,(\gamma'_N,\beta))=\frac{1}{m_0}(\overbrace{m_1,\cdots,m_1}^{n_1},\overbrace{m_2,\cdots,m_2}^{n_2},\cdots,\overbrace{m_p,\cdots,m_p}^{n_p})$$

where $m_0>0, m_1<\cdots<m_p$ are coprime integers. Since $\beta\in\mathfrak{B}$, there exists some $1\le s\le p$ such that $\frac{m_s}{m_0}=\|\beta\|^2$. Let e'_i be the coordinate vector which corresponds to γ'_i. Let V_b be the subspace spanned by

$$\{e'_i \mid (\gamma'_i,\beta)=\frac{m_b}{m_0}\}.$$

We define

$$\begin{gathered} Z_\beta=\oplus_{b=s}V_b,\; W_\beta=\oplus_{b>s}V_b,\; Y_\beta=Z_\beta\oplus W_\beta, \\ \overline{Z}_\beta=\{\pi(x)\mid x\in Z_\beta\setminus\{0\}\},\; \overline{Y}_\beta=\{\pi(x,y)\mid x\in Z_\beta\setminus\{0\},\, y\in W_\beta\}, \\ M_\beta=\{g\in G\mid \mathrm{Ad}(g)\beta=\beta\}. \end{gathered}$$

The group M_β is Stab_β in [35].

Let $p_\beta:\overline{Y}_\beta\to\overline{Z}_\beta$ be the projection map. Let ν_β be the indivisible rational character of M_β whose restriction to T is a positive multiple of β. We define $M'_\beta=\{g\in M_\beta\mid \nu_\beta(g)=1\}$. Since M_β is the Levi component of a parabolic subgroup of a connected split reductive group, it is connected. Therefore, M'_β is connected also.

The group M'_β acts on $\overline{Z}_\beta$ linearly. Let $\overline{Z}_\beta^{\mathrm{ss}}$ be the set of semi-stable points with respect to this action. Let P_β be the standard parabolic subgroup of G whose Levi component is M_β and fixes the set Y_β. Let U_β be the unipotent radical of P_β. Since ν_β is rational, all these groups are split groups over k.

Let $\overline{Y}_{\beta\bar{k}}^{\mathrm{ss}}=p_\beta^{-1}(\overline{Z}_{\beta\bar{k}}^{\mathrm{ss}})$, and $\overline{S}_{\beta\bar{k}}=G_{\bar{k}}\overline{Y}_{\beta\bar{k}}^{\mathrm{ss}}$. We define $S_{\beta\bar{k}}=\pi^{-1}(\overline{S}_{\beta\bar{k}})$ etc. Note that if $\beta=0$, $Y_\beta=Z_\beta=V$ and $G_\beta=G$.

The following inductive structure of the strata was proved by Kirwan [35] and Ness [48].

Proposition (3.2.4)
(1) $V_{\bar{k}}\setminus\{0\}=\coprod_{\beta\in\mathfrak{B}}S_{\beta\bar{k}}$.
(2) $\overline{S}_{\beta\bar{k}}\cong G_{\bar{k}}\times_{P_{\beta\bar{k}}}\overline{Y}_{\beta\bar{k}}^{\mathrm{ss}}$.

Since we are assuming that k is a perfect field, the following theorem follows from Kempf Theorem 4.2 [28] (also see Kirwan [35, pp. 150–156]).

Theorem (3.2.5) (Kempf) *Suppose that $x\in V_k\setminus\{0\}$ and $\pi(x)\in S_{\beta\bar{k}}$. Then there exists $g\in G_k$ such that $gx\in Y_{\beta k}^{\mathrm{ss}}$.*

For a more detailed survey of the Morse stratification and its rationality, see [85].

We apply (3.2.4) and (3.2.5) to G' in §3.1. Suppose that $\pi(x)$ for some $x\in V_k\setminus\{0\}$ is unstable. Then there exists $g\in G'_k$ such that $gx\in Y_{\beta k}^{\mathrm{ss}}$ for some $\beta\in\mathfrak{B}$. Clearly, points in $Y_{\beta k}^{\mathrm{ss}}$ do not satisfy the condition in (3.1.3). Therefore, $V_k^{\mathrm{s}}\subset V_k^{\mathrm{ss}}$. However, V_k^{s} can be the empty set in general. So this is not the most precise notion. But it seems that in some cases, V_k^{s} gives the largest subset L for which (3.1.2) holds.

We use the notion of the Morse stratification for the inductive computation of the principal part of the zeta function. For that purpose, we recall the notion of β-sequences introduced by Kirwan [35, p. 73].

Definition (3.2.6) *A sequence $\mathfrak{d} = (\beta_1, \cdots, \beta_a)$ of non-zero elements of $\mathfrak{t}$ is called a β-sequence if for each integer j between 1 and a,*
(1) β_j is the closest point to 0 of the convex hull

$$\mathrm{Conv}\{\gamma_i - \beta_1 - \cdots - \beta_{j-1} \mid (\gamma_i - \beta_k, \beta_k) = 0 \text{ for } 1 \leq k \leq j\}, \text{ and}$$

(2) β_j lies in the unique Weyl chamber containing $\mathfrak{t}_-$ of the subgroup

$$\bigcap\nolimits_{1 \leq i \leq j} M_{\beta_i}.$$

If $\mathfrak{d}, \mathfrak{d}'$ are β-sequences and $\mathfrak{d}'$ is an extension of $\mathfrak{d}$, we use the notation $\mathfrak{d} \prec \mathfrak{d}'$.

We identify $\mathfrak{t}$ with $\mathfrak{t}^*$ by the inner product $(\ ,\ )$, so we can consider β-sequences as sequences of elements of $\mathfrak{t}^*$ also. We call a the length of $\mathfrak{d}$ and denote it by $l(\mathfrak{d})$. We also use the notation $l(\tau)$ for the length of Weyl group elements, but the meaning of the notation will be clear from the context. We consider $\emptyset$ as a β-sequence also and we define $l(\emptyset) = 0$.

Lemma (3.2.7) *Let G be a split connected reductive group over k and χ an indivisible rational character. Let $G' = \mathrm{Ker}(\chi)$. Let $T \subset G$ be a maximal split torus. Then $G_k = G'_k T_k$.*

Proof. We only have to show that χ is indivisible as a character of T also. Since all the characters of G are rational, we assume that k is algebraically closed. Let C be the connected center, and $\overline{G} = [G, G]$. Then $G = C\overline{G}$, $\overline{G}$ is semi-simple, and $\overline{T} = T \cap \overline{G}$ is a maximal torus of $\overline{G}$. Suppose that $\chi = \widetilde{\chi}^p$ on T. Since $\overline{T}$ is connected, this implies that $\widetilde{\chi}$ is trivial on $\overline{T}$ also. Since $T/\overline{T} \cong C/C \cap \overline{G} \cong G/\overline{G}$, we can extend $\widetilde{\chi}$ to a character of G. Regular elements are dense in G, so $\chi = \widetilde{\chi}^p$. Therefore, $p = \pm 1$.

Q.E.D.

We go back to the previous situation and consider (G, V, χ_V). For a β-sequence $\mathfrak{d}$, we define subgroups $M_{\mathfrak{d}}, M'_{\mathfrak{d}}, P_{\mathfrak{d}}, U_{\mathfrak{d}} \subset G$ and subspaces $Y_{\mathfrak{d}}, Z_{\mathfrak{d}} \subset V$ inductively as follows.

If $\mathfrak{d} = \emptyset$, we define $M_{\mathfrak{d}} = P_{\mathfrak{d}} = G$, $M'_{\mathfrak{d}} = G'$, $U_{\mathfrak{d}} = \{1\}$, and $Y_{\mathfrak{d}} = Z_{\mathfrak{d}} = V$. If $\mathfrak{d} = (\beta)$, we consider the Morse stratification of V with respect to G' (not G !). Let $P'_{\mathfrak{d}} = P_\beta, M'_{\mathfrak{d}} = M'_\beta$ be the groups which we defined earlier, and $S_{\mathfrak{d}} = S_\beta$ etc. Let $U_{\mathfrak{d}}$ be the unipotent radical of $P'_{\mathfrak{d}}$. We define $P_{\mathfrak{d}} = TP'_{\mathfrak{d}}$, $M_{\mathfrak{d}} = TM'_{\mathfrak{d}}$. Then $U_{\mathfrak{d}}$ is the unipotent radical of $P_{\mathfrak{d}}$ also.

Consider β-sequences $\mathfrak{d}' = (\beta_1, \cdots, \beta_{a+1})$ and $\mathfrak{d} = (\beta_1, \cdots, \beta_a)$. If $S_{\mathfrak{d}} = \emptyset$, we do not consider $M_{\mathfrak{d}'}$ etc. Suppose $S_{\mathfrak{d}} \neq \emptyset$. Consider the representation of $M'_{\mathfrak{d}}$ on $Z_{\mathfrak{d}}$. We consider the Morse stratification for this situation. Then it is parametrized by the β_{a+1}'s for all the β-sequences $\mathfrak{d} \prec \mathfrak{d}'$'s as above. Let $P'_{\mathfrak{d}'} = P_{\beta_{a+1}}, M'_{\mathfrak{d}'} = M'_{\beta_{a+1}} \subset M'_{\mathfrak{d}}$ be the subgroups which we defined earlier. Let $P_{\mathfrak{d}'} = TP'_{\mathfrak{d}'}, M_{\mathfrak{d}} = TM'_{\mathfrak{d}}$, and $S_{\mathfrak{d}'} = S_{\beta_{a+1}} \subset Z_{\mathfrak{d}}$ etc. Let $Z^{s}_{\mathfrak{d}'k}$ be the set of k-stable points with respect to $M'_{\mathfrak{d}} \cap M_{\mathfrak{d}'}$. Let $Y^{s}_{\mathfrak{d}'k}, S^{s}_{\mathfrak{d}'k}$ be the corresponding subsets of $Z_{\mathfrak{d}k}$. Let $Z^{ss}_{\mathfrak{d}',\mathrm{st}k} = Z^{ss}_{\mathfrak{d}'k} \setminus Z^{s}_{\mathfrak{d}'k}$. We define $Y_{\mathfrak{d}',\mathrm{st}k}, S_{\mathfrak{d}',\mathrm{st}k}$ similarly. We consider $\mathfrak{d}'$'s such that $S_{\mathfrak{d}'} \neq \emptyset$. Let $U_{\mathfrak{d}'}$ be the unipotent radical of $P_{\mathfrak{d}'}$. By induction, $M'_{\mathfrak{d}}$ is connected.

Lemma (3.2.8) $M_{\mathfrak{d}k} = M'_{\mathfrak{d}k} T_k$.

Proof. We prove this lemma by induction on $l(\mathfrak{d})$. If $l(\mathfrak{d}) = 0$, this is (3.2.7). Suppose that $\mathfrak{d} \prec \mathfrak{d}'$, $l(\mathfrak{d}') = l(\mathfrak{d}) + 1$, and $\mathfrak{d}' = (\beta_1, \cdots, \beta_{a+1})$. Let ν' be the indivisible rational character of $M_{\mathfrak{d}'} \cap M'_{\mathfrak{d}}$, which is a positive multiple of β_{a+1} when restricted to $M_{\mathfrak{d}'} \cap M'_{\mathfrak{d}} \cap T$. Since $M_{\mathfrak{d}'} \cap M'_{\mathfrak{d}}$ is the Levi component of a parabolic subgroup $P_{\mathfrak{d}'} \cap M'_{\mathfrak{d}}$ of $M'_{\mathfrak{d}}$, it is connected. By definition, $M'_{\mathfrak{d}'} = \mathrm{Ker}(\nu')$. By (3.2.7), $(M_{\mathfrak{d}'} \cap M'_{\mathfrak{d}})_k = M'_{\mathfrak{d}'k}(T \cap M_{\mathfrak{d}'} \cap M'_{\mathfrak{d}})_k$.

Let $g \in M_{\mathfrak{d}'k}$. Since $M_{\mathfrak{d}'k} \subset M_{\mathfrak{d}k}$, we can write $g = g't$ for some $g' \in M'_{\mathfrak{d}k}, t \in T_k$. Then $g' \in (M_{\mathfrak{d}'} \cap M'_{\mathfrak{d}})_k$. Therefore, we can write $g' = g''t'$ for some $g'' \in M'_{\mathfrak{d}'k}, t' \in T_k$. Hence, $g = g''t't$, proving the lemma.

Q.E.D.

Now we consider the rationality of the stratification.

Proposition (3.2.9) *Suppose that $\mathfrak{d} \prec \mathfrak{d}'$ and $l(\mathfrak{d}') = l(\mathfrak{d}) + 1$. Then $S_{\mathfrak{d}k} \cong M_{\mathfrak{d}k} \times_{P_{\mathfrak{d}'k}} Y^{\mathrm{ss}}_{\mathfrak{d}'k}$.*

Proof. By (3.2.5), $M_{\mathfrak{d}k} \times_{P_{\mathfrak{d}'k}} Y^{\mathrm{ss}}_{\mathfrak{d}'k} \to S_{\mathfrak{d}k}$ is surjective.

Suppose that $g_1, g_2 \in M_{\mathfrak{d}k}, y_1, y_2 \in Y^{\mathrm{ss}}_{\mathfrak{d}'k}$, and $g_1 y_1 = g_2 y_2$. We can write $g_1 = g'_1 t_1, g_2 = g'_2 t_2$ for some $t_1, t_2 \in T_k, g'_1, g'_2 \in M'_{\mathfrak{d}k}$. Since $g'_1 \pi(t_1 y_1) = g'_2 \pi(t_2 y_2)$, there exists $p \in (P_{\mathfrak{d}'} \cap M'_{\mathfrak{d}})_k \subset P_{\mathfrak{d}'k}$ such that $g'_1 = g'_2 p$ by (3.2.4). Since $g'_2 p t_1 y_1 = g'_2 t_2 y_2$, $p t_1 y_1 = t_2 y_2$. Let $p' = t_2^{-1} p t_1$. Then $g_1 = g_2 p', p' y_1 = y_2$, and $p' \in P_{\mathfrak{d}'k}$.

Q.E.D.

Equivariant Morse theory in algebraic situations like in this section was established around 1983. However, Kempf's paper appeared in *Annals of Math.* in 1978. Therefore, he proved the rationality of the Morse stratification before equivariant Morse theory was established.

Of course, if (G, V, χ_V) is a prehomogeneous vector space over an algebraically closed field, the equivariant Morse stratification is the decomposition into G-orbits. Kimura and Ozeki determined the orbit decomposition of prehomogeneous vector spaces over algebraically closed fields (see [31]–[33]). However, the rationality question was not considered there. The reader can probably see that the construction of the equivariant Morse stratification is algorithmic and it is possible to write a computer program to determine the set $\mathfrak{B}$. It is a non-trivial task to determine which vectors in $\mathfrak{B}$ give us non-empty strata. However, this is relatively easier than just trying to determine the orbit decomposition from scratch. The author carried out this task and obtained the list of $\mathfrak{B}$ in some cases. One advantage of this approach is that one can determine the inductive structure of the strata $S_\beta \cong G \times_{P_\beta} Y^{\mathrm{ss}}_\beta$ using the vectors β. In later parts of this book, we handle the orbit decompositions of prehomogeneous vector spaces from this viewpoint.

When we consider a group of the form $G/\widetilde{T}$, where G is a product of $\mathrm{GL}(n)$'s and $\widetilde{T}$ is a split torus contained in the center of G, there is an alternative way of determining $Y^{\mathrm{ss}}_{\mathfrak{d}}$. Let $\chi_V \in X^*(G/\widetilde{T})$ be as before, and $G' = \mathrm{Ker}(\chi_V)$.

We choose a split connected reductive subgroup $G'' \subset G$ so that $G''_{\bar{k}}$ surjects to $G'_{\bar{k}}$ with finite kernel. Then we construct $M''_{\mathfrak{d}}$ using G'' instead of G'. It is easy to see that $M''_{\mathfrak{d}\bar{k}}$ surjects to $M'_{\mathfrak{d}\bar{k}}$ with finite kernel contained in $\widetilde{T}_{\bar{k}}$.

Since the notion of stability is rational over the ground field by (3.2.3) and $\widetilde{T}_{\bar{k}}$ acts trivially on $V_{\bar{k}}$, the stabilities of points in $Y_{\mathfrak{d}}$ with respect to $M''_{\mathfrak{d}}$ and $M'_{\mathfrak{d}}$ are the same.

§3.3 The paths

We introduce the notion of paths and related materials in this section. We consider groups of the form $G/\widetilde{T}$ where G is a product of $\mathrm{GL}(n)$'s and $\widetilde{T}$ is a split torus contained in the center of G. Consider $(G/\widetilde{T}, V, \chi_V)$ as in §3.1.

Definition (3.3.1)
(1) *A path of length a is a pair $\mathfrak{p} = (\mathfrak{d}, \mathfrak{s})$, where $\mathfrak{d}$ is a β-sequence of length a, and $\mathfrak{s}$ is a function from $\{1, \cdots, a\}$ to $\{0, 1\}$.*
(2) *We use the notation $\mathfrak{p} \prec \mathfrak{p}'$ if $\mathfrak{d} \prec \mathfrak{d}'$, and $\mathfrak{s}'$ is an extension of $\mathfrak{s}$.*

We define $l(\mathfrak{p}) = l(\mathfrak{d})$ and call it the length of $\mathfrak{p}$. We allow $\mathfrak{p} = (\emptyset, \emptyset)$ (which means that we do not consider the function $\mathfrak{s}$) as a path and define $l(\mathfrak{p}) = 0$ for convenience .

Let $M_{\mathfrak{d}}, S_{\mathfrak{d}}$ etc. be as in §3.2. Let $\nu_{\mathfrak{d}}$ be the longest element of the Weyl group of $M_{\mathfrak{d}}$. We define an involution of $M_{\mathfrak{d}\mathbb{A}}$ by $\theta_{\mathfrak{d}}(g) = \nu_{\mathfrak{d}}{}^t g^{-1} \nu_{\mathfrak{d}}$. We define a bilinear form $[\ ,\ ]_{\mathfrak{d}}$ on $Z_{\mathfrak{d}}$ using $\nu_{\mathfrak{d}}$ as in §3.1. We define a Fourier transform $\mathscr{F}_{\mathfrak{d}}$ on $Z_{\mathfrak{d}}$ using this bilinear form.

Suppose that $Y_{\mathfrak{d}} \subset V' \subset V$ are subspaces defined over k and $\Psi \in \mathscr{S}(V'_{\mathbb{A}})$. Then the restriction of Ψ to $\mathscr{S}(Y_{\mathfrak{d}\mathbb{A}})$ is denoted by $\widetilde{R}_{\mathfrak{d}}\Psi$. Any element $y \in Y_{\mathfrak{d}\mathbb{A}}$ can be written as $y = (y_1, y_2)$ where $y_1 \in Z_{\mathfrak{d}\mathbb{A}}, y_2 \in W_{\mathfrak{d}\mathbb{A}}$. We define $R_{\mathfrak{d}}\Psi \in \mathscr{S}(Z_{\mathfrak{d}\mathbb{A}})$ by

$$R_{\mathfrak{d}}\Psi(y_1) = \int_{W_{\mathfrak{d}\mathbb{A}}} \widetilde{R}_{\mathfrak{d}}\Psi(y_1, y_2) dy_2 \tag{3.3.2}$$

for $y_1 \in Z_{\mathfrak{d}\mathbb{A}}$.

For $\Phi \in \mathscr{S}(V_{\mathbb{A}})$, and a path $\mathfrak{p} = (\mathfrak{d}, \mathfrak{s})$, we define $\Phi_{\mathfrak{p}}$ inductively in the following manner.

If $l(\mathfrak{d}) = 0$, $\Phi_{\mathfrak{p}} = \Phi$.
If $l(\mathfrak{d}) = 1$,

$$\Phi_{\mathfrak{p}} = \begin{cases} \Phi & \mathfrak{s}(1) = 0, \\ \widehat{\Phi} & \mathfrak{s}(1) = 1. \end{cases} \tag{3.3.3}$$

Suppose that $\mathfrak{p} = (\mathfrak{d}, \mathfrak{s}) \prec \mathfrak{p}' = (\mathfrak{d}', \mathfrak{s}')$, and $l(\mathfrak{p}') = l(\mathfrak{p}) + 1 = a + 1$. Then we define

$$\Phi_{\mathfrak{p}'} = \begin{cases} R_{\mathfrak{d}}\Phi_{\mathfrak{p}} & \mathfrak{s}'(a+1) = 0, \\ \mathscr{F}_{\mathfrak{d}} R_{\mathfrak{d}}\Phi_{\mathfrak{p}} & \mathfrak{s}'(a+1) = 1. \end{cases} \tag{3.3.4}$$

If $l(\mathfrak{p}) = 1$, $\Phi_{\mathfrak{p}} \in \mathscr{S}(V_{\mathbb{A}})$. If $\mathfrak{p} \prec \mathfrak{p}'$, and $l(\mathfrak{p}') = l(\mathfrak{p}) + 1 = a + 1$, $\Phi_{\mathfrak{p}'} \in \mathscr{S}(Z_{\mathfrak{d}\mathbb{A}})$.

We also define a homomorphism $\theta_{\mathfrak{p}}$ from $M_{\mathfrak{d}}$ to G inductively in the following manner.

If $l(\mathfrak{p}) = 1$, $\theta_{\mathfrak{p}}(g) = g^{\iota}$. If $\mathfrak{p} \prec \mathfrak{p}'$, and $l(\mathfrak{p}') = l(\mathfrak{p}) + 1 = a + 1$,

$$\theta_{\mathfrak{p}'} = \begin{cases} \theta_{\mathfrak{p}} & \mathfrak{s}(a+1) = 0, \\ \theta_{\mathfrak{p}}\theta_{\mathfrak{d}} & \mathfrak{s}(a+1) = 1. \end{cases} \tag{3.3.5}$$

Definition (3.3.6) *$P_{\mathfrak{d}\#} = M_{\mathfrak{d}}U_{\mathfrak{d}\#}$ (resp. $P_{\mathfrak{p}\#} = M_{\mathfrak{p}}U_{\mathfrak{p}\#}$) is the parabolic subgroup of G which contains B and whose Levi component is $M_{\mathfrak{d}}$ (resp. $M_{\mathfrak{p}}$).*

Let $P_{\mathfrak{p}} = \theta_{\mathfrak{p}}(P_{\mathfrak{d}}), M_{\mathfrak{p}} = \theta_{\mathfrak{p}}(M_{\mathfrak{d}})$, and $U_{\mathfrak{p}} = \theta_{\mathfrak{p}}(U_{\mathfrak{d}})$. Since $\theta_{\mathfrak{p}}$ induces an automorphism of $\mathfrak{t}$ which preserves $\mathfrak{t}_-$, $\theta_{\mathfrak{p}}(P_{\mathfrak{d}\#}) = P_{\mathfrak{p}\#}$. Let $B_{\mathfrak{d}} = B \cap M_{\mathfrak{d}}, B_{\mathfrak{p}} = B \cap M_{\mathfrak{p}}$. Let $N_{\mathfrak{d}} = N \cap M_{\mathfrak{d}}, N_{\mathfrak{p}} = N \cap M_{\mathfrak{p}}$. Let $\rho_{\mathfrak{d}}$ (resp. $\rho_{\mathfrak{d}\#}$) be half the sum of the weights of $U_{\mathfrak{d}}$ (resp. $U_{\mathfrak{d}\#}$). We write elements of $U_{\mathfrak{d}\mathbb{A}}, U_{\mathfrak{d}\#\mathbb{A}}$ as $n_{\mathfrak{d}}(u_{\mathfrak{d}}), n_{\mathfrak{d}\#}(u_{\mathfrak{d}\#})$, where $u_{\mathfrak{d}}, u_{\mathfrak{d}\#}$ represent elements of some affine spaces. Let $du_{\mathfrak{d}}, du_{\mathfrak{d}\#}$ be the measures on $U_{\mathfrak{d}\mathbb{A}}, U_{\mathfrak{d}\#\mathbb{A}}$ such that the volumes of $U_{\mathfrak{d}\mathbb{A}}/U_{\mathfrak{d}k}, U_{\mathfrak{d}\#\mathbb{A}}/U_{\mathfrak{d}\#k}$ are 1. If there is no confusion, we write $n(u_{\mathfrak{d}})$ instead of $n_{\mathfrak{d}}(u_{\mathfrak{d}})$.

Let $\mathfrak{d} = (\beta_1, \cdots, \beta_a)$ be a β-sequence. Let $M_{\mathfrak{d}} = M_{\mathfrak{d}1} \times \cdots \times M_{\mathfrak{d}f}$. Let $G^1_{\mathbb{A}} = \overline{T}G^0_{\mathbb{A}}$ be as in §3.1. Let $\nu_{\mathfrak{d}1}, \cdots, \nu_{\mathfrak{d}a} : G^1_{\mathbb{A}} \cap P_{\mathfrak{d}\mathbb{A}}/P_{\mathfrak{d}k} \to \mathbb{R}_+$ be homomorphisms, which are positive multiples of $\beta_1, \cdots, \beta_a$ when restricted to T_+. We define

$$
(3.3.7)\qquad
\begin{aligned}
P^1_{\mathfrak{d}\mathbb{A}} &= \{p \in G^1_{\mathbb{A}} \cap P_{\mathfrak{d}\mathbb{A}} \mid \nu_{\mathfrak{d}1}(p) = \cdots = \nu_{\mathfrak{d}a}(p) = 1\}, \\
P^0_{\mathfrak{d}\mathbb{A}} &= \{p \in P_{\mathfrak{d}\mathbb{A}} \mid |\chi(p)| = 1 \text{ for } \chi \in X^*(P_{\mathfrak{d}})\}, \\
M^1_{\mathfrak{d}\mathbb{A}} &= \{g \in G^1_{\mathbb{A}} \cap M_{\mathfrak{d}\mathbb{A}} \mid \nu_{\mathfrak{d}1}(g) = \cdots = \nu_{\mathfrak{d}a}(g) = 1\}, \\
M^0_{\mathfrak{d}\mathbb{A}} &= \{g \in M_{\mathfrak{d}\mathbb{A}} \mid |\chi(g)| = 1 \text{ for } \chi \in X^*(M_{\mathfrak{d}})\}.
\end{aligned}
$$

If $\mathfrak{d} = \emptyset$, we define $P^1_{\mathfrak{d}\mathbb{A}} = M^1_{\mathfrak{d}\mathbb{A}} = G^1_{\mathbb{A}}$ etc.

Let $\Omega_{\mathfrak{d}} \subset B_{\mathfrak{d}\mathbb{A}}$ be a compact set, and $\eta > 0$. We define $T^0_{\mathfrak{d}\eta}$ to be the set of $t \in T_{\mathbb{A}} \cap M^0_{\mathfrak{d}\mathbb{A}}$ such that $t^\alpha \geq \eta$ for all the positive weights α of $B_{\mathfrak{d}}$. Let $T^0_{\mathfrak{d}\eta+} = T^0_{\mathfrak{d}\eta} \cap T_+$. Then for suitable Ω, η, $(K \cap M_{\mathfrak{d}\mathbb{A}})T^0_{\mathfrak{d}\eta+}\Omega_{\mathfrak{d}}$ surjects to $M^0_{\mathfrak{d}\mathbb{A}}/M_{\mathfrak{d}k}$. We call this set a Siegel domain for $M^0_{\mathfrak{d}\mathbb{A}}/M_{\mathfrak{d}k}$. It is known that there also exists a compact set $\widehat{\Omega}_{\mathfrak{d}} \subset M_{\mathfrak{d}\mathbb{A}}$ such that $(K \cap M_{\mathfrak{d}\mathbb{A}})T^0_{\mathfrak{d}\eta+}\Omega_{\mathfrak{d}} \subset \widehat{\Omega}_{\mathfrak{d}}T^0_{\mathfrak{d}\eta+}$.

Let $d_{\mathfrak{d}1}(\lambda_{\mathfrak{d}1}), \cdots, d_{\mathfrak{d}a}(\lambda_{\mathfrak{d}a})$ be 1PS's which are positive multiples of $\beta_1, \cdots, \beta_a$. Let $\lambda_{\mathfrak{d}} = d_{\mathfrak{d}1}(\lambda_{\mathfrak{d}1}) \cdots d_{\mathfrak{d}a}(\lambda_{\mathfrak{d}a})$ for $\lambda_{\mathfrak{d}1}, \cdots, \lambda_{\mathfrak{d}a} \in \mathbb{R}^a_+$. We define

$$
(3.3.8)\qquad A_{\mathfrak{d}} = \{\lambda_{\mathfrak{d}}\} \cong \mathbb{R}^a_+.
$$

Then $G^1_{\mathbb{A}} \cap M_{\mathfrak{d}\mathbb{A}} = A_{\mathfrak{d}}M^1_{\mathfrak{d}\mathbb{A}}$. Let $\lambda_{\mathfrak{d}} \in A_{\mathfrak{d}}, x \in Z_{\mathfrak{d}\mathbb{A}}$, and $\underline{\mu} = \lambda_{\mathfrak{d}}^{\beta_1+\cdots+\beta_a}$. Then $\lambda_{\mathfrak{d}}x = \underline{\mu}x$. We define homomorphisms $e_{\mathfrak{p}1}, \cdots, e_{\mathfrak{p}a} : G^1_{\mathbb{A}} \cap M_{\mathfrak{d}\mathbb{A}}/M_{\mathfrak{d}k} \to \mathbb{R}_+$ inductively in the following manner. Suppose that $\mathfrak{p} \prec \mathfrak{p}'$, and $l(\mathfrak{p}') = l(\mathfrak{p}) + 1 = a + 1$. For $i = 1, \cdots, a$, we extend $e_{\mathfrak{p}i}$ to $A_{\mathfrak{d}'}$ so that it is trivial on $d_{\mathfrak{d}'a+1}(\lambda_{\mathfrak{d}'a+1})$. Then we define

$$
(3.3.9)\qquad e_{\mathfrak{p}'i} = \begin{cases} e_{\mathfrak{p}i} & \mathfrak{s}'(a+1) = 0, \\ e_{\mathfrak{p}i}^{-1} & \mathfrak{s}'(a+1) = 1, \end{cases}
$$

and $e_{\mathfrak{p}'a+1}(\lambda_{\mathfrak{d}'}) = \lambda_{\mathfrak{d}'}^{\beta_1+\cdots+\beta_{a+1}}$.

Let $\lambda_{\mathfrak{d}} = d_{\mathfrak{d}'1}(\lambda_{\mathfrak{d}'1}) \cdots d_{\mathfrak{d}'a}(\lambda_{\mathfrak{d}'a})$. By Definition (3.2.5), $(\beta_{a+1}, \beta_k) = 0$ for $i = 1, \cdots, a$. Therefore, $\lambda_{\mathfrak{d}}^{\beta_{a+1}} = 1$. This implies that

$$
(3.3.10)\qquad e_{\mathfrak{p}'a+1}(d_{\mathfrak{d}'a+1}(\lambda_{\mathfrak{d}'a+1})) = \begin{cases} e_{\mathfrak{p}'a+1}(\lambda_{\mathfrak{d}'})e_{\mathfrak{p}'a}(\lambda_{\mathfrak{d}})^{-1} & \mathfrak{s}(a+1) = 0, \\ e_{\mathfrak{p}'a+1}(\lambda_{\mathfrak{d}'})e_{\mathfrak{p}'a}(\lambda_{\mathfrak{d}}) & \mathfrak{s}(a+1) = 1. \end{cases}
$$

Definition (3.3.11) *Let $\mathfrak{p} = (\mathfrak{d}, \mathfrak{s})$ be a path of length a. We define*

$$
\begin{aligned}
A_{\mathfrak{p}0} &= \{\lambda_{\mathfrak{d}} \in A_{\mathfrak{d}} \mid e_{\mathfrak{p}i}(\lambda_{\mathfrak{d}}) \leq 1 \textit{ for } i = 1, \cdots, a\}, \\
A_{\mathfrak{p}1} &= \{\lambda_{\mathfrak{d}} \in A_{\mathfrak{d}} \mid e_{\mathfrak{p}i}(\lambda_{\mathfrak{d}}) \leq 1 \textit{ for } i = 1, \cdots, a-1, e_{\mathfrak{p}a} \geq 1\}, \\
A_{\mathfrak{p}2} &= \{\lambda_{\mathfrak{d}} \in A_{\mathfrak{d}} \mid e_{\mathfrak{p}i}(\lambda_{\mathfrak{d}}) \leq 1 \textit{ for } i = 1, \cdots, a-1\}.
\end{aligned}
$$

We define $A^1_\mathfrak{p} = T_+ \cap M^1_{\mathfrak{d}\mathbb{A}} \cap Z(M_\mathfrak{d})_\mathbb{A}$, where $Z(M_\mathfrak{d})$ is the center of $M_\mathfrak{d}$. Let $A'_\mathfrak{d} = A_\mathfrak{d} A^1_\mathfrak{d}$. Let $A'_{\mathfrak{p}i} = A_{\mathfrak{p}i} A^1_\mathfrak{p}$ for $i = 0,1,2$. We extend $e_{\mathfrak{p}i}$ to $A'_{\mathfrak{p}i}$ so that $e_{\mathfrak{p}i}$ is trivial on $A^1_\mathfrak{p}$ for $i = 1,\cdots,a$.

We choose a Haar measure $dg_\mathfrak{d}$ on $G^1_\mathbb{A} \cap M_\mathfrak{d}$ so that if $f(g^1)$ is a K-invariant function on $G^1_\mathbb{A}/P_{\mathfrak{d}\#k}$,

$$\int_{G^1_\mathbb{A}/P_{\mathfrak{d}\#k}} f(g^1)dg^1 = \int_{P_{\mathfrak{d}\#\mathbb{A}}/P_{\mathfrak{d}\#k}} f(g_\mathfrak{d} n_{\mathfrak{d}\#}(u_{\mathfrak{d}\#}))t(g_\mathfrak{d})^{-2\rho_{\mathfrak{d}\#}} dg_\mathfrak{d} du_{\mathfrak{d}\#}.$$

Let

$$\epsilon_\mathfrak{p} = (-1)^{\#\{1\le i\le a \mid \mathfrak{s}(i)=0\}}. \tag{3.3.12}$$

We define $\omega_\mathfrak{p} = \omega$ if $\#\{1 \le i \le a \mid \mathfrak{s}(i) = 1\}$ is even and $\omega_\mathfrak{p} = \omega^{-1}$ otherwise. For $\omega, \mathfrak{d}$ and $\Psi \in \mathscr{S}(Z_{\mathfrak{d}\mathbb{A}})$, we define $M_{\mathfrak{d},\omega}\Psi \in \mathscr{S}(Z_{\mathfrak{d}\mathbb{A}})$ by

$$M_{\mathfrak{d},\omega}\Psi(x) = \int_{K\cap M_{\mathfrak{d}\mathbb{A}}} \omega(k_\mathfrak{d})\Psi(k_\mathfrak{d} x)dk_\mathfrak{d},$$

where $dk_\mathfrak{d}$ is the Haar measure such that the volume of $K \cap M_{\mathfrak{d}\mathbb{A}}$ is 1. Easy considerations show that if $\Phi = M_{V,\omega}\Phi$, $M_{\mathfrak{d},\omega_\mathfrak{p}} R_\mathfrak{d}\Phi_\mathfrak{p} = R_\mathfrak{d}\Phi_\mathfrak{p}$.

The spaces $Z_\mathfrak{d}, W_\mathfrak{d}$ are invariant under $M_\mathfrak{d}$. This implies that the determinants of $\mathrm{GL}(Z_\mathfrak{d}), \mathrm{GL}(W_\mathfrak{d})$ determine rational characters of $M_\mathfrak{d}$. Let $\bar\kappa_{\mathfrak{d}1}(g_\mathfrak{d}), \bar\kappa_{\mathfrak{d}2}(g_\mathfrak{d})$ be the rational characters of $M_\mathfrak{d}$ determined by $Z_\mathfrak{d}, W_\mathfrak{d}$ respectively. This means that if $tI \in \mathrm{GL}(Z_\mathfrak{d})$ is a scalar matrix, $\bar\kappa_{\mathfrak{d}1}(tI) = t^{\dim Z_\mathfrak{d}}$ etc. Let $\kappa_{\mathfrak{d}i}(g_\mathfrak{d}) = |\bar\kappa_{\mathfrak{d}i}(g_\mathfrak{d})|^{-1}$ for $i = 1,2$. Let $\kappa_{\mathfrak{d}3}(g_\mathfrak{d}) = \kappa_{\mathfrak{d}1}(g_\mathfrak{d})\kappa_{\mathfrak{d}2}(g_\mathfrak{d})$.

Let $g_\mathfrak{d} = \lambda_\mathfrak{d} g^1_\mathfrak{d}$, where $\lambda_\mathfrak{d} \in A_\mathfrak{d}, g^1_\mathfrak{d} \in M^1_{\mathfrak{d}\mathbb{A}}$. If $l(\mathfrak{p}) = 0$, we define $\sigma_\mathfrak{p}(g) = \sigma_\mathfrak{p}(\chi, g) = \chi(g)$. We define $\sigma_\mathfrak{p}(g_\mathfrak{d}) = \sigma_\mathfrak{p}(\chi, g_\mathfrak{d})$ for $l(\mathfrak{p}) > 0$ inductively as follows.

Definition (3.3.13)
(1) *If* $l(\mathfrak{p}) = 1$, *we define*

$$\sigma_\mathfrak{p}(g_\mathfrak{d}) = \begin{cases} \chi(g_\mathfrak{d})\kappa_{\mathfrak{d}2}(g_\mathfrak{d})t(g_\mathfrak{d})^{-2\rho_\mathfrak{d}} & \mathfrak{s}(1) = 0, \\ \chi(g_\mathfrak{d})^{-1}\kappa_V(g_\mathfrak{d})^{-1}\kappa_{\mathfrak{d}2}(g_\mathfrak{d})t(g_\mathfrak{d})^{-2\rho_\mathfrak{d}} & \mathfrak{s}(1) = 1. \end{cases}$$

(2) *Suppose that* $\mathfrak{p} \prec \mathfrak{p}'$, *and* $l(\mathfrak{p}') = l(\mathfrak{p}) + 1 = a + 1$. *Then*

$$\sigma_{\mathfrak{p}'}(g_{\mathfrak{d}'}) = \begin{cases} \sigma_\mathfrak{p}(g_{\mathfrak{d}'})\kappa_{\mathfrak{d}'2}(g_{\mathfrak{d}'})t(g_{\mathfrak{d}'})^{-2\rho_{\mathfrak{d}'}} & \mathfrak{s}(a+1) = 0, \\ \sigma_\mathfrak{p}(g_{\mathfrak{d}'})^{-1}\kappa_{\mathfrak{d}1}(g_{\mathfrak{d}'})^{-1}\kappa_{\mathfrak{d}'2}(g_{\mathfrak{d}'})t(g_{\mathfrak{d}'})^{-2\rho_{\mathfrak{d}'}} & \mathfrak{s}(a+1) = 1. \end{cases}$$

§3.4 Shintani's lemma for GL(n)

In this section, we define and prove some growth properties of the smoothed Eisenstein series for groups which are products of GL(n)'s.

Let $G = \mathrm{GL}(n_1) \times \cdots \times \mathrm{GL}(n_f)$ etc. be as in §3.2. We define

$$\mathfrak{t}^{0*}_i = \{z_i = (z_{i1}, \cdots, z_{in_i}) \in \mathbb{C}^{n_i} \mid z_{i1} + \cdots + z_{in_i} = 0\},$$
$$\mathfrak{t}^{0*}_{i-} = \{z_i \in \mathfrak{t}^{0*} \mid z_{ij} \le z_{ij+1} \text{ for all } j\}.$$

Let $\mathfrak{t}^{0*} = \mathfrak{t}_1^{0*} \times \cdots \times \mathfrak{t}_f^{0*}$, and $\mathfrak{t}_-^{0*} = \mathfrak{t}_{1-}^{0*} \times \cdots \times \mathfrak{t}_{f-}^{0*}$.

Let $E_{B_i}(g_i, z_i)$ be the Eisenstein series on $\mathrm{GL}(n_i)_{\mathbb{A}}^0$. We define

$$E_B(g^0, z) = \prod_i E_{B_i}(g_i, z_i) \tag{3.4.1}$$

for $g^0 = (g_1, \cdots, g_f) \in G_{\mathbb{A}}^0 = G_{1\mathbb{A}}^0 \times \cdots \times G_{f\mathbb{A}}^0$. If $n_i = 1$, we define $E_{B_i}(g_i, z_i) = 1$.

We define a hermitian inner product $(\ ,\ )_i$ on $\mathfrak{t}_{i\mathbb{C}}^{0*}$ by the formula $(z_i, z_i')_i = \sum_{j=1}^{n_i} z_{ij} \bar{z'_{ij}}$. Let $C_1, \cdots, C_f > 0$ be constants, and $C = (C_1, \cdots, C_f)$. We define a hermitian inner product $(\ ,\ )_0$ on $\mathfrak{t}_{\mathbb{C}}^{0*}$ by $(z, z')_0 = \sum_i C_i (z_i, z_i')_i$ for $z = (z_1, \cdots, z_f), z' = (z_1', \cdots, z_f')$. Let $\|z\|_0 = \sqrt{(z, z)_0}$. Let $\rho = (\rho_1, \cdots, \rho_f)$ be as in §1.1. For $i = 1, \cdots, f$, let

$$L_i(z_i) = 2(z_i, \rho_i)_i = \sum_{j_1 < j_2} (z_{ij_1} - z_{ij_2}). \tag{3.4.2}$$

If $n_i = 1$, we define $L_i(z_i) = 0$. We define

$$\begin{aligned} L(z) &= \sum_i C_i L_i(z_i) = 2(z, \rho)_0, \\ w_0 &= L(\rho). \end{aligned} \tag{3.4.3}$$

Lemma (3.4.4) *Let $\tau = (\tau_1, \cdots, \tau_f)$ be a Weyl group element. Then if $\tau \neq \tau_G$, $L(-\tau\rho) < L(\rho)$.*

Proof. The element τ_G is the only element of the Weyl group such that $\tau_G \rho = -\rho$. So if $\tau \neq \tau_G$, $-\tau\rho \neq \rho$. The hermitian form $(\ ,\ )_0$ is Weyl group invariant. This means that $-\tau\rho$ is on the sphere of radius $\|\rho\|_0$ in $\mathfrak{t}^{0*}$. Hence if $-\tau\rho \neq \rho$, $L(-\tau\rho) < L(\rho)$.

Q.E.D.

For a Weyl group element $\tau = (\tau_1, \cdots, \tau_f)$, we define

$$M_\tau(z) = \prod_i M_{\tau_i}(z_i). \tag{3.4.5}$$

We remind the reader that

$$M_{\tau_i}(z_i) = \prod_{\substack{j_1 > j_2 \\ \tau_i(j_1) < \tau_i(j_2)}} \phi(z_{i\tau_i(j_1)} - z_{i\tau_i(j_2)}).$$

Let $\psi(z)$ be an entire function of z which is rapidly decreasing on any vertical strip. We always assume that

$$\psi(z) = \psi(-\tau_G z),\ \psi(\rho) \neq 0.$$

We define

$$\begin{aligned} \Lambda(w; z) &= \frac{\psi(z)}{w - L(z)}, \\ \Lambda_\tau(w, z) &= M_\tau(z) \Lambda(w; z). \end{aligned} \tag{3.4.6}$$

Let $dz_i = \prod_{j=1}^{n_i-1} d(z_{ij} - z_{ij+1})$ and $dz = \prod_i dz_i$.

Definition (3.4.7) *For $g^0 \in G^0_{\mathbb{A}}, w \in \mathbb{C}$, we define*

$$\mathscr{E}(g^0, w, \psi) = \left(\frac{1}{2\pi\sqrt{-1}}\right)^{n_1+\cdots+n_f-f} \int_{\mathrm{Re}(z)=q} E_B(g^0, z)\Lambda(w; z)dz$$

where $q = (q_1, \cdots, q_f)$, $q_i = (q_{i1}, \cdots, q_{in_i})$, and $q_{ij} - q_{ij+1} > 1$ for all i, j.

We call $\mathscr{E}(g^0, w, \psi)$ the smoothed Eisenstein series. It is defined for $\mathrm{Re}(w) > L(q)$. Since we can vary the choice of q, it is holomorphic for $\mathrm{Re}(w) > w_0$. If there is no confusion, we drop ψ and write $\mathscr{E}(g^0, w)$ also. We have used a weighting of factors in the above definition. This will simplify some computations in later parts of this book. By (2.4.13) and the assumption that $\psi(-\tau_G z) = \psi(z)$, $\mathscr{E}((g^0)^\iota, w) = \mathscr{E}(g^0, w)$.

Let $\mathfrak{p}$ be a path. We define a function $\mathscr{E}_{\mathfrak{p}}(g_{\mathfrak{d}}, w)$ of $(g_{\mathfrak{d}}, w) \in (G^0_{\mathbb{A}} \cap M_{\mathfrak{d}\mathbb{A}}/M_{\mathfrak{d}k}) \times \mathbb{C}$ inductively as follows.

If $l(\mathfrak{p}) = 0$, we define $\mathscr{E}_{\mathfrak{p}}(g^0, w) = \mathscr{E}(g^0, w)$.

If $l(\mathfrak{p}) = 1$,

$$\mathscr{E}_{\mathfrak{p}}(g_{\mathfrak{d}}, w) = \begin{cases} \int_{U_{\mathfrak{d}\mathbb{A}}/U_{\mathfrak{d}k}} \mathscr{E}(g_{\mathfrak{d}} n(u_{\mathfrak{d}}), w) du_{\mathfrak{d}} & \mathfrak{s}(1) = 0, \\ \int_{U_{\mathfrak{d}\mathbb{A}}/U_{\mathfrak{d}k}} \mathscr{E}(g_{\mathfrak{d}}^\iota n(u_{\mathfrak{d}})^\iota, w) du_{\mathfrak{d}} & \mathfrak{s}(1) = 1. \end{cases} \tag{3.4.8}$$

If $\mathfrak{p} \prec \mathfrak{p}'$, and $l(\mathfrak{p}') = l(\mathfrak{p}) + 1 = a + 1$,

$$\mathscr{E}_{\mathfrak{p}'}(g_{\mathfrak{d}'}, w) = \begin{cases} \int_{U_{\mathfrak{d}'\mathbb{A}}/U_{\mathfrak{d}'k}} \mathscr{E}_{\mathfrak{p}}(g_{\mathfrak{d}'} n(u_{\mathfrak{d}'}), w) du_{\mathfrak{d}'} & \mathfrak{s}(1) = 0, \\ \int_{U_{\mathfrak{d}'\mathbb{A}}/U_{\mathfrak{d}'k}} \mathscr{E}_{\mathfrak{p}}(\theta_{\mathfrak{d}}(g_{\mathfrak{d}'} n(u_{\mathfrak{d}'})), w) du_{\mathfrak{d}'} & \mathfrak{s}(1) = 1. \end{cases} \tag{3.4.9}$$

Let

$$\widetilde{\mathscr{E}}_{\mathfrak{p}'}(g_{\mathfrak{d}'} n(u_{\mathfrak{d}'}), w) = \begin{cases} \mathscr{E}_{\mathfrak{p}}(g_{\mathfrak{d}'} n(u_{\mathfrak{d}'}), w) - \mathscr{E}_{\mathfrak{p}'}(g_{\mathfrak{d}'}, w) & \mathfrak{s}(a+1) = 0, \\ \mathscr{E}_{\mathfrak{p}}(\theta_{\mathfrak{d}}(g_{\mathfrak{d}'} n(u_{\mathfrak{d}'})), w) - \mathscr{E}_{\mathfrak{p}'}(\theta_{\mathfrak{d}}(g_{\mathfrak{d}'}), w) & \mathfrak{s}(a+1) = 1. \end{cases} \tag{3.4.10}$$

The function $\widetilde{\mathscr{E}}_{\mathfrak{p}}(g_{\mathfrak{d}} n(u_{\mathfrak{d}}), w)$ is the non-constant term with respect to $U_{\mathfrak{p}}$.

Let $P = MU \subset G$ be a parabolic subgroup of G, where M is the Levi component and U is the unipotent radical. Then $E_U(g^0, z)$ is, by definition, the constant term of $E_B(g^0, z)$ with respect to U, i.e.,

$$E_U(g^0, z) = \int_{U_{\mathbb{A}}/U_k} E_B(g^0 n(u), z) du, \tag{3.4.11}$$

where du is the measure on $U_{\mathbb{A}}$ such that the volume of $U_{\mathbb{A}}/U_k$ is 1.

Definition (3.4.12) *For $g^0 \in G^0_{\mathbb{A}}, w \in \mathbb{C}$, we define*

$$\mathscr{E}_U(g^0, w) = \left(\frac{1}{2\pi\sqrt{-1}}\right)^{n_1+\cdots+n_f-f} \int_{\mathrm{Re}(z)=q} E_U(g^0, z)\Lambda(w; z)dz,$$

where we use the same q as in (3.4.7).

The function $\mathscr{E}_U(g^0, w)$ is defined for $\mathrm{Re}(w) > L(q)$.

Let $P_{\mathfrak{d}\#}$ be as in §3.3. Then $\mathscr{E}_{\mathfrak{p}}(g_{\mathfrak{d}}, w) = \mathscr{E}_{U_{\mathfrak{p}\#}}(\theta_{\mathfrak{p}}(g_{\mathfrak{d}}), w)$ for $g_{\mathfrak{d}} \in G^0_{\mathbb{A}} \cap M_{\mathfrak{d}\mathbb{A}}$.

Suppose that $M = M_1 \times \cdots \times M_f$, $M_i = M_{i1} \times \cdots \times M_{ia_i}$, and $M_{ip} \cong \mathrm{GL}(j_{ip} - j_{ip-1})$ for all i, p ($j_{i0} = 0, j_{ia_i} = n_i$). Let $\mathfrak{J} = (\mathfrak{J}_1, \cdots, \mathfrak{J}_f)$, and $\mathfrak{J}_{ip} = \{j_{ip-1} + 1, \cdots, j_{ip}\}$ for all i, p. Let $\tau = (\tau_1, \cdots, \tau_f)$ be a Weyl group element. Consider the following condition for permutations.

Condition (3.4.13) *If $j_1, j_2 \in \mathfrak{J}_{ip}$ for some i, p and $j_1 < j_2$ then $\tau_i(j_1) < \tau_i(j_2)$.*

Let τ_P be the longest element which satisfies Condition (3.4.13). If $P = P_{\mathfrak{d}\#}$ (resp. $P = P_{\mathfrak{p}\#}$), we use the notation $\tau_{\mathfrak{d}}$ (resp. $\tau_{\mathfrak{p}}$) for $\tau_{P_{\mathfrak{d}\#}}$ (resp. $\tau_{P_{\mathfrak{p}\#}}$).

Let τ be a Weyl group element which satisfies Condition (3.4.13). For $g^0 \in G^0_{\mathbb{A}} \cap M_{\mathbb{A}}$, we define

(3.4.14) $E_{U,\tau}(g^0, z) = M_\tau(z) E_{B_M}(g^0, \tau z)$,

$$\mathscr{E}_{U,\tau}(g^0, w) = \left(\frac{1}{2\pi\sqrt{-1}}\right)^{n_1+\cdots+n_f-f} \int_{\mathrm{Re}(z)=q} E_{U,\tau}(g^0, z)\Lambda(w; z)dz,$$

where $q \in D_\tau$ and $\mathscr{E}_{U,\tau}(g, w)$ is defined for $\mathrm{Re}(w) > L(q)$

By the usual consideration,

$$\mathscr{E}_U(g^0, w) = \sum_\tau \mathscr{E}_{U,\tau}(g^0, w), \tag{3.4.15}$$

where τ runs through permutations which satisfy Condition (3.4.13).

If $U = U_{\mathfrak{p}\#}$, we also use the notation $\mathscr{E}_{\mathfrak{p},\tau}(g_{\mathfrak{d}}, w)$ for $\mathscr{E}_{U_{\mathfrak{p}\#},\tau}(\theta_{\mathfrak{p}}(g_{\mathfrak{d}}), w)$.

Suppose $l(\mathfrak{p}) = a$. For $i = 1, \cdots, a$, let $\mathfrak{p} = (\mathfrak{d}_i, \mathfrak{s}_i)$ be the unique path of length i such that $\mathfrak{p}_i \prec \mathfrak{p}$. We define two conditions for β-sequences and paths.

In (3.4.16)–(3.4.23), we assume that $G^1_{\mathbb{A}} = G^0_{\mathbb{A}}$ (i.e. (G, V) is an irreducible representation).

Condition (3.4.16)
(1) $M^1_{\mathfrak{d}_i\mathbb{A}} = M^0_{\mathfrak{d}_i\mathbb{A}}$ *for* $i = 1, \cdots, a$.
(2) $M^1_{\mathfrak{d}_i\mathbb{A}} = M^0_{\mathfrak{d}_i\mathbb{A}}$ *for* $i = 1, \cdots, a-1$.

If $\mathfrak{p} = (\mathfrak{d}, \mathfrak{s})$ is a path and $\mathfrak{d}$ satisfies (1) or (2) of Condition (3.4.16), we say that $\mathfrak{p}$ satisfies Condition (1) or (2) of Condition (3.4.16).

Suppose that $M_{\mathfrak{d}} = M_{\mathfrak{d}1} \times \cdots \times M_{\mathfrak{d}f}$, where $M_{\mathfrak{d}i} \subset \mathrm{GL}(n_i)$,

$$M_{\mathfrak{d}i} = M_{\mathfrak{d}i1} \times \cdots \times M_{\mathfrak{d}ia_i+1},$$

and

$$\begin{aligned} M_{\mathfrak{d}i1} &= \mathrm{GL}(j_{\mathfrak{d}i1}), \\ M_{\mathfrak{d}i2} &= \mathrm{GL}(j_{\mathfrak{d}i2} - j_{\mathfrak{d}i1}), \\ &\vdots \\ M_{\mathfrak{d}a_i} &= \mathrm{GL}(j_{\mathfrak{d}ia_i} - j_{\mathfrak{d}ia_i-1}), \\ M_{\mathfrak{d}ia_i+1} &= \mathrm{GL}(n_i - j_{\mathfrak{d}ia_i}) \end{aligned}$$

for some $1 \le j_{\mathfrak{d}i1} < \cdots < j_{\mathfrak{d}ia_i} \le n_i - 1$ for $i = 1, \cdots, f$. We allow some a_i to be zero, and in that case, $M_{\mathfrak{d}i} = \mathrm{GL}(n_i)$. Of course, $a = \sum_i a_i$. For convenience, let $j_{\mathfrak{d}i0} = 0, j_{\mathfrak{d}ia_i+1} = n_i$.

In this situation, $M_{\mathfrak{p}}$ is of the form $M_{\mathfrak{p}} = M_{\mathfrak{p}1} \times \cdots \times M_{\mathfrak{p}f}$, where $M_{\mathfrak{p}i} \subset \mathrm{GL}(n_i)$,

$$M_{\mathfrak{p}i} = M_{\mathfrak{p}i1} \times \cdots \times M_{\mathfrak{p}ia_i+1},$$

and

$$\begin{aligned} M_{\mathfrak{p}i1} &= \mathrm{GL}(j_{\mathfrak{p}i1}), \\ M_{\mathfrak{p}i2} &= \mathrm{GL}(j_{\mathfrak{p}i2} - j_{\mathfrak{p}i1}), \\ &\vdots \\ M_{\mathfrak{p}a_i} &= \mathrm{GL}(j_{\mathfrak{p}ia_i} - j_{\mathfrak{p}ia_i-1}), \\ M_{\mathfrak{p}ia_i+1} &= \mathrm{GL}(n_i - j_{\mathfrak{p}ia_i}) \end{aligned}$$

for some $1 \leq j_{\mathfrak{p}i1} < \cdots < j_{\mathfrak{p}ia_i} \leq n_i - 1$ for $i = 1, \cdots, f$. For convenience, let $j_{\mathfrak{p}i0} = 0, j_{\mathfrak{p}ia_i+1} = n_i$.

Let $A'_{\mathfrak{d}}$ be as in §3.3. We consider growth conditions for functions on the space $A'_{\mathfrak{p}i} M^0_{\mathfrak{d}\mathbb{A}} / M_{\mathfrak{d}k}$ for $i = 1, 2, 3$.

Let

$$\begin{aligned} r_{1ip} &= (r_{1ips})_{1 \leq s \leq j_{\mathfrak{d}ip} - j_{\mathfrak{d}ip-1} - 1} \in \mathbb{R}^{j_{\mathfrak{d}ip} - j_{\mathfrak{d}ip-1} - 1}, \\ r_{1i} &= (r_{1ip})_{1 \leq p \leq a_i}, \\ r_1 &= (r_{11}, \cdots, r_{1f}). \end{aligned}$$

Let $r_2 = (r_{21}, \cdots, r_{2a}) \in \mathbb{R}^a$ and $r = (r_1, r_2)$.

Definition (3.4.17) *Suppose that $\lambda'_{\mathfrak{d}} \in A'_{\mathfrak{d}}$ and $t^0_{\mathfrak{d}} \in M^0_{\mathfrak{d}\mathbb{A}} \cap T_{\mathbb{A}}$. Let*

$$t^0_{\mathfrak{d}} = (t_{\mathfrak{d}1}, \cdots, t_{\mathfrak{d}f}), t_{\mathfrak{d}i} = a_{n_i}(t_{\mathfrak{d}i1}, \cdots, t_{\mathfrak{d}in_i}).$$

We define ${\lambda'_{\mathfrak{d}}}^{r_2} = \prod_{i=1}^a e_{\mathfrak{p}i}(\lambda'_{\mathfrak{d}})^{r_{2i}}$, and

$$t^{0 r_1}_{\mathfrak{d}} = \prod_{i=1}^{f} \prod_{p=1}^{a_i+1} (t_{\mathfrak{d}ij_{\mathfrak{d}p-1}+1} t^{-1}_{\mathfrak{d}ij_{\mathfrak{d}p-1}+2})^{r_{1ip1}} \cdots (t_{\mathfrak{d}ij_{\mathfrak{d}p}-1} t^{-1}_{\mathfrak{d}ij_{\mathfrak{d}p}})^{r_{1ipj_{\mathfrak{d}ip}-j_{\mathfrak{d}ip-1}-1}}.$$

Let $C(A'_{\mathfrak{p}i} M^0_{\mathfrak{d}\mathbb{A}} / M_{\mathfrak{d}k}, r)$ be the set of continuous functions f satisfying the condition

$$\sup_{\substack{\lambda'_{\mathfrak{d}} \in A'_{\mathfrak{p}i} \\ k_{\mathfrak{d}} \in \widehat{\Omega}_{\mathfrak{d}} \\ t^0_{\mathfrak{d}} \in T^0_{\mathfrak{d}\eta+}}} |f(\lambda'_{\mathfrak{d}} k_{\mathfrak{d}} t^0_{\mathfrak{d}})| t^{0 r_1}_{\mathfrak{d}} {\lambda'_{\mathfrak{d}}}^{r_2} < \infty \tag{3.4.18}$$

for $i = 1, 2, 3$.

Consider elements in $\mathfrak{t}^{0*}$ of the form $\alpha = (\alpha_1, \cdots, \alpha_f)$, where

$$\alpha_i = (\overbrace{\alpha_{i1}, \cdots, \alpha_{i1}}^{j_{\mathfrak{p}i1}}, \cdots, \overbrace{\alpha_{ia_i+1}, \cdots, \alpha_{ia_i+1}}^{n - j_{\mathfrak{p}ia_i}})$$

and $\alpha_{i1}, \cdots, \alpha_{ia_i+1} \in \mathbb{R}^{a_i+1}$ for all i. We consider such α satisfying the following condition

$$\alpha_{i1} < \cdots < \alpha_{ia_i+1}, \quad \sum_p (j_{\mathfrak{p}ip} - j_{\mathfrak{p}ip-1}) \alpha_{ip} = 0 \text{ for all } i. \tag{3.4.19}$$

Proposition (3.4.20) *Let $\mathfrak{p}$ be a path of length a satisfying* Condition (3.4.16)(2). *Then there exist α satisfying* (3.4.19) *and constants $c_1, \cdots, c_a > 0$ such that*

$$\theta_{\mathfrak{p}}(\lambda'_{\mathfrak{d}})^{\alpha} = \prod_{i=1}^{a} e_{\mathfrak{p}i}(\lambda'_{\mathfrak{d}})^{c_i}$$

Proof. We prove this proposition by induction on $l(\mathfrak{p})$. Let $\mathfrak{p} = (\mathfrak{d}, \mathfrak{s}) \prec \mathfrak{p}' = (\mathfrak{d}', \mathfrak{s}')$ be paths such that $\mathfrak{p}'$ satisfies Condition (3.4.16)(2), and $l(\mathfrak{p}') = l(\mathfrak{p}) + 1 = a + 1$. This implies that $\mathfrak{p}$ satisfies Condition (3.4.16)(1). Consider $e_{\mathfrak{p}'a+1}$ on $M^0_{\mathfrak{d}\mathrm{A}} \cap M_{\mathfrak{d}'\mathrm{A}}$. $\theta_{\mathfrak{p}'}(M^0_{\mathfrak{d}\mathrm{A}} \cap M_{\mathfrak{d}'\mathrm{A}}) = M^0_{\mathfrak{p}\mathrm{A}} \cap M_{\mathfrak{p}'\mathrm{A}}$. ($\theta_{\mathfrak{d}}$ maps $M_{\mathfrak{d}}$ to $M_{\mathfrak{d}}$.)

Suppose that $M_{\mathfrak{d}ip} \cap M_{\mathfrak{d}'} = \mathrm{GL}(l_{ip1}) \times \cdots \times \mathrm{GL}(l_{ipb_{ip}})$, and $M_{\mathfrak{p}ip} \cap M_{\mathfrak{p}'} = \mathrm{GL}(m_{ip1}) \times \cdots \times \mathrm{GL}(m_{ipc_{ip}})$. Let $\mu = (\mu_1, \cdots, \mu_f)$, where

$$\mu_i = (\mu_{i1}, \cdots, \mu_{ia_i+1}),\ \mu_{ip} = a_{j_{\mathfrak{d}p} - j_{\mathfrak{d}p-1}}(\overbrace{\underline{\mu}_{ip1}, \cdots, \underline{\mu}_{ip1}}^{l_{ip1}}, \cdots, \overbrace{\underline{\mu}_{ipb_{ip}}, \cdots, \underline{\mu}_{ipb_{ip}}}^{l_{ipb_{ip}}}),$$

and $\mu_{ips} \in \mathbb{R}_+$ for all i, p, s. We consider such μ satisfying $\prod_s \mu_{ips}^{l_{ips}} = 1$ for all i, p. Then any element of $M^0_{\mathfrak{d}\mathrm{A}} \cap M_{\mathfrak{d}'\mathrm{A}}$ is of the form $\mu g^0_{\mathfrak{d}'}$ for some μ and $g^0_{\mathfrak{d}'} \in M^0_{\mathfrak{d}'\mathrm{A}}$.

Let γ be an element in $\mathfrak{t}^{0*}$ of the form $\gamma = (\gamma_1, \cdots, \gamma_f)$, where

$$\gamma_i = (\gamma_{i1}, \cdots, \gamma_{ia_i+1}),\ \gamma_{ip} = (\overbrace{\gamma_{ip1}, \cdots, \gamma_{ip1}}^{l_{ip1}}, \cdots, \overbrace{\gamma_{ipb_{ip}}, \cdots, \gamma_{ipb_{ip}}}^{l_{ipb_{ip}}}),$$

and

$$\text{(3.4.21)} \qquad \gamma_{ip1} < \cdots < \gamma_{ipb_{ip}},\ \sum_s l_{ips}\gamma_{ips} = 0 \text{ for all } i, p.$$

Also we consider an element $\widetilde{\gamma}$ in $\mathfrak{t}^{0*}$ of the form $\widetilde{\gamma} = (\widetilde{\gamma}_1, \cdots, \widetilde{\gamma}_f)$, where

$$\widetilde{\gamma}_i = (\widetilde{\gamma}_{i1}, \cdots, \widetilde{\gamma}_{ia_i+1}),\ \widetilde{\gamma}_{ip} = (\overbrace{\widetilde{\gamma}_{ip1}, \cdots, \widetilde{\gamma}_{ip1}}^{m_{ip1}}, \cdots, \overbrace{\widetilde{\gamma}_{ipc_{ip}}, \cdots, \widetilde{\gamma}_{ipc_{ip}}}^{m_{ipc_{ip}}}).$$

and

$$\text{(3.4.22)} \qquad \widetilde{\gamma}_{ip1} < \cdots < \widetilde{\gamma}_{ipc_{ip}},\ \sum_s m_{ips}\widetilde{\gamma}_{ips} = 0 \ \text{ for all } i, p.$$

Lemma (3.4.23) *There exists $\widetilde{\gamma}$ as above such that $\theta_{\mathfrak{p}'}(\mu)^{\widetilde{\gamma}} = e_{\mathfrak{p}'a+1}(\mu)$.*

Proof. By the definition of β-sequences, $e_{\mathfrak{p}'a+1}(\mu) = \mu^{\gamma}$ for some γ as in (3.4.21). Note that $\mu^{\beta_1} = \cdots = \mu^{\beta_a} = 1$. $P_{\mathfrak{p}'} = \theta_{\mathfrak{p}'}(P_{\mathfrak{d}'})$ is the unique subgroup of $M_{\mathfrak{d}}$ which contains $B \cap M_{\mathfrak{d}}$ and whose Levi component is $M_{\mathfrak{d}'}$. By $\theta_{\mathfrak{p}'}$, the positive eights with respect to $P_{\mathfrak{d}'}$ correspond to the positive weights with respect to $P_{\mathfrak{p}'}$. Therefore, the existence of such a $\widetilde{\gamma}$ follows.

Q.E.D.

We continue the proof of (3.4.20). We first assume that $\mathfrak{p}'$ also satisfies Condition (3.4.16)(1). Then $A_{\mathfrak{d}'} = A'_{\mathfrak{d}'}$. We can write $\lambda_{\mathfrak{d}'} = d_{\mathfrak{d}' a+1}(\lambda_{\mathfrak{d}' a+1})\lambda_{\mathfrak{d}}$ identifying $\lambda_{\mathfrak{d}}$ with $\prod_{i=1}^{a} d_{\mathfrak{d}i}(\lambda_{\mathfrak{d}' i})$. Since $\theta_{\mathfrak{d}}(\lambda_{\mathfrak{d}}) = \lambda_{\mathfrak{d}}^{-1}$,

$$\theta_{\mathfrak{p}'}(\lambda_{\mathfrak{d}'}) = \begin{cases} \theta_{\mathfrak{p}}(\lambda_{\mathfrak{d}})\theta_{\mathfrak{p}'}(d_{\mathfrak{d}' a+1}(\lambda_{\mathfrak{d}' a+1})) & \mathfrak{s}'(a+1) = 0, \\ \theta_{\mathfrak{p}}(\lambda_{\mathfrak{d}})^{-1}\theta_{\mathfrak{p}'}(d_{\mathfrak{d}' a+1}(\lambda_{\mathfrak{d}' a+1})) & \mathfrak{s}'(a+1) = 1. \end{cases}$$

By the above lemma ($\mu = d_{\mathfrak{d}' a+1}(\lambda_{\mathfrak{d}' a+1})$), there exists $\widetilde{\gamma}'$ satisfying the condition of the lemma for $\mathfrak{p}'$. This implies that

$$\theta_{\mathfrak{p}'}(d_{\mathfrak{d}' a+1}(\lambda_{\mathfrak{d}' a+1}))^{\widetilde{\gamma}'} = e_{\mathfrak{p}' a+1}(d_{\mathfrak{d}' a+1}(\lambda_{\mathfrak{d}' a+1})) = e_{\mathfrak{p}' a+1}(\lambda_{\mathfrak{d}'})e_{\mathfrak{p}' a}(\lambda_{\mathfrak{d}'})^{\pm 1}.$$

By induction, we choose $\alpha \in \mathfrak{t}^{0*}$ which satisfies (3.4.20) for $\mathfrak{p}$ and $\theta_{\mathfrak{p}}(\lambda_{\mathfrak{d}})^{\alpha} = \prod_{i=1}^{a} e_{\mathfrak{p}i}(\lambda_{\mathfrak{d}})^{c_i}$, where $c_i > 0$ for $i = 1, \cdots, a$. Let $\widetilde{\alpha} = m\alpha + \widetilde{\gamma}$, where m is a positive number.

If $\mathfrak{s}'(a+1) = 0$, $e_{\mathfrak{p}i}(\lambda_{\mathfrak{d}}) = e_{\mathfrak{p}' i}(\lambda_{\mathfrak{d}})$, and if $\mathfrak{s}'(a+1) = 1$, $e_{\mathfrak{p}i}(\lambda_{\mathfrak{d}}) = e_{\mathfrak{p}' i}(\lambda_{\mathfrak{d}})^{-1}$. Also $\theta_{\mathfrak{p}}(\lambda_{\mathfrak{d}})^{\widetilde{\gamma}} = 1, \theta_{\mathfrak{p}'}(d_{\mathfrak{d}' a+1}(\lambda_{\mathfrak{d}' a+1}))^{\alpha} = 1$. Therefore,

$$\theta_{\mathfrak{p}'}(\lambda_{\mathfrak{d}'})^{\widetilde{\alpha}} = e_{\mathfrak{p}' a+1}(\lambda_{\mathfrak{d}'})e_{\mathfrak{p}' a}(\lambda_{\mathfrak{d}'})^{\pm 1}\prod_{i=1}^{a} e_{\mathfrak{p}' a+1}(\lambda_{\mathfrak{d}'})^{mc_i}.$$

Hence, if m is large enough, $\widetilde{\alpha}$ satisfies the condition of (3.4.20) for $\mathfrak{p}'$.

This proves (3.4.20) for the case when $\mathfrak{p}$ satisfies Condition (3.4.16)(1). We use the same argument again, and we obtain the case when $\mathfrak{p}$ only satisfies Condition (3.4.16)(2).

Q.E.D.

Let $I_i \subset \{1, \cdots, n_i - 1\}$ be a subset and τ_i a permutation of $\{1, \cdots, n_i\}$ for $i = 1, \cdots, f$. Let ν_i be a permutation which satisfies the condition that $j_1 < j_2$, $j_2 \notin I_i$ implies $\nu_i^{-1}(j_1) < \nu_i^{-1}(j_2)$. Let $I = (I_1, \cdots, I_f)$, $\tau = (\tau_1, \cdots, \tau_f)$, and $\nu = (\nu_1, \cdots, \nu_f)$. Let I_{τ_i} be as in §2.1. Let D_{I_i,τ_i} be as in (2.3.7). Let $D_{I,\tau} = D_{I_1,\tau_1} \times \cdots \times D_{I_f,\tau_f}$. Let $\alpha_i = (\alpha_{i1}, \cdots, \alpha_{in_i-1}) \in k^{n_i-1}$ for $i = 1, \cdots, f$, and $\alpha = (\alpha_1, \cdots, \alpha_f)$. Let $u_i = (u_{ij_1j_2})_{j_1>j_2} \in \mathbb{A}^{\frac{n_i(n_i-1)}{2}}$ for $i = 1, \cdots, f$. Let $\psi_{\alpha_i}(n(u_i)) = < \alpha_{i1}u_{i21} + \cdots + \alpha_{in_i-1}u_{n_in_i-1} >$. For $g^0 = (g_1, \cdots, g_f) \in G^0_{\mathbb{A}}$, we define

$$(3.4.24) \qquad E_{B_i,\alpha_i}(g_i, z_i) = \int_{N_{i\mathbb{A}}/N_{ik}} E_{B_i,\alpha_i}(g_in_{n_i}(u_i), z_i)\psi_{\alpha_i}(n_{n_i}(u_i))du_i,$$

$$E_{B_i,\alpha_i,\tau_i}(g_i, z_i) = \int_{N^-_{\tau_i\mathbb{A}}} t(g_in_{n_i}(u_i)\tau_i)^{z_i+\rho_i}\psi_{\alpha_i}(n_{n_i}(u_i))du_i.$$

We assume that $\tau_i(j+1) < \tau_i(j)$ for $j \in I_i$ for all i, because the above integral is 0 otherwise. We define

$$(3.4.25) \qquad E_{B,\alpha,\tau}(g^0, z) = \prod_i E_{B_i,\alpha_i,\tau_i}(g_i, z_i),$$

$$E_{B,\alpha}(g^0, z) = \prod_i E_{B_i,\alpha_i}(g_i, z_i).$$

Also we define

$$(3.4.26)\quad \begin{aligned}&\mathscr{E}_\alpha(g^0,w,\psi)\\ &=\left(\frac{1}{2\pi\sqrt{-1}}\right)^{n_1+\cdots+n_f-f}\int_{\mathrm{Re}(z)=q}E_{B,\alpha}(g^0,z)\Lambda(w;z)dz,\\ &\mathscr{E}_{\alpha,\tau}(g^0,w,\psi)\\ &=\left(\frac{1}{2\pi\sqrt{-1}}\right)^{n_1+\cdots+n_f-f}\int_{\mathrm{Re}(z)=q}E_{B,\alpha,\tau}(g^0,z)\Lambda(w;z)dz,\end{aligned}$$

where we choose q as in (3.4.6) for the first integral and $q\in D_\tau$ for the second integral. These functions are defined for $\mathrm{Re}(w)>L(q)$. Let

$$(3.4.27)\quad \begin{aligned}&\mathscr{E}_{I,\tau,\nu}(g^0,w,\psi)\\ &=\sum_i\sum_{\substack{\alpha_{ij}\in k^\times \text{ if } j\in I_i\\ \alpha_{ij}=0 \text{ if } j\notin I_i}}\sum_i\sum_{\gamma_i\in N^-_{\nu_i k}}\mathscr{E}_{\alpha,\tau}(g^0(\gamma_1,\cdots,\gamma_f),w,\psi),\\ &\widetilde{\mathscr{E}}_{I,\tau,\nu}(g^0,w,\psi)\\ &=\sum_i\sum_{\substack{\alpha_{ij}\in k^\times \text{ if } j\in I_i\\ \alpha_{ij}=0 \text{ if } j\notin I_i}}\sum_i\sum_{\gamma_i\in N^-_{\nu_i k}}|\mathscr{E}_{\alpha,\tau}(g^0(\gamma_1,\cdots,\gamma_f),w,\psi)|.\end{aligned}$$

Since $D_{I,\tau}$ is contractible, $\mathscr{E}_{\alpha,I,\tau}(g^0,w,\psi)$ is well defined as long as we choose the contour defining $\mathscr{E}_{\alpha,\tau}(g(\gamma_1,\cdots,\gamma_f),w,\psi)$'s so that $q\in D_{I,\tau}$. Of course,

$$|\mathscr{E}_{I,\tau,\nu}(g^0,w,\psi)|\le\widetilde{\mathscr{E}}_{I,\tau,\nu}(g^0,w,\psi).$$

Unless the situation requires, we drop ψ, and use the notation $\mathscr{E}_\alpha(g^0,w)$ etc.

We estimate $\widetilde{\mathscr{E}}_{I,\tau,\nu}(g^0,w)$. Let l_i be a function from I_i to the set of natural numbers. If $I_i=\{j_1,\cdots,j_{N_i}\}$, $j_1<\cdots<j_{N_i}$, we define $l_{ip}=l_i(j_p)$. Let $l=(l_1,\cdots,l_f)$. Let $s_{i,I_i}(l)$ be the element of the Lie algebra determined by

$$(3.4.28)\quad t_i^{s_{i,I_i}(l)}=\prod_{j\in I_i}|t_{ij}t_{ij+1}^{-1}|^{l_i(j)}.$$

Let $s(I)=(s_{1,I_1}(l),\cdots,s_{f,I_f}(l))$, and $t^{s_I(l)}=\prod_i t_i^{s_{I_i}(l)}$. We fix τ,I. For $q\in D_{I,\tau}$, l as above, we put $q'(l)=\tau q-s_I(l)$. Consider the following conditions.

Condition (3.4.29)
(1) *For any $\delta>0$, there exists a function $c_{\delta,l}(z)$ of polynomial growth such that*

$$\sum_\alpha|W_\tau(\delta(\alpha,g^0),z)|\ll c_{\delta,l}(z)t(g^0)^{-s_I(l)}$$

for $\{z\mid \mathrm{Re}(z)\in D_{I,\tau,\delta}\}$.
$(2)_M$

$$q'(l)_{i\nu_i(j_1)}-q'(l)_{i\nu_i(j_2)}\ge M$$

for $(j_1,j_2)\in I_{\nu_i}$.

By (2.2.6), (2.4.8), (2.4.9), we get the following proposition.

Proposition (3.4.30)
(1) *There exists* $M > 0$ *independent of* q, l *such that if* q, l *satisfy the conditions* (1), $(2)_M$ *in* (3.4.29) *and* $\delta_1 > 0$, *then* $\widetilde{\mathscr{E}}_{I,\tau,\nu}(g^0, w)$ *is bounded by a finite linear combination of functions of the form*

$$t(g^0)^{\nu(\tau q+\rho-s_I(l))} \prod_i \prod_{(j_1,j_2)\in I_{\nu_i}} |t_{ij_1}(g^0)t_{ij_2}(g^0)^{-1}|^{c_{ij_1j_2}}$$

for $L(q) + \delta_1 \leq \mathrm{Re}(w)$, *where* $c_{ij_1j_2}$ *is a constant independent of* q, l *for all* $(j_1, j_2) \in I_{\nu_i}$.
(2) *Moreover, if* $q \in D_{I,\tau}$,

$$\widetilde{\mathscr{E}}_{I,\tau,\nu}(g^0, w) \ll t(g^0)^{\tau q+\rho}$$

for g^0 *in the Siegel domain and* $L(q) + \delta_1 \leq \mathrm{Re}(w)$.

Note that in (1), we are not restricting ourselves to elements of the Siegel domain. Let

$$h_c(g^0) = \prod_i \prod_{(j_1,j_2)\in I_{\nu_i}} |t_{ij_1}(g^0)t_{ij_2}(g^0)^{-1}|^{c_{ij_1j_2}}.$$

This is a slowly increasing function on $G^0_{\mathbb{A}}$ independent of q, l.

We have proved in (2.4.8) that there exists M_1 (depending on q) such that if $l_{ij} - l_{ij+1} \leq M_1$ for all i, j then $(3.4.29)(2)_M$ is satisfied for a sufficiently large M.

We are going to use (3.4.30) to estimate $\mathscr{E}_{\mathfrak{p}}(g_{\mathfrak{d}}, w), \widetilde{\mathscr{E}}_{\mathfrak{p}}(g_{\mathfrak{d}}, w)$ in this book. But our estimates are rather delicate, so the author would like to clarify the logic.

We have estimates of the form (3.4.29)(1) where the range of l does not depend on $q \in D_{I,\tau}$. But for Condition $(3.4.29)(2)_M$ to be satisfied, the range of l depends on $q \in D_{I,\tau}$.

When we use (3.4.30), we have two kinds of freedoms. One is the choice of l, and the other is the choice of q. We can always fix q, and take l which satisfies (3.4.29)(1) as long as it satisfies $(3.4.29)(2)_M$. But if we want to fix l, and vary the choice of q, we have to be a little careful. For, if $\nu_i \neq 1$, the range of l_i could depend on q. Therefore, if $\nu_i \neq 1$, we have to show that $(3.4.29)(2)_M$ does not depend on q_i. On the other hand, if $\nu = (1, \cdots, 1)$, since Condition $(3.4.29)(2)_M$ is empty, we can fix any l satisfying (3.4.29)(1) and not necessarily satisfying the condition that $l_{ij} - l_{ij+1} \ll 0$ for all i, j, and vary the choice of q. So, for example, we can choose $l_{ij} - l_{ij+1} \gg 0$ for all i, j. Also if some I_i consists of one element and $\nu_i = 1$, we can choose an arbitrary real number $l_{i1} > 1$ by (2.3.24)(2).

We go back to the previous situation where $\mathfrak{p}$ is a path as before.

Theorem (3.4.31) *Let* $C_G = (\mathfrak{V}_{n_1} \cdots \mathfrak{V}_{n_f})^{-1}$. *Suppose that* $G^1_{\mathbb{A}} = G^0_{\mathbb{A}}$ *and* $\mathfrak{p}$ *is a path satisfying* Condition (3.4.16)(2).
(1) *For any* $\epsilon > 0$, *there exist* $\delta = \delta(\epsilon) > 0$ *and a finite number of points* $c_1, \cdots, c_N \in \mathfrak{t}^{0*}$ *satisfying* $\|c_i\|_0 < \epsilon$ *such that if* $M > L(\rho)$,

$$|\mathscr{E}_{\mathfrak{p}}(g_{\mathfrak{d}}, w) - C_G\Lambda(w;\rho)| \ll \sum_i t(g_{\mathfrak{d}})^{c_i}$$

on $A'_{\mathfrak{d}}(K \cap M_{\mathfrak{d}\mathbb{A}})T^0_{\mathfrak{d}\eta+}\Omega_{\mathfrak{d}}$, *and* $w_0 - \delta \leq \mathrm{Re}(w) \leq M$.
(2) *Suppose that* $\tau = (\tau_1, \cdots, \tau_f)$, *and* $\tau \neq \tau_{\mathfrak{p}}$. *Then for any* $\epsilon > 0$, *there exist* $\delta = \delta(\epsilon) > 0$ *and* $r_i = (r_{i1}, r_{i2})$ *as in* (3.4.17), *such that* $r_{i21}, \cdots, r_{i2a} > 0$ *for all* i, $|r_{i1ps}| < \epsilon$ *for all* i, p, s, *and that if* $M > L(\rho)$,

$$|\mathscr{E}_{\mathfrak{p},\tau}(\lambda'_{\mathfrak{d}} g^0_{\mathfrak{d}}, w)| \ll \sum_i \lambda'_{\mathfrak{d}}{}^{r_{2i}} t(g^0_{\mathfrak{d}})^{r_{i1}}$$

on $A'_{\mathfrak{p}0}(K \cap M_{\mathfrak{d}\mathbb{A}})T^0_{\mathfrak{d}\eta+}\Omega_{\mathfrak{d}}$, *and* $L(\rho) - \delta \leq \mathrm{Re}(w) \leq M$.
(3) *There exists* $r = (r_1, r_2)$ *as in* (3.4.17) *such that all the entries of* r_1 *(resp.* r_2 *) are negative (resp. positive), and that if* $w_0 < M_1 < M_2$,

$$|\mathscr{E}_{\mathfrak{p},\tau}(\lambda'_{\mathfrak{d}} g^0_{\mathfrak{d}}, w)| \ll \lambda'_{\mathfrak{d}}{}^{(\mathrm{Re}(w)-w_0)r_2} t(g^0_{\mathfrak{d}})^{(\mathrm{Re}(w)-w_0)r_1},$$

on $A'_{\mathfrak{d}}(K \cap M_{\mathfrak{d}\mathbb{A}})T^0_{\mathfrak{d}\eta+}\Omega_{\mathfrak{d}}$, *and* $M_1 \leq \mathrm{Re}(w) \leq M_2$ *for all* τ.

Proof. Let $g_{\mathfrak{d}} = \lambda'_{\mathfrak{d}} g^0_{\mathfrak{d}}$. Clearly, $\theta_{\mathfrak{p}}(g_{\mathfrak{d}}) = \theta_{\mathfrak{p}}(\lambda'_{\mathfrak{d}})\theta_{\mathfrak{p}}(g^0_{\mathfrak{d}})$. Consider $I = (I_1, \cdots, I_f)$ such that $j_{\mathfrak{p}ip} \notin I_i$ for all i, p. Let $\sigma = (\sigma_1, \cdots, \sigma_f), \nu = (\nu_1, \cdots, \nu_f)$ be elements of the Weyl group, where $\sigma_i = \prod_{p=1}^{a_i+1} \sigma_{ip}, \nu_i = \prod_{p=1}^{a_i+1} \nu_{ip}$, and σ_{ip}, ν_{ip} are permutations of $\mathfrak{I}_{\mathfrak{p}ip}$ for all i, p and ν_{ip} satisfies the condition in (2.1.10). Then, by the consideration in §2.1,

$$\mathscr{E}_{\mathfrak{p},\tau}(g_{\mathfrak{d}}, w) = \sum_{I,\sigma,\nu} \mathscr{E}_{I,\sigma\tau,\nu}(\theta_{\mathfrak{p}}(g_{\mathfrak{d}}), w).$$

If $(j_1, j_2) \in I_{\nu_i}$ for some i, then $j_1, j_2 \in \mathfrak{I}_{\mathfrak{p}ip}$ for some i, p. Also if $g^0_{\mathfrak{d}}$ is in the Siegel domain, $\theta_{\mathfrak{p}}(g^0_{\mathfrak{d}})$ is contained in some Siegel domain for $M^0_{\mathfrak{p}\mathbb{A}}$ also. Therefore, by (3.4.30), if $q \in D_{I,\sigma\tau}$ and $\delta > 0$,

$$\widetilde{\mathscr{E}}_{I,\sigma\tau,\nu}(\theta_{\mathfrak{p}}(g_{\mathfrak{d}}), w) \ll t(\theta_{\mathfrak{p}}(g_{\mathfrak{d}}))^{\sigma\tau q+\rho}$$

for $g^0_{\mathfrak{d}}$ in the Siegel domain and $L(q) + \delta \leq \mathrm{Re}(w)$.

It is easy to see that

$$E_{I,\sigma,\nu}(\theta_{\mathfrak{p}}(g_{\mathfrak{d}}), z) = \theta_{\mathfrak{p}}(\lambda'_{\mathfrak{d}})^{\sigma\tau z+\rho} E_{I,\sigma,\nu}(\theta_{\mathfrak{p}}(g^0_{\mathfrak{d}}), z).$$

Clearly $\theta_{\mathfrak{p}}(\lambda'_{\mathfrak{d}})^{\sigma\tau z+\rho} = \theta_{\mathfrak{p}}(\lambda'_{\mathfrak{d}})^{\tau z+\rho}$. We choose an element $\widetilde{\gamma}$ in $\mathfrak{t}^{0*}$ of the form $\widetilde{\gamma} = (\widetilde{\gamma}_1, \cdots, \widetilde{\gamma}_f)$, where $\widetilde{\gamma}_i = (\widetilde{\gamma}_{ip}), \widetilde{\gamma}_{ip} = (\widetilde{\gamma}_{ip1}, \cdots, \widetilde{\gamma}_{ipj_{\mathfrak{p}ip}-j_{\mathfrak{p}ip-1}})$, and

$$\widetilde{\gamma}_{ip} = a_{ip1}(-1, 1, 0, \cdots, 0) + \cdots + a_{ipj_{\mathfrak{p}ip}-j_{\mathfrak{p}ip-1}-1}(0, \cdots, 0, -1, 1),$$
$$a_{ip1}, \cdots, a_{ipj_{\mathfrak{p}ip}-j_{\mathfrak{p}ip-1}-1} > 0, \ \widetilde{\gamma}_{ip1} < \cdots < \widetilde{\gamma}_{ipj_{\mathfrak{p}ip}-j_{\mathfrak{p}ip-1}}.$$

Consider α in (3.4.20). Let $h_1, h_2 > 0$ be positive numbers and $\widetilde{\alpha} = h_1\alpha + h_2\widetilde{\gamma}$. Then if $h_1 h_2^{-1} \gg 0$, $\widetilde{\alpha}$ is an interior point of $\mathfrak{t}^{0*}_-$. If h_1 is small, $\|\widetilde{\alpha}\|_0$ can be arbitrarily small.

We choose the contour in the definition of $\mathscr{E}_{I,\sigma\tau,\nu}(\theta_{\mathfrak{p}}(g^0_{\mathfrak{d}}), w)$ so that $\mathrm{Re}(z) = q$ and $q = (\sigma\tau)^{-1}(-\rho + m\widetilde{\alpha})$, where $m > 0$ is a constant. Then $\sigma\tau q + \rho = m\widetilde{\alpha}$. Since $-(\sigma\tau)^{-1}(\rho)$ is in the closure of $D_{I,\sigma\tau}$ and $\widetilde{\alpha}$ is an interior point of $\mathfrak{t}^{0*}_-$, $q \in D_{I,\sigma\tau}$.

Therefore,

$$\begin{aligned}\widetilde{\mathscr{E}}_{I,\sigma\tau,\nu}(\theta_{\mathfrak{p}}(g_{\mathfrak{d}}),w) &\ll \theta_{\mathfrak{p}}(\lambda'_{\mathfrak{d}})^{mh_1\alpha}\theta_{\mathfrak{p}}(g^0_{\mathfrak{d}})^{mh_2\widetilde{\gamma}} \\ &= \prod_{i=1}^{a} e_{\mathfrak{p}i}(\lambda'_{\mathfrak{d}})^{mh_1c_i}\theta_{\mathfrak{p}}(g^0_{\mathfrak{d}})^{mh_2\widetilde{\gamma}}.\end{aligned}$$

By the above consideration, $q \in D_{I,\sigma\tau}$ as long as $m > 0$, and $L(-(\sigma\tau)^{-1}(\rho)) \leq w_0$ for all σ, τ. Clearly,

$$L((\sigma\tau)^{-1}(m\widetilde{\alpha})) = mL((\sigma\tau)^{-1}(\widetilde{\alpha})).$$

So, if $0 < c' < |L((\sigma\tau)^{-1}(\widetilde{\alpha}))|^{-1}$ for all σ, τ, we can choose $m = c'(\mathrm{Re}(w) - w_0)$. Note that the choice of $\widetilde{\alpha}$ does not depend on $\sigma\tau$. This proves (3).

Suppose $\sigma\tau \neq \tau_G$. Since $L(-(\sigma\tau)^{-1}(\rho)) < w_0$, if $m > 0$ is sufficiently small, $L(q) < w_0$. This proves (2). Note that if $\tau \neq \tau_{\mathfrak{p}}$, $\sigma\tau \neq \tau_G$.

Next, we consider the case $\sigma\tau = \tau_G$. By the above remark, $\tau = \tau_{\mathfrak{p}}$.

Suppose $I_i \neq \emptyset$. Let $j \in I_i$. Let $\alpha' = (\overbrace{0, \cdots, 0}^{i-1}, \alpha'_i, 0, \cdots, 0) \in \mathfrak{t}^{0*}$, where $\alpha'_i = (\overbrace{-(n_i - j), \cdots, -(n_i - j)}^{j}, j, \cdots, j)$. Then if $m' > 0$ is a small number, $\rho + m'\alpha'$ is in the closure of D_{I,τ_G} and $L(\rho + m'\alpha') < w_0$. If $m'm^{-1} \gg 0$, then $L(\rho + m'\alpha' + \tau_G(m\widetilde{\alpha})) < w_0$ and $\rho + m'\alpha' + \tau_G(m\widetilde{\alpha}) \in D_{I,\tau_G}$. Let $q = \rho + m'\alpha' + \tau_G(m\widetilde{\alpha})$. Then $\tau_G q + \rho = m'\tau_G(\alpha') + m\tau_G(\widetilde{\alpha})$. Therefore, $\|\tau_G q + \rho\|_0$ can be arbitrarily small. Hence, we can choose $q = \rho + m'\alpha' + \tau_G(m\widetilde{\alpha}) \in D_{I,\tau_G}$.

Suppose that $\sigma\tau = \tau_G$ and $I = (\emptyset, \cdots, \emptyset)$. This implies that $\nu = (1, \cdots, 1)$. So

$$\mathscr{E}_{I,\tau_G,1}(g_{\mathfrak{d}}, w) = \left(\frac{1}{2\pi\sqrt{-1}}\right)^{n_1+\cdots+n_f-f} \int_{\mathrm{Re}(z)=q} M_{\tau_G}(z)t(\theta_{\mathfrak{p}}(g_{\mathfrak{d}}))^{\tau_G z+\rho}\Lambda(w;z)dz.$$

Let $\delta_1, \delta_2 > 0$. Let $q_1 = (q_{11}, \cdots, q_{1f}) \in \mathfrak{t}^{0*}$, where $q_{1i} = (q_{1i1}, \cdots, q_{1in_i})$, $q_{1ip} - q_{1ip+1} = 1 + \delta_1$ for all i, p except for $i = p = 1$, and $q_{111} - q_{112} = 1 - \delta_2$. We assume that δ_1, δ_2 are small and $\delta_2\delta_1^{-1} \gg 0$. Then $L(q) < w_0$. Let $q_2 = (q_{21}, \cdots, q_{2f}) \in \mathfrak{t}^{0*}$ such that $q_{2i} = (q_{2i1}, \cdots, q_{2in_i})$ and $q_{2ip} - q_{2ip+1} = 1 + \delta_1$ for all i, p except for $i = p = 1$ and $q_{111} - q_{112} = 1$. Let $dz' = \prod_{(i,p)\neq(1,1)} d(z_{ip} - z_{ip+1})$. Then $\mathscr{E}_{I,\tau_G,1}(g_{\mathfrak{d}}, w)$ is equal to

$$\begin{aligned}&\left(\frac{1}{2\pi\sqrt{-1}}\right)^{n_1+\cdots+n_f-f} \int_{\mathrm{Re}(z)=q_1} M_{\tau_G}(z)t(\theta_{\mathfrak{p}}(g_{\mathfrak{d}}))^{\tau_G z+\rho}\Lambda(w;z)dz \\ &+ \left(\frac{1}{2\pi\sqrt{-1}}\right)^{n_1+\cdots+n_f-f-1} \int_{\mathrm{Re}(z)=q_2} \operatorname*{Res}_{z_{11}-z_{12}=1} \left[M_{\tau_G}(z)t(\theta_{\mathfrak{p}}(g_{\mathfrak{d}}))^{\tau_G z+\rho}\Lambda(w;z)\right] dz'.\end{aligned}$$

There exists $\delta > 0$ such that the first term is holomorphic for $\mathrm{Re}(w) \geq w_0 - \delta$. We continue this process, and eventually get $C_G\Lambda(w;\rho)$. This proves (1).

Q.E.D.

Note that the right hand side of (3.4.31)(3) does not depend on M_1, M_2.

Definition (3.4.32) *Let $f_1(w), f_2(w)$ be meromorphic functions on a domain of the form $\{w \in \mathbb{C} \mid \mathrm{Re}(w) > A\}$. We use the notation $f_1 \sim f_2$ if the following two conditions are satisfied.*
(1) *There exists a constant $A' < w_0$ such that $f_1 - f_2$ can be continued meromorphically to the domain $\{w \in \mathbb{C} \mid \mathrm{Re}(w) > A'\}$.*
(2) *The function $f_1(w) - f_2(w)$ is holomorphic around $w = w_0$.*

The following corollary follows from Theorem (3.4.31)(1).

Corollary (3.4.33) *Suppose that $G^1_{\mathbb{A}} = G^0_{\mathbb{A}}$ and $\mathfrak{p}$ is a path satisfying* Condition (3.4.16)(2). *Suppose that $f(g_{\mathfrak{d}})$ is a function on $A'_{\mathfrak{p}i} M^0_{\mathfrak{d}\mathbb{A}}/M_{\mathfrak{d}k}$, where $i = 1, 2$ or 3, and that there exist finitely many points $c_1, \cdots, c_m \in \mathfrak{t}^*$ such that*

$$|f(g_{\mathfrak{d}})| \ll \inf_i((\lambda'_{\mathfrak{d}} t(g^0_{\mathfrak{d}}))^{c_i}),$$

and

$$\int_{A'_{\mathfrak{p}i}(K \cap M_{\mathfrak{d}\mathbb{A}})T^0_{\mathfrak{d}\eta}\Omega_{\mathfrak{d}}} \inf_i((\lambda'_{\mathfrak{d}} t(g^0_{\mathfrak{d}}))^{c_i}) dg_{\mathfrak{d}} < \infty.$$

Then

$$\int_{A'_{\mathfrak{p}i} M^0_{\mathfrak{d}\mathbb{A}}/M_{\mathfrak{d}k}} f(g_{\mathfrak{d}}) \mathscr{E}_{\mathfrak{p}}(g_{\mathfrak{d}}, w) dg_{\mathfrak{d}} \sim C_G \Lambda(w; \rho) \int_{A'_{\mathfrak{p}i} M^0_{\mathfrak{d}\mathbb{A}}/M_{\mathfrak{d}k}} f(g_{\mathfrak{d}}) dg_{\mathfrak{d}}.$$

In other word, if we can prove that a function is integrable by estimating on the Siegel domain, then we can get rid of the smoothed Eisenstein series in the integral.

The statement of the next theorem does not depend on the path $\mathfrak{p}$, and we do not consider Condition (3.4.16) or the assumption that $G^1_{\mathbb{A}} = G^0_{\mathbb{A}}$.

Theorem (3.4.34) (Shintani's lemma for $\mathrm{GL}(n)$)
(1) *For any $\epsilon > 0$, there exists $\delta > 0$ such that if $M > w_0$,*

$$\sup_{\substack{w_0 - \delta \le \mathrm{Re}(w) \le M \\ g^0 = ktn(u) \in \mathfrak{S}^0}} |\mathscr{E}(g^0, w) - C_G \Lambda(w; \rho)| \prod_{i=1}^{f} \prod_{j=1}^{n_i - 1} |t_{ij}(g^0) t_{ij+1}(g^0)^{-1}|^{-\epsilon} < \infty.$$

(2) *There exists a constant $c > 0$ such that if $M_1 > M_2 > w_0$,*

$$\sup_{\substack{M_2 \le \mathrm{Re}(w) \le M_1 \\ g^0 = ktn(u) \in \mathfrak{S}^0}} |\mathscr{E}(g^0, w)| \prod_{i=1}^{f} \prod_{j=1}^{n_i - 1} |t_{ij}(g^0) t_{ij+1}(g^0)^{-1}|^{c(\mathrm{Re}(w) - w_0)} < \infty.$$

Proof. The proof of this theorem is very similar to that of (3.4.31). The only place we used the assumption (3.4.16)(2) was the choice of $\widetilde{\alpha}$ in the proof of (3.4.31). This time, we just choose $\widetilde{\alpha} = h\widetilde{\gamma}$ where $\widetilde{\gamma} = (\widetilde{\gamma}_1, \cdots, \widetilde{\gamma}_f)$, and

$$\widetilde{\gamma}_i = a_{i1}(-1, 1, 0, \cdots, 0) + \cdots + a_{in_i - 1}(0, \cdots, 0, -1, 1),$$
$$a_{i1}, \cdots, a_{in_i - 1} > 0, \ \widetilde{\gamma}_{i1} < \cdots < \widetilde{\gamma}_{in_i}.$$

The rest of the proof is similar to that of (3.4.31).

Q.E.D.

Note that if c_{ij} is a constant for $i = 1, \cdots, f, j = 1, \cdots, n_i - 1$,

$$\prod_{i=1}^{f} \prod_{j=1}^{n_i-1} |t_{ij}(g^0)t_{ij+1}(g^0)^{-1}|^{c_{ij}(\mathrm{Re}(w)-w_0)} \ll \prod_{i=1}^{f} \prod_{j=1}^{n_i-1} |t_{ij}(g^0)t_{ij+1}(g^0)^{-1}|^{c(\mathrm{Re}(w)-w_0)}$$

for $g^0 \in \mathfrak{S}^0$, where $c = \inf_{i,j} c_{ij}$.

The statement (1) of (3.4.34) implies that if $f \in C(G^0_{\mathbb{A}}/G_k, r)$ for some r such that $r_{ij} > -j(n_i - j)$ for all i, j, then

$$\int_{G^0_{\mathbb{A}}/G_k} f(g^0)\mathscr{E}(g^0, w)dg^0 \sim C_G \Lambda(w; \rho) \int_{G^0_{\mathbb{A}}/G_k} f(g^0)dg^0.$$

The statement (2) of (3.4.34) implies that if f is a slowly increasing function,

$$\int_{G^0_{\mathbb{A}}/G_k} f(g^0)\mathscr{E}(g^0, w)dg^0$$

is well defined, and becomes a holomorphic function in some right half plane.

§3.5 The general process

In this section, we consider (G, V, χ_V) such that $(M_{\mathfrak{d}}, Z_{\mathfrak{d}})$ is a prehomogeneous vector space for all $\mathfrak{d} = (\beta_1, \cdots, \beta_a)$. (The character is the one which is a positive multiple of β_a.) This condition is not always satisfied, for example, $G = \mathrm{GL}(n), V = M(n, n)$ (the set of $n \times n$ matrices). In this section, we assume that (G, V) is an irreducible representation.

Let $\mathfrak{d}, \mathfrak{d}'$ be β-sequences such that $\mathfrak{d} \prec \mathfrak{d}'$ and $l(\mathfrak{d}') = l(\mathfrak{d}) + 1$. Let $\Psi_1 \in \mathscr{S}(Z_{\mathfrak{d}\mathbb{A}})$, and $\Psi_2 \in \mathscr{S}(Z_{\mathfrak{d}'\mathbb{A}})$. We define

$$\text{(3.5.1)} \qquad \begin{aligned} \Theta_{S_{\mathfrak{d}'}}(\Psi_1, g_{\mathfrak{d}}) &= \sum_{x \in S_{\mathfrak{d}' k}} \Psi_1(g_{\mathfrak{d}} x), \\ \Theta_{Y_{\mathfrak{d}'}}(\Psi_1, g_{\mathfrak{d}}) &= \sum_{x \in Y^{\mathrm{ss}}_{\mathfrak{d}' k}} \Psi_1(g_{\mathfrak{d}} x), \\ \Theta_{Z_{\mathfrak{d}'}}(\Psi_2, g_{\mathfrak{d}'}) &= \sum_{x \in Z^{\mathrm{ss}}_{\mathfrak{d}' k}} \Psi_2(g_{\mathfrak{d}'} x), \end{aligned}$$

for $g_{\mathfrak{d}} \in G^1_{\mathbb{A}} \cap M_{\mathfrak{d}\mathbb{A}}, g_{\mathfrak{d}'} \in G^1_{\mathbb{A}} \cap M_{\mathfrak{d}'\mathbb{A}}$.

We also define

$$\begin{aligned} &\Theta_{S^s_{\mathfrak{d}'}}(\Psi_1, g_{\mathfrak{d}}) = \sum_{x \in S^s_{\mathfrak{d}' k}} \Psi_1(g_{\mathfrak{d}} x), \ \Theta_{S_{\mathfrak{d}'}, \mathrm{st}}(\Psi_1, g_{\mathfrak{d}}) \sum_{x \in S_{\mathfrak{d}', \mathrm{st} k}} \Psi_1(g_{\mathfrak{d}} x), \\ &\Theta_{Y^s_{\mathfrak{d}'}}(\Psi_1, g_{\mathfrak{d}'}) = \sum_{x \in Y^s_{\mathfrak{d}' k}} \Psi_1(g_{\mathfrak{d}'} x), \ \Theta_{Y_{\mathfrak{d}'}, \mathrm{st}}(\Psi_1, g_{\mathfrak{d}'}) = \sum_{x \in Y_{\mathfrak{d}', \mathrm{st} k}} \Psi_2(g_{\mathfrak{d}'} x), \\ &\Theta_{Z^s_{\mathfrak{d}'}}(\Psi_2, g_{\mathfrak{d}'}) = \sum_{x \in Z^s_{\mathfrak{d}' k}} \Psi_2(g_{\mathfrak{d}'} x), \ \Theta_{Z_{\mathfrak{d}'}, \mathrm{st}}(\Psi_2, g_{\mathfrak{d}'}) = \sum_{x \in Z_{\mathfrak{d}', \mathrm{st} k}} \Psi_2(g_{\mathfrak{d}'} x). \end{aligned}$$

Let $\omega = (\omega_1, \cdots, \omega_f) \in \Omega(\mathbb{A}^\times/k^\times)^f$ be as before. In the following definition, we consider the case where $G^1_{\mathbb{A}} = G^0_{\mathbb{A}}$.

Definition (3.5.2) *Let* $\mathfrak{p} = (\mathfrak{d}, \mathfrak{s})$ *be a path, and* τ *a Weyl group element which satisfies* Condition (3.4.13) *for* $U_{\mathfrak{p}\#}$. *For* $\Phi \in \mathscr{S}(V_{\mathbb{A}})$, ω *as above, and* $w \in \mathbb{C}$, *we define*

$$(1)\quad \Xi_{\mathfrak{p},\tau}(\Phi,\omega,w) = \int_{A'_{\mathfrak{p}2} M^0_{\mathfrak{d}\mathbb{A}}/M_{\mathfrak{d}k}} \omega_{\mathfrak{p}}(g_{\mathfrak{d}})\sigma_{\mathfrak{p}}(g_{\mathfrak{d}})\Theta_{Z_{\mathfrak{d}}}(R_{\mathfrak{d}}\Phi_{\mathfrak{p}}, g_{\mathfrak{d}})\mathscr{E}_{\mathfrak{p},\tau}(g_{\mathfrak{d}}, w)dg_{\mathfrak{d}},$$

$$(2)\quad \Xi_{\mathfrak{p},\tau+}(\Phi,\omega,w) = \int_{A'_{\mathfrak{p}1} M^0_{\mathfrak{d}\mathbb{A}}/M_{\mathfrak{d}k}} \omega_{\mathfrak{p}}(g_{\mathfrak{d}})\sigma_{\mathfrak{p}}(g_{\mathfrak{d}})\Theta_{Z_{\mathfrak{d}}}(R_{\mathfrak{d}}\Phi_{\mathfrak{p}}, g_{\mathfrak{d}})\mathscr{E}_{\mathfrak{p},\tau}(g_{\mathfrak{d}}, w)dg_{\mathfrak{d}},$$

$$(3)\quad \widehat{\Xi}_{\mathfrak{p},\tau+}(\Phi,\omega,w) = \int_{A'_{\mathfrak{p}0} M^0_{\mathfrak{d}\mathbb{A}}/M_{\mathfrak{d}k}} \omega_{\mathfrak{p}}(g_{\mathfrak{d}})\sigma_{\mathfrak{p}}(g_{\mathfrak{d}})\kappa_{\mathfrak{d}1}(g_{\mathfrak{d}})\Theta_{Z_{\mathfrak{d}}}(\mathscr{F}_{\mathfrak{d}}R_{\mathfrak{d}}\Phi_{\mathfrak{p}}, \theta_{\mathfrak{d}}(g_{\mathfrak{d}}))\mathscr{E}_{\mathfrak{p},\tau}(g_{\mathfrak{d}}, w)dg_{\mathfrak{d}},$$

$$(4)\quad \Xi_{\mathfrak{p},\tau\#}(\Phi,\omega,w) = R_{\mathfrak{d}}\Phi_{\mathfrak{p}}(0)\int_{A'_{\mathfrak{p}0} M^0_{\mathfrak{d}\mathbb{A}}/M_{\mathfrak{d}k}} \omega_{\mathfrak{p}}(g_{\mathfrak{d}})\sigma_{\mathfrak{p}}(g_{\mathfrak{d}})\mathscr{E}_{\mathfrak{p},\tau}(g_{\mathfrak{d}}, w)dg_{\mathfrak{d}},$$

$$(5)\quad \widehat{\Xi}_{\mathfrak{p},\tau\#}(\Phi,\omega,w) = \mathscr{F}_{\mathfrak{d}}R_{\mathfrak{d}}\Phi_{\mathfrak{p}}(0)\int_{A'_{\mathfrak{p}0} M^0_{\mathfrak{d}\mathbb{A}}/M_{\mathfrak{d}k}} \omega_{\mathfrak{p}}(g_{\mathfrak{d}})\sigma_{\mathfrak{p}}(g_{\mathfrak{d}})\kappa_{\mathfrak{d}1}(g_{\mathfrak{d}})\mathscr{E}_{\mathfrak{p},\tau}(g_{\mathfrak{d}}, w)dg_{\mathfrak{d}},$$

if these integrals are well defined for $\mathrm{Re}(w) \gg 0$.

When we have to refer to the function ψ, we use the notation $\Xi_{\mathfrak{p},\tau}(\Phi,\omega,w,\psi)$ etc. We also define $\Xi^s_{\mathfrak{p},\tau+}(\Phi,\omega,w), \Xi_{\mathfrak{p},\tau,\mathrm{st}+}(\Phi,\omega,w)$ etc. using $\Theta_{Z^s_{\mathfrak{d}}}(R_{\mathfrak{d}}\Phi_{\mathfrak{p}}, g_{\mathfrak{d}})$ etc. Clearly,

$$\Xi_{\mathfrak{p},\tau+}(\Phi,\omega,w) = \Xi^s_{\mathfrak{p},\tau+}(\Phi,\omega,w) + \Xi_{\mathfrak{p},\tau,\mathrm{st}+}(\Phi,\omega,w) \text{ etc.}$$

If the distributions in (3.5.2) are well defined, we define

$$\Xi_{\mathfrak{p}}(\Phi,\omega,w) = \sum_{\tau} \Xi_{\mathfrak{p},\tau}(\Phi,\omega,w)$$

etc. Equivalently, $\Xi_{\mathfrak{p}}(\Phi,\omega,w)$ can be defined by replacing $\mathscr{E}_{\mathfrak{p},\tau}(g_{\mathfrak{d}}, w)$ in (3.5.2) by $\mathscr{E}_{\mathfrak{p}}(g_{\mathfrak{d}}, w)$. It is easy to see that

$$\Xi_{\mathfrak{p},+}(\Phi,\omega,w) = \Xi^s_{\mathfrak{p},+}(\Phi,\omega,w) + \Xi_{\mathfrak{p},\mathrm{st}+}(\Phi,\omega,w) \text{ etc.}$$

also.

We still assume that $G^1_{\mathbb{A}} = G^0_{\mathbb{A}}$. We consider a path $\mathfrak{p} = (\mathfrak{d}, \mathfrak{s})$ which satisfies Condition (3.4.16)(1), or satisfies Condition (3.4.16)(2) and there is no split torus in the center of $M'_{\mathfrak{d}}$ which fixes $Z^{\mathrm{ss}}_{\mathfrak{d}k}$. Let $x \in Z^{\mathrm{ss}}_{\mathfrak{d}k}$. This assumption implies that there is no split torus in the center of $M'_{\mathfrak{d}}$, which fixes x, because $Z^{\mathrm{ss}}_{\mathfrak{d}k}$ is a single $M'_{\mathfrak{d}\bar{k}}$-orbit.

Let $\mathfrak{t}_\mathfrak{d}, \mathfrak{t}^1_\mathfrak{d}, \mathfrak{t}'_\mathfrak{d}, \mathfrak{t}^0_\mathfrak{d}$ be the Lie algebras of $A_\mathfrak{d}, A^1_\mathfrak{d}, A'_\mathfrak{d}, T_+ \cap M^0_{\mathfrak{d}\mathbb{A}}$ respectively. Then $\mathfrak{t}'_\mathfrak{d} = \mathfrak{t}_\mathfrak{d} \oplus \mathfrak{t}^1_\mathfrak{d}$.

Let $x = (x_1, \cdots, x_N) \in Z^{ss}_{\mathfrak{d}k}$. Since x is semi-stable, it is semi-stable with respect to the action of $A^1_\mathfrak{d}$ also. Note that we are considering $A^1_\mathfrak{d}$ as a group over $\mathbb{R}$, and the notion of semi-stable points is not changed by field extensions. If $\mathfrak{p}$ satisfies Condition (3.4.16)(1), $A^1_\mathfrak{d}$ is trivial. The convex hull of $\{\gamma_i|_{\mathfrak{t}^1_\mathfrak{d}} \mid x_i \neq 0\}$ contains the origin of $\mathfrak{t}^{1*}_\mathfrak{d}$. If $\mathfrak{p}$ satisfies Condition (3.4.16)(2) and there is no split torus contained in the center of $M'_\mathfrak{d}$ which fixes $Z_\mathfrak{d}$, the origin cannot be on the boundary of this convex hull, because that implies the existence of a split torus fixing x. Therefore, it contains a neighborhood of the origin of $\mathfrak{t}^{1*}_\mathfrak{d}$.

Proposition (3.5.3) *Let $\mathfrak{p} = (\mathfrak{d}, \mathfrak{s}), \tau$ be as above.*
(1) *The distributions $\Xi_{\mathfrak{p},\tau}(\Phi, \omega, w), \Xi_{\mathfrak{p},\tau+}(\Phi, \omega, w)$, and $\widehat{\Xi}_{\mathfrak{p},\tau+}(\Phi, \omega, w)$ in* (3.5.3) *are well defined for* $\mathrm{Re}(w) \gg 0$.
(2) *Moreover if $\mathfrak{p}$ satisfies* Condition (3.4.16)(1), $\Xi_{\mathfrak{p},\tau\#}(\Phi, \omega, w), \widehat{\Xi}_{\mathfrak{p},\tau\#}(\Phi, \omega, w)$ *are also well defined for* $\mathrm{Re}(w) \gg 0$.

Proof. Let $a = l(\mathfrak{p})$. Then $\lambda_\mathfrak{d}$ acts on $Z_{\mathfrak{d}\mathbb{A}}$ by multiplication by $\underline{e_{\mathfrak{p}a}(\lambda_\mathfrak{d})}$. For $g^0_\mathfrak{d}$ in the Siegel domain, we choose $\nu(g^0_\mathfrak{d}) \in T_+$ so that $t(g^0_\mathfrak{d})^z = \nu(g^0_\mathfrak{d})^z$ for any $z \in \mathfrak{t}^*_\mathbb{C}$. We choose $0 \leq \Psi \in \mathscr{S}(Z_{\mathfrak{d}\mathbb{A}})$ so that $|R_\mathfrak{d}\Phi_\mathfrak{p}(\lambda'_\mathfrak{d} g^0_\mathfrak{d} x)| \ll \Psi(\lambda'_\mathfrak{d}\nu(g^0_\mathfrak{d})x)$ for $g^0_\mathfrak{d}$ in the Siegel domain.

For $I \subset \{1, \cdots, N\}$, let

$$\Theta_{\mathfrak{d},I}(\Psi, \lambda'_\mathfrak{d}\nu(g^0_\mathfrak{d})) = \sum_{\substack{x \in Z^{ss}_{\mathfrak{d}k} \\ x_i \in k^\times \text{ for } i \in I \\ x_i = 0 \text{ for } i \notin I}} \Psi(\lambda'_\mathfrak{d}\nu(g^0_\mathfrak{d}), x).$$

We consider such I such that the convex hull of $\{\gamma_i|_{\mathfrak{t}^1_\mathfrak{d} \oplus \mathfrak{t}^0_\mathfrak{d}} \mid i \in I\}$ contains the origin of $\mathfrak{t}^{1*}_\mathfrak{d} \oplus \mathfrak{t}^{0*}_\mathfrak{d}$, and the convex hull of $\{\gamma_i|_{\mathfrak{t}^1_\mathfrak{d}} \mid i \in I\}$ contains a neighborhood of the origin of $\mathfrak{t}^{1*}_\mathfrak{d}$.

We choose I_0 so that $\{x = (x_i)_i \mid x_i = 0 \text{ for } i \notin I_0\} = Z_\mathfrak{d}$. Then

$$|\Theta_{Z_\mathfrak{d}}(R_\mathfrak{d}\Phi_\mathfrak{p}, \lambda'_\mathfrak{d} g^0_\mathfrak{d})| \ll \sum_{I \subset I_0} \Theta_{\mathfrak{d},I}(\Psi, \lambda'_\mathfrak{d}\nu(g^0_\mathfrak{d})).$$

We fix I. We choose constants $0 < c_i$ for $i \in I$ so that $\sum_i c_i\gamma_i|_{\mathfrak{t}^1_\mathfrak{d} \oplus \mathfrak{t}^0_\mathfrak{d}} = 0$. Since $\lambda_\mathfrak{d}$ acts on $Z_{\mathfrak{d}\mathbb{A}}$ by multiplication by $\underline{e_{\mathfrak{p}a}(\lambda_\mathfrak{d})}$, this implies that

$$(\lambda'_\mathfrak{d}\nu(g^0_\mathfrak{d}))^{-\sum_{i\in I} c_i\gamma_i} = e_{\mathfrak{p}a}(\lambda'_\mathfrak{d})^{-\sum_{i \in I} c_i}.$$

Let $C = \sum_{i \in I} c_i$. Clearly, $C > 0$.

Let $d = \dim A^1_\mathfrak{d}$. We write elements of $A'_\mathfrak{d}$ in the form $\lambda'_\mathfrak{d} = \lambda_\mathfrak{d}\lambda^{(1)}_\mathfrak{d}$, where $\lambda_\mathfrak{d} \in A_\mathfrak{d}, \lambda^{(1)}_\mathfrak{d} \in A^1_\mathfrak{d}$. Since the convex hull of $\{\gamma_i|_{\mathfrak{t}^1_\mathfrak{d}} \mid i \in I\}$ contains a neighborhood of the origin of $\mathfrak{t}^{1*}_\mathfrak{d}$, as in the proof of (3.1.4), for any $M > 0$, there exist constants $c'_{M,ji}$'s for $j = 1, \cdots, h, i \in I$ and a slowly increasing function $f_M(\lambda_\mathfrak{d}, \nu(g^0_\mathfrak{d}))$ such that

$$\inf_j (\lambda'_\mathfrak{d}\nu(g^0_\mathfrak{d}))^{-\sum_{i\in I} c'_{M,ji}\gamma_i} \leq f_M(\lambda_\mathfrak{d}, \nu(g^0_\mathfrak{d}))\mathrm{rd}_{d,M}(\lambda^{(1)}_\mathfrak{d}). \tag{3.5.4}$$

By (1.2.6), for any $M_1, M_2 \geq 1$,

$$\begin{aligned}|\Theta_{Z_\mathfrak{d}}(R_\mathfrak{d}\Phi_\mathfrak{p}, \lambda'_\mathfrak{d} g^0_\mathfrak{d})| &\ll \inf_j (\lambda'_\mathfrak{d}\nu(g^0_\mathfrak{d}))^{-M_1\sum_{i\in I} c_i\gamma_i - \sum_{i\in I} c'_{M_2,ji}\gamma_i} \\ &\leq e_{\mathfrak{p}a}(\lambda'_\mathfrak{d})^{-CM_1} f_{M_2}(\lambda_\mathfrak{d}, \nu(g^0_\mathfrak{d}))\mathrm{rd}_{d,M_2}(\lambda^{(1)}_\mathfrak{d}).\end{aligned}$$

We have the estimate

$$|\mathscr{E}_{\mathfrak{p},\tau}(g_\mathfrak{d}, w)| \ll \lambda'^{(\mathrm{Re}(w)-w_0)r_2}_\mathfrak{d} \nu(g^0_\mathfrak{d})^{(\mathrm{Re}(w)-w_0)r_1},$$

where r is as in (3.4.31)(3).

We choose M_2 large enough so that the function

$$f_{M_2}(\lambda_\mathfrak{d}, \nu(g^0_\mathfrak{d}))\mathrm{rd}_{d,M_2}(\lambda^{(1)}_\mathfrak{d})$$

is integrable with respect to $\lambda^{(1)}_\mathfrak{d}$. Then the resulting integral is bounded by a slowly increasing function $f'(\lambda_\mathfrak{d}, \nu(g^0_\mathfrak{d}))$ of $\lambda_\mathfrak{d}, \nu(g^0_\mathfrak{d})$. Consider the function

$$f'(\lambda_\mathfrak{d}, \nu(g^0_\mathfrak{d}))e_{\mathfrak{p}a}(\lambda'_\mathfrak{d})^{-CM_1}\lambda'^{(\mathrm{Re}(w)-w_0)r_2}_\mathfrak{d}\nu(g^0_\mathfrak{d})^{(\mathrm{Re}(w)-w_0)r_1}.$$

Note that the functions $e_{\mathfrak{p}a}(\lambda'_\mathfrak{d}), \lambda'^{(\mathrm{Re}(w)-w_0)r_2}_\mathfrak{d}$ depend only on $\lambda_\mathfrak{d}$.

If $e_{\mathfrak{p}a}(\lambda'_\mathfrak{d}) \geq 1$, we choose $\mathrm{Re}(w) \gg 0$ so that the above function is integrable with respect to $e_{\mathfrak{p}i}(\lambda'_\mathfrak{d})$ for $i = 1, \cdots, a-1$ and $g^0_\mathfrak{d}$ in the Siegel domain. Then we take $M_1 \gg 0$ so that it is integrable with respect to $e_{\mathfrak{p}a}(\lambda'_\mathfrak{d})$. If $e_{\mathfrak{p}a}(\lambda'_\mathfrak{d}) \leq 1$, we fix M_1 and take $\mathrm{Re}(w) \gg 0$. Since there are finitely many possibilities for I, this proves that $\Xi_{\mathfrak{p},\tau}(\Phi, \omega, w), \Xi_{\mathfrak{p},\tau+}(\Phi, \omega, w)$ are well defined if $\mathrm{Re}(w) \gg 0$. We can estimate $\Theta_{Z_\mathfrak{d}}(\mathscr{F}_\mathfrak{d} R_\mathfrak{d}\Phi_\mathfrak{p}, \theta_\mathfrak{p}(\lambda'_\mathfrak{d} g^0_\mathfrak{d}))$ similarly except that the exponent of $e_{\mathfrak{p}a}(\lambda'_\mathfrak{d})$ is positive. Since we consider $A_{\mathfrak{p}0}$ for $\widehat{\Xi}_{\mathfrak{p}+}(\Phi, \omega, w)$, the proof is similar. This proves (1).

We consider (2). In this case, $A_\mathfrak{d} = A'_\mathfrak{d}, M^1_{\mathfrak{d}\mathbb{A}} = M^0_{\mathfrak{d}\mathbb{A}}$. So $(e_{\mathfrak{p}1}, \cdots, e_{\mathfrak{p}a})$ gives an isomorphism $A_\mathfrak{d} \to \mathbb{R}^a_+$.

By (3.4.31)(3), for any slowly increasing function $f(g_\mathfrak{d})$ on $A_{\mathfrak{p}0}M^0_{\mathfrak{d}\mathbb{A}}/M_{\mathfrak{d}k}$, the integral

$$\int_{A_{\mathfrak{p}0}M^0_{\mathfrak{d}\mathbb{A}}/M_{\mathfrak{d}k}} f(g_\mathfrak{d})\mathscr{E}_{\mathfrak{p},\tau}(g_\mathfrak{d}, w)dg_\mathfrak{d}$$

converges absolutely for $\mathrm{Re}(w) \gg 0$. This proves (2).

Q.E.D.

A similar proof to that in the above proposition shows the following.

Proposition (3.5.5) *Let $\mathfrak{p} = (\mathfrak{d}, \mathfrak{s})$ be as above, $\mathfrak{p} \prec \mathfrak{p}' = (\mathfrak{d}'\mathfrak{s}')$ and $l(\mathfrak{p}') = l(\mathfrak{p}) + 1$. Suppose that for any $x = (x_1, \cdots, x_N) \in S_{\mathfrak{d}'k}$, the convex hull of $\{\gamma_i|_{\mathfrak{t}^1_\mathfrak{d}} \mid x_i \neq 0\}$ contains a neighborhood of the origin of $\mathfrak{t}^{1*}_\mathfrak{d}$. Then the integrals*

$$\int_{A'_{\mathfrak{p}0}M^0_{\mathfrak{d}\mathbb{A}}/M_{\mathfrak{d}k}} \omega_\mathfrak{p}(g_\mathfrak{d})\kappa_{\mathfrak{d}1}(g_\mathfrak{d})\sigma_\mathfrak{p}(g_\mathfrak{d})\Theta_{S_{\mathfrak{d}'}}(\mathscr{F}_\mathfrak{d} R_\mathfrak{d}\Phi_\mathfrak{p}, \theta_\mathfrak{d}(g_\mathfrak{d}))\mathscr{E}_\mathfrak{p}(g_\mathfrak{d}, w)dg_\mathfrak{d},$$

$$\int_{A'_{\mathfrak{p}0}M^0_{\mathfrak{d}\mathbb{A}}/M_{\mathfrak{d}k}} \omega_\mathfrak{p}(g_\mathfrak{d})\sigma_\mathfrak{p}(g_\mathfrak{d})\Theta_{S_{\mathfrak{d}'}}(R_\mathfrak{d}\Phi_\mathfrak{p}, g_\mathfrak{d})\mathscr{E}_\mathfrak{p}(g_\mathfrak{d}, w)dg_\mathfrak{d}$$

are well defined for $\mathrm{Re}(w) \gg 0$.

If $\mathfrak{p}$ satisfies Condition (3.4.16)(1), since $A_{\mathfrak{d}}^1$ is trivial, the condition in (3.5.5) is automatically satisfied for $\mathfrak{p}'$.

Suppose that $\mathfrak{p}$ satisfies Condition (3.4.16)(1). Let

$$(3.5.6)\qquad \begin{aligned} J'_{\mathfrak{p}}(\Phi, g_{\mathfrak{d}}) = \kappa_{\mathfrak{d}1}(g_{\mathfrak{d}}) & \sum_{\substack{\mathfrak{d}\prec\mathfrak{d}' \\ l(\mathfrak{d}')=l(\mathfrak{d})+1}} \Theta_{S_{\mathfrak{d}'}}(\mathscr{F}_{\mathfrak{d}} R_{\mathfrak{d}} \Phi_{\mathfrak{p}}, \theta_{\mathfrak{d}}(g_{\mathfrak{d}})) \\ & - \sum_{\substack{\mathfrak{d}\prec\mathfrak{d}' \\ l(\mathfrak{d}')=l(\mathfrak{d})+1}} \Theta_{S_{\mathfrak{d}'}}(R_{\mathfrak{d}} \Phi_{\mathfrak{p}}, g_{\mathfrak{d}}), \\ J_{\mathfrak{p}}(\Phi, g_{\mathfrak{d}}) = J'_{\mathfrak{p}}(\Phi, g_{\mathfrak{d}}) & + \kappa_{\mathfrak{d}1}(g_{\mathfrak{d}})\mathscr{F}_{\mathfrak{d}} R_{\mathfrak{d}} \Phi_{\mathfrak{p}}(0) - R_{\mathfrak{d}} \Phi_{\mathfrak{p}}(0). \end{aligned}$$

If $\mathfrak{p} = (\emptyset, \emptyset)$, we use the notation $J(\Phi, g^0)$, $J'(\Phi, g^0)$.

By the Poisson summation formula,

$$\begin{aligned} \Xi_{\mathfrak{p}}(\Phi, \omega, w) &= \Xi_{\mathfrak{p}+}(\Phi, \omega, w) \\ &\quad + \int_{A_{\mathfrak{p}0} M_{\mathfrak{d}\mathbb{A}}^0 / M_{\mathfrak{d}k}} \omega_{\mathfrak{p}}(g_{\mathfrak{d}}) \sigma_{\mathfrak{p}}(g_{\mathfrak{d}}) \Theta_{Z_{\mathfrak{d}}}(R_{\mathfrak{d}} \Phi_{\mathfrak{p}}, g_{\mathfrak{d}}) \mathscr{E}_{\mathfrak{p}}(g_{\mathfrak{d}}, w) dg_{\mathfrak{d}} \\ &= \Xi_{\mathfrak{p}+}(\Phi, \omega, w) + \widehat{\Xi}_{\mathfrak{p}+}(\Phi, \omega, w) + \widehat{\Xi}_{\mathfrak{p}\#}(\Phi, \omega, w) - \Xi_{\mathfrak{p}\#}(\Phi, \omega, w) \\ &\quad + \int_{A_{\mathfrak{p}0} M_{\mathfrak{d}\mathbb{A}}^0 / M_{\mathfrak{d}k}} \omega_{\mathfrak{p}}(g_{\mathfrak{d}}) \sigma_{\mathfrak{p}}(g_{\mathfrak{d}}) J'_{\mathfrak{p}}(\Phi, g_{\mathfrak{d}}) \mathscr{E}_{\mathfrak{p}}(g_{\mathfrak{d}}, w) dg_{\mathfrak{d}}. \end{aligned}$$

By (3.5.3), (3.5.5), we get the following proposition.

Proposition (3.5.7) *If* $\mathfrak{p} = (\mathfrak{d}, \mathfrak{s})$ *satisfies* Condition (3.4.16)(1), *then*

$$\begin{aligned} \Xi_{\mathfrak{p}}(\Phi, \omega, w) &= \Xi_{\mathfrak{p}+}(\Phi, \omega, w) + \widehat{\Xi}_{\mathfrak{p}+}(\Phi, \omega, w) + \widehat{\Xi}_{\mathfrak{p}\#}(\Phi, \omega, w) - \Xi_{\mathfrak{p}\#}(\Phi, \omega, w) \\ &+ \sum_{\substack{\mathfrak{d}\prec\mathfrak{d}' \\ l(\mathfrak{d}')=l(\mathfrak{d})+1}} \int_{A_{\mathfrak{p}0} M_{\mathfrak{d}\mathbb{A}}^0 / M_{\mathfrak{d}k}} \omega_{\mathfrak{p}}(g_{\mathfrak{d}}) \kappa_{\mathfrak{d}1}(g_{\mathfrak{d}}) \sigma_{\mathfrak{p}}(g_{\mathfrak{d}}) \Theta_{S_{\mathfrak{d}'}}(\mathscr{F}_{\mathfrak{d}} R_{\mathfrak{d}} \Phi_{\mathfrak{p}}, \theta_{\mathfrak{d}}(g_{\mathfrak{d}})) \mathscr{E}_{\mathfrak{p}}(g_{\mathfrak{d}}, w) dg_{\mathfrak{d}} \\ &- \sum_{\substack{\mathfrak{d}\prec\mathfrak{d}' \\ l(\mathfrak{d}')=l(\mathfrak{d})+1}} \int_{A_{\mathfrak{p}0} M_{\mathfrak{d}\mathbb{A}}^0 / M_{\mathfrak{d}k}} \omega_{\mathfrak{p}}(g_{\mathfrak{d}}) \sigma_{\mathfrak{p}}(g_{\mathfrak{d}}) \Theta_{S_{\mathfrak{d}'}}(R_{\mathfrak{d}} \Phi_{\mathfrak{p}}, g_{\mathfrak{d}}) \mathscr{E}_{\mathfrak{p}}(g_{\mathfrak{d}}, w) dg_{\mathfrak{d}}. \end{aligned}$$

Let $\mathfrak{p}$ be as above, $\mathfrak{p} \prec \mathfrak{p}'$, and $l(\mathfrak{p}') = l(\mathfrak{p}) + 1 = a + 1$. For $g_{\mathfrak{d}'} \in G_{\mathbb{A}}^0 \cap M_{\mathfrak{d}'\mathbb{A}}$, $u_{\mathfrak{d}'} \in U_{\mathfrak{d}'\mathbb{A}}$, we define

$$\begin{aligned} \Sigma_{\mathfrak{p}'1}(\Phi, \omega, g_{\mathfrak{d}'}, u_{\mathfrak{d}'}) &= \omega_{\mathfrak{p}}(g_{\mathfrak{d}'}) \sigma_{\mathfrak{p}}(g_{\mathfrak{d}'}) \Theta_{Y_{\mathfrak{d}'}}(R_{\mathfrak{d}} \Phi_{\mathfrak{p}}, g_{\mathfrak{d}'} n_{\mathfrak{d}'}(u_{\mathfrak{d}'})), \\ \Sigma_{\mathfrak{p}'2}(\Phi, \omega, g_{\mathfrak{d}'}, u_{\mathfrak{d}'}) &= \omega_{\mathfrak{p}}(g_{\mathfrak{d}'}) \sigma_{\mathfrak{p}}(g_{\mathfrak{d}'}) \kappa_{\mathfrak{d}1}(g_{\mathfrak{d}'}) \Theta_{Y_{\mathfrak{d}'}}(\mathscr{F}_{\mathfrak{d}} R_{\mathfrak{d}} \Phi_{\mathfrak{p}}, \theta_{\mathfrak{d}}(g_{\mathfrak{d}'} n_{\mathfrak{d}'}(u_{\mathfrak{d}'}))). \end{aligned}$$

Definition (3.5.8)

(1) *If* $\mathfrak{s}'(a+1) = 0$, $\widetilde{\Xi}_{\mathfrak{p}'}(\Phi, \omega, w)$ *is, by definition,*

$$\int_{A_{\mathfrak{p}0} P_{\mathfrak{d}'\mathbb{A}}^0 / P_{\mathfrak{d}'k}} \Sigma_{\mathfrak{p}'1}(\Phi, \omega, g_{\mathfrak{d}'}, u_{\mathfrak{d}'}) \widetilde{\mathscr{E}}_{\mathfrak{p}}(g_{\mathfrak{d}'} n_{\mathfrak{d}'}(u_{\mathfrak{d}'}), w) t(g_{\mathfrak{d}'})^{-2\rho_{\mathfrak{d}'}} dg_{\mathfrak{d}'} du_{\mathfrak{d}'},$$

if this integral is well defined for $\mathrm{Re}(w) \gg 0$.

(2) *If* $\mathfrak{s}'(a+1)=1$, $\widetilde{\Xi}_{\mathfrak{p}'}(\Phi,\omega,w)$ *is, by definition,*

$$\int_{A_{\mathfrak{p}0}P^1_{\mathfrak{d}'\mathbb{A}}/P_{\mathfrak{d}'k}} \Sigma_{\mathfrak{p}'2}(\Phi,\omega,g_{\mathfrak{d}'},u_{\mathfrak{d}'})\widetilde{\mathscr{E}}_{\mathfrak{p}}(g_{\mathfrak{d}'}n_{\mathfrak{d}'}(u_{\mathfrak{d}'}),w)t(g_{\mathfrak{d}'})^{-2\rho_{\mathfrak{d}'}}dg_{\mathfrak{d}'}du_{\mathfrak{d}'},$$

if this integral is well defined for $\mathrm{Re}(w)\gg 0$.

We consider paths in the following three classes.

(1) The path $\mathfrak{p}$ satisfies Condition (3.4.16)(1).

(2) The path $\mathfrak{p}$ satisfies Condition (3.4.16)(2) but not (1) and $M'_{\mathfrak{d}}$ contains a split torus in the center which acts trivially on $Z_{\mathfrak{d}}$.

(3) The path $\mathfrak{p}$ satisfies Condition (3.4.16)(2) but not (1) and $M'_{\mathfrak{d}}$ does not contain a split torus in the center which acts trivially on $Z_{\mathfrak{d}}$.

Let $\mathfrak{p}'$ be a path such that $\mathfrak{p}'=(\mathfrak{d}',\mathfrak{s}')$, and $\mathfrak{s}'(a+1)=0$. Then

$$\begin{aligned}
&\int_{A_{\mathfrak{p}0}M^0_{\mathfrak{d}\mathbb{A}}/M_{\mathfrak{d}k}} \omega_{\mathfrak{p}}(g_{\mathfrak{d}})\sigma_{\mathfrak{p}}(g_{\mathfrak{d}})\Theta_{S_{\mathfrak{d}'}}(R_{\mathfrak{d}}\Phi_{\mathfrak{p}},g_{\mathfrak{d}})\mathscr{E}_{\mathfrak{p}}(g_{\mathfrak{d}},w)dg_{\mathfrak{d}}\\
&=\int_{A_{\mathfrak{p}0}M^0_{\mathfrak{d}\mathbb{A}}/P_{\mathfrak{d}'k}} \omega_{\mathfrak{p}}(g_{\mathfrak{d}})\sigma_{\mathfrak{p}}(g_{\mathfrak{d}})\Theta_{Y_{\mathfrak{d}'}}(R_{\mathfrak{d}}\Phi_{\mathfrak{p}},g_{\mathfrak{d}})\mathscr{E}_{\mathfrak{p}}(g_{\mathfrak{d}},w)dg_{\mathfrak{d}}\\
&=\int_{A_{\mathfrak{p}0}M^0_{\mathfrak{d}\mathbb{A}}\cap P_{\mathfrak{d}'\mathbb{A}}/P_{\mathfrak{d}'k}} \Sigma_{\mathfrak{p}'1}(\Phi,\omega,g_{\mathfrak{d}'},u_{\mathfrak{d}'})\mathscr{E}_{\mathfrak{p}}(g_{\mathfrak{d}'}n_{\mathfrak{d}'}(u_{\mathfrak{d}'}),w)t(g_{\mathfrak{d}'})^{-2\rho_{\mathfrak{d}'}}dg_{\mathfrak{d}'}du_{\mathfrak{d}'}.
\end{aligned}$$

The second step is because $M_{\mathfrak{d},\omega_{\mathfrak{p}}}R_{\mathfrak{d}}\Phi_{\mathfrak{p}}=R_{\mathfrak{d}}\Phi_{\mathfrak{p}}$ by the assumption in §3.1.

Since

$$\int_{U_{\mathfrak{d}'\mathbb{A}}/U_{\mathfrak{d}'k}} \Theta_{Y_{\mathfrak{d}'}}(R_{\mathfrak{d}}\Phi_{\mathfrak{p}},g_{\mathfrak{d}'}n_{\mathfrak{d}'}(u_{\mathfrak{d}'}))du_{\mathfrak{d}'}=\kappa_{\mathfrak{d}'2}(g_{\mathfrak{d}'})\Theta_{Z_{\mathfrak{d}'}}(R_{\mathfrak{d}'}\Phi_{\mathfrak{p}'},g_{\mathfrak{d}'}),$$

and $A_{\mathfrak{p}0}\{d_{\mathfrak{d}'a+1}(\lambda_{\mathfrak{d}'a+1})\}=A_{\mathfrak{p}'2}$, the above integral is equal to

$$\widetilde{\Xi}_{\mathfrak{p}'}(\Phi,\omega,w)+\Xi_{\mathfrak{p}'}(\Phi,\omega,w).$$

If $\mathfrak{p}'$ belongs to the class (3), $\Xi_{\mathfrak{p}'}(\Phi,\omega,w)$ is well defined for $\mathrm{Re}(w)\gg 0$ by (3.5.2). Therefore, $\widetilde{\Xi}_{\mathfrak{p}'}(\Phi,\omega,w)$ is well defined for $\mathrm{Re}(w)\gg 0$ also.

Similarly, let $\mathfrak{p}'=(\mathfrak{d}',\mathfrak{s}')$, and $\mathfrak{s}'(a+1)=1$. Then

$$\begin{aligned}
&\int_{A_{\mathfrak{p}0}M^0_{\mathfrak{d}\mathbb{A}}/M_{\mathfrak{d}k}} \omega_{\mathfrak{p}}(g_{\mathfrak{d}})\kappa_{\mathfrak{d}1}(g_{\mathfrak{d}})\sigma_{\mathfrak{p}}(g_{\mathfrak{d}})\Theta_{S_{\mathfrak{d}'}}(\mathscr{F}_{\mathfrak{d}}R_{\mathfrak{d}}\Phi_{\mathfrak{p}},\theta_{\mathfrak{d}}(g_{\mathfrak{d}}))\mathscr{E}_{\mathfrak{p}}(g_{\mathfrak{d}},w)dg_{\mathfrak{d}}\\
&=\widetilde{\Xi}_{\mathfrak{p}'}(\Phi,\omega,w)+\Xi_{\mathfrak{p}'}(\Phi,\omega,w).
\end{aligned}$$

Therefore, $\widetilde{\Xi}_{\mathfrak{p}'}(\Phi,\omega,w)$ is well defined for $\mathrm{Re}(w)\gg 0$ if $\mathfrak{p}'$ belongs to the class (3).

These considerations show the following proposition.

Proposition (3.5.9) *Let* $\mathfrak{p}$ *be a path in* class (1). *Then if* $\widetilde{\Xi}_{\mathfrak{p}'}(\Phi,\omega,w)$ *is well defined for* $\mathrm{Re}(w)\gg 0$ *for all* $\mathfrak{p}'$,

$$\begin{aligned}
\epsilon_{\mathfrak{p}}\Xi_{\mathfrak{p}}(\Phi,\omega,w)&=\epsilon_{\mathfrak{p}}(\Xi_{\mathfrak{p}+}(\Phi,\omega,w)+\widehat{\Xi}_{\mathfrak{p}+}(\Phi,\omega,w)+\widehat{\Xi}_{\mathfrak{p}\#}(\Phi,\omega,w)-\Xi_{\mathfrak{p}\#}(\Phi,\omega,w))\\
&\quad+\sum_{\substack{\mathfrak{d}\prec\mathfrak{d}'\\ l(\mathfrak{d}')=l(\mathfrak{d})+1}}\epsilon_{\mathfrak{p}'}\left(\widetilde{\Xi}_{\mathfrak{p}'}(\Phi,\omega,w)+\Xi_{\mathfrak{p}'}(\Phi,\omega,w)\right).
\end{aligned}$$

This proposition is the basis of the iterated use of the Poisson summation formula. We now describe how we proceed.

We assume that $(G/\widetilde{T}, V, \chi_V)$ is an irreducible representation and of complete type for simplicity. This implies that $G^1_{\mathbb{A}} = G^0_{\mathbb{A}}$. For $\Phi \in \mathscr{S}(V_{\mathbb{A}})$ and $\lambda \in \mathbb{R}_+$, we define $\Phi_\lambda(x) = \Phi(\underline{\lambda} x)$.

We want to find points $p_1, \cdots, p_{d_V} \in \mathbb{C}$ and distributions $a(\Phi, \omega), a_{ij}(\Phi, \omega)$ for $i = 1, \cdots, d_V, j = 1, \cdots, d_{V,i}$ with the following properties

$$\int_{G^0_{\mathbb{A}}/G_k} \omega(g^0) J(\Phi, g^0) \mathscr{E}(g^0, w) dg^0 \sim a(\Phi, \omega)\Lambda(w; z); \tag{3.5.10}$$

$$a(\Phi_\lambda, \omega) = \sum_{i=1}^{d_V} \sum_{j=1}^{d_{V,i}} \lambda^{-p_i} (\log \lambda)^j a_{ij}(\Phi, \omega), \tag{3.5.11}$$

where $\lambda \in \mathbb{R}_+$.

Clearly,

$$\int_0^1 \lambda^{s-p_i} (\log \lambda)^j d^\times \lambda = (-1)^j j! (s - p_i)^{-j-1}.$$

This implies that

$$\begin{aligned} Z_V(\Phi, \omega, s) &= Z_{V+}(\Phi, \omega, s) + Z_{V+}(\widehat{\Phi}, \omega^{-1}, N - s) \\ &\quad + C_G^{-1} \sum_{i=1}^{d_V} \sum_{j=1}^{d_{V,i}} (-1)^j j! \frac{a_{ij}(\Phi, \omega)}{(s - p_i)^{j+1}}, \end{aligned} \tag{3.5.12}$$

giving a principal part formula.

If (G, V, χ_V) is of incomplete type, we need 'adjusting terms' (see [86] or Chapter 4). In this book, the only such case is essentially the case of binary quadratic forms, so we do not discuss it here. For a general conjecture about prehomogeneous vector spaces of incomplete type, the reader should see [85].

In order to achieve (3.5.10) and (3.5.11), we use (3.5.9) repeatedly. If $\mathfrak{p}' = (\mathfrak{d}', \mathfrak{s}')$ in (3.5.9) belongs to class (1) and $M_{\mathfrak{d}'}$ is a torus or $\mathfrak{p}'$ belongs to class (2), then we do not consider paths $\mathfrak{p}' \prec \mathfrak{p}''$. If $\mathfrak{p}'$ belongs to class (1) and $M_{\mathfrak{d}'}$ is a torus, $\Xi_{\mathfrak{p}'}(\Phi, \omega, w)$ is essentially a contour integral of a product of Dedekind zeta functions. Therefore, we can handle such distributions. If $\mathfrak{p}'$ belongs to class (2), it is likely that we can show $\Xi_{\mathfrak{p}'}(\Phi, \omega, w), \widetilde{\Xi}_{\mathfrak{p}'}(\Phi, \omega, w) \sim 0$ as is the case in Parts III and IV.

Suppose that $\mathfrak{p}$ belongs to class (3). This implies that the representation of $M_{\mathfrak{d}}$ on $Z_{\mathfrak{d}}$ is reducible. We still do not have a general approach to handle all the reducible prehomogeneous vector spaces, but if the number of irreducible factors is two, we can use a technique to consider further paths. We discuss this feature of our program in Parts III and IV.

So we are still obliged to show that $\widetilde{\Xi}_{\mathfrak{p}}(\Phi, \omega, w)$ is well defined for $\mathrm{Re}(w) \gg 0$ for $\mathfrak{p} = (\mathfrak{d}, \mathfrak{s})$ in classes (2), (3). We will handle these distributions when we consider individual cases. Of course, we want to ignore all the $\widetilde{\Xi}_{\mathfrak{p}}(\Phi, \omega, w)$'s. However, it is too optimistic to expect that $\widetilde{\Xi}_{\mathfrak{p}}(\Phi, \omega, w) \sim 0$ for all $\mathfrak{p}$ as we will see in Part IV. For the space of pairs of ternary quadratic forms, the correct statement turns out to be

$$\sum_{\mathfrak{p}} \epsilon_{\mathfrak{p}} \widetilde{\Xi}_{\mathfrak{p}}(\Phi, \omega, w) \sim 0.$$

Also it is too optimistic to expect that for all $\mathfrak{p}$, there exists a distribution $b_{\mathfrak{p}}(\Phi,\omega)$ which satisfies the property (3.5.11) and

$$\Xi_{\mathfrak{p}}(\Phi,\omega,w) \sim b_{\mathfrak{p}}(\Phi_\lambda,\omega)\Lambda(w;z).$$

This is because some cancellations have to be established between strata. Therefore, our process is quite delicate and heavily depends on the individual cases.

The following proposition follows from (3.4.32).

Proposition (3.5.13) *Consider a path* $\mathfrak{p}$ *in* class (1). *Suppose that*

$$\sigma_{\mathfrak{p}}(\lambda_{\mathfrak{d}}) = \prod_{i=1}^{a} e_{\mathfrak{p}i}(\lambda_{\mathfrak{d}})^{\chi_{\mathfrak{p}i}},$$

and $\chi_{\mathfrak{p}i} > 0$ *for* $i = 1,\cdots,a-1$. *Then*

$$\begin{aligned}
&\Xi^s_{\mathfrak{p}+}(\Phi,\omega,w) \\
&\sim C_G\Lambda(w;\rho)\int_{A'_{\mathfrak{p}1}M^0_{\mathfrak{d}\mathbb{A}}/M_{\mathfrak{d}k}} \omega_{\mathfrak{p}}(g_{\mathfrak{d}})\sigma_{\mathfrak{p}}(g_{\mathfrak{d}})\Theta_{Z^s_{\mathfrak{d}}}(R_{\mathfrak{d}}\Phi_{\mathfrak{p}},g_{\mathfrak{d}})dg_{\mathfrak{d}}, \\
&\widehat{\Xi}^s_{\mathfrak{p}+}(\Phi,\omega,w) \\
&\sim C_G\Lambda(w;\rho)\int_{A'_{\mathfrak{p}0}M^0_{\mathfrak{d}\mathbb{A}}/M_{\mathfrak{d}k}} \omega_{\mathfrak{p}}(g_{\mathfrak{d}})\sigma_{\mathfrak{p}}(g_{\mathfrak{d}})\kappa_{\mathfrak{d}1}(g_{\mathfrak{d}})\Theta_{Z^s_{\mathfrak{d}}}(\mathscr{F}_{\mathfrak{d}}R_{\mathfrak{d}}\Phi_{\mathfrak{p}},\theta_{\mathfrak{d}}(g_{\mathfrak{d}}))dg_{\mathfrak{d}}.
\end{aligned}$$

Next, we consider paths of positive length in class (2). We are still assuming that $G^0_{\mathbb{A}} = G^1_{\mathbb{A}}$. Therefore, we do not consider χ.

Suppose that $\mathfrak{p}$ is a path which satisfies Condition (3.4.16)(2). Let $A^1_{\mathfrak{d}}$ be as before. Let $A^2_{\mathfrak{d}}$ be the connected component of $\mathrm{Ker}(A^1_{\mathfrak{d}} \to \mathrm{GL}(V)_{\mathbb{A}})$ which contains 1. Then $A^2_{\mathfrak{d}} \cong \mathbb{R}^c_+$ for some c. We choose a subgroup $A^3_{\mathfrak{d}} \subset A^1_{\mathfrak{d}}$ which is isomorphic to $\mathbb{R}^d_+$ for some d so that $A^1_{\mathfrak{d}} = A^2_{\mathfrak{d}}A^3_{\mathfrak{d}}$ and $A^2_{\mathfrak{d}} \cap A^3_{\mathfrak{d}} = \{1\}$. We write elements of $A^1_{\mathfrak{d}}$ in the form $\lambda^{(1)}_{\mathfrak{d}} = \lambda^{(2)}_{\mathfrak{d}}\lambda^{(3)}_{\mathfrak{d}}$, where $\lambda^{(i)}_{\mathfrak{d}} \in A^i_{\mathfrak{d}}$ for $i = 1,2,3$. Let $g_{\mathfrak{d}} = \lambda_{\mathfrak{d}}\lambda^{(1)}_{\mathfrak{d}}g^0_{\mathfrak{d}}$, where $g_{\mathfrak{d}} \in G^0_{\mathbb{A}} \cap M_{\mathfrak{d}\mathbb{A}}, g^0_{\mathfrak{d}} \in M^0_{\mathfrak{d}\mathbb{A}}$.

We fix an identification $A^2_{\mathfrak{d}} \cong \mathbb{R}^c_+$, and write

$$\lambda^{(2)}_{\mathfrak{d}} = (\lambda^{(2)}_{\mathfrak{d}1},\cdots,\lambda^{(2)}_{\mathfrak{d}c})$$

for $\lambda^{(2)}_{\mathfrak{d}1},\cdots,\lambda^{(2)}_{\mathfrak{d}c} \in \mathbb{R}_+$. Let $d^\times\lambda^{(2)}_{\mathfrak{d}} = \prod_i d^\times\lambda^{(2)}_{\mathfrak{d}i}$. We choose an invariant measure $d^\times\lambda^{(i)}_{\mathfrak{d}}$ on $A^i_{\mathfrak{d}}$ for $i = 1,3$ so that $d^\times\lambda^{(1)}_{\mathfrak{d}} = d^\times\lambda^{(2)}_{\mathfrak{d}}d^\times\lambda^{(3)}_{\mathfrak{d}}$, and $dg_{\mathfrak{d}} = d^\times\lambda_{\mathfrak{d}}d^\times\lambda^{(1)}_{\mathfrak{d}}dg^0_{\mathfrak{d}}$.

Suppose that τ is a Weyl group element which satisfies Condition (3.4.13) for $M_{\mathfrak{p}}$. Let $\gamma_{\mathfrak{p},\tau,i}(z)$ be a linear function for $i = 1,\cdots,c$ such that

$$\theta_{\mathfrak{p}}(\lambda^{(2)}_{\mathfrak{d}})^{\tau z} = \prod_{i=1}^{c}(\lambda^{(2)}_{\mathfrak{d}i})^{\gamma_{\mathfrak{p},\tau,i}(z)}. \tag{3.5.14}$$

We define

$$LS_{\mathfrak{p},\tau} = \{z \in \mathfrak{t}^*_{\mathbb{C}} \mid \gamma_{\mathfrak{p},\tau,i}(z) = 0 \text{ for } i = 1,\cdots,c\}. \tag{3.5.15}$$

Let $dz|_{LS_{\mathfrak{p},\tau}}$ be the differential form such that $dz = \prod_{i=1}^{c} d\gamma_{\mathfrak{p},\tau,i} dz|_{LS_{\mathfrak{p},\tau}}$.

Let $h \in \mathfrak{t}^*$ be the point such that

$$\theta_{\mathfrak{p}}(\lambda_{\mathfrak{d}}^{(2)})^{-\tau h} = \sigma_{\mathfrak{p}}(\lambda_{\mathfrak{d}}^{(2)})\theta_{\mathfrak{p}}(\lambda_{\mathfrak{d}}^{(2)})^{\rho}.$$

There exist numbers $h_1, \cdots, h_c$ such that $\theta_{\mathfrak{p}}(\lambda_{\mathfrak{d}}^{(2)})^{\tau h} = \prod_i (\lambda_{\mathfrak{d}i}^{(2)})^{h_i}$.

Consider I, σ, ν in the proof of (3.4.31). We consider the following condition.

Condition (3.5.16) *For any τ, σ, there exists $q \in (LS_{\mathfrak{p},\tau} + h) \cap D_{\sigma\tau}$ such that if $\alpha \in \mathfrak{t}^*$ and $\|\alpha\|_0$ is small, the function*

$$\sigma_{\mathfrak{p}}(\lambda_{\mathfrak{d}}\lambda_{\mathfrak{d}}^{(3)})\theta_{\mathfrak{p}}(\lambda_{\mathfrak{d}}\lambda_{\mathfrak{d}}^{(3)})^{\tau(q+\alpha)+\rho}\theta_{\mathfrak{p}}(t(g_{\mathfrak{d}}^0))^{\sigma\tau(q+\alpha)+\rho}\Theta_{Z_{\mathfrak{d}}}(R_{\mathfrak{d}}\Phi_{\mathfrak{p}}, g_{\mathfrak{d}})$$

is integrable on $A_{\mathfrak{p}2}A_{\mathfrak{p}}^3(K \cap M_{\mathfrak{d}\mathbb{A}})T_{\mathfrak{d}\eta+}^0\Omega_{\mathfrak{d}}$.

Note that $\Theta_{Z_{\mathfrak{d}}}(R_{\mathfrak{d}}\Phi_{\mathfrak{p}}, g_{\mathfrak{d}})$ does not depend on $\lambda_{\mathfrak{d}}^{(2)}$. The above condition implies that $(LS_{\mathfrak{p},\tau}+h)\cap D_{\sigma\tau} \neq \emptyset$ for all τ, σ. In particular, $(LS_{\mathfrak{p},\tau}+h)\cap D_{\tau} \neq \emptyset$. Suppose that (3.5.16) is satisfied. We choose a subspace $H_{\mathfrak{p},\tau} \subset \mathfrak{t}_{\mathbb{C}}^*$ so that $\mathfrak{t}_{\mathbb{C}}^* = LS_{\mathfrak{p},\tau} \oplus H_{\mathfrak{p},\tau}$. Let $z_1 \in (LS_{\mathfrak{p},\tau} + h)$, $z_2 \in H_{\mathfrak{p},\tau}$. Let dz_1 be the differential form on $LS_{\mathfrak{p},\tau} + h$ such that $dz_1(z-h) = dz|_{LS_{\mathfrak{p},\tau}}$. For $z_2 \in H_{\mathfrak{p},\tau}$, we define $z_{2i} = \gamma_{\mathfrak{p},\tau,i}(z_2)$ for $i = 1, \cdots, c$. We define $dz_2 = \prod_i dz_{2i}$.

Let $c' = n_1 + \cdots + n_f - f - c$. We put

$$\begin{aligned}
&f_{\mathfrak{p},\tau}(\lambda_{\mathfrak{d}}\lambda_{\mathfrak{d}}^{(3)}, g_{\mathfrak{d}}, w, z_1, z_2)\\
&= \theta_{\mathfrak{p}}(\lambda_{\mathfrak{d}}\lambda_{\mathfrak{d}}^{(3)})^{\tau(z_1+z_2)+\rho}E_{B_{\mathfrak{p}},\tau}(\theta_{\mathfrak{p}}(g_{\mathfrak{d}}^0), z_1+z_2)\Lambda(w; z_1+z_2),\\
&f_{\mathfrak{p},\tau,I,\sigma,\nu}(\lambda_{\mathfrak{d}}\lambda_{\mathfrak{d}}^{(3)}, g_{\mathfrak{d}}, w, z_1, z_2)\\
&= \theta_{\mathfrak{p}}(\lambda_{\mathfrak{d}}\lambda_{\mathfrak{d}}^{(3)})^{\tau(z_1+z_2)+\rho}E_{B_{\mathfrak{p}},I,\sigma\tau,\nu}(\theta_{\mathfrak{p}}(g_{\mathfrak{d}}^0), z_1+z_2)\Lambda(w; z_1+z_2).
\end{aligned}$$

Then we define the functions

$$\mathscr{E}'_{\mathfrak{p},\tau}(\lambda_{\mathfrak{d}}\lambda_{\mathfrak{d}}^{(3)}g_{\mathfrak{d}}^0, w, z_2),\ \mathscr{E}'_{\mathfrak{p},\tau,I,\sigma,\nu}(\lambda_{\mathfrak{d}}\lambda_{\mathfrak{d}}^{(3)}g_{\mathfrak{d}}^0, w, z_2)$$

by the integrals

$$\left(\frac{1}{2\pi\sqrt{-1}}\right)^{c'}\int_{\substack{z_1\in LS_{\mathfrak{p},\tau}+h\\ \mathrm{Re}(z_1)=q}} f_{\mathfrak{p},\tau}(\lambda_{\mathfrak{d}}\lambda_{\mathfrak{d}}^{(3)}, g_{\mathfrak{d}}, w, z_1, z_2)dz_1,$$

$$\left(\frac{1}{2\pi\sqrt{-1}}\right)^{c'}\int_{\substack{z_1\in LS_{\mathfrak{p},\tau}+h\\ \mathrm{Re}(z_1)=q}} f_{\mathfrak{p},\tau,I,\sigma,\nu}(\lambda_{\mathfrak{d}}\lambda_{\mathfrak{d}}^{(3)}, g_{\mathfrak{d}}, w, z_1, z_2)dz_1,$$

respectively, where we choose q in $(LS_{\mathfrak{p},\tau} + h) \cap D_{\tau}$, $(LS_{\mathfrak{p},\tau} + h) \cap D_{\sigma\tau}$. The above functions are well defined if $\mathrm{Re}(w) \gg 0$ and $\|\mathrm{Re}(z_2)\|_0$ is small.

Proposition (3.5.17) *Suppose that* Condition (3.5.16) *is satisfied for $\mathfrak{p}$. Let τ be as before. Then*

(1) $\Xi_{\mathfrak{p}}(\Phi, \omega, w), \Xi_{\mathfrak{p},\tau}(\Phi, \omega, w)$ *are well defined for* $\mathrm{Re}(w) \gg 0$, *and*

$$\Xi_{\mathfrak{p}}(\Phi, \omega, w) = \sum_{\tau} \Xi_{\mathfrak{p},\tau}(\Phi, \omega, w),$$

(2) $\Xi_{\mathfrak{p},\tau}(\Phi,\omega,w)$ *is equal to*

$$\int_{A_{\mathfrak{p}2}A_{\mathfrak{d}}^3 M_{\mathfrak{d}A}^0/M_{\mathfrak{d}k}} \sigma_{\mathfrak{p}}(\lambda_{\mathfrak{d}}\lambda_{\mathfrak{d}}^{(3)})\mathscr{E}'_{\mathfrak{p},\tau}(\lambda_{\mathfrak{d}}\lambda_{\mathfrak{d}}^{(3)}g_{\mathfrak{d}}^0,w,0)\Theta_{Z_{\mathfrak{d}}}(R_{\mathfrak{d}}\Phi_{\mathfrak{p}},g_{\mathfrak{d}})d^\times\lambda_{\mathfrak{d}}d^\times\lambda_{\mathfrak{d}}^{(3)}dg_{\mathfrak{d}}^0.$$

Proof. We consider subsets of $A_{\mathfrak{p}}^2$ like $\{\lambda^{(2)} \mid \lambda_{\mathfrak{d}i}^{(2)} \geq 1 \text{ for all } i\}$, $\{\lambda^{(2)} \mid \lambda_{\mathfrak{d}1}^{(2)} \leq 1, \lambda_{\mathfrak{d}2}^{(2)} \geq 1, \cdots, \lambda_{\mathfrak{d}c}^{(2)} \geq 1\}$ etc.

Let $A_{\mathfrak{d}+}^2 = \{\lambda^{(2)} \mid \lambda_{\mathfrak{d}i}^{(2)} \geq 1 \text{ for all } i\}$. Since the argument is symmetric, we consider the set $A_{\mathfrak{d}+}^2$. We choose $\alpha \in \mathfrak{t}^*$ so that $\|\alpha\|_0$ is small and

$$\theta_{\mathfrak{p}}(\lambda_{\mathfrak{d}}^{(2)})^{\tau\alpha} = (\lambda_{\mathfrak{d}1}^{(2)})^{p_1}\cdots(\lambda_{\mathfrak{d}c}^{(2)})^{p_c},$$

for some $p_1,\cdots,p_c < 0$.

By assumption,

$$\mathscr{E}'_{\mathfrak{p},\tau}(\lambda_{\mathfrak{d}}\lambda_{\mathfrak{d}}^{(3)}g_{\mathfrak{d}}^0,w,z_2) = \sum_{I,\sigma,\nu}\mathscr{E}'_{\mathfrak{p},\tau,I,\sigma,\nu}(\lambda_{\mathfrak{d}}\lambda_{\mathfrak{d}}^{(3)}g_{\mathfrak{d}}^0,w,z_2),$$

and

$$|\mathscr{E}'_{\mathfrak{p},\tau,I,\sigma,\nu}(\lambda_{\mathfrak{d}}\lambda_{\mathfrak{d}}^{(3)}g_{\mathfrak{d}}^0,w,z_2)| \ll \theta_{\mathfrak{p}}(\lambda_{\mathfrak{d}}\lambda_{\mathfrak{d}}^{(3)})^{\tau(q_\sigma+\mathrm{Re}(z_2))+\rho}t(g_{\mathfrak{d}}^0)^{\sigma\tau(q_\sigma+\mathrm{Re}(z_2))+\rho},$$

where q_σ satisfies Condition (3.5.16).

It is easy to see that

$$\sigma_{\mathfrak{p}}(\lambda_{\mathfrak{d}}^{(2)})\mathscr{E}_{\mathfrak{p},\tau}(g_{\mathfrak{d}},w) = \left(\frac{1}{2\pi\sqrt{-1}}\right)^c\int_{\mathrm{Re}(z_2)=\alpha}\mathscr{E}'_{\mathfrak{p},\tau}(\lambda_{\mathfrak{d}}\lambda_{\mathfrak{d}}^{(3)}g_{\mathfrak{d}}^0,w,z_2)\prod_i(\lambda_{\mathfrak{d}i}^{(2)})^{z_{2i}}dz_2.$$

Consider the function

$$\sigma_{\mathfrak{p}}(\lambda_{\mathfrak{d}}\lambda_{\mathfrak{d}}^{(3)})\Theta_{Z_{\mathfrak{d}}}(R_{\mathfrak{d}}\Phi_{\mathfrak{p}},g_{\mathfrak{d}})\mathscr{E}'_{\mathfrak{p},\tau}(\lambda_{\mathfrak{d}}\lambda_{\mathfrak{d}}^{(3)}g_{\mathfrak{d}}^0,w,z_2)\prod_i(\lambda_{\mathfrak{d}i}^{(2)})^{z_{2i}}. \tag{3.5.18}$$

By Condition (3.5.16) and the choice of α, the order of the integration of the function (3.5.18) with respect to the set $A_{\mathfrak{d}+}^2$ and the contour $\{\mathrm{Re}(z_2)=\alpha\}$ can be changed. Therefore, $\Xi_{\mathfrak{p},\tau}(\Phi,\omega,w)$ is well defined for all τ, and (1) follows.

The integral of the function (3.5.18) with respect to the set $A_{\mathfrak{d}+}^2$ and the contour $\{\mathrm{Re}(z_2)=\alpha\}$ is equal to

$$(-1)^c\left(\frac{1}{2\pi\sqrt{-1}}\right)^c\int_{\substack{\mathrm{Re}(z_{2i})=p_i\\ i=1,\cdots,c}}\frac{\sigma_{\mathfrak{p}}(\lambda_{\mathfrak{d}}\lambda_{\mathfrak{d}}^{(3)})\Theta_{Z_{\mathfrak{d}}}(R_{\mathfrak{d}}\Phi_{\mathfrak{p}},g_{\mathfrak{d}})\mathscr{E}'_{\mathfrak{p},\tau}(\lambda_{\mathfrak{d}}\lambda_{\mathfrak{d}}^{(3)}g_{\mathfrak{d}}^0,w,z_2)}{\prod_i z_{2i}}dz_2.$$

The integral of (3.5.18) over the set $\{\lambda^{(2)} \mid \lambda_{\mathfrak{d}1}^{(2)} \leq 1, \lambda_{\mathfrak{d}2}^{(2)} \geq 1, \cdots, \lambda_{\mathfrak{d}c}^{(2)} \geq 1\}$ etc. has a similar expression.

Let $\epsilon > 0$ be a constant . If $\psi(s)$ is holomorphic function on the set $\{s \in \mathbb{C} \mid -\epsilon < \mathrm{Re}(s) < \epsilon\}$ which is rapidly decreasing with respect to $\mathrm{Im}(s)$,

$$-\int_{\mathrm{Re}(s)=-\frac{\epsilon}{2}}\frac{\psi(s)}{s}ds + \int_{\mathrm{Re}(s)=\frac{\epsilon}{2}}\frac{\psi(s)}{s}ds = \psi(0).$$

The statement (2) of the proposition follows by applying the above relation c times.

Q.E.D.

When we apply our process later, we sometimes face strata S_β's such that there is a split torus in the center of M'_β which fixes each point in Z_β. In many cases (but not always), we can ignore such strata. This phenomenon is based on the following proposition.

Proposition (3.5.19) (The vanishing principle) *Suppose that*

$$\psi(\rho) \neq 0, \psi(-\tau_G z) = \psi(z),$$

and Condition (3.5.16) *is satisfied. Then if* $\rho \notin LS_{\mathfrak{p},\tau} + h$, *there exists a polynomial* $P(z)$ *such that* $P(\rho) \neq 0$, $P(-\tau_G z) = P(z)$, *and that if* $\psi' = \psi(z)P(z)$, *then* $\Xi_{\mathfrak{p},\tau}(\Phi, \omega, w, \psi') = 0$.

Proof. We choose a linear function $l(z)$ so that $l(z) = 0$ on $LS_{\mathfrak{p},\tau} + h$ and $l(z) \neq 0$. Let $P(z) = l(z)l(-\tau_G z)$. Since $-\tau_G \rho = \rho$, $P(\rho) \neq 0$ also. Since $\psi'(z)$ is identically zero on the contour, $\psi'(z)$ satisfies the condition of this proposition.

Q.E.D.

Proposition (3.5.20) *Let* $\delta_\#(\omega) = \delta(\omega_1) \cdots \delta(\omega_f)$. *Then*

$$\int_{G^0_{\mathbb{A}}/G_k} \omega(g^0)\mathscr{E}(g^0, w)dg^0 = \delta_\#(\omega)\Lambda(w; \rho).$$

Proof. Let db be the measure on $B_{\mathbb{A}}$ which is a product of measures we defined in §1.1. for $B_{\mathbb{A}} \cap G_{i\mathbb{A}}$ for $i = 1, \cdots, f$. Let $B^0_{\mathbb{A}} = G^0_{\mathbb{A}} \cap B_{\mathbb{A}}$. We choose a measure db^0 on $B^0_{\mathbb{A}}$ so that

$$\int_{B_{\mathbb{A}}/B_k} f(b)db = n_1 \cdots n_f \int_{\mathbb{R}^f_+} \int_{B^0_{\mathbb{A}}/B_k} f(c(\lambda)b^0) \prod_{i=1}^{f} d^\times \lambda_i db^0,$$

where $c(\lambda) = c_{n_1}(\lambda_1) \cdots c_{n_f}(\lambda_f)$ for $\lambda_1, \cdots, \lambda_f \in \mathbb{R}_+$.

Let $\mu_i = a_{n_i}(\underline{\mu}_{i1}, \underline{\mu}_{i2}\underline{\mu}_{i1}^{-1},, \cdots, \underline{\mu}_{in_i-1}^{-1})$, and $\mu = (\mu_1, \cdots, \mu_f)$. We identify

$$B^0_{\mathbb{A}} \cong \mathbb{R}_+^{n_1+\cdots+n_f-f} \times (\mathbb{A}^1/k^\times)^{n_1+\cdots+n_f},$$

and write $b^0 = \mu t^0$ where $t^0 \in (\mathbb{A}^1/k^\times)^{n_1+\cdots+n_f}$. Then, by the choice of our measure, $db = \mu^{-2\rho} \prod_{i,j} d^\times \mu_{ij} d^\times t^0$, where $d^\times t^0$ is the measure such that the volume of $(\mathbb{A}^1/k^\times)^{n_1+\cdots+n_f}$ is 1.

By the Mellin inversion formula,

$$
\begin{aligned}
&\int_{G^0_{\mathbb{A}}/G_k} \omega(g^0)\mathscr{E}(g^0,w)dg^0 \\
&= \left(\frac{1}{2\pi\sqrt{-1}}\right)^{n_1+\cdots+n_f-f} \int_{G^0_{\mathbb{A}}/G_k}\int_{\mathrm{Re}(z)=q} \omega(g^0) \sum_{\gamma\in G_k/B_k} t(g^0\gamma)^{z+\rho}\Lambda(w;z)dg^0dz \\
&= \left(\frac{1}{2\pi\sqrt{-1}}\right)^{n_1+\cdots+n_f-f} \int_{G^0_{\mathbb{A}}/G_k} \sum_{\gamma\in G_k/B_k} \int_{\mathrm{Re}(z)=q} \omega(g^0)t(g^0\gamma)^{z+\rho}\Lambda(w;z)dg^0dz \\
&= \left(\frac{1}{2\pi\sqrt{-1}}\right)^{n_1+\cdots+n_f-f} \int_{G^0_{\mathbb{A}}/B_k}\int_{\mathrm{Re}(z)=q} \omega(g^0)t(g^0)^{z+\rho}\Lambda(w;z)dg^0dz \\
&= \delta_{\#}(\omega)\left(\frac{1}{2\pi\sqrt{-1}}\right)^{n_1+\cdots+n_f-f} \\
&\quad \times \int_{\mathbb{R}_+^{n_1+\cdots+n_f-f}}\int_{\mathrm{Re}(z)=q} \prod_{i,j} \mu_{ij}^{z_{ij}-z_{ij+1}-1}\Lambda(w;z)\prod_{i,j} d^\times \mu_{ij}dz \\
&= \delta_{\#}(\omega)\Lambda(w;\rho).
\end{aligned}
$$

This proves the proposition.

Q.E.D.

Right now, (3.5.13) and (3.5.19) are the only general statements we can prove without using properties of individual representations.

§3.6 The passing principle

We have to consider various contour integrals in later chapters, and for that purpose, we prove a statement concerning a possible way to move the contour in this section. We first introduce some notation.

Let $\mathfrak{p}$ be a path, and τ a Weyl group element which satisfies Condition (3.4.13) for $M_{\mathfrak{p}}$. We define

$$
s_\tau = (s_{\tau 1}, \cdots, s_{\tau f}),\ s_{\tau i} = (s_{\tau i1}, \cdots, s_{\tau i n_i - 1}) \in \mathbb{C}^{n_i - 1}
$$

by $s_{\tau i1} = z_{i\tau(n_i)} - z_{i\tau(n_i-1)}, \cdots, s_{\tau i n_i-1} = z_{i\tau(2)} - z_{i\tau(1)}$. Since the correspondence between z and s_τ is one-to-one, any function of z can be considered as a function of s_τ. For a function $f(z)$, let $\widetilde{f}(s_\tau)$ be the corresponding function of s_τ. So for example, we consider $\widetilde{L}(s_\tau), \widetilde{\Lambda}_\tau(w;s_\tau)$ etc. Let $\rho_\tau \in \mathbb{R}^{n_1+\cdots+n_f-f}$ be the element which corresponds to ρ. Let ds_τ be the differential form which corresponds to dz.

Let $H \subset \mathbb{C}^{n_1+\cdots+n_f-f}$ be a subspace of dimension d not necessarily going through the origin. We use the letter s_H to express an element of H. We choose a differential form ds_H on H. Let $r^{(1)}, r^{(2)} \in H \cap \mathbb{R}^{n_1+\cdots+n_f-f}$. Let l_0 be the line segment joining $r^{(1)}, r^{(2)}$, and D a domain containing l_0. Let $f(s_H)$ be a meromorphic function which is at most of polynomial growth on any vertical strip contained in $\{s_H \mid \mathrm{Re}(s_H) \in D\}$.

Let

$$
\Xi(f, r^{(i)}, w, \psi) = \left(\frac{1}{2\pi\sqrt{-1}}\right)^d \int_{\mathrm{Re}(s_H)=r^{(i)}} \frac{f(s_H)\widetilde{\psi}(s_H)}{w - \widetilde{L}(s_H)} ds_H
$$

for $i = 1, 2$.

Suppose that there exists a polynomial $P(s_\tau)$ such that $P(s_H)f(s_H)$ is holomorphic on $\{s_H \mid \mathrm{Re}(s_H) \in D \cap H\}$.

Proposition (3.6.1) (The passing principle) *Suppose $P(\rho_\tau) \neq 0$. Then there exists a polynomial $\overline{P}(z)$ such that $\overline{P}(-\tau_G z) = \overline{P}(z)$, $\overline{P}(\rho) \neq 0$, and if we define $\psi'(z) = \overline{P}(z)\psi(z)$, then $\Xi(f, r^{(1)}, w, \psi') = \Xi(f, r^{(2)}, w, \psi')$.*

Proof. Let $P'(z)$ be the polynomial which corresponds to $P(s_\tau)$ by the above substitution. By assumption, $P'(\rho) \neq 0$. Let $P''(z) = P'(-\tau_G z)$. Since $-\tau_G \rho = \rho$, $P''(\rho) \neq 0$. Then $\overline{P}(z) = P'(z)P''(z)$ satisfies the condition of the proposition.

Q.E.D.

Consider the following condition

$$\psi(z) = \psi(-\tau_G z),\ \psi(\rho) \neq 0. \tag{3.6.2}$$

Consider the situation in §3.5. In later parts, we analyze the distributions $\Xi_{\mathfrak{p}}(\Phi, \omega, w)$ and $\widetilde{\Xi}_{\mathfrak{p}}(\Phi, \omega, w)$. In this process, we have to study various contour integrals. What we can do by (3.6.1) is to replace the function $\psi(z)$ if necessary to move the contour. By doing this, the condition (3.6.2) is not changed. Therefore, if for some $L \subset V_k^{\mathrm{ss}}$, we can prove that

$$\int_{G_{\mathbb{A}}^1/G_k} \omega(g^1)\Theta_L(\Phi, g^1)\mathscr{E}(g^1, w)dg^1 \sim C_G\Lambda(w; \rho)\sum_{\mathfrak{p}} a_{\mathfrak{p}}(\Phi, \omega),$$

where $a_{\mathfrak{p}}(\Phi, \omega)$ is some distribution associated with the path $\mathfrak{p}$, then we can still conclude that

$$\int_{G_{\mathbb{A}}^1/G_k} \omega(g^1)\Theta_L(\Phi, g^1)dg^1 = \sum_{\mathfrak{p}} a_{\mathfrak{p}}(\Phi, \omega).$$

Caution (3.6.3) *For the rest of this book, when we consider contour integrals of the form $\int_{\mathrm{Re}(z)=q} \cdots dz$ or $\int_{\mathrm{Re}(s_\tau)=r} \cdots ds_\tau$, we always consider $w \in \mathbb{C}$ such that* $\mathrm{Re}(w) > L(q)$ *or* $\mathrm{Re}(w) > \widetilde{L}(r)$

§3.7 Wright's principle

In this section, we introduce a technique concerning the cancelation of higher order terms of distributions.

We consider $\phi_n(s)$ for some n (see (1.1.2) for the definition).

The use of the following proposition was suggest to the author by D. Wright after he read the first manuscript of [86] (it is used in [86]) Therefore, the author calls this proposition 'Wright's principle.'

Proposition (3.7.1) (Wright's principle) *Let $A \in \mathbb{C}$, $c_1 < 1 < c_2$, and $\epsilon > 0$ be constants. Let $w \in \mathbb{C}$ be a complex variable. Let $f(s)$ be a meromorphic function on the domain $\{s \in \mathbb{C} \mid c_1 - \epsilon \leq \mathrm{Re}(s) \leq c_2 + \epsilon\}$, having no poles along the lines $\{s \mid \mathrm{Re}(s) = c_1\}, \{s \mid \mathrm{Re}(s) = c_2\}$. Suppose that $f(s)$ has a finite number of finite order poles and is rapidly decreasing with respect to the imaginary part of s. Also assume that there exists a constant A' such that*

$$\frac{1}{2\pi\sqrt{-1}}\int_{\mathrm{Re}(s)=c_2} \frac{\phi_n(s)f(s)}{w - s - A}ds - \frac{A'}{w - 1 - A}$$

is meromorphic on a domain of the form $\{w \mid \mathrm{Re}(w) > 1+A-\delta\}$ *for some* $\delta > 0$ *and is holomorphic at* $w = 1+A$. *Then* $f(s)$ *is holomorphic at* $s = 1$.

Proof. Suppose that $f(s)$ has a pole of order j at $s = 1$. Since $\phi_n(s)$ has a pole of order exactly one at $s = 1$, $\phi_n(s)f(s)$ has a pole of order $j+1$. Let

$$\phi_n(s)f(s) = \sum_{i=-j-1}^{\infty} a_i(s-1)^i$$

be the Laurent expansion at $s = 1$, where $a_j \neq 0$. Then

$$\frac{1}{2\pi\sqrt{-1}} \int_{\mathrm{Re}(s)=c_2} \frac{f(s)}{w-1-A} ds$$

$$= \frac{1}{2\pi\sqrt{-1}} \int_{\mathrm{Re}(s)=c_1} \frac{f(s)}{w-1-A} ds + h(w) + \sum_{i=j}^{-1} a_i(w-1-A)^i$$

for some rational function $h(w)$ which does not have $w-1-A$ in the denominator. Therefore, the proposition follows.

Q.E.D.

We illustrate the use of this proposition by an easy example in the next section. Even though this proposition is very easy to prove, it does save a lot of labor for us in Chapters 12 and 13.

§3.8 Examples

Before we start handling non-trivial cases, we consider two easy examples $G = \mathrm{GL}(1) \times \mathrm{GL}(2)$, $V = \mathrm{Sym}^3 k^2$ and $G = \mathrm{GL}(2) \times \mathrm{GL}(2)$, $V = k^2 \otimes k^2$, applying what we have developed in this chapter. The first case was handled by Shintani in [64] and the proof in adelic language is in [83]. The prehomogeneous vector space $G = \mathrm{GL}(2) \times \mathrm{GL}(2), V = \mathrm{M}(2,2)$ has been studied by many people, but there does not seem to be any result which uses exactly the same formulation as ours. Cogdell [8] studied this case by the same method as ours (he used the group GL(2), but it is essentially the same as our case). However, his paper is written in classical language and he also chose a particular Schwartz–Bruhat function, so his analysis superficially looks very different from our analysis. Also the use of Wright's principle in the previous section simplifies the computation significantly.

We first introduce a notation. Let $\omega = (\omega_1, \cdots, \omega_i)$ be a character of $(\mathbb{A}^\times/k^\times)^i$, and $\Psi \in \mathscr{S}(\mathbb{A}^i)$. For $t = (t_1, \cdots, t_i) \in \mathbb{A}^i$, we denote $\omega(t) = \prod_j \omega_j(t_j)$. Let $s = (s_1, \cdots, s_i) \in \mathbb{C}^i$.

Definition (3.8.1)

(1) $$\Theta_i(\Psi, t) = \sum_{x \in (k^\times)^i} \Psi(tx).$$

(2) $$\Sigma_i(\Psi, \omega, s) = \int_{(\mathbb{A}^\times/k^\times)^i} |t|^s \omega(t) \Theta_i(\Psi, t) d^\times t.$$

If ω is trivial, we drop ω, and use the notation $\Sigma_i(\Psi, s)$ also. When $i = 1$, we define an analog of $Z_{L+}(\Phi, \omega, s)$ as follows.

$$\Sigma_{1+}(\Psi, \omega, s) = \int_{\substack{\mathbb{A}^\times/k^\times \\ |t|\geq 1}} |t|^s \omega(t) \Theta_1(\Psi, t) d^\times t.$$

If ω is trivial, we may drop ω and write $\Sigma_{1+}(\Psi, s)$ also. It is well known that $\Sigma_{1+}(\Psi, \omega, s)$ is an entire function, and

$$\Sigma_1(\Psi, \omega, s) = \Sigma_{1+}(\Psi, \omega, s) + \Sigma_{1+}(\widehat{\Psi}, \omega^{-1}, 1-s) + \delta_\#(\omega) \left(\frac{\widehat{\Psi}(0)}{s-1} - \frac{\Psi(0)}{s} \right),$$

where $\widehat{\Psi}$ is the standard Fourier transform of Ψ. We refer to this formula as the 'principal part formula for the standard L-function in one variable'.

Definition (3.8.2) *For $j = (j_1, \cdots, j_i) \in \mathbb{Z}^i$, $s_0 = (s_{0,1}, \cdots, s_{0,i}) \in \mathbb{C}^i$, we use the notation $\Sigma_{i,(j_1,\cdots,j_i)}(\Psi, \omega, s_0)$ for the coefficient of $\prod_l (s_l - s_{l,0})^{j_i}$ in the Laurent expansion of $\Sigma_i(\Psi, \omega, s)$ at $s = s_0$. We also drop ω in this notation when ω is trivial.*

(a) The case $G = \mathrm{GL}(1) \times \mathrm{GL}(2)$, $V = \mathrm{Sym}^3 k^2$

Let $\widetilde{T}$ be the kernel of the homomorphism $G \to \mathrm{GL}(V)$. We use the formulation in §3.1. Since $\dim G/\widetilde{T} = \dim V$, $V_k^{\mathrm{ss}} = V_k^{\mathrm{s}}$. Therefore, this case is of complete type, and we use the notation $Z(\Phi, \omega, s), Z_+(\Phi, \omega, s)$ for the integrals defined in (3.1.8) for $L = V_k^{\mathrm{ss}}$.

In this case, we identify $\mathfrak{t}^* = \mathfrak{t}^{0*}$ with $\mathbb{R}$ so that $a_2(\underline{\lambda}^{-1}, \underline{\lambda})^a = \lambda^a$ for $\lambda \in \mathbb{R}$. We choose $\{a \in \mathbb{R} \mid a \geq 0\}$ as $\mathfrak{t}^*_-$. The Weyl group consists of two elements and the non-trivial element τ_G is the permutation $(1, 2)$. We identify V with k^4 by

$$(x_0, \cdots, x_3) \to f_x(v_1, v_2) = x_0 v_1^3 + x_1 v_1^2 v_2 + x_2 v_1 v_2^2 + x_3 v_2^3.$$

The weights of V are $-3, -1, 1, 3$. Clearly, $\mathfrak{B} \setminus \{0\}$ consists of two elements $\beta_1 = 1, \beta_2 = 3$. Let $\mathfrak{d}_1 = (\beta_1), \mathfrak{d}_2 = (\mathfrak{d}_2)$. Let $\mathfrak{p}_{1j} = (\mathfrak{d}_1, \mathfrak{s}_{1j}), \mathfrak{p}_{2j} = (\mathfrak{d}_2, \mathfrak{s}_{2j})$ be paths for $j = 1, 2$ such that $\mathfrak{s}_{ij}(1) = 0$ if $j = 1$ and $\mathfrak{s}_{ij}(1) = 1$ if $j = 2$. It is easy to see that $M_{\mathfrak{d}_1} = M_{\mathfrak{d}_2} = T$ and

$$\begin{aligned}
&Y_{\mathfrak{d}_1 k} = \{(0, 0, x_2, x_3) \mid x_2, x_3 \in k\},\ Y_{\mathfrak{d}_1 k}^{\mathrm{ss}} = \{(0, 0, x_2, x_3) \mid x_2 \in k^\times, x_3 \in k\}, \\
&Z_{\mathfrak{d}_1 k} = \{(0, 0, x_2, 0) \mid x_2 \in k\},\ Z_{\mathfrak{d}_1 k}^{\mathrm{ss}} = \{(0, 0, x_2, 0) \mid x_2 \in k^\times\}, \\
&Y_{\mathfrak{d}_2 k} = Z_{\mathfrak{d}_2 k} = \{(0, 0, 0, x_3) \mid x_3 \in k\},\ Z_{\mathfrak{d}_2 k}^{\mathrm{ss}} = \{(0, 0, 0, x_3) \mid x_3 \in k^\times\}.
\end{aligned}$$

Then by the general theory, $S_{\mathfrak{d}_i k} \cong G_k \times_{B_k} Y_{\mathfrak{d}_i k}^{\mathrm{ss}}$ for $i = 1, 2$.

We define a bilinear form $[\ ,\ ]'_V$ by

$$[x, y]'_V = x_0 y_0 + \frac{1}{3} x_1 y_1 + \frac{1}{3} x_2 y_2 + x_3 y_3$$

for $x = (x_0, \cdots, x_3), y = (y_0, \cdots, y_3)$. Then this bilinear form satisfies the property $[gx, {}^t g^{-1}]'_V = [x, y]'_V$ for all x, y. Let $[x, y]_V = [x, \tau_G y]'$. We use this bilinear form as $[\ ,\ ]_V$ in §3.1.

We choose the constant in the definition of the smoothed Eisenstein series so that $C = C_1 = 1$. This implies that $w_0 = 1$. In this case, $\tau_G = (1, (1,2))$ and $C_G = \mathfrak{V}_2^{-1}$. We assume that $\Phi \in \mathscr{S}(V_{\mathbb{A}})$ and $M_{V,\omega}\Phi = \Phi$. By (3.5.3), (3.5.5), $\Xi_{\mathfrak{p}_{ij}}(\Phi, \omega, w), \widetilde{\Xi}_{\mathfrak{p}_{1j}}(\Phi, \omega, w)$ are well defined for $\mathrm{Re}(w) \gg 0$ for $i, j = 1, 2$. Let $J(\Phi, g^0)$ be as in (3.5.6). We define

$$(3.8.3) \qquad I^0(\Phi, \omega) = \int_{G^0_{\mathbb{A}}/G_k} \omega(g^0) J(\Phi, g^0) dg^0,$$

$$I(\Phi, \omega, w) \int_{G^0_{\mathbb{A}}/G_k} \omega(g^0) J(\Phi, g^0) \mathscr{E}(g^0, w) dg^0.$$

Then by (3.4.34),

$$I(\Phi, \omega, w) \sim C_G \Lambda(w; \rho) I^0(\Phi, \omega).$$

By (3.5.9), (3.5.20),

$$\begin{aligned} I(\Phi, \omega, w) = & \sum_{\mathfrak{p}=\mathfrak{p}_{11},\mathfrak{p}_{12}} \epsilon_{\mathfrak{p}} (\Xi_{\mathfrak{p}}(\Phi, \omega, w) + \widetilde{\Xi}_{\mathfrak{p}}(\Phi, \omega, w)) \\ & + \sum_{\mathfrak{p}=\mathfrak{p}_{21},\mathfrak{p}_{22}} \epsilon_{\mathfrak{p}} \Xi_{\mathfrak{p}}(\Phi, \omega, w) + C_G \Lambda(w; \rho) \mathfrak{V}_2 \delta_{\#}(\omega)(\widehat{\Phi}(0) - \Phi(0)). \end{aligned}$$

By the remark after (3.4.7),

$$\Xi_{\mathfrak{p}_{i2}}(\Phi, \omega, w) = \Xi_{\mathfrak{p}_{i1}}(\widehat{\Phi}, \omega^{-1}, w), \; \widetilde{\Xi}_{\mathfrak{p}_{12}}(\Phi, \omega, w) = \widetilde{\Xi}_{\mathfrak{p}_{11}}(\widehat{\Phi}, \omega^{-1}, w).$$

Therefore, we only consider $\mathfrak{p}_{11}, \mathfrak{p}_{21}$. Let

$$d(\lambda_1) = (1, a_2(\underline{\lambda}_1^{-1}, \underline{\lambda}_1)), \; \widehat{t}^0 = (t, a_2(t_1, t_2)),$$

Where $t, t_1, t_2 \in \mathbb{A}^1$, $u \in \mathbb{A}$, and $t^0 = d(\lambda_1)\widehat{t}^0$. We define $d^\times \lambda_1, d^\times \widehat{t}^0$ in the usual manner. Let $d^\times t^0 = d^\times \lambda_1 d^\times \widehat{t}^0$. Then this measure satisfies the property that $dg^0 = dk d^\times t^0 du$ ($g^0 = kt^0 n_2(u) \in G^0_{\mathbb{A}}$ is the Iwasawa decomposition). Note that t^0 acts on $Y_{\mathfrak{d}_1\mathbb{A}}$ by $(x_2, x_3) \to (\underline{\lambda}_1 t t_1 t_2^2 x_2, \underline{\lambda}_1^3 t t_2^3 x_3)$, and on $Y_{\mathfrak{d}_2\mathbb{A}} = Z_{\mathfrak{d}_2\mathbb{A}}$ by $x_3 \to \underline{\lambda}_1^3 t t_2^3 x_3$.

Lemma (3.8.4) $\widetilde{\Xi}_{\mathfrak{p}_{11}}(\Phi, \omega, w) \sim 0$.

Proof. By definition,

$$\mathscr{E}(t^0, w) - \mathscr{E}_N(t^0, w) = \mathscr{E}_{\{1\}, \tau_G, 1}(t^0, w).$$

By (3.4.30), there exists a constant $\delta > 0$ such that $\mathscr{E}(t^0, w) - \mathscr{E}_N(t^0, w)$ is holomorphic for $\mathrm{Re}(w) \geq w_0 - \delta$ and if $M > w_0, l \gg 0$,

$$|\mathscr{E}(t^0, w) - \mathscr{E}_N(t^0, w)| \ll \lambda_1^l.$$

By (1.2.8), for any $N \geq 1$,

$$|\Theta_{Y_{\mathfrak{d}_1}}(\widetilde{R}_{\mathfrak{d}_1}\Phi, t^0 n_2(u))| \ll \lambda_1^{-N_1} \sup(1, \lambda_1^{-3}).$$

This implies that for any $l \gg 0, N \geq 1$,

$$|\Theta_{Y_{\mathfrak{d}_1}}(\widetilde{R}_{\mathfrak{d}_1}\Phi, t^0 n_2(u))||\mathscr{E}(t^0,w) - \mathscr{E}_N(t^0,w)|\lambda_1^2 \ll \lambda_1^{l-N+2}\sup(1,\lambda_1^{-3}).$$

Therefore, the integral defining $\widetilde{\Xi}_{\mathfrak{p}_{11}}(\Phi,\omega,w)$ converges absolutely for $\mathrm{Re}(w) \geq w_0 - \delta$.

Q.E.D.

We fix a Weyl group element τ. Let $s_\tau = z_{\tau(2)} - z_{\tau(1)}$. Then it is easy to see that

$$\sigma_{\mathfrak{p}_{11}}(t^0) = \lambda_1^2\lambda_1^{-3} = \lambda_1^{-1},\ \sigma_{\mathfrak{p}_{21}}(t^0) = \lambda_1^2,\ (t^0)^{\tau z+\rho} = \lambda_1^{s_\tau - 1}. \tag{3.8.5}$$

Easy computations show that

$$\int_{(\mathbb{A}^1/k^\times)^3} \omega(\widehat{t}^0) f(tt_1t_2^2) d^\times \widehat{t}^0 = \delta_\#(\omega)\int_{\mathbb{A}^1/k^\times} f(q)d^\times q,$$
$$\int_{(\mathbb{A}^1/k^\times)^3} \omega(\widehat{t}^0) f(tt_2^3) d^\times \widehat{t}^0 = \delta(\omega_1^3)\delta(\omega_2)\int_{\mathbb{A}^1/k^\times} \omega_1(q) f(q)d^\times q$$

for any measurable function $f(q)$ on $\mathbb{A}^1/k^\times$. Therefore,

$$\begin{aligned}
&\Xi_{\mathfrak{p}_{11},\tau}(\Phi,\omega,w)\\
&= \frac{\delta_\#(\omega)}{2\pi\sqrt{-1}}\int_{\mathrm{Re}(s_\tau)=r>3} \Sigma_1(R_{\mathfrak{d}_1}\Phi, s_\tau - 2)\widetilde{\Lambda}_\tau(w;s_\tau)ds_\tau,\\
&\Xi_{\mathfrak{p}_{21},\tau}(\Phi,\omega,w)\\
&= \frac{\delta(\omega_1^3)\delta(\omega_2)}{3}\frac{1}{2\pi\sqrt{-1}}\int_{\mathrm{Re}(s_\tau)=r>2} \Sigma_1(R_{\mathfrak{d}_2}\Phi,\omega_1,\frac{s_\tau+1}{3})\widetilde{\Lambda}_\tau(w;s_\tau)ds_\tau.
\end{aligned}$$

If $\tau = 1$, $D_\tau = \mathbb{R}$ and $\widetilde{L}(r) = -r$. Therefore, by choosing $r \gg 0$, $\Xi_{\mathfrak{p}_{i1},\tau}(\Phi,\omega,w) \sim 0$ for $i = 1,2$.

Suppose $\tau = \tau_G$. Then $\widetilde{\Lambda}_\tau(w;s_\tau) = \phi(s_\tau)\widetilde{\Lambda}(w;s_\tau)$. The point ρ corresponds to $s_\tau = 1$. The functions $\Sigma_1(R_{\mathfrak{d}_1}\Phi, s_\tau - 2), \Sigma_1(R_{\mathfrak{d}_2}\Phi, \frac{s_\tau+1}{3})$ are holomorphic at $s_\tau = 1$. The passing principle tells us that by changing ψ if necessary,

$$\begin{aligned}
\Xi_{\mathfrak{p}_{11},\tau}(\Phi,\omega,w) &= \frac{\delta_\#(\omega)}{2\pi\sqrt{-1}}\int_{\substack{\mathrm{Re}(s_\tau)=r\\ 0<r<1}} \Sigma_1(R_{\mathfrak{d}_1}\Phi, s_\tau - 2)\widetilde{\Lambda}_\tau(w;s_\tau)ds_\tau\\
&\quad + C_G\Lambda(w;\rho)\delta_\#(\omega)\Sigma_1(R_{\mathfrak{d}_1}\Phi,-1)\\
&\sim C_G\Lambda(w;\rho)\delta_\#(\omega)\Sigma_1(R_{\mathfrak{d}_1}\Phi,-1).
\end{aligned}$$

Similarly,

$$\Xi_{\mathfrak{p}_{21},\tau}(\Phi,\omega,w) \sim C_G\Lambda(w;\rho)\frac{\delta(\omega_1^3)\delta(\omega_2)}{3}\Sigma_1(R_{\mathfrak{d}_2}\Phi,\omega_1,\frac{2}{3}).$$

These considerations prove the following formula

$$\begin{aligned}
I(\Phi,\omega,w) &\sim C_G\Lambda(w;\rho)\delta_\#(\omega)\left(\Sigma_1(R_{\mathfrak{d}_1}\widehat{\Phi},-1) - \Sigma_1(R_{\mathfrak{d}_1}\Phi,-1)\right)\\
&\quad + C_G\Lambda(w;\rho)\frac{\delta(\omega_1^3)\delta(\omega_2)}{3}\left(\Sigma_1(R_{\mathfrak{d}_2}\widehat{\Phi},\omega_1^{-1},\frac{2}{3}) - \Sigma_1(R_{\mathfrak{d}_2}\Phi,\omega_1,\frac{2}{3})\right)\\
&\quad + C_G\Lambda(w;\rho)\mathfrak{V}_2\delta_\#(\omega)(\widehat{\Phi}(0) - \Phi(0)).
\end{aligned}$$

Therefore,

$$(3.8.6)\quad I^0(\Phi,\omega) = \delta_{\#}(\omega)\Big(\Sigma_1(R_{\mathfrak{d}_1}\widehat{\Phi},-1) - \Sigma_1(R_{\mathfrak{d}_1}\Phi,-1)\Big) + \frac{\delta(\omega_1^3)\delta(\omega_2)}{3}\left(\Sigma_1(R_{\mathfrak{d}_2}\widehat{\Phi},\omega_1^{-1},\frac{2}{3}) - \Sigma_1(R_{\mathfrak{d}_2}\Phi,\omega_1,\frac{2}{3})\right) + \mathfrak{V}_2\delta_{\#}(\omega)(\widehat{\Phi}(0)-\Phi(0)).$$

It is easy to see the following relations

$$\begin{aligned}
&\Sigma_1(R_{\mathfrak{d}_1}\Phi_\lambda,-1) = \Sigma_1(R_{\mathfrak{d}_1}\Phi,-1),\\
&\Sigma_1(R_{\mathfrak{d}_1}\widehat{\Phi_\lambda},-1) = \lambda^{-4}\Sigma_1(R_{\mathfrak{d}_1}\widehat{\Phi},-1),\\
&\Sigma_1(R_{\mathfrak{d}_2}\Phi_\lambda,\omega_1,\frac{2}{3}) = \lambda^{-\frac{2}{3}}\Sigma_1(R_{\mathfrak{d}_2}\Phi,\omega_1,\frac{2}{3}),\\
&\Sigma_1(R_{\mathfrak{d}_2}\widehat{\Phi_\lambda},\omega_1^{-1},\frac{2}{3}) = \lambda^{-\frac{10}{3}}\Sigma_1(R_{\mathfrak{d}_2}\widehat{\Phi},\omega_1^{-1},\frac{2}{3}),\\
&\Phi_\lambda(0) = \Phi(0),\ \widehat{\Phi_\lambda}(0) = \lambda^{-4}\widehat{\Phi}(0).
\end{aligned}$$

Therefore, by integrating $\lambda^s I^0(\Phi_\lambda,\omega)$ over the interval $[0,1]$, we get the following theorem.

Theorem (3.8.7) (Shintani) *Suppose $\Phi = M_{V,\omega}\Phi$. Then*

$$\begin{aligned}
Z(\Phi,\omega,s) &= Z_+(\Phi,\omega,s) + Z_+(\widehat{\Phi},\omega^{-1},4-s) + \mathfrak{V}_2\delta_{\#}(\omega)\left(\frac{\widehat{\Phi}(0)}{s-4} - \frac{\Phi(0)}{s}\right)\\
&\quad + \delta_{\#}(\omega)\left(\frac{\Sigma_1(R_{\mathfrak{d}_1}\widehat{\Phi},-1)}{s-4} - \frac{\Sigma_1(R_{\mathfrak{d}_1}\Phi,-1)}{s}\right)\\
&\quad + \delta(\omega_1^3)\delta(\omega_2)\left(\frac{\Sigma_1(R_{\mathfrak{d}_2}\widehat{\Phi},\omega_1^{-1},\frac{2}{3})}{3s-10} - \frac{\Sigma_1(R_{\mathfrak{d}_2}\Phi,\omega_1,\frac{2}{3})}{3s-2}\right).
\end{aligned}$$

(b) The case $G = \mathrm{GL}(2)\times\mathrm{GL}(2)$, $V = k^2\otimes k^2$

In this case, we have a canonical measure on the group $\mathrm{GL}(2)^0_{\mathbb{A}}\times\mathrm{GL}(2)^0_{\mathbb{A}}$. Let $\widetilde{T}$ be the kernel of the homomorphism $G\to\mathrm{GL}(V)$. We consider $(G/\widetilde{T},V)$ as before.

In this case, we identify $\mathfrak{t}^* = \mathfrak{t}^{0*}$ with $\mathbb{R}^2$ so that $(a_2(\underline{\lambda}_1^{-1},\underline{\lambda}_1),a_2(\underline{\lambda}_2^{-1},\underline{\lambda}_2))^{(a,b)} = \lambda_1^a\lambda_2^b$. We choose $\{(a,b)\in\mathbb{R}\mid a,b\geq 0\}$ as $\mathfrak{t}^*_-$. Any Weyl group element is of the form $\tau = (\tau_1,\tau_2)$ where τ_1,τ_2 are either 1 or $(1,2)$. For $x\in V$, let x_{ij} be the (i,j)-entry. We use $x = (x_{ij})$ as the coordinate system of V. The weights of V are $(-1,-1),(-1,1),(1,-1),(1,1)$. Let $((a_1,b_1),(a_2,b_2)) = a_1a_2+b_1b_2$. Then with the metric defined by this bilinear form, the weights of V can be identified with vertices of a square.

The index set $\mathfrak{B}\setminus\{0\}$ consists of three elements $\beta = (1,1),\beta' = (1,0),\beta'' = (0,1)$. Let $\mathfrak{d} = (\beta)$. It is easy to see that $M_{\mathfrak{d}} = T$ and $Y_{\mathfrak{d}k} = Z_{\mathfrak{d}k} = \{x\mid x_{11} = x_{12} = x_{21} = 0\}$. Also, $Z^{\mathrm{ss}}_k = \{x\in Z_{\mathfrak{d}k}\mid x_{22}\in k^\times\}$. Let $\mathfrak{p}_i = (\mathfrak{d},\mathfrak{s}_i)$ be a path for $i = 1,2$ such that $\mathfrak{s}_1(1) = 0,\mathfrak{s}_2(1) = 1$. For β',β'', $M''_{\beta'},M''_{\beta''}$ in §3.2 is SL(2) and $Z_{\beta'},Z_{\beta''}$ can be identified with the standard representation of SL(2). So, $Z^{\mathrm{ss}}_{\beta' k},Z^{\mathrm{ss}}_{\beta'' k} = \emptyset$. Therefore, there is only one unstable stratum.

We define the zeta function by the integral (3.1.8) for V_k^{ss}, and use the notation $Z(\Phi,\omega,s), Z_+(\Phi,\omega,s)$ where $\omega = (\omega_1,\omega_2)$ and ω_1,ω_2 are characters of $\mathbb{A}^\times/k^\times$. However, since $V_k^s = \emptyset$, the convergence of the integrals is not covered in (3.1.4). It easily follows from the following estimate (which is a consequence of (1.2.8)).

Lemma (3.8.8) *Let $\Phi \in \mathscr{S}(M(2,2)_\mathbb{A})$ and $g^0 = (g_1,g_2) \in \mathrm{GL}(2)^0_\mathbb{A} \times \mathrm{GL}(2)^0_\mathbb{A}$. Let $g_i = k_i a_2(\underline{\mu}_i t_{i1}, \underline{\mu}_i^{-2} t_{i2}) n_2(u_i)$ be the Iwasawa decomposition of G_i for $i=1,2$. Then for any $N_1,N_2,N_3 \geq 1$,*

$$\Theta_{M(2,2)_k^{ss}}(\Phi,g^0) \ll (\mu_1\mu_2)^{-N_1}\sup(1,\mu_1^{-1}\mu_2)\sup(1,\mu_1\mu_2^{-1})\sup(1,\mu_1\mu_2) + (\mu_1\mu_2^{-1})^{-N_2}(\mu_1^{-1}\mu_2)^{-N_3}\sup(1,\mu_1\mu_2).$$

for g^0 in the Siegel domain.

Let $[x,y]'_V = \sum_{i,j} x_{ij}y_{ij}$ for $x=(x_{ij}), y=(y_{ij})$. Then $[gx, {}^tg^{-1}y]'_V = [x,y]'_V$ for all x,y. We define $[x,y]_V = [x,\tau_G y]'_V$. Then by defining g^ι as before, $[gx,g^\iota y]_V = [x,y]_V$ for all x,y. We use this bilinear form as $[\ ,\ ]_V$ in §3.1.

We choose the constants in the definition of the smoothed Eisenstein series so that $C=(1,1)$. This implies that $w_0 = 2$. In this case, $\tau_G = ((1,2),(1,2))$ and $C_G = \mathfrak{V}_2^{-2}$. We assume that $\Phi \in \mathscr{S}(V_\mathbb{A})$ and $M_{V,\omega}\Phi = \Phi$.

By (3.5.5), $\Xi_{\mathfrak{p}_i}(\Phi,\omega,w)$ is well defined for $\mathrm{Re}(w) \gg 0$ and $\Xi_{\mathfrak{p}_2}(\Phi,\omega,w) = \Xi_{\mathfrak{p}_1}(\widehat{\Phi},\omega^{-1},w)$ as before. Let

$$d(\lambda_1,\lambda_2) = (a_2(\underline{\lambda}_1^{-1},\underline{\lambda}_1), a_2(\underline{\lambda}_2^{-1},\underline{\lambda}_2)),\ \widehat{t}^0 = (a_2(t_{11},t_{12}),a_2(t_{21},t_{22}))$$

for $\lambda_1,\lambda_2 \in \mathbb{R}_+$, $t_{ij} \in \mathbb{A}^1$ for $i,j=1,2$. Let $t^0 = d(\lambda_1,\lambda_2)\widehat{t}^0$. We define $d^\times \widehat{t}^0, d^\times t^0$ in the usual manner.

We define $I^0(\Phi,\omega), I(\Phi,\omega,w)$ by the formula (3.8.3) (the set V_k^{ss} is different of course). Then

$$I(\Phi,\omega,w) \sim C_G\Lambda(w;\rho)I^0(\Phi,\omega),$$

and by (3.5.9), (3.5.20),

$$I(\Phi,\omega,w) = \Xi_{\mathfrak{p}_2}(\Phi,\omega,w) - \Xi_{\mathfrak{p}_1}(\Phi,\omega,w) + C_G\Lambda(w;\rho)\delta_\#(\omega)\mathfrak{V}_2^2(\widehat{\Phi}(0)-\Phi(0)).$$

The element t^0 acts on $Z_{\mathfrak{d}\mathbb{A}}$ by $x_{22} \to \underline{\lambda}_1\underline{\lambda}_2 t_{12}t_{22}x_{22}$. Therefore, $\Xi_{\mathfrak{p}_2}(\Phi,\omega,w) = 0$ unless ω is trivial.

We fix a Weyl group element $\tau = (\tau_1,\tau_2)$. Let $s_{\tau i} = z_{i\tau_1(2)} - z_{i\tau_1(2)}$ for $i=1,2$. Let $ds_\tau = ds_{\tau 1}ds_{\tau 2}$. It is easy to see that

$$\sigma_{\mathfrak{p}_1}(t^0) = \lambda_1^2\lambda_2^2,\ d(\lambda_1,\lambda_2)^{\tau z+\rho} = \lambda_1^{s_{\tau 1}-1}\lambda_2^{s_{\tau 2}-1}. \tag{3.8.9}$$

We make the change of variable $\mu = \lambda_1\lambda_2, \lambda_2 = \lambda_2$ and $q = t_{12}t_{22}$. Then

$$\lambda_1^{s_{\tau 1}+1}\lambda_2^{s_{\tau 2}+1} = \lambda_2^{s_{\tau 2}-s_{\tau 1}}\mu^{s_{\tau 1}+1}.$$

Since

$$\int_{\mathbb{R}_+\times\mathbb{A}^1/k^\times} \omega(\widehat{t}^0)\mu^{s_{\tau 1}+1}\Theta_1(R_\mathfrak{d}\Phi,\underline{\mu}t_{12}t_{22})d^\times\mu d^\times\widehat{t}^0 = \delta_\#(\omega)\Sigma_1(R_\mathfrak{d}\Phi,s_{\tau 1}+1),$$

by the Mellin inversion formula,

$$\Xi_{\mathfrak{p}_1}(\Phi,\omega,w) = \frac{\delta_{\#}(\omega)}{2\pi\sqrt{-1}} \int_{\mathrm{Re}(s_{\tau 1})=r_1>1} \Sigma_1(R_{\mathfrak{d}}\Phi, s_{\tau 1}+1)\widetilde{\Lambda}_{\tau}(w; s_{\tau 1}, s_{\tau 1}) ds_{\tau 1}.$$

More precisely, we are using (3.5.17).

For $\tau = (1,1), ((1,2),1), (1,(1,2)), \widetilde{L}(r_1,r_1) = -2r_1, 0, 0$ respectively. We choose $r_1 = 1+\delta$ for a small constant $\delta > 0$. Then $\widetilde{L}(r_1,r_1) < w_0 = 2$. So $\Xi_{\mathfrak{p}_1}(\Phi,\omega,w) \sim 0$ unless $\tau = \tau_G$.

Suppose $\tau = \tau_G$. Then $\widetilde{\Lambda}_{\tau}(w; s_{\tau 1}, s_{\tau 1}) = \phi(s_{\tau 1})^2 \widetilde{\Lambda}(w; s_{\tau 1}, s_{\tau 1})$. Let

$$J(\Phi, s_{\tau 1}) = \Sigma_1(R_{\mathfrak{d}}\widehat{\Phi}, s_{\tau 1}+1) - \Sigma_1(R_{\mathfrak{d}}\Phi, s_{\tau 1}+1).$$

These considerations show that

$$\begin{aligned} I(\Phi,\omega,w) \sim{}& C_G \Lambda(w;\rho)\mathfrak{V}_2^2 \delta_{\#}(\omega)(\widehat{\Phi}(0) - \Phi(0)) \\ &+ \frac{\delta_{\#}(\omega)}{2\pi\sqrt{-1}} \int_{\mathrm{Re}(s_{\tau 1})=r_1>1} J(\Phi, s_{\tau 1})\phi(s_{\tau 1})^2 \widetilde{\Lambda}(w; s_{\tau 1}, s_{\tau 1}) ds_{\tau 1}. \end{aligned}$$

Therefore, by Wright's principle, $\Sigma_1(R_{\mathfrak{d}}\widehat{\Phi}, 2) = \Sigma_1(R_{\mathfrak{d}}\Phi, 2)$, and

$$\begin{aligned} I(\Phi,\omega,w) \sim{}& C_G \Lambda(w;\rho)\mathfrak{V}_2^2 \delta_{\#}(\omega)(\widehat{\Phi}(0) - \Phi(0)) \\ &+ C_G \Lambda(w;\rho)\delta_{\#}(\omega)\left(\Sigma_{1,(1)}(R_{\mathfrak{d}}\widehat{\Phi}, 2) - \Sigma_{1,(1)}(R_{\mathfrak{d}}\Phi, 2)\right). \end{aligned}$$

This implies that

$$\begin{aligned} I^0(\Phi,\omega) ={}& \mathfrak{V}_2^2 \delta_{\#}(\omega)(\widehat{\Phi}(0) - \Phi(0)) \\ &+ \delta_{\#}(\omega)\left(\Sigma_{1,(1)}(R_{\mathfrak{d}}\widehat{\Phi}, 2) - \Sigma_{1,(1)}(R_{\mathfrak{d}}\Phi, 2)\right). \end{aligned}$$

The following relations are easy to verify.

$$\begin{aligned} \Sigma_{1,(1)}(R_{\mathfrak{d}}\Phi_\lambda, 2) &= \lambda^{-2}\Sigma_{1,(1)}(R_{\mathfrak{d}}\Phi, 2) - \lambda^{-2}(\log\lambda)\Sigma_1(R_{\mathfrak{d}}\Phi, 2), \\ \Sigma_{1,(1)}(R_{\mathfrak{d}}\widehat{\Phi_\lambda}, 2) &= \lambda^{-2}\Sigma_{1,(1)}(R_{\mathfrak{d}}\widehat{\Phi}, 2) + \lambda^{-2}(\log\lambda)\Sigma_1(R_{\mathfrak{d}}\widehat{\Phi}, 2). \end{aligned}$$

Therefore, by integrating $\lambda^s I^0(\Phi_\lambda, \omega)$ over the interval $[0,1]$, we get the following theorem.

Theorem (3.8.10) (Weil–Cogdell) *Suppose $M_{V,\omega}\Phi = \Phi$. Then*

$$\begin{aligned} Z(\Phi,\omega,s) ={}& Z_+(\Phi,\omega,s) + Z_+(\widehat{\Phi},\omega^{-1},4-s) + \mathfrak{V}_2^2\delta_{\#}(\omega)\left(\frac{\widehat{\Phi}(0)}{s-4} - \frac{\Phi(0)}{s}\right) \\ &+ \delta_{\#}(\omega)\left(-\frac{2\Sigma_{1,(0)}(R_{\mathfrak{d}}\widehat{\Phi}, 2)}{(s-2)^2} + \frac{\Sigma_{1,(1)}(R_{\mathfrak{d}}\widehat{\Phi}, 2)}{s-2} - \frac{\Sigma_{1,(1)}(R_{\mathfrak{d}}\Phi, 2)}{s-2}\right). \end{aligned}$$

If for some $v \in \mathfrak{M}_\infty$, $\Phi_v \in C_0^\infty(V_{k_v}^{\mathrm{ss}})$, then $\Sigma_{1,(0)}(R_{\mathfrak{d}}\widehat{\Phi}, 2) = \Sigma_{1,(0)}(R_{\mathfrak{d}}\Phi, 2) = 0$. Therefore, the poles of the associated Dirichlet series are all simple. Cogdell [8] spent most of his labor trying to establish the cancellations of order two terms based on the equality $\Sigma_1(R_{\mathfrak{d}}\widehat{\Phi}, 2) = \Sigma_1(R_{\mathfrak{d}}\Phi, 2)$ explicitly. However, we used Wright's principle, and were able to avoid this cancellation.

Part II The Siegel–Shintani case

Chapter 4 The zeta function for the space of quadratic forms

We state the result of this chapter here.

Let $G = \mathrm{GL}(1) \times \mathrm{GL}(n)$, and $V_n = \mathrm{Sym}^2 k^n$, where $n \geq 1$. We consider the natural action of $\mathrm{GL}(n)$ on V_n (if $n = 1$, $\alpha \in \mathrm{GL}(1)$ acts by multiplication by α^2). We define the action of $\alpha \in \mathrm{GL}(1)$ on V_n by the ordinary multiplication by α. This defines an action of G on V_n. Let $\widetilde{T} \subset G$ be the kernel of the homomorphism $G \to \mathrm{GL}(V)$. We consider the prehomogeneous vector space $(G/\widetilde{T}, V_n)$ in this chapter. For a character $\omega = (\omega_1, \omega_2)$ of $G_{\mathbb{A}}/G_k$, $\Phi \in \mathscr{S}(V_{n\mathbb{A}})$, and $s \in \mathbb{C}$, we consider the zeta function $Z_{V_n}(\Phi, \omega, s)$ which is, by definition, the integral (3.1.8) for $L = V_{nk}^{\mathrm{ss}}$ if $n \neq 2$ and for $L = V_{2k}^{s}$ if $n = 2$. We show the convergence of those integrals in §4.1, because it is not covered by (3.1.4). The case $n = 2$ is of incomplete type and requires an 'adjusting term.' We define the adjusted zeta function $Z_{V_2,\mathrm{ad}}(\Phi, \omega, s)$ in §4.2.

Let $S_{n,i} \subset V_n$ be the set of rank $n - i$ forms. Then $S_{n,1}, \cdots, S_{n,n-1}$ are unstable strata. Let $Y_{n,i} = Z_{n,i} \subset V_n$ be the corresponding subspace for $S_{n,i}$. Then $Y_{n,i} \cong V_{n-i}$. Let $Z'_{n,n-2,0} \subset Z_{n,n-2}$ be the set of points of the form $\{(0, x_1, 0)\}$ by the above identification. We identify $Z'_{n,n-2,0}$ with the one dimensional affine space.

For $\Phi \in \mathscr{S}(V_{\mathbb{A}})$, $R_{n,i}\Phi \in \mathscr{S}(Y_{n,i\mathbb{A}}) = \mathscr{S}(Z_{n,i\mathbb{A}})$ is the restriction of Φ to $\mathscr{S}(Y_{n,i\mathbb{A}})$. For $\Phi \in \mathscr{S}(V_{\mathbb{A}})$, we define $R'_{n,n-2,0}\Phi \in \mathscr{S}(Z_{n,n-2,0\mathbb{A}})$ by the formula

$$R'_{n,n-2,0}\Phi(x_1) = \int_{\mathbb{A}} R_{n,n-2}\Phi(0, x_1, x_2)dx_2.$$

If $n = 3$, we define

$$\begin{aligned} F_3(\Phi, \omega, s) = {} & \mathfrak{V}_3 \delta_{\#}(\omega) \frac{\Phi(0)}{s} + \mathfrak{V}_2 \delta(\omega_1^2)\delta(\omega_2) \frac{\Sigma_1(R_{3,2}\Phi, \omega_1, \frac{3}{2})}{2s - 3} \\ & + \delta(\omega_1^2)\delta(\omega_2) \frac{Z_{V_2,\mathrm{ad},(0)}(R_{3,1}\Phi, (\omega_1, 1), 3)}{s - 3} \\ & + \delta_{\#}(\omega) \left(\frac{3\Sigma_1(R'_{3,1,0}\Phi, 2)}{2(s-3)^2} + \frac{\Sigma_{1,(1)}(R'_{3,1,0}\Phi, 2)}{2(s-3)} \right). \end{aligned}$$

If $n \geq 4$, we define

$$\begin{aligned} F_n(\Phi, \omega, s) = {} & \mathfrak{V}_n \delta_{\#}(\omega) \frac{\Phi(0)}{s} \\ & + \sum_{i \neq 1, n-2} (n-i) \mathfrak{V}_i \delta(\omega_1^2)\delta(\omega_2) \frac{Z_{V_{n-i}}(R_{n,i}\Phi, (\omega_1, 1), \frac{n(n-i)}{2})}{2s - n(n-i)} \\ & + \mathfrak{V}_{n-2} \delta(\omega_1^2)\delta(\omega_2) \frac{Z_{V_2,\mathrm{ad}}(R_{n,n-2}\Phi, (\omega_1, 1), n)}{s - n} \end{aligned}$$

$$+ (n-1)\delta(\omega_1^2)\delta(\omega_2)\frac{Z_{V_{n-1},(0)}(R_{n,1}\Phi, (\omega_1,1), \frac{n(n-1)}{2})}{2s - n(n-1)}$$
$$+ \mathfrak{V}_{n-2}\delta_{\#}(\omega)\left(\frac{n\Sigma_1(R'_{n,n-2,0}\Phi, n-1)}{2(s-n)^2} + \frac{\Sigma_{1,(1)}(R'_{n,n-2,0}\Phi, n-1)}{2(s-n)}\right).$$

The notation $Z_{V_{n-1},(0)}(R_{n,1}\Phi, (\omega_1,1), \frac{n(n-1)}{2})$ will be defined in §4.4.

In this chapter, we prove the following theorem.

Theorem (4.0.1) *Let $n \geq 3$. Suppose that $M_{V_n,\omega}\Phi = \Phi$. Then*

$$\begin{aligned} Z_{V_n}(\Phi,\omega,s) &= Z_{V_n+}(\Phi,\omega,s) + Z_{V_n+}(\widehat{\Phi},\omega^{-1},\frac{n(n+1)}{2} - s) \\ &\quad - F_n(\widehat{\Phi},\omega^{-1},\frac{n(n+1)}{2} - s) - F_n(\Phi,\omega,s). \end{aligned}$$

Therefore, the adelic zeta function has double poles in general. However, it turns out (see (4.7.6)) that if $\Phi = \otimes_v \Phi_v$ and the support of Φ_v is contained in $V_{k_v}^{ss}$ for some $v \in \mathfrak{M}_\infty$,

$$\delta_{\#}(\omega)\Sigma_1(R'_{n,n-2,0}\Phi, n-1) = \delta_{\#}(\omega)\Sigma_1(R'_{n,n-2,0}\widehat{\Phi}, n-1) = 0.$$

We consider the case $k = \mathbb{Q}$. The above remark implies that the poles of the associated Dirichlet series are all simple. Consider $\omega = 1$, and drop ω from the notation $Z_{V_n}(\Phi,\omega,s)$ etc. Let $V_i \subset V_{\mathbb{R}}^{ss}$ be the set of forms with signature $(i, n-i)$. We choose a measure on $(G/\widetilde{T})_{\mathbb{A}}$ so that the zeta function defined by (3.1.2) coincides with the integral in (3.1.8). Let $\mu_\infty(x)$ etc. be as in §0.3. Consider $\Phi = \Phi_\infty \otimes \otimes_p \Phi_p$ such that Φ_p is the characteristic function of $V_{\mathbb{Z}_p}$ and $\Phi_\infty \in C_0^\infty(V_i)$. Then it is easy to see that

$$X_i(\Phi, d_n) = \int_{V_i} \Phi(y_\infty)dy_\infty = \widehat{\Phi}_\infty(0).$$

Since $Z_{V_n}(\Phi,s)$ has the residue $\mathfrak{V}_n\widehat{\Phi}(0) = \mathfrak{V}_n\Phi_\infty(0)$ at $s = d_n$, the Dirichlet series

$$\sum_{x \in G_{\mathbb{Z}}\backslash V_{\mathbb{Z}}^{ss}} \frac{\mu_\infty(x)b_{x,\infty}}{o_\infty(x)} \frac{1}{|\Delta(x)|_\infty^{\frac{s}{n}}}$$

has a simple pole at $s = d_n$ with residue $\mathfrak{V}_n$. This means that the Dirichlet series

$$\sum_{x \in G_{\mathbb{Z}}\backslash V_{\mathbb{Z}}^{ss}} \frac{\mu_\infty(x)b_{x,\infty}}{o_\infty(x)} \frac{1}{|\Delta(x)|_\infty^{s}}$$

has a simple pole at $s = \frac{n+1}{2}$ with residue $\frac{\mathfrak{V}_n}{n}$. Therefore, by the Tauberian theorem, we have the following theorem.

Theorem (4.0.2) (Siegel [69]) *Let $n \geq 3$. Then*

$$\sum_{\substack{x \in G_{\mathbb{Z}} \backslash V_{\mathbb{Z}}^{ss} \cap V_i \\ |\Delta(x)|_\infty < X}} \frac{\mu_\infty(x) b_{x,\infty}}{o_\infty(x)} \sim \frac{\mathfrak{V}_n}{d_n} X^{\frac{n+1}{2}}.$$

For discussion on the error term estimate, the reader should see [65]. In order to state the principal part formula for the case $n = 2$, we need the 'adjusting term.' For this reason, we state the principal part formula of the case $n = 2$ in §4.2. In this case, the filtering process was carried out by Datskovsky, and (0.1.2) is the asymptotic formula for $G_{\mathbb{Q}}$-equivalence classes. For this, the reader should see [9].

§4.1 The space of quadratic forms

Let G, V_n be as in §4.0. Let $\widetilde{T} \subset G$ be the kernel of the homomorphism $G \to \mathrm{GL}(V_n)$. We consider V_n as the space of quadratic forms in n variables $v = (v_1, \cdots, v_n)$. Let $d_n = \frac{n(n+1)}{2}$. We identify V_n with the space of column vectors of dimension d_n by the following map

$$x = (x_{lm})_{1 \leq j_1 \leq j_2 \leq n} \to f(v) = \sum_{l,m} x_{lm} v_l v_m.$$

Let T_0 be the image of GL(1) in G. For $(t, g) \in \mathrm{GL}(1) \times \mathrm{GL}(n)$, we define $\chi_{V_n}(t, g) = t^n (\det g)^2$ if n is odd, and $\chi_{V_n}(t, g) = t^{\frac{n}{2}} (\det g)$ if n is even. We can consider χ_V as a character of $G/\widetilde{T}$. The character χ_V is indivisible. It is easy to see that $(G/\widetilde{T}, V_n, \chi_{V_n})$ is a prehomogeneous vector space and V_{nk}^{ss} consists of non-degenerate forms. For $n = 2$, V_{2k}^s consists of forms without rational factors. We use the formulation in §3.3. Since $X^*(G/\widetilde{T})$ is generated by one element, $G_{\mathbb{A}}^1 = G_{\mathbb{A}}^0$ and $\mathfrak{t} = \mathfrak{t}^0$. If there is no confusion, we identify GL(n) with its image in G by the natural inclusion map.

Let $Y_{n,i} = Z_{n,i}$ be the subspace of V_n spanned by the coordinate vectors of x_{lm}'s for $l, m > i$. We identify $Z_{n,i}$ with V_{n-i}. Let $S_{n,i}$ be the subset of V_n consisting of rank $n - i$ forms. Let $Z_{n,ik}^{ss}$ be the set of rank $n - i$ forms in $Z_{n,ik}$. Let $Z_{n,n-2k}^s$ be the set of rank 2 forms without rational factors in $Z_{n,n-2k}$. Let $Z_{n,n-2,0k} = Z_{n,n-2k}^{ss} \setminus Z_{n,n-2k}^s$. Let $Z'_{n,n-2,0k} \subset Z_{n,n-2k}$ be the set $\{(0, x_1, 0)\}$ when we identify $Z_{n,n-2} \cong V_2$. Let $Z'^{ss}_{n,n-2,0k} = \{(0, x_1, 0) \mid x_1 \in k^\times\}$. We define $S_{n,n-2,\mathrm{st}k} = G_k \cdot Z'^{ss}_{n,n-2,0k}, S_{n,n-2,k}^s = G_k \cdot Z_{n,n-2k}^s$. Let $P_{n,i} \subset G$ be the maximal parabolic subgroup whose Levi component is $\mathrm{GL}(1) \times \mathrm{GL}(i) \times \mathrm{GL}(n-i)$, where $\mathrm{GL}(i) \times \mathrm{GL}(n-i)$ is imbedded in GL(n) diagonally in that order.

Easy considerations show the following lemma.

Lemma (4.1.1)
(1) $V_{nk} \setminus \{0\} = V_{nk}^{ss} \coprod \coprod_{i=1}^{n-1} S_{n,ik}$.
(2) $S_{n,ik} \cong G_k \times_{P_{n,ik}} Z_{n,ik}^{ss}$ *for all* i.

We define a bilinear form $[\ ,\ ]'_{V_n}$ as follows.

$$[x, y]'_{V_n} = \sum_{i=1}^n x_{ii} y_{ii} + \frac{1}{2} \sum_{i<j} x_{ij} y_{ij}.$$

Then $[gx, {}^tg^{-1}y]'_{V_n} = [x,y]'_{V_n}$ for all $x, y \in V_n$. This can be proved easily by reducing to the case $n = 2$. We define $[x,y]_{V_n} = [x, \tau_G y]'_{V_n}$. We use this bilinear form as $[\ ,\ ]_V$ in §3.1. Let $\mathscr{F}_{V_n}, \widetilde{g} = (\lambda, g^1), \widetilde{g}^\iota$ etc. be as in Chapter 3. Let $\Phi \in \mathscr{S}(V_{n\mathbb{A}})$. Then the Fourier transform of the function $\Phi(\widetilde{g}\cdot\,)$ is $\lambda^{-d_n}\mathscr{F}_n\Phi(\widetilde{g}^\iota\cdot\,)$.

If $n = 2$, we use the notation $Z_{V_2}(\Phi, \omega, s)$ etc. for the integrals in (3.1.8) for $L = V_{2k}^s$. The convergence of this integral for $\mathrm{Re}(s) \gg 0$ is covered in (3.1.4).

Definition (4.1.2) *Let $n \geq 3$. For $\Phi \in \mathscr{S}(V_{n\mathbb{A}})$, $\omega = (\omega_1, \omega_2)$ as in (3.1.8), and $s \in \mathbb{C}$, we define*

$$(1) \qquad Z_{V_n}(\Phi, \omega, s) = \int_{\widetilde{G}_{\mathbb{A}}/G_k} \lambda^s \omega(g^1)\Theta_{V_{nk}^{ss}}(\Phi, \widetilde{g})d\widetilde{g},$$

$$(2) \qquad Z_{V_n+}(\Phi, \omega, s) = \int_{\substack{\widetilde{G}_{\mathbb{A}}/G_k \\ \lambda \geq 1}} \lambda^s \omega(g^1)\Theta_{V_{nk}^{ss}}(\Phi, \widetilde{g})d\widetilde{g}.$$

We prove that if $n \geq 3$, $Z_{V_n}(\Phi, \omega, s)$ is well defined for $\mathrm{Re}(s) > d_n$, and $Z_{V_n+}(\Phi, \omega, s)$ is an entire function. Note that this does not follow from (3.1.4).

Lemma (4.1.3)
(1) *Suppose $n \geq 3$. Then there exists $r = (r_1, \cdots, r_{n-1}) \in \mathbb{R}^{n-1}$ such that $r_j < j(n-j)$ for all j, and that for any $N > d_n$, there exists an estimate of the form*

$$|\Theta_{V_{nk}^{ss}}(\Phi, \underline{\lambda} kt)| \ll \lambda^{-N} \prod_{i=1}^{n-1} (t_i t_{i+1}^{-1})^{r_i}$$

for $\lambda \in \mathbb{R}_+, k \in \widehat{\Omega}, t = a_n(t_1, \cdots, t_n) \in T^0_{\eta+}$.
(2) *If $n = 2$, there exists a similar estimate as above for $\Theta_{V_{2k}^s}(\Phi, \underline{\lambda} kt)$. Also for any $N > 3$,*

$$|\Theta_{V_{2,\mathrm{st}k}^{ss}}(\Phi, \underline{\lambda} kt)| \ll \lambda^{-N} t_1 t_2^{-1}$$

for $\lambda \in \mathbb{R}_+, k \in \widehat{\Omega}, t = a_2(t_1, t_2) \in T^0_{\eta+}$.

Proof. Let $V = V_n$. We choose $0 \leq \Psi \in \mathscr{S}(V_{\mathbb{A}})$ so that $|\Phi(\underline{\lambda} kt)| \ll \Psi(\underline{\lambda} t)$ for $k \in \widehat{\Omega}, t \in T^0_+$. We consider T^0_+ as a subset of $\mathrm{GL}(n)^0_{\mathbb{A}}$. Let $\gamma_{lm} \in \mathfrak{t}^*$ be the element determined by the coordinate x_{lm}.

Let $t = a_n(\underline{t}_1, \cdots, \underline{t}_n) \in T^0_+$, and $\lambda_1 = t_1 t_2^{-1}, \cdots, \lambda_{n-1} = t_{n-1}t_n^{-1}$. Then

$$t_1 = \lambda_1^{\frac{n-1}{n}} \cdots \lambda_{n-1}^{\frac{1}{n}},$$
$$\vdots$$
$$t_j = \lambda_1^{\frac{n-1}{n}} \cdots \lambda_{j-1}^{-\frac{j-1}{n}} \lambda_i^{\frac{n-j}{n}} \cdots \lambda_{n-1}^{\frac{1}{n}},$$
$$\vdots$$
$$t_n = \lambda_1^{-\frac{1}{n}} \cdots \lambda_{n-1}^{-\frac{n-1}{n}}.$$

Suppose $t_l t_m = \prod_j \lambda_j^{\chi_{lmj}}$. Then

$$\chi_{lmj} = \begin{cases} \frac{2(n-j)}{n} & l \leq m \leq j, \\ \frac{n-2j}{n} & l \leq j < m, \\ -\frac{2j}{n} & j < l \leq m. \end{cases}$$

If $x \in V_k^{\rm ss}$, the convex hull of $\{\gamma_{lm} \mid x_{lm} \neq 0\}$ contains the origin (but not necessarily a neighborhood of the origin). For $I \subset \{(l,m) \mid 1 \leq l \leq m \leq n\}$, we define $V_I = \{x \in V_k \mid x_{lm} \neq 0 \text{ if and only if } (l,m) \in I\}$. We consider I such that the convex hull Conv_I of $\{\gamma_{lm} \mid (l,m) \in I\}$ contains the origin. We define

$$\Theta_I(\Psi, \widetilde{g}) = \sum_{x \in V_k^{\rm ss} \cap V_I} \Psi(\widetilde{g} x).$$

Then

$$\Theta_{V_k^{\rm ss}}(\Psi, \underline{\lambda} t) \leq \sum_I \Theta_I(\Psi, \underline{\lambda} t).$$

We fix I and estimate $\Theta_I(\Psi, \underline{\lambda} t)$.

Since Conv_I contains the origin, we choose $c_{lm} \geq 0$ so that $\sum_{(l,m)\in I} c_{lm}\gamma_{lm} = 0$. We define $C = \sum_{(l,m)\in I} c_{lm} > 0$. By (1.2.6), for any $M > 0$,

$$\begin{aligned} \Theta_I(\Psi, \underline{\lambda} t) &\ll \lambda^{-CM} \prod_{(l,m)\in I} t^{-M\sum_{(l,m)\in I} c_{lm}\gamma_{lm}} \prod_{(l,m)\in I} (\lambda t_l t_m)^{-1} \\ &= \lambda^{-CM-\#I} \prod_{(l,m)\in I} (t_l t_m)^{-1}. \end{aligned}$$

We consider the possibilities when

$$\sum_{(l,m)\in I} \chi_{lmj} \leq -j(n-j). \tag{4.1.4}$$

Suppose that (4.1.4) is satisfied. We first consider the case $n \geq 3$.

Case 1. $j \leq \frac{n}{2}$.

Since $n - 2j \geq 0$ and $n - j > 0$,

$$\sum_{(l,m)\in I} \chi_{lmj} \geq \sum_{\substack{j<l\leq m \\ (l,m)\in I}} \chi_{lmj} \geq -\frac{j(n-j)(n-j+1)}{n}.$$

If equality happens, then $(l,m) \in I$ for $j < l \leq m$, and $(l,m) \notin I$ for $l \leq m \leq j$. But

$$-\frac{j(n-j)(n-j+1)}{n} \geq -j(n-j),$$

and equality happens only when $j = 1$. Therefore, (4.1.4) implies that $j = 1$.

If $j = 1$, $n - 2j > 0$. So $(l,m) \notin I$ for $l = 1 < m$. Hence, the only possibility for (4.1.4) to happen is when $j = 1$, and $I = \{(l,m) \mid 2 \leq l \leq m\}$.

Case 2. $j > \frac{n}{2}$.

Since $n - 2j < 0$,

$$\sum_{(l,m)\in I} \chi_{lmj} \geq -\frac{j(n-j)(n-j+1)}{n} + \frac{(n-2j)j(n-j)}{n} = -\frac{j(n-j)(j+1)}{n}.$$

If equality happens, then $(l,m) \in I$ if and only if $m > j$. But

$$-\frac{j(n-j)(j+1)}{n} \geq -j(n-j),$$

and equality happens only when $j = n-1$. Therefore, (4.1.4) implies that $j = n-1$ and $I = \{(l,n) \mid l = 1, \cdots, n\}$.

We have proved that if $I \neq \{(l,m) \mid 2 \leq l \leq m \leq n\}$ or $\{(l,n) \mid l = 1, \cdots, n\}$,

$$\sum_{(l,m)\in I} \chi_{lmj} > -j(n-j)$$

for $j = 1, \cdots, n-1$.

We consider the case $n = 2$. Clearly, $t_1^2 = \lambda_1, t_1 t_2 = 1, t_2^2 = \lambda_1^{-1}$. Therefore, $\prod_{(l,m)\in I} t_l t_m = \lambda_1^p$ for some $p \leq 1$ if and only if $I = \{(1,2),(2,2)\}$ or $I = \{(2,2)\}$,

If $n \geq 3$ and $I = \{(l,m) \mid 2 \leq l \leq m \leq n\}$ or $\{(l,n) \mid l = 1, \cdots, n\}$, Conv_I does not contain the origin. So we do not have to consider such possibilities. Since $\#I \leq d_n$ and the choice of $M > 0$ is arbitrary, for any $N > d_n$, we can choose $M > 0$ so that $CM + \#I = N$. Let $r_j = \inf_I(\sum_{(l,m)\in I} \chi_{lmj})$. Then this r satisfies the condition of (1).

If $n = 2$, and $I = \{(2,2)\}$, then Conv_I does not contain the origin. However, Conv_I contains the origin for $I = \{(1,2),(2,2)\}$. But if $x_{11} = 0$, $x \in V^{\mathrm{ss}}_{2,\mathrm{st}k}$. So we do not have to consider $I = \{(1,2),(2,2)\}$ for the first statement of (2). Since $\prod_{(l,m)\in I}(t_l t_m)^{-1} = t_1^{-1} t_2^{-3} = t_2^{-2} = \lambda_1^{-1}$, the second statement of (2) follows also.

Q.E.D.

Proposition (4.1.5) *If $n \geq 3$, $Z_{V_n}(\Phi, \omega, s)$ converges absolutely and locally uniformly for* $\mathrm{Re}(s) > d_n$, *and $Z_{V_n+}(\Phi, \omega, s)$ is an entire function.*

Proof. Let t, λ_j be as in the proof of (4.1.3). We only have to show the convergence of

$$\int_{\mathbb{R}_+ \times T^0_{\eta+}} \lambda^{\mathrm{Re}(s)} \Theta_{V^{\mathrm{ss}}_{nk}}(\Phi, \underline{\lambda} t) t^{-2\rho} d^\times \lambda d^\times t.$$

It is easy to see that $t^{-2\rho} = \prod_j \lambda_j^{-j(n-j)}$. So if $\lambda \geq 1$, we choose $N \gg \mathrm{Re}(s)$ in (4.1.3), and this proves the convergence of $Z_{V_n+}(\Phi, \omega, s)$ for all s. If $\lambda \leq 1$, we choose $N = d_n + \delta$, where $\delta > 0$. Then the integral over the set $\{\underline{\lambda} \mid \lambda \leq 1\}$ converges absolutely if $\mathrm{Re}(s) > d_n + \delta$. Since $\delta > 0$ is arbitrary, this proves the proposition.

Q.E.D.

Let $C(G^0_{\mathbb{A}}/G_k, r)$ for $r = (r_1, \cdots, r_{n-1}) \in \mathbb{R}^{n-1}$ be as in (3.4.17) (we consider the path $(\emptyset, \emptyset)$). By considering (4.1.3) on $G^0_{\mathbb{A}}$, we get the following lemma.

Lemma (4.1.6) *If $n \geq 3$, $\Theta_{V^{\mathrm{ss}}_{nk}}(\Phi, g^0) \in C(G^0_{\mathbb{A}}/G_k, r)$ for some r such that $r_j > -j(n-j)$ for all j. Also $\Theta_{V^{s}_{2k}}(\Phi, g^0) \in C(G^0_{\mathbb{A}}/G_k, r)$ for some $r > -1$.*

§4.2 The case $n = 2$

In this section, we consider the case $n = 2$. We use the formulation in §3.1. In this case, V^s_k consists of forms without rational factors. Let $V = V_2$ in this section.

Let $\mathrm{t}^* = \mathrm{t}^{0*}$, $t^0, \widehat{t}^0$ etc. be as in §3.8(a). We identify V with k^3 by

$$(x_0, x_1, x_2) \to f_x(v_1, v_2) = x_0 v_1^2 + x_1 v_1 v_2 + x_2 v_2^2.$$

The weights of V are $-2, 0, 2$. Clearly, $\mathfrak{B} \setminus \{0\}$ consists of one element $\beta = 2$. Let $\mathfrak{d} = (\beta)$. Let $\mathfrak{p}_i = (\mathfrak{d}, \mathfrak{s}_i)$ be paths for $i = 1, 2$ such that $\mathfrak{s}_1(1) = 0, \mathfrak{s}_2(1) = 1$. It is easy to see that $M_{\mathfrak{d}} = T$ and

$$Y_{\mathfrak{d}k} = Z_{\mathfrak{d}k} = \{(0,0,x_2) \mid x_2 \in k\},\ Z^{\mathrm{ss}}_{\mathfrak{d}k} = \{(0,0,x_2) \mid x_2 \in k^\times \in k\}.$$

Then by the general theory, $S_{\mathfrak{d}k} \cong G_k \times_{B_k} Z^{\mathrm{ss}}_{\mathfrak{d}k}$. Let $V^{\mathrm{ss}}_{\mathrm{st}k}$ be the set of rational forms with no double factors but with rational factors. Then

$$V_k \setminus \{0\} = V^{\mathrm{s}}_k \coprod V^{\mathrm{ss}}_{\mathrm{st}k} \coprod S_{1k}.$$

Let $H \subset G$ be the subgroup generated by T and $(1, \tau_G)$. We define

$$\begin{aligned} Y_{0k} &= \{(0,x_1,x_2) \mid x_1, x_2 \in k\},\ Y^{\mathrm{ss}}_{0k} = \{(0,x_1,x_2) \mid x_1 \in k^\times, x_2 \in k\}, \\ Z'_{0k} &= \{(0,x_1,0) \mid x_1 \in k\},\ Z'^{\mathrm{ss}}_{0k} = \{(0,x_1,0) \mid x_1 \in k^\times\}. \end{aligned}$$

Then it is easy to see that

$$V^{\mathrm{ss}}_{\mathrm{st}k} \cong G_k \times_{H_k} Z'^{\mathrm{ss}}_{0k}.$$

Consider the function $\alpha(u)$ in §2.2.

Proposition (4.2.1) *Let $F \subset \mathbb{A}_f$ be a compact set, and $L \subset k \subset k_\infty$ a lattice. Then there exists a constant $N > 0$ such that*

$$1 \geq \alpha\left(\frac{u}{x}\right) \geq N \prod_{v \in \mathfrak{M}_\mathbb{R}} (|x|_v^2 + |u|_v^2)^{-\frac{1}{2}} \prod_{v \in \mathfrak{M}_\mathbb{C}} (|x|_v + |u|_v)^{-1},$$

for $u_f \in F$, $x \in L \setminus \{0\}$.

Proof. Clearly, $\sup(1, |\frac{u}{x}|_v) = |x|_v^{-1} \sup(|x|_v, |u|_v)$. There exists a finite set $S \subset \mathfrak{M}_f$ such that if $v \in \mathfrak{M}_f \setminus S$, then $x, u_v \in o_v$ for $x \in L \setminus \{0\}$, $u_f \in F$. So if $v \in \mathfrak{M}_f \setminus S$, then $|x|_v, |u|_v \leq 1$. Clearly,

$$\prod_{v \in S} \sup(|x|_v, |u|_v)$$

is bounded by a constant $M > 0$. So

$$\prod_{v \in \mathfrak{M}_f} \sup(1, |\frac{u}{x}|_v) \leq M \prod_{v \in \mathfrak{M}_f} |x|_v^{-1} = M|x|_\infty.$$

Therefore,

$$\begin{aligned} \alpha(\frac{u}{x}) &\geq M^{-1}|x|_\infty^{-1} \prod_{v \in \mathfrak{M}_\mathbb{R}} (1 + |\frac{u}{x}|_v^2)^{-\frac{1}{2}} \prod_{v \in \mathfrak{M}_\mathbb{C}} (1 + |\frac{u}{x}|_v)^{-1} \\ &= M^{-1} \prod_{v \in \mathfrak{M}_\mathbb{R}} (|x|_v^2 + |u|_v^2)^{-\frac{1}{2}} \prod_{v \in \mathfrak{M}_\mathbb{C}} (|x|_v + |u|_v)^{-1}. \end{aligned}$$

Since the inequality $1 \geq \alpha(\frac{u}{x})$ is clear, this proves the proposition.

Q.E.D.

Proposition (4.2.2) *Let*

$$X_V = \{g^0 = (t, g_2) \in G^0_{\mathbb{A}} \mid |t_1(g_2)| \geq \sqrt{\alpha(t_1(g_2)^{-1}t_2(g_2)u(g_2))}\}.$$

Then for any measurable function $f(g^0)$ on $G^0_{\mathbb{A}}/H_k$,

$$\int_{G^0_{\mathbb{A}}/H_k} f(g^0)dg^0 = \int_{X_V/T_k} f(g^0)dg^0,$$

if the right hand side converges absolutely.

Proof. Note that the above condition does not depend on the choice of the Iwasawa decomposition of g_2. We fix a place $\widetilde{v} \in \mathfrak{M}_\infty$. The subset $\{g^0 = (t, g_2) \in G^0_{\mathbb{A}} \mid |t_1(g_2)| = \sqrt{\alpha(t_1(g_2)^{-1}t_2(g_2)u(g_2))}\}$ is a measure zero set, because if we fix $u(g_2)$, and $|t_1(g_2)|_v$ for all $v \in \mathfrak{M} \setminus \{\widetilde{v}\}$, there are finitely many possibilities for $|t_1(g_2)|_{\widetilde{v}}$.

Consider a τ_G-orbit $\{g_2, g_2\tau_G\}$. we can assume that $|t_1(g_2)| \geq |t_1(g_2\tau_G)|$. Then if $g_2 \in \mathrm{GL}(2)^0_{\mathbb{A}}$,

$$\begin{aligned} g_2\tau_G =& k(g_2)\tau_G a(t_2(g_2), t_1(g_2))^t n(u(g_2)) \\ =& k(g_2)\tau^t_G n(t_1(g_2)^{-1}t_2(g_2)u(g_2)) \cdot a(t_2(g_2), t_1(g_2)) \\ =& k(g_2)\tau_G k_1 t({}^t n(t_1(g_2)^{-1}t_2(g_2)u(g_2))) \cdot n(u_1) \cdot a(t_2(g_2), t_1(g_2)) \end{aligned}$$

for some $k_1 \in K \cap G_{2\mathbb{A}}$, $u_1 \in \mathbb{A}$.

Therefore,

$$|t_1(g_2\tau_G)| = |t_2(g_2)|\alpha(t_1(g_2)^{-1}t_2(g_2)u(g_2)).$$

By assumption,

$$|t_1(g_2)| \geq |t_2(g_2)|\alpha(t_1(g_2)^{-1}t_2(g_2)u(g_2)), \text{ and } |t_2(g_2)| = |t_1(g_2)|^{-1}.$$

So if $g^0 = (t, g_2)$ and $|t_1(g_2)| \geq |t_1(g_2\tau_G)|$, then $g^0 \in X_V$. Hence, X_V surjects to $G^0_{\mathbb{A}}/H_k$. It is easy to see that X_V is invariant under the right action of T_k. Since $\{g^0 \mid |t_1(g_2)| = \sqrt{\alpha(t_1(g_2)^{-1}t_2(g_2)u(g_2))}\}$ is a measure zero set, this proves the proposition.

Q.E.D.

Definition (4.2.3) *Let v be a place of k. Let Ψ, Ψ_v be Schwartz–Bruhat functions on $\mathbb{A}^2, k_v^2$ respectively. For $s, s_1 \in \mathbb{C}$ and a character ω of $\mathbb{A}^\times/k^\times$, we define*

$$\text{(1)} \qquad T_{V,v}(\Psi_v, \omega, s, s_1) = \int_{k_v^\times \times k_v} |t_v|_v^s \omega(t_v)\alpha_v(u_v)^{s_1}\Psi_v(t_v, t_v u_v)d^\times t_v du_v,$$

$$\text{(2)} \qquad T_V(\Psi, \omega, s, s_1) = \int_{\mathbb{A}^\times \times \mathbb{A}} |t|^s \omega(t)\alpha(u)^{s_1}\Psi(t, tu)d^\times t du,$$

$$\text{(3)} \qquad T_{V+}(\Psi, \omega, s, s_1) = \int_{\substack{\mathbb{A}^\times \times \mathbb{A} \\ |t| \geq 1}} |t|^s \omega(t)\alpha(u)^{s_1}\Psi(t, tu)d^\times t du,$$

$$\text{(4)} \qquad T^1_V(\Psi, \omega, s_1) = \int_{\mathbb{A}^1 \times \mathbb{A}} \omega(t^1)\alpha(u)^{s_1}\Psi(t^1, t^1 u)d^\times t^1 du.$$

If $\Psi = \otimes \Psi_v$, $T_V(\Psi, \omega, s, s_1)$ has an Euler product as follows

$$T_V(\Psi, \omega, s, s_1) = |\Delta_k|^{-\frac{1}{2}} \mathfrak{C}_k^{-1} \prod_v T_{V,v}(\Psi_v, \omega, s, s_1).$$

In (4.2.4)–(4.2.6), we drop the index v from t_v, u_v, because the situation is obvious.

Proposition (4.2.4) *Let $v \in \mathfrak{M}_f$. Suppose that Ψ_v is the characteristic function of o_v^2 and ω is trivial on $k_v^\times$. Then*

$$T_{V,v}(\Psi_v, \omega, s, s_1) = \frac{1 - q_v^{-(s+s_1)}}{(1 - q_v^{-s})(1 - q_v^{-(s+s_1-1)})}.$$

Proof.

$$\begin{aligned}
T_{V,v}(\Psi_v, \omega, s, s_1) &= \int_{o_v \setminus \{0\}} \int_{o_v} |t|_v^{s-1} \sup(1, |ut^{-1}|_v)^{-s_1} d^\times t du \\
&= \sum_{a=0}^{\infty} \sum_{b=0}^{\infty} q_v^{-a(s-1)} (1 - q_v^{-1}) q_v^{-b} \sup(1, q_v^{-(b-a)})^{-s_1} \\
&= \sum_{a=0}^{\infty} q_v^{-a(s-1)} (1 - q_v^{-1}) (\sum_{b=0}^{a-1} q_v^{-as_1 + b(s_1-1)} + \sum_{b=a}^{\infty} q_v^{-b}) \\
&= \sum_{a=0}^{\infty} q_v^{-a(s-1)} (1 - q_v^{-1}) \left(\frac{q_v^{-as_1} (q_v^{a(s_1-1)} - 1)}{q_v^{s_1-1} - 1} + \frac{q_v^{-a}}{1 - q_v^{-1}} \right) \\
&= \sum_{a=0}^{\infty} \left(\frac{q_v^{-a(s+s_1-1)} (q_v^{a(s_1-1)} - 1)(1 - q_v^{-1})}{q_v^{s_1-1} - 1} + q_v^{-as} \right) \\
&= \sum_{a=0}^{\infty} \frac{q_v^{-as} (q_v^{s_1-1} - q_v^{-1}) - q_v^{-a(s+s_1-1)} (1 - q_v^{-1})}{q_v^{s_1-1} - 1} \\
&= \frac{q_v^{s_1-1} - q_v^{-1}}{(1 - q_v^{-s})(q_v^{s_1-1} - 1)} - \frac{1 - q_v^{-1}}{(1 - q_v^{-(s+s_1-1)})(q_v^{s_1-1} - 1)} \\
&= \frac{1 - q_v^{-(s+s_1)}}{(1 - q_v^{-s})(1 - q_v^{-(s+s_1-1)})}.
\end{aligned}$$

Q.E.D.

Proposition (4.2.5) *Let $v \in \mathfrak{M}_f$. Then $T_{V,v}(\Psi_v, \omega, s, s_1)$ is a rational function of $q_v^{-s}, q_v^{-s_1}$ for all v, holomorphic for* $\mathrm{Re}(s) > 0, \mathrm{Re}(s) + \mathrm{Re}(s_1) > 1$.

Proof. Choose an integer l so that $\Psi_v(x_1, x_2)$ is constant on the set $\{(x_1, x_2) \in \mathbb{A}^2 \mid x_1|_v, |x_2|_v \leq q_v^{-l}\}$. Then,

$$\begin{aligned}
&\int_{|t|_v, |u|_v \leq q_v^{-l}} \omega(t) |t|_v^{s-1} \alpha_v (\frac{u}{t})^{s_1} \Psi_v(t, u) d^\times t du \\
&= C q_v^{-ls} \int_{(o_v \setminus \{0\}) \times o_v} \omega(t) |t|_v^{s-1} \alpha_v \left(\frac{u}{t}\right) d^\times t du \\
&= C q_v^{-ls} \frac{1 - q_v^{-(s+s_1)}}{(1 - q_v^{-s})(1 - q_v^{-(s+s_1-1)})} \int_{o_v^\times} \omega(t) d^\times t
\end{aligned}$$

for some constant C.

There exist open sets $U_1, \cdots, U_N \subset k_v$ such that $U_i \subset q_v^{m_i} o_v^\times$ for some $m_i < l$, and that if (x_1, x_2) is in the support of Ψ_v and $|x_2|_v > q_v^{-l}$, then $x_2 \in U_i$ for some i. By replacing $U_1, \cdots, U_N$ if necessary, we choose an integer $l' > l$ so that Ψ_v is constant on the set $\{t \mid |t|_v \leq q_v^{-l'}\} \times U_i$ for $i = 1, \cdots, N$. Then

$$\int_{\substack{|t|_v \leq q_v^{-l'} \\ q_v^{-l} < |u|_v}} \omega(t)|t|^{s-1} \alpha_v \left(\frac{u}{t}\right)^{s_1} \Psi_v(t,u) d^\times t du$$
$$= \sum_{i=1}^{N} C_i q_v^{-m_i(s_1-1)} q_v^{-l'(s+s_1-1)} (1 - q_v^{-(s+s_1-1)})^{-1} \int_{o_v^\times} \omega(t) d^\times t$$

for some constants $C_1, \cdots, C_N$.

Clearly,

$$\int_{\substack{q_v^{-l'} < |t|_v \leq q_v^{-l} \\ q_v^{-l} < |u|_v}} \omega(t)|t|_v^{s-1} \alpha_v \left(\frac{u}{t}\right)^{s_1} \Psi_v(t,u) d^\times t du$$

is an entire function. Therefore,

$$\int_{\substack{|t|_v \leq q_v^{-l} \\ q_v^{-l} < |u|_v}} \omega(t)|t|_v^{s-1} \alpha_v \left(\frac{u}{t}\right)^{s_1} \Psi_v(t,u) d^\times t du$$

is a rational function of $q_v^{-s}, q_v^{-s_1}$, holomorphic for $\mathrm{Re}(s) + \mathrm{Re}(s_1) > 1$.

Similarly, the integral over the set $\{(t,u) \in k_v^\times \times k_v \mid q_v^{-l} < |t|_v, |u|_v \leq q_v^{-l}\}$ is a rational function of $q_v^{-s}, q_v^{-s_1}$, holomorphic for $\mathrm{Re}(s) > 0$. Since the integral over the set $\{(t,u) \in k_v^\times \times k_v \mid q_v^{-l} < |t|_v, |u|_v\}$ is an entire function, this proves the proposition.

Q.E.D.

Proposition (4.2.6) *Suppose $v \in \mathfrak{M}_\infty$. Then for any ω, $T_{V,v}(\Psi_v, \omega, s, s_1)$ is holomorphic for* $\mathrm{Re}(s) > 0, \mathrm{Re}(s) + \mathrm{Re}(s_1) > 1$, *and can be continued meromorphically to the entire* $\mathbb{C}^2$.

Proof. First we consider the case $v \in \mathfrak{M}_\mathbb{R}$. Since $\omega(t) = 1$ or $\mathrm{sign}(t)$ for $t \in \mathbb{R}^\times$, we may consider integrals of the form

$$\int_0^\infty \int_{-\infty}^\infty t^s (1+u^2)^{-\frac{s_1}{2}} \Psi_v(t, tu) d^\times t du.$$

The above integral is equal to the integral

$$\int_0^\infty \int_{-\infty}^\infty t^{s+s_1-1} (t^2+u^2)^{-\frac{s_1}{2}} \Psi_v(t,u) d^\times t du$$
$$= \int_0^\infty \int_{-\frac{\pi}{2}}^{\frac{\pi}{2}} r^{s-1} \cos^{s+s_1-2} \theta \Psi_v(r\cos\theta, r\sin\theta) dr d\theta.$$

We define

$$f_1(\Psi_v, z_1, z_2) = \int_0^\infty \int_{-\frac{\pi}{2}}^{\frac{\pi}{2}} r^{z_1} \cos^{z_2}\theta \Psi_v(r\cos\theta, r\sin\theta) dr d\theta,$$

$$f_2(\Psi_v, z_1, z_2) = \int_0^\infty \int_{-\frac{\pi}{2}}^{\frac{\pi}{2}} r^{z_1} \cos^{z_2}\theta \sin\theta \Psi_v(r\cos\theta, r\sin\theta) dr d\theta.$$

Then

$$\begin{aligned} f_1(\Psi_v, z_1, z_2) &= -\frac{1}{z_1+1}(f_1(\partial_1\Psi_v, z_1+1, z_2+1) + f_2(\partial_2\Psi_v, z_1+1, z_2)), \\ f_2(\Psi_v, z_1, z_2) &= \frac{1}{z_2+1}(f_2(\partial_1\Psi_v, z_1+1, z_2+1) - f_1(\partial_2\Psi_v, z_1+1, z_2+2)), \end{aligned}$$

where ∂_1, ∂_2 are partial derivatives with respect to the two coordinates. The meromorphic continuation of $T_v(\Psi_v, \omega, s, s_1)$ for a real place v follows from the above formula.

Next we consider an imaginary place. Suppose that $\omega(re^{2\pi\sqrt{-1}\theta}) = e^{2\pi n\sqrt{-1}\theta}$. Let $t = r\cos\theta e^{2\pi\sqrt{-1}\phi_1}, u = r\sin\theta e^{2\pi\sqrt{-1}\phi_2}$. Then

$$d^\times t du = 4r\tan\theta dr d\theta d\phi_1 d\phi_2.$$

Therefore,

$$T_v(\Psi_v, \omega, s, s_1) = 4\int_0^\infty \int_0^{\frac{\pi}{2}} \int_0^{2\pi} \int_0^{2\pi} F(r, \theta, \phi_1, \phi_2, s, s_1) dr d\theta d\phi_1 d\phi_2,$$

where

$$\begin{aligned} F(r, \theta, \phi_1, \phi_2, s, s_1) =& r^{2s-1}\cos^{2s+2s_1-3}\theta \sin\theta e^{2\pi\sqrt{-1}n\phi_1} \\ & \times \Psi_v(r\cos\theta e^{2\pi\sqrt{-1}n\phi_1}, r\cos\theta e^{2\pi\sqrt{-1}n\phi_2}). \end{aligned}$$

The meromorphic continuation of $T_v(\Psi_v, \omega, s, s_1)$ is similar to the real case, and it is holomorphic for $\mathrm{Re}(s) > 0, \mathrm{Re}(s) + \mathrm{Re}(s_1) > 1$.

Q.E.D.

Suppose $\Psi = \otimes\Psi_v$. If ω is a character of $\mathbb{A}^\times/k^\times$, ω is trivial for almost all v. Let $\mathfrak{M}_\infty \subset P \subset \mathfrak{M}$ be a finite set such that ω is trivial on $k_v^\times$ and Ψ_v is the characteristic function of the set o_v^2 for $v \notin P$. We define $\zeta_{k,P}(z) = \prod_{v\notin P}(1-q_v^{-z})^{-1}$, and $T_{V,P}(\Psi, \omega, s, s_1) = |\Delta_k|^{-\frac{1}{2}}\mathfrak{C}_k^{-1}\prod_{v\in P} T_v(\Psi_v, \omega, s, s_1)$. By (4.2.4),

$$T_V(\Psi, \omega, s, s_1) = T_{V,P}(\Psi, \omega, s, s_1)\frac{\zeta_{k,P}(s)\zeta_{k,P}(s+s_1-1)}{\zeta_{k,P}(s+s_1)}.$$

Definition (4.2.7) *For Ψ, ω, s, s_1, and P as above, we define*

(1) $\displaystyle T_V(\Psi, \omega, s) = \left.\frac{d}{ds_1}\right|_{s_1=0} T_V(\Psi, \omega, s, s_1),\ T_V^1(\Psi, \omega) = \left.\frac{d}{ds_1}\right|_{s_1=0} T_V^1(\Psi, \omega, s_1),$

(2) $\displaystyle T_{V,P}(\Psi, \omega, s) = \left.\frac{d}{ds_1}\right|_{s_1=0} T_{V,P}(\Psi, \omega, s, s_1).$

By (4.2.4)–(4.2.6), we get the following proposition.

Proposition (4.2.8) *For any ω, $T(\Psi,\omega,s,s_1)$ can be continued meromorphically to the entire $\mathbb{C}^2$ and is holomorphic for* $\mathrm{Re}(s)>1, \mathrm{Re}(s)+\mathrm{Re}(s_1)>2$

Let $\widehat{b}^{\,0}=\widehat{t}^{\,0}(1,n_2(u_0))\in \widehat{B}^0_{\mathbb{A}}$, and $b^0=d(\lambda_1)\widehat{b}^{\,0}\in B^0_{\mathbb{A}}$. We define measures $d\widehat{b}^{\,0}, db^0$ on $\widehat{B}^0_{\mathbb{A}}, B^0_{\mathbb{A}}$ in the usual manner ($db^0=d^\times\lambda_1 d^\times \widehat{t}^{\,0}du_0$).

Proposition (4.2.9) *Let $\omega=(\omega_1,\omega_2)$ be a character of $(\mathbb{A}^\times/k^\times)^2$. Then*

$$(1)\qquad \int_1^\infty\int_{\widehat{B}^0_{\mathbb{A}}/T_k}\lambda^s\omega(\widehat{b}^{\,0})\Theta_{Z_0'}(\Phi,\underline{\lambda}\widehat{b}^{\,0})\alpha(u_0)^{s_1}d^\times\lambda d\widehat{b}^{\,0}$$

converges absolutely and locally uniformly for all $s,s_1\in\mathbb{C}$,

$$(2)\qquad \int_0^1\int_{\widehat{B}^0_{\mathbb{A}}/T_k}\lambda^s\omega(\widehat{b}^{\,0})\Theta_{Z_0'}(\Phi,\underline{\lambda}\widehat{b}^{\,0})\alpha(u_0)^{s_1}d^\times\lambda d\widehat{b}^{\,0}$$

converges absolutely and locally uniformly for $\mathrm{Re}(s)>2+\epsilon, \mathrm{Re}(s_1)>-\epsilon$, *where $\epsilon>0$ is a constant, and*

$$(3)\qquad \int_{\widehat{B}^0_{\mathbb{A}}/T_k}\omega(\widehat{b}^{\,0})\Theta_{Z_0'}(\Phi,\widehat{b}^{\,0})\alpha(u_0)^{s_1}d\widehat{b}^{\,0}$$

converges absolutely and locally uniformly for all $s_1\in\mathbb{C}$.

Proof. Let $\sigma=\mathrm{Re}(s),\sigma_1=\mathrm{Re}(s_1)$. There exist Schwartz–Bruhat functions $\Phi_1,\Phi_2\geq 0$ on $\mathbb{A}$ such that

$$\Theta_{Z_0'}(\Phi,\underline{\lambda}\widehat{b}^{\,0})\ll\sum_{x_1\in k^\times}\Phi_1(\underline{\lambda}x_1)\Phi_2(\underline{\lambda}x_1u_0).$$

Since the order of the integration and the summation can be changed, we can make the change of variable $u_1=\underline{\lambda}x_1u_0$. Then (1)–(3) are bounded by constant multiples of the integrals

$$\int_1^\infty\int_{\mathbb{A}}\lambda^{\sigma-1}\sum_{x_1\in k^\times}\Phi_1(\underline{\lambda}x_1)\Phi_2(u_1)\alpha(\underline{\lambda}^{-1}x_1^{-1}u_1)^{\sigma_1}d^\times\lambda du_1,$$
$$\int_0^1\int_{\mathbb{A}}\lambda^{\sigma-1}\sum_{x_1\in k^\times}\Phi_1(\underline{\lambda}x_1)\Phi_2(u_1)\alpha(\underline{\lambda}^{-1}x_1^{-1}u_1)^{\sigma_1}d^\times\lambda du_1,$$
$$\int_{\mathbb{A}}\sum_{x_1\in k^\times}\Phi_1(x_1)\Phi_2(u_1)\alpha(x_1^{-1}u_1)^{\sigma_1}du_1,$$

respectively.

There exist a lattice $L\subset k$ and a compact set $F\subset\mathbb{A}_f$ such that $\Phi_1(\underline{\lambda}x_1)\Phi_2(u_1)=0$ unless $x\in L, u_{1f}\in F$. Since the finite part of $\underline{\lambda}u_1$ is the same as the finite part of u_1, by (4.2.1),

$$\alpha(\underline{\lambda}^{-1}x_1^{-1}u_1)\geq N\prod_{v\in\mathfrak{M}_{\mathbb{R}}}(|x_1|_v^2+|\underline{\lambda}^{-1}u_1|_v^2)^{-\frac{1}{2}}\prod_{v\in\mathfrak{M}_{\mathbb{C}}}(|x_1|_v+|\underline{\lambda}^{-1}u_1|_v)^{-1}$$

for some $N > 0$.

If $\sigma_1 \geq 0$, then $\alpha(\underline{\lambda}^{-1}x_1^{-1}u_1)^{\sigma_1} \leq 1$. If $\sigma_1 \leq 0$, since

$$(|x_1|_v^2 + |\underline{\lambda}^{-1}u_1|_v^2)^{-\frac{\sigma_1}{2}} \leq \lambda^{\frac{\sigma_1}{[k:\mathbb{Q}]}}\alpha_v(\underline{\lambda}x_1)^{\sigma_1}\alpha_v(u_1)^{\sigma_1} \text{ for } v \in \mathfrak{M}_{\mathbb{R}},$$
$$(|x_1|_v + |\underline{\lambda}^{-1}u_1|_v)^{-\sigma_1} \leq \lambda^{\frac{2\sigma_1}{[k:\mathbb{Q}]}}\alpha_v(\underline{\lambda}x_1)^{\sigma_1}\alpha_v(u_1)^{\sigma_1} \text{ for } v \in \mathfrak{M}_{\mathbb{C}},$$

we have the following inequality

$$\alpha(\underline{\lambda}^{-1}x_1^{-1}u_1)^{\sigma_1} \leq \lambda^{\sigma_1} \prod_{v \in \mathfrak{M}_\infty} \alpha_v(\underline{\lambda}x_1)^{\sigma_1}\alpha_v(u_1)^{\sigma_1}.$$

Therefore, the proposition follows from (1.2.6) (consider $\lambda = 1$ for (3)).

Q.E.D.

For later purposes, we fix some notation.

Definition (4.2.10) *For $\Phi \in V_{\mathbb{A}}$, we define $\widetilde{R}_0\Phi \in \mathscr{S}(Y_{0\mathbb{A}}), R'_0\Phi \in \mathscr{S}(Z'_{0\mathbb{A}})$ by*

$$\widetilde{R}_0\Phi(x,y) = \Phi(0,x_1,x_2),\ R'_0\Phi(x_1) = \int_{\mathbb{A}} \Phi(0,x_1,x_2)dx_2.$$

The proof of the following lemma is an easy consequence of the definition.

Lemma (4.2.11)

$$(1)\quad \int_0^\infty \int_{\widehat{B}^0_{\mathbb{A}}/T_k} \lambda^s \omega(\widehat{b}^{\,0})\Theta_{Z'_0}(\Phi,\underline{\lambda}\widehat{b}^{\,0})\alpha(u_0)^{s_1} d^\times\lambda d\widehat{b}^{\,0} = \delta(\omega_1\omega_2^{-1})T(\widetilde{R}_0\Phi,\omega_1,s,s_1),$$
$$(2)\quad \int_{\widehat{B}^0_{\mathbb{A}}/T_k} \omega(\widehat{b}^{\,0})\Theta_{Z'_0}(\Phi,\widehat{b}^{\,0})\alpha(u_0)^{s_1} d\widehat{b}^{\,0} = \delta(\omega_1\omega_2^{-1})T^1(\widetilde{R}_0\Phi,\omega_1,s_1).$$

Definition (4.2.12) *For Φ, ω as above, and $s \in \mathbb{C}$, we define*

$$\Sigma_{\mathrm{st}}(\Phi,\omega,s) = \delta(\omega_1\omega_2^{-1})T^1(\widetilde{R}_0\Phi,\omega_1,\frac{1-s}{2}).$$

Let

$$\Sigma_{\mathrm{st}}(\Phi,\omega,z) = \sum_{i=0}^\infty \Sigma_{\mathrm{st},(i)}(\Phi,\omega,1)(z-1)^i$$

be the Taylor expansion around $z = 1$. Then

$$\Sigma_{\mathrm{st},(1)}(\Phi,\omega,1) = -\frac{\delta(\omega_1\omega_2^{-1})}{2}T^1(\widetilde{R}_0\Phi,\omega_1).$$

Definition (4.2.13)

$$\Xi_{\mathrm{st}}(\Phi,\omega,w) = \int_{G^0_{\mathbb{A}}/G_k} \omega(g^0)\Theta_{V,\mathrm{st}}(\Phi,g^0)\mathscr{E}(g^0,w)dg^0.$$

Then by (3.4.31), $\Xi_{\mathrm{st}}(\Phi,\omega,w)$ is well defined for $\mathrm{Re}(w)\gg 0$. As in §3.8, $\Xi_{\mathfrak{p}_i}(\Phi,\omega,w)$ is well defined for $\mathrm{Re}(w)\gg 0$ for $i=1,2$, and

$$\Xi_{\mathfrak{p}_2}(\Phi,\omega,w)=\Xi_{\mathfrak{p}_1}(\widehat{\Phi},\omega^{-1},w).$$

We define

$$J^s(\Phi,g^0)=\sum_{x\in G_k\backslash V_k^s}\left(\widehat{\Phi}((g^0)^\iota x)-\Phi(g^0x)\right),$$

$$I^0(\Phi,\omega)=\int_{G^0_{\mathbb{A}}/G_k}\omega(g^0)J^s(\Phi,g^0)dg^0,$$

$$I(\Phi,\omega,w)=\int_{G^0_{\mathbb{A}}/G_k}\omega(g^0)J^s(\Phi,g^0)\mathscr{E}(g^0,w)dg^0.$$

Then

$$I(\Phi,\omega,w)\sim C_G(w;\rho)I^0(\Phi,\omega),$$

and by (3.5.21) and the argument of (3.5.9),

$$\begin{aligned}I(\Phi,\omega,w)=\;&\Xi_{\mathfrak{p}_1}(\widehat{\Phi},\omega^{-1},w)-\Xi_{\mathfrak{p}_1}(\Phi,\omega,w)+C_G\Lambda(w;\rho)\mathfrak{V}_2\delta_{\#}(\omega)(\widehat{\Phi}(0)-\Phi(0))\\&+\Xi_{\mathrm{st}}(\widehat{\Phi},\omega^{-1},w)-\Xi_{\mathrm{st}}(\Phi,\omega,w).\end{aligned}$$

Definition (4.2.14) *Let Φ,ω,s be as before. We define*

$$Z_{V,\mathrm{ad}}(\Phi,\omega,s)=Z_V(\Phi,\omega,s)-\frac{\delta(\omega_1\omega_2^{-1})}{2}T_V(\widetilde{R}_0\Phi,\omega_1,s).$$

We call $Z_{V,\mathrm{ad}}(\Phi,\omega,s)$ the adjusted zeta function, and $\frac{\delta(\omega_1\omega_2^{-1})}{2}T_V(\widetilde{R}_0\Phi,\omega_1,s)$ the adjusting term. We devote the rest of this section to the proof of the following theorem.

Theorem (4.2.15) (Shintani) *Suppose $\Phi=M_\omega\Phi$. Then*

$$\begin{aligned}Z_{V,\mathrm{ad}}(\Phi,\omega,s)=\;&Z_{V+}(\Phi,\omega,s)+Z_{V+}(\widehat{\Phi},\omega^{-1},3-s)\\&-\frac{\delta(\omega_1\omega_2^{-1})}{2}\left(T_{V+}(\widetilde{R}_0\Phi,\omega_1,s)+T_{V+}(\widetilde{R}_0\widehat{\Phi},\omega_1^{-1},3-s)\right)\\&+\mathfrak{V}_2\delta_{\#}(\omega)\left(\frac{\widehat{\Phi}(0)}{s-3}-\frac{\Phi(0)}{s}\right)\\&+\frac{\delta(\omega_1^2)\delta(\omega_2)}{2}\left(-\frac{\Sigma_{1,(-1)}(R_{\mathfrak{d}}\widehat{\Phi},\omega_1^{-1},1)}{(s-2)^2}+\frac{\Sigma_{1,(0)}(R_{\mathfrak{d}}\widehat{\Phi},\omega_1^{-1},1)}{s-2}\right)\\&-\frac{\delta(\omega_1^2)\delta(\omega_2)}{2}\left(\frac{\Sigma_{1,(-1)}(R_{\mathfrak{d}}\Phi,\omega_1,1)}{(s-1)^2}+\frac{\Sigma_{1,(0)}(R_{\mathfrak{d}}\Phi,\omega_1,1)}{s-1}\right).\end{aligned}$$

Since the first four terms of (4.2.15) are entire functions and we proved the meromorphic continuation of $T_V(\widetilde{R}_0\Phi,\omega_1,s)$ in (4.2.8), $Z_V(\Phi,\omega,s)$ can be continued meromorphically everywhere.

Corollary (4.2.16) *The adjusted zeta function $Z_{V,\mathrm{ad}}(\Phi,\omega,s)$ satisfies a functional equation*

$$Z_{V,\mathrm{ad}}(\Phi,\omega,s) = Z_{V,\mathrm{ad}}(\widehat{\Phi},\omega^{-1},3-s).$$

We first consider $\Xi_{\mathfrak{p}_1}(\Phi,\omega,w)$. The element t^0 acts on $Z_{0\mathbb{A}}$ by $x_2 \to \lambda_1 b^2 t t_2^2 x_2$. Let τ be a Weyl group element. Let s_τ be as in §3.8 (a). It is easy to see that $\sigma_{\mathfrak{p}_1}(t^0) = \lambda_1^2$, $(t^0)^{\tau z+\rho} = \lambda_1^{s_\tau - 1}$. Also if $f(q)$ is a function on $\mathbb{A}^1/k^\times$,

$$\int_{(\mathbb{A}^1/k^\times)^3} \omega(\widehat{t}^0) f(tt_2^2) d^\times \widehat{t}^0 = \delta(\omega_1^2)\delta(\omega_2) \int_{\mathbb{A}^1/k^\times} \omega_1(q) f(q) d^\times q.$$

Therefore,

$$\begin{aligned}
&\Xi_{\mathfrak{p}_1,\tau}(\Phi,\omega,w) \\
&= \frac{\delta(\omega_1^2)\delta(\omega_2)}{2} \frac{1}{2\pi\sqrt{-1}} \int_{\mathrm{Re}(s_\tau)=r>1} \Sigma_1\left(R_0\Phi,\omega_1,\frac{s_\tau+1}{2}\right) \widetilde{\Lambda}_\tau(w;s_\tau) ds_\tau.
\end{aligned}$$

If $\tau = 1$, we can choose $r \gg 0$ and $\widetilde{L}(r) = -r \ll 0$. So $\Xi_{\mathfrak{p}_1,\tau}(\Phi,\omega,w) \sim 0$.
Next, we consider $\Xi_{\mathrm{st}}(\Phi,\omega,w)$.

Proposition (4.2.17)

$$\Xi_{\mathrm{st}}(\Phi,\omega,w) \sim \sum_\tau \frac{1}{2\pi\sqrt{-1}} \int_{\mathrm{Re}(s_\tau)=r>1} \Sigma_{\mathrm{st}}(\Phi,\omega,z)\widetilde{\Lambda}_\tau(w;s_\tau) ds_\tau.$$

Proof. By the standard consideration,

$$\begin{aligned}
\Xi_{\mathrm{st}}(\Phi,\omega,w) &= \int_{G^0_{\mathbb{A}}/G_k} \omega(g^0)\Theta_{V,\mathrm{st}}(\Phi,g^0)\mathscr{E}(g^0,w) dg^0 \\
&= \int_{G^0_{\mathbb{A}}/H_k} \omega(g^0)\Theta_{Z_0'}(\Phi,g^0)\mathscr{E}(g^0,w) dg^0 \\
&= \int_{X_V/T_k} \omega(g^0)\Theta_{Z_0'}(\Phi,g^0)\mathscr{E}(g^0,w) dg^0 \\
&= \int_{X_V\cap B^0_{\mathbb{A}}/T_k} \omega(b^0)\Theta_{Z_0'}(\Phi,b^0)\mathscr{E}(b^0,w) db^0.
\end{aligned}$$

Lemma (4.2.18) *The integral*

$$\int_{X_V\cap B^0_{\mathbb{A}}/T_k} \omega(b^0)\Theta_{Z_0'}(\Phi,b^0)(\mathscr{E}(b^0,w) - \mathscr{E}_N(t^0,w)) db^0$$

is holomorphic for $\mathrm{Re}(w) > 0$.

Proof. As in §3.8 (a), there exists a constant $\delta > 0$ such that $\mathscr{E}(t^0,w) - \mathscr{E}_N(t^0,w)$ is holomorphic for $\mathrm{Re}(w) \geq w_0 - \delta$ and if $M > w_0, l \gg 0$,

$$|\mathscr{E}(t^0,w) - \mathscr{E}_N(t^0,w)| \ll \lambda_1^l.$$

Since $t_1, t_2 \in \mathbb{A}^1/k^\times$ which is a compact set, by (1.2.3), there exist Schwartz–Bruhat functions $\Phi_1, \Phi_2 \geq 0$ on $\mathbb{A}$ such that

$$\Theta_{Z_0'}(\Phi, b^1)\mathscr{E}_1(b^0, w) \ll \sum_{x_1 \in k^\times} \Phi_1(x_1)\Phi_2(x_1 u_0)\lambda_1^l.$$

Then the above integral is bounded by a constant multiple of

$$\begin{aligned}
&\int_{\mathbb{A}} \sum_{x_1 \in k^\times} \Phi_1(x_1)\Phi_2(x_1 u_0)\alpha(u_0)^{-\frac{l}{2}} du_0 \\
&= \sum_{x_1 \in k^\times} \int_{\mathbb{A}} \Phi_1(x_1)\Phi_2(x_1 u_0)\alpha(u_0)^{-\frac{l}{2}} du_0 \\
&= \sum_{x_1 \in k^\times} \int_{\mathbb{A}} \Phi_1(x_1)\Phi_2(u_1)\alpha(x_1^{-1} u_1)^{-\frac{l}{2}} du_1,
\end{aligned}$$

where $u_1 = x_1 u_0$. Therefore, the convergence of the above integral follows from (4.2.9). This proves the lemma.

Q.E.D.

If $f(q)$ is a function on $\mathbb{A}^1/k^\times$,

$$\int_{(\mathbb{A}^1/k^\times)^3} \omega(\widehat{t}^0) f(tt_1t_2) d^\times \widehat{t}^0 = \delta(\omega_1^2)\delta(\omega_2) \int_{\mathbb{A}^1/k^\times} \omega_1(q) f(q) d^\times q.$$

Also,

$$\int_{\lambda_1 \geq \sqrt{\alpha(u_0)}^{-1}} \lambda_1^{s_\tau - 1} d^\times \lambda_1 = \frac{\alpha(u_0)^{\frac{1-s_\tau}{2}}}{s_\tau - 1}.$$

Therefore, by the above lemma and (4.2.11),

$$\Xi_{\mathrm{st}}(\Phi, \omega, w) \sim \sum_\tau \frac{1}{2\pi\sqrt{-1}} \int_{\mathrm{Re}(s_\tau) = r > 1} \frac{\Sigma_{\mathrm{st}}(\Phi, \omega, s_\tau)}{s_\tau - 1} \widetilde{\Lambda}_\tau(w; s_\tau) ds_\tau.$$

This proves Proposition (4.2.17).

Q.E.D.

It is easy to see that

$$\frac{1}{2\pi\sqrt{-1}} \int_{\mathrm{Re}(s_\tau) = r > 1} \frac{\Sigma_{\mathrm{st}}(\Phi, \omega, s_\tau)}{s_\tau - 1} \widetilde{\Lambda}_\tau(w; s_\tau) ds_\tau \sim 0$$

if $\tau = 1$.

Let $\tau = \tau_G$, and

$$\begin{aligned}
J(\Phi, \omega, s_\tau) = {} & \frac{\delta(\omega_1\omega_2^{-1})}{2} \left(\Sigma_1\left(R_0\widehat{\Phi}, \omega_1^{-1}, \frac{s_\tau + 1}{2}\right) - \Sigma_1\left(R_0\Phi, \omega_1, \frac{s_\tau + 1}{2}\right)\right) \\
& + \frac{\Sigma_{\mathrm{st}}(\widehat{\Phi}, \omega^{-1}, s_\tau)}{s_\tau - 1} - \frac{\Sigma_{\mathrm{st}}(\Phi, \omega, s_\tau)}{s_\tau - 1}.
\end{aligned}$$

By the above considerations,

$$\begin{aligned}I^0(\Phi,\omega,w) \sim {} & C_G\Lambda(w;\rho)\mathfrak{V}_2\delta_{\#}(\omega)(\widehat{\Phi}(0)-\Phi(0)) \\ & + \frac{1}{2\pi\sqrt{-1}}\int_{\mathrm{Re}(s_\tau)=r>1} J(\Phi,\omega,s_\tau)\phi(s_\tau)\widetilde{\Lambda}(w;s_\tau)ds_\tau.\end{aligned}$$

By Wright's principle, $J(\Phi,\omega,s_\tau)$ must be holomorphic at $s_\tau=1$.
Since

$$\Sigma_1(\Psi,\omega_1,\frac{s_\tau+1}{2}) = \frac{2\Sigma_{1,(-1)}(\Psi,\omega_1,1)}{s_\tau-1} + \Sigma_{1,(0)}(\Psi,\omega_1,1)+O(s_\tau-1),$$

we may conclude that

$$\begin{aligned}I^0(\Phi,\omega,w) \sim {} & C_G\Lambda(w;\rho)\mathfrak{V}_2\delta_{\#}(\omega)(\widehat{\Phi}(0)-\Phi(0)) \\ & + C_G\Lambda(w;\rho)\frac{\delta(\omega_1^2)\delta(\omega_2)}{2}\left(\Sigma_{1,(0)}(R_{\mathfrak{d}}\widehat{\Phi},\omega_1^{-1},1)-\Sigma_{1,(0)}(R_{\mathfrak{d}}\Phi,\omega_1,1)\right) \\ & + C_G\Lambda(w;\rho)(\Sigma_{\mathrm{st},(1)}(\widehat{\Phi},\omega^{-1},1)-\Sigma_{\mathrm{st},(1)}(\Phi,\omega,1)).\end{aligned}$$

Hence,

$$\begin{aligned}I^0(\Phi,w) = {} & \mathfrak{V}_2\delta_{\#}(\omega)(\widehat{\Phi}(0)-\Phi(0)) \\ & + \frac{\delta(\omega_1^2)\delta(\omega_2)}{2}(\Sigma_{1,(0)}(R_{\mathfrak{d}}\widehat{\Phi},\omega_1^{-1},1)-\Sigma_{1,(0)}(R_{\mathfrak{d}}\Phi,\omega_1,1)) \\ & + \Sigma_{\mathrm{st},(1)}(\widehat{\Phi},\omega^{-1},1)-\Sigma_{\mathrm{st},(1)}(\Phi,\omega,1).\end{aligned}$$

It is easy to see that

$$\begin{aligned}& \Phi_\lambda(0)=\Phi(0),\ \widehat{\Phi_\lambda}(0)=\lambda^{-3}\widehat{\Phi}(0), \\ & \Sigma_{1,(0)}(R_{\mathfrak{d}}\Phi_\lambda,\omega_1,1)=\lambda^{-1}\Sigma_{1,(0)}(R_{\mathfrak{d}}\Phi,\omega_1,1)-\lambda^{-1}(\log\lambda)\Sigma_{1,(-1)}(R_{\mathfrak{d}}\Phi,\omega_1,1), \\ & \Sigma_{1,(0)}(R_{\mathfrak{d}}\widehat{\Phi_\lambda},\omega_1^{-1},1)=\lambda^{-2}\Sigma_{1,(0)}(R_{\mathfrak{d}}\widehat{\Phi},\omega_1^{-1},1)+\lambda^{-2}(\log\lambda)\Sigma_{1,(-1)}(R_{\mathfrak{d}}\widehat{\Phi},\omega_1^{-1},1).\end{aligned}$$

Also,

$$\begin{aligned}& \int_0^1 \lambda^s\Phi_\lambda(0)d^\times\lambda = \frac{\Phi(0)}{s},\ \int_0^1 \lambda^s\widehat{\Phi_\lambda}(0)d^\times\lambda = \frac{\widehat{\Phi}(0)}{s-3}, \\ & \int_0^1 \lambda^s\Sigma_{1,(0)}(R_{\mathfrak{d}}\Phi_\lambda,\omega_1,1)d^\times\lambda = \frac{\Sigma_{1,(0)}(R_{\mathfrak{d}}\Phi,\omega_1,1)}{s-1}+\frac{\Sigma_1(R_{\mathfrak{d}}\Phi,\omega_1,1)}{(s-1)^2}, \\ & \int_0^1 \lambda^s\Sigma_{1,(0)}(R_{\mathfrak{d}}\widehat{\Phi_\lambda},\omega_1^{-1},1)d^\times\lambda = \frac{\Sigma_{1,(0)}(R_{\mathfrak{d}}\widehat{\Phi},\omega_1^{-1},1)}{s-2}-\frac{\Sigma_{1,(-1)}(R_{\mathfrak{d}}\widehat{\Phi},\omega_1^{-1},1)}{(s-2)^2}, \\ & \int_0^1 \lambda^s\Sigma_{\mathrm{st},(1)}(\Phi_\lambda,\omega,1)d^\times\lambda = -\frac{\delta(\omega_1\omega_2^{-1})}{2}(T_V(\widetilde{R}_0\Phi,\omega_1,s)-T_{V+}(\widetilde{R}_0\Phi,\omega,s)), \\ & \int_0^1 \lambda^{s-3}\Sigma_{\mathrm{st},(1)}(\widehat{\Phi_\lambda},\omega_1^{-1},1)d^\times\lambda = -\frac{\delta(\omega_1\omega_2^{-1})}{2}T_{V+}(\widetilde{R}_0\widehat{\Phi},\omega_1^{-1},3-s),\end{aligned}$$

because the order of the integration with respect to λ and the differentiation with respect to s_1 can be changed by (4.2.9). This finishes the proof of (4.2.15).

§4.3 β-sequences

In this section, we introduce some notations related to β-sequences.

The unstable strata of V_n are $S_{n,1}, \cdots, S_{n,n-1}$, and $Y_{n,i} \cong V_{n-i}$ for all i. So we identify β-sequences with subsets $\mathfrak{d} = \{j_1, \cdots, j_a\} \subset \{1, \cdots, n\}$ where $1 \leq j_1 < \cdots < j_a < n$. With this identification, $Y_\mathfrak{d} \cong V_{n-j_a}$. If $\mathfrak{d} \prec \mathfrak{d}'$, and $l(\mathfrak{d}') = l(\mathfrak{d}) + 1$, $P_{\mathfrak{d}'} \subset M_\mathfrak{d}$ is a maximal parabolic subgroup. Therefore, all the β-sequences satisfy Condition (3.4.16)(1). Hence, $\Xi_\mathfrak{p}(\Phi, \omega, w)$ etc. are well defined if $\mathrm{Re}(w) \gg 0$.

We choose $d_{\mathfrak{d}1}(\lambda_{\mathfrak{d}1}), \cdots, d_{\mathfrak{d}a}(\lambda_{\mathfrak{d}a})$ in the following manner

$$d_{\mathfrak{d}1}(\lambda_{\mathfrak{d}1}) = \begin{pmatrix} \underline{\lambda}_{\mathfrak{d}1}^{-(n-j_1)} I_{j_1} & \\ & \underline{\lambda}_{\mathfrak{d}1}^{j_1} I_{n-j_1} \end{pmatrix},$$

$$d_{\mathfrak{d}2}(\lambda_{\mathfrak{d}2}) = \begin{pmatrix} I_{j_1} & & \\ & \underline{\lambda}_{\mathfrak{d}2}^{-(n-j_2)} I_{j_1} & \\ & & \underline{\lambda}_{\mathfrak{d}2}^{j_2-j_1} I_{n-j_2} \end{pmatrix},$$

$$\vdots$$

$$d_{\mathfrak{d}a}(\lambda_{\mathfrak{d}a}) = \begin{pmatrix} I_{j_{a-1}} & & \\ & \underline{\lambda}_{\mathfrak{d}a}^{-(n-j_a)} I_{j_a - j_{a-1}} & \\ & & \underline{\lambda}_{\mathfrak{d}a}^{j_a - j_{a-1}} I_{n-j_a} \end{pmatrix}.$$

Let

$$\lambda_\mathfrak{d} = d_\mathfrak{d}(\lambda_{\mathfrak{d}1}, \cdots, \lambda_{\mathfrak{d}a}) = d_{\mathfrak{d}1}(\lambda_{\mathfrak{d}1}) \cdots d_{\mathfrak{d}a}(\lambda_{\mathfrak{d}a}),$$
$$A_\mathfrak{d} = \{\lambda_\mathfrak{d} = d(\lambda_{\mathfrak{d}1}, \cdots, \lambda_{\mathfrak{d}a}) \mid \lambda_{\mathfrak{d}1}, \cdots, \lambda_{\mathfrak{d}a} \in \mathbb{R}_+\}.$$

Let $g_\mathfrak{d} \in M_{\mathfrak{d}\mathbb{A}} \cap G^0_\mathbb{A}$. Then $g_\mathfrak{d}$ can be written uniquely as $g_\mathfrak{d} = \lambda_\mathfrak{d} g^0_\mathfrak{d}$, where $\lambda_\mathfrak{d} \in A_\mathfrak{d}, g^0_\mathfrak{d} \in M^0_{\mathfrak{d}\mathbb{A}}$.

We define

$$\begin{aligned}(4.3.1) \qquad \mathfrak{F}_{\mathfrak{d}1} &= j_1(j_2 - j_1) \cdots (j_a - j_{a-1})(n - j_1) \cdots (n - j_a), \\ \mathfrak{F}_{\mathfrak{d}2} &= 2^a j_1(j_2 - j_1) \cdots (j_a - j_{a-1}), \\ \mathfrak{F}_{\mathfrak{d}3} &= \mathfrak{F}_{\mathfrak{d}1}\mathfrak{F}_{\mathfrak{d}2}^{-1} = 2^{-a}(n - j_1) \cdots (n - j_a).\end{aligned}$$

Let $d^\times \lambda_\mathfrak{d} = d^\times \lambda_{\mathfrak{d}1} \cdots d^\times \lambda_{\mathfrak{d}a}$. We define an invariant measure $dg_\mathfrak{d}$ on $M_{\mathfrak{d}\mathbb{A}} \cap G^0_\mathbb{A}$ by $dg_\mathfrak{d} = \mathfrak{F}_{\mathfrak{d}1} d^\times \lambda_\mathfrak{d} dg^0_\mathfrak{d}$. This measure satisfies the condition after (3.3.11). We define $\mathfrak{V}_\mathfrak{d} = \mathfrak{V}_{j_1} \mathfrak{V}_{j_2 - j_1} \cdots \mathfrak{V}_{j_a - j_{a-1}}$. Let $\widehat{\Omega}_\mathfrak{d}$ etc. be as in Chapter 3. If $\mathfrak{d} = \{i\}$, we also use the notation $\mathscr{F}_{n,i}$ instead of $\mathscr{F}_\mathfrak{d}$.

Let $\mathfrak{d} = \{j_1, \cdots, j_a\}$ be a β-sequence of length $a > 0$. Let $j_0 = 0, j_{a+1} = n$. We define constants $f_{\mathfrak{d}i}, h_{\mathfrak{d}i}$ in the following manner

$$(4.3.2) \qquad f_{\mathfrak{d}i} = \frac{(n - j_i)(n - j_i + 1)}{2}, \; h_{\mathfrak{d}i} = \frac{(n - j_{i-1})(n - j_i)}{2}$$

for all $i = 1, \cdots, a$. Let $f_\mathfrak{d} = f_{\mathfrak{d}a}$, $h_\mathfrak{d} = h_{\mathfrak{d}a}$.

We define a constant $c_\mathfrak{p}$ for each path $\mathfrak{p}$ of positive length inductively in the following manner.

Suppose $l(\mathfrak{p}) = 1$. Then

$$
(4.3.3) \qquad c_{\mathfrak{p}} = \begin{cases} -\frac{(n-j_1)}{2} & \mathfrak{s}(1) = 0, \\ \frac{n-j_1}{2} & \mathfrak{s}(1) = 1. \end{cases}
$$

If $\mathfrak{p} \prec \mathfrak{p}'$ and $l(\mathfrak{p}') = l(\mathfrak{p}) + 1$,

$$
(4.3.4) \qquad c_{\mathfrak{p}'} = \begin{cases} -\frac{(n-j_a)}{2(h_{\mathfrak{d}} - h_{\mathfrak{d}'})} c_{\mathfrak{p}} & \mathfrak{s}'(a+1) = 0, \\ \frac{(n-j_a)}{2(h_{\mathfrak{d}} + h_{\mathfrak{d}'} - f_{\mathfrak{d}})} c_{\mathfrak{p}} & \mathfrak{s}'(a+1) = 1. \end{cases}
$$

For $\omega = (\omega_1, \omega_2)$ and a path $\mathfrak{p} = (\mathfrak{d}, \mathfrak{s})$, we define $\delta_{\mathfrak{p}}(\omega) = \delta(\omega_1^2)\delta(\omega_2)$. Let $\omega_{\mathfrak{p}} = (\omega_1, 1)$ if $\#\{i \mid \mathfrak{s}(i) = 1\}$ is even, and $\omega_{\mathfrak{p}} = (\omega_1^{-1}, 1)$ otherwise.

Let $\mathfrak{P}_1$ be the set of $\mathfrak{p}$'s such that $\mathfrak{d} = \{1, \cdots, a-1, i\}$ for some $i \neq a, n-2$, and $l(\mathfrak{d}) > 0$. Let $\mathfrak{P}_2$ be the set of $\mathfrak{p}$'s such that $\mathfrak{d} = \{1, \cdots, a\}$ for some $a \leq n-3$, , and $l(\mathfrak{d}) > 0$. Let $\mathfrak{P}_3$ be the set of $\mathfrak{p}$'s such that $\mathfrak{d} = \{1, \cdots, a-1, n-2\}$ where $a \leq n-3$. $\mathfrak{P}_3$ is the empty set if $n = 3$. Let $\mathfrak{P}_4$ be the set of $\mathfrak{p}$'s such that $\mathfrak{d} = \{1, 2, \cdots, n-2\}$.

§4.4 An inductive formulation

We will formulate an inductive way of proving Theorem (4.0.1) in this section. We first have to introduce some notations.

Let

$$
(4.4.1) \qquad Z_{V_n}(\Phi, \omega, s) = \sum_i a_i(s - s_0)
$$

be the Laurent expansion of $Z_{V_n}(\Phi, \omega, s)$ at $s = s_0$. Then we define

$$
Z_{V_n,(i)}(\Phi, \omega, s_0) = a_i.
$$

Of course, when we consider these values for a particular n, the meromorphic continuation of $Z_{V_n}(\Phi, \omega, s)$ has to be known. However, since we prove the principal part formula by induction on n, we do not logically depend on the meromorphic continuation proved in [60], [64]. We use a similar notation $Z_{V_2,\mathrm{ad},(i)}(\Phi, \omega, s_0)$ for the adjusted zeta function for $n = 2$.

Suppose that $\mathfrak{d} = \{j_1, \cdots, j_a\}$ is a β-sequence such that $j_a = n-2$. Then we identify $Z_{\mathfrak{d}}$ with V_2.

Definition (4.4.2)
(1) We define $Z'_{\mathfrak{d},0} \subset Z_{\mathfrak{d}}$ to be the subspace which corresponds to Z'_0 in §4.2 by the above identification.
(2) For $\Psi \in \mathscr{S}(Z_{\mathfrak{d}\mathbb{A}})$, $R'_{\mathfrak{d},0}\Psi \in \mathscr{S}(\mathbb{A})$, $\widetilde{R}_0\Psi \in \mathscr{S}(\mathbb{A}^2)$ are functions which correspond to $R'_{V_2,0}\Psi \in \mathscr{S}(\mathbb{A})$, $\widetilde{R}_0\Psi \in \mathscr{S}(\mathbb{A}^2)$ in §4.2 by the above identification.

We define $Z'^{\,\mathrm{ss}}_{\mathfrak{d},0k}$ similarly.

Let $J(\Phi, g^0)$ be as in (3.5.6). Let $\mathscr{E}(g^0, w)$ be the smoothed Eisenstein series which we defined in (3.4.7). We define

$$
(4.4.3) \qquad I^0(\Phi, \omega) = \int_{G^0_{\mathbb{A}}/G_k} \omega(g^0) J(\Phi, g^0) dg^0,
$$

$$
I(\Phi, \omega, w) = \int_{G^0_{\mathbb{A}}/G_k} \omega(g^0) J(\Phi, g^0) \mathscr{E}(g^0, w) dg^0.
$$

We remind the reader that

$$Z_V(\Phi,\omega,s) = Z_{V+}(\Phi,\omega,s) + Z_{V+}(\widehat{\Phi},\omega^{-1},d_n-s) + \int_0^1 \lambda^s I^0(\Phi_\lambda,\omega)d^\times\lambda.$$

Since

$$J(\Phi,g^0) = \Theta_{V_{nk}^{ss}}(\Phi,g^0) - \Theta_{V_{nk}^{ss}}(\widehat{\Phi},(g^0)^\iota),$$

by (4.1.3), $J(\Phi,g^0) \in C(G^0_{\mathbb{A}}/G_k,r)$ for some $r=(r_1,\cdots,r_{n-1})$ such that $r_j > -j(n-j)$. Therefore, we can use Shintani's lemma for GL(n). The statement of (3.4.34) implies that

$$I(\Phi,\omega,w) \sim C_G\Lambda(w;\rho)I^0(\Phi,\omega).$$

By (3.5.9) and (3.5.20), we get the following proposition.

Proposition (4.4.4)

$$\begin{aligned} I(\Phi,\omega,w) = &\sum_{\mathfrak{p}\in\mathfrak{P}_1\cup\mathfrak{P}_3\cup\mathfrak{P}_4} \epsilon_{\mathfrak{p}}\Xi_{\mathfrak{p}}(\Phi,\omega,w) + \delta_{\#}(\omega)\Lambda(w;\rho)(\widehat{\Phi}(0)-\Phi(0)) \\ &+ \sum_{\mathfrak{p}\in\mathfrak{P}_2} \epsilon_{\mathfrak{p}}(\Xi_{\mathfrak{p}+}(\Phi,\omega,w) + \widehat{\Xi}_{\mathfrak{p}+}(\Phi,\omega,w)) \\ &+ \sum_{\mathfrak{p}\in\mathfrak{P}_2} \epsilon_{\mathfrak{p}}(\widehat{\Xi}_{\mathfrak{p}\#}(\Phi,\omega,w) - \Xi_{\mathfrak{p}\#}(\Phi,\omega,w)). \end{aligned}$$

Note that since all the paths satisfy Condition (3.4.16)(1), $\Xi_{\mathfrak{p}}(\Phi,\omega,w)$ is well defined for $\mathrm{Re}(w) \gg 0$ for all paths $\mathfrak{p}$.

Definition (4.4.5)
(1) *Let* $\mathfrak{p}\in\mathfrak{P}_1$ *and* $l(\mathfrak{p})=a$. *We define*

$$I_{\mathfrak{p}}(\Phi,\omega) = c_{\mathfrak{p}}\mathfrak{V}_{\mathfrak{d}}\delta_{\mathfrak{p}}(\omega)Z_{V_{n-j_a}}(R_{\mathfrak{d}}\Phi_{\mathfrak{p}},\omega_{\mathfrak{p}},h_{\mathfrak{d}}).$$

(2) *Let* $\mathfrak{p}\in\mathfrak{P}_2$ *and* $l(\mathfrak{p})=a$. *We define*

$$\begin{aligned} I_{\mathfrak{p}+}(\Phi,\omega) &= c_{\mathfrak{p}}\delta_{\mathfrak{p}}(\omega)\left(Z_{V_{n-a}+}(R_{\mathfrak{d}}\Phi_{\mathfrak{p}},\omega_{\mathfrak{p}},f_{\mathfrak{d}}) + Z_{V_{n-a}+}(\mathscr{F}_{\mathfrak{d}}R_{\mathfrak{d}}\Phi_{\mathfrak{p}},\omega_{\mathfrak{p}}^{-1},0)\right), \\ I_{\mathfrak{p}\#}(\Phi,\omega) &= c_{\mathfrak{p}}\delta_{\#}(\omega)\frac{\Phi_{\mathfrak{p}}(0)}{f_{\mathfrak{d}}}, \\ I_{\mathfrak{p}1}(\Phi,\omega) &= -c_{\mathfrak{p}}\mathfrak{V}_{n-a-2}\delta_{\#}(\omega)\frac{(n-a)\Sigma_1(R'_{\mathfrak{d}',0}R_{\mathfrak{d}'}\mathscr{F}_{\mathfrak{d}}\Phi_{\mathfrak{p}},n-a-1)}{2(n-a)^2} \\ &\quad + c_{\mathfrak{p}}\mathfrak{V}_{n-a-2}\delta_{\#}(\omega)\frac{\Sigma_{1,(1)}(R'_{\mathfrak{d}',0}R_{\mathfrak{d}'}\mathscr{F}_{\mathfrak{d}}\Phi_{\mathfrak{p}},n-a-1)}{2(n-a)}, \\ I_{\mathfrak{p}2}(\Phi,\omega) &= -c_{\mathfrak{p}}\mathfrak{V}_{n-a-2}\delta_{\#}(\omega)\frac{(n-a)\Sigma_1(R'_{\mathfrak{d}',0}R_{\mathfrak{d}'}\Phi_{\mathfrak{p}},n-a-1)}{2(f_{\mathfrak{d}}-(n-a))^2} \\ &\quad - c_{\mathfrak{p}}\mathfrak{V}_{n-a-2}\delta_{\#}(\omega)\frac{\Sigma_{1,(1)}(R'_{\mathfrak{d}',0}R_{\mathfrak{d}'}\Phi_{\mathfrak{p}},n-a-1)}{2(f_{\mathfrak{d}}-(n-a))}, \end{aligned}$$

where $\mathfrak{d}'=\{j_1,\cdots,j_{a+1}\}$ *is the unique* β*-sequence such that* $\mathfrak{d}\prec\mathfrak{d}', l(\mathfrak{d}')=l(\mathfrak{d})+1$ *and* $j_{a+1}=n-2$.

(3) *Let* $\mathfrak{p} \in \mathfrak{P}_3$ *and* $l(\mathfrak{p}) = a$. *We define*

$$I_{\mathfrak{p}}(\Phi, \omega) = c_{\mathfrak{p}} \mathfrak{V}_{n-a-2} \delta_{\mathfrak{p}}(\omega) Z_{V_2,\mathrm{ad}}(R_{\mathfrak{d}} \Phi_{\mathfrak{p}}, \omega_{\mathfrak{p}}, n-a+1).$$

(4) *Let* $\mathfrak{p} \in \mathfrak{P}_4$ *and* $l(\mathfrak{p}) = a$. *We define*

$$I_{\mathfrak{p}}(\Phi, \omega) = c_{\mathfrak{p}} \delta_{\mathfrak{p}}(\omega) Z_{V_2,\mathrm{ad},(0)}(R_{\mathfrak{d}} \Phi_{\mathfrak{p}}, \omega_{\mathfrak{p}}, 3).$$

Proposition (4.4.6) *Suppose* $M_{V_n,\omega}\Phi = \Phi$. *Then*

$$\begin{aligned} I^0(\Phi, \omega) &= \mathfrak{V}_n \delta_{\#}(\omega)(\widehat{\Phi}(0) - \Phi(0)) + \sum_{\mathfrak{p} \in \mathfrak{P}_1} I_{\mathfrak{p}}(\Phi, \omega) \\ &+ \frac{\mathfrak{V}_{n-2}\delta_{\#}(\omega)}{2} \left(\Sigma_{1,(1)}(R'_{n,n-2,0}\widehat{\Phi}, n-1) - \Sigma_{1,(1)}(R'_{n,n-2,0}\Phi, n-1) \right) \\ &+ \sum_{\mathfrak{p} \in \mathfrak{P}_2} (I_{\mathfrak{p}+}(\Phi, \omega) - I_{\mathfrak{p}\#}(\Phi, \omega) + I_{\mathfrak{p}1}(\Phi, \omega) + I_{\mathfrak{p}2}(\Phi, \omega)) \\ &+ \sum_{\mathfrak{p} \in \mathfrak{P}_3} I_{\mathfrak{p}}(\Phi, \omega) + \sum_{\mathfrak{p} \in \mathfrak{P}_4} I_{\mathfrak{p}}(\Phi, \omega). \end{aligned}$$

We devote §§4.5–4.7 to the proof of (4.4.6). For the rest of this section, we prove that (4.4.6) implies Theorem (4.0.1) by induction on n.

The following lemma is the basis of cancellations of various distributions in this chapter.

Lemma (4.4.7) *Let* $dx = \prod_v dx_v$ *be the ordinary measure on* $\mathbb{A}^n$, *and* $d^\times t = \prod_v d^\times t_v$ *the ordinary measure on* $\mathbb{A}^\times$. *Suppose that* $\Phi \in \mathscr{S}(\mathbb{A}^n)$ *is invariant under the action of the standard maximal compact subgroup of* $\mathrm{GL}(n)_{\mathbb{A}}$. *Then*

$$\int_{\mathbb{A}^n} \Phi(x) dx = \frac{\mathfrak{R}_k}{Z_k(n)} \int_{\mathbb{A}^\times} |t|^n \Phi(t, 0, \cdots, 0) d^\times t.$$

Proof. By the choice of our measure,

$$\int_{\mathbb{A}^n} \Phi(x) dx = |\Delta_k|^{-\frac{n}{2}} \prod_v \int_{k_v^n} \Phi(x_v) dx_v,$$

$$\int_{\mathbb{A}^\times} |t|^n \Phi(t, 0, \cdots, 0) d^\times t = \mathfrak{C}_k^{-1} \prod_v \int_{k_v^\times} |t_v|_v^n \Phi(t_v, 0, \cdots, 0) d^\times t_v.$$

Let $v \in \mathfrak{M}_f$. It is easy to see that

$$\int_{k_v^n} \Phi_v(x_v) dx_v = (1 - q_v^{-1})^n \sum_{i_1, \cdots, i_n} q_v^{-i_1 - \cdots - i_n} \Phi(\pi_v^{i_1}, \cdots, \pi_v^{i_n}).$$

Let $f_n(i) = (1 - q_v^{-1})^n \sum_{\inf(i_1, \cdots, i_n) = i} q_v^{-i_1 - \cdots - i_n}$. Then

$$\begin{aligned} f_n(i) &= (1 - q_v^{-1})^n \sum_{i_2, \cdots, i_n = i}^{\infty} q_v^{-i-i_2-\cdots-i_n} \\ &\quad + (1 - q_v^{-1})^n \sum_{i_1 = i+1}^{\infty} \sum_{\inf(i_2, \cdots, i_n) = i} q_v^{-i_1-i_2-\cdots-i_n} \\ &= (1 - q_v^{-1}) q_v^{-ni} + q_v^{-(i+1)} f_{n-1}(i). \end{aligned}$$

By induction, $f_n(i) = (1 - q_v^{-n})q_v^{-n}$. Therefore,

$$\int_{k_v^n} \Phi_v(x_v)dx = \sum_i f_n(i)\Phi(\pi_v^i, 0, \cdots, 0)$$
$$= (1 - q_v^{-n}) \int_{k_v^\times} |t_v|_v^n \Phi(t_v, 0, \cdots, 0)d^\times t_v.$$

Next, we consider the real place. Let D_n be the unit ball in $\mathbb{R}^n$. Then $\mathrm{vol}(D_n) = \pi^{\frac{n}{2}}\Gamma(\frac{n}{2}+1)^{-1}$. So,

$$\int_{\mathbb{R}^n} \Phi(x_\mathbb{R})dx_\mathbb{R} = \int_{\mathbb{R}^n} \Phi(\sqrt{x_{1\mathbb{R}}^2 + \cdots + x_{n\mathbb{R}}^2}, 0, \cdots, 0)dx_\mathbb{R}$$
$$= n\mathrm{vol}(D^n) \int_{\mathbb{R}_+} \Phi(r, 0, \cdots, 0)r^n d^\times r$$
$$= \frac{n}{2}\mathrm{vol}(D^n) \int_{\mathbb{R}^\times} |t_\mathbb{R}|_\mathbb{R}^n \Phi(t_\mathbb{R}, 0, \cdots, 0)d^\times t_\mathbb{R}$$
$$= \frac{1}{\pi^{-\frac{n}{2}}\Gamma(\frac{n}{2})} \int_{\mathbb{R}^\times} |t_\mathbb{R}|_\mathbb{R}^n \Phi(t_\mathbb{R}, 0, \cdots, 0)d^\times t_\mathbb{R}.$$

Finally, we consider the imaginary place. Note that $dx_\mathbb{C}$ is two times the usual Lebesgue measure. As in the real case,

$$\int_{\mathbb{C}^n} \Phi(x_\mathbb{C})dx_\mathbb{C} = \int_{\mathbb{C}^n} \Phi(\sqrt{|x_{1\mathbb{C}}|_\mathbb{C} + \cdots + |x_{n\mathbb{C}}|_\mathbb{C}}, 0, \cdots, 0)dx_\mathbb{C}$$
$$= 2n\mathrm{vol}(D_{2n})2^n \int_{\mathbb{R}_+} r^{2n}\Phi(r, 0, \cdots, 0)d^\times r$$
$$= \frac{(2\pi)^{-1}}{(2\pi)^{-n}\Gamma(n)} \int_{\mathbb{C}^\times} |t_\mathbb{C}|_\mathbb{C}^n \Phi(t_\mathbb{C}, 0, \cdots, 0)d^\times t_\mathbb{C}.$$

Therefore,

$$\int_{\mathbb{A}^n} \Phi(x)dx = Z_k(n)^{-1}(2\pi)^{-r_2} \prod_v \int_{k_v^\times} |t_v|_v^n \Phi(t_v, 0, \cdots, 0)d^\times t_v$$
$$= \frac{\mathfrak{R}_k}{Z_k(n)} \int_{\mathbb{A}^\times} |t|^n \Phi(t, 0, \cdots, 0)d^\times t,$$

because $\mathfrak{C}_k = (2\pi)^{r_2}\mathfrak{R}_k$.

Q.E.D.

Lemma (4.4.7) is essentially the relation between integrals with respect to the cartesian coordinate and the polar coordinate.

Easy considerations show that

$$\text{(4.4.8)}\quad \Sigma_{1,(1)}(R'_{n,n-2,0}\widehat{\Phi_\lambda}, n-1) = \lambda^{-\frac{n(n-1)}{2}}\Sigma_{1,(1)}(R'_{n,n-2,0}\widehat{\Phi}, n-1) + \lambda^{-\frac{n(n-1)}{2}}(\log\lambda)\Sigma_1(R'_{n,n-2,0}\widehat{\Phi}, n-1),$$
$$\Sigma_{1,(1)}(R'_{n,n-2,0}\Phi_\lambda, n-1) = \lambda^{-n}\Sigma_{1,(1)}(R'_{n,n-2,0}\Phi, n-1) - \lambda^{-n}(\log\lambda)\Sigma_1(R'_{n,n-2,0}\Phi, n-1),$$

$$
\begin{aligned}
&Z_{V_{n-i}}(R_{n,i}\widehat{\Phi_\lambda}, (\omega_1^{-1}, 1), \frac{n(n-i)}{2}) \\
&= \lambda^{-\frac{n(n+1)}{2}+\frac{n(n-i)}{2}} Z_{V_{n-i}}(R_{n,i}\widehat{\Phi}, (\omega_1^{-1}, 1), \frac{n(n-i)}{2}), \\
&Z_{V_{n-i}}(R_{n,i}\Phi_\lambda, (\omega_1, 1), \frac{n(n-i)}{2}) \\
&= \lambda^{-\frac{n(n-i)}{2}} Z_{V_{n-i}}(R_{n,i}\widehat{\Phi}, (\omega_1, 1), \frac{n(n-i)}{2}),
\end{aligned}
$$

for $i \neq 1, n-2$.

If $n = 3$,

$$
\begin{aligned}
(4.4.9)\quad Z_{V_2,\mathrm{ad},(0)}(R_{3,1}\widehat{\Phi_\lambda}, (\omega_1^{-1}, 1), 3) &= \lambda^{-3} Z_{V_2,\mathrm{ad},(0)}(R_{3,1}\widehat{\Phi}, (\omega_1^{-1}, 1), 3) \\
&\quad + \lambda^{-3}(\log\lambda)\mathfrak{V}_2\delta_\#(\omega)\mathscr{F}_{3,1}R_{3,1}\widehat{\Phi}(0), \\
Z_{V_2,\mathrm{ad},(0)}(R_{3,1}\Phi_\lambda, (\omega_1, 1), 3) &= \lambda^{-3} Z_{V_2,\mathrm{ad},(0)}(R_{3,1}\Phi, (\omega_1, 1), 3) \\
&\quad - \lambda^{-3}(\log\lambda)\mathfrak{V}_2\delta_\#(\omega)\mathscr{F}_{3,1}R_{3,1}\Phi(0).
\end{aligned}
$$

If $n \geq 4$,

$$
\begin{aligned}
(4.4.10)\qquad &\widehat{\Phi_\lambda}(0) = \lambda^{-d_n}\widehat{\Phi}(0),\ \Phi_\lambda(0) = \Phi(0), \\
&Z_{V_2,\mathrm{ad}}(R_{n,n-2}\widehat{\Phi_\lambda}, (\omega_1^{-1}, 1), n) \\
&= \lambda^{-\frac{n(n-1)}{2}} Z_{V_2,\mathrm{ad}}(R_{n,n-2}\widehat{\Phi}, (\omega_1^{-1}, 1), n), \\
&Z_{V_2,\mathrm{ad}}(R_{n,n-2}\Phi_\lambda, (\omega_1, 1), n) \\
&= \lambda^{-n} Z_{V_2,\mathrm{ad}}(R_{n,n-2}\Phi, (\omega_1, 1), n), \\
&Z_{V_{n-1},(0)}(R_{n,1}\widehat{\Phi_\lambda}, (\omega_1^{-1}, 1), \frac{n(n-1)}{2}) \\
&= \lambda^{-n} Z_{V_{n-1},(0)}(R_{n,1}\widehat{\Phi}, (\omega_1^{-1}, 1), \frac{n(n-1)}{2}) \\
&\quad + \lambda^{-n}(\log\lambda)\mathfrak{V}_{n-1}\delta_\#(\omega)\mathscr{F}_{n,1}R_{n,1}\widehat{\Phi}(0), \\
&Z_{V_{n-1},(0)}(R_{n,1}\Phi_\lambda, (\omega_1, 1), \frac{n(n-1)}{2}) \\
&= \lambda^{-\frac{n(n-1)}{2}} Z_{V_{n-1},(0)}(R_{n,1}\Phi, (\omega_1, 1), \frac{n(n-1)}{2}) \\
&\quad - \lambda^{-\frac{n(n-1)}{2}}(\log\lambda)\mathfrak{V}_{n-1}\delta_\#(\omega)\mathscr{F}_{n,1}R_{n,1}\Phi(0).
\end{aligned}
$$

Lemma (4.4.11) *If Φ is K-invariant,*

$$
(1)\qquad \mathfrak{V}_{n-2}\Sigma_1(R'_{n,n-2,0}\Phi, n-1) = \mathfrak{V}_{n-1}\mathscr{F}_{n,1}R_{n,1}\widehat{\Phi}(0),
$$

$$
(2)\qquad \mathfrak{V}_{n-2}\Sigma_1(R'_{n,n-2,0}\widehat{\Phi}, n-1) = \mathfrak{V}_{n-1}\mathscr{F}_{n,1}R_{n,1}\Phi(0).
$$

Proof. Note that $\mathfrak{V}_{n-2}\mathfrak{V}_{n-1}^{-1} = \frac{\mathfrak{R}_k}{Z_k(n-1)}$. We define $\Psi_1 \in \mathscr{S}(\mathbb{A}^{n-1})$ by

$$
\Psi_1(x_{12}, \cdots, x_{1n}) = \int_{\mathbb{A}} \Phi(x_{11}, x_{12}, \cdots, x_{1n}, \overbrace{0, \cdots, 0}^{\frac{n(n-1)}{2}}) dx_{11}.
$$

Let $\Psi_2(x) = \Psi_1(x, 0, \cdots, 0)$. By assumption, Ψ_1 is invariant under the action of the standard maximal compact subgroup of $\mathrm{GL}(n-1)_{\mathbb{A}}$. Easy considerations show that

$$\Sigma_1(R'_{n,n-2,0}\Phi, n-1) = \Sigma_1(\Psi_2, n-1),$$
$$\mathscr{F}_{n,1}R_{n,1}\widehat{\Phi}(0) = \int_{\mathbb{A}^{n-1}} \Psi_1(x_{12}, \cdots, x_{1n})dx_{12}\cdots dx_{1n}.$$

Then the first relation follows from (4.4.7).

The second relation is similar.

Q.E.D.

Definition (4.4.12) *For $n = 3$, we define*

$$\begin{aligned} f_3(\Phi, \omega) = {} & \mathfrak{V}_3\delta_{\#}(\omega)\Phi(0) + \frac{\mathfrak{V}_2\delta(\omega_1^2)\delta(\omega_2)}{2}\Sigma_1(R_{3,2}\Phi, \omega_1, \frac{3}{2}) \\ & + \delta(\omega_1^2)\delta(\omega_2)Z_{V_2,\mathrm{ad},(0)}(R_{3,1}\Phi, (\omega_1, 1), 3) \\ & + \frac{\delta_{\#}(\omega)}{2}\Sigma_{1,(1)}(R'_{3,1,0}\Phi, 2). \end{aligned}$$

For $n \geq 4$, we define

$$\begin{aligned} f_n(\Phi, \omega) = {} & \mathfrak{V}_n\delta_{\#}(\omega)\Phi(0) \\ & + \sum_{i \neq 1, n-2} \frac{(n-i)\mathfrak{V}_i\delta(\omega_1^2)\delta(\omega_2)}{2} Z_{V_{n-i}}(R_{n,i}\Phi, (\omega_1, 1), \frac{n(n-i)}{2}) \\ & + \mathfrak{V}_{n-2}\delta(\omega_1^2)\delta(\omega_2)Z_{V_2,\mathrm{ad}}(R_{n,n-2}\Phi, (\omega_1, 1), n) \\ & + \frac{(n-1)\delta(\omega_1^2)\delta(\omega_2)}{2} Z_{V_{n-1},(0)}(R_{n,1}\Phi, (\omega_1, 1), \frac{n(n-1)}{2}) \\ & + \frac{\mathfrak{V}_{n-2}\delta_{\#}(\omega)}{2}\Sigma_{1,(1)}(R'_{n,n-2,0}\Phi, n-1). \end{aligned}$$

Also we define

(4.4.13)
$$\begin{aligned} & I_{n,1}(\Phi, \omega) \\ & = \sum_{\substack{\mathfrak{p} \in \mathfrak{P}_1 \\ l(\mathfrak{p}) > 1 \\ s(1) = 0}} I_{\mathfrak{p}}(\Phi, \omega) + \sum_{\substack{\mathfrak{p} \in \mathfrak{P}_3 \\ l(\mathfrak{p}) > 1 \\ s(1) = 0}} I_{\mathfrak{p}}(\Phi, \omega) + \sum_{\substack{\mathfrak{p} \in \mathfrak{P}_4 \\ s(1) = 0}} I_{\mathfrak{p}}(\Phi, \omega) \\ & \quad + \sum_{\substack{\mathfrak{p} \in \mathfrak{P}_2 \\ s(1) = 0}} (I_{\mathfrak{p}+}(\Phi, \omega) - I_{\mathfrak{p}\#}(\Phi, \omega) + I_{\mathfrak{p}1}(\Phi, \omega) + I_{\mathfrak{p}2}(\Phi, \omega)), \\ & \widehat{I}_{n,1}(\Phi, \omega) \\ & = \sum_{\substack{\mathfrak{p} \in \mathfrak{P}_1 \\ l(\mathfrak{p}) > 1 \\ s(1) = 1}} I_{\mathfrak{p}}(\Phi, \omega) + \sum_{\substack{\mathfrak{p} \in \mathfrak{P}_3 \\ l(\mathfrak{p}) > 1 \\ s(1) = 1}} I_{\mathfrak{p}}(\Phi, \omega) + \sum_{\substack{\mathfrak{p} \in \mathfrak{P}_4 \\ s(1) = 1}} I_{\mathfrak{p}}(\Phi, \omega) \\ & \quad + \sum_{\substack{\mathfrak{p} \in \mathfrak{P}_2 \\ s(1) = 1}} (I_{\mathfrak{p}+}(\Phi, \omega) - I_{\mathfrak{p}\#}(\Phi, \omega) + I_{\mathfrak{p}1}(\Phi, \omega) + I_{\mathfrak{p}2}(\Phi, \omega)). \end{aligned}$$

We consider three statements $(A_n), (B_n), (C_n)$. The statement A_n is as follows.

$$(A_n) \qquad I^0(\Phi,\omega) = f_n(\widehat{\Phi},\omega^{-1}) - f_n(\Phi,\omega).$$

The statement (B_3) is as follows.

$$(B_3) \qquad \begin{aligned} I_{3,1}(\Phi,\omega) &= -\delta(\omega_1^2)\delta(\omega_2)Z_{V_2,\mathrm{ad},(0)}(R_{3,1}\Phi,(\omega_1,1),3), \\ \widehat{I}_{3,1}(\Phi,\omega) &= \delta(\omega_1^2)\delta(\omega_2)Z_{V_2,\mathrm{ad},(0)}(R_{3,1}\widehat{\Phi},(\omega_1^{-1},1),3). \end{aligned}$$

The statement (B_n) for $n \geq 4$ is as follows.

$$(B_n) \quad \begin{aligned} I_{n,1}(\Phi,\omega) &= -\frac{(n-1)\delta(\omega_1^2)\delta(\omega_2)}{2} Z_{V_{n-1},(0)}(R_{n,1}\Phi,(\omega_1,1),\frac{n(n-1)}{2}), \\ \widehat{I}_{n,1}(\Phi,\omega) &= \frac{(n-1)\delta(\omega_1^2)\delta(\omega_2)}{2} Z_{V_{n-1},(0)}(R_{n,1}\widehat{\Phi},(\omega_1^{-1},1),\frac{n(n-1)}{2}). \end{aligned}$$

The statement (C_3) is the empty statement. The statement (C_n) for $n \geq 4$ is the statement of Theorem (4.0.1) for V_{n-1}.

Proposition (4.4.14) *The statements $(A_n), (B_n), (C_n)$ are true for all n.*

Proof. Clearly, (4.4.6) and the statement (B_n) imply the statement (A_n). So we only have to deduce the statements $(B_{n+1}), (C_{n+1})$ assuming the statements $(A_n), (B_n), (C_n)$, and (4.4.6).

Consider the case $n = 3$. In this case, $\mathfrak{P}_2, \mathfrak{P}_3$ are the empty set. Therefore, (B_3) is Definition (4.4.5)(4).

Now we consider the step from n to $n+1$.

Let Φ_λ be as in Chapter 3. We multiply λ^s to $I^0(\Phi_\lambda,\omega)$ in (A_n) and integrate it over $\lambda \in [0,1]$. Then (C_{n+1}) follows by the relations (4.4.8), (4.4.9), (4.4.10).

Let $\Phi \in \mathscr{S}(V_{n+1\mathbb{A}})$, and $\Psi_1 = R_{n+1,1}\Phi, \Psi_2 = R_{n+1,1}\widehat{\Phi}$. Let $f_n = \frac{n(n+1)}{2}, h_{n,i} = \frac{n(n-i)}{2}$. By applying (C_{n+1}) to Ψ_1 and using (B_n), we get the following lemma.

Lemma (4.4.15) *If $n \geq 4$,*

$$\begin{aligned} &\delta(\omega_1^2)\delta(\omega_2)Z_{V_n,(0)}(\Psi_1,(\omega_1,1),f_n) \\ &= \delta(\omega_1^2)\delta(\omega_2)(Z_{V_n+}(\Psi_1,\omega,f_n) + Z_{V_n+}(\widehat{\Psi}_1,\omega^{-1},0)) \\ &\quad - \mathfrak{V}_n\delta_\#(\omega)\frac{\Psi_1(0)}{f_n} \\ &\quad + \sum_{i\neq 1,n-2} (n-i)\mathfrak{V}_i\delta(\omega_1^2)\delta(\omega_2)\frac{Z_{V_{n-i}}(R_{n,i}\widehat{\Psi}_1,(\omega_1^{-1},1),h_{n,i})}{2h_{n,i}} \\ &\quad - \sum_{i\neq 1,n-2} (n-i)\mathfrak{V}_i\delta(\omega_1^2)\delta(\omega_2)\frac{Z_{V_{n-i}}(R_{n,i}\Psi_1,(\omega_1,1),h_{n,i})}{2(f_n-h_{n,i})} \end{aligned}$$

$$
\begin{aligned}
&+\mathfrak{V}_{n-2}\delta(\omega_1^2)\delta(\omega_2)\frac{Z_{V_2,\mathrm{ad}}(R_{n,n-2}\widehat{\Psi}_1,(\omega_1^{-1},1),n)}{h_{n,n-2}}\\
&-\mathfrak{V}_{n-2}\delta(\omega_1^2)\delta(\omega_2)\frac{Z_{V_2,\mathrm{ad}}(R_{n,n-2}\Psi_1,(\omega_1,1),n)}{f_n-h_{n,n-2}}\\
&+\frac{\widehat{I}_{n,1}(\Psi_1,\omega)}{h_{n,1}}-\frac{I_{n,1}(\Psi_1,\omega)}{f_n-h_{n,1}}\\
&+\mathfrak{V}_{n-2}\delta_{\#}(\omega)\left(-\frac{n\Sigma_1(R'_{n,n-2,0}\widehat{\Psi}_1,n-1)}{2h_{n,n-2}^2}+\frac{\Sigma_{1,(1)}(R'_{n,n-2,0}\widehat{\Psi}_1,n-1)}{2h_{n,n-2}}\right)\\
&-\mathfrak{V}_{n-2}\delta_{\#}(\omega)\left(\frac{n\Sigma_1(R'_{n,n-2,0}\Psi_1,n-1)}{2(f_n-h_{n,n-2})^2}+\frac{\Sigma_{1,(1)}(R'_{n,n-2,0}\Psi_1,n-1)}{2(f_n-h_{n,n-2})}\right).
\end{aligned}
$$

For a path $\mathfrak{p}=(\mathfrak{d},\mathfrak{s})$ for V_n, we associate a path $\mathfrak{p}'=(\mathfrak{d}',\mathfrak{s}')$ for V_{n+1} so that $\mathfrak{d}'=\{1\}\cup\{i+1\mid i\in\mathfrak{d}\}$, $\mathfrak{s}'(1)=0$, and $\mathfrak{s}'(i+1)=\mathfrak{s}(i)$ for $i=1,\cdots,l(\mathfrak{p})$. Any path $\mathfrak{p}'$ for V_{n+1} such that $l(\mathfrak{p}')>1$ and $\mathfrak{s}'(1)=0$ can be obtained in this way. By definition, $c_{\mathfrak{p}'}=-\frac{n}{2}c_{\mathfrak{p}}$.

Let $\mathfrak{p}'_1=(\mathfrak{d}'_1,\mathfrak{s}'_1)$ be a path for V_{n+1} where $\mathfrak{d}'_1=\{1\},\mathfrak{s}'_1(1)=0$. If we multiply (4.4.13) by $-\frac{n}{2}$, the sum of the first two lines and the last two lines is

$$
I_{\mathfrak{p}'_1+}(\Phi,\omega)-I_{\mathfrak{p}'_1\#}(\Phi,\omega)+I_{\mathfrak{p}'_1 1}(\Phi,\omega)+I_{\mathfrak{p}'_1 2}(\Phi,\omega).
$$

The sum of the third and fourth lines is equal to

$$
\sum_{\substack{\mathfrak{p}'\in\mathfrak{P}_1\\ l(\mathfrak{p}')=2\\ \mathfrak{s}(1)=0}} I_{\mathfrak{p}'}(\Phi,\omega),
$$

where we consider paths for V_{n+1}. The sum of the fifth and the sixth lines is equal to

$$
\sum_{\substack{\mathfrak{p}'\in\mathfrak{P}_3\\ l(\mathfrak{p}')=2\\ \mathfrak{s}(1)=0}} I_{\mathfrak{p}'}(\Phi,\omega),
$$

where we consider paths for V_{n+1}. Finally,

$$
-\frac{n}{2}\frac{\widehat{I}_{n,1}(\Psi_1,\omega)}{h_{n,1}}+\frac{n}{2}\frac{I_{n,1}(\Psi_1,\omega)}{f_n-h_{n,1}}
$$

gives the rest of the terms in (B_{n+1}) for $I_{n+1,1}(\Phi,\omega)$. This proves the step from (B_n) to (B_{n+1}) for $n\geq 4$ and $I_{n+1,1}(\Phi,\omega)$. The argument for $\widehat{I}_{n+1,1}(\Phi,\omega)$ is similar using Ψ_2. The step from (B_3) to (B_4) is similar using $F_3(\Phi,\omega,s)$. This proves (4.4.14). Q.E.D.

Now we begin our analysis, and study the distributions $\Xi_{\mathfrak{p}}(\Phi,\omega,w)$ in the next three sections.

§4.5 Paths in $\mathfrak{P}_1$

We mainly consider paths in $\mathfrak{P}_1$ in this section. We first determine $\sigma_{\mathfrak{p}}$ for an arbitrary path $\mathfrak{p}$.

Lemma (4.5.1) *Suppose* $\sigma_{\mathfrak{p}}(\lambda_{\mathfrak{d}}) = \prod_{i=1}^{a} e_{\mathfrak{p}i}(\lambda_{\mathfrak{d}})^{\chi_{\mathfrak{p}i}}$. *Then*

$$\chi_{\mathfrak{p}i} = \begin{cases} h_{\mathfrak{d}i} - h_{\mathfrak{d}i+1} & s(i+1) = 0, \\ h_{\mathfrak{d}i} + h_{\mathfrak{d}i+1} - f_{\mathfrak{d}i} & s(i+1) = 1, \end{cases}$$

for $i = 1, \cdots, a-1$, *and* $\chi_{\mathfrak{p}a} = h_{\mathfrak{d}}$.

Proof. We prove this lemma by induction with respect to $l(\mathfrak{p})$. Suppose that $\mathfrak{p} = (\mathfrak{d}, \mathfrak{s}) \prec \mathfrak{p}' = (\mathfrak{d}', \mathfrak{s}')$ and $l(\mathfrak{p}') = l(\mathfrak{p}) + 1 = a + 1$. Let $\lambda_{\mathfrak{d}'} = d_{\mathfrak{d}'}(\lambda_{\mathfrak{d}'1}, \cdots, \lambda_{\mathfrak{d}'a+1})$. The statement (3.3.10) says

$$\lambda_{\mathfrak{d}'a+1}^{2(j_{a+1}-j_a)} = \begin{cases} e_{\mathfrak{d}'a+1}(\lambda_{\mathfrak{d}'})e_{\mathfrak{d}'a}(\lambda_{\mathfrak{d}'})^{-1} & \mathfrak{s}'(a+1) = 0, \\ e_{\mathfrak{d}'a+1}(\lambda_{\mathfrak{d}'})e_{\mathfrak{d}'a}(\lambda_{\mathfrak{d}'}) & \mathfrak{s}'(a+1) = 1. \end{cases} \tag{4.5.2}$$

Therefore,

$$d_{\mathfrak{d}'a+1}(\lambda_{\mathfrak{d}'a+1})^{-2\rho} = \begin{cases} (e_{\mathfrak{d}'a+1}(\lambda_{\mathfrak{d}'})e_{\mathfrak{d}'a}(\lambda_{\mathfrak{d}'})^{-1})^{h_{\mathfrak{d}'}} & \mathfrak{s}'(a+1) = 0, \\ (e_{\mathfrak{d}'a+1}(\lambda_{\mathfrak{d}'})e_{\mathfrak{d}'a}(\lambda_{\mathfrak{d}'}))^{h_{\mathfrak{d}'}} & \mathfrak{s}'(a+1) = 1. \end{cases} \tag{4.5.3}$$

In all the cases, $Y_{\mathfrak{d}} = Z_{\mathfrak{d}}$. So $\kappa_{\mathfrak{d}2}(\lambda_{\mathfrak{d}}) = 1$.

Suppose $\mathfrak{s}'(a+1) = 0$. Then

$$\sigma_{\mathfrak{p}'}(\lambda_{\mathfrak{d}'}) = (\prod_{i=1}^{a-1} e_{\mathfrak{p}'i}(\lambda_{\mathfrak{d}'})^{\chi_{\mathfrak{p}i}})e_{\mathfrak{p}'a}(\lambda_{\mathfrak{d}'})^{h_{\mathfrak{d}}}(e_{\mathfrak{d}'a+1}(\lambda_{\mathfrak{d}'})e_{\mathfrak{d}'a}(\lambda_{\mathfrak{d}'})^{-1})^{h_{\mathfrak{d}'}}.$$

Therefore, $\chi_{\mathfrak{p}'i} = \chi_{\mathfrak{p}i}$ for $i = 1, \cdots, a-1$, $\chi_{\mathfrak{p}'a} = h_{\mathfrak{d}} - h_{\mathfrak{d}'}$, and $\chi_{\mathfrak{p}'a+1} = h_{\mathfrak{d}'}$.

Suppose $\mathfrak{s}'(a+1) = 1$. Then $\theta_{\mathfrak{d}}(\lambda_{\mathfrak{d}'}) = \lambda_{\mathfrak{d}'}^{-1}$, and

$$\kappa_{\mathfrak{d}1}(\theta_{\mathfrak{d}}(\lambda_{\mathfrak{d}'})) = e_{\mathfrak{p}a}(\lambda_{\mathfrak{d}'})^{f_{\mathfrak{d}}} = e_{\mathfrak{p}'a}(\lambda_{\mathfrak{d}'})^{-f_{\mathfrak{d}}}.$$

Therefore,

$$\sigma_{\mathfrak{p}'}(\lambda_{\mathfrak{d}'}) = (\prod_{i=1}^{a-1} e_{\mathfrak{p}'i}(\lambda_{\mathfrak{d}'})^{\chi_{\mathfrak{p}i}})e_{\mathfrak{p}'a}(\lambda_{\mathfrak{d}'})^{h_{\mathfrak{d}}}(e_{\mathfrak{d}'a+1}(\lambda_{\mathfrak{d}'})e_{\mathfrak{d}'a}(\lambda_{\mathfrak{d}'}))^{h_{\mathfrak{d}'}}e_{\mathfrak{p}'a}^{-f_{\mathfrak{d}}}.$$

Hence, $\chi_{\mathfrak{p}'i} = \chi_{\mathfrak{p}i}$ for $i = 1, \cdots, a-1$, $\chi_{\mathfrak{p}'a} = h_{\mathfrak{d}} + h_{\mathfrak{d}'} - f_{\mathfrak{d}}$, and $\chi_{\mathfrak{p}'a+1} = h_{\mathfrak{d}'}$.

Q.E.D.

If $\mathfrak{p} \in \mathfrak{P}_1, \cdots, \mathfrak{P}_4$, $h_{\mathfrak{d}i} = f_{\mathfrak{d}i}$ for $i = 1, \cdots, a-1$. Therefore, $\chi_{\mathfrak{p}i} > 0$ for all i. The following lemma is an easy consequence of (4.5.1).

Lemma (4.5.4) $\epsilon_{\mathfrak{p}}\mathfrak{F}_{\mathfrak{d}3}\prod_{i=1}^{a-1}\chi_{\mathfrak{p}i}^{-1} = c_{\mathfrak{p}}$.

The following lemma follows from (4.1.3).

Lemma (4.5.5) *Let $\mathfrak{p} = (\mathfrak{d}, \mathfrak{s}) \in \mathfrak{P}_1$ or $\mathfrak{P}_2$. Suppose $\mathfrak{d} = \{1, \cdots, a-1, i\}$. Then for any $N_1, N_2 > \frac{(n-i)(n-i+1)}{2}$,*

$$
\begin{aligned}
(1)\qquad & \sigma_{\mathfrak{p}}(\lambda_{\mathfrak{d}})|\Theta_{Z_{\mathfrak{d}}}(R_{\mathfrak{d}}\Phi_{\mathfrak{p}}, \lambda_{\mathfrak{d}} g_{\mathfrak{d}}^0)| \\
& \ll \prod_{i=1}^{a-1} e_{\mathfrak{p}i}(\lambda_{\mathfrak{d}})^{\chi_{\mathfrak{p}i}} \inf(e_{\mathfrak{p}a}(\lambda_{\mathfrak{d}})^{h_{\mathfrak{d}}-N_1}, e_{\mathfrak{p}a}(\lambda_{\mathfrak{d}})^{h_{\mathfrak{d}}-N_2}) t(g^0)^r, \\
(2)\qquad & \sigma_{\mathfrak{p}}(\lambda_{\mathfrak{d}})\kappa_{\mathfrak{d}1}(\lambda_{\mathfrak{d}})|\Theta_{Z_{\mathfrak{d}}}(R_{\mathfrak{d}}\Phi_{\mathfrak{p}}, \theta_{\mathfrak{d}}(\lambda_{\mathfrak{d}} g_{\mathfrak{d}}^0))| \\
& \ll \prod_{i=1}^{a-1} e_{\mathfrak{p}i}(\lambda_{\mathfrak{d}})^{\chi_{\mathfrak{p}i}} e_{\mathfrak{p}a}(\lambda_{\mathfrak{d}})^{h_{\mathfrak{d}}-d_{n-i}+N_1} t(g^0)^r.
\end{aligned}
$$

for $\lambda_{\mathfrak{d}} \in A_{\mathfrak{p}2}$ and $g_{\mathfrak{d}}^0 \in (K \cap M_{\mathfrak{d}\mathbb{A}})T_{\mathfrak{d}\eta+}^0\Omega_{\mathfrak{d}}$, where $r = (r_1, \cdots, r_{n-i-1}) \in \mathbb{R}^{n-1}$, $r_j < j(n-i-j)$ for all j, and $t(g^0)^r$ is as in (3.4.17).

Proposition (4.5.6)
(1) *Suppose $\mathfrak{p} = (\mathfrak{d}, \mathfrak{s}) \in \mathfrak{P}_1$. Then*

$$\epsilon_{\mathfrak{p}}\Xi_{\mathfrak{p}}(\Phi, \omega, w) \sim C_G\Lambda(w;\rho)c_{\mathfrak{p}}\mathfrak{V}_{\mathfrak{d}}\delta_{\mathfrak{p}}(\omega)Z_{V_{n-j_a}}(R_{\mathfrak{d}}\Phi_{\mathfrak{p}}, \omega_{\mathfrak{p}}, h_{\mathfrak{d}}).$$

(2) *Suppose $\mathfrak{p} \in \mathfrak{P}_2$. Then*

$$
\begin{aligned}
&\epsilon_{\mathfrak{p}}\Xi_{\mathfrak{p}+}(\Phi, \omega, w) \sim C_G\Lambda(w;\rho)c_{\mathfrak{p}}\mathfrak{V}_{\mathfrak{d}}\delta_{\mathfrak{p}}(\omega)Z_{V_{n-j_a}+}(R_{\mathfrak{d}}\Phi_{\mathfrak{p}}, \omega_{\mathfrak{p}}, h_{\mathfrak{d}}),\\
&\epsilon_{\mathfrak{p}}\widehat{\Xi}_{\mathfrak{p}+}(\Phi, \omega, w) \sim C_G\Lambda(w;\rho)c_{\mathfrak{p}}\mathfrak{V}_{\mathfrak{d}}\delta_{\mathfrak{p}}(\omega)Z_{V_{n-a}+}(\mathscr{F}_{\mathfrak{d}}R_{\mathfrak{d}}\Phi_{\mathfrak{p}}, \omega_{\mathfrak{p}}^{-1}, 0),\\
&\epsilon_{\mathfrak{p}}\Xi_{\mathfrak{p}\#}(\Phi, \omega, w) \sim C_G\Lambda(w;\rho)c_{\mathfrak{p}}I_{\mathfrak{p}\#}(\Phi, \omega).
\end{aligned}
$$

Proof. Consider (1). We choose N_1, N_2 in (4.5.5) so that $N_1 \gg 0$ and N_2 is close to $\frac{(n-i)(n-i+1)}{2}$. Then $h_{\mathfrak{d}} - N_1 < 0$ and $h_{\mathfrak{d}} - N_2 > 0$. This is possible because $h_{\mathfrak{d}} > \frac{(n-i)(n-i+1)}{2}$ if $\mathfrak{p} \in \mathfrak{P}_1$.

This implies that the right hand side of (4.5.5)(1) is integrable on the Siegel domain. So we can apply (3.4.32), and

$$\epsilon_{\mathfrak{p}}\Xi_{\mathfrak{p}}(\Phi, \omega, w) \sim \epsilon_{\mathfrak{p}}C_G\Lambda(w;\rho)\int_{A_{\mathfrak{p}2}M_{\mathfrak{d}\mathbb{A}}^0/M_{\mathfrak{d}k}} \omega_{\mathfrak{p}}(g_{\mathfrak{d}})\sigma_{\mathfrak{p}}(\lambda_{\mathfrak{d}})\Theta_{Z_{\mathfrak{d}}}(R_{\mathfrak{d}}\Phi_{\mathfrak{p}}, \lambda_{\mathfrak{d}}g_{\mathfrak{d}}^0)dg_{\mathfrak{d}}.$$

An easy consideration shows that the right hand side is equal to the right hand side of (1).

The first two statements of (2) follow from (3.5.13).

Consider the third statement of (2). Since $h_{\mathfrak{d}} > 0$, $\prod_{i=1}^{a-1} e_{\mathfrak{p}i}(\lambda_{\mathfrak{d}})^{\chi_{\mathfrak{p}i}} e_{\mathfrak{p}a}(\lambda_{\mathfrak{d}})^{h_{\mathfrak{d}}}$ is integrable on $A_{\mathfrak{p}0}$ also. Therefore, we can apply (3.4.32) to $\Xi_{\mathfrak{p}\#}(\Phi, \omega, w)$. Thus, the proposition follows.

Q.E.D.

§4.6 Paths in $\mathfrak{P}_3, \mathfrak{P}_4$

Suppose that $\mathfrak{p} = (\mathfrak{d}, \mathfrak{s}) \in \mathfrak{P}_3$ or $\mathfrak{P}_4$, and $l(\mathfrak{p}) = a$.

Let $\mathfrak{d}' \prec \mathfrak{d}$, and $l(\mathfrak{d}') = a - 1$. For $\Psi \in \mathscr{S}(Z_{\mathfrak{d}'\mathbb{A}})$, we define

$$(4.6.1)\qquad \Theta_{Z'_{\mathfrak{d},0}}(R_{\mathfrak{d}}\Phi_{\mathfrak{p}}, g_{\mathfrak{d}}) = \sum_{x \in Z'^{ss}_{\mathfrak{d},0k}} \Psi(g_{\mathfrak{d}}x).$$

Let $\mathfrak{p}_{11} = (\mathfrak{d}_1, \mathfrak{s}_{11}), \mathfrak{p}_{12} = (\mathfrak{d}_1, \mathfrak{s}_{12})$ be paths such that $\mathfrak{p} \prec \mathfrak{p}_{11}, \mathfrak{p}_{12}, \mathfrak{d}_1 = \mathfrak{d} \cup \{n-1\}$, and $\mathfrak{s}_{11}(a+1) = 0$, $\mathfrak{s}_{12}(a+1) = 1$. Clearly, $\mathfrak{V}_\mathfrak{d} = \mathfrak{V}_{\mathfrak{d}_1} = \mathfrak{V}_{n-a-1}$.

By (3.5.9),

$$\Xi_\mathfrak{p}(\Phi, \omega, w) = \Xi_{\mathfrak{p}+}(\Phi, \omega, w) + \widehat{\Xi}_{\mathfrak{p}+}(\Phi, \omega, w) + \widehat{\Xi}_{\mathfrak{p}\#}(\Phi, \omega, w) - \Xi_{\mathfrak{p}\#}(\Phi, \omega, w) + \Xi_{\mathfrak{p}_{12}}(\Phi, \omega, w) - \Xi_{\mathfrak{p}_{11}}(\Phi, \omega, w).$$

Also

$$\Xi_{\mathfrak{p}+}(\Phi, \omega, w) = \Xi^s_{\mathfrak{p}+}(\Phi, \omega, w) + \Xi_{\mathfrak{p},\mathrm{st}+}(\Phi, \omega, w),$$
$$\widehat{\Xi}_{\mathfrak{p}+}(\Phi, \omega, w) = \widehat{\Xi}^s_{\mathfrak{p}+}(\Phi, \omega, w) + \widehat{\Xi}_{\mathfrak{p},\mathrm{st}+}(\Phi, \omega, w).$$

The following proposition follows from (3.5.13).

Proposition (4.6.2)

(1) $\epsilon_\mathfrak{p} \Xi^s_{\mathfrak{p}+}(\Phi, \omega, w) \sim C_G \Lambda(w; \rho) c_\mathfrak{p} \mathfrak{V}_\mathfrak{d} \delta_\mathfrak{p}(\omega) Z_{V_2+}(R_\mathfrak{d}\Phi_\mathfrak{p}, \omega_\mathfrak{p}, h_\mathfrak{d})$,

(2) $\epsilon_\mathfrak{p} \widehat{\Xi}^s_{\mathfrak{p}+}(\Phi, \omega, w) \sim C_G \Lambda(w; \rho) c_\mathfrak{p} \mathfrak{V}_\mathfrak{d} \delta_\mathfrak{p}(\omega) Z_{V_2+}(\mathscr{F}_\mathfrak{d} R_\mathfrak{d}\Phi_\mathfrak{p}, \omega_\mathfrak{p}^{-1}, 3 - h_\mathfrak{d})$.

We consider the contribution from $Z^{\mathrm{ss}}_{\mathfrak{d},\mathrm{st}k}$.

We define $\overline{M}_\mathfrak{d} = \prod_{l=1}^a M_{\mathfrak{d}l}$, and $\widetilde{X}_\mathfrak{d} = \mathbb{A}^1 \times \overline{M}^0_{\mathfrak{d}\mathbb{A}} \times X_{V_2}$. Let $X_+ = \{(u_0, \mu) \in \mathbb{A} \times \mathbb{R}_+ \mid \mu \geq \sqrt{\alpha(u_0)}\}$. Let $\widetilde{H}_\mathfrak{d} = \mathrm{GL}(1) \times \overline{M}_\mathfrak{d} \times H_{V_2}$ and $L_\mathfrak{d} = \mathrm{GL}(1) \times \overline{M}_\mathfrak{d} \times T_{\mathrm{GL}(2)}$, where $T_{\mathrm{GL}(2)}$ is the set of diagonal matrices in $\mathrm{GL}(2)$, and $\overline{M}_\mathfrak{d}, T_{\mathrm{GL}(2)}$ are imbedded in $\mathrm{GL}(n)$ diagonally in that order. Then $Z^{\mathrm{ss}}_{\mathfrak{d},\mathrm{st}k} = M_{\mathfrak{d}k} \times_{\widetilde{H}_{\mathfrak{d}k}} Z'^{\mathrm{ss}}_{\mathfrak{d},0k}$.

Any element $\widetilde{g}^0_\mathfrak{d}$ of $\widetilde{X}_\mathfrak{d}$ can be written in the form $\widetilde{g}^0_\mathfrak{d} = (t, \bar{g}^0_\mathfrak{d}, b^0_\mathfrak{d})$, where $\bar{g}^0_\mathfrak{d} \in \overline{M}^0_{\mathfrak{d}\mathbb{A}}$, $b^0_\mathfrak{d} = n_2(u_0) a_2(\underline{\mu} t_1, \underline{\mu}^{-1} t_2)$, $u_0 \in \mathbb{A}, \mu \in \mathbb{R}_+$, and $t, t_1, t_2 \in \mathbb{A}^1$. Let $db^0_\mathfrak{d} = du_0 d^\times \mu d^\times t_1 d^\times t_2$, and $d\widetilde{g}^0_\mathfrak{d} = d^\times t d\bar{g}^0_\mathfrak{d} db^0_\mathfrak{d}$. If $f(g^0_\mathfrak{d})$ is a measurable function on $M^0_{\mathfrak{d}\mathbb{A}}/\widetilde{H}_{\mathfrak{d}k}$, then

$$\int_{M^0_{\mathfrak{d}\mathbb{A}}/\widetilde{H}_{\mathfrak{d}k}} f(g^0_\mathfrak{d}) dg^0_\mathfrak{d} = \int_{\widetilde{X}_\mathfrak{d}/L_{\mathfrak{d}k}} f(\widetilde{g}^0_\mathfrak{d}) d\widetilde{g}^0_\mathfrak{d} \tag{4.6.3}$$

if the right hand side converges absolutely. Let $\widetilde{g}_\mathfrak{d} = \lambda_\mathfrak{d} \widetilde{g}^0_\mathfrak{d}$, and $d\widetilde{g}_\mathfrak{d} = d^\times \lambda_\mathfrak{d} d\widetilde{g}^0_\mathfrak{d}$. Let $M_\mathfrak{p}, U_\mathfrak{p}, U_{\mathfrak{p}\#}, \tau_\mathfrak{p}$ etc. be as in Chapter 3. Also let $\Theta_{Z_{\mathfrak{d},\mathrm{st}}}(R_\mathfrak{d}\Phi_\mathfrak{p}, g_\mathfrak{d})$ be as in (3.5.1).

By (4.6.3),

$$\begin{aligned}
&\int_{A_{\mathfrak{p}1} M_{\mathfrak{d}\mathbb{A}}/M_{\mathfrak{d}k}} \omega_\mathfrak{p}(g^0_\mathfrak{d}) \sigma_\mathfrak{p}(\lambda_\mathfrak{d}) \Theta_{Z_{\mathfrak{d},\mathrm{st}}}(R_\mathfrak{d}\Phi_\mathfrak{p}, g_\mathfrak{d}) \mathscr{E}_\mathfrak{p}(g_\mathfrak{d}, w) d^\times \lambda_\mathfrak{d} dg^0_\mathfrak{d} \\
&= \int_{A_{\mathfrak{p}1} M_{\mathfrak{d}\mathbb{A}}/\widetilde{H}_{\mathfrak{d}k}} \omega_\mathfrak{p}(g^0_\mathfrak{d}) \sigma_\mathfrak{p}(\lambda_\mathfrak{d}) \Theta_{Z'_{\mathfrak{d},0}}(R_\mathfrak{d}\Phi_\mathfrak{p}, g_\mathfrak{d}) \mathscr{E}_\mathfrak{p}(g_\mathfrak{d}, w) d^\times \lambda_\mathfrak{d} dg^0_\mathfrak{d} \\
&= \int_{A_{\mathfrak{p}1} \widetilde{X}_\mathfrak{d}/L_{\mathfrak{d}k}} \omega_\mathfrak{p}(\widetilde{g}^0_\mathfrak{d}) \sigma_\mathfrak{p}(\lambda_\mathfrak{d}) \Theta_{Z'_{\mathfrak{d},0}}(R_\mathfrak{d}\Phi_\mathfrak{p}, \widetilde{g}_\mathfrak{d}) \mathscr{E}_\mathfrak{p}(\widetilde{g}_\mathfrak{d}, w) d\widetilde{g}_\mathfrak{d},
\end{aligned}$$

$$\begin{aligned}
&\int_{A_{\mathfrak{p}0} M_{\mathfrak{d}\mathbb{A}}/M_{\mathfrak{d}k}} \omega_\mathfrak{p}(g^0_\mathfrak{d}) \sigma_\mathfrak{p}(\lambda_\mathfrak{d}) \kappa_{\mathfrak{d}1}(\lambda_\mathfrak{d}) \Theta_{Z_{\mathfrak{d},\mathrm{st}}}(\mathscr{F}_\mathfrak{d} R_\mathfrak{d}\Phi_\mathfrak{p}, \theta_\mathfrak{d}(g_\mathfrak{d})) \mathscr{E}_\mathfrak{p}(g_\mathfrak{d}, w) d^\times \lambda_\mathfrak{d} dg^0_\mathfrak{d} \\
&= \int_{A_{\mathfrak{p}0} M_{\mathfrak{d}\mathbb{A}}/\widetilde{H}_{\mathfrak{d}k}} \omega_\mathfrak{p}(g^0_\mathfrak{d}) \sigma_\mathfrak{p}(\lambda_\mathfrak{d}) \kappa_{\mathfrak{d}1}(\lambda_\mathfrak{d}) \Theta_{Z'_{\mathfrak{d},0}}(\mathscr{F}_\mathfrak{d} R_\mathfrak{d}\Phi_\mathfrak{p}, \theta_\mathfrak{d}(g_\mathfrak{d})) \mathscr{E}_\mathfrak{p}(g_\mathfrak{d}, w) d^\times \lambda_\mathfrak{d} dg^0_\mathfrak{d} \\
&= \int_{A_{\mathfrak{p}0} \widetilde{X}_\mathfrak{d}/L_{\mathfrak{d}k}} \omega_\mathfrak{p}(\widetilde{g}^0_\mathfrak{d}) \sigma_\mathfrak{p}(\lambda_\mathfrak{d}) \kappa_{\mathfrak{d}1}(\lambda_\mathfrak{d}) \Theta_{Z'_{\mathfrak{d},0}}(\mathscr{F}_\mathfrak{d} R_\mathfrak{d}\Phi_\mathfrak{p}, \theta_\mathfrak{d}(\widetilde{g}_\mathfrak{d})) \mathscr{E}_\mathfrak{p}(\widetilde{g}_\mathfrak{d}, w) d\widetilde{g}^0_\mathfrak{d}.
\end{aligned}$$

The group $M_{\mathfrak{p}}$ is either one of the following forms
Case 1 $\mathrm{GL}(1) \times \mathrm{GL}(1)^m \times \mathrm{GL}(n-a-1) \times \mathrm{GL}(2) \times \mathrm{GL}(1)^{a-m-1}$,
Case 2 $\mathrm{GL}(1) \times \mathrm{GL}(1)^m \times \mathrm{GL}(2) \times \mathrm{GL}(n-a-1) \times \mathrm{GL}(1)^{a-m-1}$.

Since $\widetilde{g}_{\mathfrak{d}} \in P_{\mathfrak{d}_1\mathbb{A}}$, we can consider $\widetilde{g}_{\mathfrak{d}}$ as an element of $G^0_{\mathbb{A}} \cap P_{\mathfrak{d}_1\mathbb{A}}$. Let

$$g'_{\mathfrak{d}} = \lambda_{\mathfrak{d}}(t, \bar{g}^0_{\mathfrak{d}}, a_2(\underline{\mu} t_1, \underline{\mu}^{-1} t_2^{-1})).$$

If we write $g'_{\mathfrak{d}} = \lambda_{\mathfrak{d}_1} g^0_{\mathfrak{d}_1}$ for $\lambda_{\mathfrak{d}_1} \in A_{\mathfrak{d}_1}, g^0_{\mathfrak{d}_1} \in M^0_{\mathfrak{d}_1\mathbb{A}}$, then $\lambda_{\mathfrak{d}_1} = \lambda_{\mathfrak{d}} d_{\mathfrak{d}_1 a+1}(\mu^{-1})$. Consider the function $\mathscr{E}_{\mathfrak{p}_{11}}(g'_{\mathfrak{d}}, w)$.

It is easy to see that

$$\theta_{\mathfrak{p}}(d_{\mathfrak{d}_1 a+1}(\mu^{-1})) = a_n(\overbrace{1, \cdots, 1}^{n+m-a-1}, \underline{\mu}, \underline{\mu}^{-1}, \overbrace{1, \cdots, 1}^{a-m-1})$$

for Case 1 and

$$\theta_{\mathfrak{p}}(d_{\mathfrak{d}_1 a+1}(\mu^{-1})) = a_n(\overbrace{1, \cdots, 1}^{m}, \underline{\mu}, \underline{\mu}^{-1}, \overbrace{1, \cdots, 1}^{n-m-2})$$

for Case 2.

Lemma (4.6.4)

(1)
$$\int_{A_{\mathfrak{p}1}\widetilde{X}_{\mathfrak{d}}/L_{\mathfrak{d}k}} \omega_{\mathfrak{p}}(\widetilde{g}^0_{\mathfrak{d}})\sigma_{\mathfrak{p}}(\lambda_{\mathfrak{d}})\Theta_{Z'_{\mathfrak{d},0}}(R_{\mathfrak{d}}\Phi_{\mathfrak{p}}, \widetilde{g}_{\mathfrak{d}})\mathscr{E}_{\mathfrak{p}}(\widetilde{g}_{\mathfrak{d}}, w)d\widetilde{g}_{\mathfrak{d}}$$
$$\sim \int_{A_{\mathfrak{p}1}\widetilde{X}_{\mathfrak{d}}/L_{\mathfrak{d}k}} \omega_{\mathfrak{p}}(\widetilde{g}^0_{\mathfrak{d}})\sigma_{\mathfrak{p}}(\lambda_{\mathfrak{d}})\Theta_{Z'_{\mathfrak{d},0}}(R_{\mathfrak{d}}\Phi_{\mathfrak{p}}, \widetilde{g}_{\mathfrak{d}})\mathscr{E}_{\mathfrak{p}_{11},\tau_{\mathfrak{p}_{11}}}(g'_{\mathfrak{d}}, w)d\widetilde{g}_{\mathfrak{d}},$$

(2)
$$\int_{A_{\mathfrak{p}0}\widetilde{X}_{\mathfrak{d}}/L_{\mathfrak{d}k}} \omega_{\mathfrak{p}}^{-1}(\widetilde{g}^0_{\mathfrak{d}})\sigma_{\mathfrak{p}}(\lambda_{\mathfrak{d}})\kappa_{\mathfrak{d}1}(\lambda_{\mathfrak{d}})\Theta_{Z'_{\mathfrak{d},0}}(\mathscr{F}_{\mathfrak{d}}R_{\mathfrak{d}}\Phi_{\mathfrak{p}}, \theta_{\mathfrak{d}}(\widetilde{g}_{\mathfrak{d}}))\mathscr{E}_{\mathfrak{p}}(\widetilde{g}_{\mathfrak{d}}, w)d\widetilde{g}_{\mathfrak{d}}$$
$$\sim \int_{A_{\mathfrak{p}0}\widetilde{X}_{\mathfrak{d}}/L_{\mathfrak{d}k}} \omega_{\mathfrak{p}}^{-1}(\widetilde{g}^0_{\mathfrak{d}})\sigma_{\mathfrak{p}}(\lambda_{\mathfrak{d}})\kappa_{\mathfrak{d}1}(\lambda_{\mathfrak{d}})\Theta_{Z'_{\mathfrak{d},0}}(\mathscr{F}_{\mathfrak{d}}R_{\mathfrak{d}}\Phi_{\mathfrak{p}}, \theta_{\mathfrak{d}}(\widetilde{g}_{\mathfrak{d}}))\mathscr{E}_{\mathfrak{p}_{11},\tau_{\mathfrak{p}_{11}}}(g'_{\mathfrak{p}}, w)d\widetilde{g}_{\mathfrak{d}}.$$

Proof. We first compare $\mathscr{E}_{\mathfrak{p}}(\widetilde{g}_{\mathfrak{d}}, w)$ with $\mathscr{E}_{\mathfrak{p}_{11},\tau_{\mathfrak{p}_{11}}}(g'_{\mathfrak{d}}, w)$. By definition,

$$\mathscr{E}_{\mathfrak{p}}(\widetilde{g}_{\mathfrak{d}}, w) - \mathscr{E}_{\mathfrak{p}_{11}}(g'_{\mathfrak{d}}, w) = \widetilde{\mathscr{E}}_{\mathfrak{p}_{11}}(\widetilde{g}_{\mathfrak{d}}, w).$$

Lemma (4.6.5) *For any $\epsilon > 0$ there exists $\delta = \delta(\epsilon) > 0$ and $c_1, \cdots, c_h \in \mathfrak{t}^* = \mathfrak{t}^{0*}$ such that $\|c_i\|_0 < \epsilon$ for all i and for any $l \gg 0, M > w_0$,*

$$|\widetilde{\mathscr{E}}_{\mathfrak{p}_{11}}(\widetilde{g}_{\mathfrak{d}}, w)| \ll \sum_i t(\bar{g}_{\mathfrak{d}})^{c_i}\mu^{-l}$$

for $w_0 - \delta \le \mathrm{Re}(w) \le M$, and $\bar{g}^0_{\mathfrak{d}}$ in some Siegel domain.

Proof. Let $I = \{n-m-a\}$ for Case 1, and $I = \{m+1\}$ for Case 2. Then

$$\widetilde{\mathscr{E}}_{\mathfrak{p}_{11}}(\widetilde{g}_{\mathfrak{d}}, w) = \sum_{I,\tau} \mathscr{E}_{I,\tau,1}(\theta_{\mathfrak{p}}(\widetilde{g}_{\mathfrak{d}}), w)$$

(see (3.4.27) etc. for the definition).

Since $\nu = 1$ in this case, for any $\delta > 0$ and $l \gg 0$, there exists a function $c_{l,\delta}(z)$ of polynomial growth such that

$$|E_{B,I,\tau,1}(\theta_{\mathfrak{p}}(\widetilde{g}_{\mathfrak{d}}), z)| \ll c_{l,\delta}(z) t(\theta_{\mathfrak{p}}(\widetilde{g}_{\mathfrak{d}}))^{\tau z + \rho} \mu^{-l}.$$

Since $I \neq \emptyset$, by the same argument as in the proof of (3.4.31), for any $\epsilon > 0$, we can choose $q \in D_{I,\tau}$ such that $L(q) < w_0$, and $\|\tau q + \rho\|_0 < \epsilon$. Therefore, there exist $\delta = \delta(\epsilon)$ and $c \in \mathfrak{t}^{0*} = \mathfrak{t}^*$ such that $\|c\|_0 < \epsilon$ and for any $l \gg 0, M > w_0$,

$$\widetilde{\mathscr{E}}_{I,\tau,1}(\theta_{\mathfrak{p}}(\widetilde{g}_{\mathfrak{d}}), w) \ll t(\theta_{\mathfrak{p}}(\widetilde{g}_{\mathfrak{d}}))^c \mu^{-l}$$

for $w_0 - \delta \leq \mathrm{Re}(w) \leq M$. Note that there is no condition on μ in this estimate (see the remark after (3.4.30)).

We can write $t(\theta_{\mathfrak{p}}(\widetilde{g}_{\mathfrak{d}}))^c = t(\widetilde{g}_{\mathfrak{d}}))^{c'}$ for some $c' \in \mathfrak{t}^*$, and if the norm of c is small, the norm of c' is small also.

Q.E.D.

Next, we consider the difference $\mathscr{E}_{\mathfrak{p}_{11}}(g'_{\mathfrak{d}}, w) - \mathscr{E}_{\mathfrak{p}_{11},\tau_{\mathfrak{p}_{11}}}(g'_{\mathfrak{d}}, w)$. By definition, $e_{\mathfrak{p}_{11}i}(g'_{\mathfrak{d}}) = e_{\mathfrak{p}i}(\lambda_{\mathfrak{d}})$ for $i = 1, \cdots, a$, and $e_{\mathfrak{p}_{11}a+1}(g'_{\mathfrak{d}}) = \mu^{-2} e_{\mathfrak{p}a}(\lambda_{\mathfrak{d}})$.

Therefore, (3.4.31)(2) implies that for any $\epsilon > 0$, there exist

$$\delta = \delta(\epsilon) > 0,\ r_1, \cdots, r_{a+1} > 0, \text{ and } c \in \mathfrak{t}^*$$

such that $r_i < \epsilon$ for all i, $\|c\|_0 < \epsilon$, and if $M > w_0$,

$$|\mathscr{E}_{\mathfrak{p}_{11}}(g'_{\mathfrak{d}}, w) - \mathscr{E}_{\mathfrak{p}_{11},\tau_{\mathfrak{p}_{11}}}(g'_{\mathfrak{d}}, w)| \ll \mu^{-r_{a+1}} t(\bar{g}^0_{\mathfrak{d}})^c \prod_{i=1}^{a} e_{\mathfrak{p}i}(\lambda_{\mathfrak{d}})^{r_i}$$

for $w_0 - \delta \leq \mathrm{Re}(w) \leq M$. Therefore, for any $\epsilon > 0$, there exists $\delta = \delta(\epsilon) > 0$ such that if $M > w_0$, $\mathscr{E}_{\mathfrak{p}}(\widetilde{g}_{\mathfrak{d}}, w) - \mathscr{E}_{\mathfrak{p}_{11},\tau_{\mathfrak{p}_{11}}}(g'_{\mathfrak{d}}, w)$ is bounded by a finite linear combination of functions of the form

$$\mu^{-r_{a+1}} t(\bar{g}^0_{\mathfrak{d}})^c \prod_{i=1}^{a} e_{\mathfrak{p}i}(\lambda_{\mathfrak{d}})^{r'_i}$$

for $w_0 - \delta \leq \mathrm{Re}(w) \leq M$, and $\bar{g}^0_{\mathfrak{d}}$ in some Siegel domain, where $c \in \mathfrak{t}^*$, $|r'_i|, \|c\|_0 < \epsilon$ for $i = 1, \cdots, a$, and $r_{a+1} > 0$.

We choose $0 \leq \Psi \in \mathscr{S}(\mathbb{A}^2)$ so that

$$|\Theta_{Z'_{\mathfrak{d},0}}(R_{\mathfrak{d}}\Phi_{\mathfrak{p}}, \widetilde{g}_{\mathfrak{d}})| \ll \sum_{x \in k^\times} \Psi(\underline{e_{\mathfrak{p}a}(\lambda_{\mathfrak{d}})x}, \underline{e_{\mathfrak{p}a}(\lambda_{\mathfrak{d}})x} u_0),$$

$$|\Theta_{Z'_{\mathfrak{d},0}}(\mathscr{F}_{\mathfrak{d}} R_{\mathfrak{d}}\Phi_{\mathfrak{p}}, \theta_{\mathfrak{d}}(\widetilde{g}_{\mathfrak{d}}))| \ll \sum_{x \in k^\times} \Psi(\underline{e_{\mathfrak{p}a}(\lambda_{\mathfrak{d}})^{-1}x}, \underline{e_{\mathfrak{p}a}(\lambda_{\mathfrak{d}})^{-1}x} u_0).$$

These functions do not depend on $\bar{g}^0_{\mathfrak{d}}$.

The function $t(\bar{g}^0_{\mathfrak{d}})^c$ is integrable on the Siegel domain; therefore, we can ignore this factor. By (4.5.1),

$$\sigma_{\mathfrak{p}}(\lambda_{\mathfrak{d}}) \mu^{-r_{a+1}} \prod_{i=1}^{a} e_{\mathfrak{p}i}(\lambda_{\mathfrak{d}})^{r_i} = \mu^{-r_{a+1}} \prod_{i=1}^{a} e_{\mathfrak{p}i}(\lambda_{\mathfrak{d}})^{\chi_{\mathfrak{p}i} + r_i}.$$

Since $r_{a+1} > 0$,

$$\int_{\mu \geq \sqrt{\alpha(u_0)}} \mu^{-r_{a+1}} d^\times \mu = \frac{1}{r_{a+1}} \alpha(u_0)^{-\frac{r_{a+1}}{2}}.$$

Since $\kappa_{\mathfrak{d}1}(\lambda_{\mathfrak{d}}) = e_{\mathfrak{p}a}(\lambda_{\mathfrak{d}})^{-3}$, multiplying $\kappa_{\mathfrak{d}1}(\lambda_{\mathfrak{d}})$ affects only $e_{\mathfrak{p}a}(\lambda_{\mathfrak{d}})$. Since r_i can be arbitrarily small, we can choose such r_i's that $\chi_{\mathfrak{p}i} + r_i > 0$ for $i = 1, \cdots, a-1$. Let $\mu_i = e_{\mathfrak{p}i}(\lambda_{\mathfrak{d}})$ for $i = 1, \cdots, a$. Then the integration with respect to $\mu_1, \cdots, \mu_{a-1}$ converges. So we only have to check that

$$\int_1^\infty \int_{\mathbb{A}} \mu_a^{h_{\mathfrak{d}}+r_a} \alpha(u_0)^{-\frac{r_{a+1}}{2}} \sum_{x \in k^\times} \Psi(\underline{\mu}_a x, \underline{\mu}_a x u_0) d^\times \mu_a du_0 < \infty,$$

$$\int_0^1 \int_{\mathbb{A}} \mu_a^{h_{\mathfrak{d}}-3+r_a} \alpha(u_0)^{-\frac{r_{a+1}}{2}} \sum_{x \in k^\times} \Psi(\underline{\mu}_a^{-1} x, \underline{\mu}_a^{-1} x u_0) d^\times \mu_a du_0 < \infty.$$

This follows from the fact that $T_{V_2+}(\Psi, \omega, s, s_1)$ converges absolutely and locally uniformly for all s, s_1. This proves (4.6.4).

Q.E.D.

Let $z = (z_1, \cdots, z_n) \in \mathbb{C}^n, z_1 + \cdots + z_n = 0$. Let $y_1 = z_{a-m}, y_2 = z_{a-m+1}$ for Case 1 and $y_1 = z_{n-m-1}, y_2 = z_{n-m}$ for Case 2. Let $y = y_1 - y_2$. We define $\tau_1 = (a-m, a-m+1)$ for Case 1 and $\tau_1 = (n-m-1, n-m)$ for Case 2.

In Case 1,

$$\begin{aligned}
&\tau_{\mathfrak{p}}(1) = n, \cdots, \tau_{\mathfrak{p}}(m) = n-m+1, \\
&\tau_{\mathfrak{p}}(m+1) = a-m+2, \cdots, \tau_{\mathfrak{p}}(n+m-a-1) = n-m, \\
&\tau_{\mathfrak{p}}(n+m-a) = a-m, \ \tau_{\mathfrak{p}}(n+m-a+1) = a-m+1, \\
&\tau_{\mathfrak{p}}(n+m-a+2) = a-m-1, \cdots, \tau_{\mathfrak{p}}(n) = 1.
\end{aligned}$$

In Case 2,

$$\begin{aligned}
&\tau_{\mathfrak{p}}(1) = n, \cdots, \tau_{\mathfrak{p}}(m) = n-m+1, \\
&\tau_{\mathfrak{p}}(m+1) = n-m-1, \ \tau_{\mathfrak{p}}(m+2) = n-m, \\
&\tau_{\mathfrak{p}}(m+3) = a-m, \cdots, \tau_{\mathfrak{p}}(n+m-a+1) = n-m-2, \\
&\tau_{\mathfrak{p}}(n+m-a+2) = a-m-1, \cdots, \tau_{\mathfrak{p}}(n) = 1.
\end{aligned}$$

It is easy to see that $\tau_{\mathfrak{p}_{11}} = \tau_1 \tau_{\mathfrak{p}}$. Let $\widetilde{U} = U_{\mathfrak{p}_{11}\#}, \widetilde{\tau} = \tau_{\mathfrak{p}_{11}}$. Then

$$M_{\widetilde{\tau}}(z) = M_{\tau_1}(\tau_{\mathfrak{p}} z) M_{\tau_{\mathfrak{p}}}(z) = \phi(y) M_{\tau_{\mathfrak{p}}}(z).$$

Clearly, $\theta_{\mathfrak{p}}(d_{\mathfrak{d}_1,a+1}(\mu^{-1}))^{\widetilde{\tau} z+\rho} = \mu^{1-y}$. Let $I_1 = \{(i,j) \mid n-m+1 \leq j \leq n, \ 1 \leq i < j\}$, $I_2 = \{(i,j) \mid 1 \leq i \leq a-m-1, \ a-m \leq j \leq n-m\}$, and $I_3 = \{(i,j) \mid 1 \leq i < j \leq a-m-1\}$. Then $M_{\widetilde{\tau}}(z)$ is equal to

$$\prod_{(i,j) \in I_1 \cup I_2 \cup I_3} \phi(z_i - z_j) \prod_{j=a-m+2}^{n-m} \phi(z_{a-m+1} - z_j) \prod_{j=a-m+1}^{n-m} \phi(z_{a-m} - z_j)$$

for Case 1, and

$$\prod_{(i,j)\in I_1\cup I_2\cup I_3} \phi(z_i - z_j) \prod_{i=a-m}^{n-m-2} \phi(z_i - z_{n-m-1}) \prod_{i=a-m}^{n-m-2} \phi(z_i - z_{n-m})$$

for Case 2.

Let

$$\theta_{\mathfrak{p}}(\lambda_{\mathfrak{d}})^{\widetilde{\tau}z+\rho} = \theta_{\mathfrak{p}}(\lambda_{\mathfrak{d}})^{\tau_{\mathfrak{p}}z+\rho} = e_{\mathfrak{p}1}(\lambda_{\mathfrak{d}})^{\gamma_{\mathfrak{p}1}(z)} \cdots e_{\mathfrak{p}a}(\lambda_{\mathfrak{d}})^{\gamma_{\mathfrak{p}a}(z)}.$$

Let $\widetilde{z}_i = z_i - z_{i+1}$ for $i = 1, \cdots, n-1$.

Lemma (4.6.6) *Suppose that* $\gamma_{\mathfrak{p}i}(z) = \sum_{j=1}^{n-1} c_{ij}\widetilde{z}_j + c_{i0}$, *where* c_{ij}, c_{i0} *are constants. Then*
(1) $\sum_j c_{ij} + c_{i0} = 0$,
(2) $c_{aa-m} = \frac{1}{2}$ *for* Case 1 *and* $c_{an-m-1} = \frac{1}{2}$ *for* Case 2.

Proof. In Case 1,

$$\theta_{\mathfrak{p}}(d_{\mathfrak{d}a}(\lambda_{\mathfrak{d}a})) = a_n(\overbrace{1, \cdots, 1}^{m}, \overbrace{\lambda_{\mathfrak{d}a}^{-2}, \cdots, \lambda_{\mathfrak{d}a}^{-2}}^{n-a-1}, \lambda_{\mathfrak{d}a}^{n-a-1}, \lambda_{\mathfrak{d}a}^{n-a-1}, \overbrace{1, \cdots, 1}^{a-m-1}).$$

Therefore,

$$\begin{aligned}\theta_{\mathfrak{p}}(d_{\mathfrak{d}a}(\lambda_{\mathfrak{d}a}))^{\tau_{\mathfrak{p}}z} &= \lambda_{\mathfrak{d}a}^{(n-a-1)(y_1+y_2)-2(z_{a-m+2}+\cdots+z_{n-m})} \\ &= \lambda_{\mathfrak{d}a}^{(n-a-1)y+2((y_2-z_{a-m+2})+\cdots+(y_2-z_{n-m}))}.\end{aligned}$$

In Case 2,

$$\theta_{\mathfrak{p}}(d_{\mathfrak{d}a}(\lambda_{\mathfrak{d}a})) = a_n(\overbrace{1, \cdots, 1}^{m}, \lambda_{\mathfrak{d}a}^{-(n-a-1)}, \lambda_{\mathfrak{d}a}^{-(n-a-1)}, \overbrace{\lambda_{\mathfrak{d}a}^{2}, \cdots, \lambda_{\mathfrak{d}a}^{2}}^{n-a-1}, \overbrace{1, \cdots, 1}^{a-m-1}).$$

Therefore,

$$\begin{aligned}\theta_{\mathfrak{p}}(d_{\mathfrak{d}a}(\lambda_{\mathfrak{d}a}))^{\tau_{\mathfrak{p}}z} &= \lambda_{\mathfrak{d}a}^{-(n-a-1)(y_1+y_2)+2(z_{a-m}+\cdots+z_{n-m-2})} \\ &= \lambda_{\mathfrak{d}a}^{(n-a-1)y+2((z_{a-m}-y_1)+\cdots+(z_{n-m-2}-y_1))}.\end{aligned}$$

So

$$\theta_{\mathfrak{p}}(\lambda_{\mathfrak{d}a})^{\tau_{\mathfrak{p}}z} = \prod_{i=1}^{a-1} \theta_{\mathfrak{p}}(d_{\mathfrak{d}i}(\lambda_{\mathfrak{d}i}))^{\tau_{\mathfrak{p}}z}\theta_{\mathfrak{p}}(d_{\mathfrak{d}a}(\lambda_{\mathfrak{d}a}))^{\tau_{\mathfrak{p}}z},$$

and the first factor is a product of powers of $e_{\mathfrak{p}i}(\lambda_{\mathfrak{d}})$'s for $i = 1, \cdots, a-1$. Thus, (2) follows from (4.5.2).

If $\widetilde{z}_1, \cdots, \widetilde{z}_{n-1} = 1$, then $\theta_{\mathfrak{p}}(\lambda_{\mathfrak{d}a})^{\tau_{\mathfrak{p}}z+\rho} = \theta_{\mathfrak{p}}(\lambda_{\mathfrak{d}a})^{\tau_G z+\rho} = 1$. Thus, (1) follows. Q.E.D.

We define

$$Q_{\mathfrak{p}} = \{z \mid z_i - z_{i+1} = 1 \text{ for } i = a-m+2, \cdots, n-m-1\}$$

for Case 1, and

$$Q_{\mathfrak{p}} = \{z \mid z_i - z_{i+1} = 1 \text{ for } i = a-m, \cdots, n-m-3\}$$

for Case 2.

Let $C_{\mathfrak{p}}$ be a contour of the form $\{z \in Q_{\mathfrak{p}} \mid \mathrm{Re}(z) = q\}$. Let

$$dz|_{C_{\mathfrak{p}}} = \begin{cases} \prod_{i=1}^{a-m+1} d\widetilde{z}_i \prod_{n-m}^{n-1} d\widetilde{z}_i & \text{Case 1,} \\ \prod_{i=1}^{a-m-1} d\widetilde{z}_i \prod_{n-m-2}^{n-1} d\widetilde{z}_i & \text{Case 2.} \end{cases}$$

By the Mellin inversion formula,

$$\begin{aligned} &\int_{\overline{M}^0_{\mathfrak{d}\mathbb{A}}/\overline{M}_{\mathfrak{d}k}} \omega_2(\det \bar{g}^0_{\mathfrak{d}})^{\pm 1} \mathscr{E}_{\widetilde{U},\widetilde{\tau}}(\theta_{\mathfrak{p}}(\lambda_{\mathfrak{d}}(\bar{g}^0_{\mathfrak{d}}, b^0_{\mathfrak{d}})), w) d\bar{g}^0_{\mathfrak{d}} \\ &= \delta_{\#}(\omega_2) \left(\frac{1}{2\pi\sqrt{-1}}\right)^{a+1} \int_{C_{\mathfrak{p}}} M_{\widetilde{\tau}}(z) \mu^{1-y} \theta_{\mathfrak{p}}(\lambda_{\mathfrak{d}})^{\widetilde{\tau}z+\rho} \Lambda(w; z) dz|_{C_{\mathfrak{p}}}. \end{aligned}$$

Let $F_{\mathfrak{p}1}(\Phi, t, \lambda_{\mathfrak{d}}, b^0_{\mathfrak{d}}, z), F_{\mathfrak{p}2}(\Phi, t, \lambda_{\mathfrak{d}}, b^0_{\mathfrak{d}}, z)$ be the following two functions

$$\begin{aligned} &\sigma_{\mathfrak{p}}(\lambda_{\mathfrak{d}}) \mu^{1-y} \theta_{\mathfrak{p}}(\lambda_{\mathfrak{d}})^{\widetilde{\tau}z+\rho} \Theta_{Z'_{\mathfrak{d},0}}(R_{\mathfrak{d}}\Phi_{\mathfrak{p}}, (\underline{\lambda}_0 t, 1, b^0_{\mathfrak{d}})), \\ &\sigma_{\mathfrak{p}}(\lambda_{\mathfrak{d}}) \kappa_{\mathfrak{d}1}(\lambda_{\mathfrak{d}}) \mu^{1-y} \theta_{\mathfrak{p}}(\lambda_{\mathfrak{d}})^{\widetilde{\tau}z+\rho} \Theta_{Z'_{\mathfrak{d},0}}(\mathscr{F}_{\mathfrak{d}} R_{\mathfrak{d}}\Phi_{\mathfrak{p}}, \theta_{\mathfrak{d}}(\underline{\lambda}_0 t, 1, b^0_{\mathfrak{d}})). \end{aligned}$$

Also let

$$\begin{aligned} \Sigma_{\mathfrak{p},\mathrm{st}+}(\Phi, z) &= \frac{T_{V_2+}(\widetilde{R}_{\mathfrak{d},0} R_{\mathfrak{d}}\Phi_{\mathfrak{p}}, \gamma_{\mathfrak{p}a}(z) + h_{\mathfrak{d}}, \frac{1-y}{2})}{\mathfrak{V}_{\mathfrak{d}2}(y-1) \prod_{i=1}^{a-1}(\gamma_{\mathfrak{p}i}(z) + \chi_{\mathfrak{p}i})}, \\ \widehat{\Sigma}_{\mathfrak{p},\mathrm{st}+}(\Phi, z) &= \frac{T_{V_2+}(\widetilde{R}_{\mathfrak{d},0} \mathscr{F}_{\mathfrak{d}} R_{\mathfrak{d}}\Phi_{\mathfrak{p}}, 3 - \gamma_{\mathfrak{p}a}(z) - h_{\mathfrak{d}}, \frac{1-y}{2})}{\mathfrak{V}_{\mathfrak{d}2}(y-1) \prod_{i=1}^{a-1}(\gamma_{\mathfrak{p}i}(z) + \chi_{\mathfrak{p}i})}. \end{aligned}$$

Let $\mu_i = e_{\mathfrak{p}i}(\lambda_{\mathfrak{d}})$ for $i = 1, \cdots, a$. Then $d^\times \lambda_{\mathfrak{d}} = \mathfrak{V}_{\mathfrak{d}2}^{-1} d^\times \mu_1 \cdots d^\times \mu_a$. Therefore,

$$\begin{aligned} &\int_{\mathbb{A}^1/k^\times \times A_{\mathfrak{p}1} \times X_+ \times (\mathbb{A}^1/k^\times)^2} \omega_1(tt_1t_2)^{\pm 1} F_{\mathfrak{p}1}(\Phi, t, \lambda_{\mathfrak{d}}, b^0_{\mathfrak{d}}, z) d^\times t d^\times \lambda_{\mathfrak{d}} db^0_{\mathfrak{d}} \\ &= \delta_{\#}(\omega_1) \Sigma_{\mathfrak{p},\mathrm{st}+}(\Phi, z), \\ &\int_{\mathbb{A}^1/k^\times \times A_{\mathfrak{p}0} \times X_+ \times (\mathbb{A}^1/k^\times)^2} \omega_1(tt_1t_2)^{\pm 1} F_{\mathfrak{p}2}(\Phi, t, \lambda_{\mathfrak{d}}, b^0_{\mathfrak{d}}, z) d^\times t d^\times \lambda_{\mathfrak{d}} db^0_{\mathfrak{d}} \\ &= \delta_{\#}(\omega_1) \widehat{\Sigma}_{\mathfrak{p},\mathrm{st}+}(\Phi, z). \end{aligned}$$

Hence,

$$\begin{aligned} &\int_{A_{\mathfrak{p}1} \widetilde{X}_{\mathfrak{d}}/L_{\mathfrak{d}k}} \omega_{\mathfrak{p}}(\widetilde{g}^0_{\mathfrak{d}}) \sigma_{\mathfrak{p}}(\lambda_{\mathfrak{d}}) \Theta_{Z'_{\mathfrak{d},0}}(R_{\mathfrak{d}}\Phi_{\mathfrak{p}}, \widetilde{g}_{\mathfrak{d}}) \mathscr{E}_{\widetilde{U},\widetilde{\tau}}(\widetilde{g}_{\mathfrak{p}}, w) d\widetilde{g}_{\mathfrak{d}} \\ &= \delta_{\#}(\omega) \left(\frac{1}{2\pi\sqrt{-1}}\right)^{a+1} \int_{C_{\mathfrak{p}}} \Sigma_{\mathfrak{p},\mathrm{st}+}(\Phi, z) M_{\widetilde{\tau}}(z) \Lambda(w; z) dz|_{C_{\mathfrak{p}}}, \\ &\int_{A_{\mathfrak{p}0} \widetilde{X}_{\mathfrak{d}}/L_{\mathfrak{d}k}} \omega_{\mathfrak{p}}^{-1}(\widetilde{g}^0_{\mathfrak{d}}) \sigma_{\mathfrak{p}}(\lambda_{\mathfrak{d}}) \kappa_{\mathfrak{d}1}(\lambda_{\mathfrak{d}}) \Theta_{Z'_{\mathfrak{d},0}}(\mathscr{F}_{\mathfrak{d}} R_{\mathfrak{d}}\Phi_{\mathfrak{p}}, \theta_{\mathfrak{d}}(\widetilde{g}^0_{\mathfrak{d}})) \mathscr{E}_{\widetilde{U},\widetilde{\tau}}(\widetilde{g}_{\mathfrak{p}}, w) d\widetilde{g}_{\mathfrak{d}} \\ &= \delta_{\#}(\omega) \left(\frac{1}{2\pi\sqrt{-1}}\right)^{a+1} \int_{C_{\mathfrak{p}}} \widehat{\Sigma}_{\mathfrak{p},\mathrm{st}+}(\Phi, z) M_{\widetilde{\tau}}(z) \Lambda(w; z) dz|_{C_{\mathfrak{p}}}. \end{aligned}$$

Let

$$h_{\mathfrak{p}}(z) = \prod_{(i,j)\in I_1\cup I_2\cup I_3} \phi(z_i - z_j).$$

In Case 1, we define a function $f_{\mathfrak{p}0}(y, \widetilde{z}_{a-m+1})$ by

$$f_{\mathfrak{p}0}(y, \widetilde{z}_{a-m+1}) = \operatorname*{Res}_{\widetilde{z}_{n-1}=1} \cdots \operatorname*{Res}_{\widetilde{z}_{n-m}=1} \operatorname*{Res}_{\widetilde{z}_{a-m-1}=1} \cdots \operatorname*{Res}_{\widetilde{z}_1=1} h_{\mathfrak{p}}(z)|_{Q_{\mathfrak{p}}}.$$

In Case 2, we define a function $f_{\mathfrak{p}0}(y, \widetilde{z}_{n-m-2})$ by the same formula. Let $f_{\mathfrak{p}1}(y) = f_{\mathfrak{p}0}(y, 1)$ in both cases. Let

$$f_{\mathfrak{p}2}(y) = \frac{\mathfrak{R}_k}{Z_k(n-a)} \frac{Z_k(y)}{Z_k(y+n-a)}.$$

Then

$$\operatorname*{Res}_{\widetilde{z}_{n-1}=1} \cdots \operatorname*{Res}_{\widetilde{z}_{n-m}=1} \operatorname*{Res}_{\widetilde{z}_{a-m+1}=1} \operatorname*{Res}_{\widetilde{z}_{a-m-1}=1} \cdots \operatorname*{Res}_{\widetilde{z}_1=1} M_{\mp}(z)|_{Q_{\mathfrak{p}}} = f_{\mathfrak{p}1}(y) f_{\mathfrak{p}2}(y)$$

for Case 1, and

$$\operatorname*{Res}_{\widetilde{z}_{n-1}=1} \cdots \operatorname*{Res}_{\widetilde{z}_{n-m}=1} \operatorname*{Res}_{\widetilde{z}_{n-m-2}=1} \operatorname*{Res}_{\widetilde{z}_{a-m-1}=1} \cdots \operatorname*{Res}_{\widetilde{z}_1=1} M_{\mp}(z)|_{Q_{\mathfrak{p}}} = f_{\mathfrak{p}1}(y) f_{\mathfrak{p}2}(y)$$

for Case 2. Also $\operatorname{Res}_{y=1} f_{\mathfrak{p}1}(y) f_{\mathfrak{p}2}(y) = \mathfrak{V}_{\mathfrak{d}} C_G = \mathfrak{V}_{n-a-1}\mathfrak{V}_n^{-1}$.

Let $Q'_{\mathfrak{p}} = \{z \in Q_{\mathfrak{p}} \mid \widetilde{z}_i = 1 \text{ for } i = 1, \cdots, a-m-1, a-m+1, n-m, \cdots, n-1\}$ for Case 1, and $Q'_{\mathfrak{p}} = \{z \in Q_{\mathfrak{p}} \mid \widetilde{z}_i = 1 \text{ for } i = 1, \cdots, a-m-1, n-m-2, n-m, \cdots, n-1\}$ for Case 2. Any function on $Q'_{\mathfrak{p}}$ can be considered as a function of y.

Let

$$\gamma'_{\mathfrak{p}i}(y) = \gamma_{\mathfrak{p}i}(z)|_{Q'_{\mathfrak{p}}},\ \Sigma'_{\mathfrak{p},\mathrm{st}+}(\Phi, y) = \Sigma_{\mathfrak{p},\mathrm{st}+}(\Phi, z)|_{Q'_{\mathfrak{p}}},\ \widehat{\Sigma}'_{\mathfrak{p},\mathrm{st}+}(\Phi, y) = \widehat{\Sigma}_{\mathfrak{p},\mathrm{st}+}(\Phi, z)|_{Q'_{\mathfrak{p}}}.$$

Because of (4.6.6), $\gamma'_{\mathfrak{p}a}(y) = \frac{y-1}{2}$.

Lemma (4.6.7)

(1)
$$\left(\frac{1}{2\pi\sqrt{-1}}\right)^{a+1} \int_{C_{\mathfrak{p}}} \Sigma_{\mathfrak{p},\mathrm{st}+}(\Phi, z) M_{\mp}(z) \Lambda(w; z) dz|_{C_{\mathfrak{p}}}$$
$$\sim \frac{1}{2\pi\sqrt{-1}} \int_{\mathrm{Re}(y)=1+\delta} \Sigma'_{\mathfrak{p},\mathrm{st}+}(\Phi, y) f_{\mathfrak{p}1}(y) f_{\mathfrak{p}2}(y) \left[\Lambda(w; z)\right]|_{Q'_{\mathfrak{p}}} dy,$$

(2)
$$\left(\frac{1}{2\pi\sqrt{-1}}\right)^{a+1} \int_{C_{\mathfrak{p}}} \widehat{\Sigma}_{\mathfrak{p},\mathrm{st}+}(\Phi, z) M_{\mp}(z) \Lambda(w; z) dz|_{C_{\mathfrak{p}}}$$
$$\sim \frac{1}{2\pi\sqrt{-1}} \int_{\mathrm{Re}(y)=1+\delta} \widehat{\Sigma}'_{\mathfrak{p},\mathrm{st}+}(\Phi, y) f_{\mathfrak{p}1}(y) f_{\mathfrak{p}2}(y) \left[\Lambda(w; z)\right]|_{Q'_{\mathfrak{p}}} dy,$$

where $\delta > 0$ is a small number.

Proof. Let $Q_{\mathfrak{p}1} = \{z \in Q_{\mathfrak{p}} \mid \widetilde{z}_1 = 1\}$. Let $C_{\mathfrak{p}1}$ be the contour of the form $\{z \in Q_{\mathfrak{p}1} \mid \mathrm{Re}(z) = q\}$. Let $C'_{\mathfrak{p}1}$ be the contour of the form $\{z \in Q_{\mathfrak{p}} \mid \mathrm{Re}(z) = q\}$ where $q_1 - q_2 = 1 - \delta_1$, $q_i - q_{i+1} = 1 + \delta_2$ for $i \notin \{1, a-m+2, \cdots, n-m-1\}$ or

$i \notin \{1, a-m, \cdots, n-m-3\}$, and $\delta_1\delta_2^{-1} \gg 0$. Let $dz|_{C_{\mathfrak{p}1}}$ be the differential form on $C_{\mathfrak{p}1}$ such that $dz|_{C_{\mathfrak{p}}} = d\widetilde{z}_1 dz|_{C_{\mathfrak{p}1}}$. Then

$$\left(\frac{1}{2\pi\sqrt{-1}}\right)^{a+1} \int_{C_{\mathfrak{p}}} \Sigma_{\mathfrak{p},\mathrm{st}+}(\Phi, z) M_{\widetilde{\tau}}(z) \Lambda(w; z) dz|_{C_{\mathfrak{p}}}$$
$$= \left(\frac{1}{2\pi\sqrt{-1}}\right)^{a} \int_{C_{\mathfrak{p}1}} \operatorname*{Res}_{\widetilde{z}_1=1} \left[\Sigma_{\mathfrak{p},\mathrm{st}+}(\Phi, z) M_{\widetilde{\tau}}(z) \Lambda(w; z)\right] dz|_{C_{\mathfrak{p}1}}$$
$$+ \left(\frac{1}{2\pi\sqrt{-1}}\right)^{a+1} \int_{C'_{\mathfrak{p}1}} \Sigma_{\mathfrak{p},\mathrm{st}+}(\Phi, z) M_{\widetilde{\tau}}(z) \Lambda(w; z) dz|_{C_{\mathfrak{p}}}.$$

The second term is holomorphic for $\mathrm{Re}(w) \geq w_0 - \delta$ for some $\delta > 0$. If we continue this process, we get the lemma.

Q.E.D.

These considerations show the following proposition.

Proposition (4.6.8) *Suppose* $\mathfrak{p} \in \mathfrak{P}_3 \cup \mathfrak{P}_4$. *Then*

(1)
$$\epsilon_{\mathfrak{p}} \Xi_{\mathfrak{p}+}(\Phi, \omega, w) \sim C_G \Lambda(w; \rho) c_{\mathfrak{p}} \mathfrak{V}_{\mathfrak{d}} \delta_{\mathfrak{p}}(\omega) Z_{V_2+}(R_{\mathfrak{d}} \Phi_{\mathfrak{p}}, \omega_{\mathfrak{p}}, h_{\mathfrak{d}}) + \frac{\epsilon_{\mathfrak{p}} \mathfrak{F}_{\mathfrak{d}1} \delta_{\#}(\omega)}{2\pi\sqrt{-1}} \int_{\mathrm{Re}(y)=1+\delta} \Sigma'_{\mathfrak{p},\mathrm{st}+}(\Phi, y) f_{\mathfrak{p}1}(y) f_{\mathfrak{p}2}(y) \left[\Lambda(w; z)\right]|_{Q'_{\mathfrak{p}}} dy,$$

(2)
$$\epsilon_{\mathfrak{p}} \widehat{\Xi}_{\mathfrak{p}+}(\Phi, \omega, w) \sim C_G \Lambda(w; \rho) c_{\mathfrak{p}} \mathfrak{V}_{\mathfrak{d}} \delta_{\mathfrak{p}}(\omega) Z_{V_2+}(\mathscr{F}_{\mathfrak{d}} R_{\mathfrak{d}} \Phi_{\mathfrak{p}}, \omega_{\mathfrak{p}}^{-1}, 3 - h_{\mathfrak{d}}) + \frac{\epsilon_{\mathfrak{p}} \mathfrak{F}_{\mathfrak{d}1} \delta_{\#}(\omega)}{2\pi\sqrt{-1}} \int_{\mathrm{Re}(y)=1+\delta} \widehat{\Sigma}'_{\mathfrak{p},\mathrm{st}+}(\Phi, y) f_{\mathfrak{p}1}(y) f_{\mathfrak{p}2}(y) [\Lambda(w; z)]|_{Q'_{\mathfrak{p}}} dy,$$

where $\delta > 0$ *is a small number.*

Next, we consider $\Xi_{\mathfrak{p}_{11}}(\Phi, \omega, w)$, $\Xi_{\mathfrak{p}_{12}}(\Phi, \omega, w)$.

Lemma (4.6.9) *Let* $\widetilde{\tau}$ *be as before* ($\widetilde{\tau} = \tau_{\mathfrak{p}_{12}}$ *also*). *Then*

$$\Xi_{\mathfrak{p}_{1i}}(\Phi, \omega, w) \sim \mathfrak{F}_{\mathfrak{d}1} \int_{A_{\mathfrak{p}_{1i}0} \times M^0_{\mathfrak{d}_1 \mathbb{A}}/M^0_{\mathfrak{d}_1 k}} \omega_{\mathfrak{p}_{1i}}(g^0_{\mathfrak{d}_1}) \sigma_{\mathfrak{d}_1}(\lambda_{\mathfrak{d}_1}) \Theta_{Z_{\mathfrak{d}_1}}(R_{\mathfrak{d}_1} \Phi_{\mathfrak{p}_{1i}}, g_{\mathfrak{d}_1}) \times \mathscr{E}_{\mathfrak{p}_{1i},\widetilde{\tau}}(g_{\mathfrak{d}_1}, w) d^\times \lambda_{\mathfrak{d}_1} dg_{\mathfrak{d}_1}$$

for $i = 1, 2$.

Proof. Suppose $\tau \neq \widetilde{\tau}$. By (3.4.31)(2), for any $\epsilon > 0$, there exist

$$\delta = \delta(\epsilon) > 0,\ r_1, \cdots, r_{a+1} > 0, \text{ and } c \in \mathfrak{t}^*$$

such that $r_i, \|c\|_0 < \epsilon$ for all i and if $M > w_0$,

$$|\mathscr{E}_{\mathfrak{p}_i,\tau}(g_{\mathfrak{d}_1}, w)| \ll \prod_{j=1}^{a+1} e_{\mathfrak{p}_{1i}j}(\lambda_{\mathfrak{d}_1})^{r_j} t(g^0_{\mathfrak{d}_1})^c$$

for $w_0 - \delta \leq \mathrm{Re}(w) \leq M$, and $g^0_{\mathfrak{d}_1}$ in the Siegel domain.

By definition, $h_{\mathfrak{d}_1} = 1$. Since

$$\int_{\mathbb{A}^\times} |t|^{1+r_{a+1}} \Theta_1(R_{\mathfrak{d}_1}\Phi_{\mathfrak{p}_{1i}}, t) d^\times t$$

converges absolutely for $i = 1, 2$, the lemma follows.

Q.E.D.

By (4.5.2),

$$\lambda_{\mathfrak{d}_1 a+1} = e_{\mathfrak{p}_{11}a+1}(\lambda_{\mathfrak{d}_1})^{\frac{1}{2}} e_{\mathfrak{p}_{11}a}(\lambda_{\mathfrak{d}_1})^{-\frac{1}{2}} = e_{\mathfrak{p}_{12}a+1}(\lambda_{\mathfrak{d}_1})^{\frac{1}{2}} e_{\mathfrak{p}_{12}a}(\lambda_{\mathfrak{d}_1})^{\frac{1}{2}}.$$

Therefore,

$$\begin{aligned}
\theta_{\mathfrak{p}_{11}}(\lambda_{\mathfrak{d}_1})^{\widetilde{\tau}z+\rho} &= \theta_{\mathfrak{p}}(\lambda_{\mathfrak{d}})^{\widetilde{\tau}z+\rho}\theta_{\mathfrak{p}}(d_{\mathfrak{d}_1 a+1}(\lambda_{\mathfrak{d}_1 a+1}))^{\widetilde{\tau}z+\rho} \\
&= \prod_{i=1}^{a} e_{\mathfrak{p}i}(\lambda_{\mathfrak{d}})^{\gamma_{\mathfrak{p}i}(z)} \lambda_{\mathfrak{d}_1 a+1}^{y-1} \\
&= \prod_{i=1}^{a} e_{\mathfrak{p}_{11}i}(\lambda_{\mathfrak{d}})^{\gamma_{\mathfrak{p}i}(z)} \lambda_{\mathfrak{d}_1 a+1}^{y-1} \\
&= e_{\mathfrak{p}_{11}a}(\lambda_{\mathfrak{d}_1})^{\gamma_{\mathfrak{p}a}(z)-\frac{y-1}{2}} e_{\mathfrak{p}_{11}a+1}(\lambda_{\mathfrak{d}_1})^{\frac{y-1}{2}} \prod_{i=1}^{a-1} e_{\mathfrak{p}_{11}i}(\lambda_{\mathfrak{d}_1})^{\gamma_{\mathfrak{p}i}(z)}.
\end{aligned}$$

Also

$$\begin{aligned}
\theta_{\mathfrak{p}_{12}}(\lambda_{\mathfrak{d}_1})^{\widetilde{\tau}z+\rho} &= \theta_{\mathfrak{p}}\theta_{\mathfrak{d}}(\lambda_{\mathfrak{d}})^{\widetilde{\tau}z+\rho}\theta_{\mathfrak{p}}\theta_{\mathfrak{d}}(d_{\mathfrak{d}_1 a+1}(\lambda_{\mathfrak{d}_1 a+1}))^{\widetilde{\tau}z+\rho} \\
&= \theta_{\mathfrak{p}}(\lambda_{\mathfrak{d}}^{-1})^{\widetilde{\tau}z+\rho}\theta_{\mathfrak{p}}(d_{\mathfrak{d}_1 a+1}(\lambda_{\mathfrak{d}_1 a+1}))^{\widetilde{\tau}z+\rho} \\
&= \prod_{i=1}^{a} e_{\mathfrak{p}i}(\lambda_{\mathfrak{d}})^{-\gamma_{\mathfrak{p}i}(z)} \lambda_{\mathfrak{d}_1 a+1}^{y-1} \\
&= e_{\mathfrak{p}_{12}a}(\lambda_{\mathfrak{d}_1})^{\gamma_{\mathfrak{p}a}(z)+\frac{y-1}{2}} e_{\mathfrak{p}_{12}a+1}(\lambda_{\mathfrak{d}_1})^{\frac{y-1}{2}} \prod_{i=1}^{a-1} e_{\mathfrak{p}_{12}i}(\lambda_{\mathfrak{d}_1})^{\gamma_{\mathfrak{p}i}(z)}.
\end{aligned}$$

By definition, $\chi_{\mathfrak{p}_{11}a} = h_{\mathfrak{d}} - 1$, $\chi_{\mathfrak{p}_{11}a+1} = 1$, $\chi_{\mathfrak{p}_{12}a} = h_{\mathfrak{d}} - 2$, $\chi_{\mathfrak{p}_{12}a+1} = 1$. We define

$$\begin{aligned}
\Sigma_{\mathfrak{p}_{11}}(\Phi, \omega, z) &= \frac{\delta_{\mathfrak{p}}(\omega)\Sigma_1(R_{\mathfrak{d}_1}\Phi_{\mathfrak{p}_{11}}, \omega_{\mathfrak{p}}, \frac{y+1}{2})}{2\mathfrak{F}_{\mathfrak{d}2}(h_{\mathfrak{d}} - 1 + \gamma_{\mathfrak{p}a}(z) - \frac{y-1}{2})\prod_{i=1}^{a-1}(\gamma_{\mathfrak{p}i}(z) + \chi_{\mathfrak{p}i})}, \\
\Sigma_{\mathfrak{p}_{12}}(\Phi, \omega, z) &= \frac{\delta_{\mathfrak{p}}(\omega)\Sigma_1(R_{\mathfrak{d}_1}\Phi_{\mathfrak{p}_{12}}, \omega_{\mathfrak{p}}^{-1}, \frac{y+1}{2})}{2\mathfrak{F}_{\mathfrak{d}2}(h_{\mathfrak{d}} - 2 + \gamma_{\mathfrak{p}a}(z) + \frac{y-1}{2})\prod_{i=1}^{a-1}(\gamma_{\mathfrak{p}i}(z) + \chi_{\mathfrak{p}i})},
\end{aligned}$$

where we consider $\omega_{\mathfrak{p}}, \omega_{\mathfrak{p}}^{-1}$ as characters of $\mathbb{A}^\times/k^\times$ by restricting to elements of the form $(t, 1)$. Let $\mu_j = e_{\mathfrak{p}_{1i}j}(\lambda_{\mathfrak{d}_1})$ for $i = 1, 2, j = 1, \cdots, a+1$. Then

$$d^\times \lambda_{\mathfrak{d}} = \mathfrak{F}_{\mathfrak{d}_1 2}^{-1} d^\times \mu_1 \cdots d^\times \mu_{a+1} = 2^{-1}\mathfrak{F}_{\mathfrak{d}2}^{-1} d^\times \mu_1 \cdots d^\times \mu_{a+1}.$$

Also $\mathfrak{F}_{\mathfrak{d}1} = \mathfrak{F}_{\mathfrak{d}_1 1}$. Therefore, we get the following proposition.

Proposition (4.6.10)

$$\Xi_{\mathfrak{p}_{1i}}(\Phi,\omega,w) \sim \mathfrak{F}_{\mathfrak{d}1}\left(\frac{1}{2\pi\sqrt{-1}}\right)^{a+1}\int_{C_\mathfrak{p}} \Sigma_{\mathfrak{p}_{1i}}(\Phi,\omega,z)M_{\mp}(z)\Lambda(w;z)dz|_{C_\mathfrak{p}}$$

for $i=1,2$.

We define $\Sigma'_{\mathfrak{p}_{1i}}(\Phi,\omega,y) = \Sigma_{\mathfrak{p}_{1i}}(\Phi,\omega,z)|_{Q'_\mathfrak{p}}$ for $i=1,2$. We can get the following lemma as in (4.6.7) and the proof is left to the reader.

Lemma (4.6.11)

$$\Xi_{\mathfrak{p}_{1i}}(\Phi,\omega,w) \sim \frac{\mathfrak{F}_{\mathfrak{d}1}}{2\pi\sqrt{-1}}\int_{\mathrm{Re}(y)=1+\delta} \Sigma'_{\mathfrak{p}_{1i}}(\Phi,\omega,y)f_{\mathfrak{p}1}(y)f_{\mathfrak{p}2}(y)\left[\Lambda(w;z)\right]|_{Q'_\mathfrak{p}}dy$$

for $i=1,2$, *where* $\delta>0$ *is a small number.*

The proof of the following proposition is similar to that of (4.5.6)(2), and is left to the reader.

Proposition (4.6.12)
(1) *Suppose* $\mathfrak{p}\in\mathfrak{P}_3\cup\mathfrak{P}_4$. *Then*

$$\epsilon_\mathfrak{p}\Xi_{\mathfrak{p}\#}(\Phi,\omega,w) \sim C_G\Lambda(w;\rho)c_\mathfrak{p}\mathfrak{V}_\mathfrak{d}\delta_\#(\omega)\frac{\Phi_\mathfrak{p}(0)}{h_\mathfrak{d}}.$$

(2) *Suppose* $\mathfrak{p}\in\mathfrak{P}_3$. *Then*

$$\epsilon_\mathfrak{p}\widehat{\Xi}_{\mathfrak{p}\#}(\Phi,\omega,w) \sim C_G\Lambda(w;\rho)c_\mathfrak{p}\mathfrak{V}_\mathfrak{d}\delta_\#(\omega)\frac{\mathscr{F}_\mathfrak{d}\Phi_\mathfrak{p}(0)}{h_\mathfrak{d}-3}.$$

Now we are going to establish a cancellation of distributions. This cancellation basically comes from the case $n=2$. In this process, we reconstruct the order two pole of the adjusted zeta function for the case $n=2$.

Let

$$\begin{aligned}
\Sigma^{(1)}_\mathfrak{p}(\Phi,y) &= T_{V_2+}(\widetilde{R}_{\mathfrak{d},0}R_\mathfrak{d}\Phi_\mathfrak{p},\frac{y-1}{2}+h_\mathfrak{d},\frac{1-y}{2}) \\
&\quad + T_{V_2+}(\widetilde{R}_{\mathfrak{d},0}\mathscr{F}_\mathfrak{d}R_\mathfrak{d}\Phi_\mathfrak{p},3-\frac{y-1}{2}-h_\mathfrak{d},\frac{1-y}{2}), \\
\Sigma^{(2)}_\mathfrak{p}(\Phi,\omega,y) &= \frac{\delta_\mathfrak{p}(\omega)\Sigma_1(R_{\mathfrak{d}_1}\Phi_{\mathfrak{p}_{12}},\omega_\mathfrak{p}^{-1},\frac{y+1}{2})}{2(h_\mathfrak{d}-2+y-1)} - \frac{\delta_\mathfrak{p}(\omega)\Sigma_1(R_{\mathfrak{d}_1}\Phi_{\mathfrak{p}_{11}},\omega_\mathfrak{p},\frac{y+1}{2})}{2(h_\mathfrak{d}-1)}, \\
\Sigma^{(3)}_\mathfrak{p}(\Phi,\omega,y) &= \frac{\delta_\#(\omega)\Sigma^{(1)}_\mathfrak{p}(\Phi,y)}{y-1} - \frac{\delta_\#(\omega)\Sigma_1(R'_{\mathfrak{d},0}R_\mathfrak{d}\Phi_\mathfrak{p},\frac{y-1}{2}+h_\mathfrak{d}-1)}{y-1} \\
&\quad + \Sigma^{(2)}_\mathfrak{p}(\Phi,\omega,y).
\end{aligned}$$

We consider the Laurent expansion of $\Sigma^{(3)}_\mathfrak{p}(\Phi,\omega,y)$ at $y=1$.

Since

$$T_{V_2+}(\widetilde{R}_{\mathfrak{d},0}R_\mathfrak{d}\Phi_\mathfrak{p},s,0) = \Sigma_{1+}(R'_{\mathfrak{d},0}R_\mathfrak{d}\Phi_\mathfrak{p},s-1) \text{ etc.},$$

$$T_{V_2+}(\widetilde{R}_{\mathfrak{d},0}R_{\mathfrak{d}}\Phi_{\mathfrak{p}}, \frac{y-1}{2}+h_{\mathfrak{d}}, 0) + T_{V_2+}(\widetilde{R}_{\mathfrak{d},0}\mathscr{F}_{\mathfrak{d}}R_{\mathfrak{d}}\Phi_{\mathfrak{p}}, 3-\frac{y-1}{2}-h_{\mathfrak{d}}, 0)$$
$$= \Sigma_1(R'_{\mathfrak{d},0}R_{\mathfrak{d}}\Phi_{\mathfrak{p}}, \frac{y-1}{2}+h_{\mathfrak{d}}-1) + \frac{R'_{\mathfrak{d},0}R_{\mathfrak{d}}\Phi_{\mathfrak{p}}(0)}{\frac{y-1}{2}+h_{\mathfrak{d}}-1} + \frac{R'_{\mathfrak{d},0}R_{\mathfrak{d}}\mathscr{F}_{\mathfrak{d}}\Phi_{\mathfrak{p}}(0)}{2-\frac{y-1}{2}-h_{\mathfrak{d}}}.$$

Clearly,

$$\frac{1}{\frac{y-1}{2}+h_{\mathfrak{d}}-1} = \frac{1}{h_{\mathfrak{d}}-1} - \frac{1}{2(h_{\mathfrak{d}}-1)^2}(y-1) + O((y-1)^2),$$
$$\frac{1}{2-\frac{y-1}{2}-h_{\mathfrak{d}}} = \frac{1}{2-h_{\mathfrak{d}}} + \frac{1}{2(2-h_{\mathfrak{d}})^2}(y-1) + O((y-1)^2),$$

and

$$\frac{\delta_{\mathfrak{p}}(\omega)\Sigma_1(R_{\mathfrak{d}_1}\Phi_{\mathfrak{p}_{11}}, \omega_{\mathfrak{p}}, \frac{y+1}{2})}{2(h_{\mathfrak{d}}-1)} = \frac{\delta_{\#}(\omega)\Sigma_{1,(-1)}(R_{\mathfrak{d}_1}\Phi_{\mathfrak{p}_{11}}, 1)}{(h_{\mathfrak{d}}-1)(y-1)} + \frac{\delta_{\mathfrak{p}}(\omega)\Sigma_{1,(0)}(R_{\mathfrak{d}_1}\Phi_{\mathfrak{p}_{11}}, \omega_{\mathfrak{p}}, 1)}{2(h_{\mathfrak{d}}-1)} + O(y-1),$$
$$\frac{\delta_{\mathfrak{p}}(\omega)\Sigma_1(R_{\mathfrak{d}_1}\Phi_{\mathfrak{p}_{12}}, \omega_{\mathfrak{p}}^{-1}, \frac{y+1}{2})}{2(h_{\mathfrak{d}}-2+y-1)} = \frac{\delta_{\#}(\omega)\Sigma_{1,(-1)}(R_{\mathfrak{d}_1}\Phi_{\mathfrak{p}_{12}}, 1)}{(h_{\mathfrak{d}}-2)(y-1)} - \frac{\delta_{\#}(\omega)\Sigma_{1,(-1)}(R_{\mathfrak{d}_1}\Phi_{\mathfrak{p}_{12}}, 1)}{(h_{\mathfrak{d}}-2)^2} + \frac{\delta_{\mathfrak{p}}(\omega)\Sigma_{1,(0)}(R_{\mathfrak{d}_1}\Phi_{\mathfrak{p}_{12}}, \omega_{\mathfrak{p}}^{-1}, 1)}{2(h_{\mathfrak{d}}-2)} + O(y-1).$$

If $\delta_{\#}(\omega)=1$, Φ is K-invariant. This implies that

$$R'_{\mathfrak{d},0}R_{\mathfrak{d}}\Phi_{\mathfrak{p}}(0) = \Sigma_{1,(-1)}(R_{\mathfrak{d}_1}\Phi_{\mathfrak{p}_{11}}, 1),$$
$$R'_{\mathfrak{d},0}\mathscr{F}_{\mathfrak{d}}R_{\mathfrak{d}}\Phi_{\mathfrak{p}}(0) = \Sigma_{1,(-1)}(R_{\mathfrak{d}_1}\Phi_{\mathfrak{p}_{12}}, 1).$$

Therefore, $\Sigma_{\mathfrak{p}}^{(3)}(\Phi,\omega,y)$ is holomorphic at $y=1$ and

$$\Sigma_{\mathfrak{p}}^{(3)}(\Phi,\omega,1) = -\frac{\delta_{\#}(\omega)}{2}(T_{V_2+}(\widetilde{R}_{\mathfrak{d},0}R_{\mathfrak{d}}\Phi_{\mathfrak{p}}, h_{\mathfrak{d}}) + T_{V_2+}(\widetilde{R}_{\mathfrak{d},0}\mathscr{F}_{\mathfrak{d}}R_{\mathfrak{d}}\Phi_{\mathfrak{p}}, 3-h_{\mathfrak{d}})) - \frac{\delta_{\#}(\omega)\Sigma_{1,(-1)}(R_{\mathfrak{d}_1}\Phi_{\mathfrak{p}_{12}}, 1)}{2(h_{\mathfrak{d}}-2)^2} + \frac{\delta_{\mathfrak{p}}(\omega)\Sigma_{1,(0)}(R_{\mathfrak{d}_1}\Phi_{\mathfrak{p}_{12}}, \omega_{\mathfrak{p}}^{-1}, 1)}{2(h_{\mathfrak{d}}-2)} - \frac{\delta_{\#}(\omega)\Sigma_{1,(-1)}(R_{\mathfrak{d}_1}\Phi_{\mathfrak{p}_{11}}, 1)}{2(h_{\mathfrak{d}}-1)^2} - \frac{\delta_{\mathfrak{p}}(\omega)\Sigma_{1,(0)}(R_{\mathfrak{d}_1}\Phi_{\mathfrak{p}_{11}}, \omega_{\mathfrak{p}}, 1)}{2(h_{\mathfrak{d}}-1)}. \tag{4.6.13}$$

Let

$$\Sigma_{\mathfrak{p}}^{(4)}(\Phi,\omega,w) = \delta_{\#}(\omega)\Sigma'_{\mathfrak{p},\mathrm{st}+}(\Phi,w) + \delta_{\#}(\omega)\widehat{\Sigma}'_{\mathfrak{p},\mathrm{st}+}(\Phi,w) + \Sigma'_{\mathfrak{p}_{12}}(\Phi,\omega,y) - \Sigma'_{\mathfrak{p}_{11}}(\Phi,\omega,y). \tag{4.6.14}$$

Also let $\Xi'_{\mathfrak{p},\mathrm{st}}(\Phi,\omega,w)$ be the following integral

(4.6.15)

$$\frac{\mathfrak{F}_{\mathfrak{d}3}\delta_{\#}(\omega)}{2\pi\sqrt{-1}}\int_{\mathrm{Re}(y)=1+\delta}\frac{\Sigma_1(R'_{\mathfrak{d},0}R_{\mathfrak{d}}\Phi_{\mathfrak{p}},\frac{y-1}{2}+h_{\mathfrak{d}}-1)}{(y-1)\prod_{i=1}^{a-1}(\gamma'_{\mathfrak{p}i}(y)+\chi_{\mathfrak{p}i})}f_{\mathfrak{p}1}(y)f_{\mathfrak{p}2}(y)[\Lambda(w;z)]|_{Q_{\mathfrak{p}'}}dy.$$

Then

(4.6.16)

$$\begin{aligned}
&\frac{\mathfrak{F}_{\mathfrak{d}}}{2\pi\sqrt{-1}}\int_{\mathrm{Re}(y)=1+\delta}\Sigma^{(4)}_{\mathfrak{p}}(\Phi,\omega,w)f_{\mathfrak{p}1}(y)f_{\mathfrak{p}2}(y)[\Lambda(w;z)]|_{Q_{\mathfrak{p}'}}dy\\
&\sim\frac{\mathfrak{F}_{\mathfrak{d}3}}{2\pi\sqrt{-1}}\int_{\mathrm{Re}(y)=1+\delta}\frac{\Sigma^{(3)}_{\mathfrak{p}}(\Phi,\omega,y)}{\prod_{i=1}^{a-1}(\gamma'_{\mathfrak{p}i}(y)+\chi_{\mathfrak{p}i})}f_{\mathfrak{p}1}(y)f_{\mathfrak{p}2}(y)[\Lambda(w;z)]|_{Q_{\mathfrak{p}'}}dy\\
&\quad+\Xi'_{\mathfrak{p},\mathrm{st}}(\Phi,\omega,w).
\end{aligned}$$

Since $\Sigma^{(3)}_{\mathfrak{p}}(\Phi,\omega,y)$ is holomorphic at $y=1$, and $\chi_{\mathfrak{p}i}>0$ for $i=1,\cdots,a-1$,

(4.6.17)

$$\begin{aligned}
&\frac{\epsilon_{\mathfrak{p}}\mathfrak{F}_{\mathfrak{d}3}}{2\pi\sqrt{-1}}\int_{\mathrm{Re}(y)=1+\delta}\frac{\Sigma^{(3)}_{\mathfrak{p}}(\Phi,\omega,y)}{\prod_{i=1}^{a-1}(\gamma'_{\mathfrak{p}i}(y)+\chi_{\mathfrak{p}i})}f_{\mathfrak{p}1}(y)f_{\mathfrak{p}2}(y)[\Lambda(w;z)]|_{Q_{\mathfrak{p}'}}dy\\
&\sim c_{\mathfrak{p}}\mathfrak{V}_{\mathfrak{d}}\Sigma^{(3)}_{\mathfrak{p}}(\Phi,\omega,1)\Lambda(w;z)\\
&\quad+\frac{\epsilon_{\mathfrak{p}}\mathfrak{F}_{\mathfrak{d}3}}{2\pi\sqrt{-1}}\int_{\mathrm{Re}(y)=1-\delta}\frac{\Sigma^{(3)}_{\mathfrak{p}}(\Phi,\omega,y)}{\prod_{i=1}^{a-1}(\gamma'_{\mathfrak{p}i}(y)+\chi_{\mathfrak{p}i})}f_{\mathfrak{p}1}(y)f_{\mathfrak{p}2}(y)[\Lambda(w;z)]|_{Q_{\mathfrak{p}'}}dy\\
&\sim c_{\mathfrak{p}}\mathfrak{V}_{\mathfrak{d}}\Sigma^{(3)}_{\mathfrak{p}}(\Phi,\omega,1)\Lambda(w;z).
\end{aligned}$$

Note that $\epsilon_{\mathfrak{p}}\mathfrak{F}_{\mathfrak{d}3}\prod_{i=1}^{a-1}\chi_{\mathfrak{p}i}^{-1}=c_{\mathfrak{p}}$.

By (4.2.15), (4.6.8), (4.6.12), (4.6.17), we get the following proposition.

Proposition (4.6.18)
(1) *If* $\mathfrak{p}\in\mathfrak{P}_3$,

$$\epsilon_{\mathfrak{p}}\Xi_{\mathfrak{p}}(\Phi,\omega,w)\sim C_G\Lambda(w;\rho)I_{\mathfrak{p}}(\Phi,\omega)+\epsilon_{\mathfrak{p}}\Xi'_{\mathfrak{p},\mathrm{st}}(\Phi,\omega,w).$$

(2) *If* $\mathfrak{p}\in\mathfrak{P}_4$,

$$\epsilon_{\mathfrak{p}}\Xi_{\mathfrak{p}}(\Phi,\omega,w)\sim C_G\Lambda(w;\rho)I_{\mathfrak{p}}(\Phi,\omega)+\epsilon_{\mathfrak{p}}\Xi'_{\mathfrak{p},\mathrm{st}}(\Phi,\omega,w)+\epsilon_{\mathfrak{p}}\widehat{\Xi}_{\mathfrak{p}\#}(\Phi,\omega,w).$$

§4.7 The cancellations

In the last section, we established the cancellation of distributions based on the case $n=2$. In this section, we will establish another kind of cancellation of distributions. This cancellation is established between $\Xi'_{\mathfrak{p}_1,\mathrm{st}}(\Phi,\omega,w)$ and $\widehat{\Xi}_{\mathfrak{p}_2\#}(\Phi,\omega,w)$ for $\mathfrak{p}_1\in\mathfrak{P}_3,\mathfrak{p}_2\in\mathfrak{P}_2$ or $\mathfrak{p}_1,\mathfrak{p}_2\in\mathfrak{P}_4$.

Let $\mathfrak{p}_1=(\mathfrak{d}_1,\mathfrak{s}_1)$, $\mathfrak{p}_2=(\mathfrak{d}_2,\mathfrak{s}_2)$ be paths such that $\mathfrak{d}_1=\{1,\cdots,a-1,n-2\}$, $\mathfrak{d}_2=\{1,\cdots,a\}$, $\mathfrak{s}_1(i)=\mathfrak{s}_2(i)$ for $i=1,\cdots,a-1$, and $\mathfrak{s}_1(a)+\mathfrak{s}_2(a)=1$.

The correspondence between $\mathfrak{p}_1$ and $\mathfrak{p}_2$ is one-to-one. Therefore, if $\mathfrak{p}_1$ is of Case 1 (resp. Case 2) in §4.6, then we say $\mathfrak{p}_2$ is of Case 1 (resp. Case 2). If $\mathfrak{p}_2$ is of Case 1, $M_{\mathfrak{p}_2} \cong \mathrm{GL}(1) \times \mathrm{GL}(1)^m \times \mathrm{GL}(n-a) \times \mathrm{GL}(1)^{a-m}$. If $\mathfrak{p}_2$ is of Case 2, $M_{\mathfrak{p}_2} \cong \mathrm{GL}(1) \times \mathrm{GL}(1)^{m+1} \times \mathrm{GL}(n-a) \times \mathrm{GL}(1)^{a-m-1}$.

Proposition (4.7.1) *If $a > 1$, let $\mathfrak{p}_3$ be the unique path such that $\mathfrak{p}_3 \prec \mathfrak{p}_1, \mathfrak{p}_2$ and $l(\mathfrak{p}_3) = a-1$.*
(1) *If $a = 1$,*

$$\begin{aligned}&\epsilon_{\mathfrak{p}_1}\Xi'_{\mathfrak{p}_1,\mathrm{st}}(\Phi,\omega,w) + \epsilon_{\mathfrak{p}_2}\widehat{\Xi}_{\mathfrak{p}_2\#}(\Phi,\omega,w)\\ &\sim C_G\Lambda(w;\rho)\frac{\epsilon_{\mathfrak{p}_1}\mathfrak{V}_{n-2}\delta_{\#}(\omega)}{2}\Sigma_{1,(1)}(R'_{\mathfrak{d}_1,0}R_{\mathfrak{d}_1}\Phi_{\mathfrak{p}_1}, n-1).\end{aligned}$$

(2) *If $a > 1$ and $\mathfrak{s}_1(a) = 1$,*

$$\epsilon_{\mathfrak{p}_1}\Xi'_{\mathfrak{p}_1,\mathrm{st}}(\Phi,\omega,w) + \epsilon_{\mathfrak{p}_2}\widehat{\Xi}_{\mathfrak{p}_2\#}(\Phi,\omega,w) \sim C_G\Lambda(w;\rho)I_{\mathfrak{p}_3 1}(\Phi,\omega).$$

(3) *If $a > 1$ and $\mathfrak{s}_1(a) = 0$,*

$$\epsilon_{\mathfrak{p}_1}\Xi'_{\mathfrak{p}_1,\mathrm{st}}(\Phi,\omega,w) + \epsilon_{\mathfrak{p}_2}\widehat{\Xi}_{\mathfrak{p}_2\#}(\Phi,\omega,w) \sim C_G\Lambda(w;\rho)I_{\mathfrak{p}_3 2}(\Phi,\omega).$$

We devote the rest of this section to the proof of this proposition. This will complete the proof of (4.4.6).

Let I_1, I_2, I_3 be the subsets of $\{1,\cdots,n-1\}$ which we defined for $\mathfrak{p}_1$ in §4.6. By definition,

$$M_{\tau_{\mathfrak{p}_2}}(z) = \prod_{(i,j)\in I_1\cup I_2\cup I_3}\phi(z_i - z_j)\prod_{j=a-m+1}^{n-m}\phi(z_{a-m} - z_j)$$

for Case 1, and

$$M_{\tau_{\mathfrak{p}_2}}(z) = \prod_{(i,j)\in I_1\cup I_2\cup I_3}\phi(z_i - z_j)\prod_{i=a-m}^{n-m-2}\phi(z_i - z_{n-m})$$

for Case 2. Also by definition, $\chi_{\mathfrak{p}_2 a} = h_{\mathfrak{d}_2} = f_{\mathfrak{d}_2}$, and $\kappa_{\mathfrak{d}_2 1}(\lambda_{\mathfrak{d}_2}) = e_{\mathfrak{p}_2 a}(\lambda_{\mathfrak{d}_2})^{-f_{\mathfrak{d}_2}}$. Therefore, $\sigma_{\mathfrak{p}_2}(\lambda_{\mathfrak{d}_2})\kappa_{\mathfrak{d}_2 1}(\lambda_{\mathfrak{d}_2}) = \prod_{i=1}^{a-1} e_{\mathfrak{p}_2 i}(\lambda_{\mathfrak{d}_2})^{\chi_{\mathfrak{p}_2 i}}$.

Lemma (4.7.2)

$$\begin{aligned}&\widehat{\Xi}_{\mathfrak{p}_2\#}(\Phi,\omega,w)\\ &\sim \mathfrak{F}_{\mathfrak{d}_2 1}\int_{A_{\mathfrak{p}_2 0}M^0_{\mathfrak{d}_2\mathbb{A}}/M_{\mathfrak{d}_2 k}}\omega_{\mathfrak{p}_2}(g^0)_{\mathfrak{d}_2}\sigma_{\mathfrak{p}_2}(\lambda_{\mathfrak{d}_2})\kappa_{\mathfrak{d}_2 1}(\lambda_{\mathfrak{d}_2})\mathscr{E}_{\mathfrak{p}_2,\tau_{\mathfrak{p}_2}}(\theta_{\mathfrak{p}_2}(g_{\mathfrak{d}_2}),w)d^\times\lambda_{\mathfrak{d}_2}dg^0_{\mathfrak{d}_2}.\end{aligned}$$

Proof. Suppose $\tau \neq \tau_{\mathfrak{p}_2}$. By (3.4.31)(2), for any $\epsilon > 0$, there exist $\delta = \delta(\epsilon) > 0$ $r_1,\cdots,r_{a+1} > 0$, $c \in \mathfrak{t}^*$ such that $r_i, \|c\|_0 < \epsilon$ for all i and if $M > w_0$,

$$|\mathscr{E}_{\mathfrak{p}_2,\bar{\tau}}(g_{\mathfrak{d}_2},w)| \ll \prod_{j=1}^{a} e_{\mathfrak{p}_2 j}(\lambda_{\mathfrak{d}_2})^{r_j}t(g^0_{\mathfrak{d}_2})^c,$$

for $w-\delta \le \mathrm{Re}(w) \le M$ and $g^0_{\mathfrak{d}_2}$ in the Siegel domain.

Since $\kappa_{\mathfrak{d}_2}(\lambda_{\mathfrak{d}_2}) = e_{\mathfrak{p}_2 a}(\lambda_{\mathfrak{d}_2})^{-h_\flat}$, $\sigma_{\mathfrak{p}_2}(\lambda_{\mathfrak{d}_2})\kappa_{\mathfrak{d}_2}(\lambda_{\mathfrak{d}_2}) = \prod_{i=1}^{a-1} e_{\mathfrak{p}_2 i}(\lambda_{\mathfrak{d}_2})^{\chi_{\mathfrak{p}_2 i}}$. Since $r_i + \chi_{\mathfrak{p}_2} > 0$ for $i = 1, \cdots, a-1$, and $r_a > 0$, the lemma follows.

Q.E.D.

Let

$$Q_{\mathfrak{p}_2} = \{z \mid z_i - z_{i+1} = 1 \text{ for } i = a-m+1, \cdots, n-m-1\}$$

for Case 1, and

$$Q_{\mathfrak{p}_2} = \{z \mid z_i - z_{i+1} = 1 \text{ for } i = a-m, \cdots, n-m-2\}$$

for Case 2.

Let $C_{\mathfrak{p}_2}$ be a contour of the form $\{z \in Q_{\mathfrak{p}_2} \mid \mathrm{Re}(z) = q\}$. We use the same definition of y_1, y_2, y as in §4.6. We define $Q'_{\mathfrak{p}_2}$ similarly as in §4.6.

Lemma (4.7.3) *Let*

$$\theta_{\mathfrak{p}_2}(\lambda_{\mathfrak{d}_2})^{\tau_{\mathfrak{p}_2} z+\rho} = \prod_{i=1}^{a} e_{\mathfrak{p}_2 i}(\lambda_{\mathfrak{d}_2})^{\gamma_{\mathfrak{p}_2 i}(z)}.$$

Then $\gamma_{\mathfrak{d}_2 a}(z) = \frac{(n-a)(y-1)}{2}$ *on* $Q_{\mathfrak{p}_2}$.

Proof. In Case 1,

$$\theta_{\mathfrak{p}_2}(d_{\mathfrak{d}_2 a}(\lambda_{\mathfrak{d}_2 a}))^{\tau_{\mathfrak{p}_2} z+\rho} = \lambda_{\mathfrak{d}_2 a}^{(n-a)z_{a-m} - (z_{a-m+1}+\cdots+z_{n-m}) - \frac{(n-a+1)(n-a)}{2}}.$$

On $Q_{\mathfrak{p}_2}$, $\widetilde{z}_i = 1$ for $i = a-m+1, \cdots, n-m-1$. Therefore,

$$\begin{aligned}
&(n-a)z_{a-m} - (z_{a-m+1}+\cdots+z_{n-m}) - \frac{(n-a+1)(n-a)}{2} \\
&= (n-a)y + \sum_{i=1}^{n-a-1} i = (n-a)y + \frac{(n-a-1)(n-a)}{2} - \frac{(n-a+1)(n-a)}{2} \\
&= (n-a)(y-1).
\end{aligned}$$

Since $\lambda_{\mathfrak{d}_2 a}^2 = e_{\mathfrak{p}_2 a}(\lambda_{\mathfrak{d}_2})e_{\mathfrak{p}_2 a-1}(\lambda_{\mathfrak{d}_2})^{\pm 1}$, we get the lemma for Case 1.

Case 2 is similar.

Q.E.D.

Let

$$f_{\mathfrak{p}_2 1}(y) = f_{\mathfrak{p}_1 1}(y),\ f_{\mathfrak{p}_2 2}(y) = \frac{Z_k(y)}{Z_k(y+n-a)}.$$

Note that

$$f_{\mathfrak{p}_2 2}(y) = \frac{Z_k(n-a)}{\mathfrak{R}_k} f_{\mathfrak{p}_1 2}(y).$$

Let

$$\widehat{\Sigma}_{\mathfrak{p}_2 \#}(\Phi, \omega, z) = \frac{2\delta_\#(\omega)\mathscr{F}_{\mathfrak{d}_2}\Phi_{\mathfrak{p}_2}(0)}{(n-a)(y-1)\prod_{i=1}^{a-1}(\gamma_{\mathfrak{p}_2 i}(z) + \chi_{\mathfrak{p}_2 i})}.$$

We define $\widehat{\Sigma}'_{\mathfrak{p}_2\#}(\Phi,\omega,y)=\widehat{\Sigma}_{\mathfrak{p}_2\#}(\Phi,\omega,z)|_{Q'_{\mathfrak{p}_2}}$. By (4.7.3) and a similar computation to that in §4.6,

$$\widehat{\Xi}_{\mathfrak{p}_2\#}(\Phi,\omega,w)\sim\mathfrak{F}_{\mathfrak{d}_2 3}\left(\frac{1}{2\pi\sqrt{-1}}\right)^a\int_{C_{\mathfrak{p}_2}}\widehat{\Sigma}_{\mathfrak{p}_2\#}(\Phi,\omega,z)M_{\tau_{\mathfrak{p}_2}}(z)[\Lambda(w;z)dz]|_{C_\mathfrak{p}}$$
$$\sim\frac{\mathfrak{F}_{\mathfrak{d}_2 3}}{2\pi\sqrt{-1}}\int_{\mathrm{Re}(y)=1+\delta}\widehat{\Sigma}'_{\mathfrak{p}_2\#}(\Phi,\omega,y)f_{\mathfrak{p}_2 1}(y)f_{\mathfrak{p}_2 2}(y)[\Lambda(w;z)]|_{Q'_{\mathfrak{p}_2}}dy.$$

Clearly, $\epsilon_{\mathfrak{p}_1}=-\epsilon_{\mathfrak{p}_2}$. If $a=1$, we define

$$\Sigma_{\mathrm{canc}}(\Phi,y)=\frac{\Sigma_1(R'_{\mathfrak{d}_1,0}R_{\mathfrak{d}_1}\Phi_{\mathfrak{p}_1},\frac{y-1}{2}+n-a)}{y-1}-\frac{Z_k(n-a)}{\mathfrak{R}_k}\frac{\mathscr{F}_{\mathfrak{d}_2}R_{\mathfrak{d}_2}\Phi_{\mathfrak{p}_2}(0)}{y-1}.$$

If $a>1$, we define

$$\Sigma_{\mathrm{canc}}(\Phi,y)=\frac{\Sigma_1(R'_{\mathfrak{d}_1,0}R_{\mathfrak{d}_1}\Phi_{\mathfrak{p}_1},\frac{y-1}{2}+n-a)}{(y-1)(\gamma'_{\mathfrak{p}_1 a-1}(y)+\chi_{\mathfrak{p}_2 a-1})}-\frac{Z_k(n-a)}{\mathfrak{R}_k}\frac{\mathscr{F}_{\mathfrak{d}_2}R_{\mathfrak{d}_2}\Phi_{\mathfrak{p}_2}(0)}{(y-1)(\gamma'_{\mathfrak{p}_2 a-1}(y)+\chi_{\mathfrak{p}_2 a-1})}.$$

Lemma (4.7.4) *We have the equality* $\gamma'_{\mathfrak{p}_1 i}(y)=\gamma'_{\mathfrak{p}_2 i}(y)$ *for* $i=1,\cdots,a-2$, *and*
(1) $\gamma'_{\mathfrak{p}_1 a-1}(y)=\gamma'_{\mathfrak{p}_2 a-1}(y)-\frac{(n-a+1)(y-1)}{2}$ *if* $\mathfrak{s}_1(a)=0$,
(2) $\gamma'_{\mathfrak{p}_1 a-1}(y)=\gamma'_{\mathfrak{p}_2 a-1}(y)+\frac{(n-a+1)(y-1)}{2}$ *if* $\mathfrak{s}_1(a)=1$.

Proof. If $\mathfrak{s}_1(a)=0$,

$$\lambda_{\mathfrak{d}_1 a}=e_{\mathfrak{p}_1 a}(\lambda_{\mathfrak{d}_1})^{\frac12}e_{\mathfrak{p}_1 a-1}(\lambda_{\mathfrak{d}_1})^{-\frac12},$$
$$\lambda_{\mathfrak{d}_2 a}=e_{\mathfrak{p}_2 a}(\lambda_{\mathfrak{d}_2})^{\frac12}e_{\mathfrak{p}_2 a-1}(\lambda_{\mathfrak{d}_2})^{\frac12}.$$

The lemma follows by the proofs of (4.6.5) and (4.7.3). The argument is similar for the case $\mathfrak{s}_1(a)=1$.

Q.E.D.

Since $\frac{2}{n-a}\mathfrak{F}_{\mathfrak{d}_2 3}=\mathfrak{F}_{\mathfrak{d}_1 3}$,

$$\epsilon_{\mathfrak{p}_1}\Xi'_{\mathfrak{p}_1,\mathrm{st}}(\Phi,\omega,w)+\epsilon_{\mathfrak{p}_2}\widehat{\Xi}_{\mathfrak{p}_2\#}(\Phi,\omega,w)$$
$$\sim\frac{\epsilon_{\mathfrak{p}_1}\mathscr{F}_{\mathfrak{d}_1 3}\delta_\#(\omega)}{2\pi\sqrt{-1}}\int_{\mathrm{Re}(y)=1+\delta}\frac{\Sigma_{\mathrm{canc}}(\Phi,y)}{\prod_{i=1}^{a-2}(\gamma'_{\mathfrak{p}_1 i}(y)+\chi_{\mathfrak{p}_1 i})}f_{\mathfrak{p}_1 1}(y)f_{\mathfrak{p}_1 2}(y)[\Lambda(w;z)]|_{Q'_{\mathfrak{p}_1}}dy.$$

(If $a=1$, we do not consider the denominator in the above integral.)

We consider $\Sigma_{\mathrm{canc}}(\Phi,y)$ only when ω is trivial. This implies that Φ is K-invariant. Therefore, $\Phi_{\mathfrak{p}_2}$ is invariant under the action of the standard maximal compact subgroup of $\mathrm{GL}(n-a)_\mathbb{A}$. So by (4.4.7), $\Sigma_{\mathrm{canc}}(\Phi,y)$ is holomorphic at $y=1$.

We consider the value $\Sigma_{\mathrm{canc}}(\Phi,1)$. If $a=1$,

$$\Sigma_{\mathrm{canc}}(\Phi,1)=\frac12\Sigma_{1,(1)}(R'_{\mathfrak{d}_1,0}R_{\mathfrak{d}_1}\Phi_{\mathfrak{p}_1},n-1).$$

Therefore, in this case,

$$\begin{aligned}&\epsilon_{\mathfrak{p}_1}\Xi'_{\mathfrak{p}_1,\mathrm{st}}(\Phi,\omega,w)+\epsilon_{\mathfrak{p}_2}\widehat{\Xi}_{\mathfrak{p}_2\#}(\Phi,\omega,w)\\ &\quad\sim C_G\Lambda(w;\rho)\frac{\epsilon_{\mathfrak{p}_1}\mathfrak{V}_{n-2}\delta_{\#}(\omega)}{2}\Sigma_{1,(1)}(R'_{\mathfrak{d}_1,0}R_{\mathfrak{d}_1}\Phi_{\mathfrak{p}_1},n-1).\end{aligned}$$

Suppose $a>1$. There are two possibilities for $\mathfrak{s}_1,\mathfrak{s}_2$, namely, $\mathfrak{s}_1(a)=0,\mathfrak{s}_2(a)=1$ or $\mathfrak{s}_1(a)=1,\mathfrak{s}_2(a)=0$.

The following lemma is an easy consequence of (4.7.4).

Lemma (4.7.5) *Suppose* $a>1$. *Then*
(1) *If* $\mathfrak{s}_1(a)=0$,

$$\Sigma_{\mathrm{canc}}(\Phi,1)=\frac{(n-a+1)\Sigma_1(R'_{\mathfrak{d}_1,0}R_{\mathfrak{d}_1}\Phi_{\mathfrak{p}_1},n-a)}{2\chi^2_{\mathfrak{p}_1a-1}}+\frac{\Sigma_{1,1}(R'_{\mathfrak{d}_1,0}R_{\mathfrak{d}_1}\Phi_{\mathfrak{p}_1},n-a)}{2\chi_{\mathfrak{p}_1a-1}},$$

(2) *If* $\mathfrak{s}_1(a)=1$,

$$\Sigma_{\mathrm{canc}}(\Phi,1)=-\frac{(n-a+1)\Sigma_1(R'_{\mathfrak{d}_1,0}R_{\mathfrak{d}_1}\Phi_{\mathfrak{p}_1},n-a)}{2\chi^2_{\mathfrak{p}_1a-1}}+\frac{\Sigma_{1,1}(R'_{\mathfrak{d}_1,0}R_{\mathfrak{d}_1}\Phi_{\mathfrak{p}_1},n-a)}{2\chi_{\mathfrak{p}_1a-1}}.$$

Clearly, $\mathfrak{F}_{\mathfrak{d}_1 3}=\mathfrak{F}_{\mathfrak{d}_3 3}$ and $\mathfrak{V}_{\mathfrak{d}_1}=\mathfrak{V}_{n-a-1}$. Also

$$\epsilon_{\mathfrak{p}_3}\mathfrak{F}_{\mathfrak{d}_1 3}\prod_{i=1}^{a-2}\chi_{\mathfrak{p}_1 i}^{-1}=\epsilon_{\mathfrak{p}_3}\mathfrak{F}_{\mathfrak{d}_3 3}\prod_{i=1}^{a-2}\chi_{\mathfrak{p}_3 i}^{-1}=c_{\mathfrak{p}_3}.$$

Clearly, $\epsilon_{\mathfrak{p}_1}=-\epsilon_{\mathfrak{p}_3}$ if $\mathfrak{s}_1(a)=0$, and $\epsilon_{\mathfrak{p}_1}=\epsilon_{\mathfrak{p}_3}$ if $\mathfrak{s}_1(a)=1$. Therefore,

$$\begin{aligned}&\frac{\epsilon_{\mathfrak{p}_1}\mathscr{F}_{\mathfrak{d}_1 3}\delta_{\#}(\omega)}{2\pi\sqrt{-1}}\int_{\mathrm{Re}(y)=1+\delta}\frac{\Sigma_{\mathrm{canc}}(\Phi,y)}{\prod_{i=1}^{a-2}(\gamma'_{\mathfrak{p}_1 i}(y)+\chi_{\mathfrak{p}_1 i})}f_{\mathfrak{p}_1 1}(y)f_{\mathfrak{p}_1 2}(y)[\Lambda(w;z)]|_{Q'_{\mathfrak{p}_1}}dy\\ &=\frac{\epsilon_{\mathfrak{p}_1}\mathscr{F}_{\mathfrak{d}_1 3}\mathfrak{V}_{n-a-1}\delta_{\#}(\omega)}{\prod_{i=1}^{a-2}\chi_{\mathfrak{p}_3 i}}\Sigma_{\mathrm{canc}}(\Phi,1)\Lambda(w;\rho)\\ &\quad+\frac{\epsilon_{\mathfrak{p}_1}\mathscr{F}_{\mathfrak{d}_1 3}\delta_{\#}(\omega)}{2\pi\sqrt{-1}}\int_{\mathrm{Re}(y)=1-\delta}\frac{\Sigma_{\mathrm{canc}}(\Phi,y)}{\prod_{i=1}^{a-2}(\gamma'_{\mathfrak{p}_1 i}(y)+\chi_{\mathfrak{p}_1 i})}f_{\mathfrak{p}_1 1}(y)f_{\mathfrak{p}_1 2}(y)[\Lambda(w;z)]|_{Q'_{\mathfrak{p}_1}}dy\\ &\sim(-1)^{\mathfrak{s}_1(a)+1}c_{\mathfrak{p}_3}\mathfrak{V}_{n-a-1}\delta_{\#}(\omega)\Sigma_{\mathrm{canc}}(\Phi,1)\Lambda(w;\rho).\end{aligned}$$

By definition,

$$\chi_{\mathfrak{p}_1 a-1}=\begin{cases}\frac{(n-a+1)(n-a)}{2}=f_{\mathfrak{d}_3}-(n-a+1) & \mathfrak{s}_1(a)=0,\\ n-a+1 & \mathfrak{s}_1(a)=1.\end{cases}$$

Let $a'=a-1$. Then $l(\mathfrak{p}_3)=a'$, and $n-a+1=n-a'$ etc. Therefore, this completes the proof of (4.7.2), and hence (4.4.6).

Remark (4.7.6) The order two terms of the poles of the zeta function are constant multiples of

$$\delta_{\#}(\omega)\Sigma_1(R'_{n,n-2,0}\Phi,n-1),\ \delta_{\#}(\omega)\Sigma_1(R'_{n,n-2,0}\widehat{\Phi},n-1).$$

These distributions are zero if ω is not trivial. If ω is is trivial and $\Phi = M_{V,\omega}\Phi$, Φ is K-invariant. Therefore, by (4.4.7),

$$\Sigma_1(R'_{n,n-2,0}\widehat{\Phi}, n-1) = \frac{Z_k(n-1)}{\mathfrak{R}_k}\mathscr{F}_{n,1}R_{n,1}\Phi(0).$$

So, if $\Phi = \otimes_v \Phi_v$ and Φ_v is supported on $V_{k_v}^{ss}$ for some $v \in \mathfrak{M}_\infty$,

$$\Sigma_1(R'_{n,n-2,0}\Phi, n-1) = \mathscr{F}_{n,1}R_{n,1}\Phi(0) = 0.$$

This implies that in the case $k = \mathbb{Q}$, all the poles the associated Dirichlet series are simple as was proved by Shintani [65].

§4.8 The work of Siegel and Shintani

Both Siegel and Shintani worked on the space of quadratic forms. We consider their work from our viewpoint. Even though I was an undergraduate student at the University of Tokyo and took some courses from Shintani, I was not personally acquainted with him nor was I aware of the theory of prehomogeneous vector spaces at that time. So the greater part of this section is my speculations based on papers by Siegel and Shintani.

Siegel did at least two kinds of work on quadratic forms. One is the proof of what we call Siegel–Weil formula [67], and the other is the proof of the average density theorem for the equivalence classes of integral quadratic forms as in this chapter [69].

His argument in [69] was clever, and essentially determined the rightmost pole of the zeta function even though he used the argument based on his Siegel–Weil formula and did not use the zeta function explicitly. Shintani was apparently influenced by Siegel's paper and tried to interpret the content of Siegel's paper from the viewpoint of prehomogeneous vector spaces. Here, we try to trace what Shintani might have thought in pursuing this goal.

We assume that $k = \mathbb{Q}$ for simplicity. One difficulty of the space of quadratic forms comes from the fact that we have to adjust the zeta function for the case $n = 2$ as we saw in §4.2. If we use this formulation, the case $n = 2$ is more or less included in the general case. However, Siegel and Shintani handled the cases $n = 2$ and $n \geq 3$ quite differently.

We used the smoothed Eisenstein series for all the cases, but Shintani used the ordinary Eisenstein series for the case $n = 2$. The function he considered was the following integral

$$\int_{\widetilde{G}_\mathbb{A}/G_k} |\chi_V(\widetilde{g})|^s \Theta_{V_k^s}(\Phi, \widetilde{g}) E(g^0, z) d\widetilde{g}, \tag{4.8.1}$$

where V is the space of binary quadratic forms, $\widetilde{g} = (t, g^0) \in \mathbb{A}^\times \times \mathrm{GL}(2)^0_\mathbb{A}$, and $E(g^0, z)$ is the Eisenstein series for the Borel subgroup of $\mathrm{GL}(2)$. This function can be considered as a two variable zeta function associated with the action of the Borel subgroup on V. As in the case of the smoothed Eisenstein series, the pole of (4.8.1) at $z = \rho$ gives us the one variable zeta function. Shintani himself was aware of the approach in this book, but chose the above formulation. He might have chosen the ordinary Eisenstein series, maybe because he wanted to achieve the cancellation in §4.2 explicitly. Then why didn't he consider the generalization of the above function for the general case?

Let us consider the following integral

$$\int_{\widetilde{G}_{\mathbb{A}}/G_k} |\chi_V(\widetilde{g})|^s \Theta_{V_k^{ss}}(\Phi, \widetilde{g}) E(g^0, z) d\widetilde{g}, \tag{4.8.2}$$

where V is the space of quadratic forms in $n \geq 3$ variables, $\widetilde{g} = (t, g^0) \in \mathbb{A}^\times \times \mathrm{GL}(n)^0_{\mathbb{A}}$, and $E(g^0, z)$ is the Eisenstein series for the Borel subgroup of $\mathrm{GL}(n)$. This would be the natural generalization of the integral (4.8.1). However, there is a slight difference here, i.e. we have to consider all the points in V_k^{ss}. The effect of this is that the theta series $\Theta_{V_k^{ss}}(\Phi, g^0)$ is not rapidly decreasing with respect to the diagonal part of g^0 any more, whereas it is true for $\Theta_{V_k^s}(\Phi, g^0)$ for the case $n = 2$. In fact the function $\Theta_{V_k^{ss}}(\Phi, g^0)$ is barely integrable on $G^0_{\mathbb{A}}/G_k$, and it is already non-trivial to show the convergence of the integral (4.8.2) in a certain domain. Moreover, we do not know if (4.8.2) can be continued meromorphically to a domain which contains ρ. This means that we cannot recover the one variable zeta function from (4.8.2).

Shintani must have considered integrals of the form (4.8.2), and faced this difficulty. My guess is that this is the reason why he chose a different approach for the case $n \geq 3$. However, he could not compute the residue of the associated Dirichlet series for indefinite quadratic forms in general. So it was natural to look at Siegel's original approach and see how he handled this difficulty.

Siegel's approach for $n \geq 3$ was quite different from the case $n = 2$ also. Roughly speaking we can summarize his argument as follows. First, he used his Mass formula to describe $\mu_\infty(x)$ (see §0.3 for the definition) for this case. Then in order to estimate

$$\sum_{|\det(x)|_\infty < N} \frac{\mu_\infty(x) b_{x,\infty}}{o_\infty}, \tag{4.8.3}$$

he considered integral quadratic forms with two different signatures $(m, n-m), (m-2, n-m+2)$ such that the discriminant is less than or equal to

$$Q = \prod_p p^{3\left[\frac{\log N}{\log p}\right]}.$$

For $0 \leq S \leq N$, let $C_m(S)$ be the number of residue classes modulo Q which contain an integral quadratic form with the signature $(m, n-m)$. Let $D(S)$ be the number of modulo Q congruent classes such that $\det x = (-1)^{n-m} S$ for x in such classes. Siegel related (4.8.3) with $C_m(S)$ by the Mass formula, and showed that $C_m(S) + C_{m-2}(S) \leq D(S)$. By estimating $D(S)$, he obtained an upper bound for $C_m(S) + C_{m-2}(S)$. The lower bound is relatively easy, and this argument was the point of his argument. But why two different signatures $(m, n-m), (m-2, n-m+2)$?

His argument was very clever, and I spent a long time trying to see from our viewpoint what he was trying to accomplish. My conclusion was that Siegel was trying to accomplish cancellations of divergent integrals one way or another. After investigating which distributions $\Xi_{\mathfrak{p}}(\Phi, \omega, w)$ may give such cancellations for some time, I reached the cancellations of distributions as in (4.6.18), (4.7.5). In a way, his two different methods for $n = 2$ and $n \geq 3$ correspond to two different cancellations (4.6.18), (4.7.5) in this chapter. Siegel did not have a fully developed theory of Eisenstein series or adelic language to hand, and the fact that he was able to handle the case $n \geq 3$ is due to his genius.

We now go back to Shintani's approach for the case $n \geq 3$. Recognizing the difficulty of using (4.8.2), Shintani used the approach essentially based on micro-local analysis. If we summarize his approach, we may say, that he considered the relative invariant polynomial as a hyperfunction, and tried to compute its Fourier transform. As I look back, he was unlucky in choosing this approach. Shintani was basically considering the associated Dirichlet series $\xi_i(\Phi, s)$ (see §0.3 for the definition). However, while the adelic zeta function has order two poles, the associated Dirichlet series only have simple poles. Nevertheless, the difficulty based on the order two poles of the adelic zeta function existed. In our approach, we indirectly reconstructed the order two part of the adelic zeta function in the inductive computation as in (4.7.5). But, since the associated Dirichlet series only have simple poles, Shintani was, in a sense, fighting against an invisible enemy. It is an interesting question to see whether it is possible to reconstruct the proof of Theorem (4.0.1) by micro-local analysis.

Shintani passed away about three years before equivariant Morse theory was established. Nevertheless, he managed to produce all the essential ideas in this book in his single paper [64]. He had almost all the right tools except for the uniform estimate of Whittaker functions at infinite places. In this sense, there is no doubt that the content of this book is what Shintani would have done by himself.

The purpose of this section is not to subordinate the contribution of Shintani concerning his work on the space of quadratic forms, but rather to point out how a true pioneer sometimes has to go through a tremendous struggle and yet may not achieve all his goals. And Shintani was a true pioneer of the subject.

Part III Preliminaries for the quartic case

In the next three chapters, we consider the following three prehomogeneous vector spaces

(1) $G = \mathrm{GL}(2) \times \mathrm{GL}(2), V = \mathrm{Sym}^2 k^2 \otimes k^2$,

(2) $G = \mathrm{GL}(2) \times \mathrm{GL}(1)^2, V = \mathrm{Sym}^2 k^2 \oplus k$,

(3) $G = \mathrm{GL}(2) \times \mathrm{GL}(1)^2, V = \mathrm{Sym}^2 k^2 \oplus k^2$.

Case (3) was handled by F. Sato [55] in a slightly different formulation. These are rather easy cases, and are essentially the same as the space of binary quadratic forms. However, we need the principal part formula for the zeta function for these cases, because these representations appear as unstable strata for the prehomogeneous vector space $G = \mathrm{GL}(3) \times \mathrm{GL}(2)$, $V = \mathrm{Sym}^2 k^3 \otimes k^2$ which we call the quartic case.

Coefficients of the Laurent expansions of the zeta functions for (2) and (3) appear in the Laurent expansion at the rightmost pole of the zeta function for the quartic case, and therefore they are particularly important. The principal part formulas which we are going to prove are (5.6.4), (6.3.11), (7.3.7).

Chapter 5 The case G=GL(2)×GL(2), V=Sym^{2}k^2 ⊗ k^2

§5.1 The space $\mathrm{Sym}^2 k^2 \otimes k^2$

We consider the prehomogeneous vector space $G = \mathrm{GL}(2) \times \mathrm{GL}(2)$, $V = \mathrm{Sym}^2 k^2 \otimes k^2$ in this chapter. We will use a new technique concerning the choice of the constants in the definition of the smoothed Eisenstein series in this chapter. Generally speaking, if G is a product of more than one $\mathrm{GL}(n)$, it sometimes makes the computation easier to choose the constants in the definition of the smoothed Eisenstein series in a certain way. The reader can see the effect of this technique in the proofs of (5.3.9), (5.4.10), and (5.5.6).

For the rest of this book, W is the space of binary quadratic forms in two variables $v = (v_1, v_2)$. We can identify W with k^3 by the map

$$f_x(v) = x_0 v_1^2 + x_1 v_1 v_2 + x_2 v_2^2 \to (x_0, x_1, x_2).$$

The vector space W is V_2 in §4.2.

Let $V = \mathrm{Sym}^2 k^2 \otimes k^2 = W \oplus W$. Any element of V is of the form

$$f = (f_1, f_2),\ f_1 = f_{x_1},\ f_2 = f_{x_2},$$

where $x_1 = (x_{10}, x_{11}, x_{12}), x_2 = (x_{20}, x_{21}, x_{22})$. We use

$$x = (x_1, x_2) = (x_{10}, x_{11}, x_{12}, x_{20}, x_{21}, x_{22})$$

as the coordinate system of V.

We define $G_1 = G_2 = \mathrm{GL}(2)$, and $G = G_1 \times G_2$. The group G acts on V by the formula

$$(g_1, g_2)(f_{x_1}, f_{x_2}) = (a f_{g_1 x_1} + b f_{g_1 x_2}, c f_{g_1 x_1} + d f_{g_1 x_2}) \text{ for } g_2 = \begin{pmatrix} a & b \\ c & d \end{pmatrix}.$$

Let $\widetilde{T} \subset G$ be the kernel of the homomorphism $G \to \mathrm{GL}(V)$, and $\chi_V(g_1, g_2) = (\det g_1)^2 \det g_2$. Then χ_V can be considered as a character of $G/\widetilde{T}$ and it is indivisible.

Let

$$p_1(x) = \det \begin{pmatrix} x_{10} & x_{11} \\ x_{20} & x_{21} \end{pmatrix}, \; p_2(x) = \det \begin{pmatrix} x_{11} & x_{12} \\ x_{21} & x_{22} \end{pmatrix}, \; p_3(x) = \det \begin{pmatrix} x_{12} & x_{10} \\ x_{22} & x_{20} \end{pmatrix}.$$

Let $U = \{x \in V \mid p_i(x) \neq 0 \text{ for some } i\}$. Consider the map

$$V_k \ni x \to (p_1(x), p_2(x), p_3(x)) \in k^3.$$

Then $G_{2k} \setminus U_k \cong \mathbb{P}^2_k$, and the action of G_{1k} can be identified with the action of $\mathrm{GL}(2)_k$ on $\mathbb{P}(\mathrm{Sym}^2 k^2)$. Therefore, this is a prehomogeneous vector space which is essentially the same as $\mathrm{Sym}^2 k^2$.

By the above consideration, $V_k^s \neq \emptyset$. Let $G^0_{\mathbb{A}}$ etc. be as in §3.1. In this case, $G^1_{\mathbb{A}} = G^0_{\mathbb{A}}$, so $\mathfrak{t} = \mathfrak{t}^0$. We identify $\mathfrak{t}$ with $\{(z_1, z_2) = (z_{11}, z_{12}, z_{21}, z_{22}) \in \mathbb{R}^4 \mid z_{11} + z_{12} = z_{21} + z_{22} = 0\}$. We define a Weyl group invariant inner product by $(z, z') = \sum_{i,j} z_{ij} z'_{ij}$ for $z = (z_{ij}), z' = (z'_{ij})$. Let $\gamma_{ij} \in \mathfrak{t}^*$ be the weight of x_{ij} for all i, j. We use the notation $\gamma_{ij}(t)$ for the rational character of T determined by γ_{ij}. This should not be confused with $t^{\gamma_{ij}}$ which is a positive real number. We identify γ_{ij}'s with elements of $\mathbb{R}^2$ as follows.

$\bullet \quad \bullet \quad \bullet \; (\sqrt{2}, \frac{1}{\sqrt{2}})$

$\bullet \quad \bullet \quad \bullet \; (\sqrt{2}, -\frac{1}{\sqrt{2}})$

Let $\mathfrak{B}$ be the parametrizing set of the Morse stratification. Elements of $\mathfrak{B} \setminus \{0\}$ are as follows.

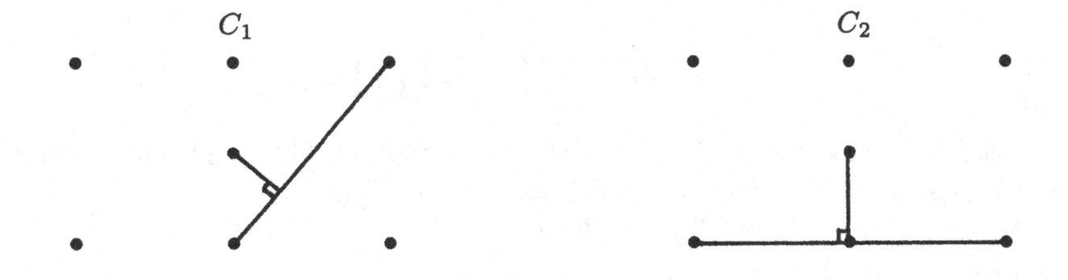

Let $\beta_1, \cdots, \beta_4$ be the closest points to the origin from the convex hulls $C_1, \cdots, C_4$ respectively. We choose $G'' = \mathrm{SL}(2) \times \mathrm{SL}(2)$ as G'' in §3.1. Then Z_{β_i} etc. for $i = 1, 2, 3$ are as follows.

Table (5.1.1)

β	Z_β	W_β	M''_β
$\beta_1 = (-\frac{1}{4}, \frac{1}{4}; -\frac{1}{4}, \frac{1}{4})$	x_{12}, x_{21}	x_{22}	$\{(a_2(t^{-1}, t), a_2(t, t^{-1}))\}$
$\beta_2 = (0, 0; -\frac{1}{2}, \frac{1}{2})$	x_{20}, x_{21}, x_{22}	—	$\mathrm{SL}(2) \times \{1\}$
$\beta_3 = (-1, 1; -\frac{1}{2}, \frac{1}{2})$	x_{22}	—	$\{(a_2(t^{-1}, t), a_2(t^2, t^{-2}))\}$
$\beta_4 = (-1, 1; 0, 0)$	x_{12}, x_{22}	—	$\mathrm{SL}(2) \times \{1\}$

Let $\mathfrak{d}_i = (\beta_i)$ for $i = 1, 2, 3$. By the above table, $Z^{\mathrm{ss}}_{\mathfrak{d}_1 k} = \{x \in Z_{\beta_1 k} \mid x_{12}, x_{21} \in k^\times\}$. We identify $Z_{\mathfrak{d}_2}$ with W and consider $Z^{\mathrm{ss}}_{\mathfrak{d}_2 k}, Z'^{\mathrm{ss}}_{\mathfrak{d}_2,0k}$ etc. Since $M''_{\mathfrak{d}_3}$ acts trivially on $Z_{\mathfrak{d}_3}$, $Z^{\mathrm{ss}}_{\mathfrak{d}_3 k} = Z_{\mathfrak{d}_3 k} \setminus \{0\}$. The vector space $Z_{\mathfrak{d}_4}$ is a standard representation of $M''_{\mathfrak{d}_4}$. Therefore, $Z^{\mathrm{ss}}_{\mathfrak{d}_4 k} = \emptyset$.

There is one β-sequence of length 2 which is $\mathfrak{d}_4 = (\beta_2, \beta_3 - \beta_2)$. Clearly, $Z_{\mathfrak{d}_4}$ is the subspace spanned by $\{e_{22}\}$, and $Z^{\mathrm{ss}}_{\mathfrak{d}_4 k} = Z_{\mathfrak{d}_4 k} \setminus \{0\}$.

It is easy to see that V^{s}_k is the set of $x \in V_k$ which are not conjugate to elements of the form $(0, 0, x_{12}, x_{20}, 0, 0)$. This set corresponds to the set of forms without rational factors by the map $U \to k^3$. Let $Y_{V,0}$ (resp. $Z_{V,0}$) be the subspace spanned by $\{e_{12}, e_{20}, e_{21}, e_{22}\}$ (resp. $\{e_{12}, e_{20}, e_{21}\}$). Let $Z'_{V,0}$ be the subspace spanned by $\{e_{12}, e_{20}\}$. Let $Z'^{\mathrm{ss}}_{V,0k} = \{x \in Z'_{V,0k} \mid x_{12}, x_{20} \in k^\times\}$. Then $V^{\mathrm{ss}}_{\mathrm{st}k} = G_k Z'^{\mathrm{ss}}_{V,0k}$. Therefore,

$$V_k \setminus \{0\} = V^{\mathrm{s}}_k \coprod V^{\mathrm{ss}}_{\mathrm{st}k} \coprod \coprod_{i=1}^{3} S_{\beta_i k}.$$

Let $Z'_{\mathfrak{d}_2,0} \subset Z_{\mathfrak{d}_2}$ (resp. $Z_{\mathfrak{d}_2,0}$) be the subspace spanned by $\{e_{21}\}$ (resp. $\{e_{21}, e_{22}\}$). Let $Z'^{\mathrm{ss}}_{\mathfrak{d}_2,0k} = Z'_{\mathfrak{d}_2,0k} \setminus \{0\}$. Then $Z^{\mathrm{ss}}_{\mathfrak{d}_2,\mathrm{st}k} = M_{\mathfrak{d}_2 k} Z'^{\mathrm{ss}}_{\mathfrak{d}_2,0k}$.

We write elements of $T^0_{\mathbb{A}}$ in the form

$$\begin{aligned} \text{(5.1.2)} \qquad & \widehat{t}^0 = (a_2(t_{11}, t_{12}), a_2(t_{21}, t_{22})), \\ & \lambda = d(\lambda_1, \lambda_2) = (a_2(\underline{\lambda_1}^{-1}, \underline{\lambda_1}), a_2(\underline{\lambda_2}^{-1}, \underline{\lambda_2})),\ t^0 = \lambda \widehat{t}^0, \end{aligned}$$

where $\lambda_1, \lambda_2 \in \mathbb{R}^2_+$, $t_{ij} \in \mathbb{A}^1$ for $i, j = 1, 2$. We define

$$d^\times \widehat{t}^0 = \prod_{i,j} d^\times t_{ij},\ d^\times \lambda = d^\times \lambda_1 d^\times \lambda_2,\ d^\times t^0 = d^\times \lambda d^\times \widehat{t}^0.$$

The inductive structures of $V^{\mathrm{ss}}_{\mathrm{st}k}$ and $Z^{\mathrm{ss}}_{\mathfrak{d}_2,\mathrm{st}k}$ can be described in the following manner.

Lemma (5.1.3) *Let $H_{\mathfrak{d}_2} \subset G$ be the subgroup generated by T and $(\tau_{\mathrm{GL}(2)}, 1)$. Let $H_V \subset G$ be the subgroup generated by T and τ_G. Then*
(1) $V^{\mathrm{ss}}_{\mathrm{st},k} \cong G_k \times_{H_{Vk}} Z'^{\mathrm{ss}}_{V,0k}$,
(2) $Z^{\mathrm{ss}}_{\mathfrak{d}_2,\mathrm{st}k} \cong M_{\mathfrak{d}_2 k} \times_{H_{\mathfrak{d}_2 k}} Z'^{\mathrm{ss}}_{\mathfrak{d}_2,0k}$.

The proof of the above lemma is easy and is left to the reader.

Let $\omega = (\omega_1, \omega_2)$ be as in §3.1. We use the alternative definition (3.1.8) of the zeta function using $L = V^{\mathrm{s}}_k$, and use the notation $Z_V(\Phi, \omega, s), Z_{V+}(\Phi, \omega, s)$.

Let $[\ ,\ ]'_V$ be the bilinear form such that

$$[x, y]'_V = x_{10}y_{10} + x_{12}y_{12} + x_{20}y_{20} + x_{22}y_{22} + \frac{1}{2}x_{11}y_{11} + \frac{1}{2}x_{21}y_{21}$$

for $x = (x_{ij}), y = (y_{ij}) \in V$. Then it is easy to see that this bilinear form satisfies the property $[gx, {}^t g^{-1} y]'_V = [x, y]'_V$. We define $[x, y]_V = [x, \tau_G y]'_V$, and use this bilinear form as $[\ ,\ ]_V$ in §3.1.

Let

$$(5.1.4) \qquad J^s(\Phi, g^0) = \sum_{x \in V_k \setminus V^s_k} \Phi((g^0)^\iota x) - \sum_{x \in V_k \setminus V^s_k} \Phi(g^0 x),$$

$$I^0(\Phi, \omega) = \int_{G^0_{\mathbb{A}}/G_k} \omega(g^0) J^s(\Phi, g^0) dg^0.$$

For $\lambda \in \mathbb{R}_+$, let Φ_λ be as in §3.5. By the Poisson summation formula,

$$Z_V(\Phi, \omega, s) = Z_{V+}(\Phi, \omega, s) + Z_{V+}(\widehat{\Phi}, \omega^{-1}, 6 - s) + \int_{\mathbb{R}_+} \lambda^s I^0(\Phi_\lambda, \omega) d^\times \lambda.$$

We will study the third term in §§5.2–5.5.

Throughout this chapter we assume that $\Phi = M_{V,\omega}\Phi$ (see (3.1.11)).

Let $\mathscr{E}(g^0, w)$ be the smoothed Eisenstein series which we introduced in §3.4. We choose the constants in (3.4.3) so that $C_1 = 1, C = C_2 > 4$. This choice of constants C_1, C_2 will have some effect later as we mentioned at the beginning of this chapter.

We define

$$(5.1.5) \qquad I(\Phi, \omega, w) = \int_{G^0_{\mathbb{A}}/G_k} \omega(g^0) J^s(\Phi, g^0) \mathscr{E}(g^0, w, \psi) dg^0.$$

Since $J^s(\Phi, g^0) = \Theta_{V^s_k}(\Phi, g^0) - \Theta_{V^s_k}(\widehat{\Phi}, (g^0)^\iota)$, by (3.1.6), (3.4.34),

$$I(\Phi, \omega, w) \sim C_G \Lambda(w; \rho) I^0(\Phi, \omega).$$

We define

$$(5.1.6) \qquad \Xi_{V,\mathrm{st}}(\Phi, \omega, w) = \int_{G^0_{\mathbb{A}}/G_k} \omega(g^0) \Theta_{V^{\mathrm{ss}}_{\mathrm{st}k}}(\Phi, g^0) \mathscr{E}(g^0, w) dg^0;$$

$$\widehat{\Xi}_{V,\mathrm{st}}(\Phi, \omega, w) = \int_{G^0_{\mathbb{A}}/G_k} \omega(g^0) \Theta_{V^{\mathrm{ss}}_{\mathrm{st}k}}(\widehat{\Phi}, (g^0)^\iota) \mathscr{E}(g^0, w) dg^0.$$

The functions

$$\Theta_{V^{ss}_{stk}}(\Phi, g^0),\ \Theta_{V^{ss}_{stk}}(\widehat{\Phi}, (g^0)^\iota)$$

are slowly increasing by (1.2.6). Therefore, by (3.4.34), the above distributions are well defined for $\mathrm{Re}(w) \gg 0$. Since $\mathscr{E}((g^0)^\iota, w) = \mathscr{E}(g^0, w)$,

$$\widehat{\Xi}_{V,\mathrm{st}}(\Phi, \omega, w) = \Xi_{V,\mathrm{st}}(\widehat{\Phi}, \omega^{-1}, w).$$

Let $\delta_{\#}(\omega)$ be as in §3.6. By (3.5.9) and (3.5.20),

$$(5.1.7) \qquad \begin{aligned} I(\Phi, \omega, w) = {} & \delta_{\#}(\omega)\Lambda(w; \rho)(\widehat{\Phi}(0) - \Phi(0)) \\ & + \sum_{\mathfrak{p}, l(\mathfrak{p})=1} \epsilon_{\mathfrak{p}}(\widetilde{\Xi}_{\mathfrak{p}}(\Phi, \omega, w) + \Xi_{\mathfrak{p}}(\Phi, \omega, w)) \\ & + \widehat{\Xi}_{V,\mathrm{st}}(\Phi, \omega, w) - \Xi_{V,\mathrm{st}}(\Phi, \omega, w). \end{aligned}$$

If $\mathfrak{p} = (\mathfrak{d}, \mathfrak{s})$, and $\mathfrak{d} = \mathfrak{d}_1$, $\mathfrak{p}$ belongs to class (3) in §3.5. Therefore, $\Xi_{\mathfrak{p}}(\Phi, \omega, w)$ is well defined for $\mathrm{Re}(w) \gg 0$. This implies that $\widetilde{\Xi}_{\mathfrak{p}}(\Phi, \omega, w)$ is well defined for $\mathrm{Re}(w) \gg 0$ also. If $\mathfrak{d} = \mathfrak{d}_2, \mathfrak{d}_3$, $\widetilde{\Xi}_{\mathfrak{p}}(\Phi, \omega, w) = 0$, and $\Xi_{\mathfrak{p}}(\Phi, \omega, w)$ is well defined for $\mathrm{Re}(w) \gg 0$ by (3.5.5). Therefore, all the distributions in (5.1.7) are well defined for $\mathrm{Re}(w) \gg 0$. If $\mathfrak{d} = \mathfrak{d}_4$, $\mathfrak{d}$ satisfies Condition (3.4.16)(1), and therefore, $\Xi_{\mathfrak{p}}(\Phi, \omega, w)$ is well defined for $\mathrm{Re}(w) \gg 0$ also.

Let τ be a Weyl group element as in (3.4.13). We use the same notation as in §3.6 in this chapter. So $s_\tau = (s_{\tau 1}, s_{\tau 2}) \in \mathbb{C}^2$ where $s_{\tau i} = z_{i\tau_1(2)} - z_{i\tau_1(1)}$ for $i = 1, 2$. We define $ds_\tau = ds_{\tau 1} ds_{\tau 2}$. For a similar reason to that above, $\Xi_{\mathfrak{p},\tau}(\Phi, \omega, w)$ is well defined for $\mathrm{Re}(w) \gg 0$ in the same cases as $\Xi_{\mathfrak{p}}(\Phi, \omega, w)$

§5.2 The adjusting term

In this section, we define the adjusting term $T_V(\Psi, \omega, s, s_1)$. This distribution is required in order to describe the contribution from V^{ss}_{stk}.

Let $\Psi \in \mathscr{S}(Z_{V,0\mathbb{A}})$. Let $q = (q_1, q_2) \in (\mathbb{A}^1)^2$, $u_0 \in \mathbb{A}$, $\mu = (\mu_1, \mu_2) \in \mathbb{R}^2_+$, $d^\times q = d^\times q_1 d^\times q_2$, and $d^\times \mu = d^\times \mu_1 d^\times \mu_2$. We consider the function $\alpha(u_0)$ in §2.2. Let ω_1, ω_2 be characters of $\mathbb{A}^\times / k^\times$, and $\omega = (\omega_1, \omega_2)$. We define $\omega(q) = \omega_1(q_1)\omega_2(q_2)$. Let

$$f_V(\Psi, \mu, q, u_0, s_1, s_2) = \mu_2^{s_1} \alpha(u_0)^{s_2} \Psi(\underline{\mu}_1 \underline{\mu}_2^{-1} q_1, \underline{\mu}_1 \underline{\mu}_2 q_2, 2\underline{\mu}_1 \underline{\mu}_2 q_2 u_0).$$

Note '2' in the third coordinate.

Definition (5.2.1) *For complex variables s, s_1, s_2 and Ψ, ω as above, we define*

(1) $$T_V(\Psi, \omega, s, s_1, s_2) = \int_{\mathbb{R}^2_+ \times (\mathbb{A}^1)^2 \times \mathbb{A}} \omega(q) \mu_1^s f_V(\Psi, \mu, q, u_0, s_1, s_2) d^\times \mu d^\times q du_0,$$

(2) $$T_{V+}(\Psi, \omega, s, s_1, s_2) = \int_{\substack{\mathbb{R}^2_+ \times (\mathbb{A}^1)^2 \times \mathbb{A} \\ \mu_1 \geq 1}} \omega(q) \mu_1^s f_V(\Psi, \mu, q, u_0, s_1, s_2) d^\times \mu d^\times q du_0,$$

(3) $$T_V^1(\Psi, \omega, s_1, s_2) = \int_{\mathbb{R}_+ \times (\mathbb{A}^1)^2 \times \mathbb{A}} \omega(q) f_V(\Psi, 1, \mu_2, q, u_0, s_1, s_2) d^\times \mu_2 d^\times q du_0.$$

Proposition (5.2.2) *The distribution $T_V(\Psi, \omega, s, s_1, s_2)$ can be continued meromorphically everywhere, and is holomorphic for $\frac{\mathrm{Re}(s-s_1)}{2} > 1$, $\frac{\mathrm{Re}(s+s_1)}{2} > 1$, $\frac{\mathrm{Re}(s+s_1)}{2} + \mathrm{Re}(s_2) > 2$. The functions* (2), (3) *are entire functions.*

Proof. The convergence of (2), (3) is easy, so we briefly discuss (1). Suppose that there exist Schwartz–Bruhat functions $\Psi_1 \in \mathscr{S}(\mathbb{A}), \Psi_2 \in \mathscr{S}(\mathbb{A}^2)$, such that $\Psi(y_1, y_2, 2y_3) = \Psi_1(y_1)\Psi_2(y_2, y_3)$. Then

$$T_V(\Psi, \omega, s, s_1, s_2) = \frac{1}{2}\Sigma_1(\Psi_1, \omega_1, \frac{s-s_1}{2})T_W(\Psi_2, \omega_2, \frac{s+s_1}{2}, s_2),$$

where the last factor is as in §4.2. So $T_V(\Psi, \omega, s, s_1, s_2)$ can be continued meromorphically everywhere. In general, we can assume that $\Psi = \otimes_v \Psi_v$, and the finite part Ψ_f of Ψ has a similar decomposition. So in order to consider the meromorphic continuation of $T_V(\Psi, \omega, s, s_1, s_2)$, we can assume that $T_V(\Psi, \omega, s, s_1, s_2)$ has an Euler product and its finite part can be continued meromorphically everywhere and is holomorphic for $\frac{\mathrm{Re}(s-s_1)}{2} > 1$, $\frac{\mathrm{Re}(s+s_1)}{2} > 1$, $\frac{\mathrm{Re}(s+s_1)}{2} + \mathrm{Re}(s_2) > 2$. One can prove that the infinite part of $T_V(\Psi, \omega, s, s_1, s_2)$ can be continued meromorphically everywhere and is holomorphic for $\frac{\mathrm{Re}(s-s_1)}{2} > 0$, $\frac{\mathrm{Re}(s+s_1)}{2} > 0$, $\frac{\mathrm{Re}(s+s_1)}{2} + \mathrm{Re}(s_2) > 1$ using the polar coordinate as in §4.2.

Q.E.D.

Definition (5.2.3) *For $\Psi, \omega, s, s_1, s_2$ as above, we define*

(1) $$T_V(\Psi, \omega, s, s_1) = \left.\frac{d}{ds_2}\right|_{s_2=0} T_V(\Psi, \omega, s, s_1, s_2),$$

(2) $$T_{V^+}(\Psi, \omega, s, s_1) = \left.\frac{d}{ds_2}\right|_{s_2=0} T_{V^+}(\Psi, \omega, s, s_1, s_2),$$

(3) $$T_V^1(\Psi, \omega, s_1) = \left.\frac{d}{ds_2}\right|_{s_2=0} T_V^1(\Psi, \omega, s_1, s_2).$$

Definition (5.2.4)
(1) *Let $\Psi_1 \in \mathscr{S}(Z_{V,0\mathbb{A}})$. Let μ, q, u_0 be as in the beginning of this section. We define*

$$\widetilde{\Theta}_{Z_{V,0}}(\Psi_1, \mu, q, u_0) = \sum_{x,y \in k^\times} \Psi_1(\underline{\mu}_1 \underline{\mu}_2^{-1} q_1 x, \underline{\mu}_1 \underline{\mu}_2 q_2 y, 2\underline{\mu}_1 \underline{\mu}_2 q_2 y u_0).$$

Note '2' in the third coordinate.
(2) *Let $\Psi_2 \in \mathscr{S}(Z_{\mathfrak{d}_2,0\mathbb{A}})$. For $t' \in \mathbb{A}^\times$, $u_0 \in \mathbb{A}$, we define*

$$\widetilde{\Theta}_{Z_{\mathfrak{d}_2,0}}(\Psi_2, t', u_0) = \sum_{x \in k^\times} \Psi_2(t'x, t'xu_0).$$

(3) *For $\Psi_3 \in \mathscr{S}(Z_{\mathfrak{d}_2\mathbb{A}})$ and $g_{\mathfrak{d}_2} \in G^0_{\mathbb{A}} \cap M_{\mathfrak{d}_2\mathbb{A}}$, we define*

$$\Theta_{Z'_{\mathfrak{d}_2,0}}(\Psi_3, g_{\mathfrak{d}_2}) = \sum_{x \in Z'_{\mathfrak{d}_2,0k}} \Psi_3(g_{\mathfrak{d}_2} x).$$

The following lemma is clear from the above definition.

Lemma (5.2.5) *Let ω_1, ω_2 be characters of $\mathbb{A}^\times/k^\times$, and $\omega = (\omega_1, \omega_2)$. Let s_1, s_2, q, μ be as in (5.2.1). Then for $\Psi \in \mathscr{S}(Z_{V,0\mathbb{A}})$,*

$$\begin{aligned} & T_V^1(\Psi, \omega, s_1, s_2) \\ & = \int_{\mathbb{R}_+ \times (\mathbb{A}^1/k^\times)^2 \times \mathbb{A}} \omega(q) \mu_2^{s_1} \alpha(u_0)^{s_2} \widetilde{\Theta}_{Z_{V,0}}(\Psi, 1, \underline{\mu}_2, q, u_0) d^\times \mu_2 d^\times q du_0. \end{aligned}$$

§5.3 Contributions from $\mathfrak{d}_1, \mathfrak{d}_3$

In this section, we consider paths $\mathfrak{p} = (\mathfrak{d}, \mathfrak{s}), \mathfrak{p}' = (\mathfrak{d}, \mathfrak{s}')$ such that $\mathfrak{d} = \mathfrak{d}_1$ or $\mathfrak{d}_3$ and $\mathfrak{s}(1) = 0, \mathfrak{s}'(1) = 1$. Since $\mathscr{E}(g^0, w) = \mathscr{E}((g^0)^\iota, w)$, $\Xi_{\mathfrak{p}'}(\Phi, \omega, w) = \Xi_{\mathfrak{p}}(\widehat{\Phi}, \omega^{-1}, w)$. So we only consider $\mathfrak{p}$. For β-sequences $\mathfrak{d} = \mathfrak{d}_1, \mathfrak{d}_3$, $M_\mathfrak{d} = T_k$. Since $l(\mathfrak{d}) = 1$, we do not have to worry too much about $A_{\mathfrak{p}2}$ etc. for these β-sequences. So, instead of the the general notation $g_\mathfrak{d} \in G^0_\mathbb{A} \cap M_{\mathfrak{d}\mathbb{A}}$, we just use t^0 in (5.1.2) to describe elements of $G^0_\mathbb{A} \cap M_{\mathfrak{d}\mathbb{A}}$. When we consider $\mathbb{A}, \mathbb{A}^\times$ in the next three sections, we consider the standard measures on them.

We start with some definitions.

Definition (5.3.1) *Let $\omega = (\omega_1, \omega_2)$ be as before. We define*
(1) $\delta_{V,\mathrm{st}}(\omega) = \delta(\omega_1\omega_2^{-2})$, $\omega_{V,\mathrm{st}} = (\omega_2, \omega_2)$,
(2) $\delta_{\mathfrak{d}_1}(\omega) = \delta(\omega_1\omega_2^{-1})\delta(\omega_2^2)$, $\omega_{\mathfrak{d}_1} = (\omega_2, \omega_2)$,
(3) $\delta_{\mathfrak{d}_2}(\omega) = \delta(\omega_2)$, $\omega_{\mathfrak{d}_2} = (1, \omega_1)$,
(4) $\delta_{\mathfrak{d}_2,\mathrm{st}}(\omega) = \delta_{\mathfrak{d}_3}(\omega) = \delta_{\mathfrak{d}_4}(\omega) = \delta_\#(\omega) = \delta(\omega_1)\delta(\omega_2)$.

Definition (5.3.2) *For $\Phi_1 \in \mathscr{S}(V_\mathbb{A})$, $\Phi_2 \in \mathscr{S}(Z_{\mathfrak{d}_2\mathbb{A}})$, we define $R_{V,0}\Phi_1 \in \mathscr{S}(Z_{V,0\mathbb{A}})$ and $\widetilde{R}_{\mathfrak{d}_2,0}\Phi_2 \in \mathscr{S}(Z_{\mathfrak{d}_2,0\mathbb{A}})$ by*

$$(1) \qquad R_{V,0}\Phi(x_{12}, x_{20}, x_{21}) = \int_\mathbb{A} \Phi(0, 0, x_{12}, x_{20}, x_{21}, x_{22}) dx_{22},$$

$$(2) \qquad \widetilde{R}_{\mathfrak{d}_2,0}\Phi(x_{21}, x_{22}) = \Phi(0, x_{21}, x_{22}).$$

Definition (5.3.3) *For $\Phi \in \mathscr{S}(V_\mathbb{A})$, and $s = (s_1, s_2) \in \mathbb{C}^2$, we define*

$$\Sigma_{\mathfrak{d}_1}(\Phi, \omega, s) = \frac{\delta_{\mathfrak{d}_1}(\omega)}{2} \Sigma_2(R_{\mathfrak{d}_1}\Phi, \omega_{\mathfrak{d}_1}, \frac{s_1 - 1}{2}, \frac{s_1 + 2s_2 - 1}{2}),$$

where the first (resp. second) coordinate of $R_{\mathfrak{d}_1}\Phi$ corresponds to x_{12} (resp. x_{21}).

For $s \in \mathbb{C}$, we define $\widetilde{\Sigma}_{\mathfrak{d}_1}(\Phi, \omega, s) = \Sigma_{\mathfrak{d}_1}(\Phi, \omega, s, 1)$. Let

$$(5.3.4) \qquad \widetilde{\Sigma}_{\mathfrak{d}_1}(\Phi, \omega, s) = \sum_{j=-2}^{\infty} \widetilde{\Sigma}_{\mathfrak{d}_1,(j)}(\Phi, \omega, s_0)(s - s_0)^j$$

be the Laurent expansion at $s = s_0$.

For the rest of this chapter, $r = (r_1, r_2)$ is a point in $\mathbb{R}^2$.

First we consider the case $\mathfrak{d} = \mathfrak{d}_1$.

Lemma (5.3.5) $\widetilde{\Xi}_\mathfrak{p}(\Phi, \omega, w) \sim 0.$

Proof. Let $\alpha = (\alpha_1, \alpha_2) \in k^2$. For $u = (u_1, u_2) \in \mathbb{A}^2$, we define

$$< \alpha n(u) > = < \alpha_1 u_1 >< \alpha_2 u_2 > .$$

Let

$$f_{1,\alpha}(t^0) = \int_{N_{\mathbb{A}}/N_k} \Theta_{Y_{\mathfrak{d}_1}}(\Phi, t^0 n(u)) < \alpha u > du,$$

$$f_{2,\alpha}(w, t^0) = \int_{N_{\mathbb{A}}/N_k} \mathscr{E}(t^0 n(u), w) < \alpha u > du.$$

Then by the Parseval formula,

$$\widetilde{\Xi}_{\mathfrak{p}}(\Phi, \omega, w) = \int_{T^0_{\mathbb{A}}/T_k} \omega(t^0) \sum_{\alpha \in k^2 \setminus \{0\}} f_{1,\alpha}(t^0) f_{2,-\alpha}(w, t^0)(t^0)^{-2\rho} d^\times t^0.$$

Lemma (5.3.6) *For any $N_1, N_2, N_3 \geq 1$,*

$$\sum_{\alpha \in k^2 \setminus \{0\}} |f_{1,\alpha}(t^0)| \ll \lambda_1^{-2} \lambda_2^{-1} (\lambda_1^2 \lambda_2^{-1})^{-N_1} \lambda_2^{-N_2} (\lambda_1^2 \lambda_2)^{N_3}.$$

Proof. Let $u'_1 = u'_1(x, u) = x_{21} u_1 + x_{12} u_2$, $u'_2 = u'_2(x, u) = x_{12} u_2$. Then $u_1 = x_{21}^{-1}(u'_1 - u'_2)$, $u_2 = x_{12}^{-1} u'_2$. Since $\mathbb{A}/k$ is compact, $f_{1,\alpha}(t^0)$ is equal to

$$\int_{\mathbb{A}^2/k^2} \sum_{\substack{x_{12}, x_{21} \in k^\times \\ x_{22} \in k}} \widetilde{R}_{\mathfrak{d}_1} \Phi(\gamma_{12}(t^0) x_{12}, \gamma_{21}(t^0) x_{21}, \gamma_{22}(t^0)(x_{22} + u'_1)) < \alpha u > du$$

$$= \sum_{x_{12}, x_{21} \in k^\times} \int_{\mathbb{A}^2/k^2} \sum_{x_{22} \in k} \widetilde{R}_{\mathfrak{d}_1} \Phi(\gamma_{12}(t^0) x_{12}, \gamma_{21}(t^0) x_{21}, \gamma_{22}(t^0)(x_{22} + u'_1))$$
$$\times < \alpha u > du.$$

Since

$$< \alpha u > = < \alpha_1 x_{21}^{-1} u'_1 >< (\alpha_2 x_{12}^{-1} - \alpha_1 x_{21}^{-1}) u'_2 >,$$

we only have to consider terms which satisfy the condition $\alpha_2 x_{12}^{-1} = \alpha_1 x_{21}^{-1}$. This implies that α_1, α_2 are both non-zero.

Let Φ_1 be the partial Fourier transform of $\widetilde{R}_{\mathfrak{d}_1} \Phi$ with respect to the third coordinate and the character $< >$. Then it is easy to see that

$$f_{1,\alpha}(t^0) = \lambda_1^{-2} \lambda_2^{-1} \sum_{x_{12} \in k^\times} \Phi_1(\gamma_{12}(t^0) x_{12}, \gamma_{21}(t^0) \alpha_1 \alpha_2^{-1} x_{12}, \gamma_{22}(t^0)^{-1} \alpha_2 x_{12}^{-1}).$$

Therefore, if we choose $\Phi_2 \in \mathscr{S}(\mathbb{A}^3)$ so that $|\Phi_1| \leq \Phi_2$,

$$\sum_{\alpha \in k^2 \setminus \{0\}} |f_{1,\alpha}(t^0)| \leq \lambda_1^{-2} \lambda_2^{-1} \Theta_3(\Phi_2, \gamma_{12}(t^0), \gamma_{21}(t^0), \gamma_{22}(t^0)^{-1}).$$

Hence, the lemma follows from (1.2.6).

Q.E.D.

Let $M > 1 + C$. Clearly,

$$\sum_{\alpha \in (k^\times)^2} |f_{2,\alpha}(w, t^0)| = \widetilde{\mathscr{E}}_{I,\tau_G,1}(t^0, w),$$

where $I = (\{1\}, \{1\})$. Let $\delta > 0$. By (2.4.10), for any $l_1, l_2 \gg 0$,

$$|E_{B,I,\tau_G,1}(t^0, z)| \ll c_{l,\delta}(z)(t^0)^{\tau_G \mathrm{Re}(z)+\rho} \lambda_1^{l_1} \lambda_2^{l_2},$$

for $\mathrm{Re}(z) \in D_{I,\tau_G,\delta}$, where $c_{l,\delta}(z)$ is a function of polynomial growth. We can choose $q \in D_{I,\tau,\delta}$ so that $L(q) < w_0$. Therefore, there exists $\delta' > 0$ such that for any $l_1, l_2 \gg 0$,

$$\widetilde{\mathscr{E}}_{I,\tau_G,1}(t^0, w) \ll \lambda_1^{l_1} \lambda_2^{l_2}$$

for $w_0 - \delta' \le \mathrm{Re}(w) \le M$.

This implies that for any $N_1, N_2, N_3 \ge 1$ and $l_1, l_2 \gg 0$,

$$\sum_{\alpha \in k^2 \setminus \{0\}} |f_{1,\alpha}(t^0) f_{2,\alpha}(w, t^0)| \ll \lambda_1^{l_1} \lambda_2^{l_2} \lambda_1^{-2} \lambda_2^{-1} (\lambda_1^2 \lambda_2^{-1})^{-N_1} \lambda_2^{-N_2} (\lambda_1^2 \lambda_2)^{N_3}.$$

Since the convex hull of $\{(2, -1), (0, 1), (-2, -1)\}$ contains a neighborhood of the origin of $\mathbb{R}^2$, the above function is bounded by a constant multiple of $\mathrm{rd}_{2,N}(\lambda_1, \lambda_2)$ for any $N > 0$. This proves (5.3.5).

Q.E.D.

Lemma (5.3.7)

$$\Xi_{\mathfrak{p},\tau}(\Phi, \omega, w) = \left(\frac{1}{2\pi\sqrt{-1}}\right)^2 \int_{\substack{\mathrm{Re}(s_\tau) = r \\ r_1 > 3,\ r_2 > 1}} \Sigma_{\mathfrak{d}}(\Phi, \omega, s_\tau) \widetilde{\Lambda}_\tau(w; s_\tau) ds_\tau.$$

Proof. Suppose that $f(q)$ is a function of $q = (q_1, q_2) \in (\mathbb{A}^1/k^\times)^2$. Then

$$\int_{(\mathbb{A}^1/k^\times)^4} \omega(t^0) f(t_{12}^2 t_{21}, t_{11} t_{12} t_{22}) d^\times t^0 = \delta_{\mathfrak{d}}(\omega) \int_{(\mathbb{A}^1/k^\times)^2} \omega_{\mathfrak{d}}(q) f(q) d^\times q$$

after the change of variable $q_1 = t_{12}^2 t_{21}$, $q_2 = t_{11} t_{12} t_{22}$. Let $t' = (t'_1, t'_2) = (\underline{\mu}_1 q_1, \underline{\mu}_2 q_2) \in (\mathbb{A}^\times)^2$ where $\mu_1, \mu_2 \in \mathbb{R}_+^2$, $q_1, q_2 \in \mathbb{A}^1$. Let $d^\times t' = d^\times t'_1 d^\times t'_2$. We make the change of variable $\mu_1 = \lambda_1^{-2} \lambda_2$, $\mu_2 = \lambda_2^{-1}$. Then $d^\times \mu_1 d^\times \mu_2 = 2 d^\times \lambda_1 d^\times \lambda_2$, and $\lambda_1 = \mu_1^{-\frac{1}{2}} \mu_2^{-\frac{1}{2}}$, $\lambda_2 = \mu_2^{-1}$. It is easy to see that $\sigma_{\mathfrak{p}}(t^0)(t^0)^{\tau z + \rho} = \mu_1^{\frac{s_{\tau 1} - 1}{2}} \mu_2^{\frac{s_{\tau 1} + 2 s_{\tau 2} - 1}{2}}$.

Therefore,

$$\frac{\delta_{\mathfrak{d}}(\omega)}{2} \int_{(\mathbb{A}^\times/k^\times)^2} \omega_2(t'_1 t'_2) \Theta_2(R_{\mathfrak{d}_1} \Phi, t') |t'_1|^{\frac{s_{\tau 1} - 1}{2}} |t'_2|^{\frac{s_{\tau 1} + 2 s_{\tau 2} - 1}{2}} d^\times t' = \Sigma_{\mathfrak{d}}(\Phi, \omega, s_\tau).$$

Hence,

$$\Xi_{\mathfrak{p},\tau}(\Phi, \omega, w) = \left(\frac{1}{2\pi\sqrt{-1}}\right)^2 \int_{\substack{\mathrm{Re}(s_\tau) = r \\ r_1 > 3,\ r_2 > 1}} \Sigma_{\mathfrak{d}}(\Phi, \omega, s_\tau) \widetilde{\Lambda}_\tau(w; s_\tau) ds_\tau. \qquad \text{Q.E.D.}$$

Lemma (5.3.8) $\Xi_{\mathfrak{p},\tau}(\Phi, \omega, w) \sim 0$ *unless* $\tau = \tau_G$.

Proof. We choose $r_1, r_2 \gg 0$ for $\tau = (1,1)$, r_1 fixed and $r_2 \gg 0$ for $\tau = (\tau_{\mathrm{GL}(2)}, 1)$, and $r_2 \gg 0$ and r_2 fixed for $\tau = (1, \tau_{\mathrm{GL}(2)})$. Then $\widetilde{L}(r) < w_0$. This proves the lemma.

Q.E.D.

If $\tau = \tau_G$, then $s_{\tau 1} = z_{11} - z_{12}$ and $s_{\tau 2} = z_{21} - z_{22}$.

Proposition (5.3.9) *Let* $\tau = \tau_G$ *and* $s = s_{\tau 1}$. *Then*

$$\Xi_{\mathfrak{p}}(\Phi, \omega, w) \sim \frac{\varrho}{2\pi\sqrt{-1}} \int_{\substack{\mathrm{Re}(s)=r_1 \\ r_1 > 3}} \widetilde{\Sigma}_{\mathfrak{d}}(\Phi, \omega, s, 1)\phi(s)\widetilde{\Lambda}(w; s, 1)ds.$$

Proof. By (5.3.8),

$$\begin{aligned}
\Xi_{\mathfrak{p}}(\Phi, \omega, w) &\sim \left(\frac{1}{2\pi\sqrt{-1}}\right)^2 \int_{\substack{\mathrm{Re}(s_\tau)=r \\ r_1>3,\ r_2>1}} \Sigma_{\mathfrak{d}}(\Phi, \omega, s_\tau)\widetilde{\Lambda}_\tau(w; s_\tau)ds_\tau \\
&= \left(\frac{1}{2\pi\sqrt{-1}}\right)^2 \int_{\substack{\mathrm{Re}(s_\tau)=r \\ r_1>3,\ 1>r_2>0}} \Sigma_{\mathfrak{d}}(\Phi, \omega, s_\tau)\widetilde{\Lambda}_\tau(w; s_\tau)ds_\tau \\
&\quad + \frac{\varrho}{2\pi\sqrt{-1}} \int_{\substack{\mathrm{Re}(s)=r_1 \\ r_1>3}} \widetilde{\Sigma}_{\mathfrak{d}}(\Phi, \omega, s)\phi(s)\widetilde{\Lambda}(w; s, 1)ds.
\end{aligned}$$

If r_1 is close to 3 and r_2 is close to 0, then $r_1 + Cr_2 < 1 + C$, because $C > 4$. Therefore, we can ignore the first term. This proves (5.3.9).

Q.E.D.

Next, we consider a path $\mathfrak{p} = (\mathfrak{d}, \mathfrak{s})$ such that $\mathfrak{d} = \mathfrak{d}_3$.

Proposition (5.3.10) *By changing* ψ *if necessary,*

$$\Xi_{\mathfrak{p}}(\Phi, \omega, w) \sim 0.$$

Proof. If $f(q)$ is a function of $q \in \mathbb{A}^1/k^\times$,

$$\int_{(\mathbb{A}^1/k^\times)^4} \omega(\widehat{t}^0) f(t_{12}^2 t_{22}) d^\times \widehat{t}^0 = \delta_\#(\omega) \int_{\mathbb{A}^1/k^\times} f(q)dq.$$

Consider the situation in (3.1.18). In this case, $A_{\mathfrak{d}} = \{d(\mu_1^2, \mu_1) \mid \mu_1 \in \mathbb{R}_+\}$ and $A_{\mathfrak{d}}^1 = A_{\mathfrak{d}}^2 = \{d(\mu_2, \mu_2^{-2}) \mid \mu_2 \in \mathbb{R}_+\}$. We make the change of variable $\lambda_1 = \mu_1^2\mu_2$, $\lambda_2 = \mu_1\mu_2^{-1}$. Then $d^\times\lambda = 5d^\times\mu_1 d^\times\mu_2$. Let τ, s_τ be as before. It is easy to see that $e_{\mathfrak{p}1}(\lambda) = \mu_1^5$ and and $\lambda^{\tau z - \rho} = \mu_1^{2s_{\tau 1}+s_{\tau 2}-3}\mu_2^{s_{\tau 1}-2s_{\tau 2}-1}$. Also $LS_{\mathfrak{p},\tau} + h$ in (3.5.17) is $\{s_\tau \mid s_{\tau 1} = 2s_{\tau 2} + 1\}$.

Therefore, by (3.5.17), $\Xi_{\mathfrak{p},\tau}(\Phi, \omega, w)$ is well defined for $\mathrm{Re}(w) \gg 0$ for all τ, and

$$\Xi_{\mathfrak{p},\tau}(\Phi, \omega, w) = \frac{\delta_{\mathfrak{d}}(\omega)}{2\pi\sqrt{-1}} \int_{\substack{\mathrm{Re}(s_{\tau 2})=r_2 \\ r_2>1}} \Sigma_1(R_{\mathfrak{d}}\Phi, s_{\tau 2}+1)\widetilde{\Lambda}_\tau(w; 2s_{\tau 2}+1, s_{\tau 2})ds_{\tau 2}.$$

Let ρ_τ be the point in $\mathbb{C}^2$ which corresponds to ρ by the substitution in §3.6. Then it is easy to see that $\rho_\tau = (1,-1)$ for $\tau = (\tau_{\mathrm{GL}(2)}, 1)$, $\rho_\tau = (-1,1)$ for $\tau = (1, \tau_{\mathrm{GL}(2)})$, and $\rho_\tau = (1,1)$ for $\tau = \tau_G$. For these cases, ρ_τ is not on the line $s_{\tau 1} = 2s_{\tau 2}+1$. Therefore, by (3.5.19), we can change ψ if necessary and assume that $\Xi_{\mathfrak{p},\tau}(\Phi,\omega,w) = 0$. For $\tau = (1,1)$, we choose $r_1 = 2r_2+1 \gg 0$. Then $\widetilde{L}_\tau(r) = -r_1 - Cr_2 < w_0$. Therefore, $\Xi_{\mathfrak{p},\tau}(\Phi,\omega,w) \sim 0$ for this case also. This proves the proposition.

Q.E.D.

§5.4 Contributions from $\mathfrak{d}_2, \mathfrak{d}_4$

In this section, we consider paths which start with β_2. We do not consider paths $\mathfrak{p} = (\mathfrak{d}, \mathfrak{s})$ such that $\mathfrak{s}(1) = 1$ for a similar reason as in §5.3. Let $\mathfrak{p}_2 = (\mathfrak{d}_2, \mathfrak{s}_2)$ be a path such that $\mathfrak{s}_2(1) = 0$. Let $\mathfrak{p}_{4i} = (\mathfrak{d}_4, \mathfrak{s}_{4i})$ be a path for $i = 1,2$ such that $\mathfrak{s}_{4i}(1) = 0$ and $\mathfrak{s}_{41}(2) = 0, \mathfrak{s}_{42}(2) = 1$.

We use the notation

$$\Sigma_{\mathfrak{d}_2}(\Psi,\omega,s),\ \Sigma_{\mathfrak{d}_2,\mathrm{ad}}(\Psi,\omega,s)$$

for the zeta function, the adjusted zeta function for the space $Z_{\mathfrak{d}}$ respectively. Also we use the notation $T_{\mathfrak{d}}(\widetilde{R}_{\mathfrak{d}_2,0}\Psi,\omega,s)$ etc. for the adjusting term etc. We define $\Sigma_{\mathfrak{d}_2,\mathrm{ad},(j)}(\Psi,\omega,s_0)$ similarly as in (5.3.4).

Since $\mathfrak{d}_2, \mathfrak{d}_4$ satisfy the condition (3.4.16) (1),

$$\begin{aligned}(5.4.1)\qquad \Xi_{\mathfrak{p}_2}(\Phi,\omega,w) &= \Xi_{\mathfrak{p}_2+}(\Phi,\omega,w) + \widehat{\Xi}_{\mathfrak{p}_2+}(\Phi,\omega,w)\\ &\quad + \widehat{\Xi}_{\mathfrak{p}_2\#}(\Phi,\omega,w) - \Xi_{\mathfrak{p}_2\#}(\Phi,\omega,w)\\ &\quad + \Xi_{\mathfrak{p}_{42}}(\Phi,\omega,w) - \Xi_{\mathfrak{p}_{41}}(\Phi,\omega,w).\end{aligned}$$

Also

$$\begin{aligned}\Xi_{\mathfrak{p}_2+}(\Phi,\omega,w) &= \Xi^s_{\mathfrak{p}_2+}(\Phi,\omega,w) + \Xi_{\mathfrak{p}_2,\mathrm{st}+}(\Phi,\omega,w),\\ \widehat{\Xi}_{\mathfrak{p}_2+}(\Phi,\omega,w) &= \widehat{\Xi}^s_{\mathfrak{p}_2+}(\Phi,\omega,w) + \widehat{\Xi}_{\mathfrak{p}_2,\mathrm{st}+}(\Phi,\omega,w).\end{aligned}$$

Let f be a function on $Z_{\mathfrak{d}_2\mathbb{A}}$. Then

$$\begin{aligned}&\int_{(\mathbb{A}^1/k^\times)^2} \omega_2(t_{21}t_{22})\omega_1(\det g_1) f(t_{22}g_1x)d^\times t_{21}d^\times t_{22}\\ &= \delta(\omega_2)\int_{\mathbb{A}^1/k^\times} \omega_1(\det g_1)f(t_{22}g_1x)d^\times t_{22},\end{aligned}$$

where $g_1 \in G_{1\mathbb{A}}, x \in Z_{\mathfrak{d}_2\mathbb{A}}$. It is easy to see that $\sigma_{\mathfrak{p}_2}(a_2(\underline{\lambda}_1^{-1}, \underline{\lambda}_1), 1) = \lambda_1^2$.

Therefore, by (3.5.13),

$$\begin{aligned}(5.4.2)\qquad \Xi^s_{\mathfrak{p}_2+}(\Phi,\omega,w) &\sim C_G\Lambda(w;\rho)\delta_{\mathfrak{d}_2}(\omega)\Sigma_{\mathfrak{d}_2+}(R_{\mathfrak{d}_2}\Phi,\omega_{\mathfrak{p}_2},2),\\ \widehat{\Xi}^s_{\mathfrak{p}_2+}(\Phi,\omega,w) &\sim C_G\Lambda(w;\rho)\delta_{\mathfrak{d}_2}(\omega)\Sigma_{\mathfrak{d}_2+}(\mathscr{F}_{\mathfrak{d}_2}R_{\mathfrak{d}_2}\Phi,\omega_{\mathfrak{p}_2}^{-1},1).\end{aligned}$$

Let $\widetilde{g}_{\mathfrak{d}_2} = (n_2(u_0),1)t^0$. We define

$$X_{\mathfrak{d}_2} = \{k\widetilde{g}_{\mathfrak{d}_2} \mid k \in K, \lambda_1 \le \alpha(u_0)^{-\frac{1}{2}}\}.$$

As in the case $W = \mathrm{Sym}^2 k^2$, $X_{\mathfrak{d}_2}/T_k$ is the fundamental domain for $G^0_{\mathbb{A}}/H_{\mathfrak{d}_2 k}$. Let $d\widetilde{g}_{\mathfrak{d}_2} = d^\times t^0 du_0$. Then

$$\Xi_{\mathfrak{p}_2,\mathrm{st}+}(\Phi,\omega,w) = \int_{\substack{X_{\mathfrak{d}_2}\cap B_{\mathbb{A}}/T_k \\ \lambda_1 \geq 1}} \omega(\widetilde{g}_{\mathfrak{d}_2})\Theta_{Z'_{\mathfrak{d}_2}}(R_{\mathfrak{d}_2}\Phi,\widetilde{g}_{\mathfrak{d}_2})\lambda_2^{-2}\mathscr{E}_{\mathfrak{p}_2}(\widetilde{g}_{\mathfrak{d}_2},w)d\widetilde{g}_{\mathfrak{d}_2},$$

$$\widehat{\Xi}_{\mathfrak{p}_2,\mathrm{st}+}(\Phi,\omega,w) = \int_{\substack{X_{\mathfrak{d}_2}\cap B_{\mathbb{A}}/T_k \\ \lambda_1 \leq 1}} \omega(\widetilde{g}_{\mathfrak{d}_2})\Theta_{Z'_{\mathfrak{d}_2}}(R_{\mathfrak{d}_2}\Phi,\theta_{\mathfrak{d}_2}(\widetilde{g}_{\mathfrak{d}_2}))\lambda_2\mathscr{E}_{\mathfrak{p}_2}(\widetilde{g}_{\mathfrak{d}_2},w)d\widetilde{g}_{\mathfrak{d}_2}.$$

Lemma (5.4.3)

(1)
$$\int_{X_{\mathfrak{d}_2}\cap B_{\mathbb{A}}/T_k} \omega(\widetilde{g}_{\mathfrak{d}_2})\Theta_{Z'_{\mathfrak{d}_2}}(R_{\mathfrak{d}_2}\Phi,\widetilde{g}_{\mathfrak{d}_2})\lambda_2^{-2}\mathscr{E}_{\mathfrak{p}_2}(\widetilde{g}_{\mathfrak{d}_2},w)d\widetilde{g}_{\mathfrak{d}_2}$$
$$\sim \int_{X_{\mathfrak{d}_2}\cap B_{\mathbb{A}}/T_k} \omega(\widetilde{g}_{\mathfrak{d}_2})\Theta_{Z'_{\mathfrak{d}_2}}(R_{\mathfrak{d}_2}\Phi,\widetilde{g}_{\mathfrak{d}_2})\lambda_2^{-2}\mathscr{E}_N(t^0,w)d\widetilde{g}_{\mathfrak{d}_2}.$$

(2)
$$\int_{X_{\mathfrak{d}_2}\cap B_{\mathbb{A}}/T_k} \omega(\widetilde{g}_{\mathfrak{d}_2})\Theta_{Z'_{\mathfrak{d}_2}}(R_{\mathfrak{d}_2}\Phi,\theta_{\mathfrak{d}_2}(\widetilde{g}_{\mathfrak{d}_2}))\lambda_2\mathscr{E}_{\mathfrak{p}_2}(\widetilde{g}_{\mathfrak{d}_2},w)d\widetilde{g}_{\mathfrak{d}_2}$$
$$\sim \int_{X_{\mathfrak{d}_2}\cap B_{\mathbb{A}}/T_k} \omega(\widetilde{g}_{\mathfrak{d}_2})\Theta_{Z'_{\mathfrak{d}_2}}(R_{\mathfrak{d}_2}\Phi,\theta_{\mathfrak{d}_2}(\widetilde{g}_{\mathfrak{d}_2}))\lambda_2\mathscr{E}_N(t^0,w)d\widetilde{g}_{\mathfrak{d}_2}.$$

Proof. Let $I = (\{1\},\emptyset)$. Then

$$|\mathscr{E}_{\mathfrak{p}_2}(\widetilde{g}_{\mathfrak{d}_2},w) - \mathscr{E}_N(t^0,w)| \ll \sum_\tau \widetilde{\mathscr{E}}_{I,\tau,1}(\widetilde{g}_{\mathfrak{d}_2},w).$$

Let $M > w_0$. By (3.4.30), for any $\epsilon > 0$, there exist $\delta = \delta(\epsilon) > 0$ and $c \in \mathfrak{t}^*$ such that $\|c\|_0 < \epsilon$ and for any $l \gg 0$,

$$\widetilde{\mathscr{E}}_{I,\tau,1}(\widetilde{g}_{\mathfrak{d}_2},w) \ll (t^0)^c\lambda_1^l$$

for $w_0 - \delta \leq \mathrm{Re}(w) \leq M$.

We choose $0 \leq \Psi \in \mathscr{S}(Z_{\mathfrak{d}_2,0\mathbb{A}})$ so that $|\widetilde{R}_{\mathfrak{d}_2,0}\Phi| \leq \Psi$. Then the above integral is bounded by a constant multiple of

$$\int_{\substack{\mathbb{R}_+\times\mathbb{A}^\times/k^\times\times\mathbb{A} \\ \lambda_1\leq\sqrt{\alpha(u_0)}^{-1},|t'|\geq 1}} \lambda_1^{l'}|t'|^{c'}\widetilde{\Theta}_{Z_{\mathfrak{d}_2,0}}(\Psi,t',u_0)d^\times\lambda_1 d^\times t' du_0$$

where c' is some constant, and l' can be arbitrarily large. Since $T_{\mathfrak{d}_2+}(\Psi,s,s_1)$ converges absolutely for all s,s_1, the above integral converges absolutely.

Q.E.D.

Let τ, s_τ be as before. We define

(5.4.4)
$$\Sigma_{\mathfrak{p}_2,\mathrm{st}+}(\Phi,s_\tau) = \frac{T_{\mathfrak{d}_2+}(\widetilde{R}_{\mathfrak{d}_2,0}R_{\mathfrak{d}_2}\Phi,1+s_{\tau 2},\frac{1-s_{\tau 1}}{2})}{s_{\tau 1}-1},$$
$$\widehat{\Sigma}_{\mathfrak{p}_2,\mathrm{st}+}(\Phi,s_\tau) = \frac{T_{\mathfrak{d}_2+}(\widetilde{R}_{\mathfrak{d}_2,0}\mathscr{F}_{\mathfrak{d}_2}R_{\mathfrak{d}_2}\Phi,2-s_{\tau 2},\frac{1-s_{\tau 1}}{2})}{s_{\tau 1}-1}.$$

Lemma (5.4.5)

(1)
$$\int_{\substack{X_{\mathfrak{d}_2}\cap B_{\mathbb{A}}/T_k\\ \lambda_1\geq 1}} \omega(\widetilde{g}_{\mathfrak{d}_2})\Theta_{Z'_{\mathfrak{d}_2,0}}(R_{\mathfrak{d}_2}\Phi,\widetilde{g}_{\mathfrak{d}_2})\lambda_2^2\mathscr{E}_{N,\tau}(t^0,w)d\widetilde{g}_{\mathfrak{d}_2}$$
$$= \delta_{\mathfrak{d}_2,\mathrm{st}}(\omega)\left(\frac{1}{2\pi\sqrt{-1}}\right)^2\int_{\substack{\mathrm{Re}(s_\tau)=r\\ r_1>1,\ r_2>1}} \Sigma_{\mathfrak{p}_2,\mathrm{st}+}(\Phi,s_\tau)\widetilde{\Lambda}_\tau(w;s_\tau)ds_\tau.$$

(2)
$$\int_{\substack{X_{\mathfrak{d}_2}\cap B_{\mathbb{A}}/T_k\\ \lambda_1\leq 1}} \omega(\widetilde{g}_{\mathfrak{d}_2})\Theta_{Z'_{\mathfrak{d}_2,0}}(R_{\mathfrak{d}_2}\Phi,\theta_{\mathfrak{d}_2}(\widetilde{g}_{\mathfrak{d}_2}))\lambda_2^{-1}\mathscr{E}_{N,\tau}(t^0,w)d\widetilde{g}_{\mathfrak{d}_2}$$
$$= \delta_{\mathfrak{d}_2,\mathrm{st}}(\omega)\left(\frac{1}{2\pi\sqrt{-1}}\right)^2\int_{\substack{\mathrm{Re}(s_\tau)=r\\ r_1>1,\ r_2>1}} \widehat{\Sigma}_{\mathfrak{p}_2,\mathrm{st}+}(\Phi,s_\tau)\widetilde{\Lambda}_\tau(w;s_\tau)ds_\tau.$$

In particular, these integrals are well defined.

Proof. Suppose that $f(q)$ is a function of $q\in(\mathbb{A}^1/k^\times)^2$. Then
$$\int_{(\mathbb{A}^1/k^\times)^4}\omega(t^0)f(t_{11}t_{12}t_{22})d^\times t^0 = \delta_{\mathfrak{d}_2,\mathrm{st}}(\omega)\int_{\mathbb{A}^1/k^\times} f(q)d^\times q.$$

Therefore, (1) is equal to
$$\delta_{\mathfrak{d}_2,\mathrm{st}}(\omega)\int_{\substack{\mathbb{R}_+\times\mathbb{A}^\times/k^\times\times\mathbb{A}\\ \lambda_1\leq\sqrt{\alpha(u_0)}^{-1},|t'|\geq 1}} \widetilde{\Theta}_{Z_{\mathfrak{d}_2,0}}(\widetilde{R}_{\mathfrak{d}_2,0}R_{\mathfrak{d}_2}\Phi,t',u_0)\lambda_2^2\mathscr{E}_{N,\tau}(t^0,w)d^\times\lambda_2 d^\times t' du_0,$$

where $t'=\lambda_2 t_{11}t_{12}t_{22}$. Since $(t^0)^{\tau z+\rho}=|t'|^{s_{\tau 2}+1}\lambda_1^{s_{\tau 1}-1}$, the lemma follows for (1). The proof of (2) is similar.

Q.E.D.

Lemma (5.4.6) *If* $\tau\neq\tau_G$,

(1)
$$\left(\frac{1}{2\pi\sqrt{-1}}\right)^2\int_{\substack{\mathrm{Re}(s_\tau)=r\\ r_1>1,\ r_2>1}} \Sigma_{\mathfrak{p}_2,\mathrm{st}+}(\Phi,s_\tau)\widetilde{\Lambda}_\tau(w;s_\tau)ds_\tau\sim 0,$$

(2)
$$\left(\frac{1}{2\pi\sqrt{-1}}\right)^2\int_{\substack{\mathrm{Re}(s_\tau)=r\\ r_1>1,\ r_2>1}} \widehat{\Sigma}_{\mathfrak{p}_2,\mathrm{st}+}(\Phi,s_\tau)\widetilde{\Lambda}_\tau(w;s_\tau)ds_\tau\sim 0.$$

Proof. Since $T_{\mathfrak{d}_2+}(\Psi,s,s_1)$ is an entire function for $\Psi=\widetilde{R}_{\mathfrak{d}_2,0}R_{\mathfrak{d}_2}\Phi,\widetilde{R}_{\mathfrak{d}_2,0}\mathscr{F}_{\mathfrak{d}_2}R_{\mathfrak{d}_2}\Phi$, we can choose $r_1,r_2\gg 0$ for $\tau=1$, r_1 fixed and $r_2\gg 0$ for $\tau=(\tau_{\mathrm{GL}(2)},1)$, and $r_1\gg 0$ and r_2 fixed for $\tau=(1,\tau_{\mathrm{GL}(2)})$. Then $\widetilde{L}(r)<w_0$. This proves the lemma. Q.E.D.

By moving the contour in the above integrals so that $r_2<1$, we get the following proposition.

Proposition (5.4.7) *Let* $\tau=\tau_G$, *and* $s=s_{\tau 1}$. *Then*

(1)
$$\Xi_{\mathfrak{p}_2,\mathrm{st}+}(\Phi,\omega,w)\sim\frac{\varrho\delta_{\mathfrak{d}_2,\mathrm{st}}(\omega)}{2\pi\sqrt{-1}}\int_{\substack{\mathrm{Re}(s)=r_1\\ r_1>1}} \Sigma_{\mathfrak{p}_2,\mathrm{st}+}(\Phi,s,1)\phi(s)\widetilde{\Lambda}(w;s,1)ds,$$

(2)
$$\widehat{\Xi}_{\mathfrak{p}_2,\mathrm{st}+}(\Phi,\omega,w)\sim\frac{\varrho\delta_{\mathfrak{d}_2,\mathrm{st}}(\omega)}{2\pi\sqrt{-1}}\int_{\substack{\mathrm{Re}(s)=r_1\\ r_1>1}} \widehat{\Sigma}_{\mathfrak{p}_2,\mathrm{st}+}(\Phi,s,1)\phi(s)\widetilde{\Lambda}(w;s,1)ds.$$

Next, we consider $\mathfrak{p}_{41}, \mathfrak{p}_{42}$. Let τ, s_τ be as before. We consider the domain D_τ (2.4.7) for $r = \mathrm{Re}(s_\tau)$ also.

Lemma (5.4.8)

$$(1)\ \Xi_{\mathfrak{p}_{41},\tau}(\Phi,\omega,w) = \frac{\varrho\delta_{\mathfrak{d}_4}(\omega)}{2\pi\sqrt{-1}} \int_{\substack{\mathrm{Re}(s_\tau)=r \\ r\in D_\tau,\ r_1<2r_2+1}} \frac{\Sigma_1(R_{\mathfrak{d}_4}\Phi_{\mathfrak{p}_{41}}, \frac{s_{\tau 1}+1}{2})}{-s_{\tau 1}+2s_{\tau 2}+1} \widetilde{\Lambda}_\tau(w;s_\tau)ds_\tau.$$

$$(2)\ \Xi_{\mathfrak{p}_{42},\tau}(\Phi,\omega,w) = \frac{\varrho\delta_{\mathfrak{d}_4}(\omega)}{2\pi\sqrt{-1}} \int_{\substack{\mathrm{Re}(s_\tau)=r \\ r\in D_\tau,\ r_1>3-2r_2}} \frac{\Sigma_1(R_{\mathfrak{d}_4}\Phi_{\mathfrak{p}_{42}}, \frac{s_{\tau 1}+1}{2})}{s_{\tau 1}+2s_{\tau 2}-3} \widetilde{\Lambda}_\tau(w;s_\tau)ds_\tau.$$

Proof. Consider s_τ such that $r = \mathrm{Re}(s_\tau)$ satisfies the conditions in (1) and (2). It is easy to see that $A_{\mathfrak{p}_{41}2}T^0_{\mathbb{A}} = \{t^0 \mid \lambda_2 \le 1\}$, and $A_{\mathfrak{p}_{42}2}T^0_{\mathbb{A}} = \{t^0 \mid \lambda_2 \ge 1\}$. Easy computations show that

$$\sigma_{\mathfrak{p}_{41}}(t^0) = \lambda_1^2\lambda_2^2,\ \sigma_{\mathfrak{p}_{42}}(t^0) = \lambda_1^2\lambda_2,$$
$$(t^0)^{\tau z+\rho} = \lambda_1^{s_{\tau 1}-1}\lambda_2^{s_{\tau 2}-1},\ \theta_{\mathfrak{d}_2}(t^0)^{\tau z+\rho} = \lambda_1^{s_{\tau 1}-1}\lambda_2^{1-s_{\tau 2}}.$$

Suppose that $f(q)$ is a function of $q \in \mathbb{A}^\times/k^\times$. Then

$$\int_{(\mathbb{A}^1/k^\times)^4} f(t_{12}^2 t_{22})d^\times t^0 = \delta_{\mathfrak{d}_4}(\omega)\int_{\mathbb{A}^1/k^\times} f(q)d^\times q$$

after the change of variable $q = t_{12}^2 t_{22}$. Let $t' = \underline{\mu} q$ where $\mu \in \mathbb{R}_+$. We make the change of variable $\mu = \lambda_1^2\lambda_2$, $\lambda_2 = \lambda_2$. Then $d^\times\mu d^\times\lambda_2 = 2d^\times\lambda_1 d^\times\lambda_2$, and $\lambda_1 = \mu^{\frac12}\lambda_2^{-\frac12}$. So

$$\sigma_{\mathfrak{p}_{41}}(t^0)(t^0)^{\tau z+\rho} = \mu^{\frac{1+s_{\tau 1}}{2}}\lambda_2^{\frac{-s_{\tau 1}+2s_{\tau 2}+1}{2}},\ \sigma_{\mathfrak{p}_{41}}(t^0)\theta_{\mathfrak{d}_2}(t^0)^{\tau z+\rho} = \mu^{\frac{1+s_{\tau 1}}{2}}\lambda_2^{-\frac{s_{\tau 1}+2s_{\tau 2}-3}{2}}.$$

Since

$$\begin{aligned}
&\frac12\int_{\substack{\mathbb{R}_+^2\times\mathbb{A}^1/k^\times \\ \lambda_2\le 1}} \mu^{\frac{1+s_{\tau 1}}{2}}\lambda_2^{\frac{-s_{\tau 1}+2s_{\tau 2}+1}{2}}\Theta_{Z_{\mathfrak{d}_4}}(R_{\mathfrak{d}_4}\Phi_{\mathfrak{p}_{41}},\underline{\mu}q)d^\times\lambda_2 d^\times\mu d^\times q \\
&= \frac{\Sigma_1(R_{\mathfrak{d}_4}\Phi_{\mathfrak{p}_{41}}, \frac{s_{\tau 1}+1}{2})}{-s_{\tau 1}+2s_{\tau 2}+1}, \\
&\frac12\int_{\substack{\mathbb{R}_+^2\times\mathbb{A}^1/k^\times \\ \lambda_2\ge 1}} \mu^{\frac{1+s_{\tau 1}}{2}}\lambda_2^{-\frac{s_{\tau 1}+2s_{\tau 2}-3}{2}}\Theta_{Z_{\mathfrak{d}_4}}(R_{\mathfrak{d}_4}\Phi_{\mathfrak{p}_{42}},\underline{\mu}q)d^\times\lambda_2 d^\times\mu d^\times q \\
&= \frac{\Sigma_1(R_{\mathfrak{d}_4}\Phi_{\mathfrak{p}_{42}}, \frac{s_{\tau 1}+1}{2})}{s_{\tau 1}+2s_{\tau 2}-3},
\end{aligned}$$

the lemma follows.

Q.E.D.

Lemma (5.4.9) *If $\tau \ne \tau_G$, $\Xi_{\mathfrak{p}_{4i},\tau}(\Phi,\omega,w) \sim 0$ for $i = 1, 2$.*

Proof. If $\tau = 1$ or $(\tau_{\mathrm{GL}(2)}, 1)$, we fix r_1 and choose $r_2 \gg 0$ for $\mathfrak{p}_{41}, \mathfrak{p}_{42}$. Then $\widetilde{L}(r) < w_0$ and $r_1 < 2r_2+1, r_1 > 3-2r_2$. If r_1, r_2 are close to 1, then $\widetilde{L}(r) < w_0$

and $r_1 < 2r_2 + 1$ for $\tau = (1, \tau_{\mathrm{GL}(2)})$ and $\mathfrak{p}_{41}$. If $r_1 \gg 0$, then $r_1 > 3 - 2r_2$, and $\widetilde{L}(r) < w_0$ for $\tau = (1, \tau_{\mathrm{GL}(2)})$ and $\mathfrak{p}_{42}$. This proves the lemma.

Q.E.D.

Proposition (5.4.10) *Let $\tau = \tau_G$, and $s = s_{\tau 1}$. Then*

$$(1) \qquad \Xi_{\mathfrak{p}_{41}}(\Phi, \omega, w) \sim \frac{\varrho\delta_{\mathfrak{d}_4}(\omega)}{2\pi\sqrt{-1}} \int_{\substack{\mathrm{Re}(s)=r_1 \\ 1<r_1<3}} \frac{\Sigma_1(R_{\mathfrak{d}_4}\Phi_{\mathfrak{p}_{41}}, \frac{s+1}{2})}{3-s} \phi(s)\widetilde{\Lambda}(w; s, 1)ds,$$

$$(2) \qquad \Xi_{\mathfrak{p}_{42}}(\Phi, \omega, w) \sim \frac{\varrho\delta_{\mathfrak{d}_4}(\omega)}{2\pi\sqrt{-1}} \int_{\substack{\mathrm{Re}(s)=r_1 \\ 1<r_1}} \frac{\Sigma_1(R_{\mathfrak{d}_4}\Phi_{\mathfrak{p}_{42}}, \frac{s+1}{2})}{s-1} \phi(s)\widetilde{\Lambda}(w; s, 1)ds.$$

Proof.

$$\begin{aligned}
\Xi_{\mathfrak{p}_{41}}(\Phi, \omega, w) &\sim \int_{\substack{\mathrm{Re}(s_\tau)=r \\ r_1=r_2=2}} \frac{\Sigma_1(R_{\mathfrak{d}_4}\Phi_{\mathfrak{p}_{41}}, \frac{s_{\tau 1}+1}{2})}{-s_{\tau 1} + 2s_{\tau 2} + 1} \widetilde{\Lambda}_\tau(w; s_\tau)ds_\tau \\
&= \int_{\substack{\mathrm{Re}(s_\tau)=r \\ r_1=2,\ r_2=\frac{3}{4}}} \frac{\Sigma_1(R_{\mathfrak{d}_4}\Phi_{\mathfrak{p}_{41}}, \frac{s_{\tau 1}+1}{2})}{-s_{\tau 1} + 2s_{\tau 2} + 1} \widetilde{\Lambda}_\tau(w; s_\tau)ds_\tau \\
&\quad + \frac{\varrho\delta_{\mathfrak{d}_4}(\omega)}{2\pi\sqrt{-1}} \int_{\substack{\mathrm{Re}(s)=r_1 \\ 1<r_1<3}} \frac{\Sigma_1(R_{\mathfrak{d}_4}\Phi_{\mathfrak{p}_{41}}, \frac{s+1}{2})}{3-s} \phi(s)\widetilde{\Lambda}(w; s, 1)ds \\
&\sim \frac{\varrho\delta_{\mathfrak{d}_4}(\omega)}{2\pi\sqrt{-1}} \int_{\substack{\mathrm{Re}(s)=r_1 \\ 1<r_1<3}} \frac{\Sigma_1(R_{\mathfrak{d}_4}\Phi_{\mathfrak{p}_{41}}, \frac{s+1}{2})}{3-s} \phi(s)\widetilde{\Lambda}(w; s, 1)ds,
\end{aligned}$$

because $\widetilde{L}(2, \frac{3}{4}) = 2 + \frac{3}{4}C < 1 + C$ $(C > 4)$. This proves the first statement.

$$\begin{aligned}
\Xi_{\mathfrak{p}_{42}}(\Phi, \omega, w) &\sim \frac{\varrho\delta_{\mathfrak{d}_4}(\omega)}{2\pi\sqrt{-1}} \int_{\mathrm{Re}(s_\tau)=(2,\frac{3}{4})} \frac{\Sigma_1(R_{\mathfrak{d}_4}\Phi_{\mathfrak{p}_{41}}, \frac{s_{\tau 1}+1}{2})}{s_{\tau 1} + 2s_{\tau 2} - 3} \widetilde{\Lambda}_\tau(w; s_\tau)ds_\tau \\
&\quad + \frac{\varrho\delta_{\mathfrak{d}_4}(\omega)}{2\pi\sqrt{-1}} \int_{\substack{\mathrm{Re}(s)=r_1 \\ 1<r_1}} \frac{\Sigma_1(R_{\mathfrak{d}_4}\Phi_{\mathfrak{p}_{42}}, \frac{s+1}{2})}{s-1} \phi(s)\widetilde{\Lambda}(w; s, 1)ds \\
&\sim \frac{\varrho\delta_{\mathfrak{d}_4}(\omega)}{2\pi\sqrt{-1}} \int_{\substack{\mathrm{Re}(s)=r_1 \\ 1<r_1}} \frac{\Sigma_1(R_{\mathfrak{d}_4}\Phi_{\mathfrak{p}_{42}}, \frac{s+1}{2})}{s-1} \phi(s)\widetilde{\Lambda}(w; s, 1)ds
\end{aligned}$$

for the same reason.

Q.E.D.

Proposition (5.4.11)

$$(1) \qquad \Xi_{\mathfrak{p}_{2\#}}(\Phi, \omega, w) \sim \Lambda(w; \rho) \frac{\varrho\delta_{\#}(\omega)\Phi(0)}{2}.$$

$$(2) \qquad \widehat{\Xi}_{\mathfrak{p}_{2\#}}(\Phi, \omega, w) \sim -\Lambda(w; \rho)\varrho\delta_{\#}(\omega)\mathscr{F}_{\mathfrak{d}_2}R_{\mathfrak{d}_2}\Phi(0).$$

Proof. Let τ_2 be a permutation of $\{1, 2\}$. By the Mellin inversion formula,

$$\begin{aligned}
&\left(\frac{1}{2\pi\sqrt{-1}}\right)^2 \int_{\substack{G^0_{1\mathbb{A}}/G_{1k} \times T^0_{2\mathbb{A}}/T_{2k} \\ \lambda_2 \le 1}} \omega(g_1, t_2)\lambda_2^{-1}\mathscr{E}_{\{1\}\times N_2,(1,\tau_2)}((g_1, t_2), w)dg_1 d^\times t_2 \\
&= \frac{\delta_{\#}(\omega)}{2\pi\sqrt{-1}} \int_{\substack{\mathrm{Re}(s)=r_1 \\ r_1>2}} \frac{\widetilde{\Lambda}_{(1,\tau_2)}(w; -1, s)}{s-2} ds.
\end{aligned}$$

If $\tau_2 \neq \tau_{\mathrm{GL}(2)}$, we can choose $r \gg 0$, and ignore this function. If $\tau_2 = \tau_{\mathrm{GL}(2)}$,

$$\frac{1}{2\pi\sqrt{-1}} \int_{\substack{\mathrm{Re}(s)=r_1 \\ r_1>2}} \frac{\widetilde{\Lambda}_{(1,\tau_2)}(w;-1,s)}{s-2} ds \sim -\varrho\delta_{\#}(\omega)\Lambda(w;\rho).$$

Similarly,

$$\begin{aligned}
&\left(\frac{1}{2\pi\sqrt{-1}}\right)^2 \int_{\substack{G^0_{1\mathbb{A}}/G_{1k} \times T^0_{2\mathbb{A}}/T_{2k} \\ \lambda_2 \leq 1}} \omega(g_1,t_2)\lambda^2 \mathscr{E}_{\{1\}\times N_2,(1,\tau_2)}((g_1,t_2),w) dg_1 d^\times t_2 \\
&= \frac{\delta_{\#}(\omega)}{2\pi\sqrt{-1}} \int_{\substack{\mathrm{Re}(s)=r_1 \\ r_1>1}} \frac{\widetilde{\Lambda}_{(1,\tau_2)}(w;-1,s)}{s+1} ds \\
&\sim \frac{\varrho\delta_{\#}(\omega)}{2}\Lambda(w;\rho).
\end{aligned}$$

This proves (5.4.11).

Q.E.D.

§5.5 The contribution from $V^{\mathrm{ss}}_{\mathrm{st}k}$

In this section, we consider the contribution from $V^{\mathrm{ss}}_{\mathrm{st}k}$. Let

$$b = (n_2(u_0),1)t^0(1,n_2(u_2)),\ db = \lambda_2^2 d^\times t^0 du_0 du_2,$$

where $t^0 \in T^0_{\mathbb{A}}, u_0, u_2 \in \mathbb{A}$. We define

$$X_V = \{kb \mid k \in K, \lambda_1 \leq \alpha(u_0)^{-\frac{1}{2}}\}.$$

By a similar consideration as before, X_V/T_k is the fundamental domain for $G^0_{\mathbb{A}}/H_{Vk}$.

Proposition (5.5.1) *Let $\tau = \tau_G$, and $s = s_{\tau 1}$. Then*

$$\Xi_{V,\mathrm{st}}(\Phi,\omega,w) \sim \frac{\varrho\delta_{V,\mathrm{st}}(\omega)}{2\pi\sqrt{-1}} \int_{\substack{\mathrm{Re}(s)=r_1 \\ r_1>1}} \frac{T^1_V(R_{V,0}\Phi,\omega_{V,\mathrm{st}},1,\frac{1-s}{2})}{s-1} \phi(s)\widetilde{\Lambda}(w;s,1)ds.$$

Proof. By (5.1.3),

$$\begin{aligned}
\Xi_{V,\mathrm{st}}(\omega,\Phi,w) &= \int_{G^0_{\mathbb{A}}/G_k} \omega(g^0)\Theta_{V^{\mathrm{ss}}_{\mathrm{st}k}}(\Phi,g^0)\mathscr{E}(g^0,w)dg^0 \\
&= \int_{G^0_{\mathbb{A}}/H_{Vk}} \omega(g^0)\Theta_{Z'_{V,0}}(\Phi,g^0)\mathscr{E}(g^0,w)dg^0 \\
&= \int_{X_V \cap B_{\mathbb{A}}/T_k} \omega(b)\Theta_{Z'_{V,0}}(\Phi,b)\mathscr{E}(b,w)db.
\end{aligned}$$

Lemma (5.5.2) *Let c_1, c_2 be any constants such that $c_1+2c_2 > 0$. Then the integral*

$$\int_{X_V \cap B_{\mathbb{A}}/T_k} \lambda_1^{c_1}\lambda_2^{c_2}\Theta_{Z'_{V,0}}(\Phi,b)db$$

converges absolutely.

Proof. Let $\mu = \lambda_1^{-2}\lambda_2$, and $\Psi = R_{Z_{V,0}}\Phi$. It is easy to see that

$$\int_{\mathbb{A}} \Theta_{Z'_{V,0}}(\Phi, b)du_2 = \lambda_1^{-2}\lambda_2^{-1}\widetilde{\Theta}_{Z_{V,0}}(\Psi, 1, \mu, q, u_0),$$

where $q = (q_1, q_2)$, $q_1 = t_{12}^2 t_{21}, q_2 = t_{11}^2 t_{22}$.

Therefore, the above integral is equal to

$$\begin{aligned}
&\int_{\substack{\mathbb{R}_+^2 \times (\mathbb{A}^1/k^\times)^2 \times \mathbb{A} \\ \lambda_1 \le \sqrt{\alpha(u_0)}^{-1}}} \lambda_1^{c_1-2}\lambda_2^{c_2+1}\widetilde{\Theta}_{Z_{V,0}}(\Psi, 1, \mu, q, u_0)d^\times\lambda_1 d^\times\mu d^\times q du_0 \\
&= \int_{\substack{\mathbb{R}_+^2 \times (\mathbb{A}^1/k^\times)^2 \times \mathbb{A} \\ \lambda_1 \le \sqrt{\alpha(u_0)}^{-1}}} \lambda_1^{c_1+2c_2}\mu^{c_2+1}\widetilde{\Theta}_{Z_{V,0}}(\Psi, 1, \mu, q, u_0)d^\times\lambda_1 d^\times\mu d^\times q du_0 \\
&= \int_{\mathbb{R}_+ \times (\mathbb{A}^1/k^\times)^2 \times \mathbb{A}} \frac{\mu^{c_2+1}\alpha(u_0)^{-\frac{c_1+2c_2}{2}}}{c_1+2c_2}\widetilde{\Theta}_{Z_{V,0}}(\Psi, 1, \mu, q, u_0)d^\times\mu d^\times t^0 du_0,
\end{aligned}$$

which converges absolutely.

Q.E.D.

Lemma (5.5.3)

$$\int_{X_V \cap B_{\mathbb{A}}/T_k} \omega(b)\Theta_{Z'_{V,0}}(\Phi, b)\mathscr{E}(b, w)db \sim \int_{X_V \cap B_{\mathbb{A}}/T_k} \omega(b)\Theta_{Z'_{V,0}}(\Phi, b)\mathscr{E}_N(t^0, w)db.$$

Proof. Let $M > w_0$. By (3.4.30), there exists a slowly increasing function $h(\lambda_1, \lambda_2)$ and a constant $\delta > 0$ such that for any $l_1, l_2 \gg 0$,

$$|\mathscr{E}(b, w) - \mathscr{E}_N(t^0, w)| \ll h(\lambda_1, \lambda_2) \sup(\lambda_1^{l_1}, \lambda_2^{l_2}, \lambda_1^{l_1}\lambda_2^{l_2})$$

for $w_0 - \delta \le \mathrm{Re}(w) \le M$. Therefore this lemma follows from (5.5.2).

Q.E.D.

Suppose that $f(q)$ is a function of $q \in \mathbb{A}^\times/k^\times$. Then

$$\int_{(\mathbb{A}^1/k^\times)^4} f(t_{12}^2 t_{21}, t_{11}^2 t_{22})d^\times\widehat{t}^0 = \delta_{V,\mathrm{st}}(\omega)\int_{(\mathbb{A}^1/k^\times)^2} \omega_{V,\mathrm{st}}(q)f(q)d^\times q$$

after the change of variable $q_1 = t_{12}^2 t_{21}, q_2 = t_{11}^2 t_{12}$. We make the change of variable $\mu = \lambda_1^{-2}\lambda_2, \lambda_1 = \lambda_1$. Then $d^\times\mu d^\times\lambda_1 = d^\times\lambda_1 d^\times\lambda_2$, and $\lambda_2 = \lambda_1^2\mu$.

Clearly,

$$\int_{\mathbb{A}} \Theta_{Z'_{V,0}}(\Phi, (n_2(u_0), 1)t^0(1, n_2(u_1)))du_0 = \lambda_1^{-2}\lambda_2^{-1}\widetilde{\Theta}_{Z_{V,0}}(\Psi, 1, \mu, q, u_0).$$

Therefore,

$$\begin{aligned}
&\Xi_{V,\mathrm{st}}(\Phi, \omega, w) \\
&\sim \delta_{V,\mathrm{st}}(\omega)\int_{\substack{\mathbb{R}_+^2 \times (\mathbb{A}^1/k^\times)^2 \times \mathbb{A} \\ \lambda_1 \le \sqrt{\alpha(u_0)}^{-1}}} \omega_{V,\mathrm{st}}(q)\widetilde{\Theta}_{Z_{V,0}}(\Psi, 1, \mu, q, u_0)\mu\mathscr{E}_N(t^0, w)d^\times\lambda_1 d^\times\mu d^\times q du_0.
\end{aligned}$$

Let τ, s_τ be as before. We define $\Xi_{V,\mathrm{st},\tau}(\Phi, \omega, w)$ by the following integral

(5.5.4)

$$\delta_{V,\mathrm{st}}(\omega) \int_{\substack{\mathbb{R}_+^2 \times (\mathbb{A}^1/k^\times)^2 \times \mathbb{A} \\ \lambda_1 \le \sqrt{\alpha(u_0)}^{-1}}} \omega_{V,\mathrm{st}}(q) \widetilde{\Theta}_{Z_{V,0}}(\Psi, 1, \mu, q, u_0) \mu \mathscr{E}_{N,\tau}(t^0, w) d^\times \lambda_1 d^\times \mu d^\times q du_0,$$

if this integral is well defined.

Let

$$\Sigma_{V,\mathrm{st}}(\Phi, \omega, s_\tau) = \delta_{V,\mathrm{st}}(\omega) \frac{T_V^1(R_{V,0}\Phi, \omega_{V,\mathrm{st}}, s_{\tau 2}, \frac{-s_{\tau 1} - 2s_{\tau 2} + 3}{2})}{s_{\tau 1} + 2s_{\tau 2} - 3}. \tag{5.5.5}$$

It is easy to see that $(t^0)^{\tau z + \rho} = \lambda_1^{s_{\tau 1} + 2s_{\tau 2} - 3} \mu^{s_{\tau 2} - 1}$. Therefore, we get the following proposition.

Proposition (5.5.6) *The distribution $\Xi_{V,\mathrm{st},\tau}(\Phi, \omega, w)$ is well defined for all τ and*

$$\Xi_{V,\mathrm{st},\tau}(\Phi, \omega, w) \sim \left(\frac{1}{2\pi\sqrt{-1}}\right)^2 \int_{\substack{\mathrm{Re}(s_\tau) = r \\ r \in D_\tau,\ r_1 > 3 - 2r_2}} \Sigma_{V,\mathrm{st}}(\Phi, \omega, s_\tau) \widetilde{\Lambda}_\tau(w, s_\tau) ds_\tau.$$

If $\tau \neq \tau_G$, we can choose r_1 or $r_2 \gg 0$, and $\Xi_{V,\mathrm{st},\tau}(\Phi, \omega, w) \sim 0$. Let $\tau = \tau_G$. Then

$$\begin{aligned} \Xi_{V,\mathrm{st},\tau}(\Phi, \omega, w) &= \left(\frac{1}{2\pi\sqrt{-1}}\right)^2 \int_{\mathrm{Re}(s_\tau) = (\frac{3}{2}, \frac{4}{5})} \Sigma_{V,\mathrm{st}}(\Phi, \omega, s_\tau) \widetilde{\Lambda}_\tau(w, s_\tau) ds_\tau \\ &\quad + \frac{\varrho \delta_{V,\mathrm{st}}(\omega)}{2\pi\sqrt{-1}} \int_{\substack{\mathrm{Re}(s) = r_1 \\ r_1 > 1}} \frac{T_V^1(R_{V,0}\Phi, \omega_{V,\mathrm{st}}, 1, \frac{1-s}{2})}{s - 1} \phi(s) \widetilde{\Lambda}(w, s, 1) ds. \end{aligned}$$

Since $C > 4$, $\frac{3}{2} + \frac{4}{5}C < 1 + C$. Therefore, we can ignore the first term. This completes the proof of (5.5.1).

Q.E.D.

§5.6 The principal part formula

In this section, we prove the principal part formula for the adjusted zeta function. Throughout this section, we assume that $\Phi = M_{V,\omega}\Phi$. This implies that $M_{\mathfrak{d},\omega_\mathfrak{p}} R_\mathfrak{d}\Phi = R_\mathfrak{d}\Phi$ for all the path $\mathfrak{p}$.

We define

$$\begin{aligned} J_1(\Phi, \omega, s) = &\frac{\delta_{\mathfrak{d}_1}(\omega)}{2} \Sigma_2(R_{\mathfrak{d}_1}\Phi, \omega_{\mathfrak{d}_1}, \frac{s-1}{2}, \frac{s+1}{2}) \\ &+ \delta_{\mathfrak{d}_2,\mathrm{st}}(\omega) \frac{T_{\mathfrak{d}_2+}(\widetilde{R}_{\mathfrak{d}_2,0} R_{\mathfrak{d}_2}\Phi, 2, \frac{1-s}{2})}{s-1} \\ &+ \delta_{\mathfrak{d}_2,\mathrm{st}}(\omega) \frac{T_{\mathfrak{d}_2+}(\widetilde{R}_{\mathfrak{d}_2,0} \mathscr{F}_{\mathfrak{d}_2} R_{\mathfrak{d}_2}\Phi, 1, \frac{1-s}{2})}{s-1} \\ &+ \delta_{\mathfrak{d}_3}(\omega) \left(\frac{\Sigma_1(R_{\mathfrak{d}_4}\Phi_{\mathfrak{p}_{42}}, \frac{s+1}{2})}{s-1} + \frac{\Sigma_1(R_{\mathfrak{d}_4}\Phi_{\mathfrak{p}_{41}}, \frac{s+1}{2})}{s-3} \right) \\ &+ \delta_{V,\mathrm{st}}(\omega) \frac{T_V^1(R_{V,0}\Phi, \omega_{V,\mathrm{st}}, 1, \frac{1-s}{2})}{s-1}, \end{aligned}$$

$$J_2(\Phi,\omega)=\delta_{\mathfrak{d}_2}(\omega)\left(\Sigma_{\mathfrak{d}_2+}(R_{\mathfrak{d}_2}\Phi,\omega_{\mathfrak{d}_2},2)+\Sigma_{\mathfrak{d}_2+}(\mathscr{F}_{\mathfrak{d}_2}R_{\mathfrak{d}_2}\Phi,\omega_{\mathfrak{d}_2}^{-1},1)\right)$$
$$-\mathfrak{V}_2\delta_{\#}(\omega)(\mathscr{F}_{\mathfrak{d}_2}\Phi(0)+\frac{1}{2}\Phi(0))+\mathfrak{V}_2^2\delta_{\#}(\omega)\Phi(0).$$

Then, by (5.3.9), (5.3.10), (5.4.2), (5.4.7), (5.4.10), (5.4.11), (5.5.1),

$$I(\Phi,\omega,w)\sim\frac{\varrho}{2\pi\sqrt{-1}}\int_{\substack{\mathrm{Re}(s)=r_1\\ r_1>1}}(J_1(\widehat{\Phi},\omega^{-1},s)-J_1(\Phi,\omega,s))\phi(s)\widetilde{\Lambda}(w;s,1)ds$$
$$+C_G\Lambda(w;\rho)(J_2(\widehat{\Phi},\omega^{-1})-J_2(\Phi,\omega)).$$

We define

$$\begin{aligned}J_3(\Phi,\omega)=&\ \delta_{\mathfrak{d}_1}(\omega)\widetilde{\Sigma}_{\mathfrak{d}_1,(0)}(\Phi,\omega,1)\\&-\frac{\delta_{\mathfrak{d}_2,\mathrm{st}}(\omega)}{2}(T_{\mathfrak{d}_2+}(\widetilde{R}_{\mathfrak{d}_2,0}R_{\mathfrak{d}_2}\Phi,2)+T_{\mathfrak{d}_2+}(\widetilde{R}_{\mathfrak{d}_2,0}R_{\mathfrak{d}_2}\mathscr{F}_{\mathfrak{d}_2}\Phi,1))\\&+\frac{\delta_{\mathfrak{d}_4}(\omega)}{2}\Sigma_{1,(1)}(R_{\mathfrak{d}_4}\Phi_{\mathfrak{p}_{42}},1)\\&-\frac{\delta_{\mathfrak{d}_4}(\omega)}{2}(\Sigma_{1,(-1)}(R_{\mathfrak{d}_4}\Phi_{\mathfrak{p}_{41}},1)+\Sigma_{1,(0)}(R_{\mathfrak{d}_4}\Phi_{\mathfrak{p}_{41}},1))\\&-\frac{\delta_{V,\mathrm{st}}(\omega)}{2}T_V^1(R_{V,0}\Phi,\omega_{V,\mathrm{st}},1)\\&+\delta_{\mathfrak{d}_2}(\omega)(\Sigma_{\mathfrak{d}_2+}(R_{\mathfrak{d}_2}\Phi,\omega_{\mathfrak{d}_2},2)+\Sigma_{\mathfrak{d}_2+}(R_{\mathfrak{d}_2}\mathscr{F}_{\mathfrak{d}_2}\Phi,\omega_{\mathfrak{d}_2}^{-1},1))\\&-\mathfrak{V}_2\delta_{\#}(\omega)\left(\mathscr{F}_{\mathfrak{d}_2}\Phi(0)+\frac{1}{2}\Phi(0)\right)+\mathfrak{V}_2^2\delta_{\#}(\omega)\Phi(0).\end{aligned}$$

Then, by Wright's principle, $J_1(\widehat{\Phi},\omega^{-1},s)-J_1(\Phi,\omega,s)$ must be holomorphic at $s=1$, and the following proposition follows.

Proposition (5.6.1) $I^0(\Phi,\omega)=J_3(\widehat{\Phi},\omega^{-1})-J_3(\Phi,\omega)$.

Let $\Psi=R_{\mathfrak{d}_2}\Phi$. Then, by the principal part formula (4.2.15),

$$\begin{aligned}\Sigma_{\mathfrak{d}_2,\mathrm{ad},(0)}(\Psi,\omega_{\mathfrak{d}_2},2)=&\Sigma_{\mathfrak{d}_2+}(\Psi,\omega_{\mathfrak{d}_2},2)+\Sigma_{\mathfrak{d}_2+}(\widehat{\Psi},\omega_{\mathfrak{d}_2}^{-1},1)\\&-\frac{\delta(\omega_1)}{2}(T_{\mathfrak{d}_2+}(\widetilde{R}_{\mathfrak{d}_2,0}\Psi,2)+T_{\mathfrak{d}_2+}(\widetilde{R}_{\mathfrak{d}_2,0}\widehat{\Psi},1))\\&-\mathfrak{V}_2\delta(\omega_1)\left(\mathscr{F}_{\mathfrak{d}_2}\Psi(0)+\frac{1}{2}\Psi(0)\right)\\&-\frac{\delta(\omega_1)}{2}(\Sigma_{1,(-1)}(R_{\mathfrak{d}_4}\Phi_{\mathfrak{p}_{41}},1)+\Sigma_{1,(0)}(R_{\mathfrak{d}_4}\Phi_{\mathfrak{p}_{41}},1)).\end{aligned}$$

Therefore,

$$\begin{aligned}J_3(\Phi,\omega)=&\ \delta_{\mathfrak{d}_1}(\omega)\widetilde{\Sigma}_{\mathfrak{d}_1,(0)}(\Phi,\omega,1)+\frac{\delta_{\mathfrak{d}_4}(\omega)}{2}\Sigma_{1,(1)}(R_{\mathfrak{d}_4}\Phi_{\mathfrak{p}_{42}},1)+\mathfrak{V}_2^2\delta_{\#}(\omega)\Phi(0)\\&+\delta_{\mathfrak{d}_2}(\omega)\Sigma_{\mathfrak{d}_2,\mathrm{ad},(0)}(R_{\mathfrak{d}_2}\Phi,\omega_{\mathfrak{d}_2},2)-\frac{\delta_{V,\mathrm{st}}(\omega)}{2}T_V^1(R_{V,0}\Phi,\omega_{V,\mathrm{st}},1).\end{aligned}$$

The following relations are easy to prove, and the proof is left to the reader.

Lemma (5.6.2)

(1)
$$\begin{aligned}\widetilde{\Sigma}_{\mathfrak{d}_1,(0)}(\Phi_\lambda,\omega,1) = &\frac{1}{2}\lambda^{-2}(\log\lambda)^2\widetilde{\Sigma}_{\mathfrak{d}_1,(-2)}(\Phi,\omega,1)\\ &-\lambda^{-2}(\log\lambda)\widetilde{\Sigma}_{\mathfrak{d}_1,(-1)}(\Phi,\omega,1)\\ &+\lambda^{-2}\widetilde{\Sigma}_{\mathfrak{d}_1,(0)}(\Phi,\omega,1).\end{aligned}$$

(2)
$$\begin{aligned}\widetilde{\Sigma}_{\mathfrak{d}_1,(0)}(\widehat{\Phi_\lambda},\omega^{-1},1) = &\frac{1}{2}\lambda^{-4}(\log\lambda)^2\widetilde{\Sigma}_{\mathfrak{d}_1,(-2)}(\widehat{\Phi},\omega^{-1},1)\\ &+\lambda^{-4}(\log\lambda)\widetilde{\Sigma}_{\mathfrak{d}_1,(-1)}(\widehat{\Phi},\omega^{-1},1)\\ &+\lambda^{-4}\widetilde{\Sigma}_{\mathfrak{d}_1,(0)}(\widehat{\Phi},\omega^{-1},1).\end{aligned}$$

(3)
$$\begin{aligned}\Sigma_{\mathfrak{d}_2,\mathrm{ad},(0)}(R_{\mathfrak{d}_2}\Phi_\lambda,\omega_{\mathfrak{d}_2},2) = &\frac{1}{2}\lambda^{-2}(\log\lambda)^2\Sigma_{\mathfrak{d}_2,(-2)}(R_{\mathfrak{d}_2}\Phi,\omega_{\mathfrak{d}_2},2)\\ &-\lambda^{-2}(\log\lambda)\Sigma_{\mathfrak{d}_2,(-1)}(R_{\mathfrak{d}_2}\Phi,\omega_{\mathfrak{d}_2},2)\\ &+\lambda^{-2}\Sigma_{\mathfrak{d}_2,\mathrm{ad},(0)}(R_{\mathfrak{d}_2}\Phi,\omega_{\mathfrak{d}_2},2).\end{aligned}$$

(4)
$$\begin{aligned}\Sigma_{\mathfrak{d}_2,\mathrm{ad},(0)}(R_{\mathfrak{d}_2}\widehat{\Phi_\lambda},\omega_{\mathfrak{d}_2}^{-1},2) = &\frac{1}{2}\lambda^{-4}(\log\lambda)^2\Sigma_{\mathfrak{d}_2,\mathrm{ad},(-2)}(R_{\mathfrak{d}_2}\widehat{\Phi},\omega_{\mathfrak{d}_2}^{-1},2)\\ &+\lambda^{-4}(\log\lambda)\Sigma_{\mathfrak{d}_2,\mathrm{ad},(-1)}(R_{\mathfrak{d}_2}\widehat{\Phi},\omega_{\mathfrak{d}_2}^{-1},2)\\ &+\lambda^{-4}\Sigma_{\mathfrak{d}_2,\mathrm{ad},(0)}(\widehat{\Phi},\omega_{\mathfrak{d}_2}^{-1},2).\end{aligned}$$

(5)
$$\begin{aligned}\Sigma_{1,(1)}(R_{\mathfrak{d}_4}\Phi_{\lambda\mathfrak{p}_{42}},1) = &\frac{1}{2}\lambda^{-2}(\log\lambda)^2\Sigma_{1,(-1)}(R_{\mathfrak{d}_4}\Phi_{\mathfrak{p}_{42}},1)\\ &+\lambda^{-2}(\log\lambda)\Sigma_{1,(0)}(R_{\mathfrak{d}_4}\Phi_{\mathfrak{p}_{42}},1)\\ &+\lambda^{-2}\Sigma_{1,(1)}(R_{\mathfrak{d}_4}\Phi_{\mathfrak{p}_{42}},1).\end{aligned}$$

(6)
$$\begin{aligned}\Sigma_{1,(1)}(R_{\mathfrak{d}_4}\widehat{\Phi}_{\lambda\mathfrak{p}_{42}},1) = &\frac{1}{2}\lambda^{-4}(\log\lambda)^2\Sigma_{1,(-1)}(R_{\mathfrak{d}_3}\mathscr{F}_{\mathfrak{d}_2}\widehat{\Phi},1)\\ &-\lambda^{-4}(\log\lambda)\Sigma_{1,(0)}(R_{\mathfrak{d}_3}\mathscr{F}_{\mathfrak{d}_2}\widehat{\Phi},1)\\ &+\lambda^{-4}\Sigma_{1,(1)}(R_{\mathfrak{d}_3}\mathscr{F}_{\mathfrak{d}_2}\widehat{\Phi},1).\end{aligned}$$

Since

$$\Sigma_{\mathfrak{d}_2,\mathrm{ad},(-2)}(\Phi,\omega_{\mathfrak{d}_2},2) = -\frac{\delta(\omega_1)}{2}\Sigma_{1,(-1)}(R_{\mathfrak{d}_4}\Phi_{\mathfrak{p}_{42}},1),$$
$$\Sigma_{\mathfrak{d}_2,\mathrm{ad},(-1)}(\Phi,\omega_{\mathfrak{d}_2},2) = \frac{\delta(\omega_1)}{2}\Sigma_{1,(0)}(R_{\mathfrak{d}_4}\Phi_{\mathfrak{p}_{42}},1),$$

etc. and $\delta_{\mathfrak{d}_4}(\omega) = \delta(\omega_1)\delta_{\mathfrak{d}_2}(\omega)$,

$$\begin{aligned}&\delta_{\mathfrak{d}_2}(\omega)\Sigma_{\mathfrak{d}_2,\mathrm{ad},(0)}(R_{\mathfrak{d}_2}\Phi_\lambda,\omega_{\mathfrak{d}_2},2) + \frac{\delta_{\mathfrak{d}_4}(\omega)}{2}\Sigma_{1,(1)}(R_{\mathfrak{d}_4}\Phi_{\lambda\mathfrak{p}_{42}},1)\\ &= \delta_{\mathfrak{d}_2}(\omega)\lambda^{-2}\Sigma_{\mathfrak{d}_2,\mathrm{ad},(0)}(R_{\mathfrak{d}_2}\Phi,\omega_{\mathfrak{d}_2},2) + \frac{\delta_{\mathfrak{d}_4}(\omega)}{2}\lambda^{-2}\Sigma_{1,(1)}(R_{\mathfrak{d}_4}\Phi_{\mathfrak{p}_{42}},1),\\ &\delta_{\mathfrak{d}_2}(\omega)\Sigma_{\mathfrak{d}_2,\mathrm{ad},(0)}(\widehat{\Phi}_{\lambda\mathfrak{p}_{42}},\omega_{\mathfrak{d}_2}^{-1},2) + \frac{\delta_{\mathfrak{d}_4}(\omega)}{2}\Sigma_{1,(1)}(R_{\mathfrak{d}_4}\widehat{\Phi}_{\lambda\mathfrak{p}_{42}},1)\\ &= \delta_{\mathfrak{d}_2}(\omega)\lambda^{-4}\Sigma_{\mathfrak{d}_2,\mathrm{ad},(0)}(R_{\mathfrak{d}_2}\widehat{\Phi},\omega^{-1}) + \frac{\delta_{\mathfrak{d}_4}(\omega)}{2}\lambda^{-4}\Sigma_{1,(1)}(R_{\mathfrak{d}_4}\widehat{\Phi}_{\mathfrak{p}_{42}},1).\end{aligned}$$

Definition (5.6.3)

$$Z_{V,\mathrm{ad}}(\Phi,\omega,s)=Z_V(\Phi,\omega,s)-\frac{\delta_{V,\mathrm{st}}(\omega)}{2}T_V(R_{V,0}\Phi,\omega_{V,\mathrm{st}},s-1,1).$$

We call $Z_{V,\mathrm{ad}}(\Phi,\omega,s)$ the adjusted zeta function, and $\frac{\delta_{V,\mathrm{st}}(\omega)}{2}T_V(R_{V,0}\Phi,\omega_{V,\mathrm{st}},s-1,1)$ the adjusting term.

It is easy to see that

$$\begin{aligned}\int_0^1\lambda^sT_V^1(R_{V,0}\Phi_\lambda,\omega_{V,\mathrm{st}},1)d^\times\lambda&=T_V(R_{V,0}\Phi,\omega_{V,\mathrm{st}},s-1,1)\\&\quad-T_{V+}(R_{V,0}\Phi,\omega_{V,\mathrm{st}},s-1,1),\\\int_0^1\lambda^sT_V^1(R_{V,0}\widehat{\Phi}_\lambda,\omega_{V,\mathrm{st}}^{-1},1)d^\times\lambda&=T_{V+}(R_{V,0}\widehat{\Phi},\omega_{V,\mathrm{st}}^{-1},5-s,1).\end{aligned}$$

We define

$$\widetilde{\Sigma}(\Phi,\omega)=\delta_{\mathfrak{d}_2}(\omega)\Sigma_{\mathfrak{d}_2,\mathrm{ad},(0)}(R_{\mathfrak{d}_2}\Phi,\omega_{\mathfrak{d}_2},2)+\frac{\delta_{\mathfrak{d}_4}(\omega)}{2}\Sigma_{1,(1)}(R_{\mathfrak{d}_4}\Phi_{\mathfrak{p}_{42}},1).$$

Then, by the above considerations, $Z_{V,\mathrm{ad}}(\Phi,\omega,s)$ satisfies the following principal part formula.

Theorem (5.6.4)

$$\begin{aligned}Z_{V,\mathrm{ad}}(\Phi,\omega,s)&=Z_{V+}(\Phi,\omega,s)+Z_{V+}(\widehat{\Phi},\omega^{-1},6-s)\\&\quad-\frac{\delta_{V,\mathrm{st}}(\omega)}{2}T_{V+}(R_{V,0}\Phi,\omega_{V,\mathrm{st}},s-1,1)\\&\quad-\frac{\delta_{V,\mathrm{st}}(\omega)}{2}T_{V+}(R_{V,0}\widehat{\Phi},\omega_{V,\mathrm{st}}^{-1},5-s,1)\\&\quad+\mathfrak{V}_2^2\delta_\#(\omega)\left(\frac{\widehat{\Phi}(0)}{s-6}-\frac{\Phi(0)}{s}\right)+\frac{\widetilde{\Sigma}(\widehat{\Phi},\omega^{-1})}{s-4}-\frac{\widetilde{\Sigma}(\Phi,\omega)}{s-2}\\&\quad+\sum_{j=1}^3\left((-1)^{j+1}\frac{\widetilde{\Sigma}_{\mathfrak{d}_1,(-j+1)}(\widehat{\Phi},\omega^{-1})}{(s-4)^j}-\frac{\widetilde{\Sigma}_{\mathfrak{d}_1,(-j+1)}(\Phi,\omega)}{(s-2)^j}\right)\end{aligned}$$

The following functional equation follows from the above formula.

Corollary (5.6.5) $Z_{V,\mathrm{ad}}(\Phi,\omega,s)=Z_{V,\mathrm{ad}}(\widehat{\Phi},\omega^{-1},6-s)$.

Also it is easy to see that

$$\begin{aligned}\widetilde{\Sigma}_{\mathfrak{d}_1,(-2)}(\Phi,\omega)&=2\delta_\#(\omega)\Sigma_{2,(-1,-1)}(R_{\mathfrak{d}_1}\Phi,0,1),\\\widetilde{\Sigma}_{\mathfrak{d}_1,(-1)}(\Phi,\omega)&=\delta_{\mathfrak{d}_1}(\omega)(\Sigma_{2,(-1,0)}(R_{\mathfrak{d}_1}\Phi,\omega_{\mathfrak{d}_1},0,1)+\Sigma_{2,(-0,-1)}(R_{\mathfrak{d}_1}\Phi,\omega_{\mathfrak{d}_1},0,1)),\\\widetilde{\Sigma}_{\mathfrak{d}_1,(0)}(\Phi,\omega)&=\frac{\delta_{\mathfrak{d}_1}(\omega)}{2}(\Sigma_{2,(0,0)}(R_{\mathfrak{d}_1}\Phi,\omega_{\mathfrak{d}_1},0,1)+\Sigma_{2,(-1,1)}(R_{\mathfrak{d}_1}\Phi,\omega_{\mathfrak{d}_1},0,1)\\&\quad+\frac{\delta_{\mathfrak{d}_1}(\omega)}{2}\Sigma_{2,(1,-1)}(R_{\mathfrak{d}_1}\Phi,\omega_{\mathfrak{d}_1},0,1).\end{aligned}$$

Chapter 6 The case G=GL(2)× GL(1)^2, V=Sym^2 k^2⊕ k

§6.1 Reducible prehomogeneous vector spaces with two irreducible factors

We consider two prehomogeneous vector spaces $G = \mathrm{GL}(2) \times \mathrm{GL}(1)^2, V = \mathrm{Sym}^2 k^2 \oplus k, \mathrm{Sym}^2 k^2 \oplus k^2$ in this chapter and the next chapter. In both these cases, V is of the form $V = V_1 \oplus V_2$ for some irreducible representations V_1, V_2. Therefore, we consider such representations in general. We assume that $V_k^{\mathrm{s}} \neq \emptyset$. We choose $\overline{T}_+$ as in §3.1. If we restrict the weights γ_i to $\overline{T}_+$, the convex hull should contain a neighborhood of the origin. This can happen only if $\dim \overline{T}_+ \leq 1$. If $\dim \overline{T}_+ = 0$, $G_{\mathbb{A}}^1 = G_{\mathbb{A}}^0$, and we considered such cases. Therefore, we assume that $\dim \overline{T}_+ = 1$. We choose an isomorphism $a : \mathbb{R}_+ \to \overline{T}_+$. Then any element of $G_{\mathbb{A}}^1$ is of the form $g^1 = a(\lambda_1) g^0$, where $\lambda_1 \in \mathbb{R}_+, g^0 \in G_{\mathbb{A}}^0$.

We define $\bar{\kappa}_{V_1}, \bar{\kappa}_{V_2}$ to be the rational characters defined by the determinants of $\mathrm{GL}(V_1), \mathrm{GL}(V_2)$ respectively. Let $\kappa_{V_i}(g^1) = |\bar{\kappa}_{V_i}(g^1)|^{-1}$ for $i = 1, 2$. We assume that $a(\lambda_1)$ acts on V_1, V_2 by multiplication by $\underline{\lambda_1^a}, \underline{\lambda_1^b}$ for some $a > 0, b < 0$. This means that

$$\kappa_{V_1}(a(\lambda_1)) = \lambda_1^{-a \dim V_1}, \quad \kappa_{V_2}(a(\lambda_1)) = \lambda_1^{b \dim V_2}.$$

For $i = 1, 2$, let $[\ ,\ ]_{V_i}$ be a bilinear form on V_i such that $[g x_i, g^\iota y_i]_{V_i} = [x_i, y_i]_{V_i}$ for all x, y, where $g^\iota = \tau_G {}^t g^{-1} \tau_G$ as before. Let $[\ ,\ ]_V$ be a bilinear form on V defined by the formula $[(x_1, x_2), (y_1, y_2)]_V = [x_1, y_1]_{V_1} + [x_2, y_2]_{V_2}$. For $\Phi \in \mathscr{S}(V_{\mathbb{A}})$, let $R_{V_i}\Phi \in \mathscr{S}(V_{i\mathbb{A}})$ be the restriction for $i = 1, 2$. For $\Phi \in \mathscr{S}(V_{1\mathbb{A}})$ (resp. $\mathscr{S}(V_{2\mathbb{A}})$), we define a Fourier transform $\mathscr{F}_{V_1}\Phi$ (resp. $\mathscr{F}_{V_1}\Phi$) by $[\ ,\ ]_{V_1}$ (resp. $[\ ,\ ]_{V_2}$). We also define partial Fourier transforms $\mathscr{F}_{V_1}\Phi, \mathscr{F}_{V_2}\Phi$ for $\Phi \in \mathscr{S}(V_{\mathbb{A}})$. Since the restriction to V_1 or V_2 gives the same result, we use the same notation.

If $g^1 \in G_{\mathbb{A}}^1$,

$$\begin{aligned}
(6.1.1)\quad \Theta_{V_k^{\mathrm{ss}}}(\Phi, g^1) &= \kappa_V(g^1)\Theta_{V_k^{\mathrm{ss}}}(\mathscr{F}_V\Phi, (g^1)^\iota) + \kappa_V(g^1)\mathscr{F}_V\Phi(0) - \Phi(0) \\
&\quad + \sum_{S_\beta \not\subset V_1, V_2} \left(\kappa_V(g^1)\Theta_{S_\beta}(\mathscr{F}_V\Phi, (g^1)^\iota) - \Theta_{S_\beta}(\Phi, g^1)\right) \\
&\quad + \sum_{S_\beta \subset V_1 \text{ or } V_2} \left(\kappa_V(g^1)\Theta_{S_\beta}(\mathscr{F}_V\Phi, (g^1)^\iota) - \Theta_{S_\beta}(\Phi, g^1)\right).
\end{aligned}$$

Lemma (6.1.2)

$$\begin{aligned}
&\sum_{S_\beta \subset V_1 \text{ or } V_2} \left(\kappa_V(g^1)\Theta_{S_\beta}(\mathscr{F}_V\Phi, (g^1)^\iota) - \Theta_{S_\beta}(\Phi, g^1)\right) + \kappa_V(g^1)\mathscr{F}_V\Phi(0) - \Phi(0) \\
&= \kappa_{V_2}(g^1) \sum_{S_\beta \subset V_1} \Theta_{S_\beta}(\mathscr{F}_{V_1}\mathscr{F}_V\Phi, g^1) + \kappa_V(g^1) \sum_{S_\beta \subset V_2} \Theta_{S_\beta}(\mathscr{F}_V\Phi, (g^1)^\iota) \\
&\quad - \sum_{S_\beta \subset V_1} \Theta_{S_\beta}(\Phi, g^1) - \kappa_{V_2}(g^1) \sum_{S_\beta \subset V_2} \Theta_{S_\beta}(\mathscr{F}_{V_2}\Phi, (g^1)^\iota)
\end{aligned}$$

$$= \kappa_V(g^1) \sum_{S_\beta \subset V_1} \Theta_{S_\beta}(\mathscr{F}_V \Phi, (g^1)^\iota) + \kappa_{V_1}(g^1) \sum_{S_\beta \subset V_2} \Theta_{S_\beta}(\mathscr{F}_{V_2}\mathscr{F}_V \Phi, g^1)$$

$$- \kappa_{V_1}(g^1) \sum_{S_\beta \subset V_1} \Theta_{S_\beta}(\mathscr{F}_{V_1}\Phi, (g^1)^\iota) - \sum_{S_\beta \subset V_2} \Theta_{S_\beta}(\Phi, g^1).$$

Proof. This lemma follows from the following relation

(6.1.3)

$$\kappa_V(g^1) \sum_{x_1 \in V_{1k}\setminus\{0\}} \mathscr{F}_V\Phi((g^1)^\iota x_1, 0) + \kappa_V(g^1) \sum_{x_1 \in V_{1k}\setminus\{0\}} \mathscr{F}_V\Phi(0, (g^1)^\iota x_2)$$

$$- \sum_{x_1 \in V_{1k}\setminus\{0\}} \Phi(g^1 x_1, 0) - \sum_{x_1 \in V_{2k}\setminus\{0\}} \Phi(0, g^1 x_2) + \kappa_V(g^1)\mathscr{F}_V\Phi(0) - \Phi(0)$$

$$= \kappa_{V_2}(g^1) \sum_{x_1 \in V_{1k}\setminus\{0\}} \mathscr{F}_{V_1}\mathscr{F}_V\Phi(g^1 x_1, 0) + \kappa_V(g^1) \sum_{x_1 \in V_{1k}\setminus\{0\}} \mathscr{F}_V\Phi(0, (g^1)^\iota x_2)$$

$$- \sum_{x_1 \in V_{1k}\setminus\{0\}} \Phi(g^1 x_1, 0) - \kappa_{V_2}(g^1) \sum_{x_1 \in V_{2k}\setminus\{0\}} \mathscr{F}_{V_2}\Phi(0, (g^1)^\iota x_2)$$

$$= \kappa_V(g^1) \sum_{x_1 \in V_{1k}\setminus\{0\}} \mathscr{F}_V\Phi((g^1)^\iota x_1, 0) + \kappa_{V_1}(g^1) \sum_{x_2 \in V_{2k}\setminus\{0\}} \mathscr{F}_{V_2}\mathscr{F}_V\Phi(0, g^1 x_2)$$

$$- \kappa_{V_1}(g^1) \sum_{x_1 \in V_{1k}\setminus\{0\}} \mathscr{F}_{V_1}\Phi((g^1)^\iota x_1, 0) - \sum_{x_2 \in V_{2k}\setminus\{0\}} \Phi(0, g^1 x_2).$$

Q.E.D.

Consider a β-sequence $\mathfrak{d}$ such that $l(\mathfrak{d}) = 1$ and $S_\mathfrak{d} \not\subset V_1, V_2$.

Lemma (6.1.4) *Let $g^1 = a(\lambda_1)g^0$ be as before, and χ as in §3.1. Then the integral*

$$\int_{G^1_\mathbb{A}/G_k} \omega(g^1)\chi(g^1)\Theta_{S_\mathfrak{d}}(\Phi, g^1)\mathscr{E}(g^0, w)d^\times g^1$$

is well defined for $\mathrm{Re}(w) \gg 0$.

Proof. Since $S_\mathfrak{d} \not\subset V_1, V_2$, for any $N_1, N_2 \gg 0$, there exists a slowly increasing function $h_{N_1,N_2}(g^0)$ of g^0 such that

$$\Theta_{S_\mathfrak{d}}(\Phi, g^1) \ll h_{N_1,N_2}(g^0)\inf(\lambda_1^{N_1}\lambda_1^{-N_2}).$$

We choose N_1, N_2 large enough so that the function $\inf(\lambda_1^{N_1}\lambda_1^{-N_2})\chi(a(\lambda_1))$ is integrable on $\overline{T}_+$. We fix such N_1, N_2. If $\mathrm{Re}(w) \gg 0$, the function $h_{N_1,N_2}(g^0)\mathscr{E}(g^0, w)$ is integrable on the Siegel domain. Therefore, the convergence of the above integral follows.

Q.E.D.

The following lemma follows from the same argument as the above lemma.

Lemma (6.1.5) *Let χ be as in §3.1. Then the integral*

$$\Xi_{V,\mathrm{st}}(\Phi, \omega, \chi, w) = \int_{G^1_\mathbb{A}/G_k} \omega(g^1)\chi(g^1)\Theta_{V^{ss}_{\mathrm{st}k}}(\Phi, g^1)\mathscr{E}(g^0, w)d^\times g^1$$

is well defined for $\mathrm{Re}(w) \gg 0$.

Definition (6.1.6) *Let* $\mathfrak{d} = (\beta)$ *be a* β*-sequence such that* $S_\beta \subset V_1$ *or* V_2. *Let* χ *be as in* §3.1. *We define*

$$\Xi_{\mathfrak{d},\geq}(\Phi,\omega,\chi,w) = \begin{cases} \int_{\substack{G^1_{\mathbb{A}}/G_k \\ \lambda_1 \geq 1}} \omega(g^1)\chi(g^1)\Theta_{S_\beta}(\Phi,g^1)\mathscr{E}(g^0,w)dg^1 & S_\beta \subset V_1, \\ \int_{\substack{G^1_{\mathbb{A}}/G_k \\ \lambda_1 \geq 1}} \omega(g^1)\chi(g^1)\Theta_{S_\beta}(\Phi,(g^1)^\iota)\mathscr{E}(g^0,w)dg^1 & S_\beta \subset V_2, \end{cases}$$

$$\Xi_{\mathfrak{d},\leq}(\Phi,\omega,\chi,w) = \begin{cases} \int_{\substack{G^1_{\mathbb{A}}/G_k \\ \lambda_1 \leq 1}} \omega(g^1)\chi(g^1)\Theta_{S_\beta}(\Phi,(g^1)^\iota)\mathscr{E}(g^0,w)dg^1 & S_\beta \subset V_1, \\ \int_{\substack{G^1_{\mathbb{A}}/G_k \\ \lambda_1 \leq 1}} \omega(g^1)\chi(g^1)\Theta_{S_\beta}(\Phi,g^1)\mathscr{E}(g^0,w)dg^1 & S_\beta \subset V_2. \end{cases}$$

If $S_\beta \subset V_1$, we define

$$\begin{aligned} \Xi_{\mathfrak{d},1}(\Phi,\omega,\chi,w) &= \Xi_{\mathfrak{d},\geq}(\mathscr{F}_{V_1}\mathscr{F}_V\Phi,\omega,\kappa_{V_2}\chi,w), \\ \Xi_{\mathfrak{d},2}(\Phi,\omega,\chi,w) &= \Xi_{\mathfrak{d},\geq}(\Phi,\omega,\chi,w), \\ \Xi_{\mathfrak{d},3}(\Phi,\omega,\chi,w) &= \Xi_{\mathfrak{d},\leq}(\mathscr{F}_V\Phi,\omega,\kappa_V\chi,w), \\ \Xi_{\mathfrak{d},4}(\Phi,\omega,\chi,w) &= \Xi_{\mathfrak{d},\leq}(\mathscr{F}_{V_1}\Phi,\omega,\kappa_{V_1}\chi,w). \end{aligned} \tag{6.1.7}$$

Also if $S_\beta \subset V_2$, we define

$$\begin{aligned} \Xi_{\mathfrak{d},1}(\Phi,\omega,\chi,w) &= \Xi_{\mathfrak{d},\geq}(\mathscr{F}_V\Phi,\omega,\kappa_V\chi,w), \\ \Xi_{\mathfrak{d},2}(\Phi,\omega,\chi,w) &= \Xi_{\mathfrak{d},\geq}(\mathscr{F}_{V_2}\Phi,\omega,\kappa_{V_2}\chi,w), \\ \Xi_{\mathfrak{d},3}(\Phi,\omega,\chi,w) &= \Xi_{\mathfrak{d},\leq}(\mathscr{F}_{V_2}\mathscr{F}_V\Phi,\omega,\kappa_{V_1}\chi,w), \\ \Xi_{\mathfrak{d},4}(\Phi,\omega,\chi,w) &= \Xi_{\mathfrak{d},\leq}(\Phi,\omega,\chi,w). \end{aligned} \tag{6.1.8}$$

We also define

$$\Xi^s_{\mathfrak{d},i}(\Phi,\omega,\chi,w), \Xi_{\mathfrak{d},\mathrm{st},i}(\Phi,\omega,\chi,w)$$

similarly by considering $S^s_{\beta k}, S_{\beta,\mathrm{st}k}$.

Proposition (6.1.9) *The distribution* $\Xi_{\mathfrak{d},i}(\Phi,\omega,\chi,w)$ *is well defined for all* i *if* $\mathrm{Re}(w) \gg 0$.

Proof. Since the argument is similar, we prove this proposition for $\Xi_{\mathfrak{d},1}(\Phi,\omega,\chi,w)$ for $S_{\mathfrak{d}} \subset V_1$. Let $g^1 = a(\lambda_1)g^0$. By (1.2.6), for any $N \gg 0$, there exists a slowly increasing function $h_N(g^0)$ such that

$$\Theta_{S_{\mathfrak{d}}}(\Phi,g^1) \ll \lambda_1^{-N} h_N(g^0)$$

for $g^0 \in \mathfrak{S}^0$. We fix such N so that $\lambda_1^{-N}\chi(a(\lambda_1))$ is integrable on the set $\{\lambda_1 \mid \lambda_1 \geq 1\}$. Since $h_N(g^0)\mathscr{E}(g^0,w)$ is integrable on $\mathfrak{S}^0$ if $\mathrm{Re}(w) \gg 0$, this proves the proposition.

Q.E.D.

An easy consideration shows that

$$\Xi_{\mathfrak{d},1}(\Phi,\omega,\chi,w) = \Xi_{\mathfrak{d},4}(\mathscr{F}_V\Phi,\omega^{-1},\kappa_V^{-1}\chi^{-1},w).$$

Therefore, $\Xi_{\mathfrak{d},i}(\Phi,\omega,\chi,w)$ for $i=1,\cdots,4$ can be constructed from any one of them.

Let $S_{\mathfrak{d}}\subset V_1$, and $\mathfrak{d}=(\beta)$. Suppose that $Y_{\mathfrak{d}}=Z_{\mathfrak{d}}$ for all $S_{\mathfrak{d}}\subset V_1$ or V_2. Notice that this condition is satisfied by our two representations.

It is easy to see that

(6.1.10)
$$\Xi_{\mathfrak{d},1}(\Phi,\omega,\chi,w)=\int_{\substack{G^1_{\mathbb{A}}\cap M_{\mathfrak{d}\mathbb{A}}/M_{\mathfrak{d}k}\\ \lambda_1\geq 1}}\chi(g_{\mathfrak{d}})t(g_{\mathfrak{d}})^{-2\rho_{\mathfrak{d}}}\Theta_{Z_{\mathfrak{d}}}(R_{\mathfrak{d}}\Phi,g_{\mathfrak{d}})\mathscr{E}_{U_{\mathfrak{d}}}(g_{\mathfrak{d}},w)dg_{\mathfrak{d}}.$$

We define $J^s(\Phi,g^1)$ similarly as in (5.1.4)

Let

$$\text{(6.1.11)}\qquad \Xi_{\text{tot}}(\Phi,\omega,\chi,w)=\sum_{S_{\mathfrak{d}}\subset V_1,V_2}\sum_{i=1}^{4}(-1)^{i+1}\Xi_{(\beta),i}(\Phi,\omega,\chi,w),$$
$$I(\Phi,\omega,\chi,w)=\int_{G^1_{\mathbb{A}}/G_k}\omega(g^1)\chi(g^1)J^s(\Phi,g^1)\mathscr{E}(g^0,w)dg^1.$$

We have proved the following proposition.

Proposition (6.1.12) *Suppose that $Y_{\mathfrak{d}}=Z_{\mathfrak{d}}$ for all $S_{\mathfrak{d}}\subset V_1$ or V_2. Then*

$$\begin{aligned}I(\Phi,\omega,\chi,w)=&\sum_{l(\mathfrak{d})=1,S_{\mathfrak{d}}\not\subset V_1,V_2}\int_{G^1_{\mathbb{A}}/G_k}\omega(g^1)\chi(g^1)\kappa_V(g^1)\Theta_{S_{\mathfrak{d}}}(\widehat{\Phi},(g^1)^{\iota})\mathscr{E}(g^0,w)d^{\times}g^1\\&-\sum_{l(\mathfrak{d})=1,S_{\mathfrak{d}}\not\subset V_1,V_2}\int_{G^1_{\mathbb{A}}/G_k}\omega(g^1)\chi(g^1)\Theta_{S_{\mathfrak{d}}}(\Phi,g^1)\mathscr{E}(g^0,w)d^{\times}g^1\\&+\Xi_{\text{tot}}(\Phi,\omega,\chi,w)+\Xi_{V,\text{st}}(\widehat{\Phi},\omega^{-1},\kappa_V^{-1}\chi^{-1},w)-\Xi_{V,\text{st}}(\Phi,\omega,\chi,w).\end{aligned}$$

§6.2 The spaces $\mathrm{Sym}^2k^2\oplus k,\mathrm{Sym}^2k^2\oplus k^2$

Let $V_1=W=\mathrm{Sym}^2k^2$, $V_2=k$ or $V_2=k^2$ and $V=V_1\oplus V_2$. We consider these cases in this chapter and the next chapter.

We write elements of V in the form

$$x=(x_1,x_2)=(x_{10},x_{11},x_{12},x_{20}),\ \text{where}\ x_1\in V_1,\ x_2\in V_2,$$

or

$$x=(x_1,x_2)=(x_{10},x_{11},x_{12},x_{20},x_{21}),\ \text{where}\ x_1\in V_1,\ x_2\in V_2.$$

Let $G_1=\mathrm{GL}(2)$, $G_i=T_i=\mathrm{GL}(1)$ for $i=2,3$, and $G=G_1\times G_2\times G_3$. The group G acts on V by

$$(g_1,t_2,t_3)x=(t_2g_1x_1,t_3x_2)\ \text{or}\ (t_2g_1x_1,t_3g_1x_2)$$

for $g=(g_1,t_2,t_3)\in G$.

Let $\widetilde{T}\subset G$ be the kernel of the homomorphism $G\to\mathrm{GL}(V)$. Consider a rational character $\chi_V(g_1,t_2,t_3)=(\det g_1)^{a_1}t_2^{a_2}t_3^{a_3}$ of G. This character is trivial on $\widetilde{T}$ if and only if $a_1 = a_2$ or $2a_1 - 2a_2 - a_3 = 0$. We choose such $a_1, a_2, a_3 > 0$ so that they are coprime integers. Then we can consider χ_V as a character of $G/\widetilde{T}$ and it is indivisible. We consider $(G/\widetilde{T}, V)$. We fix $a_1 = a_2 = 5, a_3 = 3$ for $V = \mathrm{Sym}^2 k^2 \oplus k$, and $a_1 = 5, a_2 = 3, a_3 = 4$ for $V = \mathrm{Sym}^2 k^2 \oplus k^2$, because these are the normalizations which we will use in later chapters.

We have to choose $\overline{T}_+ \subset T_+$ (see §3.1) so that $\overline{T}_+ \cong T_+^1/\widetilde{T}_+$. We can choose $\overline{T}_+ = \{(1, t_2, t_3) \in T_{2+} \times T_{3+} \mid t_2^5 t_3^3 = 1\}$ or $\overline{T}_+ = \{(1, t_2, t_3) \in T_{2+} \times T_{3+} \mid t_2^3 t_3^4 = 1\}$.

In Chapters 6 and 7, t_{11}, t_{12}, t_2, t_3 are elements of $\mathbb{A}^1$. We define $t_1(\lambda_1) = a_2(\underline{\lambda}_1^{-1}, \underline{\lambda}_1)$ and $t_2(\lambda_2) = (\underline{\lambda}_2^{3}, \underline{\lambda}_2^{-5})$ or $(\underline{\lambda}_2^{4}, \underline{\lambda}_2^{-3})$ for $\lambda_1, \lambda_2 \in \mathbb{R}_+$. Let

$$d(\lambda_1, \lambda_2) = (t_1(\lambda_1), t_2(\lambda_2)).$$

Then $\overline{T}_+ = \{(1, t_2(\lambda_2)) \mid \lambda_2 \in \mathbb{R}_+\}$. Let $t_1 = a_2(t_{11}, t_{12})$.

We define

$$\widehat{t}^0 = (t_1, t_2, t_3),\ t^0 = (t_1(\lambda_1), 1)\widehat{t}^0,\ t^1 = d(\lambda_1, \lambda_2)\widehat{t}^0. \tag{6.2.1}$$

We use this notation in Chapters 6 and 7.

Let

$$d^\times \widehat{t}^0 = d^\times t_{11} d^\times t_{12} d^\times t_2 d^\times t_3,\ d^\times t^0 = d^\times \lambda_1 d^\times \widehat{t}^0,\ d^\times t^1 = d^\times \lambda_1 d^\times \lambda_2 d^\times \widehat{t}^0.$$

Let $\widetilde{G}_{\mathbb{A}}$, $\widetilde{g}$ be as in Chapter 3. Any element of $G_{\mathbb{A}}^1$ is of the form $g^1 = t_2(\lambda_2) g^0$ where $g^0 \in G_{\mathbb{A}}^0$. We use $d^\times t^1$ for the diagonal part to define a measure dg^1 on $G_{\mathbb{A}}^1$. Let $d\widetilde{g} = d^\times \lambda dg^1$. We define an action of $\widetilde{G}_{\mathbb{A}}$ on $V_{\mathbb{A}}$ by $\widetilde{g}x = \underline{\lambda} g^1 x$, where we consider the ordinary multiplication by $\underline{\lambda}$.

Let e_{ij} be the coordinate vector of x_{ij}. Let γ_{ij} be the weight of e_{ij} i.e. $te_{ij} = \gamma_{ij}(t)e_{ij}$ for $t \in T$. We identify $\mathfrak{t}$ with $\{(z_1, z_2) \in \mathbb{R}^2 \mid z_1 + z_2 = 0\} \times \mathbb{R}$ by

$$d(\lambda_1, \lambda_2)^{(-a, a; b)} = \lambda_1^{2a} \lambda_2^{b}.$$

Let

$$((a, -a; b), (a', -a'; b')) = \begin{cases} 2aa' + \frac{1}{8}bb' & \mathrm{Sym}^2 k^2 \oplus k, \\ 2aa' + \frac{1}{14}bb' & \mathrm{Sym}^2 k^2 \oplus k^2. \end{cases}$$

Then this is a Weyl group invariant inner product on $\mathfrak{t}^*$. We use this particular inner product, because it is the inner product which we will consider in Part IV.

We can identify γ_{ij}'s with elements of $\mathbb{R}^2$ as follows.

• • • $(\sqrt{2}, 3\sqrt{8})$ • • • $(\sqrt{2}, \frac{4}{\sqrt{14}})$

• $(0, -\frac{5}{\sqrt{8}})$ • • $(\frac{1}{\sqrt{2}}, -\frac{3}{\sqrt{14}})$

Note that the origin is above the line segment joining the two points $(\sqrt{2}, \frac{4}{\sqrt{14}})$ and $(-\frac{1}{\sqrt{2}}, -\frac{3}{\sqrt{14}})$.

Let $\mathfrak{B}$ be the parametrizing set of the Morse stratification. Then elements in $\mathfrak{B} \setminus \{0\}$ for $V = \mathrm{Sym}^2 k^2 \oplus k$ are as follows.

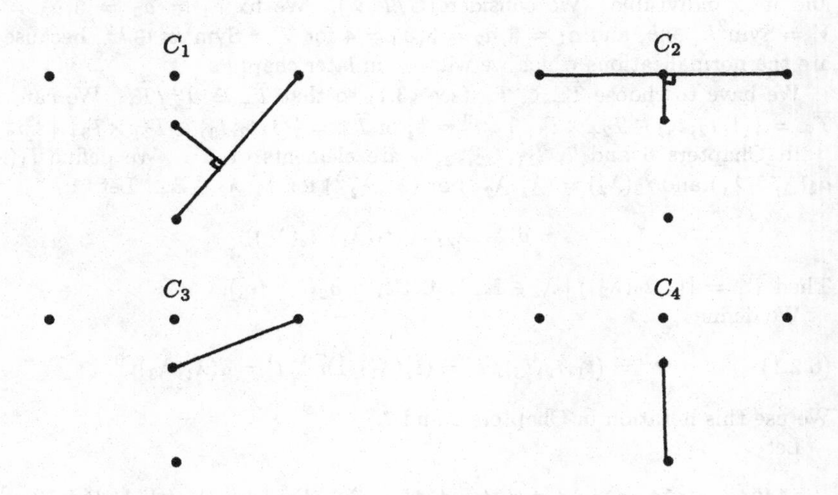

Elements in $\mathfrak{B} \setminus \{0\}$ for $V = \mathrm{Sym}^2 k^2 \oplus k^2$ are as follows.

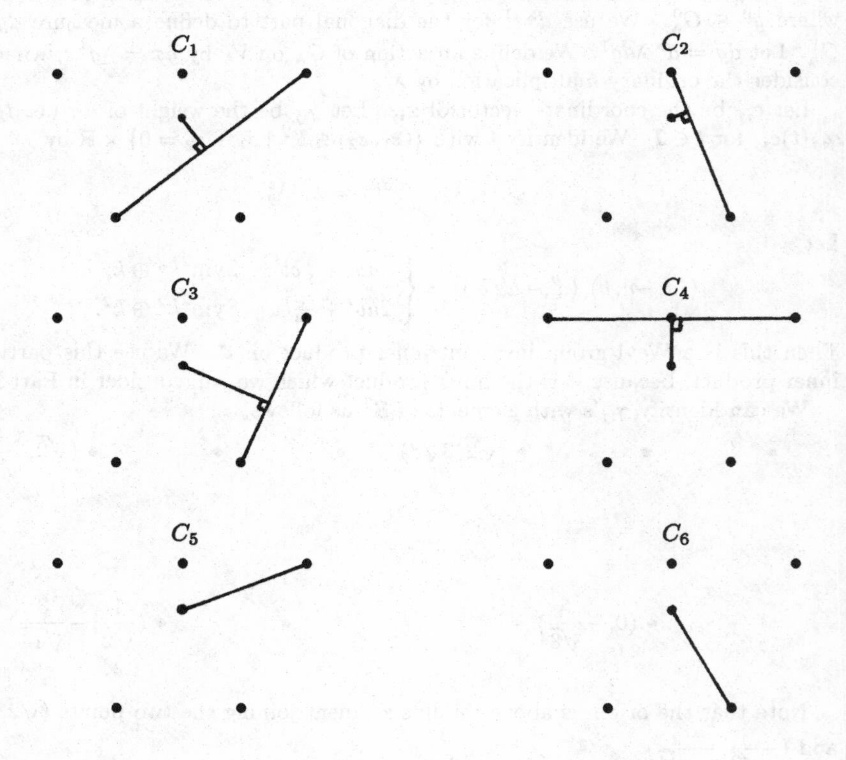

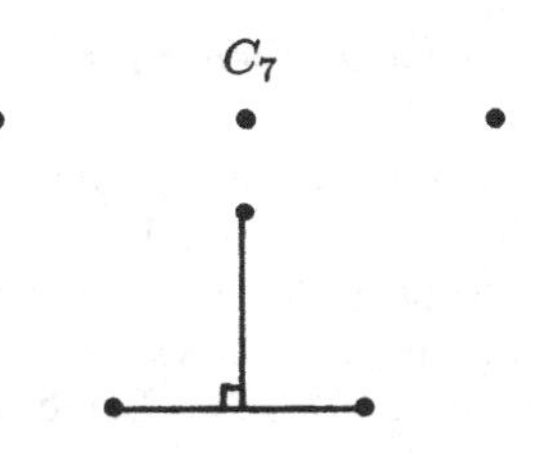

We choose $\mathrm{SL}(2) \times \{(t^3, t^{-5})\}$ or $\mathrm{SL}(2) \times \{(t^4, t^{-3})\}$ as G'' in §3.2. Then Z_β etc. are as follows.

Table (6.2.2) (for $V = \mathrm{Sym}^2 k^2 \oplus k$)

β	Z_β	W_β	M''_β
$\beta_1 = \frac{1}{2}(-1, 1; -2)$	x_{12}, x_{20}	–	$\{(a_2(t^{-1}, t), t^3, t^{-5})\}$
$\beta_2 = (0, 0; 3)$	x_{10}, x_{11}, x_{12}	–	$\mathrm{SL}(2) \times \{1\}$
$\beta_3 = (-1, 1; 3)$	x_{12}	–	$\{(a_2(t^{-3}, t^3), t^{-6}, t^{10})\}$
$\beta_4 = (0, 0; 5)$	x_{20}	–	$\mathrm{SL}(2) \times \{1\}$

Table (6.2.3) (for $V = \mathrm{Sym}^2 k^2 \oplus k^2$)

β	Z_β	W_β	M''_β
$\beta_1 = \frac{1}{16}(-1, 1; -6)$	x_{12}, x_{20}	x_{21}	$\{(a_2(t^{-3}, t^3), t^4, t^{-3})\}$
$\beta_2 = \frac{1}{4}(-1, 1; 2)$	x_{11}, x_{21}	x_{12}	$\{(a_2(t^{-1}, t), t^{-4}, t^3)\}$
$\beta_3 = \frac{1}{10}(-7, 7; -2)$	x_{12}, x_{21}	–	$\{(a_2(t^{-2}, t^2), t^{28}, t^{-21})\}$
$\beta_4 = (0, 0; 4)$	x_{10}, x_{11}, x_{12}	–	$\mathrm{SL}(2) \times \{1\}$
$\beta_5 = (-1, 1; 4)$	x_{12}	–	$\{(a_2(t^2, t^{-2}), t^4, t^{-3})\}$
$\beta_6 = (-\frac{1}{2}, \frac{1}{2}; -3)$	x_{21}	–	$\{(a_2(t^3, t^{-3}), t^{-4}, t^3)\}$
$\beta_7 = (0, 0; -3)$	x_{20}, x_{21}	–	$\mathrm{SL}(2) \times \{1\}$

Consider $V = \mathrm{Sym}^2 k^2 \oplus k$. It is easy to see that $x \in V$ is k-stable if and only if (x_{10}, x_{11}, x_{12}) is k-stable with respect to the action of $\mathrm{GL}(2)$ and $x_{20} \neq 0$. In particular, $V_k^{\mathrm{s}} \neq \emptyset$. Let $\mathfrak{d}_i = (\beta_i)$ for $i = 1, \cdots, 4$. By Table (6.2.2), the weights of x_{12}, x_{20} with respect to M''_{β_1} are t^5, t^{-6} respectively. Since there are both positive and negative weights, $Z^{\mathrm{ss}}_{\mathfrak{d}_1 k} = \{x \in Z_{\mathfrak{d}_1 k} \mid x_{12}, x_{20} \neq 0\}$. We identify $Z_{\mathfrak{d}_2}$ with W, and consider $Z^{\mathrm{ss}}_{\mathfrak{d}_2 k}, Z'_{\mathfrak{d}_2,0}$ etc. Let $\Sigma_{\mathfrak{d}_2}(\Psi, \omega, s), \Sigma_{\mathfrak{d}_2,\mathrm{ad}}(\Psi, \omega, s)$ be the zeta function, the adjusted zeta function respectively. We also use the notation $T_{\mathfrak{d}_2}(\widetilde{R}_{\mathfrak{d}_2,0}\Psi, \omega, s)$ etc. for the adjusting term etc. Let $Z_{V,0}$ (resp. $Z'_{V,0}$) be the subspace spanned by $\{e_{11}, e_{12}, e_{20}\}$ (resp. $\{e_{11}, e_{20}\}$). Let $Z'^{\mathrm{ss}}_{V,0k} = \{x \in Z'_{V,0k} \mid x_{11}, x_{20} \neq 0\}$. Let $H_V = H_{\mathfrak{d}_2}$ be the subgroup generated by τ_G and T_k. Then $V^{\mathrm{ss}}_{\mathrm{st}k} \cong G_k \times_{H_{Vk}} Z'^{\mathrm{ss}}_{V,0k}$, and $S_{\mathfrak{d}_2,\mathrm{st}k} \cong G_k \times_{H_{\mathfrak{d}_2 k}} Z'^{\mathrm{ss}}_{\mathfrak{d}_2,0k}$. Since $M''_{\mathfrak{d}}$ acts trivially on $Z_{\mathfrak{d}}$ for $\mathfrak{d} = \mathfrak{d}_3, \mathfrak{d}_4$, $Z^{\mathrm{ss}}_{\mathfrak{d}k} =$

$Z_{\mathfrak{d}k} \setminus \{0\}$ for $\mathfrak{d} = \mathfrak{d}_3, \mathfrak{d}_4$. We do not consider β-sequences of length ≥ 2. Let $R_{V,0}\Phi$ be the restriction to $Z_{V,0}$.

Consider $V = \mathrm{Sym}^2 k^2 \oplus k^2$. Note that if a convex hull of a subset of γ_{ij}'s contains the origin, it contains a neighborhood of the origin. Therefore, $V_k^{\mathrm{ss}} = V_k^{\mathrm{s}}$. Let Q be the binary quadratic form which corresponds to (x_{10}, x_{11}, x_{12}), and l the linear function which corresponds to (x_{20}, x_{21}). Then it is easy to see that x is semi-stable if and only l does not divide Q. In particular, $V_k^{\mathrm{ss}} \neq \emptyset$. Let $\mathfrak{d}_i = (\beta_i)$ for $i = 1, \cdots, 6$. By the above table, the weights of x_{12}, x_{20} with respect to M''_{β_1} are t^{10}, t^{-6}, the weights of x_{11}, x_{21} with respect to M''_{β_2} are t^{-4}, t^4, and the weights of x_{12}, x_{21} with respect to M''_{β_3} are t^{32}, t^{-19}. For these cases, there are both positive and negative weights. Therefore,

$$\begin{aligned}
Z_{\mathfrak{d}_1 k}^{\mathrm{ss}} &= \{x \in Z_{\mathfrak{d}_1 k} \mid x_{12}, x_{20} \neq 0\}, \\
Z_{\mathfrak{d}_2 k}^{\mathrm{ss}} &= \{x \in Z_{\mathfrak{d}_2 k} \mid x_{11}, x_{21} \neq 0\}, \\
Z_{\mathfrak{d}_3 k}^{\mathrm{ss}} &= \{x \in Z_{\mathfrak{d}_3 k} \mid x_{12}, x_{21} \neq 0\}.
\end{aligned}$$

We identify $Z_{\mathfrak{d}_4}$ with W, and use similar notations to $\mathfrak{d}_2$ of the previous case. Let $H_{\mathfrak{d}_4}$ be the subgroup generated by τ_G and T_k. Then $S_{\mathfrak{d}_4,\mathrm{st}k} \cong G_k \times_{H_{\mathfrak{d}_4 k}} Z'^{\mathrm{ss}}_{\mathfrak{d}_4,0k}$. Since $M''_{\mathfrak{d}}$ acts trivially on $Z_{\mathfrak{d}}$ for $\mathfrak{d} = \mathfrak{d}_5, \mathfrak{d}_6$, $Z_{\mathfrak{d}k}^{\mathrm{ss}} = Z_{\mathfrak{d}k} \setminus \{0\}$ for $\mathfrak{d} = \mathfrak{d}_5, \mathfrak{d}_6$. As in Chapter 5, $Z_{\beta_7 k}^{\mathrm{ss}} = \emptyset$. We do not consider β-sequences of length ≥ 2.

In both cases,

$$V_k \setminus \{0\} = V_k^{\mathrm{ss}} \coprod \coprod_i S_{\mathfrak{d}_i k}.$$

We define $\widetilde{\Theta}_{Z_{\mathfrak{d},0}}(\Psi, t', u_0)$ etc. for $\mathfrak{d} = \mathfrak{d}_2$ in the first case and for $\mathfrak{d} = \mathfrak{d}_4$ in the second case as in (5.2.4).

Let $\omega = (\omega_1, \omega_2, \omega_3)$ be a character of $(\mathbb{A}^\times/k^\times)^3$. We define $\omega(g^1)$ as in §3.1. It is easy to see that $\kappa_V(g^1) = \lambda_2^{-4}$ $\kappa_{V_1}(g^1) = \lambda_2^{-9}$ and $\kappa_{V_2}(g^1) = \lambda_2^5$, for $V = \mathrm{Sym}^2 k^2 \oplus k$. Also $\kappa_V(g^1) = \lambda_2^{-6}$, $\kappa_{V_1}(g^1) = \lambda_2^{-12}$, and $\kappa_{V_2}(g^1) = \lambda_2^6$ for $V = \mathrm{Sym}^2 k^2 \oplus k^2$. For $V_2 = k^2$ or k, let $[\ ,\ ]'_{V_2}$ be the standard inner product, and $[x,y]_{V_2} = [x, \tau_G y]'_{V_2}$. We defined a bilinear form $[\ ,\ ]_V$ for $V = \mathrm{Sym}^2 k^2$ in §4.1. We use this bilinear form as $[\ ,\ ]_{V_1}$ in §6.1.

We define $J^s(\Phi, g^1)$ similarly as in (5.1.4). Let

$$I^1(\Phi, \omega, \chi) = \int_{G^1_{\mathbb{A}}/G_k} \omega(g^1)\chi(g^1) J^s(\Phi, g^1)\, dg^1, \tag{6.2.4}$$

$$I(\Phi, \omega, \chi, w) = \int_{G^1_{\mathbb{A}}/G_k} \omega(g^1)\chi(g^1) J^s(\Phi, g^1) \mathscr{E}(g^0, w)\, dg^1.$$

($J^s(\Phi, g^1) = J(\Phi, g^1)$ for $\mathrm{Sym}^2 k^2 \oplus k^2$.)

We define the zeta function using (3.1.8) for $L = V_k^{\mathrm{s}}$ for the first case, and $L = V_k^{\mathrm{ss}}$ for the second case, and use the notation $Z_V(\Phi, \omega, \chi, s)$ etc. Let $\chi(t_2(\lambda_2)) = \lambda_2^c$ for some $c \in \mathbb{C}$. By the Poisson summation formula,

$$\begin{aligned}
Z_V(\Phi, \omega, s) = {} & Z_{V+}(\Phi, \omega, \chi, s) + Z_{V+}(\widehat{\Phi}, \omega^{-1}, \kappa_V^{-1}\chi^{-1}, N - s) \\
& + \int_{\mathbb{R}_+} \lambda^s I^1(\Phi_\lambda, \omega, \chi)\, d^\times\lambda,
\end{aligned}$$

where $N = 4$ for the first case, and $N = 5$ for the second case.

By (3.4.34),

$$I(\Phi, \omega, \chi, w) \sim C_G \Lambda(w; \rho) I^1(\Phi, \omega, \chi).$$

As in Chapter 5, we assume that $\Phi = M_{V,\omega}\Phi$ in this chapter and the next chapter.

§6.3 The principal part formula

In this section, we prove a principal part formula for the zeta function for the space $\mathrm{Sym}^2 k^2 \oplus k$. Since the zeta function is almost a product of the Riemann zeta function and the zeta function for the space of binary quadratic forms, we only outline the proof.

We assume that $c \neq -5, -2, 0, 1, 3, 4, 9$.

Let $\Psi \in \mathscr{S}(Z_{V,0\mathbb{A}})$. Suppose that $q = (q_1, q_2) \in (\mathbb{A}^1)^2$, $u_0 \in \mathbb{A}$, and $\mu = (\mu_1, \mu_2) \in \mathbb{R}_+^2$. Let $\alpha(u_0)$ be as in §2.2. Let $d^\times q = d^\times q_1 d^\times q_2$, and $d^\times \mu = d^\times \mu_1 d^\times \mu_2$. Let ω_1, ω_2 be characters of $\mathbb{A}^\times / k^\times$, and $\omega = (\omega_1, \omega_2)$. We define $\omega(q) = \omega_1(q_1)\omega_2(q_2)$. Let

$$f_V(\Psi, \mu, q; u_0, s_1, s_2) = \mu_2^{s_1} \alpha(u_0)^{s_2} \Psi(\underline{\mu}_1 \underline{\mu}_2^3 q_1, \underline{\mu}_1 \underline{\mu}_2^3 q_1 u_0, \underline{\mu}_1 \underline{\mu}_2^{-5} q_2).$$

Definition (6.3.1) *For complex variables s, s_1, s_2 and Ψ, ω as above, we define*

(1) $$T_V(\Psi, \omega, s, s_1, s_2) = \int_{\mathbb{R}_+^2 \times (\mathbb{A}^1)^2 \times \mathbb{A}} \omega(q) \mu_1^s f_V(\Psi, \mu, q, u_0, s_1, s_2) d^\times \mu d^\times q du_0,$$

(2) $$T_{V+}(\Psi, \omega, s, s_1, s_2) = \int_{\substack{\mathbb{R}_+^2 \times (\mathbb{A}^1)^2 \times \mathbb{A} \\ \mu_1 \geq 1}} \omega(q) \mu_1^s f_V(\Psi, \mu, q, u_0, s_1, s_2) d^\times \mu d^\times q du_0,$$

(3) $$T_V^1(\Psi, \omega, s_1, s_2) = \int_{\mathbb{R}_+ \times (\mathbb{A}^1)^2 \times \mathbb{A}} \omega(q) f_V(\Psi, 1, \mu_2, q, u_0, s_1, s_2) d^\times \mu_2 d^\times q du_0.$$

As in (5.2.2), $T_V(\Psi, \omega, s, s_1, s_2)$ has a meromorphic continuation.

We define

$$T_V(\Psi, \omega, s, s_1) = \left.\frac{d}{ds_2}\right|_{s_2=0} T_V(\Psi, \omega, s, s_1, s_2),$$

$$T_{V+}(\Psi, \omega, s, s_1) = \left.\frac{d}{ds_2}\right|_{s_2=0} T_{V+}(\Psi, \omega, s, s_1, s_2),$$

$$T_V^1(\Psi, \omega, s_1) = \left.\frac{d}{ds_2}\right|_{s_2=0} T_V^1(\Psi, \omega, s_1, s_2).$$

In our case, $Y_\mathfrak{d} = Z_\mathfrak{d}$ for all $\mathfrak{d}$. Therefore, all the distributions in (6.1.12) are well defined. Let

$$\Sigma_{V,\mathrm{st}}(\Phi, \omega, \chi, s) = \delta(\omega_1 \omega_2^{-1}) \delta(\omega_3) \frac{T_V^1(R_{V,0}\Phi, (\omega_1, 1), c, \frac{1}{2}(1-s))}{s-1}.$$

Then by considering $\tau = \tau_G$, $s = s_\tau$,

(6.3.2) $$\Xi_{V,\mathrm{st}}(\Phi, \omega, \chi, w) \sim \frac{1}{2\pi\sqrt{-1}} \int_{\mathrm{Re}(s)=r>1} \Sigma_{V,\mathrm{st}}(\Phi, \omega, \chi, s) \phi(s) \widetilde{\Lambda}(w; s) ds.$$

Let

$$\Sigma_{\mathfrak{d}_1}(\Phi,\omega,\chi,s)=\frac{\delta(\omega_1)\delta(\omega_2^2)}{10}\Sigma_2(R_{\mathfrak{d}_1}\Phi,(\omega_2,\omega_3),\frac{1}{2}(s+1),\frac{1}{10}(3s+3-2c)).$$

Then similarly as in Chapter 5, by considering $\tau=\tau_G$, $s=s_\tau$,

$$(6.3.3)\qquad \int_{G^1_{\mathbb{A}}/G_k}\omega(g^1)\chi(g^1)\Theta_{S_{\mathfrak{d}_1}}(\Phi,g^1)\mathscr{E}(g^0,w)d^\times g^1 \sim \frac{1}{2\pi\sqrt{-1}}\int_{\mathrm{Re}(s)=r>1}\Sigma_{\mathfrak{d}_1}(\Phi,\omega,\chi,s)\phi(s)\widetilde{\Lambda}(w;s)ds.$$

Let

$$J_2(\Phi,\omega,\chi)=\frac{\varrho\delta(\omega_3)}{3}(\Sigma_{\mathfrak{d}_2+}(R_{\mathfrak{d}_2}\Phi,\omega_2,\omega_1,\frac{c}{3})+\Sigma_{\mathfrak{d}_2+}(R_{\mathfrak{d}_2}\mathscr{F}_{V_1}\Phi,(\omega_2^{-1},\omega_1^{-1}),3-\frac{c}{3}).$$

By (3.4.31),

$$(6.3.4)\qquad \Xi^s_{\mathfrak{d}_2,2}(\Phi,\omega,\chi,w)+\Xi^s_{\mathfrak{d}_2,4}(\Phi,\omega,\chi,w)\sim C_G\Lambda(w;\rho)J_2(\Phi,\omega,\chi).$$

Let

$$\Sigma_{\mathfrak{d}_2,\mathrm{st}}(\Phi,\omega,\chi,s)=\frac{\delta(\omega_3)\delta(\omega_1\omega_2^{-1})}{3}\frac{T_{\mathfrak{d}_2+}(\widetilde{R}_{\mathfrak{d}_2,0}R_{\mathfrak{d}_2}\Phi,\omega_1,\frac{c}{3},\frac{1}{2}(1-s))}{s-1} + \frac{\delta(\omega_3)\delta(\omega_1\omega_2^{-1})}{3}\frac{T_{\mathfrak{d}_2+}(\widetilde{R}_{\mathfrak{d}_2,0}\mathscr{F}_{\mathfrak{d}_2}R_{\mathfrak{d}_2}\Phi,\omega_1^{-1},3-\frac{c}{3},\frac{1}{2}(1-s))}{s-1}.$$

Then similarly as in Chapter 5, by considering $\tau=\tau_G$, $s=s_\tau$,

$$(6.3.5)\qquad \Xi_{\mathfrak{d}_2,\mathrm{st},2}(\Phi,\omega,\chi,w)+\Xi_{\mathfrak{d}_2,\mathrm{st},4}(\Phi,\omega,\chi,w) \sim\frac{1}{2\pi\sqrt{-1}}\int_{\mathrm{Re}(s)=r>1}\Sigma_{\mathfrak{d}_2,\mathrm{st}}(\Phi,\omega,\chi,s)\phi(s)\widetilde{\Lambda}(w;s)ds.$$

Let $\Sigma_{\mathfrak{d}_3}(\Phi,\omega,\chi,s)$ be the following function

$$\delta(\omega_1)\delta(\omega_2^2)\delta(\omega_3)\left(\frac{\Sigma_1(R_{\mathfrak{d}_3}\Phi,\omega_2,\frac{1}{2}(s+1))}{3s-2c+3}+\frac{\Sigma_1(R_{\mathfrak{d}_3}\mathscr{F}_{V_1}\Phi,\omega_2^{-1},\frac{1}{2}(s+1))}{3s+2c-15}\right).$$

Then similarly as in Chapter 5, by considering $\tau=\tau_G$, $s=s_\tau$,

$$(6.3.6)\qquad \Xi_{\mathfrak{d}_3,2}(\Phi,\omega,\chi,w)+\Xi_{\mathfrak{d}_3,4}(\Phi,\omega,\chi,w) \sim\frac{1}{2\pi\sqrt{-1}}\int_{\mathrm{Re}(s)=r>1}\Sigma_{\mathfrak{d}_3}(\Phi,\omega,\chi,s)\phi(s)\widetilde{\Lambda}(w;s)ds.$$

Let

$$J_4(\Phi,\omega,\chi)=\frac{\mathfrak{V}_2\delta(\omega_1)\delta(\omega_2)}{5}\Sigma_{1+}(R_{\mathfrak{d}_4}\Phi,\omega_3,-\frac{c}{5}) +\frac{\mathfrak{V}_2\delta(\omega_1)\delta(\omega_2)}{5}\Sigma_{1+}(\mathscr{F}_{\mathfrak{d}_4}R_{\mathfrak{d}_4}\Phi,\omega_3^{-1},\frac{1}{5}(c+5)).$$

Then by using the Mellin inversion formula,

$$\Xi_{\mathfrak{d}_4,2}(\Phi,\omega,\chi,w) + \Xi_{\mathfrak{d}_4,4}(\Phi,\omega,\chi,w) \sim C_G \Lambda(w;\rho) J_4(\Phi,\omega,\chi), \tag{6.3.7}$$

where

$$\begin{aligned} J_4(\Phi,\omega,\chi) = {} & \frac{\mathfrak{V}_2 \delta(\omega_1)\delta(\omega_2)}{5} \Sigma_1(R_{\mathfrak{d}_4}\Phi, \omega_3, -\frac{c}{5}) \\ & + \mathfrak{V}_2 \delta(\omega_1)\delta(\omega_2)\delta(\omega_3) \left(-\frac{\Phi(0)}{c} + \frac{\mathscr{F}_{V_2}\Phi(0)}{c+5} \right). \end{aligned}$$

All these computations are valid replacing Φ by $\widehat{\Phi}$, χ by $\kappa_V^{-1}\chi^{-1}$ etc. by the assumption on c.

We define $\Sigma_{\mathfrak{d}_1,(j)}(\Phi,\omega,\chi,s_0)$ etc. similarly as before.

Easy considerations show that

$$\begin{aligned} \Sigma_{V,\mathrm{st},(0)}(\Phi,\omega,\chi,1) &= -\frac{\delta(\omega_1\omega_2^{-1})\delta(\omega_3)}{2} T_V^1(R_{V,0}\Phi, (\omega_1,1), c), \\ \Sigma_{\mathfrak{d}_1,(0)}(\Phi,\omega,\chi,1) &= \frac{3\delta(\omega_1)\delta(\omega_2^2)}{50} \Sigma_{2,(1,-1)}(R_{\mathfrak{d}_1}\Phi, (\omega_2,\omega_3), 1, \frac{1}{5}(3-c)) \\ & \quad - \frac{\delta(\omega_1)\delta(\omega_2^2)}{10} \Sigma_{2,(0,0)}(R_{\mathfrak{d}_1}\Phi, (\omega_2,\omega_3), 1, \frac{1}{5}(3-c)). \end{aligned} \tag{6.3.8}$$

Also

$$\begin{aligned} & J_2(\widehat{\Phi},\omega^{-1},\kappa_V^{-1}\chi^{-1}) + J_4(\widehat{\Phi},\omega^{-1},\kappa_V^{-1}\chi^{-1}) \\ & \quad + \Sigma_{\mathfrak{d}_2,\mathrm{st},(0)}(\widehat{\Phi},\omega^{-1},\kappa_V^{-1}\chi^{-1},1) + \Sigma_{\mathfrak{d}_3,(0)}(\widehat{\Phi},\omega^{-1},\kappa_V^{-1}\chi^{-1},1) \\ & \quad - J_2(\Phi,\omega,\chi) - J_4(\Phi,\omega,\chi) \\ & \quad - \Sigma_{\mathfrak{d}_2,\mathrm{st},(0)}(\Phi,\omega,\chi,1) - \Sigma_{\mathfrak{d}_3,(0)}(\Phi,\omega,\chi,1) \\ & = \frac{\delta(\omega_3)}{3} \left(\Sigma_{\mathfrak{d}_2,\mathrm{ad}}(R_{\mathfrak{d}_2}\widehat{\Phi}, (\omega_2^{-1},\omega_1^{-1}), \frac{4-c}{3}) - \Sigma_{\mathfrak{d}_2,\mathrm{ad}}(R_{\mathfrak{d}_2}\Phi, (\omega_2,\omega_1), \frac{c}{3}) \right) \\ & \quad + \frac{\mathfrak{V}_2\delta(\omega_1)\delta(\omega_2)}{5} \left(\Sigma_1(R_{\mathfrak{d}_4}\widehat{\Phi}, \omega_3^{-1}, \frac{c-4}{5}) - \Sigma_1(R_{\mathfrak{d}_4}\Phi, \omega_3, -\frac{c}{5}) \right). \end{aligned} \tag{6.3.9}$$

Definition (6.3.10)

$$Z_{V,\mathrm{ad}}(\Phi,\omega,\chi,s) = Z_V(\Phi,\omega,\chi,s) - \frac{\delta(\omega_1\omega_2^{-1})\delta(\omega_3)}{2} T_V(R_{V,0}\Phi, (\omega_1,1), s, c).$$

We use the terminology 'adjusted zeta function', 'adjusting term' for this case also.

Let

$$\begin{aligned} F(\Phi,\omega,\chi,s) = {} & \frac{\delta(\omega_1)\delta(\omega_2^2)\Sigma_{2,(0,0)}(R_{\mathfrak{d}_1}\Phi, (\omega_2,\omega_3), 1, \frac{1}{5}(3-c))}{2(5s+c-8)} \\ & + \frac{3\delta(\omega_1)\delta(\omega_2^2)\Sigma_{2,(1,-1)}(R_{\mathfrak{d}_1}\Phi, (\omega_2,\omega_3), 1, \frac{1}{5}(3-c))}{2(25s+5c-40)} \\ & + \frac{4\delta(\omega_1)\delta(\omega_2^2)\Sigma_{2,(0,-1)}(R_{\mathfrak{d}_1}\Phi, (\omega_2,\omega_3), 1, \frac{1}{5}(3-c))}{(5s+c-8)^2} \end{aligned}$$

$$+\delta(\omega_3)\frac{\Sigma_{\mathfrak{d}_2,\mathrm{ad}}(R_{\mathfrak{d}_2}\Phi,(\omega_2,\omega_1),\frac{c}{3})}{3s-c}$$
$$+\mathfrak{V}_2\delta(\omega_1)\delta(\omega_2)\frac{\Sigma_1(R_{\mathfrak{d}_4}\Phi,\omega_3,-\frac{c}{5})}{5s+c}.$$

Then by considering relations as in (5.6.2), we get the following proposition.

Proposition (6.3.11) *Suppose* $\Phi = M_{V,\omega}\Phi$. *We also assume that*

$$c \neq -5,-2,0,1,3,4,9.$$

Then $Z_{V,\mathrm{ad}}(\Phi,\omega,\chi,s)$ *satisfies the following principal part formula*

$$\begin{aligned}Z_{V,\mathrm{ad}}(\Phi,\omega,\chi,s) &= Z_{V+}(\Phi,\omega,\chi,s)+Z_{V+}(\widehat{\Phi},\omega^{-1},\kappa_V^{-1}\chi^{-1},4-s)\\&\quad-\frac{\delta(\omega_1\omega_2^{-1})\delta(\omega_3)}{2}T_{V+}(R_{V,0}\Phi,(\omega_1,1),s,c)\\&\quad-\frac{\delta(\omega_1\omega_2^{-1})\delta(\omega_3)}{2}T_{V+}(R_{V,0}\widehat{\Phi},(\omega_1^{-1},1),4-s,4-c)\\&\quad-F(\widehat{\Phi},\omega^{-1},\kappa_V^{-1}\chi^{-1},4-s)-F(\Phi,\omega,\chi,s).\end{aligned}$$

When we apply this result in Chapter 13, we consider the situation $s=-3, c=-1$.

Chapter 7 The case G=GL(2)×GL(1)^2, V=Sym^{2}k^2⊕ k^2

In this chapter, we consider the case $G = \mathrm{GL}(2) \times \mathrm{GL}(1)^2$, $V = \mathrm{Sym}^2 k^2 \oplus k^2$ using the formulation in Chapter 6. We use the notation in §6.2.

§7.1 Unstable distributions

In this section, we define distributions which arise from unstable strata.

Definition (7.1.1) *Let ω be as before. We define*
(1) $\delta_{\mathfrak{d}_1}(\omega) = \delta(\omega_1\omega_2^{-2})\delta(\omega_1\omega_3^{-1})$, $\omega_{\mathfrak{d}_1} = (\omega_2, \omega_3)$,
(2) $\delta_{\mathfrak{d}_2}(\omega) = \delta(\omega_1\omega_2^{-1})\delta(\omega_3)$, $\omega_{\mathfrak{d}_2} = (\omega_2, 1)$,
(3) $\delta_{\mathfrak{d}_3}(\omega) = \delta(\omega_1)\delta(\omega_2^2\omega_3)$, $\omega_{\mathfrak{d}_3} = (\omega_2, \omega_3)$,
(4) $\delta_{\mathfrak{d}_4}(\omega) = \delta(\omega_3)$, $\delta_{\mathfrak{d}_4,\mathrm{st}}(\omega) = \delta(\omega_3)\delta(\omega_1\omega_2^{-1})$, $\omega_{\mathfrak{d}_4} = (\omega_2, \omega_1)$, $\omega_{\mathfrak{d},\mathrm{st}} = \omega_2$,
(5) $\delta_{\mathfrak{d}_5}(\omega) = \delta(\omega_1)\delta(\omega_2^2)\delta(\omega_3)$, $\omega_{\mathfrak{d}_5} = \omega_2$,
(6) $\delta_{\#}(\omega) = \delta_{\mathfrak{d}_6}(\omega) = \delta(\omega_1)\delta(\omega_2)\delta(\omega_3)$.

Definition (7.1.2) *Let Φ, ω, χ be as in §6.1. For a complex variable $s \in \mathbb{C}$, we define*

$$(1) \qquad \Sigma_{\mathfrak{d}_1}(\Phi,\omega,\chi,s) = \frac{\delta_{\mathfrak{d}_1}(\omega)}{2}\Sigma_2(R_{\mathfrak{d}_1}\Phi, \omega_{\mathfrak{d}_1}, \frac{3s-3-c}{2}, 2s-3-c),$$

$$(2) \qquad \Sigma_{\mathfrak{d}_2}(\Phi,\omega,\chi,s) = \frac{\delta_{\mathfrak{d}_2}(\omega)}{4}\Sigma_2(R_{\mathfrak{d}_2}\Phi, \omega_{\mathfrak{d}_2}, \frac{3s-7+c}{4}, s-1),$$

$$(3) \qquad \Sigma_{\mathfrak{d}_3}(\Phi,\omega,\chi,s) = \frac{\delta_{\mathfrak{d}_3}(\omega)}{10}\Sigma_2(R_{\mathfrak{d}_3}\Phi, \omega_{\mathfrak{d}_3}, \frac{3s+3+c}{10}, \frac{2s+2-c}{5}),$$

$$(4) \qquad \widetilde{\Sigma}_{\mathfrak{d}_4}(\Phi,\omega,\chi,s) = \frac{\delta_{\mathfrak{d}_4}(\omega)}{4}\Sigma_{\mathfrak{d}_4,\mathrm{ad}}(R_{\mathfrak{d}_4}\Phi, \omega_{\mathfrak{d}_4}, \frac{c}{4}+s-1),$$

$$(5) \qquad \Sigma_{\mathfrak{d}_5,1}(\Phi,\omega,\chi,s) = \delta_{\mathfrak{d}_5}(\omega)\frac{\Sigma_1(R_{\mathfrak{d}_5}\Phi, \omega_{\mathfrak{d}_5}, \frac{s+1}{2})}{4(s+1)-2c},$$

$$\Sigma_{\mathfrak{d}_5,2}(\Phi,\omega,\chi,s) = \delta_{\mathfrak{d}_5}(\omega)\frac{\Sigma_1(R_{\mathfrak{d}_5}\mathscr{F}_{V_1}\Phi, \omega_{\mathfrak{d}_5}^{-1}, \frac{s+1}{2})}{4(s-5)+2c},$$

$$(6) \qquad \Sigma_{\mathfrak{d}_6,1}(\Phi,\chi,\omega,s) = \delta_{\mathfrak{d}_6}(\omega)\frac{\Sigma_1(R_{\mathfrak{d}_6}\Phi, s+1)}{3s+3+c},$$

$$\Sigma_{\mathfrak{d}_6,2}(\Phi,\chi,\omega,s) = \delta_{\mathfrak{d}_6}(\omega)\frac{\Sigma_1(R_{\mathfrak{d}_6}\mathscr{F}_{V_2}\Phi, s+1)}{3s-3-c}.$$

We define $\widetilde{\Sigma}_{\mathfrak{d}_4,\mathrm{ad}}(\Phi,\omega,\chi,s)$, $\widetilde{\Sigma}_{\mathfrak{d}_4+}(\Phi,\omega,\chi,s)$ similarly.

The distributions $\Sigma_{\mathfrak{d}_i}(\Phi,\omega,\chi,s)$ for $i = 1,\cdots,3$ and $\widetilde{\Sigma}_{\mathfrak{d}_4}(\Phi,\omega,\chi,s)$ have at most a double pole at $s = 1$. The distribution $\Sigma_{\mathfrak{d}_5,l}(\Phi,\omega,\chi,s)$ has at most a double pole at $s = 1$ for $l = 1, 2$. The distribution $\Sigma_{\mathfrak{d}_6,l}(\Phi,\omega,\chi,s)$ has at most a simple pole at $s = 1$ for $l = 1, 2$.

We define $\Sigma_{\mathfrak{d}_i,(j)}(\Phi,\omega,\chi,1)$ etc. similarly as in (5.3.4).

§7.2 Contributions from unstable strata

We consider $I(\Phi, \omega, \chi, w)$ in this section. Let $\mathfrak{p}_i = (\mathfrak{d}_i, \mathfrak{s}_i)$ be a path such that $\mathfrak{s}_i(1) = 0$, and $\mathfrak{p}'_i = (\mathfrak{d}_i, \mathfrak{s}'_i)$ a path such that $\mathfrak{s}_i(1) = 1$ for $i = 1, 2, 3$.

Let $\mathfrak{d} = \mathfrak{d}_1, \mathfrak{d}_2$ or $\mathfrak{d}_3$. It is easy to see that

$$\begin{aligned}&\int_{G^1_{\mathbb{A}}/G_k} \omega(g^1)\chi(g^1)\kappa_V(g^1)\Theta_{S_\mathfrak{d}}(\widehat{\Phi}, (g^1)^\iota)\mathscr{E}(g^0, w)d^\times g^1 \\ &= \int_{G^1_{\mathbb{A}}/G_k} \omega^{-1}(g^1)\chi^{-1}(g^1)\kappa_V^{-1}(g^1)\Theta_{S_\mathfrak{d}}(\widehat{\Phi}, g^1)\mathscr{E}(g^0, w)d^\times g^1\end{aligned}$$

because $\mathscr{E}((g^0)^\iota, w) = \mathscr{E}(g^0, w)$. Therefore, we only consider

$$\int_{G^1_{\mathbb{A}}/G_k} \omega(g^1)\chi(g^1)\Theta_{S_\mathfrak{d}}(\Phi, g^1)\mathscr{E}(g^0, w)d^\times g^1.$$

Lemma (7.2.1) *For* $\mathfrak{d} = \mathfrak{d}_1, \mathfrak{d}_2, \mathfrak{d}_3$,

$$\begin{aligned}&\int_{G^1_{\mathbb{A}}/G_k} \omega(g^1)\chi(g^1)\Theta_{S_\mathfrak{d}}(\Phi, g^1)\mathscr{E}(g^0, w)d^\times g^1 \\ &\sim \int_{G^1_{\mathbb{A}}\cap T_{\mathbb{A}}/T_k} \omega(t^1)\chi(t^1)(t^0)^{-2\rho}\kappa_{\mathfrak{d}1}(t^1)\Theta_{Z_\mathfrak{d}}(R_\mathfrak{d}\Phi, t^1)\mathscr{E}_N(t^0, w)d^\times t^1.\end{aligned}$$

Proof. Clearly,

$$\begin{aligned}&\int_{G^1_{\mathbb{A}}/G_k} \omega(g^1)\chi(g^1)\Theta_{S_\mathfrak{d}}(\Phi, g^1)\mathscr{E}(g^0, w)d^\times g^1 \\ &= \int_{G^1_{\mathbb{A}}\cap B_{\mathbb{A}}/B_k} \omega(t^1)\chi(t^1)(t^0)^{-2\rho}\Theta_{Y_\mathfrak{d}}(\widetilde{R}_\mathfrak{d}\Phi, t^1 n(u))\mathscr{E}(t^0 n(u), w)d^\times t^1 du.\end{aligned}$$

For $\alpha \in k$, we define

$$\begin{aligned}f_{1,\alpha}(t^1) &= \int_{N_{\mathbb{A}}/N_k} \Theta_{Y_{\mathfrak{d}_1}}(\Phi, t^1(n_2(u), 1)) < \alpha u > du, \\ f_{2,\alpha}(t^1) &= \int_{N_{\mathbb{A}}/N_k} \Theta_{Y_{\mathfrak{d}_2}}(\Phi, t^1(n_2(u), 1)) < \alpha u > du, \\ f_{3,\alpha}(w, t^1) &= \int_{N_{\mathbb{A}}/N_k} \mathscr{E}(t^0 n_2(u), w) < \alpha u > du.\end{aligned}$$

Then, by the Parseval formula, we only have to show that

$$\int_{T^1_{\mathbb{A}}/T_k} \omega(t^1)\lambda_2^c \sum_{\alpha\in k^\times} f_{i,\alpha}(t^1)f_{3,-\alpha}(w, t^1)\lambda_1^2 d^\times t^1 \sim 0$$

for $i = 1, 2$.

We fix a constant $M > w_0$. By (3.4.3), there exists $\delta > 0$ such that if $l \gg 0$,

$$|f_{3,-\alpha}(w, t^1)| \ll \lambda_1^l$$

for $w_0 - \delta \le \mathrm{Re}(w) \le M$. Let Φ_1 be the partial Fourier transform of $\widetilde{R}_{\mathfrak{d}_1}\Phi$ with respect to x_{21} and the character $<>$, and Φ_2 the partial Fourier transform of $\widetilde{R}_{\mathfrak{d}_2}\Phi$ with respect to x_{12} and the character $<>$. Then

$$
\begin{aligned}
& f_{1,\alpha}(t^1) \\
&= \int_{\mathbb{A}/k} \sum_{x_{12},x_{20}\in k^\times, x_{21}\in k} \Phi(\gamma_{12}(t^1)x_{12}, \gamma_{20}(t^1)x_{20}, \gamma_{21}(t^1)(x_{21}+ux_{20})) < \alpha u > du \\
&= \sum_{x_{12},x_{20}\in k^\times} \int_{\mathbb{A}/k} \sum_{x_{21}\in k} \Phi(\gamma_{12}(t^1)x_{12}, \gamma_{20}(t^1)x_{20}, \gamma_{21}(t^1)(x_{21}+ux_{20})) < \alpha u > du \\
&= \sum_{x_{12},x_{20}\in k^\times} \int_{\mathbb{A}} \Phi(\gamma_{12}(t^1)x_{12}, \gamma_{20}(t^1)x_{20}, \gamma_{21}(t^1)x_{20}u) < \alpha u > du \\
&= \lambda_1^{-1}\lambda_2^3 \sum_{x_{12},x_{20}\in k^\times} \Phi_1(\gamma_{12}(t^1)x_{12}, \gamma_{20}(t^1)x_{20}, \gamma_{21}(t^1)^{-1}x_{20}^{-1}\alpha).
\end{aligned}
$$

Similarly,

$$
f_{2,\alpha}(t^1) = \lambda_1^{-2}\lambda_2^{-4} \sum_{x_{11},x_{21}\in k^\times} \Phi_2(\gamma_{11}(t^1)x_{11}, \gamma_{21}(t^1)x_{21}, \gamma_{12}(t^1)^{-1}x_{12}^{-1}\alpha).
$$

Therefore, by (1.2.6), for any $N_1, N_2, N_3 \ge 1$,

$$
\begin{aligned}
&\sum_{\alpha\in k^\times} |f_{1,\alpha}(t^1)| \ll \lambda_1^{-1}\lambda_2^3(\lambda_1^2\lambda_2^4)^{-N_1}(\lambda_1^{-1}\lambda_2^{-3})^{-N_2}(\lambda_1^{-1}\lambda_2^3)^{-N_3}, \\
&\sum_{\alpha\in k^\times} |f_{2,\alpha}(t^1)| \ll \lambda_1^{-2}\lambda_2^{-4}\lambda_2^{-4N_1}(\lambda_1\lambda_2^{-3})^{-N_2}(\lambda_1^{-2}\lambda_2^{-4})^{-N_3}.
\end{aligned}
\tag{7.2.2}
$$

Since the convex hull of each of $\{(2,4),(-1,-3),(-1,3)\}$, $\{(0,4),(1,-3),(-2,-4)\}$ contains a neighborhood of the origin of $\mathbb{R}^2$, for any $N \gg 0$,

$$
\sum_{\alpha\in k^\times} |f_{1,\alpha}(t^1)|, \ \sum_{\alpha\in k^\times} |f_{2,\alpha}(t^1)| \ll \mathrm{rd}_{2,N}(\lambda_1,\lambda_2).
$$

We fix l. Then

$$
\int_{T_{\mathbb{A}}^1/T_k} \omega(t^1)\lambda_2^c \sum_{\alpha\in k^\times} f_{i,\alpha}(t^1) f_{3,-\alpha}(w,t^1)\lambda_1^2 d^\times t^1
$$

is bounded by a constant multiple of

$$
\int_{\mathbb{R}_+^2} \mathrm{rd}_{2,N}(\lambda_1,\lambda_2)\lambda_1^{2+l}\lambda_2^c d^\times\lambda_1 d^\times\lambda_2
$$

for $i = 1, 2$. This proves (7.2.1).

Q.E.D.

Let τ be any Weyl group element. For $i = 1, 2, 3$, we define

$$
\Xi_{\mathfrak{p}_i,\tau}(\Phi,\omega,\chi,w) = \int_{T_{\mathbb{A}}^1/T_k} \omega(t^1)\sigma_{\mathfrak{p}_i}(t^1)\Theta_{Z_{\mathfrak{d}}}(R_{\mathfrak{d}}\Phi,t^1)\mathscr{E}_{N,\tau}(t^0,w)d^\times t^1.
\tag{7.2.3}
$$

By a similar proof as in (6.1.4), $\Xi_{\mathfrak{p}_i}(\Phi, \omega, \chi, w)$ is well defined for $\mathrm{Re}(w) \gg 0$. By (7.2.1), the first two terms of (6.1.12)

$$\sim \sum_{i=1}^{3} \sum_{\tau} (\Xi_{\mathfrak{p}_i,\tau}(\widehat{\Phi}, \omega^{-1}, \kappa_V^{-1}\chi^{-1}, w) - \Xi_{\mathfrak{p}_i,\tau}(\Phi, \omega, \chi, w)).$$

Easy computations show that $\sigma_{\mathfrak{p}_1}(t^1) = \lambda_1\lambda_2^{c+3}$, $\sigma_{\mathfrak{p}_2}(t^1) = \lambda_2^{c-4}$, and $\sigma_{\mathfrak{p}_3}(t^1) = \lambda_1^2\lambda_2^c$.

Proposition (7.2.4) *Let $\tau, s = s_\tau$ be as before. Then*

$$\Xi_{\mathfrak{p}_i,\tau}(\Phi, \omega, \chi, w) \sim \frac{1}{2\pi\sqrt{-1}} \int_{\substack{\mathrm{Re}(s)=r \\ r \gg 0}} \Sigma_{\mathfrak{d}_i}(\Phi, \omega, \chi, s)\widetilde{\Lambda}_\tau(w; s)ds.$$

Proof. Let $t' = (t'_1, t'_2) = (\underline{\mu}_1 q_1, \underline{\mu}_2 q_2) \in (\mathbb{A}^\times)^2$, where $\mu_1, \mu_2 \in \mathbb{R}_+$, $q_1, q_2 \in \mathbb{A}^1$. Let $d^\times t' = d^\times t'_1 d^\times t'_2$.

Suppose that f is a function on $(\mathbb{A}^1/k^\times)^2$. It is easy to see that

$$\int_{(\mathbb{A}^1/k^\times)^4} \omega(\widehat{t}^0) f(t_{12}^2 t_2, t_{11}t_3) d^\times \widehat{t}^0 = \delta_{\mathfrak{d}_1}(\omega) \int_{(\mathbb{A}^1/k^\times)^2} \omega_{\mathfrak{d}_1}(q) f(q) d^\times q,$$
$$\int_{(\mathbb{A}^1/k^\times)^4} \omega(\widehat{t}^0) f(t_{11}t_{12}t_2, t_{12}t_3) d^\times \widehat{t}^0 = \delta_{\mathfrak{d}_2}(\omega) \int_{(\mathbb{A}^1/k^\times)^2} \omega_{\mathfrak{d}_2}(q) f(q) d^\times q,$$
$$\int_{(\mathbb{A}^1/k^\times)^4} \omega(\widehat{t}^0) f(t_{12}^2 t_2, t_{12}t_3) d^\times \widehat{t}^0 = \delta_{\mathfrak{d}_3}(\omega) \int_{(\mathbb{A}^1/k^\times)^2} \omega_{\mathfrak{d}_3}(q) f(q) d^\times q,$$

after the change of variable $(q_1, q_2) = (t_{12}^2 t_2, t_{11}t_3)$, $(t_{11}t_{12}t_2, t_{12}t_3)$, $(t_{12}^2 t_2, t_{12}t_3)$ respectively.

We make the change of variable $\mu_1 = \lambda_1^2\lambda_2^4, \mu_2 = \lambda_1^{-1}\lambda_2^{-3}$ for $\mathfrak{p}_1$, $\mu_1 = \lambda_2^4$, $\mu_2 = \lambda_1\lambda_2^{-3}$ for $\mathfrak{p}_2$, and $\mu_1 = \lambda_1^2\lambda_2^4$, $\mu_2 = \lambda_1\lambda_2^{-3}$ for $\mathfrak{p}_3$. Then

$$d^\times \mu_1 d^\times \mu_2 = 2d^\times \lambda_1 d^\times \lambda_2, \ \lambda_1 = \mu_1^{\frac{3}{2}} \mu_2^2, \ \lambda_2 = \mu_1^{-\frac{1}{2}} \mu_2^{-1}$$

for $\mathfrak{p}_1$,

$$d^\times \mu_1 d^\times \mu_2 = 4d^\times \lambda_1 d^\times \lambda_2, \ \lambda_1 = \mu_1^{\frac{3}{4}} \mu_2, \ \lambda_2 = \mu_1^{\frac{1}{4}}$$

for $\mathfrak{p}_2$, and

$$d^\times \mu_1 d^\times \mu_2 = 10d^\times \lambda_1 d^\times \lambda_2, \ \lambda_1 = \mu_1^{\frac{3}{10}} \mu_2^{\frac{4}{10}}, \ \lambda_2 = \mu_1^{\frac{1}{10}} \mu_2^{-\frac{2}{10}}$$

for $\mathfrak{p}_3$.

Clearly, $\mathscr{E}_{\mathfrak{p}_i,\tau}(t^0, w) = \mathscr{E}_{N,\tau}(t^0, w)$. It is easy to see that

$$\sigma_{\mathfrak{p}_i}(t^1)(t^0)^{\tau z+\rho} = \begin{cases} \mu_1^{\frac{3s-3-c}{2}} \mu_2^{2s-3-c} & i = 1, \\ \mu_1^{\frac{3s-7+c}{4}} \mu_2^{s-1} & i = 2, \\ \mu_1^{\frac{3s+3+c}{10}} \mu_2^{\frac{4s+4-2c}{10}} & i = 3. \end{cases}$$

Therefore,

$$\int_{T^1_{\mathbb{A}}/T_k} \omega(t^1)\sigma_{\mathfrak{p}_i}(t^1)\Theta_{Z_{\mathfrak{d}_i}}(R_{\mathfrak{d}_i}\Phi, t^1)\mathscr{E}_{N,\tau}(t^0, w)(t^0)^{-2\rho} d^\times t^1 = \Sigma_{\mathfrak{d}_i}(\Phi, \omega, \chi, s)$$

for $i = 1, 2, 3$. This proves the proposition.

Q.E.D.

Proposition (7.2.5) $\Xi_{\mathfrak{p}_i,\tau}(\Phi, \omega, \chi, w) \sim 0$ *if* $\tau \neq \tau_G$ *for* $i = 1, 2, 3$.

Proof. If $\tau \neq \tau_G$, we can choose $r \gg 0$. Then $\widetilde{L}(r) < w_0$.

Q.E.D.

Next, we consider $S_{\mathfrak{d}_i}$ for $i = 4, 5, 6$. Since

$$\Xi_{\mathfrak{d}_i,2}(\Phi, \omega, \chi, w) = \Xi_{\mathfrak{d}_i,3}(\widehat{\Phi}, \omega^{-1}, \kappa_V^{-1}\chi^{-1}, w),$$
$$\Xi_{\mathfrak{d}_i,4}(\Phi, \omega, \chi, w) = \Xi_{\mathfrak{d}_i,1}(\widehat{\Phi}, \omega^{-1}, \kappa_V^{-1}\chi^{-1}, w),$$

we only consider $\Xi_{\mathfrak{d}_i,j}(\Phi, \omega, \chi, w)$ for $j = 2, 4$.

The following proposition is a direct consequence of (3.4.31)(1).

Proposition (7.2.6)

(1) $$\Xi^s_{\mathfrak{d}_4,2}(\Phi, \omega, \chi, w) \sim C_G\Lambda(w; \rho)\frac{\delta_{\mathfrak{d}_4}(\omega)}{4}\Sigma_{\mathfrak{d}_4+}(R_{\mathfrak{d}_4}\Phi, \omega_{\mathfrak{d}_4}, \frac{c}{4}).$$

(2) $$\Xi^s_{\mathfrak{d}_4,4}(\Phi, \omega, \chi, w) \sim C_G\Lambda(w; \rho)\frac{\delta_{\mathfrak{d}_4}(\omega)}{4}\Sigma_{\mathfrak{d}_4+}(\mathscr{F}_{\mathfrak{d}_4}R_{\mathfrak{d}_4}\Phi, \omega_{\mathfrak{d}_4}^{-1}, 3 - \frac{c}{4}).$$

Let

$$\Sigma_{\mathfrak{d}_4,\mathrm{st},2}(\Phi, \omega, \chi, s) = \delta_{\mathfrak{d}_4,\mathrm{st}}(\omega)\frac{T_{\mathfrak{d}_4+}(\widetilde{R}_{\mathfrak{d}_4,0}R_{\mathfrak{d}_4}\Phi, \omega_{\mathfrak{d}_4,\mathrm{st}}, \frac{c}{4}, \frac{1-s}{2})}{4(s-1)},$$

$$\Sigma_{\mathfrak{d}_4,\mathrm{st},4}(\Phi, \omega, \chi, s) = \delta_{\mathfrak{d}_4,\mathrm{st}}(\omega)\frac{T_{\mathfrak{d}_4+}(\widetilde{R}_{\mathfrak{d}_4,0}\mathscr{F}_{\mathfrak{d}_4}R_{\mathfrak{d}_4}\Phi, \omega_{\mathfrak{d}_4,\mathrm{st}}^{-1}, 3 - \frac{c}{4}, \frac{1-s}{2})}{4(s-1)}.$$

The proof of the following proposition is similar to (5.4.3)–(5.4.7), and is left to the reader.

Proposition (7.2.7) *Let* $\tau = \tau_G$, *and* $s = s_{\tau_G}$. *Then*

(1) $$\Xi_{\mathfrak{d}_4,\mathrm{st},2}(\Phi, \omega, \chi, w) \sim \frac{1}{2\pi\sqrt{-1}}\int_{\substack{\mathrm{Re}(s)=r\\ r>1}} \Sigma_{\mathfrak{d}_4,\mathrm{st},2}(\Phi, \omega, \chi, s)\phi(s)\widetilde{\Lambda}(w; s)ds,$$

(2) $$\Xi_{\mathfrak{d}_4,\mathrm{st},4}(\Phi, \omega, \chi, w) \sim \frac{1}{2\pi\sqrt{-1}}\int_{\substack{\mathrm{Re}(s)=r\\ r>1}} \Sigma_{\mathfrak{d}_4,\mathrm{st},4}(\Phi, \omega, \chi, s)\phi(s)\widetilde{\Lambda}(w; s)ds.$$

Since $Y_{\mathfrak{d}_i} = Z_{\mathfrak{d}_i}$ and $M_{\mathfrak{d}_i} = T$ for $i = 5, 6$,

$$\Xi_{\mathfrak{d}_i,2}(\Phi, \omega, \chi, w) = \int_{\substack{T^1_{\mathbb{A}}/T_k\\ \lambda_1 \geq 1}} \omega(t^1)\chi(t^1)\Theta_{Z_{\mathfrak{d}_i}}(\Phi, t^1)\mathscr{E}_N(t^0, w)(t^0)^{-2\rho}d^\times t^1,$$

$$\Xi_{\mathfrak{d}_i,4}(\Phi, \omega, \chi, w) = \int_{\substack{T^1_{\mathbb{A}}/T_k\\ \lambda_1 \leq 1}} \omega(t^1)\chi(t^1)\Theta_{Z_{\mathfrak{d}_i}}(\Phi, t^1)\mathscr{E}_N(t^0, w)(t^0)^{-2\rho}d^\times t^1,$$

for $i = 5, 6$.

Let τ be a Weyl group element. We define

$$(7.2.8)\quad \Xi_{\mathfrak{d}_5,2,\tau}(\Phi,\omega,\chi,w) = \int_{\substack{T^1_{\mathbb{A}}/T_k \\ \lambda_1 \geq 1}} \omega(t^1)\chi(t^1)\Theta_{Z_{\mathfrak{d}_5}}(\Phi,t^1)\mathscr{E}_{N,\tau}(t^0,w)(t^0)^{-2\rho}d^\times t^1.$$

We define $\Xi_{\mathfrak{d}_5,i,\tau}(\Phi,\omega,\chi,w)$ for $i \neq 2$, and $\Xi_{\mathfrak{d}_6,i,\tau}(\Phi,\omega,\chi,w)$ for $i = 1, \cdots, 4$ similarly.

Proposition (7.2.9) *The distribution* $\Xi_{\mathfrak{d}_5,i,\tau}(\Phi,\omega,\chi,w)$ *is well defined for* $i = 1, \cdots, 4$ *if* $\mathrm{Re}(w) \gg 0$, *and*

$$(1)\quad \Xi_{\mathfrak{d}_5,2,\tau}(\Phi,\omega,\chi,w) \sim \frac{\delta_{\mathfrak{d}_5}(\omega)}{2\pi\sqrt{-1}} \int_{\substack{\mathrm{Re}(s)=r \\ r \gg 0}} \Sigma_{\mathfrak{d}_5,1}(\Phi,\omega,\chi,s)\widetilde{\Lambda}_\tau(w;s)ds,$$

$$(2)\quad \Xi_{\mathfrak{d}_5,4,\tau}(\Phi,\omega,\chi,w) \sim \frac{\delta_{\mathfrak{d}_5}(\omega)}{2\pi\sqrt{-1}} \int_{\substack{\mathrm{Re}(s)=r \\ r \gg 0}} \Sigma_{\mathfrak{d}_5,2}(\Phi,\omega,\chi,s)\widetilde{\Lambda}_\tau(w;s)ds.$$

Proof. Let $t' = \underline{\mu}q$, where $\mu \in \mathbb{R}_+$, $q \in \mathbb{A}^1$. Suppose that f is a function on $\mathbb{A}^1/k^\times$. It is easy to see that

$$\int_{(\mathbb{A}^1/k^\times)^4} \omega(\widehat{t}^0)f(t_{12}^2t_2)d^\times\widehat{t}^0 = \delta_{\mathfrak{d}_5}(\omega)\int_{\mathbb{A}^1/k^\times} \omega_{\mathfrak{p}_5}(q)f(q)d^\times q$$

after the change of variable $q = t_{12}^2t_2$. We make the change of variable $\mu = \lambda_1^2\lambda_2^4$, $\lambda_2 = \lambda_2$. Then $d^\times\mu d^\times\lambda_2 = 2d^\times\lambda_1 d^\times\lambda_2$, $\lambda_1 = \mu^{\frac{1}{2}}\lambda_2^{-2}$.

It is easy to see that

$$\chi(t^1)(t^0)^{\tau z-\rho} = \mu^{\frac{s+1}{2}}\lambda_2^{-2(s+1)+c}.$$

Therefore,

$$\int_{\substack{T^1_{\mathbb{A}}/T_k \\ \lambda_2 \geq 1}} \omega(t^1)\chi(t^1)\Theta_{Z_{\mathfrak{d}_5}}(R_{\mathfrak{d}_5}\Phi,t^1)(t^0)^{\tau z-\rho}d^\times t^1 = \Sigma_{\mathfrak{d}_5,1}(\Phi,\omega,\chi,s).$$

This proves the first statement. By the remark after (6.1.9), the second statement can be obtained by replacing Φ by $\mathscr{F}_V\Phi$, $\omega_{\mathfrak{d}}$ by $\omega_{\mathfrak{d}}^{-1}$, and c by $6-c$.

Q.E.D.

Proposition (7.2.10) $\Xi_{\mathfrak{d}_6,i,\tau}(\omega,\Phi,\chi,w)$ *is well defined for* $i = 1, \cdots, 4$ *if* $\mathrm{Re}(w) \gg 0$, *and*

$$(1)\quad \Xi_{\mathfrak{d}_6,4,\tau}(\omega,\Phi,\chi,w) \sim \frac{1}{2\pi\sqrt{-1}} \int_{\substack{\mathrm{Re}(s)=r \\ r>1}} \Sigma_{\mathfrak{d}_6,1}(\Phi,\chi,s)\widetilde{\Lambda}_\tau(w;s)ds,$$

$$(2)\quad \Xi_{\mathfrak{d}_6,2,\tau}(\omega,\Phi,\chi,w) \sim \frac{1}{2\pi\sqrt{-1}} \int_{\substack{\mathrm{Re}(s)=r \\ r>1}} \Sigma_{\mathfrak{d}_6,2}(\Phi,\chi,s)\widetilde{\Lambda}_\tau(w;s)ds.$$

Proof. Let $t' = \underline{\mu}q$, where $\mu \in \mathbb{R}_+$, $q \in \mathbb{A}^1$. Suppose that f is a function on $\mathbb{A}^1/k^\times$. It is easy to see that

$$\int_{(\mathbb{A}^1/k^\times)^4} \omega(\widehat{t}^0)f(t_{12}t_{22})d^\times\widehat{t}^0 = \delta_{\#}(\omega)\int_{\mathbb{A}^1/k^\times} f(q)d^\times q$$

after the change of variable $q = t_{12}t_{22}$. We make the change of variable $\mu = \lambda_1\lambda_2^{-3}$, $\lambda_2 = \lambda_2$. Then $d^\times \mu d^\times \lambda_2 = d^\times \lambda_1 d^\times \lambda_2$, $\lambda_1 = \mu\lambda_2^3$.

It is easy to see that $\chi(t^1)(t^0)^{\tau z-\rho} = \mu^{s+1}\lambda_2^{3s+3+c}$.

Therefore,

$$\int_{\substack{T_{\mathbb{A}}^1/T_k \\ \lambda_2 \leq 1}} \omega(t^1)\chi(t^1)\Theta_{Z_{\mathfrak{d}_6}}(R_{\mathfrak{d}_6}\Phi, t^1)(t^0)^{\tau z-\rho} dt^1 = \Sigma_{\mathfrak{d}_6,1}(\Phi, \chi, s).$$

This proves the first statement. The second statement can be obtained similarly.

Q.E.D.

Proposition (7.2.11) *If $\tau \neq \tau_G$,*

$$\Xi_{\mathfrak{d}_5,i,\tau}(\Phi, \omega, \chi, w),\ \Xi_{\mathfrak{d}_6,i,\tau}(\Phi, \omega, \chi, w) \sim 0$$

for $i = 1, \cdots, 4$.

Proof. We only have to choose $r \gg 0$.

Q.E.D.

§7.3 The principal part formula

We define

$$\begin{aligned}
J_1(\Phi, \omega, \chi, s) = & \sum_{i=1}^{3} \Sigma_{\mathfrak{d}_i}(\Phi, \omega, \chi, s) \\
& + \frac{\delta_{\mathfrak{d}_4,\mathrm{st}}(\omega)}{4(s-1)} T_{\mathfrak{d}_4+}(\widetilde{R}_{\mathfrak{d}_4,0} R_{\mathfrak{d}_4}\Phi, \omega_{\mathfrak{d}_4,\mathrm{st}}, \frac{c}{4}, \frac{1-s}{2}) \\
& + \frac{\delta_{\mathfrak{d}_4,\mathrm{st}}(\omega)}{4(s-1)} T_{\mathfrak{d}_4+}(\widetilde{R}_{\mathfrak{d}_4,0} \mathscr{F}_{\mathfrak{d}_4} R_{\mathfrak{d}_4}\Phi, \omega_{\mathfrak{d}_4,\mathrm{st}}^{-1}, 3 - \frac{c}{4}, \frac{1-s}{2}) \\
& + \Sigma_{\mathfrak{d}_5,1}(\Phi, \omega, \chi, s) + \Sigma_{\mathfrak{d}_5,2}(\Phi, \omega, \chi, s) \\
& + \Sigma_{\mathfrak{d}_6,1}(\Phi, \omega, \chi, s) + \Sigma_{\mathfrak{d}_6,2}(\Phi, \omega, \chi, s), \\
J_2(\Phi, \omega, \chi) = & \frac{\delta_{\mathfrak{d}_4}(\omega)}{4} \left(\Sigma_{\mathfrak{d}_4+}(R_{\mathfrak{d}_4}\Phi, \omega_{\mathfrak{p}_4}, \frac{c}{4}) + \Sigma_{\mathfrak{d}_4+}(\mathscr{F}_{\mathfrak{d}_4} R_{\mathfrak{d}_4}\Phi, \omega_{\mathfrak{p}_4}^{-1}, 3 - \frac{c}{4}) \right), \\
J_3(\Phi, \omega, \chi) = & - \frac{\delta_{\mathfrak{d}_4,\mathrm{st}}(\omega)}{8} T_{\mathfrak{d}_4+}(\widetilde{R}_{\mathfrak{d}_4,0} R_{\mathfrak{d}_4}\Phi, \omega_{\mathfrak{d}_4,\mathrm{st}}, \frac{c}{4}) \\
& - \frac{\delta_{\mathfrak{d}_4,\mathrm{st}}(\omega)}{8} T_{\mathfrak{d}_4+}(\widetilde{R}_{\mathfrak{d}_4,0} \mathscr{F}_{\mathfrak{d}_4} R_{\mathfrak{d}_4}\Phi, \omega_{\mathfrak{d}_4,\mathrm{st}}^{-1}, 3 - \frac{c}{4}).
\end{aligned}$$

Let $\tau = \tau_G$, and $s = s_\tau$. Then by (7.2.1), (7.2.4)–(7.2.11),

$$\begin{aligned}
& I(\Phi, \omega, \chi, w) \\
& \sim \frac{1}{2\pi\sqrt{-1}} \int_{\substack{\mathrm{Re}(s)=r \\ r \gg 0}} (J_1(\widehat{\Phi}, \omega^{-1}, \kappa_V^{-1}\chi^{-1}, s) - J_1(\Phi, \omega, \chi, s))\phi(s)\widetilde{\Lambda}(w; s) ds \\
& \quad + C_G \Lambda(w; \rho)(J_2(\widehat{\Phi}, \omega^{-1}, \kappa_V^{-1}\chi^{-1}) - J_2(\Phi, \omega, \chi)).
\end{aligned}$$

By Wright's principle, $J_1(\widehat{\Phi},\omega^{-1},\kappa_V^{-1}\chi^{-1},s)-J_1(\Phi,\omega,\chi,s)$ must be holomorphic at $s=1$. Therefore,

$$\begin{aligned}
&\frac{1}{2\pi\sqrt{-1}}\int_{\substack{\mathrm{Re}(s)=r\\ r\gg 0}}(J_1(\widehat{\Phi},\omega^{-1},\kappa_V^{-1}\chi^{-1},s)-J_1(\Phi,\omega,\chi,s))\phi(s)\widetilde{\Lambda}(w;s)ds\\
&\sim C_G\Lambda(w;\rho)\sum_{i=1,2,3}(\Sigma_{\mathfrak{d}_i,(0)}(\widehat{\Phi},\omega^{-1},\kappa_V^{-1}\chi^{-1},1)-\Sigma_{\mathfrak{d}_i,(0)}(\Phi,\omega,\chi,1))\\
&\quad+C_G\Lambda(w;\rho)\sum_{\substack{i=5,6\\ l=1,2}}(\Sigma_{\mathfrak{d}_i,l,(0)}(\widehat{\Phi},\omega^{-1},\kappa_V^{-1}\chi^{-1},1)-\Sigma_{\mathfrak{d}_i,l,(0)}(\Phi,\omega,\chi,1))\\
&\quad+C_G\Lambda(w;\rho)(J_3(\widehat{\Phi},\omega^{-1},\kappa_V^{-1}\chi^{-1})-J_3(\Phi,\omega,\chi)).
\end{aligned}$$

This implies that

$$\begin{aligned}
I^1(\Phi,\omega,\chi)=&\sum_{i=1,2,3}\left(\Sigma_{\mathfrak{d}_i,(0)}(\widehat{\Phi},\omega^{-1},\kappa_V^{-1}\chi^{-1},1)-\Sigma_{\mathfrak{d}_i,(0)}(\Phi,\omega,\chi,1)\right)\\
&+\sum_{\substack{i=5,6\\ l=1,2}}\left(\Sigma_{\mathfrak{d}_i,l,(0)}(\widehat{\Phi},\omega^{-1},\kappa_V^{-1}\chi^{-1},1)-\Sigma_{\mathfrak{d}_i,l,(0)}(\Phi,\omega,\chi,1)\right)\\
&+\sum_{i=2,3}(J_i(\widehat{\Phi},\omega^{-1},\kappa_V^{-1}\chi^{-1})-J_i(\Phi,\omega,\chi)).
\end{aligned}$$

Since the reason for studying this case is to apply it to the quartic case, we restrict ourselves to χ satisfying the condition $\frac{c}{4},\frac{6-c}{4}\neq 0,2,3$. In Part IV, we will consider the situation $c=2$, so this condition is satisfied.

By assumption, $\Sigma_{\mathfrak{d}_6,j}(\Phi,\omega,\chi,s),\Sigma_{\mathfrak{d}_6,j}(\widehat{\Phi},\omega^{-1},\kappa_V^{-1}\chi^{-1},s)$ for $i=1,2$ are holomorphic at $s=1$, and their values are

$$\begin{aligned}
&\delta_{\mathfrak{d}_6}(\omega)\frac{\Sigma_1(R_{\mathfrak{d}_6}\Phi,2)}{6+c},\ -\delta_{\mathfrak{d}_6}(\omega)\frac{\Sigma_1(R_{\mathfrak{d}_6}\mathscr{F}_{V_2}\Phi,2)}{c},\\
&\delta_{\mathfrak{d}_6}(\omega)\frac{\Sigma_1(R_{\mathfrak{d}_6}\widehat{\Phi},2)}{12-c},\ -\delta_{\mathfrak{d}_6}(\omega)\frac{\Sigma_1(R_{\mathfrak{d}_6}\mathscr{F}_{V_2}\widehat{\Phi},2)}{6-c},
\end{aligned}$$

respectively. The following lemma is an easy consequence of (4.4.7).

Lemma (7.3.1) *If Φ is K-invariant,*

$$\Sigma_1(R_{\mathfrak{d}_6}\Phi,2)=\frac{Z_k(2)}{\mathfrak{R}_k}\mathscr{F}_{\mathfrak{d}_4}\widehat{\Phi}(0),\ \Sigma_1(R_{\mathfrak{d}_6}\mathscr{F}_{V_2}\Phi,2)=\frac{Z_k(2)}{\mathfrak{R}_k}\Phi(0).$$

The above relations hold by replacing Φ by $\widehat{\Phi}$. Also by assumption,

$$\begin{aligned}
\Sigma_{\mathfrak{d}_5,2,(0)}(\Phi,\omega,\chi,1)=&-\frac{\delta_{\mathfrak{d}_5}(\omega)}{8}\frac{\Sigma_{1,(-1)}(R_{\mathfrak{d}_5}\mathscr{F}_{V_1}\Phi,\omega_{\mathfrak{d}_5}^{-1},1)}{(\frac{c}{4}-2)^2}\\
&+\frac{\delta_{\mathfrak{d}_5}(\omega)}{8}\frac{\Sigma_{1,(0)}(R_{\mathfrak{d}_5}\mathscr{F}_{V_1}\Phi,\omega_{\mathfrak{d}_5}^{-1},1)}{(\frac{c}{4}-2)}.
\end{aligned}$$

The above relation holds by replacing Φ,ω,χ by $\widehat{\Phi},\omega^{-1},\kappa_V^{-1}\chi^{-1}$ respectively.

If $c \neq 4$,

$$\Sigma_{\mathfrak{d}_5,1,(0)}(\Phi,\omega,\chi,1) = -\frac{\delta_{\mathfrak{d}_5}(\omega)}{8}\left(\frac{\Sigma_{1,(-1)}(R_{\mathfrak{d}_5}\Phi,\omega_{\mathfrak{d}_5},1)}{(\frac{c}{4}-1)^2} + \frac{\Sigma_{1,(0)}(R_{\mathfrak{d}_5}\Phi,\omega_{\mathfrak{d}_5},1)}{(\frac{c}{4}-1)}\right).$$

If $c = 4$,

$$\Sigma_{\mathfrak{d}_5,1,(0)}(\Phi,\omega,\chi,1) = \frac{\delta_{\mathfrak{d}_5}(\omega)}{8}\Sigma_{1,(1)}(R_{\mathfrak{d}_5}\Phi,\omega_{\mathfrak{d}_5},1).$$

If $c \neq 4$, we define

$$J_4(\Phi,\omega,\chi) = \frac{\delta_{\mathfrak{d}_4}(\omega)}{4}\Sigma_{\mathfrak{d}_4,\mathrm{ad}}(R_{\mathfrak{d}_4}\Phi,\omega_{\mathfrak{d}_4},\frac{c}{4}).$$

If $c = 4$, we define

$$J_4(\Phi,\omega,\chi) = \frac{\delta_{\mathfrak{d}_4}(\omega)}{4}\Sigma_{\mathfrak{d}_4,\mathrm{ad},(0)}(R_{\mathfrak{d}_4}\Phi,\omega_{\mathfrak{d}_4},1) + \frac{\delta_{\mathfrak{d}_5}(\omega)}{8}\Sigma_{1,(1)}(R_{\mathfrak{d}_5}\Phi,\omega_{\mathfrak{d}_5},1).$$

By the principal part formula for $\Sigma_{\mathfrak{d}_4,\mathrm{ad}}(R_{\mathfrak{d}_4}\Phi,\omega_{\mathfrak{d}_4},s)$ and the above considerations, we get following proposition.

Proposition (7.3.2)

$$\begin{aligned} I^1(\Phi,\omega,\chi) = &\sum_{i=1,2,3}\left(\Sigma_{\mathfrak{d}_i,(0)}(\widehat{\Phi},\omega^{-1},\kappa_V^{-1}\chi^{-1},1) - \Sigma_{\mathfrak{d}_i,(0)}(\Phi,\omega,\chi,1)\right) \\ &+ J_4(\widehat{\Phi},\omega^{-1},\kappa_V^{-1}\chi^{-1}) - J_4(\Phi,\omega,\chi). \end{aligned}$$

If $c \neq 4$, $J_4(\Phi_\lambda,\omega,\chi) = \lambda^{-\frac{c}{4}}J_4(\Phi,\omega,\chi)$. If $c = 4$, $J_4(\Phi_\lambda,\omega,\chi) = \lambda^{-1}J_4(\Phi,\omega,\chi)$ as in §1.7.

Definition (7.3.3)

(1) $$p_1(\chi) = -\frac{3c}{2},\ p_2(\chi) = \frac{c}{4},\ p_3(\chi) = \frac{14-c}{10},\ p_4(\chi) = \frac{c}{4},$$

(2) $$q_1 = \frac{7}{2},\ q_2 = \frac{7}{4},\ q_4 = 1.$$

Easy computations show the following relations and the proof is left to the reader.

Lemma (7.3.4) *For* $i = 1,2,3$,

(1) $$\Sigma_{\mathfrak{d}_i}(\Phi_\lambda,\omega,\chi,s) = \lambda^{-q_i(s-1)-p_i(\chi)}\Sigma_{\mathfrak{d}_i}(\Phi,\omega,\chi,s),$$

(2) $$\Sigma_{\mathfrak{d}_i}(\widehat{\Phi_\lambda},\omega^{-1},\kappa_V^{-1}\chi^{-1},s) = \lambda^{q_i(s-1)-(5-p_i(\kappa_V^{-1}\chi^{-1}))}\Sigma_{\mathfrak{d}_i}(\widehat{\Phi},\omega^{-1},\kappa_V^{-1}\chi^{-1},s).$$

The following proposition is an easy consequence of (7.3.4).

Proposition (7.3.5) *For* $i = 1,2,3$,

(1) $$\begin{aligned} &\Sigma_{\mathfrak{d}_i,(0)}(\Phi_\lambda,\omega,\chi,1) \\ &= \sum_{j=-2}^{0}\frac{1}{(-j)!}(-q_i\log\lambda)^{-j}\lambda^{-p_i(\chi)}\Sigma_{\mathfrak{d}_i,(j)}(\Phi,\omega,\chi,1), \end{aligned}$$

(2) $$\begin{aligned} &\Sigma_{\mathfrak{d}_i,(0)}(\Phi_\lambda,\omega^{-1},\kappa_V^{-1}\chi^{-1},1) \\ &= \sum_{j=-2}^{0}\frac{1}{(-j)!}(q_i\log\lambda)^{-j}\lambda^{-(5-p_i(\kappa_V^{-1}\chi^{-1}))}\Sigma_{\mathfrak{d}_i,(j)}(\Phi,\omega^{-1},\kappa_V^{-1}\chi^{-1},1). \end{aligned}$$

Definition (7.3.6)

$$F_1(\Phi,\omega,\chi,s)=\sum_{i=1}^{3}\sum_{j=-2}^{0}\frac{(q_i)^{-j}\delta_{\mathfrak{d}_i}(\omega)\Sigma_{\mathfrak{d}_i,(j)}(\Phi,\omega,\chi,1)}{(s-p_i)^{1-j}}.$$

$$F_2(\Phi,\omega,\chi,s)=\frac{J_4(\Phi,\omega,\chi)}{s-\frac{c}{4}}.$$

These considerations imply the following principal part formula.

Theorem (7.3.7) (F. Sato) *Suppose $\Phi=M_\omega\Phi$. Also assume that $\frac{c}{4},\frac{6-c}{4}\neq 0,2,3$. Then $Z_V(\Phi,\omega,\chi,s)$ satisfies the following principal part formula*

$$\begin{aligned}Z_V(\Phi,\omega,\chi,s)&=Z_{V+}(\Phi,\omega,\chi,s)+Z_{V+}(\widehat{\Phi},\omega^{-1},\kappa_V^{-1}\chi^{-1},5-s)\\&\quad-\sum_{i=1,2}(F_i(\Phi,\omega,s)+F_i(\widehat{\Phi},\omega^{-1},\kappa_V^{-1}\chi^{-1},5-s)).\end{aligned}$$

Because of the above formula, the following functional equation follows.

Corollary (7.3.8) $Z_V(\Phi,\omega,\chi,s)=Z_V(\widehat{\Phi},\omega^{-1},\kappa_V^{-1}\chi^{-1},5-s)$.

Since we want to use (7.3.7) for the quartic case, we describe $F_1(\Phi,\omega,\chi,s)$ explicitly here. We assume that $c=2$. So $6-c=4$.

We define

$$F_{\mathfrak{d}_1}(\Phi,\omega,s)=\frac{1}{2}\frac{\Sigma_2(R_{\mathfrak{d}_1}\Phi,\omega_{\mathfrak{d}_1},-1,-3)}{s+3},$$

$$\begin{aligned}F_{\mathfrak{d}_2}(\Phi,\omega,s)&=\frac{1}{4}\frac{\Sigma_{2,(0,0)}(R_{\mathfrak{d}_2}\Phi,\omega_{\mathfrak{d}_2},-\frac{1}{2},0)}{s-\frac{1}{2}}+\frac{3}{16}\frac{\Sigma_{2,(1,-1)}(R_{\mathfrak{d}_2}\Phi,\omega_{\mathfrak{d}_2},-\frac{1}{2},0)}{s-\frac{1}{2}}\\&\quad+\frac{7}{16}\frac{\Sigma_{2,(0,-1)}(R_{\mathfrak{d}_2}\Phi,\omega_{\mathfrak{d}_2},-\frac{1}{2},0)}{(s-\frac{1}{2})^2},\end{aligned}$$

$$F_{\mathfrak{d}_3}(\Phi,\omega,s)=\frac{1}{10}\frac{\Sigma_2(R_{\mathfrak{d}_3}\Phi,\omega_{\mathfrak{d}_3},\frac{4}{5},\frac{2}{5})}{s-\frac{6}{5}},$$

$$\widehat{F}_{\mathfrak{d}_1}(\Phi,\omega,s)=\frac{1}{2}\frac{\Sigma_2(R_{\mathfrak{d}_1}\widehat{\Phi},\omega_{\mathfrak{d}_1}^{-1},-2,-5)}{11-s},$$

$$\begin{aligned}\widehat{F}_{\mathfrak{d}_2}(\Phi,\omega,s)&=\frac{1}{4}\frac{\Sigma_{2,(0,0)}(R_{\mathfrak{d}_2}\widehat{\Phi},\omega_{\mathfrak{d}_2}^{-1},0,0)}{4-s}+\frac{3}{16}\frac{\Sigma_{2,(1,-1)}(R_{\mathfrak{d}_2}\widehat{\Phi},\omega_{\mathfrak{d}_2}^{-1},0,0)}{4-s}\\&\quad+\frac{1}{3}\frac{\Sigma_{2,(-1,1)}(R_{\mathfrak{d}_2}\widehat{\Phi},\omega_{\mathfrak{d}_2}^{-1},0,0)}{4-s}+\frac{7}{16}\frac{\Sigma_{2,(0,-1)}(R_{\mathfrak{d}_2}\widehat{\Phi},\omega_{\mathfrak{d}_2}^{-1},0,0)}{(4-s)^2}\\&\quad+\frac{7}{12}\frac{\Sigma_{2,(-1,0)}(R_{\mathfrak{d}_2}\widehat{\Phi},\omega_{\mathfrak{d}_2}^{-1},0,0)}{(4-s)^2}+\frac{49}{48}\frac{\Sigma_{2,(-1,-1)}(R_{\mathfrak{d}_2}\widehat{\Phi},\omega_{\mathfrak{d}_2}^{-1},0,0)}{(4-s)^3},\end{aligned}$$

$$\begin{aligned}\widehat{F}_{\mathfrak{d}_3}(\Phi,\omega,s)&=\frac{1}{10}\frac{\Sigma_{2,(0,0)}(R_{\mathfrak{d}_3}\widehat{\Phi},\omega_{\mathfrak{d}_3}^{-1},1,0)}{4-s}+\frac{3}{40}\frac{\Sigma_{2,(1,-1)}(R_{\mathfrak{d}_3}\widehat{\Phi},\omega_{\mathfrak{d}_3}^{-1},1,0)}{4-s}\\&\quad+\frac{2}{15}\frac{\Sigma_{2,(-1,1)}(R_{\mathfrak{d}_3}\widehat{\Phi},\omega_{\mathfrak{d}_3}^{-1},1,0)}{4-s}+\frac{7}{40}\frac{\Sigma_{2,(0,-1)}(R_{\mathfrak{d}_3}\widehat{\Phi},\omega_{\mathfrak{d}_3}^{-1},1,0)}{(4-s)^2}\\&\quad+\frac{7}{30}\frac{\Sigma_{2,(-1,0)}(R_{\mathfrak{d}_3}\widehat{\Phi},\omega_{\mathfrak{d}_3}^{-1},1,0)}{(4-s)^2}+\frac{49}{120}\frac{\Sigma_{2,(-1,-1)}(R_{\mathfrak{d}_3}\widehat{\Phi},\omega_{\mathfrak{d}_3}^{-1},1,0)}{(4-s)^3}.\end{aligned}$$

Then

$$F_1(\Phi, \omega, \chi, s) + F_1(\widehat{\Phi}, \omega^{-1}, \kappa_V^{-1}\chi^{-1}, 5-s) = \sum_{i=1,2,3} \delta_{\mathfrak{d}_i}(\omega)(F_{\mathfrak{d}_i}(\Phi, \omega, s) + \widehat{F}_{\mathfrak{d}_i}(\Phi, \omega, s)). \tag{7.3.9}$$

Therefore, the poles of $Z_V(\Phi, \omega, \chi, s)$ are $s = -3, \frac{1}{2}, \frac{6}{5}, 4, 11$. In Chapter 12,13, we show that the Laurent expansion of $Z_V(\Phi, \omega, \chi, s)$ at $s = -3$ will contribute to the rightmost pole of the zeta function for the quartic case. The function $Z_V(\Phi, \omega, \chi, s)$ has a simple pole at $s = -3$, and the coefficient of the order two term of the principal part of the quartic case turns out to be a constant multiple of $\Sigma_2(R_{\mathfrak{d}_1}\Phi, \omega_{\mathfrak{d}_1}, -1, -3)$.

Part IV The quartic case

In the next six chapters, we consider the quartic case $G = \mathrm{GL}(3) \times \mathrm{GL}(2)$, $V = \mathrm{Sym}^2 k^3 \otimes k^2$. In Chapter 8, we study the stability and the Morse stratification of our case. In particular, we explicitly describe all the β-sequences we need.

In Chapter 3, we proved that the distributions associated with certain paths are well defined. However, we have some paths which are not covered there. The representations Z_{β_6}, Z_{β_8} of M_{β_6}, M_{β_8} (in §8.2) are reducible and require a handling similar to Chapters 6 and 7. We prove that certain distributions associated with β_6, β_8 are well defined in §9.1. In §9.2, we prove some special estimates of the smoothed Eisenstein series which are required in Chapter 10. In Chapter 10, we prove that we can ignore the non-constant terms associated with unstable strata. We will prove that the distribution $\widetilde{\Xi}_{\mathfrak{p}}(\Phi, \omega, w)$ is well defined for all $\mathfrak{p}$ and

$$\sum_{\mathfrak{p}} \epsilon_{\mathfrak{p}} \widetilde{\Xi}_{\mathfrak{p}}(\Phi, \omega, w) \sim 0.$$

For this purpose, some cancellations between different paths have to established as in §10.4, 10.7. In Chapters 11–13, we compute the constant terms $\Xi_{\mathfrak{p}}(\Phi, \omega, w)$ associated with paths explicitly, and prove a principal part formula for the zeta function of our case. In our case, we can fortunately use Wright's principle.

This case is of complete type. Let $Z_V(\Phi, \omega, s)$ be the zeta function defined by (3.1.8) for $L = V_k^{\mathrm{ss}} = V_k^{\mathrm{s}}$. The principal part formula for the zeta function for this case is (13.2.2). The location of the poles is $s = 0, 2, 3, 9, 10, 12$. The orders of the poles at $s = 2, 10$ do not coincide with the multiplicities of the corresponding roots of the b-function for this case.

Chapter 8 Invariant theory of pairs of ternary quadratic forms

§8.1 The space of pairs of ternary quadratic forms

Let $\widetilde{V}$ be the space of quadratic forms in three variables $v = (v_1, v_2, v_3)$. We identify $\widetilde{V}$ with k^6 as follows:

$$\begin{aligned} Q_x(v) &= x_{11}v_1^2 + x_{12}v_1v_2 + x_{13}v_1v_3 + x_{22}v_2^2 + x_{23}v_2v_3 + x_{33}v_3^2 \\ &\to x = (x_{11}, x_{12}, x_{13}, x_{22}, x_{23}, x_{33}). \end{aligned}$$

The group $G_1 = \mathrm{GL}(3)$ acts on $\widetilde{V}$ in the following way:

$$Q_{g_1 \cdot x}(v) = Q_x(vg_1),$$

for $g_1 \in G_1$.

Consider $V = \mathrm{Sym}^2 k^3 \otimes k^2 = \widetilde{V} \oplus \widetilde{V}$. Any element of V is of the form

$$Q = (Q_{x_1}, Q_{x_2}),$$

where

$$\begin{aligned} x_1 &= (x_{1,11}, x_{1,12}, x_{1,13}, x_{1,22}, x_{1,23}, x_{1,33}), \\ x_2 &= (x_{2,11}, x_{2,12}, x_{2,13}, x_{2,22}, x_{2,23}, x_{2,33}). \end{aligned}$$

We choose $x = (x_1, x_2)$ as the coordinate system of V.

Let $G_2 = \mathrm{GL}(2)$, and $G = G_1 \times G_2$. The group G acts on V as follows. Let $g = (g_1, g_2)$ where $g_1 \in G_1$, and $g_2 = \begin{pmatrix} a & b \\ c & d \end{pmatrix} \in G_2$. We define

$$g \cdot Q = (aQ_{g_1 \cdot x_1} + bQ_{g_1 \cdot x_2}, cQ_{g_1 \cdot x_1} + dQ_{g_1 \cdot x_2}).$$

Let $\widetilde{T}$ be the kernel of the homomorphism $G \to \mathrm{GL}(V)$. We define

$$\chi_V(g) = (\det g_1)^4 (\det g_2)^3.$$

Then χ_V can be considered as a character of $G/\widetilde{T}$, and it is indivisible. In the next six chapters, we consider $(G/\widetilde{T}, V)$.

Throughout the next six chapters, we use the notation

$$a(t_{11}, t_{12}, t_{13}; t_{21}, t_{22}) = (a_3(t_{11}, t_{12}, t_{13}), a_2(t_{21}, t_{22})). \tag{8.1.1}$$

Let $T \subset G, G', \mathfrak{t}^*, \mathfrak{t}^*_-$ etc. be as in Chapter 3. In our case, $G^0_{\mathbb{A}} = G^1_{\mathbb{A}}$, and therefore, $\mathfrak{t}^* = \mathfrak{t}^{0*}$. We identify $\mathfrak{t}^*$ with

$$\{z = (z_{11}, z_{12}, z_{13}; z_{21}, z_{22}) \in \mathbb{R}^5 \mid z_{11} + z_{12} + z_{13} = 0, \ z_{21} + z_{22} = 0\}.$$

We use the notation $z_1 = (z_{11}, z_{12}, z_{13}), z_2 = (z_{21}, z_{22})$, and write $z = (z_1, z_2)$.

For $z = (z_{11}, z_{12}, z_{13}; z_{21}, z_{22})$, $z' = (z'_{11}, z'_{12}, z'_{13}; z'_{21}, z'_{22})$, we define

$$(z, z') = z_{11}z'_{11} + z_{12}z'_{12} + z_{13}z'_{13} + z_{21}z'_{21} + z_{22}z'_{22}.$$

This inner product is Weyl group invariant, and we use this inner product to determine the Morse stratification in the next section. Let $\| \ \|$ be the metric defined by this bilinear form. We recall that $G' = \mathrm{Ker}(\chi_V)$. We choose $G'' = \mathrm{SL}(3) \times \mathrm{SL}(2)$ for G'' in §3.1.

The weights of $x_{1,ij}$, $x_{2,ij}$ for $1 \le i \le j \le 3$ are as follows:

$$\begin{array}{llll}
x_{1,11} & (\frac{4}{3}, -\frac{2}{3}, -\frac{2}{3}; \frac{1}{2}, -\frac{1}{2}) & x_{2,11} & (\frac{4}{3}, -\frac{2}{3}, -\frac{2}{3}; -\frac{1}{2}, \frac{1}{2}) \\
x_{1,12} & (\frac{1}{3}, \frac{1}{3}, -\frac{2}{3}; \frac{1}{2}, -\frac{1}{2}) & x_{2,12} & (\frac{1}{3}, \frac{1}{3}, -\frac{2}{3}; -\frac{1}{2}, \frac{1}{2}) \\
x_{1,13} & (\frac{1}{3}, -\frac{2}{3}, \frac{1}{3}; \frac{1}{2}, -\frac{1}{2}) & x_{2,13} & (\frac{1}{3}, -\frac{2}{3}, \frac{1}{3}; -\frac{1}{2}, \frac{1}{2}) \\
x_{1,22} & (-\frac{2}{3}, \frac{4}{3}, -\frac{2}{3}; \frac{1}{2}, -\frac{1}{2}) & x_{2,22} & (-\frac{2}{3}, \frac{4}{3}, -\frac{2}{3}; -\frac{1}{2}, \frac{1}{2}) \\
x_{1,23} & (-\frac{2}{3}, \frac{1}{3}, \frac{1}{3}; \frac{1}{2}, -\frac{1}{2}) & x_{2,23} & (-\frac{2}{3}, \frac{1}{3}, \frac{1}{3}; -\frac{1}{2}, \frac{1}{2}) \\
x_{1,33} & (-\frac{2}{3}, -\frac{2}{3}, \frac{4}{3}; \frac{1}{2}, -\frac{1}{2}) & x_{2,33} & (-\frac{2}{3}, -\frac{2}{3}, \frac{4}{3}; -\frac{1}{2}, \frac{1}{2})
\end{array}$$

Therefore, with our metric, the weights of coordinates look as in the picture on the next page.

Let $\gamma_{1,ij}, \gamma_{2,ij}$ be the weights of $x_{1,ij}, x_{2,ij}$ respectively. If $t \in T$, we denote the value of the rational character determined by $\gamma_{1,ij}$ (resp. $\gamma_{2,ij}$) by $\gamma_{1,ij}(t)$ (resp. $\gamma_{2,ij}(t)$). This should not be confused with $t^{\gamma_{1,ij}}$ etc. which is the adelic absolute value of $\gamma_{1,ij}(t)$ etc. We use this notation throughout the next six chapters.

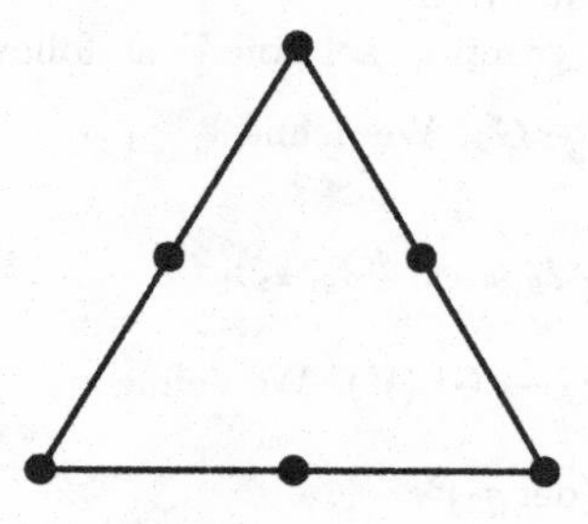

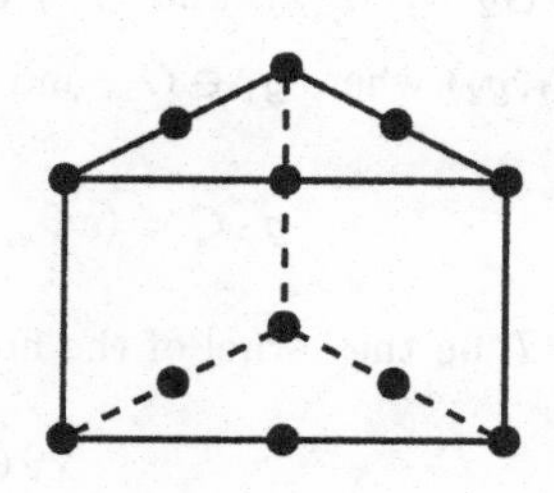

§8.2 The Morse stratification

We study the stability over $\bar{k}$. Consider a point $x = (x_1, x_2) \in V_{\bar{k}}$ as before. Let $Q_1 = Q_{x_1}, Q_2 = Q_{x_2}$. The quadratic forms Q_1, Q_2 define subschemes $V(Q_1), V(Q_2)$ of $\mathbb{P}^2_{\bar{k}}$ respectively. We define $\mathrm{Zero}(x) = V(Q_1) \cap V(Q_2)$ and call it the zero set of x. The stability with respect to G', G'' are the same.

We proved the following proposition in [84] ((1) was proved in [59]). Here, we give a proof based on geometric invariant theory.

Proposition (8.2.1)
(1) $\mathbb{P}(V)^{ss}_{\bar{k}} = \mathbb{P}(V)^{s}_{(0)\bar{k}}$ *and is a single* $G'_{\bar{k}}$ *orbit.*
(2) *If* $x \in V_{\bar{k}} \setminus \{0\}$, $\pi(x) \in \mathbb{P}(V)^{ss}_{\bar{k}}$ *if and only if* $V(Q_1) \cap V(Q_2)$ *consists of four points.*

Proof. Let $\mathfrak{B}$ be the set of minimal combination of weights. Table (8.2.2)–(8.2.4) are the list of $\beta \in \mathfrak{B} \setminus \{0\}$.

We first show that $V^{ss}_{\bar{k}} \neq \emptyset$. It is easy to verify that if $x \in V_{\bar{k}}$ and $x \in S_{\beta_i}$ for some $1 \leq i \leq 10$, then $\mathrm{Zero}(x)$ has at least a double point or contains a line or a conic. Therefore, if we can show that $S_{\beta_i} = \emptyset$ for $i = 11, 12$, any $x \in V_{\bar{k}}$ such that $\mathrm{Zero}(x)$ consists of four distinct points should be semi-stable. Such x clearly exists, so $V^{ss} \neq \emptyset$. So we consider $S_{\beta_{11}}, S_{\beta_{12}}$.

Consider $S_{\beta_{11}}$. Let $a_{\beta_{11}}(\alpha) = a(\alpha^{-1}, \alpha^{-1}, \alpha^2; \alpha, \alpha^{-1})$ for $\alpha \in \mathrm{GL}(1)$. Then by Table (8.2.3),

$$M''_{\beta_{11}} = \left\{ \left(\begin{pmatrix} g_{1,12} & \\ & 1 \end{pmatrix}, I_2 \right) a_{\beta_{11}}(\alpha) \,\middle|\, g_{1,12} \in \mathrm{SL}(2), \alpha \in \mathrm{GL}(1) \right\}.$$

Elements of the form $a_{\beta_{11}}(\alpha)$ act trivially on $Z_{\beta_{11}}$, and the action of $g_{1,12}$ on $Z_{\beta_{11}}$ can be identified with the standard representation of SL(2). Therefore, $Z^{ss}_{\beta_{11}k} = \emptyset$.

Consider $S_{\beta_{12}}$. By Table (8.2.3),

$$M''_{\beta_{12}} = \left\{ \left(\begin{pmatrix} g_{1,12} & \\ & 1 \end{pmatrix}, g_2 \right) \,\middle|\, g_{1,12}, g_2 \in \mathrm{SL}(2) \right\}.$$

Elements of the form $\left(\begin{pmatrix} g_{1,12} & \\ & 1 \end{pmatrix}, 1 \right)$ act on $Z_{\beta_{12}}$ trivially, and the action of g_2 on $Z_{\beta_{12}}$ can be identified with the standard representation of SL(2). Therefore, $Z^{ss}_{\beta_{12}k} = \emptyset$. This proves that $V^{ss} \neq \emptyset$. Note that we did not use a relative invariant polynomial.

Table (8.2.2)

Strata	Convex hull	Conics
S_{β_1}	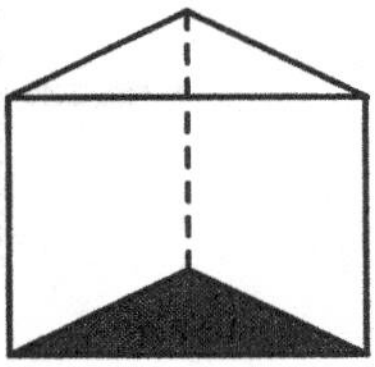	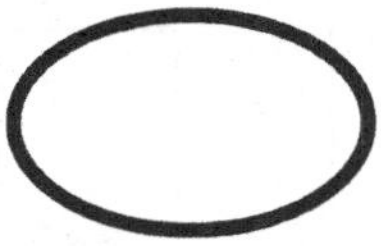 2 identical non-singular conics
S_{β_2}	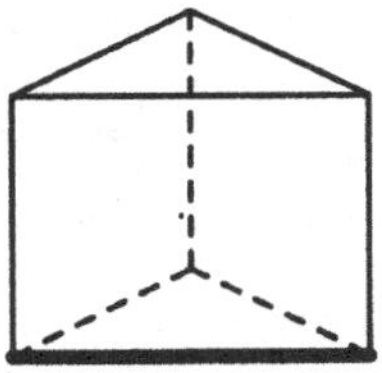	 2 identical reducible conics
S_{β_3}	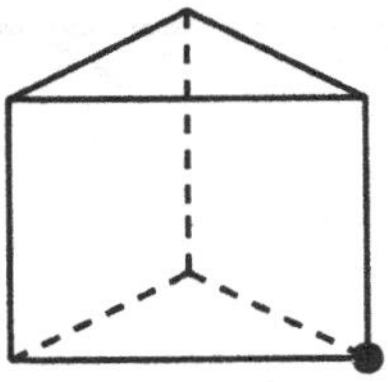	 2 identical double lines
S_{β_4}	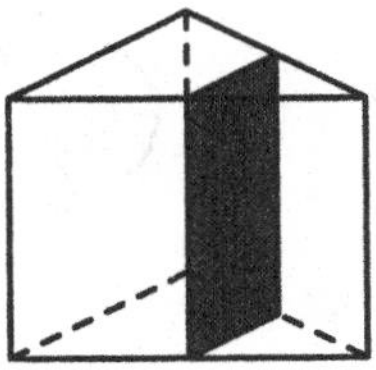	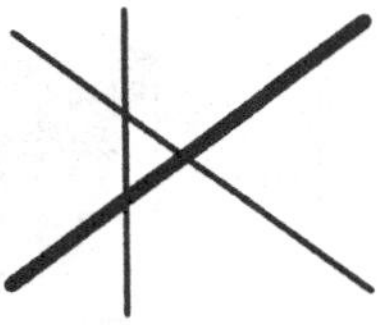 1 common component

Strata	Convex hull	Conics
S_{β_5}	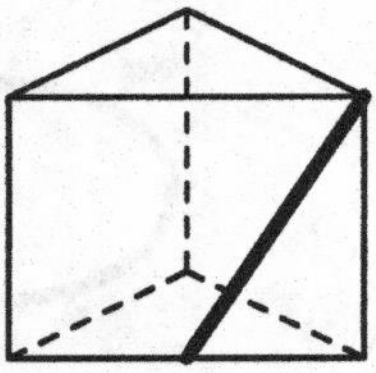	 1 common double line
S_{β_6}	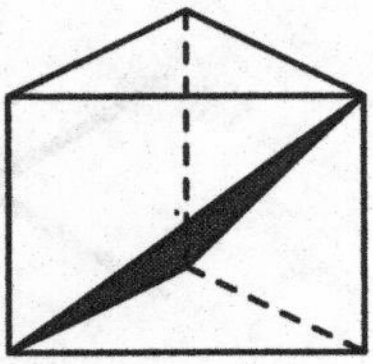	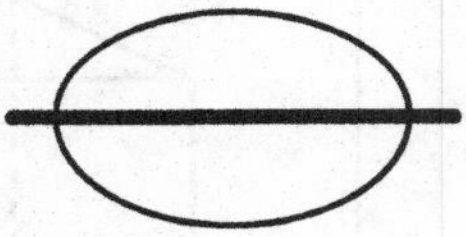 2 multiplicity 2 points
S_{β_7}	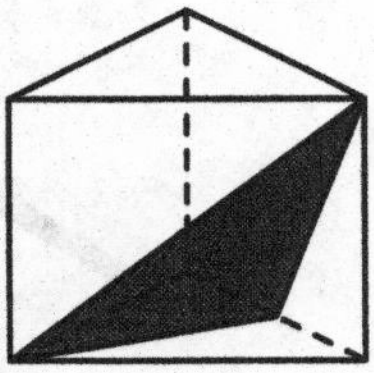	 The local ring $\cong \bar{k}[\epsilon]/(\epsilon^4)$
S_{β_8}	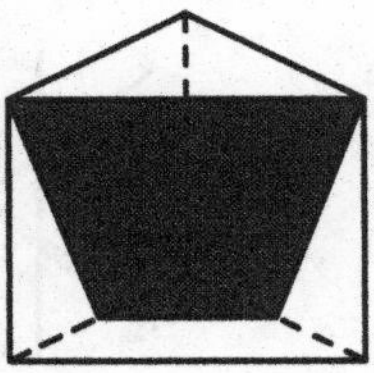	 1 multiplicity 2 point

Strata	Convex hull	Conics
S_{β_9}	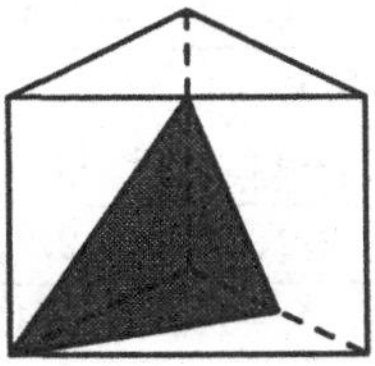	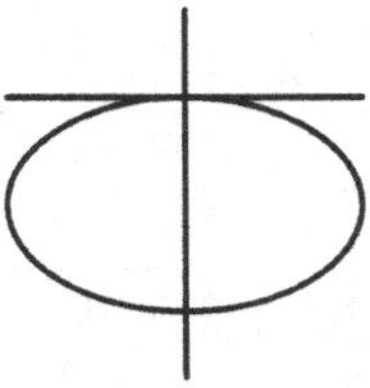 1 multiplicity 3 point
$S_{\beta_{10}}$	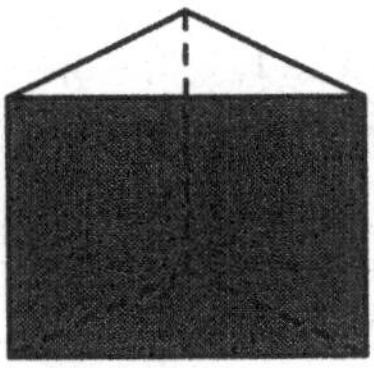	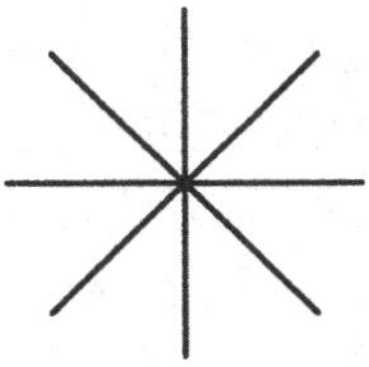 The local ring $\cong \bar{k}[\epsilon_1, \epsilon_2]/(\epsilon_1^2, \epsilon_2^2)$
$S_{\beta_{11}}$	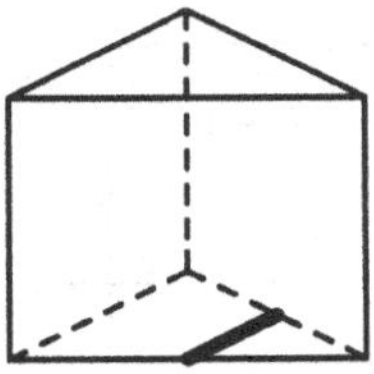	
$S_{\beta_{12}}$	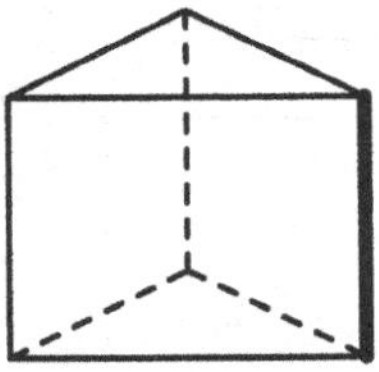	

Table (8.2.3)

β	$((\gamma_{1,11}\cdot\beta),(\gamma_{1,12}\cdot\beta),\cdots,(\gamma_{2,33}\cdot\beta))$
$\beta_1=(0,0,0;-\frac{1}{2},\frac{1}{2})$	$\frac{1}{2}(-1,-1,-1,-1,-1,-1,1,1,1,1,1,1)$
$\beta_2=(-\frac{2}{3},\frac{1}{3},\frac{1}{3};-\frac{1}{2},\frac{1}{2})$	$\frac{1}{6}(-11,-5,-5,1,1,1,-5,1,1,7,7,7)$
$\beta_3=(-\frac{2}{3},-\frac{2}{3},\frac{4}{3};-\frac{1}{2},\frac{1}{2})$	$\frac{1}{6}(-11,-11,1,-11,1,11,-5,-5,7,-5,7,17)$
$\beta_4=(-\frac{1}{6},-\frac{1}{6},\frac{1}{3};0,0)$	$\frac{1}{6}(-2,-2,1,-2,1,4,-2,-2,1,-2,1,4)$
$\beta_5=(-\frac{2}{3},\frac{1}{12},\frac{7}{12};-\frac{1}{4},\frac{1}{4})$	$\frac{1}{12}(-19,-10,-4,-1,5,11,-13,-4,2,5,11,17)$
$\beta_6=(-\frac{1}{24},-\frac{1}{24},\frac{1}{12};-\frac{1}{8},\frac{1}{8})$	$\frac{1}{24}(-5,-5,-1,-5,-1,1,1,1,5,1,5)$
$\beta_7=(-\frac{1}{4},0,\frac{1}{4};-\frac{1}{4},\frac{1}{4})$	$\frac{1}{4}(-3,-2,-1,-1,0,1,-1,0,1,1,2,3)$
$\beta_8=(-\frac{2}{21},\frac{1}{21},\frac{1}{21};-\frac{1}{14},\frac{1}{14})$	$\frac{1}{42}(-11,-5,-5,1,1,1,-5,1,1,7,7,7)$
$\beta_9=(-\frac{1}{5},0,\frac{1}{5};-\frac{1}{10},\frac{1}{10})$	$\frac{1}{10}(-5,-3,-1,-1,1,3,-3,-1,1,1,3,1)$
$\beta_{10}=(-\frac{2}{3},\frac{1}{3},\frac{1}{3};0,0)$	$\frac{1}{3}(-4,-1,-1,2,2,2,-4,-1,-1,2,2,2)$
$\beta_{11}=(-\frac{1}{6},-\frac{1}{6},\frac{1}{3};-\frac{1}{2},\frac{1}{2})$	$\frac{1}{6}(-5,-5,-2,-5,-2,1,1,1,4,1,4,17)$
$\beta_{12}=(-\frac{2}{3},-\frac{2}{3},\frac{4}{3};-\frac{1}{2},\frac{1}{2})$	$\frac{1}{3}(-4,-4,2,-4,2,8,-4,-4,2,-4,2,8)$

Table (8.2.4)

β	$\|\beta\|^2$	Z_β	W_β
β_1	$\frac{1}{2}$	x_{2,j_1j_2} for $j_1,j_2=1,2,3$	—
β_2	$\frac{7}{6}$	$x_{2,22},x_{2,23},x_{2,33}$	—
β_3	$\frac{17}{6}$	$x_{2,33}$	—
β_4	$\frac{1}{6}$	$x_{1,13},x_{1,23},x_{2,13},x_{2,23}$	$x_{1,33},x_{2,33}$
β_5	$\frac{11}{12}$	$x_{1,33},x_{2,23}$	$x_{2,33}$
β_6	$\frac{1}{24}$	$x_{1,33},x_{2,j_1,j_2}$ for $j_1,j_2=1,2,3$	$x_{2,13},x_{2,23},x_{2,33}$
β_7	$\frac{1}{4}$	$x_{1,33},x_{2,13},x_{2,22}$	$x_{2,23},x_{2,33}$
β_8	$\frac{1}{42}$	$x_{1,22},x_{1,23},x_{1,33},x_{2,12},x_{2,13}$	$x_{2,22},x_{2,23},x_{2,33}$
β_9	$\frac{1}{10}$	$x_{1,23},x_{2,13},x_{2,22}$	$x_{1,33},x_{2,23},x_{2,33}$
β_{10}	$\frac{2}{3}$	x_{i,j_1j_2} for $i=1,2,\ j_1,j_2=2,3$	—
β_{11}	$\frac{2}{3}$	$x_{2,13},x_{2,23}$	$x_{2,33}$
β_{12}	$\frac{8}{3}$	$x_{1,33},x_{2,33}$	—

Let $w = (Q_1, Q_2)$ where

$$Q_1 = v_2v_3 - v_1v_3, \; Q_2 = v_1v_2 - v_2v_3.$$

Easy computations show that

$$\mathrm{Zero}(w) = \{(1,0,0),(0,1,0),(0,0,1),(1,1,1)\}.$$

If a conic in $\mathbb{P}^2$ contains three points on the same line, it contains the line. Therefore, if $\mathrm{Zero}(x)$ consists of four points, they are in general position. There exists an element of $\mathrm{GL}(3)_{\bar{k}}$ which sends $\mathrm{Zero}(x)$ to $\mathrm{Zero}(w)$. Therefore, we may assume that $x = (Q_1, Q_2)$ where

$$Q_1 = x_{1,12}v_1v_2 + x_{1,23}v_2v_3 + x_{1,13}v_1v_3,$$
$$Q_2 = x_{2,12}v_1v_2 + x_{2,23}v_2v_3 + x_{2,13}v_1v_3,$$

and

$$x_{1,12} + x_{1,23} + x_{1,13} = x_{2,12} + x_{2,23} + x_{2,13} = 0.$$

Without loss of generality, we can assume that $x_{1,12} \neq 0$. By applying a lower triangular matrix in $\mathrm{GL}(2)_{k'}$, we can assume that $x_{2,12} = 0$. After a coordinate change, we can assume that Q_2 is a constant multiple of $v_2v_3 - v_1v_3$. Then by applying an upper triangular matrix, we can assume that $Q_1 = v_1v_2 - v_2v_3$. Thus, the set of x such that $\mathrm{Zero}(x)$ consists of four points is a single $G_{\bar{k}}$-orbit. Since the set of such points is an open set, this proves (8.2.1).

Q.E.D.

A straightforward argument shows that $x \in S_{\beta_i}$ for some $1 \leq i \leq 10$ if and only if $\mathrm{Zero}(x)$ has the corresponding geometric property. However, we do not logically depend on this statement. All we have to know is the fact that $S_{\beta_i} \neq \emptyset$ for $i = 1, \cdots, 10$ and the inductive structure of S_{β_i} for $i = 1, \cdots, 10$. So the verification of the geometric interpretation of Table (8.2.2) is left to the reader.

We now consider the strata S_{β_i}. In our situation, there are only a few possibilities for parabolic subgroups of G. Let

$$P_1 = \left\{ \begin{pmatrix} g_{1,12} & 0 \\ 0 & t_{13} \end{pmatrix} \middle| \; g_{1,12} \in \mathrm{GL}(2), \; t_{13} \in \mathrm{GL}(1) \right\},$$
$$P_2 = \left\{ \begin{pmatrix} t_{11} & 0 \\ 0 & g_{1,23} \end{pmatrix} \middle| \; t_{11} \in \mathrm{GL}(1), \; g_{1,23} \in \mathrm{GL}(2) \right\}.$$

Standard parabolic subgroups of G_1 are G_1, P_1, P_2, and B_1. Standard parabolic subgroups of G_2 are G_2 and B_2.

We consider M''_β in §3.2.

(a) S_{β_1}

We can identify the vector space Z_{β_1} with the space of quadratic forms in three variables, and

$$M''_{\beta_1} \cong \mathrm{SL}(3) \cong \{(g_1, I_2) \mid g_1 \in \mathrm{SL}(3)\}.$$

We discussed this case in §4.1, and $Z^{\mathrm{ss}}_{\beta_1 k}$ consists of non-degenerate quadratic forms.

(b) S_{β_2}

We can identify the vector space Z_{β_2} with the space of binary quadratic forms. Let $a_{\beta_2}(\alpha) = a(\alpha^2, \alpha^{-1}, \alpha^{-1}; \alpha^{-2}, \alpha^2)$ for $\alpha \in \mathrm{GL}(1)$. Then

$$M''_{\beta_2} = \left\{ \left(\begin{pmatrix} 1 & 0 \\ 0 & g_{1,23} \end{pmatrix}, I_2 \right) a_{\beta_2}(\alpha) \,\middle|\, g_{1,23} \in \mathrm{SL}(2),\ \alpha \in \mathrm{GL}(1) \right\}.$$

Since elements of the form $a_{\beta_2}(\alpha)$ act trivially on Z_β, a point in Z_β is semi-stable if and only if it is semi-stable under the action of

$$\mathrm{SL}(2) \cong \left\{ \left(\begin{pmatrix} 1 & 0 \\ 0 & g_{1,23} \end{pmatrix}, I_2 \right) \,\middle|\, \det g_{1,23} \in \mathrm{SL}(2) \right\}.$$

Therefore, $Z^{\mathrm{ss}}_{\beta_2 k}$ consists of non-degenerate forms again.

(c) S_{β_3}

In this case, we can identify Z_{β_3} with the one dimensional affine space. Let

$$a_{\beta_3}(\alpha) = a(\alpha, \alpha, \alpha^{-2}; \alpha^{-2}, \alpha^2)$$

for $\alpha \in \mathrm{GL}(1)$. Then

$$M''_{\beta_3} = \left\{ \left(\begin{pmatrix} g_{1,12} & 0 \\ 0 & 1 \end{pmatrix}, I_2 \right) a_{\beta_3}(\alpha) \,\middle|\, g_{1,12} \in \mathrm{SL}(2),\ \alpha \in \mathrm{GL}(1) \right\}.$$

Since M''_{β_3} acts trivially on Z_{β_3}, $Z^{\mathrm{ss}}_{\beta_3 k} = Z_{\beta_3 k} \setminus \{0\}$.

(d) S_{β_4}

We can identify the vector space Z_{β_4} with $\mathrm{M}(2,2)$, i.e. the vector space of 2×2 matrices. It is easy to see that

$$M''_{\beta_4} = \mathrm{SL}(2) \times \mathrm{SL}(2) \cong \left\{ \left(\begin{pmatrix} g_{1,12} & 0 \\ 0 & 1 \end{pmatrix}, g_2 \right) \,\middle|\, g_{1,12}, g_2 \in \mathrm{SL}(2) \right\}.$$

Therefore, $Z^{\mathrm{ss}}_{\beta_4 k}$ is the set of rank two matrices.

(e) S_{β_5}

The vector space Z_{β_5} is spanned by coordinate vectors of $x_{1,33}, x_{2,23}$. Let

$$a_{\beta_5}(\alpha_1, \alpha_2) = a(\alpha_1^2, \alpha_2, \alpha_1^{-2}\alpha_2^{-1}; \alpha_1^{-5}\alpha_2^{-1}, \alpha_1^5\alpha_2)$$

for $\alpha_1, \alpha_2 \in \mathrm{GL}(1)$. Then $M''_{\beta_5} = \{a_{\beta_5}(\alpha_1, \alpha_2) \,|\, \alpha_1, \alpha_2 \in \mathrm{GL}(1)\}$.

The weights of the coordinates $x_{1,33}, x_{2,23}$ determine the characters $\alpha_1^{-9}\alpha_2^{-3}, \alpha_1^3\alpha_2$ of $a_{\beta_5}(\alpha_1, \alpha_2)$ respectively. These characters depend only on $\alpha_3 = \alpha_1^3\alpha_2$, and there are both positive and negative weights of α_3. Therefore, $Z^{\mathrm{ss}}_{\beta_5 k} = \{(x_{1,33}, x_{2,23}) \in (k^\times)^2\}$.

(f) S_{β_6}

Let $a_{\beta_6}(\alpha) = a(\alpha, \alpha, \alpha^{-2}; \alpha^{-1}, \alpha)$ for $\alpha \in \mathrm{GL}(1)$. Then

$$M''_{\beta_6} = \left\{ \left(\begin{pmatrix} g_{1,12} & 0 \\ 0 & 1 \end{pmatrix}, g_2 \right) a_{\beta_6}(\alpha) \,\middle|\, g_{1,12}, g_2 \in \mathrm{SL}(2),\ \alpha \in \mathrm{GL}(1) \right\}.$$

Let V_1 (resp. V_2) be the subspaces spanned by coordinate vectors of

$$\{x_{2,11}, x_{2,12}, x_{2,22}\} \text{ (resp. } \{x_{1,33}\}).$$

Clearly $Z_{\beta_6} = V_1 \oplus V_2$. Elements of the form $a_{\beta_6}(\alpha)$ act on V_1, V_2 by multiplication by α^3, α^{-5} respectively. Therefore, we have already discussed this case in Chapter 6. So $Z^{\text{ss}}_{\beta_6 k} \neq \emptyset$. Let $Z^{s}_{\beta_6 k}$ be the set of k-stable points, and $Z^{\text{ss}}_{\beta_6,\text{st}k} = Z^{\text{ss}}_{\beta_6 k} \setminus Z^{s}_{\beta_6 k}$. As we defined in Chapter 3, $S^{s}_{\beta_6 k} = G_k Z^{s}_{\beta_6 k}$, $S_{\beta_6,\text{st}k} = G_k Z^{\text{ss}}_{\beta_6,\text{st}k}$.

(g) S_{β_7}

Let $a_{\beta_7}(\alpha_1, \alpha_2) = a(\alpha_1, \alpha_1^{-2}\alpha_2^{-2}, \alpha_1\alpha_2^2; \alpha_2, \alpha_2^{-1})$ for $\alpha_1, \alpha_2 \in \text{GL}(1)$. Then

$$M''_{\beta_7} = \{a_{\beta_7}(\alpha_1, \alpha_2) \mid \alpha_1, \alpha_2 \in \text{GL}(1)\}.$$

The coordinates $x_{1,33}, x_{2,13}, x_{2,22}$ determine characters $\alpha_1^2\alpha_2^5, \alpha_1^2\alpha_3, \alpha_1^{-4}\alpha_2^{-3}$ respectively. The convex hull of $\{(2,5),(2,1),(-4,-3)\}$ contains a neighborhood of the origin of $\mathbb{R}^2$. Therefore, $Z^{\text{ss}}_{\beta_7 k} = Z^{s}_{\beta_7 k} = \{(x_{1,33}, x_{2,13}, x_{2,22}) \in (k^\times)^3\}$.

(h) S_{β_8}

Let $a_{\beta_8}(\alpha) = a(\alpha^{-2}, \alpha, \alpha; \alpha^2, \alpha^{-2})$ for $\alpha \in \text{GL}(1)$. Then

$$M''_{\beta_8} = \left\{ \left(\begin{pmatrix} 1 & 0 \\ 0 & g_{1,23} \end{pmatrix}, I_2 \right) a_{\beta_8}(\alpha) \,\middle|\, g_{1,23} \in \text{SL}(2),\ \alpha \in \text{GL}(1) \right\}.$$

Let V_1 (resp. V_2) be the subspace spanned by coordinate vectors of

$$\{x_{1,22}, x_{1,23}, x_{1,33}\} \text{ (resp. } \{x_{2,12}, x_{2,13}\}).$$

Clearly, $Z_{\beta_8} = V_1 \oplus V_2$. Elements of the form $a_{\beta_8}(\alpha)$ act on V_1, V_2 by multiplication by α^4, α^{-3} respectively. Therefore, we have already discussed this case in Chapter 6. In this case, $Z^{\text{ss}}_{\beta_8 k} \neq \emptyset$, and $Z^{\text{ss}}_{\beta_8 k} = Z^{s}_{\beta_8 k}$.

(i) S_{β_9}

Let $a_{\beta_9}(\alpha_1, \alpha_2) = a(\alpha_1, \alpha_2, \alpha_1^{-1}\alpha_2^{1}; \alpha_1^{-2}\alpha_2^{-1}, \alpha_1^2\alpha_2)$ for $\alpha_1, \alpha_2 \in \text{GL}(1)$. Then

$$M''_{\beta_9} = \{a_{\beta_9}(\alpha_1, \alpha_2) \mid \alpha_1, \alpha_2 \in \text{GL}(1)\}.$$

The coordinates $x_{1,23}, x_{2,13}, x_{2,22}$ determine characters $\alpha_1^{-3}\alpha_2, \alpha_1^2, \alpha_1^2\alpha_2^2$ respectively. The convex hull of $\{(-3,-1),(2,0),(2,2)\}$ contains a neighborhood of the origin of $\mathbb{R}^2$. Therefore, $Z^{\text{ss}}_{\beta_9 k} = Z^{s}_{\beta_9 k} = \{(x_{1,23}, x_{2,13}, x_{2,22}) \in (k^\times)^3\}$.

(j) $S_{\beta_{10}}$

We can identify $Z_{\beta_{10}}$ with $\text{Sym}^2 k^2 \otimes k^2$. It is easy to see that

$$M''_\beta = \text{SL}(2) \times \text{SL}(2) \cong \left\{ \left(\begin{pmatrix} 1 & 0 \\ 0 & g_{1,23} \end{pmatrix}, g_2 \right) \,\middle|\, g_{1,23}, g_2 \in \text{SL}(2) \right\}.$$

Therefore, we have already discussed this case in Chapter 5. So $Z^{\text{ss}}_{\beta_{10} k} \neq \emptyset$

Let $Z^{s}_{\beta_{10},k}$ be the set of k-stable points, and $Z^{\text{ss}}_{\beta_{10},\text{st}k} = Z^{\text{ss}}_{\beta_{10} k} \setminus Z^{s}_{\beta_{10},k}$. As we defined in Chapter 3, $S^{s}_{\beta_{10},k} = G_k Z^{s}_{\beta_{10},k}$, $S_{\beta_{10},\text{st}k} = G_k Z^{\text{ss}}_{\beta_{10},\text{st}k}$.

§8.3 β-sequences of lengths ≥ 2

In this section, we describe β-sequences of lengths ≥ 2 for our representation explicitly. Let $\beta_1, \cdots, \beta_{10}$ be as in §8.2.

Let $\beta_{1,1} = \beta_2 - \beta_1, \beta_{1,2} = \beta_3 - \beta_1$. Then $(\beta_1, \beta_{1,1}), (\beta_1, \beta_{1,2})$ are the β-sequences of length 2 which start with β_1. Let $\beta_{1,1,1} = \beta_3 - \beta_2$. Then $(\beta_1, \beta_{1,1}, \beta_{1,1,1})$ is the only β-sequence of length 3 which starts with β_1.

Let $\beta_{2,1} = \beta_3 - \beta_2$. Then $(\beta_2, \beta_{1,2})$ is the only β-sequence of length 2 which starts with β_2.

There is no β-sequence of length ≥ 2 which starts with β_3.

Let $\beta_{4,1} = \beta_2 - \beta_4$. Then $(\beta_4, \beta_{4,2})$ is the only β-sequence of length 2 which starts with β_4.

We do not consider β-sequences of lengths ≥ 2 which start with β_5.

We can write any element of T_+ in the form

$$\nu_{\beta_6}(\underline{\lambda}_1)a_{\beta_6}(\underline{\lambda}_2)(a_3(\underline{\lambda}_3^{-1}, \underline{\lambda}_3, 1), 1),$$

where ν_{β_6} is a 1PS proportional to β_6 and $\lambda_1, \lambda_2, \lambda_3 \in \mathbb{R}_+$. If we identify $\nu_{\beta_6}(\underline{\lambda}_1)$, $a_{\beta_6}(\underline{\lambda}_2)$, $(a_3(\underline{\lambda}_3^{-1}, \underline{\lambda}_3, 1), 1)$ with 1PS's, these are orthogonal to each other. Let $\mathfrak{t}^*_{\beta_6} = \{z \in \mathfrak{t} \mid (\beta_6, z) = 0\}$. We identify $\mathfrak{t}^*_{\beta_6}$ with $\{(-a, a; b) \mid a, b \in \mathbb{R}\} \cong \mathbb{R}^2$ so that

$$(\nu_{\beta_6}(\underline{\lambda}_1)a_{\beta_6}(\underline{\lambda}_2)(a_3(\underline{\lambda}_3^{-1}, \underline{\lambda}_3, 1), 1))^{(-a,a;b)} = \lambda_2^b\lambda_3^{2a}.$$

Since $\|(1, 1, -2; -1, 1)\|^2 = 8$ and $\|(-1, 1, 0; 0, 0)\|^2 = 2$, $\|(-a, a; b)\|^2 = 2a^2 + \frac{1}{8}b^2$.

If (β_6, β) is a β-sequence, β is an element of $\mathfrak{t}^*_{\beta_6}$. Since the metric on $\mathfrak{t}^*_{\beta_6}$ is the same metric as we considered in §6.2, β corresponds to β-sequences which we considered in §6.2.

Let $\beta_{6,1}, \cdots, \beta_{6,4}$ be the elements which correspond to $\beta_1, \cdots, \beta_4$ in §6.2. Then β-sequences of length 2 which start with β_6 are $(\beta_6, \beta_{6,1}), \cdots, (\beta_6, \beta_{6,4})$.

We do not have to know $\beta_{6,1}$ for our purpose. We can describe $\beta_{6,2}, \cdots, \beta_{6,4}$ explicitly as follows:

$$\beta_{6,2} = (\frac{1}{3}, \frac{1}{3}, -\frac{2}{3}; -\frac{1}{2}, \frac{1}{2}) - \beta_6,$$
$$\beta_{6,3} = (-\frac{2}{3}, \frac{4}{3}, -\frac{2}{3}; -\frac{1}{2}, \frac{1}{2}) - \beta_6,$$
$$\beta_{6,4} = (-\frac{2}{3}, -\frac{2}{3}, \frac{4}{3}; \frac{1}{2}, -\frac{1}{2}) - \beta_6.$$

We do not consider β-sequences of length 3 which start with β_6.

We do not consider β-sequences of length ≥ 2 which start with β_7.

We define $\mathfrak{t}^*_{\beta_8}$ similarly as in the case S_{β_6}. Since $\|(-2, 1, 1; 2, -2)\|^2 = 14$, the metric on $\mathfrak{t}^*_{\beta_8}$ coincides with the metric we considered in §6.2.

Let $\beta_{8,1}, \cdots, \beta_{8,6}$ be the elements which correspond to $\beta_1, \cdots, \beta_6$ in §6.2. Then β-sequences of length 2 which start with β_8 are $(\beta_8, \beta_{8,1}), \cdots, (\beta_8, \beta_{8,6})$.

We do not have to know $\beta_{8,1}, \beta_{8,2}, \beta_{8,3}$ for our purpose. We can describe $\beta_{8,4}, \cdots,$ $\beta_{8,6}$ explicitly as follows:

$$\beta_{8,4} = (-\frac{2}{3}, \frac{1}{3}, \frac{1}{3}; \frac{1}{2}, -\frac{1}{2}) - \beta_8,$$
$$\beta_{8,5} = (-\frac{2}{3}, -\frac{2}{3}, \frac{4}{3}; \frac{1}{2}, -\frac{1}{2}) - \beta_8,$$
$$\beta_{8,6} = (\frac{1}{3}, -\frac{2}{3}, \frac{1}{3}; -\frac{1}{2}, \frac{1}{2}) - \beta_8.$$

We do not consider β-sequences of length 3 which start with β_8.

We do not consider β-sequences of lengths ≥ 2 which start with β_9.

We define

$$\beta_{10,1} = (0, -\frac{1}{4}, \frac{1}{4}; -\frac{1}{4}, \frac{1}{4}),$$
$$\beta_{10,2} = \beta_2 - \beta_{10},$$
$$\beta_{10,3} = \beta_3 - \beta_{10}.$$

Then $(\beta_{10}, \beta_{10,1}), (\beta_{10}, \beta_{10,2}), (\beta_{10}, \beta_{10,3})$ are the β-sequences of length 2 which start with β_{10}. Let $\beta_{10,2,1} = \beta_{10,3}$. Then $(\beta_{10}, \beta_{10,2}, \beta_{10,2,1})$ is the only β-sequence of length 3 which starts with β_{10}.

Chapter 9 Preliminary estimates

§9.1 Distributions associated with paths

Let $\Phi, \omega, s, \widetilde{G}_{\mathbb{A}}, \widetilde{g}$ etc. be as in §3.1. In our case, the condition $\dim G/\widetilde{T} = \dim V$ is satisfied. Therefore, $(G/\widetilde{T}/V)$ is of complete type. We consider the integrals in (3.1.8) and use the notation $Z_V(\Phi, \omega, s)$ etc. for $L = V_k^{\rm ss}$. Note that since $G_{\mathbb{A}}^1 = G_{\mathbb{A}}^0$, we have a canonical measure on $G_{\mathbb{A}}^0$.

Let $[\ ,\ ]_{\widetilde{V}}$ be the bilinear form for $\widetilde{V} = \mathrm{Sym}^2 k^3$ which we defined in §4.1. We define a bilinear form on V by

$$[(x_1, x_2), (y_1, y_2)]_V = [x_1, y_2]_{\widetilde{V}} + [x_2, y_1]_{\widetilde{V}}.$$

Then this bilinear form satisfies the property $[gx, g^{\iota}y]_V = [x, y]_V$ for all $x, y \in V$ where $g^{\iota} = \tau_G {}^t g^{-1} \tau_G$. We use this bilinear form as $[\ ,\ ]_V$ in §3.1.

Let $J(\Phi, g^0)$ be as in (3.5.6). We define

$$(9.1.1) \qquad I^0(\Phi, \omega) = \int_{G_{\mathbb{A}}^0/G_k} \omega(g^0) J(\Phi, g^0) dg^0;$$

$$I(\Phi, \omega, w) = \int_{G_{\mathbb{A}}^0/G_k} \omega(g^0) J(\Phi, g^0) \mathscr{E}(g^0, w) dg^0.$$

The Poisson summation formula implies that

$$(9.1.2) \quad Z_V(\Phi, \omega, s) = Z_{V+}(\Phi, \omega, s) + Z_{V+}(\widehat{\Phi}, \omega^{-1}, 12 - s) + \int_0^1 \lambda^s I^0(\Phi_\lambda, \omega) d^\times \lambda.$$

Since the first two terms are entire functions, the last term is the issue. As in previous parts, we assume that $\Phi = M_{V,\omega}\Phi$. Also we assume that $\psi(z) = \psi(-\tau_G z)$ and $\psi(\rho) \neq 0$, so $\mathscr{E}(g^0, w) = \mathscr{E}((g^0)^{\iota}, w)$.

By (3.4.34),

$$(9.1.3) \qquad I(\Phi, \omega, w) \sim C_G \Lambda(w; \rho) I^0(\Phi, \omega) \quad (C_G = \mathfrak{V}_2^{-1} \mathfrak{V}_3^{-1}).$$

So we study $I(\Phi, \omega, w)$ as a function of w. By (3.5.9) and (3.5.20),

$$(9.1.4) \qquad I(\Phi, \omega, w) = \delta_{\#}(\omega) \Lambda(w; \rho)(\widehat{\Phi}(0) - \Phi(0)) + \sum_{\mathfrak{p}, l(\mathfrak{p})=1} \epsilon_{\mathfrak{p}} (\widetilde{\Xi}_{\mathfrak{p}}(\Phi, \omega, w) + \Xi_{\mathfrak{p}}(\Phi, \omega, w)).$$

We know that $\widetilde{\Xi}_{\mathfrak{p}}(\Phi, \omega, w) + \Xi_{\mathfrak{p}}(\Phi, \omega, w)$ is well defined for $\mathrm{Re}(w) \gg 0$. If $\mathfrak{p} = (\mathfrak{d}, \mathfrak{s})$ and $\mathfrak{d} = (\beta_1), (\beta_2), (\beta_3)$ or (β_{10}), $\widetilde{\Xi}_{\mathfrak{p}}(\Phi, \omega, w) = 0$. If $\mathfrak{p}$ belongs to class (1) or (3) in §3.5, $\Xi_{\mathfrak{p}}(\Phi, \omega, w)$ is well defined for $\mathrm{Re}(w) \gg 0$. This applies to $\mathfrak{p} = (\mathfrak{d}, \mathfrak{s})$ where $\mathfrak{d} = (\beta_4), (\beta_6), (\beta_7), (\beta_8)$ or (β_9). Therefore, all the distributions in (9.1.4) are well defined for $\mathrm{Re}(w) \gg 0$ except for $\mathfrak{p} = (\mathfrak{d}, \mathfrak{s})$ such that $\mathfrak{d} = (\beta_5)$. We prove that $\widetilde{\Xi}_{\mathfrak{p}}(\Phi, \omega, w)$ is well defined for $\mathrm{Re}(w) \gg 0$ and $\widetilde{\Xi}_{\mathfrak{p}}(\Phi, \omega, w) \sim 0$ for this case in §10.3.

Since the rank of our group is 3, we have to consider paths of lengths up to 3. It turns out that if $\mathfrak{d}$ is a β-sequence of length ≥ 2 which starts with $\beta_1, \beta_2, \beta_4, \beta_{10}$, we can apply our results in Chapter 3. However, if $\mathfrak{d}$ is a β-sequence of length ≥ 2 which starts with β_6 or β_8, we have to use a method similar to Chapters 6 and 7. For the rest of this section, we define certain distributions associated with β_6, β_8, and prove that they are well defined.

Let $\mathfrak{p} = (\mathfrak{d}, \mathfrak{s})$ be a path such that $\mathfrak{d} = (\beta_6)$ or (β_8). In both cases, $Z_{\mathfrak{d}} = V_{\mathfrak{d}1} \oplus V_{\mathfrak{d}2}$ where $V_{\mathfrak{d}1k} = \mathrm{Sym}^2 k^2$, $V_{\mathfrak{d}2k} = k$ for β_6, and $V_{\mathfrak{d}1k} = \mathrm{Sym}^2 k^2$, $V_{\mathfrak{d}2k} = k^2$ for β_8. We define a bilinear form $[\ ,\ ]_{V_{\mathfrak{d}i}}$ on $V_{\mathfrak{d}i}$ for $i = 1, 2$ similarly as in §6.1, using the longest element of the Weyl group of $M_{\mathfrak{d}}$. Let $\mathscr{F}_{V_{\mathfrak{d}i}}$ be the partial Fourier transform with respect to $[\ ,\ ]_{V_{\mathfrak{d}i}}$ for $i = 1, 2$. In both cases, $A_{\mathfrak{p}2} = A_{\mathfrak{d}}$ (see (3.3.11)).

By Table (8.2.3), $A_{\mathfrak{d}} = \{d_{\mathfrak{d}}(\lambda_1) \mid \lambda_1 \in \mathbb{R}_+\}$, where

$$d_{\mathfrak{d}}(\lambda_1) = \begin{cases} a(\underline{\lambda}_1^{-6}, \underline{\lambda}_1^{3}, \underline{\lambda}_1^{3}; \underline{\lambda}_1^{-2}, \underline{\lambda}_1^{2}) & \mathfrak{d} = (\beta_6), \\ a(\underline{\lambda}_1^{-1}, \underline{\lambda}_1^{-1}, \underline{\lambda}_1^{2}; \underline{\lambda}_1^{-3}, \underline{\lambda}_1^{3}) & \mathfrak{d} = (\beta_8). \end{cases} \tag{9.1.5}$$

Let $\bar{\kappa}_{V_{\mathfrak{d}i}}$ be the rational character of $M_{\mathfrak{d}}$ defined by the determinant of $\mathrm{GL}(V_{\mathfrak{d}i})$ for $i = 1, 2$. For $g_{\mathfrak{d}} \in G^0_{\mathbb{A}} \cap M_{\mathfrak{d}\mathbb{A}}$, we define $\kappa_{V_{\mathfrak{d}1}}(g_{\mathfrak{d}}) = |\bar{\kappa}_{V_{\mathfrak{d}1}}(g_{\mathfrak{d}})|^{-1}$. for $i = 1, 2$.

By (6.1.1), (6.1.2),

$$\begin{aligned}
\Theta_{Z_{\mathfrak{d}}}(R_{\mathfrak{d}}\Phi_{\mathfrak{p}}, g_{\mathfrak{d}}) = \kappa_{\mathfrak{d}1}(g_{\mathfrak{d}})\Theta_{Z_{\mathfrak{d}}}(\mathscr{F}_{\mathfrak{d}}R_{\mathfrak{d}}\Phi_{\mathfrak{p}}, \theta_{\mathfrak{d}}(g_{\mathfrak{d}})) &+ \mathscr{F}_{\mathfrak{d}}R_{\mathfrak{d}}\Phi_{\mathfrak{p}}(0) - R_{\mathfrak{d}}\Phi_{\mathfrak{p}}(0) \\
+ \sum_{Z_{\mathfrak{d}'} \not\subset V_{\mathfrak{d}1}, V_{\mathfrak{d}2}} &\left(\kappa_{\mathfrak{d}1}(g_{\mathfrak{d}})\Theta_{S_{\mathfrak{d}'}}(\mathscr{F}_{\mathfrak{d}}R_{\mathfrak{d}}\Phi_{\mathfrak{p}}, \theta_{\mathfrak{d}}(g_{\mathfrak{d}})) - \Theta_{S_{\mathfrak{d}'}}(R_{\mathfrak{d}}\Phi_{\mathfrak{p}}, g_{\mathfrak{d}})\right) \\
+ \sum_{Z_{\mathfrak{d}'} \subset V_{\mathfrak{d}1}} &\left(\kappa_{\mathfrak{d}1}(g_{\mathfrak{d}})\Theta_{S_{\mathfrak{d}'}}(\mathscr{F}_{\mathfrak{d}}R_{\mathfrak{d}}\Phi_{\mathfrak{p}}, \theta_{\mathfrak{d}}(g_{\mathfrak{d}})) - \Theta_{S_{\mathfrak{d}'}}(R_{\mathfrak{d}}\Phi_{\mathfrak{p}}, g_{\mathfrak{d}})\right) \\
+ \sum_{Z_{\mathfrak{d}'} \subset V_{\mathfrak{d}2}} &\left(\kappa_{\mathfrak{d}1}(g_{\mathfrak{d}})\Theta_{S_{\mathfrak{d}'}}(\mathscr{F}_{\mathfrak{d}}R_{\mathfrak{d}}\Phi_{\mathfrak{p}}, \theta_{\mathfrak{d}}(g_{\mathfrak{d}})) - \Theta_{S_{\mathfrak{d}'}}(R_{\mathfrak{d}}\Phi_{\mathfrak{p}}, g_{\mathfrak{d}})\right),
\end{aligned}$$

and

$$\begin{aligned}
&\sum_{Z_{\mathfrak{d}'} \subset V_{\mathfrak{d}1}, V_{\mathfrak{d}2}} \left(\kappa_{\mathfrak{d}1}(g_{\mathfrak{d}})\Theta_{S_{\mathfrak{d}'}}(\mathscr{F}_{\mathfrak{d}}R_{\mathfrak{d}}\Phi_{\mathfrak{p}}, \theta_{\mathfrak{d}}(g_{\mathfrak{d}})) - \Theta_{S_{\mathfrak{d}'}}(R_{\mathfrak{d}}\Phi_{\mathfrak{p}}, g_{\mathfrak{d}})\right) \\
&\quad + \mathscr{F}_{\mathfrak{d}}R_{\mathfrak{d}}\Phi_{\mathfrak{p}}(0) - R_{\mathfrak{d}}\Phi_{\mathfrak{p}}(0) \\
&= \kappa_{V_{\mathfrak{d}2}}(g_{\mathfrak{d}}) \sum_{Z_{\mathfrak{d}'} \subset V_{\mathfrak{d}1}} \Theta_{S_{\mathfrak{d}'}}(\mathscr{F}_{V_{\mathfrak{d}1}}\mathscr{F}_{\mathfrak{d}}R_{\mathfrak{d}}\Phi_{\mathfrak{p}}, g_{\mathfrak{d}}) \\
&\quad + \kappa_{\mathfrak{d}1}(g_{\mathfrak{d}}) \sum_{Z_{\mathfrak{d}'} \subset V_{\mathfrak{d}2}} \Theta_{S_{\mathfrak{d}'}}(\mathscr{F}_{\mathfrak{d}}R_{\mathfrak{d}}\Phi_{\mathfrak{p}}, \theta_{\mathfrak{d}}(g_{\mathfrak{d}})) \\
&\quad - \sum_{Z_{\mathfrak{d}'} \subset V_{\mathfrak{d}1}} \Theta_{S_{\mathfrak{d}'}}(R_{\mathfrak{d}}\Phi_{\mathfrak{p}}, g_{\mathfrak{d}}) - \kappa_{V_{\mathfrak{d}2}}(g_{\mathfrak{d}}) \sum_{Z_{\mathfrak{d}'} \subset V_{\mathfrak{d}2}} \Theta_{S_{\mathfrak{d}'}}(\mathscr{F}_{V_{\mathfrak{d}2}}R_{\mathfrak{d}}\Phi_{\mathfrak{p}}, \theta_{\mathfrak{d}}(g_{\mathfrak{d}})) \\
&= \kappa_{\mathfrak{d}1}(g_{\mathfrak{d}}) \sum_{Z_{\mathfrak{d}'} \subset V_{\mathfrak{d}1}} \Theta_{S_{\mathfrak{d}'}}(\mathscr{F}_{\mathfrak{d}}R_{\mathfrak{d}}\Phi_{\mathfrak{p}}, \theta_{\mathfrak{d}}(g_{\mathfrak{d}})) \\
&\quad + \kappa_{V_{\mathfrak{d}1}}(g_{\mathfrak{d}}) \sum_{Z_{\mathfrak{d}'} \subset V_{\mathfrak{d}2}} \Theta_{S_{\mathfrak{d}'}}(\mathscr{F}_{V_{\mathfrak{d}2}}\mathscr{F}_{\mathfrak{d}}R_{\mathfrak{d}1}\Phi_{\mathfrak{p}}, g_{\mathfrak{d}}) \\
&\quad - \kappa_{V_{\mathfrak{d}1}}(g_{\mathfrak{d}}) \sum_{Z_{\mathfrak{d}'} \subset V_{\mathfrak{d}1}} \Theta_{S_{\mathfrak{d}'}}(\mathscr{F}_{V_{\mathfrak{d}1}}R_{\mathfrak{d}}\Phi_{\mathfrak{p}}, \theta_{\mathfrak{d}}(g_{\mathfrak{d}})) - \sum_{Z_{\mathfrak{d}'} \subset V_{\mathfrak{d}2}} \Theta_{S_{\mathfrak{d}'}}(R_{\mathfrak{d}}\Phi_{\mathfrak{p}}, g_{\mathfrak{d}}),
\end{aligned}$$

where $\mathfrak{d}'$ runs through β-sequences of length 2 and $\mathfrak{d} \prec \mathfrak{d}'$.

Let χ be a principal quasi-character of $g_{\mathfrak{d}} \in G^0_{\mathbb{A}} \cap M_{\mathfrak{d}\mathbb{A}}/M_{\mathfrak{d}k}$, and $\Psi \in \mathscr{S}(Z_{\mathfrak{d}\mathbb{A}})$. We define

$$\Xi_{\mathfrak{p},\mathfrak{d}',\geq}(\Psi,\omega,\chi,w) = \begin{cases} \int_{\substack{A_{\mathfrak{d}0}M^1_{\mathfrak{d}\mathbb{A}}/M_{\mathfrak{d}k} \\ \kappa_{V_{\mathfrak{d}1}}(g_{\mathfrak{d}})^{-1}\geq 1}} \chi(g_{\mathfrak{d}})\Theta_{S_{\mathfrak{d}'}}(\Psi,g_{\mathfrak{d}})\mathscr{E}_{\mathfrak{p}}(g_{\mathfrak{d}},w)dg_{\mathfrak{d}} & S_{\mathfrak{d}'} \subset V_{\mathfrak{d}1}, \\ \int_{\substack{A_{\mathfrak{d}0}M^1_{\mathfrak{d}\mathbb{A}}/M_{\mathfrak{d}k} \\ \kappa_{V_{\mathfrak{d}1}}(g_{\mathfrak{d}})^{-1}\geq 1}} \chi(g_{\mathfrak{d}})\Theta_{S_{\mathfrak{d}'}}(\Psi,\theta_{\mathfrak{d}}(g_{\mathfrak{d}}))\mathscr{E}_{\mathfrak{p}}(g_{\mathfrak{d}},w)dg_{\mathfrak{d}} & S_{\mathfrak{d}'} \subset V_{\mathfrak{d}2}, \end{cases}$$

$$\Xi_{\mathfrak{p},\mathfrak{d}',\leq}(\Psi,\omega,\chi,w) = \begin{cases} \int_{\substack{A_{\mathfrak{d}0}M^1_{\mathfrak{d}\mathbb{A}}/M_{\mathfrak{d}k} \\ \kappa_{V_{\mathfrak{d}1}}(g_{\mathfrak{d}})^{-1}\leq 1}} \chi(g_{\mathfrak{d}})\Theta_{S_{\mathfrak{d}'}}(\Psi,\theta_{\mathfrak{d}}(g_{\mathfrak{d}}))\mathscr{E}_{\mathfrak{p}}(g_{\mathfrak{d}},w)dg_{\mathfrak{d}} & S_{\mathfrak{d}'} \subset V_{\mathfrak{d}1}, \\ \int_{\substack{A_{\mathfrak{d}0}M^1_{\mathfrak{d}\mathbb{A}}/M_{\mathfrak{d}k} \\ \kappa_{V_{\mathfrak{d}1}}(g_{\mathfrak{d}})^{-1}\leq 1}} \chi(g_{\mathfrak{d}})\Theta_{S_{\mathfrak{d}'}}(\Psi,g_{\mathfrak{d}})\mathscr{E}_{\mathfrak{p}}(g_{\mathfrak{d}},w)dg_{\mathfrak{d}} & S_{\mathfrak{d}'} \subset V_{\mathfrak{d}2}. \end{cases}$$

Lemma (9.1.6) *The distributions* $\Xi_{\mathfrak{p},\mathfrak{d}',\geq}(\Psi,\omega,\chi,w), \Xi_{\mathfrak{p},\mathfrak{d}',\leq}(\Psi,\omega,\chi,w)$ *are well defined if* $\mathrm{Re}(w) \gg 0$.

Proof. Let $g_{\mathfrak{d}} = \lambda_{\mathfrak{d}} g^1_{\mathfrak{d}}$, where $\lambda_{\mathfrak{d}} = d_{\mathfrak{d}}(\lambda_1)$ is as in (9.1.5), and $g^1_{\mathfrak{d}} \in M^1_{\mathfrak{d}\mathbb{A}}$. Let $a_{\mathfrak{d}}(\lambda_2) = a_{\beta_6}(\underline{\lambda_2})$ or $a_{\beta_8}(\underline{\lambda_2})$ for $\lambda_2 \in \mathbb{R}_+$. Then we can write $g^1_{\mathfrak{d}} = a_{\mathfrak{d}}(\lambda_2)g^0_{\mathfrak{d}}$, where $g^0_{\mathfrak{d}} \in M^0_{\mathfrak{d}\mathbb{A}}$.

The element $\lambda_{\mathfrak{d}} a_{\mathfrak{d}}(\lambda_2)$ acts on $V_{\mathfrak{d}1\mathbb{A}}, V_{\mathfrak{d}2\mathbb{A}}$ by multiplication by

$$\underline{e_{\mathfrak{p}1}(\lambda_{\mathfrak{d}})\lambda_2^a}, \ \underline{e_{\mathfrak{p}1}(\lambda_{\mathfrak{d}})\lambda_2^{-b}},$$

respectively, where $a, b > 0$ are constants. Then $\kappa_{V_{\mathfrak{d}1}}(g_{\mathfrak{d}})^{-1} = (e_{\mathfrak{p}1}(\lambda_{\mathfrak{d}})\lambda_2^a)^{\dim V_{\mathfrak{d}2}}$.

By (1.2.6), for any $N \gg 0$, there exists a slowly increasing function $h_N(g^0_{\mathfrak{d}})$ of $g^0_{\mathfrak{d}}$ such that

$$\begin{aligned} &\Theta_{S_{\mathfrak{d}'}}(\Psi,g_{\mathfrak{d}}) \ll (e_{\mathfrak{p}1}(\lambda_{\mathfrak{d}})\lambda_2^a)^{-N}h_N(g^0_{\mathfrak{d}}) && \text{for } Z_{\mathfrak{d}'} \subset V_{\mathfrak{d}1}, \\ &\Theta_{S_{\mathfrak{d}'}}(\Psi,\theta_{\mathfrak{d}}(g_{\mathfrak{d}})) \ll (e_{\mathfrak{p}1}(\lambda_{\mathfrak{d}})^{-1}\lambda_2^b)^{-N}h_N(g^0_{\mathfrak{d}}) && \text{for } Z_{\mathfrak{d}'} \subset V_{\mathfrak{d}2}, \\ &\Theta_{S_{\mathfrak{d}'}}(\Psi,\theta_{\mathfrak{d}}(g_{\mathfrak{d}})) \ll (e_{\mathfrak{p}1}(\lambda_{\mathfrak{d}})\lambda_2^a)^{N}h_N(g^0_{\mathfrak{d}}) && \text{for } Z_{\mathfrak{d}'} \subset V_{\mathfrak{d}1}, \\ &\Theta_{S_{\mathfrak{d}'}}(\Psi,g_{\mathfrak{d}}) \ll (e_{\mathfrak{p}1}(\lambda_{\mathfrak{d}})^{-1}\lambda_2^b)^{N}h_N(g^0_{\mathfrak{d}}) && \text{for } Z_{\mathfrak{d}'} \subset V_{\mathfrak{d}2}. \end{aligned}$$

Let $\mu_1 = e_{\mathfrak{p}1}(\lambda_{\mathfrak{d}})$, $\mu_2 = \kappa_{V_{\mathfrak{d}1}}(g_{\mathfrak{d}})^{-1}$. Then $\lambda_2 = \mu_1^{-\frac{1}{a}}\mu_2^{\frac{1}{a\dim V_{\mathfrak{d}2}}}$. Therefore if $N \gg 0$, there exists a slowly increasing function $h'_N(\mu_1, g^0_{\mathfrak{d}})$ such that

$$\begin{aligned} &\Theta_{S_{\mathfrak{d}'}}(\Psi,g_{\mathfrak{d}}) \ll \mu_2^{-N}h'_N(\mu_1,g^0_{\mathfrak{d}}) && \text{for } Z_{\mathfrak{d}'} \subset V_{\mathfrak{d}1}, \\ &\Theta_{S_{\mathfrak{d}'}}(\Psi,\theta_{\mathfrak{d}}(g_{\mathfrak{d}})) \ll \mu_2^{-N}h'_N(\mu_1,g^0_{\mathfrak{d}}) && \text{for } Z_{\mathfrak{d}'} \subset V_{\mathfrak{d}2}, \\ &\Theta_{S_{\mathfrak{d}'}}(\Psi,\theta_{\mathfrak{d}}(g_{\mathfrak{d}})) \ll \mu_2^{N}h'_N(\mu_1,g^0_{\mathfrak{d}}) && \text{for } Z_{\mathfrak{d}'} \subset V_{\mathfrak{d}1}, \\ &\Theta_{S_{\mathfrak{d}'}}(\Psi,g_{\mathfrak{d}}) \ll \mu_2^{N}h'_N(\mu_1,g^0_{\mathfrak{d}}) && \text{for } Z_{\mathfrak{d}'} \subset V_{\mathfrak{d}2}. \end{aligned}$$

Let

$$g^0_{\mathfrak{d}} = \left(\begin{pmatrix} g_{1,12} & \\ & t_{13} \end{pmatrix}, a_2(t_{21},t_{22})\right) \text{ or } \left(\begin{pmatrix} t_{11} & \\ & g_{1,23} \end{pmatrix}, a_2(t_{21},t_{22})\right)$$

for $\mathfrak{d} = (\beta_6), (\beta_8)$ respectively, where $t_{11}, t_{13}, t_{21}, t_{22} \in \mathbb{A}^1, g_{1,12}, g_{1,23} \in \mathrm{GL}(2)^0_{\mathbb{A}}$. By (3.4.31), there exist constants $r_1, r_2 > 0$ such that if $M_1, M_2 > w_0$,

$$\mathscr{E}_{\mathfrak{p}}(g_{\mathfrak{d}},w) \ll \begin{cases} \mu_1^{r_1(\mathrm{Re}(w)-w_0)}|t_{11}(g_{1,12})t_{12}(g_{1,12})^{-1}|^{-r_2(\mathrm{Re}(w)-w_0)} & \mathfrak{d} = (\beta_6), \\ \mu_1^{r_1(\mathrm{Re}(w)-w_0)}|t_{12}(g_{1,23})t_{13}(g_{1,23})^{-1}|^{-r_2(\mathrm{Re}(w)-w_0)} & \mathfrak{d} = (\beta_8), \end{cases}$$

for $M_1 \le \mathrm{Re}(w) \le M_2$ and $g_{\mathfrak{d}}^0$ in the Siegel domain.

We choose $\mathrm{Re}(w) \gg N \gg 0$. Then the function

$$\chi(g_{\mathfrak{d}})\mu_2^{-N} h'_N(\mu_1, g_{\mathfrak{d}}^0)\mu_1^{r_1(\mathrm{Re}(w)-w_0)}|t_{11}(g_{1,12})t_{12}(g_{1,12})^{-1}|^{-r_2(\mathrm{Re}(w)-w_0)}$$

is integrable for $\mu_1 \le 1, \mu_2 \ge 1$, and $g_{\mathfrak{d}}^0$ in the Siegel domain. Other cases are similar. This proves the lemma.

Q.E.D.

Let $\Psi = R_{\mathfrak{d}}\Phi_{\mathfrak{p}}$. If $S_{\mathfrak{d}'} \subset V_{\mathfrak{d}1}$, we define

$$\begin{aligned}
\Xi_{\mathfrak{p},\mathfrak{d}',1}(\Phi,\omega,w) &= \Xi_{\mathfrak{p},\mathfrak{d}',\ge}(\mathscr{F}_{V_{\mathfrak{d}1}}\mathscr{F}_{\mathfrak{d}}\Psi,\omega,\kappa_{V_{\mathfrak{d}2}}\sigma_{\mathfrak{p}},w),\\
\Xi_{\mathfrak{p},\mathfrak{d}',2}(\Phi,\omega,w) &= \Xi_{\mathfrak{p},\mathfrak{d}',\ge}(\Psi,\omega,\sigma_{\mathfrak{p}},w),\\
\Xi_{\mathfrak{p},\mathfrak{d}',3}(\Phi,\omega,w) &= \Xi_{\mathfrak{p},\mathfrak{d}',\le}(\mathscr{F}_{\mathfrak{d}}\Psi,\omega,\kappa_{\mathfrak{d}1}\sigma_{\mathfrak{p}},w),\\
\Xi_{\mathfrak{p},\mathfrak{d}',4}(\Phi,\omega,w) &= \Xi_{\mathfrak{p},\mathfrak{d}',\le}(\mathscr{F}_{V_{\mathfrak{d}1}}\Psi,\omega,\kappa_{V_{\mathfrak{d}1}}\sigma_{\mathfrak{p}},w).
\end{aligned} \tag{9.1.7}$$

Also if $S_{\mathfrak{d}'} \subset V_{\mathfrak{d}2}$, we define

$$\begin{aligned}
\Xi_{\mathfrak{p},\mathfrak{d}',1}(\Phi,\omega,w) &= \Xi_{\mathfrak{p},\mathfrak{d}',\ge}(\mathscr{F}_{\mathfrak{d}}\Psi,\omega,\kappa_{\mathfrak{d}1}\sigma_{\mathfrak{p}},w),\\
\Xi_{\mathfrak{p},\mathfrak{d}',2}(\Phi,\omega,w) &= \Xi_{\mathfrak{p},\mathfrak{d}',\ge}(\mathscr{F}_{V_{\mathfrak{d}2}}\Psi,\omega,\kappa_{V_{\mathfrak{d}2}}\sigma_{\mathfrak{p}},w),\\
\Xi_{\mathfrak{p},\mathfrak{d}',3}(\Phi,\omega,w) &= \Xi_{\mathfrak{p},\mathfrak{d}',\le}(\mathscr{F}_{V_{\mathfrak{d}2}}\mathscr{F}_{\mathfrak{d}}\Psi,\omega,\kappa_{V_{\mathfrak{d}1}}\sigma_{\mathfrak{p}},w),\\
\Xi_{\mathfrak{p},\mathfrak{d}',4}(\Phi,\omega,w) &= \Xi_{\mathfrak{p},\mathfrak{d}',\le}(\Psi,\omega,\sigma_{\mathfrak{p}},w).
\end{aligned} \tag{9.1.8}$$

We define

$$\Xi_{\mathfrak{p},\mathrm{tot}}(\Phi,\omega,w) = \sum_{Z_{\mathfrak{d}'} \subset V_{\mathfrak{d}1},V_{\mathfrak{d}2}} \sum_{i=1}^{4} (-1)^{i+1}\Xi_{\mathfrak{p},\mathfrak{d}',i}(\Phi,\omega,w). \tag{9.1.9}$$

For $\mathfrak{d}'$ such that $Z_{\mathfrak{d}'} \not\subset V_{\mathfrak{d}1}, V_{\mathfrak{d}2}$, we define $\widetilde{\Xi}_{\mathfrak{p}'}(R_{\mathfrak{d}'}\Phi_{\mathfrak{p}'},\omega,w), \Xi_{\mathfrak{p}'}(R_{\mathfrak{d}'}\Phi_{\mathfrak{p}'},\omega,w)$ in exactly the same manner as in (3.5.8). Then by (3.5.5),

$$\widetilde{\Xi}_{\mathfrak{p}'}(R_{\mathfrak{d}'}\Phi_{\mathfrak{p}'},\omega,w) + \Xi_{\mathfrak{p}'}(R_{\mathfrak{d}'}\Phi_{\mathfrak{p}'},\omega,w)$$

is well defined for $\mathrm{Re}(w) \gg 0$.

These considerations show that

$$\begin{aligned}
\Xi_{\mathfrak{p}}(\Phi,\omega,w) &= \Xi_{\mathfrak{p}+}(\Phi,\omega,w) + \widehat{\Xi}_{\mathfrak{p}+}(\Phi,\omega,w)\\
&\quad + \sum_{\substack{\mathfrak{p} \prec \mathfrak{p}'=(\mathfrak{d}',\mathfrak{s}')\\ l(\mathfrak{p}')=2\\ Z_{\mathfrak{d}'} \not\subset V_{\mathfrak{d}1},V_{\mathfrak{d}2}}} (\widetilde{\Xi}_{\mathfrak{p}'}(R_{\mathfrak{d}'}\Phi_{\mathfrak{p}'},\omega,w) + \Xi_{\mathfrak{p}'}(R_{\mathfrak{d}'}\Phi_{\mathfrak{p}'},\omega,w))\\
&\quad + \Xi_{\mathfrak{p},\mathrm{tot}}(\Phi,\omega,w).
\end{aligned} \tag{9.1.10}$$

We will prove in Chapter 10 that the distributions

$$\widetilde{\Xi}_{\mathfrak{p}'}(R_{\mathfrak{d}'}\Phi_{\mathfrak{p}'},\omega,w),\ \Xi_{\mathfrak{p}'}(R_{\mathfrak{d}'}\Phi_{\mathfrak{p}'},\omega,w)$$

are well defined for $\mathrm{Re}(w) \gg 0$.

§9.2 The smoothed Eisenstein series

In this section, we prove some estimates of the smoothed Eisenstein series which are not covered in Chapter 3. We will use these estimates in Chapter 10.

We first fix some notation.

For $u = (u_1, u_2) \in \mathbb{A}^2$, let

$$(9.2.1) \qquad n_{P_1}(u) = \left(\begin{pmatrix} I_2 & \\ u & 1 \end{pmatrix}, I_2 \right), \quad n_{P_2}(u) = \left(\begin{pmatrix} 1 & \\ {}^t u & I_2 \end{pmatrix}, I_2 \right).$$

For $u = (u_1, u_2, u_3) \in \mathbb{A}^3$, let

$$(9.2.2) \qquad n_{B_1}(u) = \left(\begin{pmatrix} 1 & & \\ u_1 & 1 & \\ u_2 & u_3 & 1 \end{pmatrix}, I_2 \right).$$

For $u \in \mathbb{A}$, let

$$(9.2.3) \qquad n'(u) = \left(\begin{pmatrix} 1 & & \\ u & 1 & \\ & & 1 \end{pmatrix}, I_2 \right), \; n''(u) = \left(\begin{pmatrix} 1 & & \\ & 1 & \\ & u & 1 \end{pmatrix}, I_2 \right), \quad n_{B_2}(u) = \left(I_3, \begin{pmatrix} 1 & \\ u & 1 \end{pmatrix} \right).$$

For $\lambda_1 \in \mathbb{R}$, $t_{11}, \cdots, t_{22} \in \mathbb{A}^1$, let

$$(9.2.4) \qquad \begin{aligned} d_{P_1}(\lambda_1) &= a(\underline{\lambda}_1^{-1}, \underline{\lambda}_1^{-1}, \underline{\lambda}_1^2; 1, 1), \\ d_{P_2}(\lambda_1) &= a(\underline{\lambda}_1^{-2}, \underline{\lambda}_1, \underline{\lambda}_1; 1, 1), \\ t^0 &= a(t_{11}, t_{12}, t_{13}; t_{21}, t_{22}). \end{aligned}$$

For $u = (u_1, \cdots, u_4)$, let $n_B(u) = (n_{B_1}(u_1, u_2, u_3), n_{B_2}(u_4))$. Also let $d^\times \widehat{t}^0 = \prod d^\times t_{ij}$.

In our situation, the Eisenstein series $E_B(g^0, z)$ is a function of $g^0 \in G^0_{\mathbb{A}}$, $z = (z_1, z_2)$, where $z_1 = (z_{11}, z_{12}, z_{13}) \in \mathbb{C}^3$, $z_2 = (z_{21}, z_{22}) \in \mathbb{C}^2$ and $z_{11} + z_{12} + z_{13} = 0$, $z_{21} + z_{22} = 0$. Let $\mathscr{E}(g^0, w)$ be the smoothed Eisenstein series. We choose the constants (C_1, C_2) in (3.4.3) so that $C_1 = 1, C_2 = C > 100$. Let $I = (I_1, I_2), \tau = (\tau_1, \tau_2), \nu = (\nu_1, \nu_2), s_I(l) = (s_{1,I_1}(l), s_{2,I_2}(l))$ be as in (3.4.24)–(3.4.28). Note that ν_2 is always 1.

Consider an element of the form

$$\lambda = d_{P_1}(\lambda_1) a(\underline{\mu}_1, \underline{\mu}_1^{-1}, 1; \underline{\mu}_2, \underline{\mu}_2^{-1}),$$

where $\lambda_1, \mu_1, \mu_2 \in \mathbb{R}_+$. We can write any $g^0 \in G^0_{\mathbb{A}}$ in the form $g^0 = k\lambda \widehat{t}^0 n_B(u)$, where $k \in K$ and $u = (u_1, \cdots, u_4) \in \mathbb{A}^4$. We write $\lambda \widehat{t}^0 = (t_1, t_2)$, where $t_1 \in T^0_{1\mathbb{A}}, t_2 \in T^0_{2\mathbb{A}}$. For the rest of this section, we estimate $\widetilde{\mathscr{E}}_{I,\tau,\nu}(g^0, w)$ for various cases. We recall that by (3.4.30), if $q \in \mathfrak{t}^*$, for any $\delta > 0, M > w_0$,

$$\widetilde{\mathscr{E}}_{I,\tau,\nu}(g^0, w) \ll \lambda^{\nu(\tau q + \rho - s_I(l))} h(\lambda)$$

for $L(q)+\delta \le \mathrm{Re}(w) \le M$, where $h(\lambda)$ is a slowly increasing function independent of δ, M. Moreover, if $\nu = 1$, $h(\lambda) = 1$.

Let $q = (q_1, q_2) = (q_{11}, q_{12}, q_{13}; q_{21}, q_{22}) \in D_{I,\tau}$. We define $r_1 = q_{11} - q_{12}$, $r_2 = q_{12} - q_{13}$, and $r_3 = q_{21} - q_{22}$. Then

$$q_{11} = \frac{2r_1 + r_2}{3}, \quad q_{12} = \frac{-r_1 + r_2}{3}, \quad q_{13} = -\frac{r_1 + 2r_2}{3}. \tag{9.2.5}$$

The following table shows the condition for q to belong to the domain $D_{I,\tau}$.

Table (9.2.6)

τ_1	$I_1 = \emptyset$	$I_1 = \{1\}$	$I_1 = \{2\}$	$I_1 = \{1,2\}$
1	no condition	—	—	—
(1,2)	$r_1 > 1$	$r_1 > 0$	—	—
(1,3)	$r_1>1$, $r_2>1$	$r_1>1$, $r_2>0$	$r_1>0$, $r_2>1$	$r_1>0$, $r_2>0$
(2,3)	$r_2 > 1$	—	$r_2 > 0$	—
(1,2,3)	$r_1>1$, $r_1+r_2>1$	—	$r_1>1$, $r_1+r_2>0$	—
(1,3,2)	$r_2>1$, $r_1+r_2>1$	$r_2>1$, $r_1+r_2>0$	—	—

It is easy to see that

$$t_1^{\tau_1 q_1 + \rho_1} = \lambda_1^{-3}\mu_1 \times \begin{cases} \lambda_1^{-r_1-2r_2}\mu_1^{r_1} & \tau_1 = 1, \\ \lambda_1^{-r_1-2r_2}\mu_1^{-r_1} & \tau_1 = (1,2), \\ \lambda_1^{2r_1+r_2}\mu_1^{-r_2} & \tau_1 = (1,3), \\ \lambda_1^{-r_1+r_2}\mu_1^{r_1+r_2} & \tau_1 = (2,3), \\ \lambda_1^{2r_1+r_2}\mu_1^{r_2} & \tau_1 = (1,2,3), \\ \lambda_1^{-r_1+r_2}\mu_1^{-r_1-r_2} & \tau_1 = (1,3,2). \end{cases} \tag{9.2.7}$$

Also

$$t_1^{-s_{1,I_1}(l)} = \begin{cases} \mu_1^{-2l_{11}} & I_1 = \{1\}, \\ (\lambda_1^{-3}\mu_1^{-1})^{-l_{11}} = \lambda_1^{3l_{11}}\mu_1^{l_{11}} & I_1 = \{2\}, \\ \mu_1^{-2l_{11}}(\lambda_1^{-3}\mu_1^{-1})^{-l_{12}} = \lambda_1^{3l_{12}}\mu_1^{l_{12}-2l_{11}} & I_1 = \{1,2\}, \end{cases} \tag{9.2.8}$$

$$t_2^{-s_{2,I_2}(l)} = \mu_2^{-2l_2}.$$

Suppose $\nu_1 = (1,2)$. Then

$$t_1^{-\nu_1(s_{1,I_1}(l))} = \begin{cases} (\lambda_1^{-3}\mu_1)^{-l_{11}} = \lambda_1^{3l_{11}}\mu_1^{-l_{11}} & I_1 = \{2\}, \\ \mu_1^{2l_{11}}(\lambda_1^{-3}\mu_1)^{-l_{12}} = \lambda_1^{3l_{12}}\mu_1^{2l_{11}-l_{12}} & I_1 = \{1,2\}. \end{cases} \tag{9.2.9}$$

Proposition (9.2.10) *Suppose $I \neq (\emptyset, \emptyset)$.*
(1) *Suppose that $\tau_1 \neq (1,3), (1,3,2)$, or $\tau = (1,3)$ and $I_1 = \{1,2\}$, or $\tau = (1,3,2)$ and $I_1 = \{1\}$. Then for any $N > 0, \epsilon > 0$, there exists a constant $\delta > 0$ such that if $M > w_0$,*

$$\widetilde{\mathscr{E}}_{I,\tau,1}(g^0, w) \ll \lambda_1^{c_1} \mu_1^{c_2} \mu_2^{c_3}$$

for $w_0 - \delta \leq \mathrm{Re}(w) \leq M$, where $c_1 \geq N$, $c_2 \leq -N$, $c_3 < \epsilon$.
(2) *Suppose that $I_2 = \{1\}$, and $\tau_1 = (1,3)$ or $(1,3,2)$. Then for any $\epsilon > 0$ there exists a constant $\delta > 0$ such that if $M > w_0$,*

$$\widetilde{\mathscr{E}}_{I,\tau,1}(g^0, w) \ll \lambda_1^{c_1} \mu_1^{c_2} \mu_2^{c_3}$$

for $w_0 - \delta \leq \mathrm{Re}(w) \leq M$, where $c_1 \geq -2 + \frac{1}{8}C$, $c_2 \leq -(-2 + \frac{1}{16}C)$, and $c_3 < \epsilon$.
(3) *Suppose that $(I_1, \tau_1) = (\{2\}, (1,3))$ and $I_2 = \emptyset$. Then for any $\epsilon > 0$ there exists a constant $\delta > 0$ such that if $M > w_0$,*

$$\widetilde{\mathscr{E}}_{I,\tau,1}(g^0, w) \ll \lambda_1^{2+c_1} \mu_1^{c_2} \mu_2^{c_3}$$

for $w_0 - \delta \leq \mathrm{Re}(w) \leq M$, where $|c_1|, |c_2|, |c_3| < \epsilon$.

Proof. If $I_2 = \emptyset$,

$$t_2^{\tau_2 z_2 + \rho_2} = \begin{cases} \mu_2^{1+r_3} & \tau_2 = 1, \\ \mu_2^{1-r_3} & \tau_2 = (1,2). \end{cases}$$

If $\tau_2 = 1$, we choose $r_3 = -1$, and if $\tau_2 = (1,2)$, we choose $r_3 > 1$ close to 1. If $I_2 = \{1\}$,

$$t_2^{\tau_2 z_2 + \rho_2 - s_{2,I_2}(l)} = \mu_2^{1-r_3-2l_2}.$$

We choose $r_3 = \frac{1}{2}$ for this case. In all the cases, $L_1(q_2) \leq 1 + \epsilon_1$, where $\epsilon_1 > 0$ is a small number. Also $t_2^{\tau_2 z_2 + \rho_2} = \mu_2^{c_3}$ for some c_3, where $c_3 < \epsilon$ in all the cases and $|c_3| < \epsilon$ if $I_2 = \emptyset$.

Consider (1). Let r_1, r_2, r_3 be as in (9.2.5). If $\tau_1 = 1$, we choose $r_1, r_2 \ll 0$. If $\tau_1 = (1,2)$, we choose $r_1 = -r_2 \gg 0$. If $\tau_1 = (2,3)$, we fix r_2, and choose $r_1 \ll 0$. If $\tau_1 = (1,2,3)$, we choose $r_1 = \frac{3}{2} - r_2 \gg 0$. Then $L(q) = 3 + (1+\epsilon_1)C < w_0 = 4 + C$ if ϵ_1 is sufficiently small. Then $t_1^{\tau_1 q_1 + \rho_1}$ satisfies the condition of (1). Since $\nu = 1$, we can fix l. This proves (1) for $\tau_1 \neq (1,3), (1,3,2)$.

If $I_1 = \{1,2\}$, we choose $r_1 = r_2 = \frac{1}{2}$, and $l_{11} \gg l_{12} \gg 0$ in (9.2.8). Note that we can do this because $\nu = 1$.

Consider the case $\tau_1 = (1,3,2), I_1 = \{1\}$. We choose $r_2 \gg 0$ and $r_1 = 1 - r_2$. Then $L(q) < w_0$, and

$$t_1^{\tau_1 q_1 + \rho_1 - s_{1,I_1}(l)} = \lambda_1^{2r_2 - 1} \mu_1^{-1-2l_{11}}.$$

Therefore, if we choose $l_{11} \gg 0$, we have an estimate of the form (1).

Next, we consider (2). If $\tau_1 = (1,3)$, we choose $r_1 = r_2 = 1 + \frac{1}{16}C$ and $r_3 = \frac{1}{2}$. Then $L(q) = 4 + \frac{3}{4}C < w_0$, and

$$t_1^{\tau_1 q_1 + \rho_1} = \lambda_1^{\frac{3}{16}C} \mu_1^{-\frac{1}{16}C}.$$

We have already considered the case $I_1 = \{1,2\}$ in (1). If $I_1 = \{1\}$ or $\{2\}$, we choose $l_{11} = 2$ in (9.2.8). Note that we can choose any $l_{11} > 1$ in this case. This consideration shows that we have an estimate of the form (2).

If $\tau_1 = (1,3,2)$, we choose $r_1 = 0, r_2 = 1 + \frac{1}{8}C$, and $r_3 = \frac{1}{2}$. Then $L(q) < w_0$, and

$$t_1^{\tau_1 q_1 + \rho_1} = \lambda_1^{-2+\frac{1}{8}C} \mu_1^{-\frac{1}{8}C}.$$

We have already considered the case $I_1 = \{1\}$ in (1). Therefore, we have an estimate of the form (2).

Finally, we consider (3). We choose $r_1 = \epsilon_2, r_2 = 2 - \epsilon_3,\ l_{11} = 1 + \epsilon_4$, where $\epsilon_1, \cdots, \epsilon_4 > 0$. Then $L(q) = w_0 + 2(\epsilon_2 - \epsilon_3) + C\epsilon_1$, and

$$\widetilde{\mathscr{E}}_{I,\tau,1}(g^0, w) \ll \lambda_1^{2+2\epsilon_2-\epsilon_3+3\epsilon_4} \mu_1^{\epsilon_3+\epsilon_4} \mu_2^{c_3},$$

where $|c_3| < \epsilon$.

Since we can choose $\epsilon_1, \cdots, \epsilon_4$ so that $L(q) < w_0$, we have an estimate of the form (3).

Q.E.D.

Proposition (9.2.11)
(1) *Suppose $I_2 = \{1\}$. Then there exist a constant $\delta > 0$, a slowly increasing function $h(\lambda_1, \mu_2)$, and a constant $c > 1$ such that if $M > w_0$,*

$$\widetilde{\mathscr{E}}_{I,\tau,\nu}(g^0, w) \ll \mu_1^{-c} h(\lambda_1, \mu_2)$$

for $w_0 - \delta \leq \mathrm{Re}(w) \leq M$.
(2) *Suppose that $I_1 \ni 2$ and $\nu_1 = (1,2)$. Then there exist a constant $\delta > 0$ and a slowly increasing function $h(\lambda)$ such that if $M > w_0$, for any $l \gg 0$,*

$$\widetilde{\mathscr{E}}_{I,\tau,\nu}(g^0, w) \ll h(\lambda) \lambda_1^{3l} \mu_1^{-l}$$

for $w_0 - \delta \leq \mathrm{Re}(w) \leq M$.

Proof. Since we assumed that $C > 100$, we covered the case when $\nu = 1$ in (9.2.9). Therefore, we assume that $\nu_1 = (1,2)$ ($\nu_2 = 1$) and prove (2).

If $I_1 = \{1,2\}$, by (3.4.30),

$$\widetilde{\mathscr{E}}_{I,\tau,\nu}(g^0, w) \ll \lambda_1^{3l_{12}} \mu_1^{2l_{11}-l_{12}} \lambda^{\nu(\tau q+\rho)} h(\lambda),$$

where $h(\lambda)$ is a slowly increasing function. So we choose any $q \in D_{I,\tau}$ so that $L(q) < w_0$, fix l_{11}, and choose $l_{12} \gg 0$.

If $I_1 = \{2\}$, by (3.4.30),

$$\widetilde{\mathscr{E}}_{I,\tau,\nu}(g^0, w) \ll \lambda_1^{3l_{11}} \mu_1^{-l_{11}} \lambda^{\nu(\tau q+\rho)} h(\lambda),$$

where $h(\lambda)$ is a slowly increasing function. So we choose any $q \in D_{I,\tau}$ so that $L(q) < w_0$, and choose $l_{11} \gg 0$.

Q.E.D.

Chapter 10 The non-constant terms associated with unstable strata

In this chapter, we prove the following two theorems.

Theorem (10.0.1) *Let* $\mathfrak{p} = (\mathfrak{d}, \mathfrak{s})$ *be a path. Then* $\widetilde{\Xi}_{\mathfrak{p}}(\Phi, \omega, w) \sim 0$ *for*
(1) $\mathfrak{d} = (\beta_4)$,
(2) $\mathfrak{d} = (\beta_5)$,
(3) $\mathfrak{d} = (\beta_7)$,
(4) $\mathfrak{d} = (\beta_8)$,
(5) $\mathfrak{d} = (\beta_8, \beta_{8,1})$, *and* $\mathfrak{s}(2) = 1$,
(6) $\mathfrak{d} = (\beta_8, \beta_{8,2})$, *and* $\mathfrak{s}(2) = 1$,
(7) $\mathfrak{d} = (\beta_{10}, \beta_{10,1})$.

Theorem (10.0.2) *Let* $\mathfrak{p}_1 = (\mathfrak{d}_1, \mathfrak{s}_1), \mathfrak{p}_2 = (\mathfrak{d}_2, \mathfrak{s}_2)$ *be paths. Then*

$$\epsilon_{\mathfrak{p}_1} \widetilde{\Xi}_{\mathfrak{p}_1}(\Phi, \omega, w) + \epsilon_{\mathfrak{p}_2} \widetilde{\Xi}_{\mathfrak{p}_2}(\Phi, \omega, w) \sim 0 \quad \textit{for}$$

(1) $\mathfrak{d}_1 = (\beta_6), \mathfrak{d}_2 = (\beta_8, \beta_{8,1})$, *and* $\mathfrak{s}_1(1) = \mathfrak{s}_2(1)$, $\mathfrak{s}_2(2) = 0$,
(2) $\mathfrak{d}_1 = (\beta_8, \beta_{8,2}), \mathfrak{d}_2 = (\beta_9)$, *and* $\mathfrak{s}_1(1) = \mathfrak{s}_2(1)$, $\mathfrak{s}_1(2) = 0$.

We will prove (10.0.1) (1) in §10.1, (2) and (7) in §10.2, (3) in §10.4, (4) in §10.5, (5) in §10.3, and (6) in §10.6. Also we will prove (10.0.2) (1) in §10.3, and (2) in §10.6.

Corollary (10.0.3)

$$\sum_{\mathfrak{p}} \epsilon_{\mathfrak{p}} \widetilde{\Xi}_{\mathfrak{p}}(\Phi, \omega, w) \sim 0.$$

In this chapter, $g_{\mathfrak{d}}$ always denotes an element of $G^0_{\mathbb{A}} \cap M_{\mathfrak{d}\mathbb{A}}$. We use the measure $dg_{\mathfrak{d}}$ on $G^0_{\mathbb{A}} \cap M_{\mathfrak{d}\mathbb{A}}$ which we defined in §3.3 (just after (3.3.11)). Let $d^\times t^0$ be the measure on $T^0_{\mathbb{A}}$ such that $dg^0 = dk d^\times t^0 du$, where dk, du are the measures on $K, N_{\mathbb{A}}$ which we defined before.

§10.1 The case $\mathfrak{d} = (\beta_4)$

In this section, we prove the following proposition.

Proposition (10.1.1) *Let* $\mathfrak{p} = (\mathfrak{d}, \mathfrak{s})$ *be a path such that* $\mathfrak{d} = \mathfrak{d}_4$. *Then* $\widetilde{\Xi}_{\mathfrak{p}}(\Phi, \omega, w) \sim 0$.

Proof. Let $n(u_{\mathfrak{d}}) = n_{P_1}(u_{\mathfrak{d}})$ for $u_{\mathfrak{d}} = (u_2, u_3) \in \mathbb{A}^2$. Consider

$$(10.1.2) \qquad \int_{(\mathbb{A}/k)^2} \Theta_{Y_{\mathfrak{d}}}(\Psi, g_{\mathfrak{d}} n(u_{\mathfrak{d}})) \mathscr{E}(g_{\mathfrak{d}} n(u_{\mathfrak{d}}), w) du_{\mathfrak{d}}$$

for $\Psi = \widetilde{R}_{\mathfrak{d}} \Phi_{\mathfrak{p}}$.

Let $d_{\mathfrak{d}}(\lambda_1) = d_{P_1}(\lambda_1)$ for $\lambda_1 \in \mathbb{R}_+$ (see (9.2.4)). Let

$$g^0_{\mathfrak{d}} = (g_1, g_2),\ g_1 = \begin{pmatrix} g_{1,12} & \\ & t_{13} \end{pmatrix},$$

where $|\det g_{1,12}| = |t_{13}| = |\det g_2| = 1$. Let $g_\mathfrak{d} = d(\lambda_1) g_\mathfrak{d}^0$.

For $\alpha = (\alpha_1, \alpha_2) \in k^2$, let $< \alpha \cdot u_\mathfrak{d} > = < \alpha_1 u_2 + \alpha_2 u_3 >$. We define

$$(10.1.3) \qquad f_{1,\alpha}(g_\mathfrak{d}) = \int_{(\mathbb{A}/k)^2} \Theta_{Y_\mathfrak{d}}(\Psi, g_\mathfrak{d} n(u_\mathfrak{d})) < \alpha \cdot u_\mathfrak{d} > du_\mathfrak{d},$$

$$f_{2,\alpha}(g_\mathfrak{d}, w) = \int_{(\mathbb{A}/k)^2} \mathscr{E}(g_\mathfrak{d} n(u_\mathfrak{d}), w) < \alpha \cdot u_\mathfrak{d} > du_\mathfrak{d}.$$

By the Parseval formula, (10.1.2) is equal to

$$\sum_{\alpha \in k^2} f_{1,\alpha}(g_\mathfrak{d}) f_{2,-\alpha}(g_\mathfrak{d}, w).$$

We consider $g_{1,12}, g_2$ in the Siegel domain as follows:

$$(10.1.4) \qquad g_{1,12} = k_1 \begin{pmatrix} \underline{\mu}_1 t_{11} & \\ & \underline{\mu}_1^{-1} t_{12} \end{pmatrix} \begin{pmatrix} 1 & \\ u_1 & 1 \end{pmatrix},$$

$$g_2 = k_2 \begin{pmatrix} \underline{\mu}_2 t_{21} & \\ & \underline{\mu}_2^{-1} t_{22} \end{pmatrix} \begin{pmatrix} 1 & \\ u_4 & 1 \end{pmatrix},$$

where t_{11} etc. are chosen from a fixed compact set in $\mathbb{A}^1$, and μ_1, μ_2 are bounded below.

We can write any element $y \in Y_{\mathfrak{d}k}^{ss}$ in the form $y = (y_1, y_2)$ where $y_1 \in Z_{\mathfrak{d}k}^{ss}, y_2 \in W_{\mathfrak{d}k}$ and

$$y_1 = \begin{pmatrix} x_{1,13} & x_{1,23} \\ x_{2,13} & x_{2,23} \end{pmatrix}, \quad y_2 = \begin{pmatrix} x_{1,33} \\ x_{2,33} \end{pmatrix}.$$

Then

$$g_\mathfrak{d} n(u_\mathfrak{d}) y = (\underline{\lambda}_1 t_{13} g_2 y_1 {}^t g_{1,12}, \underline{\lambda}_1^4 t_{13}^2 g_2 (y_2 + y_1 {}^t u_\mathfrak{d})).$$

Therefore,

$$\begin{aligned} f_{1,\alpha}(g_\mathfrak{d}) &= \int_{\mathbb{A}^2} \sum_{y_1 \in Z_{\mathfrak{d}k}^{ss}} \Psi(\underline{\lambda}_1 t_{13} g_2 y_1 {}^t g_{1,12}, \underline{\lambda}_1^4 t_{13}^2 g_2 y_1 {}^t u_\mathfrak{d}) < \alpha \cdot u_\mathfrak{d} > du_\mathfrak{d} \\ &= \sum_{y_1 \in Z_{\mathfrak{d}k}^{ss}} \int_{\mathbb{A}^2} \Psi(\underline{\lambda}_1 t_{13} g_2 y_1 {}^t g_{1,12}, \underline{\lambda}_1^4 t_{13}^2 g_2 y_1 {}^t u_\mathfrak{d}) < \alpha \cdot u_\mathfrak{d} > du_\mathfrak{d}. \end{aligned}$$

For $y_2 = {}^t(x_{1,33}, x_{2,33}), y_2' = {}^t(x_{1,33}', x_{2,33}')$, we define $[y_2, y_2'] = x_{1,33} x_{1,33}' + x_{2,33} x_{2,33}'$. Let Ψ_1 be the partial Fourier transform of Ψ with respect to y_2 and the above bilinear form. Then

$$f_{1,\alpha}(g_\mathfrak{d}) = \lambda_1^{-8} \sum_{y_1 \in Z_{\mathfrak{d}k}^{ss}} \Psi_1(\underline{\lambda}_1 t_{13} g_2 y_1 {}^t g_{1,12}, \underline{\lambda}_1^{-4} t_{13}^{-2} {}^t g_2^{-1} {}^t y_1^{-1} {}^t \alpha).$$

Let

$$(10.1.5) \qquad f_{3,\alpha}(g_\mathfrak{d}) = \lambda_1^{-8} \sum_{\substack{y_1 \in Z_{\mathfrak{d}k}^{ss} \\ x_{1,13} \neq 0}} \Psi_1(\underline{\lambda}_1 t_{13} g_2 y_1 {}^t g_{1,12}, \underline{\lambda}_1^{-4} t_{13}^{-2} {}^t g_2^{-1} {}^t y_1^{-1} {}^t \alpha),$$

$$f_{4,\alpha}(g_\mathfrak{d}) = \lambda_1^{-8} \sum_{\substack{y_1 \in Z_{\mathfrak{d}k}^{ss} \\ x_{1,13} = 0}} \Psi_1(\underline{\lambda}_1 t_{13} g_2 y_1 {}^t g_{1,12}, \underline{\lambda}_1^{-4} t_{13}^{-2} {}^t g_2^{-1} {}^t y_1^{-1} {}^t \alpha).$$

Then $f_{1,\alpha}(g_{\mathfrak{d}}) = f_{3,\alpha}(g_{\mathfrak{d}}) + f_{4,\alpha}(g_{\mathfrak{d}})$.

Lemma (10.1.6) *Suppose that μ_1, μ_2 are bounded below. Then for any $N \geq 1$,*

$$\sum_{\alpha \in k^2 \setminus \{0\}} |f_{3,\alpha}(g_{\mathfrak{d}})| \ll \mathrm{rd}_{3,N}(\lambda_1, \mu_1, \mu_2).$$

Proof. Let $\alpha' = (\alpha'_1, \alpha'_2) = \alpha y_1^{-1}$. Then

$$\sum_{\alpha \in k^2 \setminus \{0\}} f_{3,\alpha}(g_{\mathfrak{d}}) = \lambda_1^{-8} \sum_{\substack{y_1 \in Z^{\mathrm{ss}}_{\mathfrak{d}k} \\ \alpha' \in k^2 \setminus \{0\} \\ x_{1,13} \neq 0}} \Psi_1(\underline{\lambda}_1 t_{13} g_2 y_1 {}^t g_{1,12}, \underline{\lambda}_1^{-4} t_{13}^{-2t} g_2^{-1t} \alpha').$$

An easy consideration shows that

$$\underline{\lambda}_1^{-4} t_{13}^{-2t} g_2^{-1t} \alpha' = {}^t k_2^{-1} \underline{\lambda}_1^{-4} t_{13}^{-2} \begin{pmatrix} \mu_2^{-1} t_{21}^{-1} (\alpha'_1 - u_3 \alpha'_2) \\ \mu_2 t_{22}^{-1} \alpha'_2 \end{pmatrix}.$$

Let

$$f_5(g_{\mathfrak{d}}) = \sum_{\alpha'_1 \in k, \alpha'_2 \in k^{\times}} |f_{3,\alpha'}(g_{\mathfrak{d}})|, \quad f_6(g_{\mathfrak{d}}) = \sum_{\alpha'_2 = 0, \alpha'_1 \in k^{\times}} |f_{3,\alpha'}(g_{\mathfrak{d}})|.$$

We choose $0 \leq \Psi_2 \in \mathscr{S}(Z_{\mathfrak{d}\mathbb{A}}), 0 \leq \Psi_3 \in \mathscr{S}(W_{\mathfrak{d}\mathbb{A}})$ so that

$$|\Psi_1(y_1, y_2)| \ll \Psi_2(y_1) \Psi_3(y_2).$$

By (1.2.8), for any $N \geq 1$,

$$\sum_{\substack{\alpha'_1 \in k \\ \alpha'_2 \in k^{\times}}} \Psi_3(\underline{\lambda}_1^{-4} t_{13}^{-2t} g_2^{-1t} \alpha') \ll \sup(1, \lambda_1^4 \mu_2)(\lambda_1^{-4} \mu_2)^{-N},$$

$$\sum_{\substack{\alpha'_1 \in k^{\times} \\ \alpha'_2 = 0}} \Psi_3(\underline{\lambda}_1^{-4} t_{13}^{-2t} g_2^{-1t} \alpha') \ll (\lambda_1^4 \mu_2)^N,$$

$$\sum_{\substack{y_1 \in Z^{\mathrm{ss}}_{\mathfrak{d}k} \\ x_{1,13} \neq 0}} \Psi_2(\underline{\lambda}_1 t_{13} g_2 y_1 {}^t g_{1,12}) \ll h_1(\lambda_1, \mu_1, \mu_2)(\lambda_1 \mu_1 \mu_2)^{-N},$$

where

$$h_1(\lambda_1, \mu_1, \mu_2) = \sup(1, \lambda_1^{-1} \mu_1^{-1} \mu_2) \sup(1, \lambda_1^{-1} \mu_1 \mu_2^{-1}) \sup(1, \lambda_1^{-1} \mu_1 \mu_2).$$

So for any $N_1, N_2 \geq 1$,

$$f_5(g_{\mathfrak{d}}) \ll \lambda_1^{-8} h_1(\lambda_1, \mu_1, \mu_2) \sup(1, \lambda_1^4 \mu_2)(\lambda_1 \mu_1 \mu_2)^{-N_1} (\lambda_1^{-4} \mu_2)^{-N_2},$$
$$f_6(g_{\mathfrak{d}}) \ll \lambda_1^{-8} h_1(\lambda_1, \mu_1, \mu_2)(\lambda_1 \mu_1 \mu_2)^{-N_1} (\lambda_1^{-4} \mu_2^{-1})^{-N_2}.$$

Therefore, for any $N_1, N_2, N_3 \geq 1$,

$$\sum_{\alpha \in k^2 \setminus \{0\}} f_{3,\alpha}(g_{\mathfrak{d}}) \ll h_2(\lambda_1, \mu_1, \mu_2)(\lambda_1 \mu_1 \mu_2)^{-N_1} ((\lambda_1^{-4} \mu_2)^{-N_2} + (\lambda_1^{-4} \mu_2^{-1})^{-N_3}),$$

where

$$h_2(\lambda_1,\mu_1,\mu_2)=\lambda_1^{-8}\sup(1,\lambda_1^4\mu_2)h_1(\lambda_1,\mu_1,\mu_2).$$

If $\lambda_1\geq 1$, we choose $N_1\gg 0$, and if $\lambda_1\leq 1$, we choose $N_1=2N_2=2N_3\gg 0$. Since μ_1,μ_2 are bounded below, we get the proposition (10.1.6).

Q.E.D.

Next, we consider $f_{4,\alpha}(g_{\mathfrak{d}})$. Consider $y_1\in Z_{\mathfrak{d}k}^{\mathrm{ss}}$ such that $x_{1,13}=0$. This implies that $x_{1,23}\neq 0$ and $x_{2,13}\neq 0$. An easy computation shows that

$${}^t g_2^{-1t}y_1^{-1t}\alpha=\det x^{-1}\begin{pmatrix}\underline{\mu}_2^{-1}t_{21}^{-1}((x_{2,23}+x_{2,13}u_3)\alpha_1-x_{1,23}\alpha_2)\\ -\underline{\mu}_2 t_{22}^{-1}x_{2,13}\alpha_1\end{pmatrix}.$$

Let

$$f_7(g_{\mathfrak{d}})=\sum_{\alpha_1\in k^\times,\alpha_2\in k}|f_{4,\alpha}(g_{\mathfrak{d}})|,\ f_8(g_{\mathfrak{d}})=\sum_{\alpha_1=0,\alpha_2\in k^\times}|f_{4,\alpha}(g_{\mathfrak{d}})|.$$

Lemma (10.1.7) *Suppose that μ_1,μ_2 are bounded below. Then we have the following estimates.*
(1) *For any $N\geq 1$, $f_7(g_{\mathfrak{d}})\ll \mathrm{rd}_{3,N}(\lambda_1,\mu_1,\mu_2)$.*
(2) *For any $N\geq 0$, $f_8(g_{\mathfrak{d}})\ll \lambda_1^{-\frac{13}{2}-2N}\sup(1,\lambda_1^{-1}\mu_1\mu_2)\mu_1^{\frac12}\mu_2^{\frac12}$.*

Proof. By (1.2.8), for any $N_1,N_2,N_3\geq 1$,

$$\begin{aligned}f_7(g_{\mathfrak{d}})\ll&\lambda_1^{-8}\sup(1,\lambda_1^{-1}\mu_1\mu_2)\sup(1,\lambda_1^4\mu_2)\\ &\times(\lambda_1\mu_1\mu_2^{-1})^{-N_1}(\lambda_1\mu_1^{-1}\mu_2)^{-N_2}(\lambda_1^{-4}\mu_2)^{-N_3}.\end{aligned}$$

We choose $N_1=N_4+\frac12 N_3$, and $N_2=N_4$, where $N_3,N_4\geq 1$. Then

$$f_7(g_{\mathfrak{d}})\ll\lambda_1^{-8}\sup(1,\lambda_1^{-1}\mu_1\mu_2)\sup(1,\lambda_1^4\mu_2)\lambda_1^{-2N_4}(\lambda_1^{-\frac72}\mu_1^{\frac12}\mu_1^{\frac12})^{-N_3}.$$

Since μ_1,μ_2 are bounded below, we get (1).
Consider $f_8(g_{\mathfrak{d}})$. By (1.2.8), for any $N_1,N_2,N_3\geq 1$,

$$f_8(g_{\mathfrak{d}})\ll\lambda_1^{-8}\sup(1,\lambda_1^{-1}\mu_1\mu_2)(\lambda_1\mu_1\mu_2^{-1})^{-N_1}(\lambda_1\mu_1^{-1}\mu_2)^{-N_2}(\lambda_1^{-4}\mu_2^{-1})^{-N_3}.$$

We choose $N_1=1+N, N_2=\frac32+N, N_3=1$, where $N\geq 0$. Then

$$f_8(g_{\mathfrak{d}})\ll\lambda_1^{-\frac{13}{2}-2N}\sup(1,\lambda_1^{-1}\mu_1\mu_2)\mu_1^{\frac12}\mu_2^{\frac12}.$$

This proves (2).

Q.E.D.

We continue the proof of (10.1.1). By the consideration as in §2.1,

$$\sum_{\alpha\in k^2\setminus\{0\}}|f_{2,\alpha}(g_{\mathfrak{d}},w)|\ll\sum\widetilde{\mathscr{E}}_{I,\tau,\nu}(g_{\mathfrak{d}},w),$$

$$\sum_{\alpha_1=0,\alpha_2\in k^\times}|f_{2,\alpha}(g_{\mathfrak{d}},w)|\ll\sum\widetilde{\mathscr{E}}_{I,\tau,1}(g_{\mathfrak{d}},w),$$

where $I=(I_1,I_2)$ and either I_1 or I_2 is non-empty. Therefore, there exists $\delta>0$ such that if $M>w_0$, $\sum_{\alpha\in k^2\setminus\{0\}}|f_{2,\alpha}(g_{\mathfrak{d}},w)|$ is bounded by a slowly increasing function for $w_0-\delta\le \mathrm{Re}(w)\le M$. This implies that

$$\sum_{\alpha\in k^2\setminus\{0\}}|f_{3,\alpha}(g_{\mathfrak{d}})f_{2,\alpha}(g_{\mathfrak{d}})|t(g_{\mathfrak{d}})^{-2\rho_{\mathfrak{d}}},\quad \sum_{\alpha_1\in k^\times,\alpha_2\in k}|f_{4,\alpha}(g_{\mathfrak{d}})f_{2,\alpha}(g_{\mathfrak{d}})|t(g_{\mathfrak{d}})^{-2\rho_{\mathfrak{d}}}$$

are integrable for $g_{\mathfrak{d}}$ in the Siegel domain.

By (9.2.10), there exists $\delta>0$ such that if $M>w_0$,

$$\widetilde{\mathscr{E}}_{I,\tau,1}(g_{\mathfrak{d}},w)\ll \sup(\lambda_1^{c_1}\mu_1^{c_2}\mu_2^{c_3},\lambda_1^{2+c_1'}\mu_1^{c_2'}\mu_2^{c_3'}),$$

where $c_1\ge 8, c_2\le -4$ and $|c_3|,|c_1'|,|c_2'|,|c_3'|$ are small. Note that we are restricting ourselves to the Siegel domain, so we can assume that $|c_3|,|c_3'|$ are small.

It is easy to see that $t(g_{\mathfrak{d}})^{-2\rho}=\lambda_1^6\mu_1^{-2}\mu_2^{-2}$. Therefore, for any $N\ge 0$,

$$t(g_{\mathfrak{d}})^{-2\rho}f_8(g_{\mathfrak{d}})\sum_{\alpha_1=0,\alpha_2\in k^\times}|f_{2,-\alpha}(g_{\mathfrak{d}},w)|$$

is bounded by a constant multiple of

$$\sup(1,\lambda_1^{-1}\mu_1\mu_2)\sup(\lambda_1^{-\frac{1}{2}-2N+c_1}\mu_1^{-\frac{3}{2}+c_2}\mu_2^{-\frac{3}{2}+c_3},\lambda_1^{\frac{3}{2}-2N+c_1'}\mu_1^{-\frac{3}{2}+c_2'}\mu_2^{-\frac{3}{2}+c_3'}).$$

If $\lambda_1\ge 1$, we choose $N\gg 0$, and if $\lambda_1\le 1$, we choose $N=0$. Hence,

$$\int_{A_{\mathfrak{d}}M^1_{\mathfrak{d}\mathbb{A}}/M_{\mathfrak{d}k}}\omega_{\mathfrak{p}}(g_{\mathfrak{d}})\sum_{\alpha\in k^2\setminus\{0\}}f_{1,\alpha}(g_{\mathfrak{d}})f_{2,-\alpha}(g_{\mathfrak{d}})t(g_{\mathfrak{d}})^{-2\rho}dg_{\mathfrak{d}}\sim 0.$$

This proves (10.1.1).

Q.E.D.

§10.2 The cases $\mathfrak{d}=(\beta_5),(\beta_{10},\beta_{10,1})$

Let $\mathfrak{d}_1=(\beta_5)$, $\mathfrak{d}_2=(\beta_{10},\beta_{10,1})$, and $\mathfrak{d}_3=(\beta_{10})$. We consider paths $\mathfrak{p}_1=(\mathfrak{d}_1,\mathfrak{s}_1),\mathfrak{p}_2=(\mathfrak{d}_2,\mathfrak{s}_2)$, and $\mathfrak{p}_3=(\mathfrak{d}_3,\mathfrak{s}_3)$ such that $\mathfrak{s}_1(1)=\mathfrak{s}_2(1)=\mathfrak{s}_3(1)$ (i.e. $\mathfrak{p}_3\prec\mathfrak{p}_2$). For $\mathfrak{d}=\mathfrak{d}_1,\mathfrak{d}_2$, $M_{\mathfrak{d}}=T$. Therefore, we use the notation $g_{\mathfrak{d}}=t^0=(t_1,t_2)$ where $t_1\in T^0_{1\mathbb{A}},t_2\in T^0_{2\mathbb{A}}$.

We first consider $\mathfrak{p}_1$. Let $\mathfrak{p}=\mathfrak{p}_1,\mathfrak{d}=\mathfrak{d}_1$.

Proposition (10.2.1) *The distribution* $\widetilde{\Xi}_{\mathfrak{p}}(\Phi,\omega,w)$ *is well defined for* $\mathrm{Re}(w)\gg 0$, *and* $\widetilde{\Xi}_{\mathfrak{p}}(\Phi,\omega,w)\sim 0$.

Proof. Let $\Psi=\widetilde{R}_{\mathfrak{d}}\Phi_{\mathfrak{p}}$, and $n(u_{\mathfrak{d}})=n_B(u_{\mathfrak{d}})$ for $u_{\mathfrak{d}}=(u_1,u_2,u_3,u_4)\in\mathbb{A}^4$. It is easy to see that $\Theta_{Y_{\mathfrak{d}}}(\Psi,t^0n(u_{\mathfrak{d}}))$ does not depend on u_1,u_2. For $\alpha=(\alpha_1,\alpha_2)\in k^2$, let $<\alpha\cdot u_{\mathfrak{d}}>=<\alpha_1u_3+\alpha_2u_4>$.

We define

$$f_{1,\alpha}(t^0)=\int_{(\mathbb{A}/k)^4}\Theta_{Y_{\mathfrak{d}}}(\Psi,t^0n(u_{\mathfrak{d}}))<\alpha\cdot u_{\mathfrak{d}}>du_{\mathfrak{d}},\tag{10.2.2}$$

$$f_{2,\alpha}(t^0,w)=\int_{(\mathbb{A}/k)^4}\mathscr{E}(t^0n(u_{\mathfrak{d}}),w)<\alpha\cdot u_{\mathfrak{d}}>du_{\mathfrak{d}}.$$

By the Parseval formula,

$$\int_{(\mathbb{A}/k)^4} \Theta_{Y_\mathfrak{d}}(\Psi, t^0 n(u_\mathfrak{d}))\mathscr{E}(t^0 n(u_\mathfrak{d}), w) du_\mathfrak{d} = \sum_{\alpha \in k^2} f_{1,\alpha}(t^0) f_{2,-\alpha}(t^0, w).$$

We can write any element $y \in Y^{\rm ss}_{\mathfrak{d}k}$ as $y = (y_1, y_2)$ where $y_1 = (x_{1,33}, x_{2,23}) \in Z^{\rm ss}_{\mathfrak{d}k}, y_2 = x_{2,33} \in W_{\mathfrak{d}k}$. Then

$$f_{1,\alpha}(t^0) = \sum_{y_1 \in Z^{\rm ss}_{\mathfrak{d}k}} \int_{(\mathbb{A}/k)^4} \sum_{y_2 \in k} \Psi(t^0 n(u_\mathfrak{d}) y) < \alpha \cdot u_\mathfrak{d} > du_\mathfrak{d}.$$

An easy computation shows that

$$t^0 n(u_\mathfrak{d}) y = (\gamma_{1,33}(t^0) x_{1,33}, \gamma_{2,23}(t^0) x_{2,23}, \gamma_{2,33}(t^0)(x_{2,33} + x_{1,33} u_4 + x_{2,23} u_3)).$$

Let $u_4' = x_{1,33} u_4 + x_{2,23} u_3$, $u_i' = u_i$ for $i = 1, 2, 3$. Then $u_4 = x_{1,33}^{-1}(u_4' - x_{2,23} u_3')$. We define

$$\Psi'(t^0, u_4', y) = \Psi(\gamma_{1,33}(t^0) x_{1,33}, \gamma_{2,23}(t^0) x_{2,23}, \gamma_{2,33}(t^0)(x_{2,33} + u_4')). \tag{10.2.3}$$

Since $t^0 n(u_\mathfrak{d}) y$ does not depend on u_1', u_2',

$$f_{1,\alpha}(t^0) = \sum_{y_1 \in Z^{\rm ss}_{\mathfrak{d}k}} \int_{(\mathbb{A}/k)^2} \sum_{y_2 \in k} \Psi'(t^0, u_4', y) < \alpha_1 u_3' + \alpha_2 x_{1,33}^{-1}(u_4' - x_{2,23} u_3') > du_3' du_4'.$$

Since

$$< \alpha_1 u_3' + \alpha_2 x_{1,33}^{-1}(u_4' - x_{2,23}, u_3') > = < (\alpha_1 - \alpha_2 x_{1,33}^{-1} x_{2,23}) u_3' + \alpha_2 x_{1,33}^{-1} u_4' >,$$

the above integral is 0 unless $x_{1,33} = \alpha_1^{-1}\alpha_2 x_{2,23}$. Therefore,

$$f_{1,\alpha}(t^0) = \sum_{\substack{y_1 \in Z^{\rm ss}_{\mathfrak{d}k} \\ x_{1,33} = \alpha_1^{-1}\alpha_2 x_{2,23}}} \int_{\mathbb{A}} \Psi'(t^0, u_4', y_1) < \alpha_2 x_{1,33}^{-1} u_4' > du_4'.$$

Let Ψ_1 be the partial Fourier transform of Ψ with respect to the coordinate $x_{2,33}$ and $< >$. Then $f_{1,\alpha}(t^0)$ is equal to

$$|\gamma_{2,33}(t^0)|^{-1} \sum_{\substack{y_1 \in Z^{\rm ss}_{\mathfrak{d}k} \\ x_{1,33} = \alpha_1^{-1}\alpha_2 x_{2,23}}} \Psi_1(\gamma_{1,33}(t^0) x_{1,33}, \gamma_{2,23}(t^0) x_{2,23}, \gamma_{2,33}(t^0)^{-1} \alpha_2 x_{1,33}^{-1}).$$

Since $x_{1,33}, x_{2,23} \neq 0$ for $y_1 \in Z^{\rm ss}_{\mathfrak{d}k}$, the above integral is 0 unless $\alpha_1, \alpha_2 \neq 0$ or $\alpha_1 = \alpha_2 = 0$. We choose $0 \leq \Psi_2 \in \mathscr{S}(Y_{\mathfrak{d}\mathbb{A}})$ so that $|\Psi_1| \leq \Psi_2$. Then

$$\sum_{\alpha \in k^2 \setminus \{0\}} |f_{1,\alpha}(t^0)| \leq |\gamma_{2,33}(t^0)|^{-1} \Theta_3(\Psi_2, \gamma_{1,33}(t^0), \gamma_{2,23}(t^0), \gamma_{2,33}(t^0)^{-1}).$$

Let $\mu_1 = |\gamma_{1,33}(t^0)|, \mu_2 = |\gamma_{2,23}(t^0)|, \mu_3 = |\gamma_{2,33}(t^0)|^{-1}$, and

$$d(\mu_1, \mu_2, \mu_3) = a(\underline{\mu}_1^{-\frac{1}{2}} \underline{\mu}_2^{-1} \underline{\mu}_3^{-\frac{1}{2}}, \underline{\mu}_1^{\frac{1}{4}} \underline{\mu}_2 \underline{\mu}_3^{\frac{3}{4}}, \underline{\mu}_1^{\frac{1}{4}} \underline{\mu}_3^{-\frac{1}{4}}; \underline{\mu}_1^{\frac{1}{2}} \underline{\mu}_3^{\frac{1}{2}}, \underline{\mu}_1^{-\frac{1}{2}} \underline{\mu}_3^{-\frac{1}{2}}). \tag{10.2.4}$$

Then t^0 can be written in the form $t^0 = d(\mu_1, \mu_2, \mu_3)\widehat{t}^{\,0}$.
By (1.2.6), for any $N_1, N_2, N_3 \geq 1$,

$$\sum_{\alpha \in k^2 \setminus \{0\}} |f_{1,\alpha}(t^0)| \ll \mu_1^{-N_1} \mu_2^{-N_2} \mu_3^{1-N_3}. \tag{10.2.5}$$

By the consideration as in §2.1,

$$\sum_{\alpha \in (k^\times)^2} |f_{2,\alpha}(t^0, w)| \leq \sum_{\tau} \widetilde{\mathscr{E}}_{I,\tau,1}(t^0, w)$$

where $I_1 = \{2\}, I_2 = \{1\}$.

Lemma (10.2.6) *Suppose that $I_1 = \{2\}$ and $I_2 = \{1\}$.*
(1) If $\tau_1 = (2,3)$ or $(1,2,3)$, there exists a constant $\delta > 0$ such that for any $N > 0, M > w_0$ there exist $c_1, c_2, c_3 > N$ satisfying

$$\widetilde{\mathscr{E}}_{I,\tau,1}(t^0, w) \ll \mu_1^{c_1} \mu_2^{c_2} \mu_3^{c_3}$$

for $w_0 - \delta \leq \mathrm{Re}(w) \leq M$.
(2) If $\tau_1 = (1,3)$, there exists a constant $\delta > 0$ such that if $M > w_0$,

$$\widetilde{\mathscr{E}}_{I,\tau,1}(t^0, w) \ll \mu_1^{\frac{11}{4}} \mu_2^{7} \mu_3^{\frac{3}{4}}$$

for $w_0 - \delta \leq \mathrm{Re}(w) \leq M$.

Proof. Let $q \in D_{I,\tau}$. Let r_1, r_2, r_3 be as in (9.2.5). By (3.4.30), for any $\delta > 0, M > w_0, l_{11}, l_{21} > 1$,

$$\widetilde{\mathscr{E}}_{I,\tau,1}(t^0, w) \ll (\mu_2\mu_3)^{-l_{11}} (\mu_1\mu_3)^{\frac{1}{2} - \frac{1}{2}r_3 - l_{21}} \times \begin{cases} t_1^{(q_{11}, q_{13}, q_{12}) + \rho_1} & \tau_1 = (2,3), \\ t_1^{(q_{12}, q_{13}, q_{11}) + \rho_1} & \tau_1 = (1,2,3), \\ t_1^{(q_{13}, q_{12}, q_{11}) + \rho_1} & \tau_1 = (1,3) \end{cases}$$

for $L(q) + \delta \leq \mathrm{Re}(w) \leq M$.

An easy computation shows that $t_1^{\rho_1} = \mu_1^{-\frac{3}{4}} \mu_2^{-1} \mu_3^{-\frac{1}{4}}$. We fix $r_3 = \frac{1}{2}$, $l_{11} = l_{21} = 2$. Then

$$\widetilde{\mathscr{E}}_{I,\tau,1}(t^0, w) \ll \mu_1^{-\frac{5}{2}} \mu_2^{-3} \mu_3^{-4} \times \begin{cases} \mu_1^{-\frac{1}{2}r_1 - \frac{1}{4}r_2} \mu_2^{-r_1 - r_2} \mu_3^{-\frac{1}{2}r_1 - \frac{3}{4}r_2} & \tau_1 = (2,3), \\ \mu_1^{\frac{1}{4}r_1 - \frac{1}{4}r_2} \mu_2^{-r_2} \mu_3^{-\frac{1}{4}r_1 - \frac{3}{4}r_2} & \tau_1 = (1,2,3), \\ \mu_1^{\frac{1}{4}r_1 + \frac{1}{2}r_2} \mu_2^{r_2} \mu_3^{-\frac{1}{4}r_1 + \frac{1}{2}r_2} & \tau_1 = (1,3). \end{cases}$$

In the first case, we fix r_2 and choose $r_1 \ll 0$. In the second case, we choose $r_1 = -r_2 + \delta \gg 0$ where $\delta > 0$ is a small number. Then it is easy to see that the statement of (1) is satisfied. In the third case, By choosing q so that $r_2 = 10,\ r_1 = 1,$

we get the estimate of (2). Since $L(q) = 22 + \frac{1}{2}C$, $w_0 = 4 + C$, and $C > 100$, $L(q) < w_0$. This proves (2).

Q.E.D.

By (10.2.5), (10.2.6), for any $N, N_1, N_2, N_3 \geq 1$,

$$\sum_{\alpha \in k^2 \setminus \{0\}} |f_{1,\alpha}(t^0) f_{2,-\alpha}(t^0, w)|(t^0)^{-2\rho} \ll \mathrm{rd}_{3,N}(\mu_1, \mu_2, \mu_3) + \mu_1^{\frac{17}{4}-N_1} \mu_2^{9-N_2} \mu_3^{\frac{9}{4}-N_3}.$$

Therefore,

$$\int_{T_{\mathbb{A}}^0/T_k} \omega_{\mathfrak{p}}(t^0) \sum_{\alpha \in k^2 \setminus \{0\}} f_{1,\alpha}(t^0) f_{2,-\alpha}(t^0, w)(t^0)^{-2\rho} d^\times t^0 \sim 0.$$

This proves (10.2.1) (We have also showed the absolute convergence for $\mathrm{Re}(w) \gg 0$).

Q.E.D.

Next, we consider $\mathfrak{p}_2$ such that $\mathfrak{s}_2(2) = 0$.

Proposition (10.2.7) $\widetilde{\Xi}_{\mathfrak{p}_2}(\Phi, \omega, w) \sim 0$.

Proof. Let $\Psi = \widetilde{R}_{\mathfrak{d}_2} \Phi_{\mathfrak{p}_2}$. Let $n_{\mathfrak{d}_2}(u_3, u_4) = (n''(u_3), n_{B_2}(u_4))$ for $u_3, u_4 \in \mathbb{A}$. Then

$$\int_{(\mathbb{A}/k)^2} \mathscr{E}_{\mathfrak{p}_3}(t^0 n_{\mathfrak{d}_2}(u_3, u_4), w) < \alpha_1 u_3 + \alpha_4 u_4 > du_3 du_4 = f_{2,\alpha}(t^0, w).$$

Since $e_{\mathfrak{p}_2 1}(t^0) = (t^0)^{\beta_{10}} = \mu_1^{\frac{1}{2}} \mu_2 \mu_3^{\frac{1}{2}}$ and $\mathfrak{s}_2(1) = \mathfrak{s}_1(1)$, we are considering the integral

$$\int_{\substack{T_{\mathbb{A}}^0/T_k \\ \mu_1 \mu_2^2 \mu_3 \leq 1}} \omega_{\mathfrak{p}_2}(t^0) \sum_{\alpha \in k^2 \setminus \{0\}} f_{1,\alpha}(t^0) f_{2,-\alpha}(t^0, w)(t^0)^{-2\rho} d^\times t^0.$$

But we have already estimated this integral in a larger domain. Therefore,

$$\widetilde{\Xi}_{\mathfrak{p}_2}(\Phi, \omega, w) \sim 0.$$

Q.E.D.

Next we consider $\mathfrak{p}_2$ such that $\mathfrak{s}_2(2) = 1$.

Proposition (10.2.8) $\widetilde{\Xi}_{\mathfrak{p}_2}(\Phi, \omega, w) \sim 0$.

Proof. Let

$$\text{(10.2.9)} \quad f_{1,\alpha}(t^0) = \int_{(\mathbb{A}/k)^2} \Theta_{Y_{\mathfrak{d}_2}}(\Psi, t^0 \theta_{\mathfrak{d}_3}(n_{\mathfrak{d}_2}(u_3, u_4))) < \alpha_1 u_3 + \alpha_2 u_4 > du_3 du_4,$$

$$f_{2,\alpha}(t^0, w) = \int_{(\mathbb{A}/k)^2} \mathscr{E}_{\mathfrak{p}_3}(\theta_{\mathfrak{d}_3}(t^0) n_{\mathfrak{d}_2}(u_3, u_4), w) < \alpha_1 u_3 + \alpha_4 u_4 > du_3 du_4.$$

Then, by the Parseval formula, $\widetilde{\Xi}_{\mathfrak{p}_2}(\Phi, \omega, w)$ is equal to

$$\int_{\substack{T_{\mathbb{A}}^0/T_k \\ \mu_1 \mu_2^2 \mu_3 \geq 1}} \omega_{\mathfrak{p}_2}(t^0) \sum_{\alpha \in k^2 \setminus \{0\}} f_{1,\alpha}(t^0) f_{2,-\alpha}(t^0, w)((t^0)^{-2\rho_{\mathfrak{d}_3}} \kappa_{\mathfrak{d}_3 1}(t^0))^{-1}(t^0)^{-2\rho_{\mathfrak{d}_2}} d^\times t^0.$$

We can estimate $\sum_{\alpha\in k^2\setminus\{0\}} |f_{1,\alpha}(t^0)|$ in exactly the same way as in (10.2.5). Let $\mu_4 = \mu_1\mu_2^2\mu_3$. Then

$$\mu_1^{-N_1}\mu_2^{-N_2}\mu_3^{-N_3} = \mu_1^{\frac{N_2}{2}-N_1}\mu_4^{-\frac{N_2}{2}}\mu_3^{\frac{N_2}{2}-N_3}.$$

Therefore, for any $N > 0$,

$$\sum_{\alpha\in k^\times} |f_{1,\alpha}(t^0)| \ll \mu_4^{-N}\mathrm{rd}_{2,N}(\mu_1,\mu_2).$$

Since $\sum_{\alpha\in k^\times} |f_{2,-\alpha}(t^0,w)|$ is a slowly increasing function, well defined for $\mathrm{Re}(w) \geq w_0 - \delta$ for some $\delta > 0$,

$$\widetilde{\Xi}_{\mathfrak{p}_2}(\Phi,\omega,w) \sim 0.$$

Q.E.D.

§10.3 The cases $\mathfrak{d} = (\beta_6), (\beta_8, \beta_{8,1})$

In this section, we consider the cases $\mathfrak{d}_1 = (\beta_6)$, $\mathfrak{d}_2 = (\beta_8, \beta_{8,1})$. Let $\mathfrak{d}_3 = (\beta_8)$. Consider paths $\mathfrak{p}_1 = (\mathfrak{d}_1, \mathfrak{s}_1), \mathfrak{p}_2 = (\mathfrak{d}_2, \mathfrak{s}_2)$, and $\mathfrak{p}_3 = (\mathfrak{d}_3, \mathfrak{s}_3)$ such that $\mathfrak{s}_1(1) = \mathfrak{s}_2(1) = \mathfrak{s}_3(1)$.

We first consider $\mathfrak{d} = \mathfrak{d}_1, \mathfrak{s} = \mathfrak{s}_1$. In this case, $P_{\mathfrak{d}} = P_1 \times B_2$. The proof of the following lemma is easy, and is left to the reader.

Lemma (10.3.1) *The convex hull of each of the sets*
(1) $\{\gamma_{1,33}, \gamma_{2,11}, \gamma_{2,22}, -\gamma_{2,33}\}$,
(2) $\{\gamma_{1,33}, \gamma_{2,11}, \gamma_{2,22}, -\gamma_{2,13}\}$,
(3) $\{\gamma_{1,33}, \gamma_{2,11}, \gamma_{2,22}, -\gamma_{2,23}\}$,
(4) $\{\gamma_{1,33}, \gamma_{2,12}, -\gamma_{2,13}, -\gamma_{2,23}\}$
contains a neighborhood of the origin.

For $u_{\mathfrak{d}} = (u_2, u_3, u_4) \in \mathbb{A}^3$, let $n(u_{\mathfrak{d}}) = (n_{P_1}(u_2, u_3), n_{B_2}(u_4))$. Let

$$g_1 = \begin{pmatrix} g_{1,12} & 0 \\ 0 & t_{13} \end{pmatrix},\ t_2 = \begin{pmatrix} t_{21} & 0 \\ 0 & t_{22} \end{pmatrix},$$

where $|\det g_{1,12}| = |t_{13}| = |t_{21}| = |t_{22}| = 1$. Let $g_{\mathfrak{d}} = d_{P_1}(\lambda_1)a_{\mathfrak{s}_6}(\underline{\mu}_2)(g_1, t_2)$, where $a_{\mathfrak{s}_6}(\underline{\mu}_2)$ is as in §1.2. Let $g_{1,12}, u_1$ etc. be as in (10.1.4), and $g_{\mathfrak{d}} = kt^0n$ the Iwasawa decomposition of $g_{\mathfrak{d}}$. We can write any element $y \in Y_{\mathfrak{d}k}^{\mathrm{ss}}$ in the form $y = (y_1, y_2)$ where $y_1 = (x_{1,33}, x_{2,11}, x_{2,12}, x_{2,22}) \in Z_{\mathfrak{d}k}^{\mathrm{ss}}$ and $y_2 = (x_{2,13}, x_{2,23}, x_{2,33}) \in W_{\mathfrak{d}k}$. Let $\Psi = \widetilde{R}_{\mathfrak{d}}\Phi_{\mathfrak{p}}$. Then

$$\Theta_{Y_{\mathfrak{d}}}(\Psi, g_{\mathfrak{d}}n(u_{\mathfrak{d}})) = \Theta_{Y_{\mathfrak{d}}^s}(\Psi, g_{\mathfrak{d}}n(u_{\mathfrak{d}})) + \Theta_{Y_{\mathfrak{d},\mathrm{st}}}(\Psi, g_{\mathfrak{d}}n(u_{\mathfrak{d}})).$$

We recall that these are theta series associated with subsets of $\{(y_1, y_2) \in Z_{\mathfrak{d}k}, y_2 \in W_{\mathfrak{d}k}\}$, where $y_1 \in Z_{\mathfrak{d}k}^s, Z_{\mathfrak{d},\mathrm{st}k}^{\mathrm{ss}}$ respectively. By assumption,

$$\Theta_{Y_{\mathfrak{d}}}(\Psi, g_{\mathfrak{d}}n(u_{\mathfrak{d}})) = \omega^{\pm 1}(k)\Theta_{Y_{\mathfrak{d}}}(\Psi, t^0nn(u_{\mathfrak{d}})).$$

So when we estimate various functions, we assume that $k = 1$.

Let

$$<\alpha\cdot u_{\mathfrak{d}}> = <\alpha_1 u_2+\alpha_2 u_3+\alpha_3 u_4>.$$

We define

$$f_{1,\alpha}(g_{\mathfrak{d}}) = \int_{(\mathbb{A}/k)^3} \Theta_{Y_{\mathfrak{d}}^s}(\Psi, g_{\mathfrak{d}} n(u_{\mathfrak{d}})) <\alpha\cdot u_{\mathfrak{d}}> du_{\mathfrak{d}}, \tag{10.3.2}$$

$$f_{2,\alpha}(g_{\mathfrak{d}}, w) = \int_{(\mathbb{A}/k)^3} \mathscr{E}(g_{\mathfrak{d}} n(u_{\mathfrak{d}}), w) <\alpha\cdot u_{\mathfrak{d}}> du_{\mathfrak{d}}.$$

Proposition (10.3.3)

$$\int_{A_{\mathfrak{d}} M^1_{\mathfrak{d}\mathbb{A}}/M_{\mathfrak{d}k}} \omega_{\mathfrak{p}}(g_{\mathfrak{d}}) \sum_{\alpha\in k^3\setminus\{0\}} f_{1,\alpha}(g_{\mathfrak{d}}) f_{2,-\alpha}(g_{\mathfrak{d}}, w) t(g_{\mathfrak{d}})^{-2\rho_{\mathfrak{d}}} dg_{\mathfrak{d}} \sim 0.$$

Proof. We define

$$Q_{y_1}(u_2,u_3) = x_{2,11}u_2^2 + x_{2,12}u_2u_3 + x_{2,22}u_3^2,\ l_{y_2}(u_2,u_3) = x_{2,13}u_2 + x_{2,23}u_3.$$

Then

$$n(u_{\mathfrak{d}})y = (y_1, x_{2,13}+\partial_{u_2}Q_{y_1}, x_{2,23}+\partial_{u_3}Q_{y_1}, x_{2,33}+l_{y_2}+Q_{y_1}+x_{1,33}u_4),$$

where ∂_{u_2} etc. are the partial derivatives. Let $u_2'=\partial_{u_2}Q_{y_1}, u_3'=\partial_{u_3}Q_{y_1}$. Then

$$\begin{pmatrix} u_2' \\ u_3' \end{pmatrix} = \begin{pmatrix} 2x_{2,11} & x_{2,12} \\ x_{2,12} & 2x_{2,22} \end{pmatrix} \begin{pmatrix} u_2 \\ u_3 \end{pmatrix}.$$

So if $\Delta(y_1)$ is the determinant of the above matrix,

$$\begin{pmatrix} u_2 \\ u_3 \end{pmatrix} = \Delta(y_1)^{-1} \begin{pmatrix} 2x_{2,22} & -x_{2,12} \\ -x_{2,12} & 2x_{2,11} \end{pmatrix} \begin{pmatrix} u_2' \\ u_3' \end{pmatrix}.$$

Note that if $y_1\in Z^{\mathrm{ss}}_{\mathfrak{d}k}$, $\Delta(y_1)\neq 0$.

As before,

$$f_{1,\alpha}(g_{\mathfrak{d}}) = \sum_{y_1\in Z^s_{\mathfrak{d}k}} \int_{\mathbb{A}^3} \Psi(g_{\mathfrak{d}} n(u_{\mathfrak{d}})y_1) <\alpha\cdot u_{\mathfrak{d}}> du_{\mathfrak{d}}.$$

Let

$$\begin{aligned} A_{2,13} &= A_{2,13}(g_{\mathfrak{d}}, y_1, u_1, u_{\mathfrak{d}}) = \gamma_{2,13}(t^0)\partial_{u_2}Q_{y_1}, \\ A_{2,23} &= A_{2,13}(g_{\mathfrak{d}}, y_1, u_1, u_{\mathfrak{d}}) = \gamma_{2,23}(t^0)(\partial_{u_3}Q_{y_1} + u_1(x_{2,13}+\partial_{u_2}Q_{y_1})), \\ A_{2,33} &= A_{2,13}(g_{\mathfrak{d}}, y_1, u_1, u_{\mathfrak{d}}) = \gamma_{2,33}(t^0)(l_{y_2}+Q_{y_1}+x_{1,33}u_4). \end{aligned}$$

Then

$$f_{1,\alpha}(g_{\mathfrak{d}}) = \sum_{y_1\in Z^s_{\mathfrak{d}k}} \int_{\mathbb{A}^3} \Psi(g_{\mathfrak{d}} y_1, A_{2,13}, A_{2,23}, A_{2,33}) <\alpha\cdot u_{\mathfrak{d}}> du_{\mathfrak{d}}.$$

Lemma (10.3.4) *There exists a constant $\delta > 0$ such that for any $N > 0, M > w_0$,*

$$\sum_{\alpha \in k^3 \setminus \{0\}} |f_{1,\alpha}(g_{\mathfrak{d}}) f_{2,\alpha}(g_{\mathfrak{d}}, w)| \ll \mathrm{rd}_{3,N}(\lambda_1, \mu_1, \mu_2)$$

for $w_0 - \delta \leq \mathrm{Re}(w) \leq M$.

Proof. Let $\Psi_1 \in \mathscr{S}(Y_{\mathfrak{d}\mathbb{A}})$ be the partial Fourier transform with respect to the coordinate $x_{2,33}$ and $< >$. Then

$$\begin{aligned} &\int_{\mathbb{A}} \Psi(g_{\mathfrak{d}} n(u_{\mathfrak{d}}) y_1) < \alpha_3 u_4 > du_4 \\ &= \Psi_1(g_{\mathfrak{d}} y_1, A_{2,13}, A_{2,23}, \gamma_{2,33}(t^0)^{-1} \alpha_3 x_{1,33}^{-1}) < -\alpha_3(l_{y_2} + Q_{y_1}) > . \end{aligned}$$

We choose $0 \leq \Psi_2 \in \mathscr{S}(Z_{\mathfrak{d}\mathbb{A}} \times \mathbb{A})$ so that

$$\int_{\mathbb{A}^2} |\Psi_1(y_1, x_{2,13}, x_{2,23}, x_{1,33})| dx_{2,13} dx_{2,23} \ll \Psi_2(y_1, x_{1,33}).$$

Then

$$\sum_{\alpha_3 \in k^\times} |f_{1,\alpha}(g_{\mathfrak{d}})| \ll \lambda_1^{-6} \mu_2^3 \sum_{y_1 \in Z_{\mathfrak{d}k}^s} \Psi_2(g_{\mathfrak{d}} y_1, \gamma_{2,33}(t^0)^{-1} \alpha_3 x_{1,33}^{-1}).$$

Note that this bound does not depend on α_1, α_2. By (10.3.1), for any $N > 0$,

$$\sum_{\alpha_3 \in k^\times} |f_{1,\alpha}(g_{\mathfrak{d}})| \ll \mathrm{rd}_{3,N}(\lambda_1, \mu_1, \mu_2).$$

By the consideration as in §2.1,

$$\sum_{\substack{\alpha_1, \alpha_2 \in k \\ \alpha_3 \in k^\times}} |f_{2,\alpha}(g_{\mathfrak{d}}, w)| \ll \sum_{I, \tau, \nu} \widetilde{\mathscr{E}}_{I,\tau,\nu}(g_{\mathfrak{d}}, w),$$

where $I = (I_1, I_2)$ and $I_2 = \{1\}$.

Therefore, there exists $\delta > 0$ such that if $M > w_0$, $\sum_{\alpha_1, \alpha_2 \in k, \alpha_3 \in k^\times} |f_{2,\alpha}(g_{\mathfrak{d}}, w)|$ is bounded by a slowly increasing function for $w_0 - \delta \leq \mathrm{Re}(w) \leq M$. So for any $N > 0$,

$$\sum_{\substack{\alpha_1, \alpha_2 \in k \\ \alpha_3 \in k^\times}} |f_{1,\alpha}(g_{\mathfrak{d}}) f_{2,\alpha}(g_{\mathfrak{d}}, w)| \ll \mathrm{rd}_{3,N}(\lambda_1, \mu_1, \mu_2) \tag{10.3.5}$$

for $w_0 - \delta \leq \mathrm{Re}(w) \leq M$.

We consider α such that $\alpha_3 = 0$. Let

$$\alpha_1' = \Delta(y_1)^{-1}(2\alpha_1 x_{2,22} - \alpha_2 x_{2,12}), \; \alpha_2' = \Delta(y_1)^{-1}(-\alpha_1 x_{2,12} + 2\alpha_2 x_{2,11}).$$

Then $\alpha_1 u_2 + \alpha_2 u_3 = \alpha_1' u_2' + \alpha_2' u_3'$. The condition $\alpha \neq 0$ is equivalent to the condition $\alpha' = (\alpha_1', \alpha_2') \neq 0$.

We define

$$\begin{aligned}&\Psi_3(y_1, x_{2,13}, x_{2,23})\\&= \int_{\mathbb{A}^2} \Psi(y_1, x'_{2,13}, x'_{2,23}, x'_{2,33}) < x_{2,13}x'_{2,13} + x_{2,23}x'_{2,23} > dx'_{2,13}dx'_{2,23}dx'_{2,33}.\end{aligned}$$

Then

$$f_{1,\alpha}(g_{\mathfrak{d}}) = |\gamma_{2,33}(t^0)|^{-1} \sum_{y_1 \in Z^s_{\mathfrak{d}k}} \Psi_3(g_{\mathfrak{d}}y_1, \gamma_{2,13}(t^0)^{-1}\alpha'_1 - \alpha'_2 u_1, \gamma_{2,23}(t^0)^{-1}\alpha'_2).$$

We choose $0 \le \Psi_4 \in \mathscr{S}(Z_{\mathfrak{d}\mathbb{A}}), 0 \le \Psi_5 \in \mathscr{S}(\mathbb{A}^2)$ so that

$$|\Psi_3(y_1, x_{2,13}, x_{2,23})| \le \Psi_4(y_1)\Psi_5(x_{2,13}, x_{2,23}).$$

By (1.2.6), for any $N_1, N_2, N_3 \ge 1$,

$$\begin{aligned}&\sum_{y_1 \in Z^s_{\mathfrak{d}k}} \Psi_4(g_{\mathfrak{d}}y_1) \ll |\gamma_{1,33}(t^0)|^{-N_1}|\gamma_{2,11}(t^0)|^{-N_2}|\gamma_{2,22}(t^0)|^{-N_3},\\&\sum_{\substack{\alpha'_1 \in k\\ \alpha'_2 \in k^\times}} \Psi_5(\gamma_{2,13}(t^0)^{-1}\alpha'_1 - \alpha'_2 u_1, \gamma_{2,23}(t^0)^{-1}\alpha'_2) \ll \sup(1, |\gamma_{2,13}(t^0)|)|\gamma_{2,23}(t^0)|^{N_1},\\&\sum_{\alpha'_1 \in k^\times} \Psi_5(\gamma_{2,13}(t^0)^{-1}\alpha'_1, 0) \ll |\gamma_{2,13}(t^0)|^{N_1}.\end{aligned}$$

Therefore, by (10.3.1), for any $N > 0$,

$$\sum_{\substack{(\alpha_1,\alpha_2)\in k^2\setminus\{0\}\\ \alpha_3=0}} |f_{1,\alpha}(g_{\mathfrak{d}})f_{2,-\alpha}(g_{\mathfrak{d}}, w)| \ll \mathrm{rd}_{3,N}(\lambda_1, \mu_1, \mu_2). \tag{10.3.6}$$

Hence, (10.3.4) follows from (10.3.5) and (10.3.6).

Q.E.D.

Now Proposition (10.3.3) is an easy consequence of (10.3.4).

Q.E.D.

Next, we consider

$$\int_{(\mathbb{A}/k)^3} \Theta_{Y_{\mathfrak{d},st}}(\Psi, g_{\mathfrak{d}}n(u_{\mathfrak{d}}))\mathscr{E}(g_{\mathfrak{d}}n(u_{\mathfrak{d}}), w)du_{\mathfrak{d}}.$$

Let

$$d(\lambda_1, \mu_1, \mu_2) = a(\underline{\lambda}_1^{-1}\underline{\mu}_1, \underline{\lambda}_1^{-1}\underline{\mu}_1^{-1}, \underline{\lambda}_1^2; \underline{\mu}_2, \underline{\mu}_2^{-1}), \tag{10.3.7}$$

and $t^0 = d(\lambda_1, \mu_1, \mu_2)\widehat{t}^0$. Let $g_{\mathfrak{d}} = n'(u_0)t^0$ for $u_0 \in \mathbb{A}$.

Let $H_{\mathfrak{d}}$ be the subgroup generated by T and the element $((1,2),1)$. We identify $Z_{\mathfrak{d}k}$ with the space $\mathrm{Sym}^2 k^2 \oplus k$ in Chapter 6. We define $Z'_{\mathfrak{d},0k}$ to be the subspace which corresponds to $Z'_{V,0k}$ in Chapter 6. We define $Z'^{ss}_{\mathfrak{d},0k}$ similarly. Then $Z^{ss}_{\mathfrak{d},stk} \cong M_{\mathfrak{d}k} \times_{H_{\mathfrak{d}k}} Z'^{ss}_{\mathfrak{d},0k}$. We define

$$(10.3.8) \qquad X_{\mathfrak{d}} = \{kn'(u_0)t^0 \mid k \in K \cap M_{\mathfrak{d}\mathbb{A}},\ \mu_1 \geq \sqrt{\alpha(u_0)}\}.$$

Then as in §4.2, if $f(g_{\mathfrak{d}})$ is a function on $G^0_{\mathbb{A}} \cap M_{\mathfrak{d}\mathbb{A}}/H_{\mathfrak{d}k}$,

$$\int_{G^0_{\mathbb{A}} \cap M_{\mathfrak{d}\mathbb{A}}/H_{\mathfrak{d}k}} f(g_{\mathfrak{d}})dg_{\mathfrak{d}} = \int_{X_{\mathfrak{d}} \cap B_{\mathbb{A}}/T_k} f(g_{\mathfrak{d}})(t^0)^{-2\rho}\mu_1^2 d^\times t^0 du_0.$$

We define

$$(10.3.9) \quad \widetilde{\Theta}_{Z'_{\mathfrak{d},0}}(\Psi, g_{\mathfrak{d}}n(u_{\mathfrak{d}})) = \sum_{\substack{y_1=(x_{1,33},0,x_{2,12},0),\, y_2 \in W_{\mathfrak{d}k} \\ x_{1,33},x_{2,12} \in k^\times}} \Psi(g_{\mathfrak{d}}n(u_{\mathfrak{d}})(y_1,y_2)),$$

$$f_{3,\alpha}(g_{\mathfrak{d}}) = \int_{(\mathbb{A}/k)^3} \widetilde{\Theta}_{Z'_{\mathfrak{d},0}}(\Psi, g_{\mathfrak{d}}n(u_{\mathfrak{d}})) < \alpha \cdot u_{\mathfrak{d}} > du_{\mathfrak{d}}.$$

Then

$$\begin{aligned} &\int_{A_{\mathfrak{d}} M^1_{\mathfrak{d}\mathbb{A}}/M_{\mathfrak{d}k}} \int_{(\mathbb{A}/k)^3} \Theta_{Y_{\mathfrak{d},st}}(\Psi, g_{\mathfrak{d}}n(u_{\mathfrak{d}}))\mathscr{E}(g_{\mathfrak{d}}n(u_{\mathfrak{d}}), w) dg_{\mathfrak{d}} du_{\mathfrak{d}} \\ &= \int_{X_{\mathfrak{d}} \cap B_{\mathbb{A}}/T_k} \sum_{\alpha \in k^3 \setminus \{0\}} f_{3,\alpha}(g_{\mathfrak{d}}) f_{2,-\alpha}(g_{\mathfrak{d}}, w)(t^0)^{-2\rho}\mu_1^2 d^\times t^0 du_0. \end{aligned}$$

By a similar consideration as before,

$$f_{3,\alpha}(g_{\mathfrak{d}}) = \sum_{\substack{y_1=(x_{1,33},0,x_{2,12},0) \\ x_{1,33},x_{2,12} \in k^\times}} \int_{\mathbb{A}^3} \Psi(g_{\mathfrak{d}}n(u_{\mathfrak{d}})y_1) du_{\mathfrak{d}}.$$

As in the proof of (10.3.4),

$$|f_{3,\alpha}(g_{\mathfrak{d}})| \ll \lambda_1^{-6}\mu_2^3 \sum_{\substack{y_1=(x_{1,33},0,x_{2,12},0) \\ x_{1,33},x_{2,12} \in k^\times}} \Psi_2(g_{\mathfrak{d}}y_1, \gamma_{2,33}(t^0)^{-1}x_{1,33}^{-1}\alpha_3).$$

This bound does not depend on α_1, α_2.

We choose $0 \leq \Psi_6, \Psi_7 \in \mathscr{S}(\mathbb{A}^2)$ so that

$$\begin{aligned} &\Psi_2(g_{\mathfrak{d}}y_1, \gamma_{2,33}(t^0)^{-1}x_{1,33}^{-1}\alpha_3) \\ &\ll \Psi_6(\underline{\lambda_1}^{-2}\underline{\mu_2}^{-1}x_{2,12}, \underline{\lambda_1}^{-2}\underline{\mu_2}^{-1}x_{2,12}u_0)\Psi_7(\underline{\lambda_1}^4\underline{\mu_2}x_{1,33}, \underline{\lambda_1}^{-4}\underline{\mu_2}x_{1,33}^{-1}\alpha_3). \end{aligned}$$

For $\mu \in \mathbb{R}_+, u_0 \in \mathbb{A}$, let

$$h_1(\mu, u_0) = \sum_{x_{2,12} \in k^\times} \Psi_6(\underline{\mu}x_{2,12}, \underline{\mu}x_{2,12}u_0).$$

We also define

$$h_2(\lambda_1, \mu_2) = \sum_{x_{1,33},\alpha_3 \in k^\times} \Psi_7(\underline{\lambda_1}^4\underline{\mu_2}x_{1,33}, \underline{\lambda_1}^{-4}\underline{\mu_2}x_{1,33}^{-1}\alpha_3).$$

Then

$$\sum_{\alpha_3\in k^\times} |f_{3,\alpha}(g_{\mathfrak{d}})| \ll \lambda_1^{-6}\mu_2^3 h_1(\lambda_1^{-2}\mu_2^{-1}, u_0)h_2(\lambda_1,\mu_2). \tag{10.3.10}$$

Lemma (10.3.11) *For any $s_1, s_2, s_3 \in \mathbb{C}$, the integral*

$$\int_{\mathbb{R}_+^2\times\mathbb{A}} \lambda_1^{s_1}\mu_2^{s_2}\alpha(u_0)^{s_3}h_1(\lambda_1^{-2}\mu_2^{-1}, u_0)h_2(\lambda_1,\mu_2)d^\times\lambda_1 d^\times\mu_2 du_0$$

converges absolutely.

Proof. Let $\mu_3 = \lambda_1^{-2}\mu_2^{-1}$, $\mu_4 = \lambda_1^4\mu_2$. Then $\lambda_1 = \mu_3^{\frac{1}{2}}\mu_4^{\frac{1}{2}}$, $\mu_2 = \mu_3^{-2}\mu_4^{-1}$. By (1.2.6), for any $N_1, N_2 \geq 1$, $h_2(\lambda_1,\mu_2) \ll \mu_4^{-N_1}(\mu_3^{-4}\mu_4^{-3})^{-N_2}$. Therefore, our integral is bounded by a constant multiple of

$$\int_{\mathbb{R}_+^2\times\mathbb{A}} \mu_3^{4N_2+s_1'}\mu_4^{3N_2-N_1+s_2'}\alpha(u_0)^{\mathrm{Re}(s_3)}h_1(\mu_3,u_0)d^\times\mu_3 d^\times\mu_4 du_0,$$

where $s_1' = \mathrm{Re}(\frac{1}{2}s_1 - 2s_2)$, $s_2' = \mathrm{Re}(\frac{1}{2}s_1 - s_2)$.

Let $r_1, r_2 \in \mathbb{R}$, and $\epsilon > 0$. Then the integral

$$\int_{\mathbb{R}_+\times\mathbb{A}} \mu_3^{r_1}\alpha(u_0)^{r_2}h_1(\mu_3,u_0)d^\times\mu_3 du_0$$

converges absolutely and locally uniformly on the domain of the form $\{(r_1,r_2) \in \mathbb{R}^2 \mid r_1 > 2+\epsilon, r_2 > -\epsilon\}$ by (4.2.9). So if $\mu_4 \geq 1$, we choose $N_1 \gg N_2 \gg 0$, and if $\mu_4 \leq 1$, we choose $N_2 \gg N_1 \gg 0$. This proves (10.3.11).

Q.E.D.

By the consideration as in §2.1,

$$\sum_{\substack{\alpha_3\in k^\times\\ \alpha_1,\alpha_2\in k}} |f_{2,\alpha}(g_{\mathfrak{d}},w)| \ll \sum \widetilde{\mathscr{E}}_{I,\tau,\nu}(g_{\mathfrak{d}},w),$$

where $I_2 = \{1\}$.

By (3.4.30), and (9.2.11)(1), (10.3.10), (10.3.11), we get the following proposition.

Proposition (10.3.12)

$$\int_{X_{\mathfrak{d}}\cap B_{\mathbb{A}}/T_k} \omega_{\mathfrak{p}}(t^0) \sum_{\substack{\alpha_1,\alpha_2\in k\\ \alpha_3\in k^\times}} f_{3,\alpha}(n'(u_0)t^0)f_{2,-\alpha}(n'(u_0)t^0,w)(t^0)^{-2\rho}\mu_1^2 d^\times t^0 du_0 \sim 0.$$

Note that

$$\sum_{\substack{\alpha_1,\alpha_2\in k\\ \alpha_3\in k^\times}} f_{3,\alpha}(n'(u_0)t^0)f_{2,-\alpha}(n'(u_0)t^0,w)$$

is invariant under the action of T_k.

We consider α such that $\alpha_3 = 0$.

Proposition (10.3.13)

$$\int_{X_\mathfrak{d}\cap B_\mathbb{A}/T_k} \omega_\mathfrak{p}(g_\mathfrak{d}) \sum_{\substack{\alpha_1\in k^\times \\ \alpha_2\in k \\ \alpha_3=0}} f_{3,\alpha}(g_\mathfrak{d}) f_{2,\alpha}(g_\mathfrak{d}, w)(t^0)^{-2\rho}\mu_1^2 d^\times t^0 du_0 \sim 0.$$

Proof. It is easy to see that $(t^0)^{-2\rho}\mu_1^2 = \lambda_1^6\mu_2^{-2}, \kappa_{\mathfrak{d}2}(t^0) = \lambda_1^{-6}\mu_2^3$. Let $\alpha_1' = -\alpha_2 x_{2,12}^{-1}, \alpha_2' = -\alpha_1 x_{2,12}^{-1}$. Then $f_{3,\alpha}(g_\mathfrak{d})$ is equal to

$$\lambda_1^{-6}\mu_2^3 \sum_{\substack{y_1=(x_{1,33},0,x_{2,12},0) \\ x_{1,33},x_{2,12}\in k^\times}} \Psi_3(g_\mathfrak{d} y_1, \gamma_{2,13}(t^0)^{-1}\alpha_1' - \gamma_{2,23}(t^0)^{-1}\alpha_2' u_0, \gamma_{2,23}(t^0)^{-1}\alpha_2').$$

We choose $0\le \Psi_8 \in \mathscr{S}(\mathbb{A}^2), 0\le\Psi_9\in\mathscr{S}(\mathbb{A}^3)$ so that

$$\begin{aligned}&\Psi_3(g_\mathfrak{d} y_1, x_{2,13}, x_{2,23}) \\ &\ll \Psi_8(\underline{\lambda}_1^{-2}\underline{\mu}_2^{-1}x_{2,12}, \underline{\lambda}_1^{-2}\underline{\mu}_2^{-1}x_{2,12}u_0)\Psi_9(\underline{\lambda}_1^4\underline{\mu}_2 x_{1,33}, x_{2,13}, x_{2,23}).\end{aligned}$$

Let $h_3(\mu, u_0)$, $h_4(t^0, u_0)$ be the following functions

$$\sum_{x_{2,12}\in k^\times} \Psi_8(\underline{\mu}x_{2,12}, \underline{\mu}x_{2,12}u_0),$$

$$\sum_{\substack{x_{1,33}\in k^\times \\ \alpha_1'\in k \\ \alpha_2'\in k^\times}} \Psi_9(\underline{\lambda}_1^4\underline{\mu}_2 x_{1,33}, \gamma_{2,13}(t^0)^{-1}\alpha_1' - \gamma_{2,23}(t^0)^{-1}\alpha_2' u_0, \gamma_{2,23}(t^0)^{-1}\alpha_2').$$

The condition $\alpha_1\ne 0$ is equivalent to the condition $\alpha_2'\ne 0$. So

$$\sum_{\substack{\alpha_1\in k^\times \\ \alpha_2\in k}} |f_{3,\alpha}(g_\mathfrak{d})| \ll \lambda_1^{-6}\mu_2^3 h_3(\lambda_1^{-2}\mu_2^{-1}, u_0) h_4(t^0, u_0).$$

By (1.2.8), for any $N_1, N_2\ge 1$,

$$h_4(t^0, u_0) \ll \sup(1, \lambda_1\mu_1\mu_2^{-1})(\lambda_1^4\mu_2)^{-N_1}(\lambda_1^{-1}\mu_1\mu_2)^{-N_2}.$$

By the consideration as in §2.1,

$$\sum_{\substack{\alpha_1\in k^\times \\ \alpha_2\in k \\ \alpha_3=0}} |f_{2,\alpha}(g_\mathfrak{d}, w)| \le \sum_{I,\tau,\nu} \widetilde{\mathscr{E}}_{I,\tau,\nu}(g_\mathfrak{d}, w),$$

where $I_1 \ni 2, \nu_1 = (1,2)$. Therefore, by (3.4.30), there exists a slowly increasing function $h_5(t^0)$ and a constant $\delta > 0$ such that if $M > w_0$,

$$\sum_{\substack{\alpha_1\in k^\times \\ \alpha_2\in k \\ \alpha_3=0}} |f_{2,\alpha}(g_\mathfrak{d}, w)| \ll h_5(t^0)$$

for $w_0 - \delta \leq \mathrm{Re}(w) \leq M$.

Let

$$h_6(t^0) = \lambda_1^{-6}\mu_2^3 \sup(1, \lambda_1\mu_1\mu_2^{-1})h_5(t^0).$$

Then, for any $N_1, N_2 \geq 1$,

$$\sum_{\substack{\alpha_1 \in k^\times \\ \alpha_2 \in k \\ \alpha_3 = 0}} |f_{3,\alpha}(g_{\mathfrak{d}})f_{2,-\alpha}(g_{\mathfrak{d}}, w)|$$
$$\ll h_3(\lambda_1^{-2}\mu_2^{-1}, u_0)h_6(t^0)\lambda_1^{3N_2-2N_1}(\lambda_1^{-2}\mu_2^{-1})^{N_1+N_2}\mu_1^{-N_2}.$$

If $\lambda_1 \geq 1$, we choose $N_1 \gg N_2 \gg 0$, and if $\lambda_1 \leq 1$, we choose $N_2 \gg N_1 \gg 0$. Then the right hand side is integrable on $X_{\mathfrak{d}} \cap B_{\mathbb{A}}/T_k$. This proves the proposition.

Q.E.D.

We consider α such that $\alpha_1 = \alpha_3 = 0$. This implies that $\alpha_2' = 0$. Let

$$h_7(t^0) = \sum_{x_{1,33}, \alpha_1' \in k^\times} \Psi_9(\underline{\lambda_1^4\mu_2} x_{1,33}, \gamma_{2,13}(t^0)^{-1}\alpha_1', 0).$$

Then for any $N_1, N_2 \geq 1$,

$$h_7(t^0) \ll (\lambda_1^4\mu_2)^{-N_1}(\lambda_1^{-1}\mu_1^{-1}\mu_2)^{-N_2},$$

and

$$\sum_{\substack{\alpha_1 = \alpha_3 = 0 \\ \alpha_2 \in k^\times}} |f_{3,\alpha}(g_{\mathfrak{d}})| \ll \lambda_1^{-6}\mu_2^3 h_3(\lambda_1^{-2}\mu_2^{-1}, u_0)h_7(t^0).$$

By the consideration as in §2.1,

$$\sum_{\substack{\alpha_1 = \alpha_3 = 0 \\ \alpha_2 \in k^\times}} |f_{2,\alpha}(g_{\mathfrak{d}})| \leq \widetilde{\mathscr{E}}_{I,\tau,1}(g_{\mathfrak{d}}, w),$$

where $I_1 \ni 2$. Therefore, by (3.4.30), there exists a slowly increasing function $h_8(t^0)$, and a constant $\delta > 0$ such that if $M > w_0$ and $l \gg 0$,

$$\sum_{\substack{\alpha_1 = \alpha_3 = 0 \\ \alpha_2 \in k^\times}} |f_{2,\alpha}(g_{\mathfrak{d}})| \ll h_8(t^0)(\lambda_1^3\mu_1)^l$$

for $w_0 - \delta \leq \mathrm{Re}(w) \leq M$.

Let $h_9(t^0) = \lambda_1^{-6}\mu_2^3 h_8(t^0)$. Then for any $N_1, N_2 \geq 1$, $l \gg 0$,

$$\text{(10.3.14)} \quad \sum_{\substack{\alpha_1 = \alpha_3 = 0 \\ \alpha_2 \in k^\times}} |f_{3,\alpha}(g_{\mathfrak{d}})f_{2,-\alpha}(g_{\mathfrak{d}})|$$
$$\ll h_9(t^0)h_3(\lambda_1^{-2}\mu_2^{-1}, u_0)(\lambda_1^4\mu_2)^{-N_1}(\lambda_1^{-1}\mu_1^{-1}\mu_2)^{-N_2}(\lambda_1^3\mu_1)^l$$
$$= h_9(t^0)h_3(\lambda_1^{-2}\mu_2^{-1}, u_0)\lambda_1^{3l-2N_1+3N_2}\mu_1^{l+N_2}(\lambda_1^{-2}\mu_2^{-1})^{N_1+N_2}.$$

Lemma (10.3.15) The integral

$$\int_{T^0_{\mathbb{A}}/T_k\times\mathbb{A}} \omega_{\mathfrak{p}}(t^0)\sum_{\substack{\alpha_1=\alpha_3=0\\ \alpha_2\in k^\times}} f_{3,\alpha}(n'(u_0)t^0)f_{2,-\alpha}(n'(u_0)t^0,w)(t^0)^{-2\rho}\mu_1^2 d^\times t^0 du_0$$

is well defined if $\mathrm{Re}(w)\gg 0$.

Proof. By (3.4.30), there exists a slowly increasing function $h_{10}(t^0)$ such that

$$\widetilde{\mathscr{E}}_{I,\tau,1}(g_{\mathfrak{d}},w)\ll h_{10}(t^0)\begin{cases}\lambda_1^{-r_1+r_2}\mu_1^{r_1+r_2} & \tau_1=(2,3),\\ \lambda_1^{2r_1+r_2}\mu_1^{r_2} & \tau_1=(1,2,3),\\ \lambda_1^{-r_1+r_2}\mu_1^{-r_2} & \tau_1=(1,3).\end{cases}$$

This is possible because ν in (3.4.30) is 1 for our case.

If $\tau_1=(2,3)$, we fix r_2 and consider $r_1\ll 0$. If $\tau_1=(1,2,3)$, we fix r_1+r_2 and consider $r_1\gg 0$. If $\tau_1=(1,3)$, we fix r_1 and consider $r_2\gg 0$. It is easy to see that the convex hull of the set

$$\{(-2,0,-1),(4,0,1),(-1,-1,1),(-1,1,0)\}$$

contains a neighborhood of the origin.

Let $t^0=(t_1,t_2)$ where $t_1\in T^0_{1\mathbb{A}}, t_2\in T^0_{2\mathbb{A}}$. In all the cases, there are a finite number of points $q_1,\cdots,q_j\in D_{I_1,\tau_1}$ such that

$$\sum_{\substack{\alpha_1=\alpha_3=0\\ \alpha_2\in k^\times}} |f_{3,\alpha}(n'(u_0)t^0)|\inf_i((t_1)^{\tau_1 q_i})(t^0)^{-2\rho}\mu_1^2 h_{10}(t^0)$$

is integrable on $T^0_{\mathbb{A}}/T_k\times\mathbb{A}$. Then our integral is well defined if $\mathrm{Re}(w)>L(q_i)$ for $i=1,\cdots,j$.

Q.E.D.

Lemma (10.3.16)

$$\widetilde{\Xi}_{\mathfrak{p}_1}(\Phi,\omega,w)\sim\int_{T^0_{\mathbb{A}}/T_k\times\mathbb{A}}\omega_{\mathfrak{p}_1}(t^0)\sum_{\substack{\alpha_1=\alpha_3=0\\ \alpha_2\in k^\times}} f_{3,\alpha}(g_{\mathfrak{d}})f_{2,-\alpha}(g_{\mathfrak{d}})(t^0)^{-2\rho}\mu_1^2 d^\times t^0 du_0.$$

Proof. Consider the integral

$$\int_{\substack{T^0_{\mathbb{A}}/T_k\times\mathbb{A}\\ \mu_1\le\sqrt{\alpha(u_0)}}}\omega_{\mathfrak{p}_1}(t^0)\sum_{\substack{\alpha_1=\alpha_3=0\\ \alpha_2\in k^\times}} f_{3,\alpha}(g_{\mathfrak{d}})f_{2,-\alpha}(g_{\mathfrak{d}})(t^0)^{-2\rho}\mu_1^2 d^\times t^0 du_0.$$

We use the estimate (10.3.14). If $\lambda_1\ge 1$, we choose $N_1\gg N_2,l$, and if $\lambda_1\le 1$, we choose $l\gg N_1\gg N_2$. Then the above integral converges absolutely. Therefore,

$$\int_{\substack{T^0_{\mathbb{A}}/T_k\times\mathbb{A}\\ \mu_1\le\sqrt{\alpha(u_0)}}}\omega_{\mathfrak{p}_1}(t^0)\sum_{\substack{\alpha_1=\alpha_3=0\\ \alpha_2\in k^\times}} f_{3,\alpha}(g_{\mathfrak{d}})f_{2,-\alpha}(g_{\mathfrak{d}})(t^0)^{-2\rho}\mu_1^2 d^\times t^0 du_0\sim 0. \qquad \text{Q.E.D.}$$

We define

$$f_{4,\alpha_2,\alpha_4}(t^0) = \int_{\mathbb{A}} f_{3,(0,\alpha_2,0)}(t^0 n'(u_1)) < \alpha_4 u_1 > du_1,$$

$$f_{5,\alpha_2,\alpha_4}(t^0, w) = \int_{\mathbb{A}/k} f_{2,(0,\alpha_2,0)}(t^0 n'(u_1), w) < \alpha_4 u_1 > du_1.$$

Then by the Parseval formula,

$$\int_{T^0_{\mathbb{A}}/T_k \times \mathbb{A}} \omega_{\mathfrak{p}_1}(t^0) \sum_{\substack{\alpha_1=\alpha_3=0 \\ \alpha_2 \in k^\times}} f_{3,\alpha}(g_{\mathfrak{d}}) f_{2,-\alpha}(g_{\mathfrak{d}})(t^0)^{-2\rho} \mu_1^2 d^\times t^0 du_0$$

$$= \int_{T^0_{\mathbb{A}}/T_k} \omega_{\mathfrak{p}_1}(t^0) \sum_{\substack{\alpha_2 \in k^\times \\ \alpha_4 \in k}} f_{4,\alpha_2,\alpha_4}(t^0) f_{5,\alpha_2,\alpha_4}(t^0, w)(t^0)^{-2\rho} \mu_1^2 d^\times t^0.$$

By (10.3.1)(4), for any $N > 0$,

$$\sum_{\alpha_2,\alpha_4 \in k^\times} |f_{4,\alpha_2,\alpha_4}(t^0)| \ll \mathrm{rd}_{3,N}(\lambda_1, \mu_1, \mu_2).$$

Since we have an estimate of $\sum_{\alpha_2,\alpha_4 \in k^\times} |f_{5,\alpha_2,\alpha_4}(t^0, w)|$ by a slowly increasing function as before, we get the following lemma.

Lemma (10.3.17)

$$\int_{T^0_{\mathbb{A}}/T_k} \omega_{\mathfrak{p}_1}(t^0) \sum_{\substack{\alpha_2 \in k^\times \\ \alpha_4 \in k^\times}} f_{4,\alpha_2,\alpha_4}(t^0) f_{5,\alpha_2,\alpha_4}(t^0, w)(t^0)^{-2\rho} d^\times t^0 \sim 0.$$

Let $f_{6,\alpha_2}(t^0) = f_{4,\alpha_2,0}(t^0)$, $f_{7,\alpha_2}(t^0, w) = f_{5,\alpha_2,0}(t^0, w)$. Then

$$f_{6,\alpha_2}(t^0) = \int_{\mathbb{A}/k} \Theta_{Y_{\mathfrak{d}_2}}(\widetilde{R}_{\mathfrak{d}_2} \Phi_{\mathfrak{p}_2}, t^0 n''(u_3)) < \alpha_2 u_3 > du_3,$$

$$f_{7,\alpha_2}(t^0, w) = \int_{\mathbb{A}/k} \mathscr{E}_{\mathfrak{p}_3}(t^0 n''(u_3), w) < \alpha_2 u_3 > du_3.$$

Consider $\mathfrak{p}_2$ such that $\mathfrak{s}_2(2) = 0$. It is easy to see that $e_{\mathfrak{p}_2 1}(t^0) = (t^0)^{\beta_8} = \lambda_1^{\frac{1}{7}} \mu_1^{-\frac{1}{7}} \mu_2^{-\frac{1}{7}}$. So $A_{\mathfrak{p}_2 0} M_{\mathfrak{d}_2 \mathbb{A}} = \{t^0 \in T^0_{\mathbb{A}} \mid \lambda_1 \mu_1^{-1} \mu_2^{-1} \le 1\}$. Therefore,

$$\widetilde{\Xi}_{\mathfrak{p}_2}(\Phi, \omega, w) = \int_{\substack{T^0_{\mathbb{A}}/T_k \\ \lambda_1 \mu_1^{-1} \mu_2^{-1} \le 1}} \omega_{\mathfrak{p}_2}(t^0) \sum_{\alpha_2 \in k^\times} f_{6,\alpha_2}(t^0) f_{7,-\alpha}(t^0, w)(t^0)^{-2\rho} d^\times t^0.$$

Hence,

$$\epsilon_{\mathfrak{p}_1} \widetilde{\Xi}_{\mathfrak{p}_1}(\Phi, \omega, w) + \epsilon_{\mathfrak{p}_2} \widetilde{\Xi}_{\mathfrak{p}_2}(\Phi, \omega, w) \tag{10.3.18}$$

$$\sim \epsilon_{\mathfrak{p}_1} \int_{\substack{T^0_{\mathbb{A}}/T_k \\ \lambda_1 \mu_1^{-1} \mu_2^{-1} \ge 1}} \omega_{\mathfrak{p}_1}(t^0) \sum_{\alpha_2 \in k^\times} f_{6,\alpha_2}(t^0) f_{7,-\alpha_2}(t^0, w)(t^0)^{-2\rho} d^\times t^0.$$

Note that $\epsilon_2 = -\epsilon_1$.

Lemma (10.3.19) *There exists a constant $\delta > 0$ such that for any $N > 0, M > w_0$,*

$$\sum_{\alpha_2 \in k^\times} |f_{6,\alpha_2}(t^0) f_{7,-\alpha_2}(t^0, w)| \ll (\lambda_1 \mu_1^{-1} \mu_2^{-1})^{-N} \mathrm{rd}_{2,N}(\mu_1, \mu_2)$$

for $w_0 - \delta \leq \mathrm{Re}(w) \leq M$.

Proof. There exists a slowly increasing function $h_{11}(t^0)$ such that for any $N_1, N_2, N_3 \geq 1$, $l > 1$,

$$\begin{aligned} &\sum_{\substack{\alpha_1=\alpha_3=0 \\ \alpha_2 \in k^\times}} |f_{6,\alpha_2}(t^0) f_{7,-\alpha_2}(t^0, w)| \\ &\ll h_{11}(t^0)(\lambda_1^4 \mu_2)^{-N_1} (\lambda_1^{-1} \mu_1^{-1} \mu_2)^{-N_2} (\lambda_1^{-2} \mu_2^{-1})^{-N_3} (\lambda_1^3 \mu_1)^l. \end{aligned}$$

Let $\mu_3 = \lambda_1 \mu_1^{-1} \mu_2^{-1}$. Then the right hand side of the above inequality is

$$h_{11}(t^0) \mu_1^{-(4N_1 - 2N_2 - 2N_3 - 4l)} \mu_2^{-(5N_1 - 3N_3 - 3l)} \mu_3^{-(4N_1 - N_2 - 2N_3 - 3l)}. \tag{10.3.20}$$

We choose N_1, N_2, N_3, l in the following manner:

$$\begin{cases} N_2, N_3, l \text{ fixed, and } N_1 \gg 0 & \mu_1 \geq 1, \mu_2 \geq 1, \\ N_2, l \text{ fixed, and } N_1 = \frac{4}{7} N_3 \gg 0 & \mu_1 \geq 1, \mu_2 \leq 1, \\ N_2, N_3 \text{ fixed, and } l = \frac{5}{4} N_1 \gg 0 & \mu_1 \leq 1, \mu_2 \geq 1, \\ N_2 \text{ fixed, and } N_3 = 4l, N_1 = \frac{23}{8} l \gg 0 & \mu_1 \leq 1, \mu_2 \leq 1. \end{cases}$$

This gives us the bound of the lemma.

Q.E.D.

The following proposition follows from (10.3.18), (10.3.19).

Proposition (10.3.21) *If $\mathfrak{s}_1(1) = \mathfrak{s}_2(1)$ and $\mathfrak{s}_2(2) = 0$,*

$$\epsilon_{\mathfrak{p}_1} \widetilde{\Xi}_{\mathfrak{p}_1}(\Phi, \omega, w) + \epsilon_{\mathfrak{p}_2} \widetilde{\Xi}_{\mathfrak{p}_2}(\Phi, \omega, w) \sim 0.$$

Consider $\mathfrak{p}_2$ such that $\mathfrak{s}_2(2) = 1$. Then $\widetilde{\Xi}_{\mathfrak{p}_2}(\Phi, \omega, w)$ is equal to

$$\int_{\substack{T^0_{\mathbb{A}}/T_k \\ \lambda_1 \mu_1^{-1} \mu_2^{-1} \geq 1}} \omega_{\mathfrak{p}_2}(t^0) \sum_{\alpha_2 \in k^\times} f_{6,\alpha_2}(t^0) f_{7,-\alpha}(t^0, w)((t^0)^{-2\rho_{\mathfrak{d}_3}} \kappa_{\mathfrak{d}_3 1}(t^0))^{-1} (t^0)^{-2\rho_{\mathfrak{d}_2}} d^\times t^0.$$

By (10.3.19), we get the following proposition.

Proposition (10.3.22) *If $\mathfrak{s}_2(2) = 1$, $\widetilde{\Xi}_{\mathfrak{p}_2}(\Phi, \omega, w) \sim 0$*

§10.4 The case $\mathfrak{d} = (\beta_7)$

In this section, we consider a path $\mathfrak{p} = (\mathfrak{d}, \mathfrak{s})$ such that $\mathfrak{d} = (\beta_7)$.

An easy consideration shows the following lemma and we omit the proof.

Lemma (10.4.1) *The convex hull of each of the set*
(1) $\{\gamma_{1,33}, \gamma_{2,22}, \gamma_{2,13}, -\gamma_{2,23}\}$,
(2) $\{\gamma_{1,33}, \gamma_{2,22}, \gamma_{2,13}, -\gamma_{2,33}\}$
contains a neighborhood of the origin.

Let $\Psi = \widetilde{R}_{\mathfrak{d}}\Phi_{\mathfrak{p}}$. In this case, $P_{\mathfrak{d}} = B$, so we write $g_{\mathfrak{d}} = t^0 = (t_1, t_2)$ where $t_1 \in T^0_{1\mathbb{A}}, t_2 \in T^0_{2\mathbb{A}}$. We choose an isomorphism $d: \mathbb{R}^3_+ \to T^0_+$. Therefore, we can write $t^0 = d(\lambda_1, \lambda_2, \lambda_3)\widehat{t}^0$, where $\lambda_1, \lambda_2, \lambda_3 \in \mathbb{R}_+$. Let $n(u_{\mathfrak{d}}) = (n_{P_1}(u_2, u_3), n_{B_2}(u_4))$ for $u_{\mathfrak{d}} = (u_2, u_3, u_4) \in \mathbb{A}^3$. For fixed t^0 and $u_1 \in \mathbb{A}$, $\Theta_{Y_{\mathfrak{d}}}(\Psi, t^0 n'(u_1)n(u_{\mathfrak{d}}))$, $\mathscr{E}(t^0 n'(u_1)n(u_{\mathfrak{d}}), w)$ are functions on $(\mathbb{A}/k)^3$.

For $\alpha = (\alpha_1, \alpha_2, \alpha_3) \in k^3$, let $< \alpha \cdot u_{\mathfrak{d}} > = < \alpha_1 u_2 + \alpha_2 u_3 + \alpha_3 u_4 >$. By the Parseval formula,

$$\begin{aligned}&\int_{(\mathbb{A}/k)^3} \Theta_{Y_{\mathfrak{d}}}(\Psi, t^0 n'(u_1)n(u_{\mathfrak{d}}))\mathscr{E}(t^0 n'(u_1)n(u_{\mathfrak{d}}), w)du_{\mathfrak{d}} \\ &= \sum_{\alpha\in k^3} f_{1,\alpha}(t^0 n'(u_1)) f_{2,-\alpha}(t^0 n'(u_1), w),\end{aligned}$$

where

$$f_{1,\alpha}(t^0 n'(u_1)) = \int_{(\mathbb{A}/k)^3} \Theta_{Y_{\mathfrak{d}}}(\Psi, t^0 n'(u_1)n(u_{\mathfrak{d}})) < \alpha \cdot u_{\mathfrak{d}} > du_{\mathfrak{d}},$$
$$f_{2,\alpha}(t^0 n'(u_1), w) = \int_{(\mathbb{A}/k)^3} \mathscr{E}(t^0 n'(u_1)n(u_{\mathfrak{d}}), w) < \alpha \cdot u_{\mathfrak{d}} > du_{\mathfrak{d}}.$$

Proposition (10.4.2) *There exists a constant $\delta > 0$ such that for any $N > 0, M > w_0$,*

$$\sum_{\alpha\in k^3\setminus\{0\}} |f_{1,\alpha}(t^0 n'(u_1)) f_{2,-\alpha}(t^0 n'(u_1), w)| \ll \mathrm{rd}_{3,N}(\lambda_1, \lambda_2, \lambda_3)$$

for $w_0 - \delta \leq \mathrm{Re}(w) \leq M$.

Proof. We can write any element $y \in Y^{ss}_{\mathfrak{d}k}$ in the form $y = (y_1, y_2)$ where

$$y_1 = (x_{1,33}, x_{2,22}, x_{2,13}) \in Z^{ss}_{\mathfrak{d}k},\ y_2 = (x_{2,23}, x_{2,33}) \in W_{\mathfrak{d}k}.$$

By definition,

$$n'(u_1)n(u_{\mathfrak{d}})y = (y_1, x_{2,23} + x_{2,13}u_1 + 2x_{2,22}u_3, x_{2,33} + x_{2,22}u_3^2 + x_{2,13}u_2 + x_{1,33}u_4).$$

Let

$$\begin{aligned}u_2' &= x_{2,13}u_2 + x_{2,22}u_3^2 + x_{1,33}u_4, \\ u_3' &= x_{2,13}u_1 + 2x_{2,22}u_3, \\ u_4' &= u_4.\end{aligned}$$

Then

$$\begin{aligned}u_2 &= x_{2,13}^{-1}u_2' - 4^{-1}x_{2,13}^{-1}x_{2,22}^{-1}u_3'^2 + 2^{-1}x_{2,22}^{-1}u_1u_3' - 4^{-1}x_{2,22}^{-1}x_{2,13}u_1^2 - x_{2,13}^{-1}x_{1,33}u_4', \\ u_3 &= 2^{-1}x_{2,22}^{-1}u_3' - 2^{-1}x_{2,22}^{-1}x_{2,13}u_1, \\ u_4 &= u_4'.\end{aligned}$$

Clearly, $n'(u_1)n(u_\eth)y$ does not depend on u_4'. Therefore,

$$f_{1,\alpha}(t^0 n'(u_1)) = \sum_{y_1 \in Z_{\eth k}^{ss}} \int_{\mathbb{A}^2} \Psi(t^0 n'(u_1)n(u_\eth)y_1)du_2'du_3'.$$

Let

$$\begin{aligned}
\psi_1(\alpha, u_1, u') &= < \alpha_1 x_{2,13}^{-1} u_2' + 2^{-1}\alpha_2 x_{2,22}^{-1} u_3' > \\
&\quad \times < \alpha_1(-4^{-1} x_{2,13}^{-1} x_{2,22}^{-1} u_3'^2 + 2^{-1} x_{2,22}^{-1} u_1 u_3') > \\
&\quad \times < (\alpha_3 - \alpha_1 x_{2,13}^{-1} x_{1,33}) u_4' >, \\
\psi_2(\alpha, u_1) &= < -4^{-1}\alpha_1 x_{2,22}^{-1} x_{2,13} u_1^2 - \alpha_2 x_{2,22}^{-1} x_{2,13} u_1 >, \\
\psi_3(\alpha, u_1, u_3') &= < 2^{-1}\alpha_2 x_{2,22}^{-1} u_3' + \alpha_1(-4^{-1} x_{2,13}^{-1} x_{2,22}^{-1} u_3'^2 + 2^{-1} x_{2,22}^{-1} u_1 u_3') > .
\end{aligned}$$

Then $f_{1,\alpha}(t^0 n'(u_1))$ is equal to

$$\sum_{y_1 \in Z_{\eth k}^{ss}} \int_{\mathbb{A}^2} \Psi(t^0 y_1, \gamma_{2,23}(t^0)u_3', \gamma_{2,33}(t^0)u_2')\psi_1(\alpha, u_1, u')du_2'du_3'\psi_2(\alpha, u_1).$$

The above integral is zero unless $\alpha_3 - \alpha_1 x_{2,13}^{-1} x_{1,33} = 0$, because $n'(u_1)n(u_\eth)y$ does not depend on u_4'.

Let Ψ_1 be the partial Fourier transform of Ψ with respect to the coordinate $x_{2,33}$ and $< >$. Then

$$\begin{aligned}
&\int_{\mathbb{A}^2} \Psi(t^0 y_1, \gamma_{2,23}(t^0)u_3', \gamma_{2,33}(t^0)u_2')\psi_1(\alpha, u_1, u')du_2'du_3' \\
&= |\gamma_{2,33}(t^0)|^{-1} \int_{\mathbb{A}} \Psi_1(t^0 y_1, \gamma_{2,23}(t^0)u_3', \alpha_1 x_{2,13}^{-1}\gamma_{2,33}(t^0)^{-1})\psi_3(\alpha, u_1, u_3')du_3'.
\end{aligned}$$

We choose $0 \leq \Psi_2 \in \mathscr{S}(Z_{\eth\mathbb{A}} \times \mathbb{A})$ so that

$$\int_{\mathbb{A}} |\Psi_1(y_1, x_{2,23}, x_{2,33})| dx_{2,23} \leq \Psi_2(y_1, x_{2,33}).$$

Since $|\psi_2(\alpha, u_1)| = |\psi_3(\alpha, u_1, u_3')| = 1$,

$$|f_{1,\alpha}(t^0 n'(u_1)) \ll \kappa_{\eth 2}(t^0) \sum_{\substack{y_1 \in Z_{\eth k}^{ss} \\ x_{2,13} = \alpha_1 \alpha_3^{-1} x_{1,33}}} \Psi_2(t^0 y_1, \alpha_1 x_{2,13}^{-1}\gamma_{2,33}(t^0)^{-1}).$$

If the condition $x_{2,13} = \alpha_1 \alpha_3^{-1} x_{1,33}$ is satisfied, $\alpha_1 \neq 0$ implies $\alpha_3 \neq 0$. So

$$\sum_{\substack{\alpha_1 \in k^\times \\ \alpha_3 \in k}} |f_{1,\alpha}(t^0)| \ll \kappa_{\eth 2}(t^0) \sum_{\substack{y_1 \in Z_{\eth k}^{ss} \\ \alpha' \in k^\times}} \Psi_2(t^0 y_1, \alpha' \gamma_{2,33}(t^0)^{-1}).$$

This bound does not depend on α_2. By (10.4.1), for any $N > 0$,

$$\sum_{\alpha_1, \alpha_3 \in k^\times} |f_{1,\alpha}(t^0)| \ll \mathrm{rd}_{3,N}(\lambda_1, \lambda_2, \lambda_3).$$

There exist a slowly increasing function $h(t^0)$ and a constant $\delta > 0$ such that if $M > w_0$,

$$\sum_{\substack{\alpha_1 \in k^\times \\ \alpha_2, \alpha_3 \in k}} |f_{2,\alpha}(t^0, w)| \ll h(t^0)$$

for $w_0 - \delta \leq \mathrm{Re}(w) \leq M$. Therefore, for any $N > 0$,

$$\sum_{\substack{\alpha_1 \in k^\times \\ \alpha_2, \alpha_3 \in k}} |f_{1,\alpha}(t^0) f_{2,-\alpha}(t^0, w)| \ll \mathrm{rd}_{3,N}(\lambda_1, \lambda_2, \lambda_3).$$

Next, we consider terms such that $\alpha_1 = \alpha_3 = 0,\ \alpha_2 \neq 0$. Let

$$\Psi_3(y_1, x_{2,23}) = \int_{\mathbb{A}^2} \Psi(y_1, x'_{2,23}, x'_{2,33}) < x_{2,23} x'_{2,23} > dx'_{2,23} dx'_{2,33}.$$

Then $f_{1,\alpha}(t^0 n'(u_1))$ is equal to

$$\kappa_{\mathfrak{d}2}(t^0) \sum_{y_1 \in Z^{ss}_{\mathfrak{d}k}} \Psi_3(t^0 y_1, 2^{-1} \alpha_2 x_{2,22}^{-1} \gamma_{2,23}(t^0)^{-1}) < -2^{-1} \alpha_2 x_{2,22}^{-1} x_{2,13} u_1 > .$$

By (10.4.1), for any $N > 0$,

$$\sum_{\substack{\alpha_2 \in k^\times \\ \alpha_1 = \alpha_3 = 0}} |f_{1,\alpha}(t^0 n'(u_1))| \ll \mathrm{rd}_{3,N}(\lambda_1, \lambda_2, \lambda_3).$$

The rest of the argument is similar. This proves (10.4.2).

Q.E.D.

The following proposition follows from (10.4.2).

Proposition (10.4.3) $\widetilde{\Xi}_{\mathfrak{p}}(\Phi, \omega, w) \sim 0$.

§10.5 The case $\mathfrak{d} = (\beta_8)$

In this section, we consider a path $\mathfrak{p} = (\mathfrak{d}, \mathfrak{s})$ such that $\mathfrak{d} = (\beta_8)$.

Proposition (10.5.1) $\widetilde{\Xi}_{\mathfrak{p}}(\Phi, \omega, w) \sim 0$.

We devote the rest of this section to the proof of (10.5.1).

For $\lambda_1, \mu_1, \mu_2 \in \mathbb{R}_+$, we define

$$d(\lambda_1, \mu_1, \mu_2) = a(\underline{\lambda}_1^{-2}, \underline{\lambda}_1 \underline{\mu}_1, \underline{\lambda}_1 \underline{\mu}_1^{-1}; \underline{\mu}_2, \underline{\mu}_2^{-1}). \tag{10.5.2}$$

We write $g_{\mathfrak{d}} = k d(\lambda_1, \mu_1, \mu_2) \widehat{t}^0 n''(u_3)$ where $k \in K \cap M_{\mathfrak{d}\mathbb{A}}$. As in §3.4, when we estimate various functions, we assume that $k = 1$. Let $t^0 = d(\lambda_1, \mu_1, \mu_2) \widehat{t}^0$. Let $n(u_{\mathfrak{d}}) = (n_{P_2}(u_1, u_2), n_{B_2}(u_4))$ for $u_{\mathfrak{d}} = (u_1, u_2, u_4) \in \mathbb{A}^3$.

For $N = (N_1, N_2, N_3)$, we define

$$\begin{aligned} h_{1,N}(g_{\mathfrak{d}}) &= |\gamma_{1,22}(t^0)|^{-N_1} |\gamma_{2,13}(t^0)|^{-N_2} |\gamma_{2,23}(t^0)|^{N_3}, \\ h_{2,N}(g_{\mathfrak{d}}) &= |\gamma_{1,22}(t^0)|^{-N_1} |\gamma_{2,13}(t^0)|^{-N_2} |\gamma_{2,33}(t^0)|^{N_3}, \\ h_{3,N}(g_{\mathfrak{d}}) &= |\gamma_{1,23}(t^0)|^{-N_1} |\gamma_{2,12}(t^0)|^{-N_2} |\gamma_{2,23}(t^0)|^{N_3}, \\ h_{4,N}(g_{\mathfrak{d}}) &= |\gamma_{1,23}(t^0)|^{-N_1} |\gamma_{2,12}(t^0)|^{-N_2} |\gamma_{2,33}(t^0)|^{N_3}. \end{aligned} \tag{10.5.3}$$

Then

$$h_{1,N}(g_\mathfrak{d}) = \lambda_1^{-(2N_1-N_2-2N_3)}\mu_1^{-(2N_1-N_2)}\mu_2^{-(N_1-N_2+N_3)},$$
$$h_{2,N}(g_\mathfrak{d}) = \lambda_1^{-(2N_1-N_2-2N_3)}\mu_1^{-(2N_1-N_2+2N_3)}\mu_2^{-(N_1-N_2+N_3)},$$
$$h_{3,N}(g_\mathfrak{d}) = \lambda_1^{-(2N_1-N_2-2N_3)}\mu_1^{-N_2}\mu_2^{-(N_1-N_2+N_3)},$$
$$h_{4,N}(g_\mathfrak{d}) = \lambda_1^{-(2N_1-N_2-2N_3)}\mu_1^{-(N_2+2N_3)}\mu_2^{-(N_1-N_2+N_3)}.$$

Consider the condition

$(10.5.4)_i$ For any $N=(N_1,N_2,N_3)$ satisfying $N_1,N_2,N_3\geq 1$, $|f(g_\mathfrak{d})|\ll h_{i,N}(g_\mathfrak{d})$.

Lemma (10.5.5) *Suppose that $f(g_\mathfrak{d})$ is a function on $G^0_\mathbb{A}\cap M_{\mathfrak{d}\mathbb{A}}/M_{\mathfrak{d}k}$ which satisfies* Condition $(10.5.4)_i$ *for some $1\leq i\leq 4$. Then for any $N>0$,*

$$|f(g_\mathfrak{d})|\ll \mu_1^{-N}\mathrm{rd}62, N(\lambda_1,\mu_2)$$

for $g_\mathfrak{d}$ in the Siegel domain.

Proof. Since $|\gamma_{2,23}(t^0)| = \mu_1^2|\gamma_{2,33}(t^0)| \gg |\gamma_{2,33}(t^0)|$, we only have to consider $i=1,3$.

Suppose that $(10.5.4)_1$ is satisfied. Then we choose N_1,N_2,N_3 in the following manner:

$$\begin{cases} N_2, N_3 \text{ fixed and } N_1\gg 0, \gg 0 & \text{if } \lambda_1\geq 1, \mu_2\geq 1,\\ N_3 \text{ fixed and } N_1=\frac{3}{4}N_2\gg 0 & \text{if } \lambda_1\geq 1, \mu_2\leq 1,\\ N_3\gg N_1\gg N_2 & \text{if } \lambda_1\leq 1, \mu_2\geq 1,\\ N_1=\frac{2}{3}N_2, N_3=\frac{1}{4}N_2\gg 0 & \text{if } \lambda_1\leq 1, \mu_2\leq 1.\end{cases}$$

Suppose that $(10.5.4)_3$ is satisfied. Then we choose N_1,N_2,N_3 in the following manner:

$$\begin{cases} N_3 \text{ fixed and } N_1\gg N_2\gg 0 & \text{if } \lambda_1\geq 1, \mu_2\geq 1,\\ N_3 \text{ fixed and } N_1=\frac{3}{4}N_2\gg 0 & \text{if } \lambda_1\geq 1, \mu_2\leq 1,\\ N_1 \text{ fixed and } N_3\gg N_2 & \text{if } \lambda_1\leq 1, \mu_2\geq 1,\\ N_1 \text{ fixed and } N_3=\frac{2}{3}N_2 & \text{if } \lambda_1\leq 1, \mu_2\leq 1.\end{cases}$$

Then we have the estimate of the lemma.

Q.E.D.

Let $\Psi=\widetilde{R}_\mathfrak{d}\Phi_\mathfrak{p}$. As usual, we can write any element $y\in Y^{\mathrm{ss}}_{\mathfrak{d}k}$ in the form $y=(y_1,y_2)$ where

$$y_1=(x_{1,22},x_{1,23},x_{1,33},x_{2,12},x_{2,13})\in Z^{\mathrm{ss}}_{\mathfrak{d}k} \text{ and } y_2=(x_{2,22},x_{2,23},x_{2,33})\in W_{\mathfrak{d}k}.$$

For $\alpha=(\alpha_1,\alpha_2,\alpha_3)\in k^3$, we define $<\alpha\cdot u_\mathfrak{d}> = <\alpha_1u_1+\alpha_2u_2+\alpha_3u_4>$. Let

$$(10.5.6)\qquad f_{1,\alpha}(g_\mathfrak{d}) = \int_{(\mathbb{A}/k)^3}\Theta_{Y_\mathfrak{d}}(\Psi, g_\mathfrak{d}n(u_\mathfrak{d}))<\alpha\cdot u_\mathfrak{d}>du_\mathfrak{d},$$
$$f_{2,\alpha}(g_\mathfrak{d},w) = \int_{(\mathbb{A}/k)^3}\mathscr{E}(g_\mathfrak{d}n(u_\mathfrak{d}),w)<\alpha\cdot u_\mathfrak{d}>du_\mathfrak{d}.$$

The proof of the following lemma is easy, and is left to the reader.

Lemma (10.5.7) *Let $P(y_1) = x_{2,12}^2 x_{1,33} - x_{2,13}x_{2,12}x_{1,23} + x_{2,13}^2 x_{1,22}$. Then $P(y_1)$ is a relative invariant polynomial under the action of $M_{\mathfrak{d}}$.*

Let

$$A(y_1) = \begin{pmatrix} 2x_{2,12} & 0 & x_{2,22} \\ x_{2,13} & x_{2,12} & x_{1,23} \\ 0 & 2x_{2,13} & x_{1,33} \end{pmatrix}, \quad \begin{pmatrix} u_1' \\ u_2' \\ u_4' \end{pmatrix} = A(y_1) \begin{pmatrix} u_2 \\ u_2 \\ u_4 \end{pmatrix}.$$

Then

$$n(u_{\mathfrak{d}})y = (y_1, x_{2,22} + u_1', x_{2,23} + u_2', x_{2,33} + u_4')$$

Therefore,

$$f_{1,\alpha}(g_{\mathfrak{d}}) = \sum_{y_1 \in Z_{\mathfrak{d}k}^{ss}} \int_{\mathbb{A}^3} \Psi(g_{\mathfrak{d}} n(u_{\mathfrak{d}})y_1) < \alpha \cdot u_{\mathfrak{d}} > du_{\mathfrak{d}}.$$

Let $(\alpha_1', \alpha_2', \alpha_3') = (\alpha_1, \alpha_2, \alpha_3)A(y_1)^{-1}$ ($\det A(y_1) = P(y_1) \neq 0$). Then

$$< \alpha_1' u_1' + \alpha_2' u_2' + \alpha_3' u_4' >=< \alpha_1 u_1 + \alpha_2 u_2 + \alpha_3 u_4 > .$$

It is easy to see that

$$g_{\mathfrak{d}} n(u_{\mathfrak{d}})y_1 = (g_{\mathfrak{d}} y_1, u_1', u_2' + 2u_1 u_1', u_4' + u_1^2 u_1' + u_1 u_2').$$

For $u' = (u_1', u_2', u_4')$, we define

$$\Psi'(g_{\mathfrak{d}}, y_1, u') = \Psi(g_{\mathfrak{d}} y_1, u_1', u_2' + 2u_1 u_1', u_4' + u_1^2 u_1' + u_1 u_2').$$

Also we define

$$\begin{aligned} u_3'' &= u_2' + 2u_1 u_1', \\ u_4'' &= u_4' + u_1^2 u_1' + u_1 u_2'. \end{aligned}$$

Then

$$\begin{aligned} u_2' &= u_3'' - 2u_3 u_1', \\ u_4' &= u_4'' - u_3^2 u_1' - u_3(u_3'' - 2u_3 u_1') = u_4'' + u_3^2 u_1' - u_3 u_3''. \end{aligned}$$

Therefore,

$$< \alpha' \cdot u' >=< (\alpha_1' - 2\alpha_2' u_3 + \alpha_3' u_3^2)u_1' + (\alpha_2' - \alpha_3 u_3)u_2'' + \alpha_3' u_4'' > .$$

Let

$$\psi_{\alpha'}'(u_1', u_2') =< (\alpha_1' - 2\alpha_2' u_3 + \alpha_3' u_3^2)u_1' + (\alpha_2' - \alpha_3 u_3)u_2'' > .$$

Let Ψ_1 be the partial Fourier transform of Ψ with respect to the coordinate $x_{2,33}$ and $< >$. Then

$$\begin{aligned} &\sum_{y_1 \in Z_{\mathfrak{d}k}^{ss}} \int_{\mathbb{A}^3} \Psi(g_{\mathfrak{d}} n(u_{\mathfrak{d}})y_1) < \alpha \cdot u_{\mathfrak{d}} > du_{\mathfrak{d}} \\ &= |\gamma_{2,33}(t^0)|^{-1} \sum_{y_1 \in Z_{\mathfrak{d}k}^{ss}} \int_{\mathbb{A}^2} \Psi_1(g_{\mathfrak{d}} y_1, u_1', u_3'', \alpha_3' \gamma_{2,33}(t^0)^{-1}) \psi_{\alpha'}'(u_1', u_2') du_1' du_3''. \end{aligned}$$

We choose $0 \le \Psi_2 \in \mathscr{S}(Z_{\mathfrak{d}\mathbb{A}} \times \mathbb{A})$ so that

$$\int_{\mathbb{A}^2} |\Psi_1(y_1, u_1', u_3'', x_{2,23})| du_1' du_3'' \le \Psi_2(y_1, x_{2,23}).$$

Then

$$|f_{1,\alpha}(g_{\mathfrak{d}})| \ll \kappa_{\mathfrak{d}2}(g_{\mathfrak{d}}) \sum_{y_1 \in Z_{\mathfrak{d}k}^{\mathrm{ss}}} \Psi_2(g_{\mathfrak{d}} y_1, \alpha_3' \gamma_{2,33}(t^0)^{-1}).$$

By (1.2.8), (10.5.5), for any $N > 0$,

$$\sum_{\alpha_3' \in k^\times} |f_{1,\alpha}(g_{\mathfrak{d}})| \ll \mu_1^{-N} \mathrm{rd}_{2,N}(\lambda_1, \mu_2)$$

for $g_{\mathfrak{d}}$ in the Siegel domain. This bound does not depend on α_1', α_2'.

Therefore, by a similar argument as before, there exists a constant $\delta > 0$ such that if $M > w_0$,

$$\sum_{\substack{\alpha_1', \alpha_2' \in k \\ \alpha_3' \in k^\times}} |f_{1,\alpha}(g_{\mathfrak{d}}) f_{2,-\alpha}(g_{\mathfrak{d}}, w)| \ll \mu_1^{-N} \mathrm{rd}_{2,N}(\lambda_1, \mu_2) \tag{10.5.8}$$

for $g_{\mathfrak{d}}$ in the Siegel domain and $w_0 - \delta \le \mathrm{Re}(w) \le M$.

Next, we consider α such that $\alpha_3' = 0$.

Let

$$\Psi_3(y_1, x_{2,22}, x_{2,23}) = \int_{\mathbb{A}^2} \Psi(y_1, x_{2,22}, x_{2,23}', x_{2,33}') < x_{2,23} x_{2,23}' > dx_{2,23}' dx_{2,33}'.$$

We choose $0 \le \Psi_4 \in \mathscr{S}(Z_{\mathfrak{d}\mathbb{A}} \times \mathbb{A})$ so that

$$\int_{\mathbb{A}} |\Psi_3(y_1, x_{2,22}, x_{2,23})| dx_{2,22} \le \Psi_4(y_1, x_{2,23}).$$

Then

$$\sum_{\substack{\alpha_3'=0 \\ \alpha_2' \in k^\times}} |f_{1,\alpha}(g_{\mathfrak{d}})| \ll \kappa_{\mathfrak{d}2}(g_{\mathfrak{d}2}) \sum_{\substack{\alpha_3'=0 \\ \alpha_2' \in k^\times}} \sum_{y_1 \in Z_{\mathfrak{d}k}^{\mathrm{ss}}} \Psi_4(g_{\mathfrak{d}} y_1, \alpha_2' \gamma_{2,23}(t^0)^{-1}).$$

By (1.2.8), (10.5.5), for any $N > 0$,

$$\sum_{\substack{\alpha_3'=0 \\ \alpha_2' \in k^\times}} |f_{1,\alpha}(g_{\mathfrak{d}})| \ll \mu_1^{-N} \mathrm{rd}_{2,N}(\lambda_1, \mu_2)$$

for $g_{\mathfrak{d}}$ in the Siegel domain. Therefore, there exists a constant $\delta > 0$ such that if $M > w_0$,

$$\sum_{\substack{\alpha_3'=0 \\ \alpha_2' \in k^\times \\ \alpha_1' \in k}} |f_{1,\alpha}(g_{\mathfrak{d}}) f_{2,-\alpha}(g_{\mathfrak{d}}, w)| \ll \mu_1^{-N} \mathrm{rd}_{2,N}(\lambda_1, \mu_2) \tag{10.5.9}$$

for $g_{\mathfrak{d}}$ in the Siegel domain and $w_0 - \delta \leq \mathrm{Re}(w) \leq M$.

Finally, we consider α' such that $\alpha'_2 = \alpha'_3 = 0$. Let

$$\Psi_5(y_1, x_{2,22}) = \int_{\mathbb{A}^3} \Psi(y_1, x'_{2,22}, x'_{2,23} x'_{2,33}) < x_{2,22} x'_{2,22} > dx'_{2,22} dx'_{2,23} dx'_{2,33}.$$

We choose $0 \leq \Psi_6 \in \mathscr{S}(Z_{\mathfrak{d}\mathbb{A}} \times \mathbb{A})$ so that $|\Psi_5(y_1, x_{2,22})| \leq \Psi_6(y_1, x_{2,22})$. Then

$$\sum_{\substack{\alpha'_2=\alpha'_3=0 \\ \alpha'_1 \in k^\times}} |f_{1,\alpha}(g_{\mathfrak{d}})| \ll \kappa_{\mathfrak{d}2}(g_{\mathfrak{d}}) \sum_{y_1 \in Z^{\mathrm{ss}}_{\mathfrak{d}k}} \Psi_6(g_{\mathfrak{d}} y_1, \gamma_{2,22}(t^0)^{-1} \alpha'_1).$$

Clearly, $\alpha = (\alpha'_1, 0, 0) A(y_1) = (2x_{2,12}\alpha'_1, 0, x_{1,22}\alpha'_1)$. If $y_1 \in Z^{\mathrm{ss}}_{\mathfrak{d}k}$, $x_{2,12} \neq 0$ or $x_{1,22} \neq 0$.

Suppose $\alpha_1 \neq 0$. Then

$$\sum_{\substack{\alpha'_2=\alpha'_3=0 \\ \alpha'_1 \in k^\times \\ \alpha_1 \neq 0}} |f_{2,\alpha}(g_{\mathfrak{d}}, w)| \leq \sum_{I,\tau} \widetilde{\mathscr{E}}_{I,\tau,1}(g_{\mathfrak{d}}, w)$$

where $I = (I_1, I_2)$ and $I_1 \ni 2$. So by (3.4.30), there exists a constant $\delta > 0$ and a slowly increasing function $h(t^0)$ such that if $l \gg 0, M > w_0$,

$$(10.5.10) \qquad \sum_{\substack{\alpha'_2=\alpha'_3=0 \\ \alpha'_1 \in k^\times \\ \alpha_1 \neq 0}} |f_{2,\alpha}(g_{\mathfrak{d}}, w)| \ll h(t^0)(\lambda_1^3 \mu_1)^l.$$

Consider the following conditions

$(10.5.11)_1$ For any $N_1, N_2 \geq 1, l \gg 0$,

$$f(g_{\mathfrak{d}}) \ll |\gamma_{1,22}(t^0)|^{-N_1} |\gamma_{2,13}(t^0)|^{-N_2} (\lambda_1^3 \mu_1)^l;$$

$(10.5.11)_2$ For any $N_1, N_2 \geq 1, l \gg 0$,

$$f(g_{\mathfrak{d}}) \ll |\gamma_{1,23}(t^0)|^{-N_1} |\gamma_{2,12}(t^0)|^{-N_2} (\lambda_1^3 \mu_1)^l.$$

Lemma (10.5.12) *Suppose that $f(g_{\mathfrak{d}})$ is a function on $G^0_{\mathbb{A}} \cap M_{\mathfrak{d}\mathbb{A}}/M_{\mathfrak{d}k}$ which satisfies either $(10.5.11)_1$ or $(10.5.11)_2$. Then for any $N > 0$,*

$$|f(g_{\mathfrak{d}})| \ll \mu_1^{-N} \mathrm{rd}_{2,N}(\lambda_1, \mu_2)$$

for $g_{\mathfrak{d}}$ in the Siegel domain.

Proof. If $f(g_{\mathfrak{d}})$ satisfies $(10.5.11)_1$, we choose

$$\begin{cases} N_1 \gg N_2 \gg l \gg 0 & \lambda_1 \geq 1, \mu_2 \geq 1, \\ N_1 = \frac{3}{4} N_2 \gg 0, \ l \text{ fixed} & \lambda_1 \leq 1, \mu_2 \geq 1, \\ N_1 = \frac{3}{2} N_2, \ l = N_2, \ N_2 \gg 0 & \lambda_1 \geq 1, \mu_2 \leq 1, \\ N_1 = \frac{2}{3} N_2, \ l = \frac{1}{4} N_2, \ N_2 \gg 0 & \lambda_1 \leq 1, \mu_2 \leq 1. \end{cases}$$

If $f(g_{\mathfrak{d}})$ satisfies $(10.5.11)_2$, we choose

$$\begin{cases} N_1 \gg N_2 \gg l \gg 0 & \lambda_1 \geq 1, \mu_2 \geq 1, \\ N_1 = \frac{3}{4}N_2 \gg 0,\ l \text{ fixed} & \lambda_1 \leq 1, \mu_2 \geq 1, \\ N_1 = \frac{3}{2}N_2,\ l = \frac{3}{4}N_2,\ N_2 \gg 0 & \lambda_1 \geq 1, \mu_2 \leq 1, \\ N_1 = \frac{2}{3}N_2,\ l = \frac{1}{4}N_2,\ N_2 \gg 0 & \lambda_1 \leq 1, \mu_2 \leq 1. \end{cases}$$

Then we have the estimate of the lemma.

Q.E.D.

By (10.5.10), (10.5.12), for any $N > 0$,

$$\sum_{\substack{\alpha_2' = \alpha_3' = 0 \\ \alpha_1' \in k^\times \\ \alpha_1 \neq 0}} |f_{1,\alpha}(g_{\mathfrak{d}}) f_{2,-\alpha}(g_{\mathfrak{d}}, w)| \ll \mu_1^{-N} \mathrm{rd}_{2,N}(\lambda_1, \mu_2) \tag{10.5.13}$$

for $g_{\mathfrak{d}}$ in the Siegel domain.

Suppose that $\alpha_1 = 0$, $\alpha_3 \neq 0$. If $\alpha_2' = \alpha_3' = 0$, $\alpha_2 = 0$. So

$$\sum_{\substack{\alpha_1 = \alpha_2' = \alpha_3' = 0 \\ \alpha_1' \in k^\times \\ \alpha_3 \neq 0}} |f_{2,\alpha}(g_{\mathfrak{d}}, w)| \leq \sum_{I,\tau} \widetilde{\mathscr{E}}_{I,\tau,1}(g_{\mathfrak{d}}, w)$$

where $I = (I_1, I_2)$ and $I_2 = \{1\}, I_1 \not\ni 1$. Consider $\widetilde{\mathscr{E}}_{I,\tau,1}(g_{\mathfrak{d}}, w)$ for such I.

Let $\mu_3 = |\gamma_{2,13}(t^0)|, \mu_4 = |\gamma_{2,22}(t^0)|$. Then $\mu_3 = \lambda_1^{-1}\mu_1^{-1}\mu_2^{-1}, \mu_4 = \lambda_1^2\mu_1^2\mu_2^{-1}$, and $\lambda_1 = \mu_1^{-1}\mu_3^{-\frac{1}{3}}\mu_4^{\frac{1}{3}}, \mu_2 = \mu_3^{-\frac{2}{3}}\mu_4^{-\frac{1}{3}}$. Let

$$d(\mu_1, \mu_3, \mu_4) = a(\underline{\mu}_1^2 \underline{\mu}_3^{\frac{2}{3}} \underline{\mu}_4^{-\frac{2}{3}}, \underline{\mu}_3^{-\frac{1}{3}} \underline{\mu}_4^{\frac{1}{3}}, \underline{\mu}_1^{-2} \underline{\mu}_3^{-\frac{1}{3}} \underline{\mu}_4^{\frac{1}{3}}; \underline{\mu}_3^{-\frac{2}{3}} \underline{\mu}_4^{-\frac{1}{3}}, \underline{\mu}_3^{\frac{2}{3}} \underline{\mu}_4^{\frac{1}{3}}). \tag{10.5.14}$$

Then $t^0 = d(\mu_1, \mu_3, \mu_4)\widehat{t}^0$.

By (1.2.6), for any $N_1, N_2, N_3 \geq 1$,

$$\sum_{\substack{y_1 \in Z^{ss}_{\mathfrak{d}k} \\ x_{1,22}, x_{2,13} \in k^\times \\ \alpha_1' \in k^\times}} |\Psi_6(g_{\mathfrak{d}} y_1, \gamma_{2,22}(t^0)^{-1}\alpha_1')| \ll \mu_3^{2N_1 - N_2} \mu_4^{-N_4}. \tag{10.5.15}$$

Lemma (10.5.16) *Suppose that $I_1 \not\ni 1$, $I_2 = \{1\}$. Then there exist a constant $\delta > 0$ and a slowly increasing function $h(\mu_3, \mu_4)$ such that if $M > w_0, l > 1$,*

$$\widetilde{\mathscr{E}}_{I,\tau,1}(g_{\mathfrak{d}}, w) \ll \mu_1^{-6} h(\mu_3, \mu_4)(\mu_3^{\frac{4}{3}}\mu_4^{\frac{2}{3}})^l$$

for $g_{\mathfrak{d}}$ in the Siegel domain and $w_0 - \delta \leq \mathrm{Re}(w) \leq M$.

Proof. Let $q \in D_{I,\tau}, r_1, r_2, r_3$ be as in (9.2.5). There exists a slowly increasing function $h_{\tau,r}(\mu_3, \mu_4)$ for each τ such that

$$\widetilde{\mathscr{E}}_{I,\tau,1}(g_{\mathfrak{d}}, w) \ll \mu_1^4 (\mu_3^{\frac{4}{3}}\mu_4^{\frac{2}{3}})^l \times \begin{cases} \mu_1^{2(r_1+r_2)} h_{\tau,r}(\mu_3, \mu_4) & \tau_1 = 1, \\ \mu_1^{2r_2} h_{\tau,r}(\mu_3, \mu_4) & \tau_1 = (1,2), \\ \mu_1^{-2(r_1+r_2)} h_{\tau,r}(\mu_3, \mu_4) & \tau_1 = (1,3), \\ \mu_1^{2r_1} h_{\tau,r}(\mu_3, \mu_4) & \tau_1 = (2,3), \\ \mu_1^{-2r_1} h_{\tau,r}(\mu_3, \mu_4) & \tau_1 = (1,2,3), \\ \mu_1^{-2r_2} h_{\tau,r}(\mu_3, \mu_4) & \tau_1 = (1,3,2). \end{cases}$$

Note that if $I_1 = \{2\}$, we have an extra $\mu_1^{-2l'}$ factor for some $l' \ll 0$, but since μ_1 is bounded below, for $g_{\mathfrak{d}}$ in the Siegel domain, we can ignore this factor.

We choose $r_3 = \frac{1}{2}$. If $\tau_1 = 1$, we choose $r_1, r_2 \ll 0$. If $\tau_1 = (1,2)$, we fix r_1 and choose $r_2 \ll 0$. If $\tau_1 = (2,3)$, we fix r_2 and choose $r_1 \ll 0$. If $\tau_1 = (1,2,3)$, we choose $r_1 = 1 - r_2 \gg 0$. If $\tau_1 = (1,3,2)$, we choose $r_2 = 1 - r_1 \gg 0$. Then $L(q) < w_0$, and the estimate of the lemma is satisfied. If $\tau_1 = (1,3)$, we can choose $r_1 = r_2 = 2$. Then $L(q) = 8 + \frac{1}{2}C < 4 + C$ since $C > 100$.

Q.E.D.

It is easy to see that $(t^0)^{-2\rho} = \mu_1^{-8}\mu_3^{-\frac{1}{3}}\mu_4^{-\frac{4}{3}}$ and $\kappa_{\mathfrak{d}2}(g_{\mathfrak{d}}) = \mu_1^6\mu_4^{-3}$. Therefore, by (10.5.15), (10.5.16), for any $N > 0$,

(10.5.17)
$$(t^0)^{-2\rho}\kappa_{\mathfrak{d}2}(g_{\mathfrak{d}}) \sum_{\substack{y_1 \in Z_{\mathfrak{d}k}^{ss} \\ x_{1,22}, x_{2,13} \in k^\times \\ \alpha_1' \in k^\times}} |\Psi_6(g_{\mathfrak{d}}y_1, \gamma_{2,22}(t^0)^{-1}\alpha_1')|\widetilde{\mathscr{E}}_{I,\tau,1}(g_{\mathfrak{d}}, w)$$
$$\ll \mu_1^{-8}\mathrm{rd}_{2,N}(\mu_3, \mu_4).$$

Proposition (10.5.1) now follows from (10.5.8), (10.5.9), (10.5.13), (10.5.17).

§10.6 The cases $\mathfrak{d} = (\beta_8, \beta_{8,2}), (\beta_9)$

Let $\mathfrak{d}_1 = (\beta_8, \beta_{8,2})$, $\mathfrak{d}_2 = (\mathfrak{d}_9)$, and $\mathfrak{d}_3 = (\beta_8)$. Consider paths $\mathfrak{p}_1 = (\mathfrak{d}_1, \mathfrak{s}_1), \mathfrak{p}_2 = (\mathfrak{d}_2, \mathfrak{s}_2)$, and $\mathfrak{p}_3 = (\mathfrak{d}_3, \mathfrak{s}_3)$ such that $\mathfrak{s}_1(1) = \mathfrak{s}_2(1) = \mathfrak{s}_3(1)$. This means that $\mathfrak{p}_3 \prec \mathfrak{p}_1$. Let

(10.6.1)
$$d(\lambda_1, \mu_1, \mu_2) = a(\underline{\lambda}_1^{-4}\underline{\mu}_1, \underline{\lambda}_1^2\underline{\mu}_2, \underline{\lambda}_1^2\underline{\mu}_1^{-1}\underline{\mu}_2^{-1}; \underline{\lambda}_1^{-3}\underline{\mu}_1^{-1}, \underline{\lambda}_1^3\underline{\mu}_1),$$

and $t^0 = d(\lambda_1, \mu_1, \mu_2)\widehat{t}^0$. For $u_2, u_3, u_4 \in \mathbb{A}, \alpha = (\alpha_1, \alpha_2, \alpha_3) \in k^3$, Let $n(u_{\mathfrak{d}}), < \alpha \cdot u_{\mathfrak{d}} >$ be as in §10.4.

We first consider $\mathfrak{p}_1$ such that $\mathfrak{s}_1(2) = 1$. It is easy to see that $e_{\mathfrak{p}_1 1}(t^0) = (t^0)^{-\beta_8} = \lambda_1^{-1}$ and $(\kappa_{\mathfrak{d}_3 2}(t^0)(t^0)^{-2\rho_{\mathfrak{d}_3}})^{-1}(t^0)^{-2\rho_{\mathfrak{d}_1}} = \lambda_1^8\mu_1^{-2}\mu_2^{-2}$. Clearly, $\mathscr{E}_{\mathfrak{p}_3}(t^0, w)$ is the constant term with respect to $n_{P_2}(u_2, u_2)$.

Let $\Psi = \widetilde{R}_{\mathfrak{d}_1}\Phi_{\mathfrak{p}_1}$. We consider

$$\int_{\mathbb{A}/k} \Theta_{Y_{\mathfrak{d}_1}}(\Psi, t^0 n''(u_3))\mathscr{E}_{\mathfrak{p}_3}(t^0 n''(u_3), w)du_3.$$

For $\alpha \in k$, we define

(10.6.2)
$$f_{1,\alpha}(t^0) = \int_{\mathbb{A}/k} \Theta_{Y_{\mathfrak{d}_1}}(\Psi, t^0 n''(u_3)) < \alpha u_3 > du_3,$$
$$f_{2,\alpha}(t^0, w) = \int_{\mathbb{A}/k} \mathscr{E}_{\mathfrak{p}_3}(t^0 n''(u_3), w) < \alpha u_3 > du_3.$$

Lemma (10.6.3) *Let $f(t^0)$ be a function on $T_{\mathbb{A}}^0/T_k$. Suppose that for any $N_1, N_2, N_3 \geq 1$,*

$$f(t^0) \ll |\gamma_{1,23}(t^0)|^{-N_1}|\gamma_{2,13}(t^0)|^{-N_2}|\gamma_{1,33}(t^0)|^{N_3}.$$

Then for any $N > 0$,

$$f(t^0) \ll \lambda_1^{-N} \mathrm{rd}_{2,N}(\mu_1, \mu_2).$$

Proof. Easy computations show that

$$\begin{aligned}&|\gamma_{1,23}(t^0)|^{-N_1}|\gamma_{2,13}(t^0)|^{-N_2}|\gamma_{1,33}(t^0)|^{N_3}\\ &= \lambda_1^{-(N_1+N_2-N_3)}\mu_1^{-(-2N_1+N_2+3N_3)}\mu_2^{-(-N_2+2N_3)}.\end{aligned}$$

We just have to choose N_1, N_2, N_3 in the following manner:

$$\begin{cases} N_1 = N_2 = N_3 \gg 0 & \mu_1 \geq 1, \mu_2 \geq 1, \\ N_1, N_3 \text{ fixed and } N_2 \gg 0 & \mu_1 \geq 1, \mu_2 \leq 1, \\ N_2 \text{ fixed and } N_1 \gg N_3 \gg 0 & \mu_1 \leq 1, \mu_2 \geq 1, \\ N_3 \text{ fixed and } N_1 \gg N_2 \gg 0 & \mu_1 \leq 1, \mu_2 \leq 1. \end{cases}$$

Q.E.D.

The following proposition is an easy consequence of the above lemma.

Proposition (10.6.4) *Suppose* $\mathfrak{s}_1(2) = 1$. *Then*

$$\widetilde{\Xi}_{\mathfrak{p}_1}(\Phi, \omega, w) = \int_{\substack{T_{\mathbb{A}}^0/T_k \\ \lambda_1 \geq 1}} \omega_{\mathfrak{p}_1}(t^0) \sum_{\alpha \in k^\times} \lambda_1^8 \mu_1^{-2} \mu_2^{-2} f_{1,\alpha}(t^0) f_{2,-\alpha}(t^0, w) d^\times t^0 \sim 0.$$

Next, we consider $\mathfrak{p}_2$ and $\mathfrak{p}_1$ such that $\mathfrak{s}_1(2) = 0$. The proof of the following lemma is easy, and is left to the reader.

Lemma (10.6.5) *The convex hull of each of the sets*
(1) $\{\gamma_{1,23}, \gamma_{2,22}, \gamma_{2,13}, -\gamma_{2,33}\}$,
(2) $\{\gamma_{1,23}, \gamma_{2,22}, \gamma_{2,13}, -\gamma_{2,23}\}$
contains a neighborhood of the origin.

Proposition (10.6.6)

$$\epsilon_{\mathfrak{p}_1} \widetilde{\Xi}_{\mathfrak{p}_1}(\Phi, \omega, w) + \epsilon_{\mathfrak{p}_2} \widetilde{\Xi}_{\mathfrak{p}_2}(\Phi, \omega, w) \sim 0.$$

We devote the rest of this section to the proof of the above proposition.

Let $\Psi = \widetilde{R}_{\mathfrak{d}_2} \Phi_{\mathfrak{p}_2}$. Any element $y \in Y_{\mathfrak{d}_2 k}^{\mathrm{ss}}$ can be written in the form $y = (y_1, y_2)$ where $y_1 = (x_{1,23}, x_{2,13}, x_{2,22}) \in Z_{\mathfrak{d}_2 k}^{\mathrm{ss}}$ and $y_2 = (x_{1,33}, x_{2,23}, x_{2,33}) \in W_{\mathfrak{d}_2 k}$. Then

$$n'(u_1) n(u_{\mathfrak{d}}) y = (y_1, x_{1,33} + u_3', x_{2,13} + u_4' + x_{2,13} u_1, x_{2,33} + u_2'),$$

where

$$\begin{aligned} u_3' &= x_{1,23} u_3, \\ u_4' &= 2x_{2,22} u_3 + x_{1,23} u_4, \\ u_2' &= x_{2,13} u_2 + x_{2,22} u_3^2 + u_3' u_4. \end{aligned}$$

Let

$$(10.6.7) \quad f_{3,\alpha}(t^0 n'(u_1)) = \sum_{y_1 \in Z_{\mathfrak{d}_2 k}^{\mathrm{ss}}} \int_{\mathbb{A}^3} \Psi(t^0 n'(u_1) n(u_{\mathfrak{d}})(y_1, 0)) < \alpha \cdot u_{\mathfrak{d}} > du_{\mathfrak{d}};$$

$$f_{4,\alpha}(t^0 n'(u_1), w) = \int_{(\mathbb{A}/k)^3} \mathscr{E}(t^0 n'(u_1) n(u_{\mathfrak{d}}), w) < \alpha \cdot u_{\mathfrak{d}} > du_{\mathfrak{d}}.$$

An easy computation shows that

$$<\alpha\cdot u_{\mathfrak{d}}>=<\alpha_1 x_{2,13}^{-1}u_2' - \alpha_1(x_{2,13}^{-1}x_{1,23}^{-2}x_{2,22}u_3'^{\,2} - x_{2,13}^{-1}u_3'(u_4' - 2x_{1,23}^{-1}x_{2,22}u_3'))> \\ \times <\alpha_2 x_{1,23}^{-1}u_3' + \alpha_3(x_{1,23}^{-1}u_4' - 2x_{1,23}^{-2}x_{2,22}u_3')>.$$

Let

$$\psi_\alpha'(y_1,u_3',u_4') = <-\alpha_1(x_{2,13}^{-1}x_{1,23}^{-2}x_{2,22}u_3'^{\,2} - x_{2,13}^{-1}u_3'(u_4' - 2x_{1,23}^{-1}x_{2,22}u_3'))> \\ \times <\alpha_2 x_{1,23}^{-1}u_3' + \alpha_3(x_{1,23}^{-1}u_4' - 2x_{1,23}^{-2}x_{2,22}u_3')>.$$

We also define

$$\Psi'(t^0,y_1,u_1,u_2',u_3',u_4') \\ = \Psi(t^0n'(u_1)y_1,\gamma_{1,33}(t^0)u_3',\gamma_{2,23}(t^0)(u_4'+x_{2,13}u_1),\gamma_{2,33}(t^0)u_2').$$

Then $f_{3,\alpha}(t^0n'(u_1))$ is equal to

$$\sum_{y_1\in Z_{\mathfrak{d}_2 k}^{ss}} \int_{\mathbb{A}^3} \Psi'(t^0,y_1,u_1,u_2',u_3',u_4') <\alpha_1 x_{2,13}^{-1}u_2'> \psi_\alpha'(y_1,u_3',u_4')du'.$$

Let Ψ_1 be the partial Fourier transform of Ψ with respect to the coordinate $x_{2,33}$ and $< >$. We choose $0\leq\Psi_2\in\mathscr{S}(Z_{\mathfrak{d}_2\mathbb{A}}\times\mathbb{A})$ so that

$$\int_{\mathbb{A}^2} |\Psi_1(y_1,x_{1,33},x_{2,23},x_{2,33})|dx_{1,33}dx_{2,23}\leq\Psi_2(y_1,x_{2,33}).$$

Then

$$|f_{1,\alpha}(t^0n'(u_1))|\ll\kappa_{\mathfrak{d}_2 2}(t^0)\sum_{y_1\in Z_{\mathfrak{d}_2 k}^{ss}}\Psi_2(t^0y_1,\gamma_{2,33}(t^0)^{-1}\alpha_1x_{2,13}^{-1}).$$

This bound does not depend on α_2,α_3. By (10.6.6), for any $N>0$,

$$\sum_{\alpha_1\in k^\times}|f_{3,\alpha}(t^0n'(u_1))|\ll \mathrm{rd}_{3,N}(\lambda_1,\mu_1,\mu_3).$$

Therefore, there exists a constant $\delta>0$ such that if $M>w_0$,

$$\sum_{\substack{\alpha_1\in k^\times\\ \alpha_2,\alpha_3\in k}}|f_{3,\alpha}(t^0n'(u_1))f_{4,-\alpha}(t^0n'(u_1),w)|\ll \mathrm{rd}_{3,N}(\lambda_1,\mu_1,\mu_3) \tag{10.6.8}$$

for $w_0-\delta\leq\mathrm{Re}(w)\leq M$.

Next, we consider α such that $\alpha_1=0$, $(\alpha_2,\alpha_3)\neq 0$.

Let

$$\Psi_3(y_1,x_{1,33},x_{2,23}) = \int_{\mathbb{A}^2}\Psi(y_1,x_{1,33},x_{2,23}',x_{2,33}')<x_{2,23}x_{2,23}'>dx_{2,23}'dx_{2,33}'.$$

We choose $0\leq\Psi_4\in\mathscr{S}(Z_{\mathfrak{d}\mathbb{A}}\times\mathbb{A})$ so that

$$\int_{\mathbb{A}}|\Psi_3(y_1,x_{1,33},x_{2,23})|dx_{1,33}\leq\Psi_4(y_1,x_{2,23}).$$

Then

$$|f_{3,\alpha}(t^0 n'(u_1))| \ll \kappa_{\mathfrak{d}_2 2}(t^0) \sum_{y_1 \in Z^{\rm ss}_{\mathfrak{d}_2 k}} \Psi_4(t^0 y_1, \gamma_{2,23}(t^0)^{-1} \alpha_3 x_{1,23}^{-1}).$$

By (10.6.5), for any $N > 0$,

$$\sum_{\substack{\alpha_3 \in k^\times \\ \alpha_1 = 0}} |f_{3,\alpha}(t^0 n'(u_1))| \ll \mathrm{rd}_{3,N}(\lambda_1, \mu_1, \mu_2).$$

Therefore, there exists a constant $\delta > 0$ such that if $M > w_0$,

$$\sum_{\substack{\alpha_3 \in k^\times \\ \alpha_1 = 0 \\ \alpha_2 \in k}} |f_{3,\alpha}(t^0 n'(u_1)) f_{4,-\alpha}(t^0 n'(u_1), w)| \ll \mathrm{rd}_{3,N}(\lambda_1, \mu_1, \mu_2) \tag{10.6.9}$$

for $w_0 - \delta \leq \mathrm{Re}(w) \leq M$.

Finally, we consider α such that $\alpha_1 = \alpha_3 = 0, \alpha_2 \in k^\times$. Let Ψ_5 be the partial Fourier transform of Ψ_3 with respect to the coordinate $x_{1,33}$ and $< >$. Then

$$f_{3,\alpha}(t^0 n'(u_1)) = \kappa_{\mathfrak{d}_2 2}(t^0) \sum_{y_1 \in Z^{\rm ss}_{\mathfrak{d}_2 k}} \Psi_5(t^0 y_1, \gamma_{1,33}(t^0)^{-1} \alpha_2 x_{1,23}^{-1}).$$

Thus $f_{3,\alpha}(t^0 n'(u_1))$ does not depend on u_1 any more. Therefore,

$$\begin{aligned} &\int_{\mathbb{A}/k} \sum_{\substack{\alpha_1 = \alpha_3 = 0 \\ \alpha_2 \in k^\times}} f_{3,\alpha}(t^0 n'(u_1)) f_{4,-\alpha}(t^0 n'(u_1), w) du_1 \\ &\quad = \sum_{\substack{\alpha_1 = \alpha_3 = 0 \\ \alpha_2 \in k^\times}} f_{3,\alpha}(t^0) \int_{\mathbb{A}/k} f_{4,-\alpha}(t^0 n'(u_1), w) du_1, \end{aligned}$$

and

$$\int_{\mathbb{A}/k} f_{4,-\alpha}(t^0 n'(u_1), w) du_1 = f_{2,-\alpha_2}(t^0, w). \tag{10.6.10}$$

Let Ψ_6 be the partial Fourier transform of Ψ_5 with respect to the coordinate $x_{2,22}$ and $< >$. Let Ψ_7, Ψ_8 be restrictions of Ψ_5, Ψ_6 to the first three coordinates. Let

$$\begin{aligned} &A(t^0, x_{1,23}, x_{2,13}, x_{2,22}, \alpha_2) \\ &= (\gamma_{1,23}(t^0) x_{1,23}, \gamma_{2,13}(t^0) x_{2,13}, \gamma_{2,22}(t^0)^{-1} x_{2,22}, \gamma_{1,33}(t^0)^{-1} \alpha_2 x_{1,23}^{-1}). \end{aligned}$$

We define

$$\begin{aligned} f_{5,\alpha_2}(t^0) &= \kappa_{\mathfrak{d}_2 2}(t^0) |\gamma_{2,22}(t^0)|^{-1} \sum_{y_1 \in Z^{\rm ss}_{\mathfrak{d}_2 k}} \Psi_5(A(t^0, x_{1,23}, x_{2,13}, x_{2,22}, \alpha_2)), \\ f_{6,\alpha_2}(t^0) &= \kappa_{\mathfrak{d}_2 2}(t^0) |\gamma_{2,22}(t^0)|^{-1} \sum_{x_{1,23}, x_{2,13} \in k^\times} \Psi_8(A(t^0, x_{1,23}, x_{2,13}, 0, \alpha_2)), \\ f_{7,\alpha_2}(t^0) &= \kappa_{\mathfrak{d}_2 2}(t^0) \sum_{x_{1,23}, x_{2,13} \in k^\times} \Psi_7(A(t^0, x_{1,23}, x_{2,13}, 0, \alpha_2)). \end{aligned}$$

Then $f_{3,(0,\alpha_2,0)}(t^0) = f_{5,\alpha_2}(t^0) + f_{6,\alpha_2}(t^0) - f_{7,\alpha_2}(t^0)$. We choose Schwartz–Bruhat functions $0 \le \Psi_9, \Psi_{10}$ on $\mathbb{A}^4, \mathbb{A}^3$ so that $|\Psi_5| \le \Psi_9, |\Psi_6|, |\Psi_7| \le \Psi_{10}$. Then

$$\sum_{\alpha_2 \in k^\times} |f_{5,\alpha_2}(t^0)|, \sum_{\alpha_2 \in k^\times} |f_{6,\alpha_2}(t^0)|, \sum_{\alpha_2 \in k^\times} |f_{7,\alpha_2}(t^0)|$$

are bounded by

$$\begin{aligned}
&\kappa_{\mathfrak{d}_2 2}(t^0)|\gamma_{2,22}(t^0)|^{-1}\Theta_4(\Psi_9, \gamma_{1,23}(t^0), \gamma_{2,13}(t^0), \gamma_{2,22}(t^0)^{-1}, \gamma_{1,33}(t^0)^{-1}),\\
&\kappa_{\mathfrak{d}_2 2}(t^0)|\gamma_{2,22}(t^0)|^{-1}\Theta_3(\Psi_{10}, \gamma_{1,23}(t^0), \gamma_{2,13}(t^0), \gamma_{1,33}(t^0)^{-1}),\\
&\kappa_{\mathfrak{d}_2 2}(t^0)\Theta_3(\Psi_{10}, \gamma_{1,23}(t^0), \gamma_{2,13}(t^0), \gamma_{1,33}(t^0)^{-1}),
\end{aligned}$$

respectively.

Proposition (10.6.11)
(1) *Suppose that $f(t^0)$ is a function such that for any $N_1, N_2, N_3, N_4 \ge 1$,*

$$f(t^0) \ll |\gamma_{1,23}(t^0)|^{-N_1}|\gamma_{2,22}(t^0)|^{-N_2}|\gamma_{2,13}(t^0)|^{-N_3}|\gamma_{1,33}(t^0)|^{N_4}.$$

Then for any $N > 0$, $|f(t^0)| \ll \lambda_1^{-N}\mathrm{rd}_{2,N}(\mu_1, \mu_2)$ for $\lambda_1 \ge 1$.
(2) *Suppose that $f(t^0)$ is a function such that for any $N_1, N_2, N_3, N_4 \ge 1$,*

$$f(t^0) \ll |\gamma_{1,23}(t^0)|^{-N_1}|\gamma_{2,22}(t^0)|^{N_2}|\gamma_{2,13}(t^0)|^{-N_3}|\gamma_{1,33}(t^0)|^{N_4}.$$

Then for any $N > 0$, $|f(t^0)| \ll \lambda_1^{N}\mathrm{rd}_{2,N}(\mu_1, \mu_2)$ for $\lambda_1 \le 1$.

Proof. Easy computations show that

$$\begin{aligned}
&|\gamma_{1,23}(t^0)|^{-N_1}|\gamma_{2,22}(t^0)|^{-N_2}|\gamma_{2,13}(t^0)|^{-N_3}|\gamma_{1,33}(t^0)|^{N_4}\\
&= \lambda_1^{-(N_1+7N_2+N_3-N_4)}\mu_1^{-(-2N_1+N_2+N_3+3N_4)}\mu_2^{-(2N_1-N_3+2N_4)},\\
&|\gamma_{1,23}(t^0)|^{-N_1}|\gamma_{2,22}(t^0)|^{N_2}|\gamma_{2,13}(t^0)|^{-N_3}|\gamma_{1,33}(t^0)|^{N_4}\\
&= \lambda_1^{-(N_1-7N_2+N_3-N_4)}\mu_1^{-(-2N_1-N_2+N_3+3N_4)}\mu_2^{-(2N_1-N_3+2N_4)}.
\end{aligned}$$

We choose $N_1, \cdots, N_4$ in the following manner.

$$\begin{cases}
N_1, N_3 \text{ fixed, and } N_2 \gg N_4 \gg 0 & \mu_1 \ge 1, \mu_2 \ge 1,\\
N_1, N_2, N_4 \text{ fixed, and } N_3 \gg 0 & \mu_1 \ge 1, \mu_2 \le 1,\\
N_3, N_4 \text{ fixed, and } N_1 \gg N_2 \gg 0 & \mu_1 \le 1, \mu_2 \ge 1,\\
N_2, N_4 \text{ fixed, and } N_1 \gg N_3 \gg 0 & \mu_1 \le 1, \mu_2 \le 1.
\end{cases}$$

This proves (1).

We choose $N_1, \cdots, N_4$ in the following manner.

$$\begin{cases}
N_1, N_4 \text{ fixed, and } N_3 = 2N_2 \gg 0 & \mu_1 \ge 1, \mu_2 \ge 1,\\
N_1, N_2, N_4 \text{ fixed, and } N_3 \gg 0 & \mu_1 \ge 1, \mu_2 \le 1,\\
N_3, N_4 \text{ fixed, and } N_2 \gg N_1 \gg 0 & \mu_1 \le 1, \mu_2 \ge 1,\\
N_1, N_4 \text{ fixed, and } N_2 \gg N_3 \gg 0 & \mu_1 \le 1, \mu_2 \le 1.
\end{cases}$$

This proves (2).

Q.E.D.

By (10.6.11)(1),

$$\int_{\substack{T^0_{\mathbb{A}}/T_k \\ \lambda_1\geq 1}} \omega_{\mathfrak{p}_2}(t^0) \sum_{\alpha_2\in k^\times} f_{3,(0,\alpha_2,0)}(t^0) f_{2,-\alpha_2}(t^0,w)(t^0)^{-2\rho} d^\times t^0 \sim 0, \tag{10.6.12}$$

and by (10.6.11)(2),

$$\int_{\substack{T^0_{\mathbb{A}}/T_k \\ \lambda_1\leq 1}} \omega_{\mathfrak{p}_2}(t^0) \sum_{\alpha_2\in k^\times} f_{5,\alpha_2}(t^0) f_{2,-\alpha_2}(t^0,w)(t^0)^{-2\rho} d^\times t^0 \sim 0. \tag{10.6.13}$$

The following lemma is easy to prove, and we omit the proof.

Lemma (10.6.14) *The convex hull of each of the sets*
(1) $\{(1,-2,0),(1,1,-1),(-1,3,2),(-4,1,0)\}$,
(2) $\{(1,-2,0),(1,1,-1),(-1,3,2),(-2,1,1)\}$
contains a neighborhood of the origin.

Now we want to cancel out $\widetilde{\Xi}_{\mathfrak{p}_1}(\Phi,\omega,w)$ with $\widetilde{\Xi}_{\mathfrak{p}_2}(\Phi,\omega,w)$. However, we are still obliged to prove that $\widetilde{\Xi}_{\mathfrak{p}_1}(\Phi,\omega,w)$ is well defined for $\mathrm{Re}(w)\gg 0$. Consider $\widetilde{\mathscr{E}}_{I,\tau,1}(t^0,w)$ such that $I=(I_1,I_2)$ and $I_2=\emptyset, I_1=\{2\}$.

Lemma (10.6.15)
(1) *If* $\tau_1=(2,3)$ *or* $(1,2,3)$, *there exists a constant* $\delta>0$ *and a slowly increasing function* $h(t^0)$ *such that for any* $N\gg 0, M>w_0$,

$$\widetilde{\mathscr{E}}_{I,\tau,1}(t^0,w)\ll h(t^0)(\lambda_1^4\mu_1^{-1})^N$$

for $w_0-\delta\leq \mathrm{Re}(w)\leq M$.
(2) *If* $\tau_1=(1,3)$, *there exists a slowly increasing function* $h(t^0)$ *and a constant* $c>0$ *such that if* $M_2>M_1>w_0$,

$$\widetilde{\mathscr{E}}_{I,\tau,1}(t^0,w)\ll h(t^0)(\lambda_1^2\mu_1^{-1}\mu_2^{-1})^{c\mathrm{Re}(w)}$$

for $M_1\leq \mathrm{Re}(w)\leq M_2$.

Proof. Let $q\in D_{I,\tau}$. Let $r=(r_1,r_2,r_3)$ be as in (9.2.5). By (3.4.30), there exists a constant $\delta>0$ and a slowly increasing function $h(t^0)$ such that if $M>w_0$,

$$\widetilde{\mathscr{E}}_{I,\tau,1}(t^0,w)\ll h(t^0)\times\begin{cases}\lambda_1^{-4r_1-2r_2}\mu_1^{r_1}\mu_2^{-r_2} & \tau_1=(2,3),\\ \lambda_1^{2r_1-2r_2}\mu_1^{-r_1}\mu_2^{-(r_1+r_2)} & \tau_1=(1,2,3),\\ \lambda_1^{2r_1+r_2}\mu_1^{-(r_1+r_2)}\mu_2^{-r_1} & \tau_1=(1,3)\end{cases}$$

for $L(q)+\delta\leq \mathrm{Re}(w)\leq M$ where $\delta>0, M>w_0$ are constants. If $\tau_1=(2,3)$, we fix r_2 and choose $r_1\ll 0$. If $\tau_1=(1,2,3)$, we choose $r_1=1-r_2\gg 0$. Then we get an estimate of the form (1). If $\tau_1=(2,3)$, we fix r_2 and choose $r_1\gg 0$. Then $L(q)$ depends linearly on r_1. Therefore, we get an estimate of the form (2).

Q.E.D.

Since $e_{\mathfrak{p}_1 1}(t^0) = (t^0)^{\beta_8} = \lambda_1$, by the Parseval formula,

$$\widetilde{\Xi}_{\mathfrak{p}_1}(\Phi, \omega, w) = \int_{\substack{T^0_{\mathbb{A}}/T_k \\ \lambda_1 \leq 1}} \omega_{\mathfrak{p}_1}(t^0) \sum_{\alpha_2 \in k^\times} f_{6,\alpha_2}(t^0) f_{2,-\alpha_2}(t^0, w)(t^0)^{-2\rho} d^\times t^0.$$

By the consideration as in §2.1,

$$\sum_{\alpha_2 \in k^\times} |f_{2,\alpha_2}(t^0, w)| \ll \sum_{\tau} \widetilde{\mathscr{E}}_{I,\tau,1}(t^0, w),$$

where $I_1 = \{2\}, I_2 = \emptyset$. Therefore, the following proposition follows from (10.6.14), (10.6.15).

Lemma (10.6.16) *The distribution* $\widetilde{\Xi}_{\mathfrak{p}_1}(\Phi, \omega, w)$ *is well defined if* $\mathrm{Re}(w) \gg 0$.

Therefore, by (10.6.10), (10.6.16),

$$\begin{aligned} &\epsilon_{\mathfrak{p}_1} \widetilde{\Xi}_{\mathfrak{p}_1}(\Phi, \omega, w) + \epsilon_{\mathfrak{p}_2} \widetilde{\Xi}_{\mathfrak{p}_2}(\Phi, \omega, w) \\ &\sim \epsilon_{\mathfrak{p}_1} \int_{\substack{T^0_{\mathbb{A}}/T_k \\ \lambda_1 \leq 1}} \omega_{\mathfrak{p}_1}(t^0) \sum_{\alpha_2 \in k^\times} f_{7,\alpha_2}(t^0) f_{2,-\alpha_2}(t^0, w)(t^0)^{-2\rho} d^\times t^0. \end{aligned}$$

It is easy to see that $(t^0)^{-2\rho} = \lambda_1^{18}\mu_1^{-2}\mu_2^{-2}$ and $\kappa_{\mathfrak{d}_2 2}(t^0) = \lambda_1^{-15}\mu_1^4\mu_2^4$. Therefore, for any $N_1, N_2, N_3 \geq 1$,

$$\lambda_1^{18}\mu_1^{-2}\mu_2^{-2} \sum_{\alpha_2 \in k^\times} |f_{7,\alpha_2}(t^0)| \ll \lambda_1^3\mu_1^2\mu_2^2(\lambda_1\mu_1^{-2})^{-N_1}(\lambda_1\mu_2\mu_2^{-1})^{-N_2}(\lambda_1\mu_1^3\mu_2^2)^{-N_3}.$$

We have already estimated

$$\sum_{\alpha_2 \in k^\times} |f_{7,\alpha_2}(t^0)| \widetilde{\mathscr{E}}_{I,\tau,1}(t^0, w)$$

for $\tau_1 \neq (1,3)$, so we assume that $\tau_1 = (1,3)$. By (9.2.10), for any $\epsilon > 0$, there exists a constant $\delta > 0$ such that if $M > w_0$,

$$\widetilde{\mathscr{E}}_{I,\tau,1}(t^0, w) \ll \lambda_1^{2+c_1}\mu_1^{c_2}\mu_2^{c_3}$$

for $w_0 - \delta \leq \mathrm{Re}(w) \leq M$ where $|c_1|, |c_2|, |c_3| < \epsilon$.

Therefore,

$$\text{(10.6.17)} \quad \begin{aligned} &(t^0)^{-2\rho} \sum_{\alpha_2 \in k^\times} |f_{7,\alpha_2}(t^0)| \widetilde{\mathscr{E}}_{I,\tau,1}(t^0, w) \\ &\ll \lambda_1^{5+c_1}\mu_1^{2+c_2}\mu_2^{2+c_3}(\lambda_1\mu_1^{-2})^{-N_1}(\lambda_1\mu_1\mu_2^{-1})^{-N_2}(\lambda_1^{-1}\mu_1^3\mu_2^2)^{-N_3}. \end{aligned}$$

We fix N_1, N_2 and choose $N_3 \gg 0$. Then there exists a slowly increasing function $h_2(t^0)$ such that for any $N > 0$, (10.6.17) is bounded by a constant multiple of $h_2(t^0)(\lambda_1\mu_1^{-3}\mu_2^{-2})^N$. (10.6.17) is also bounded by a constant multiple of

$$\inf(\lambda_1^{2+c_1}\mu_1^{-2+c_2}\mu_2^{3+c_3}, \lambda_1^{3+c_1}\mu_1^{1+c_2}\mu_2^{-1+c_3}, \lambda_1^{\frac{1}{2}+c_1}\mu_1^{\frac{5}{2}+c_3}\mu_2^{\frac{5}{2}+c_3})$$

by choosing $(N_1, N_2, N_3) = (1,3,1), (3,1,2), (3,\frac{5}{2},1)$.

Therefore,

$$\int_{\substack{T^0_{\mathbb{A}}/T_k \\ \lambda_1 \leq 1}} \omega_{\mathfrak{p}_2}(t^0) \sum_{\alpha_2 \in k^\times} f_{7,\alpha_2}(t^0) f_{2,-\alpha_2}(t^0)(t^0)^{-2\rho} d^\times t^0 \sim 0.$$

This completes the proof of (10.6.6). Hence, we finished the proof of Theorem (10.0.1), (10.0.2). This implies that we can ignore the non-constant terms associated with unstable strata.

Chapter 11 Unstable distributions

In the next three chapters, we consider distributions $\Xi_{\mathfrak{p}}(\Phi, \omega, w)$ and compute the principal part of the zeta function for the quartic case.

§11.1 Unstable distributions

In this section, we define certain distributions associated with unstable strata. We remind the reader that we use the notation in previous chapters like $n'(u), n''(u)$ in (9.2.3) or $\widehat{t}^0$ in (9.2.4).

Definition (11.1.1) *For a character* $\omega = (\omega_1, \omega_2)$ of $(\mathbb{A}^\times/k^\times)^2$, *we define*
(1) $\delta_{\#}(\omega) = \delta(\omega_1)\delta(\omega_2)$,
(2) $\delta_{\mathfrak{d}}(\omega) = \delta(\omega_2)$, $\omega_{\mathfrak{d}} = (1, \omega_1)$ *for* $\mathfrak{d} = (\beta_1)$,
(3) $\delta_{\mathfrak{d}}(\omega) = \delta(\omega_1)$, $\omega_{\mathfrak{d}} = (1, \omega_2)$ *for* $\mathfrak{d} = (\beta_4)$,
(4) $\delta_{\mathfrak{d}}(\omega) = \delta(\omega_1^2)$, $\omega_{\mathfrak{d}} = (\omega_1, \omega_2, \omega_2)$ *for* $\mathfrak{d} = (\beta_6)$,
(5) $\delta_{\mathfrak{d}}(\omega) = \delta(\omega_1^3)\delta(\omega_2^2)$, $\omega_{\mathfrak{d}} = (\omega_2, \omega_1, \omega_1^{-1}\omega_2)$ *for* $\mathfrak{d} = (\beta_7)$,
(6) $\delta_{\mathfrak{d}}(\omega) = \delta(\omega_1)$, $\omega_{\mathfrak{d}} = (\omega_1, \omega_2, \omega_2)$ *for* $\mathfrak{d} = (\beta_8)$,
(7) $\delta_{\mathfrak{d}}(\omega) = \delta(\omega_1^3)\delta(\omega_2)$, $\omega_{\mathfrak{d}} = (1, \omega_1^{-1}, \omega_1)$ *for* $\mathfrak{d} = (\beta_9)$,
(8) $\delta_{\mathfrak{d}}(\omega) = \delta(\omega_1), \delta_{\mathfrak{d},\mathrm{st}}(\omega) = \delta(\omega_1)\delta(\omega_2^2)$, $\omega_{\mathfrak{d}} = (1, \omega_2), \omega_{\mathfrak{d},\mathrm{st}} = (\omega_2, \omega_2)$ *for* $\mathfrak{d} = (\beta_{10})$.

Definition (11.1.2) *We define subsets* $\mathfrak{W}_1, \mathfrak{W}_2$ *of permutations of* $\{1,2,3\}$ *as follows.*
(1) $\mathfrak{W}_1 = \{1, (2,3), (1,2,3)\}$.
(2) $\mathfrak{W}_2 = \{1, (1,2), (1,3,2)\}$.

Let $\mathfrak{d}$ be a β-sequence for which $P_{\mathfrak{d}\#} = P_1 \times B_2$ or $P_2 \times B_2$. The double coset $P_{\mathfrak{d}\# k} \setminus G_k / B_k$ is represented by $\mathfrak{W}_1 \times \{1, (1,2)\}$ if $P_{\mathfrak{d}\#} = P_1 \times B_2$, and by $\mathfrak{W}_2 \times \{1, (1,2)\}$ if $P_{\mathfrak{d}\#} = P_2 \times B_2$. The reductive part $M_{\mathfrak{d}}$ of $P_{\mathfrak{d}\#}$ is of the form $M_{\mathfrak{d}} = M_{\mathfrak{d}1} \times M_{\mathfrak{d}2}$ and $M_{\mathfrak{d}1} = M_{\mathfrak{d}11} \times M_{\mathfrak{d}12}$ where $M_{\mathfrak{d}11} \cong \mathrm{GL}(2)$, $M_{\mathfrak{d}12} \cong \mathrm{GL}(1)$ if $P_{\mathfrak{d}\#} = P_1 \times B_2$ and $M_{\mathfrak{d}11} \cong \mathrm{GL}(1)$, $M_{\mathfrak{d}12} \cong \mathrm{GL}(2)$ if $P_{\mathfrak{d}\#} = P_2 \times B_2$.

By a standard argument, $P_{1k}\tau_1 B_{1k}/B_{1k}, P_{2k}\tau_1 B_{1k}/B_{1k}$ are represented by the sets

$$\{\gamma_1\gamma_2 \mid \gamma_1 \in M_{\mathfrak{d}11k}/M_{\mathfrak{d}11k} \cap B_{1k}, \gamma_2 \in N^-_{\tau_1 k}\},$$
$$\{\gamma_1\gamma_2 \mid \gamma_1 \in M_{\mathfrak{d}12k}/M_{\mathfrak{d}12k} \cap B_{1k}, \gamma_2 \in N^-_{\tau_1 k}\},$$

respectively.

Let $d_{P_i}(\lambda_1)$ be as in (9.2.4), and $d_i(\lambda_1, \lambda_2) = (d_{P_i}(\lambda_1), a_2(\underline{\lambda_2}, \underline{\lambda_2}^{-1}))$ for $i = 1, 2$. Let $\lambda'_{\mathfrak{d}} = d_1(\lambda_1, \lambda_2)$ if $P_{\mathfrak{d}\#} = P_1 \times B_2$, and $\lambda'_{\mathfrak{d}} = d_2(\lambda_1, \lambda_2)$ if $P_{\mathfrak{d}\#} = P_2 \times B_2$. Any element $g_{\mathfrak{d}}$ can be written in the form $g_{\mathfrak{d}} = \lambda'_{\mathfrak{d}} g^0_{\mathfrak{d}}$ where $g^0_{\mathfrak{d}} \in M^0_{\mathfrak{d}\mathbb{A}}$. The element $g^0_{\mathfrak{d}}$ is of the form

$$(11.1.3) \qquad \left(\begin{pmatrix} g_{1,12} & \\ & t_{13} \end{pmatrix}\begin{pmatrix} t_{21} & \\ & t_{22} \end{pmatrix}\right) \text{ or } \left(\begin{pmatrix} t_{11} & \\ & g_{1,23} \end{pmatrix}\begin{pmatrix} t_{21} & \\ & t_{22} \end{pmatrix}\right),$$

where $g_{1,12}, g_{1,23} \in \mathrm{GL}(2)^0_{\mathbb{A}}$ and $|t_{13}| = |t_{21}| = |t_{22}| = |t_{11}| = 1$. The following lemma can be obtained by a standard argument and the proof is left to the reader.

Lemma (11.1.4) *In the following two statements, $\tau = (\tau_1, \tau_2)$.*
(1) *If $P_{\mathfrak{d}\#} = P_1 \times B_2$,*

$$\int_{U_{\mathfrak{d}\#\mathbb{A}}/U_{\mathfrak{d}\#k}} E_B(g_{\mathfrak{d}} n(u_{\mathfrak{d}}), z) du_{\mathfrak{d}} = \sum_{\substack{\tau_1 \in \mathfrak{W}_1 \\ \tau_2 = 1,(1,2)}} M_\tau(z) \lambda'^{\tau z + \rho}_{\mathfrak{d}} E_{M_{\mathfrak{d}11} \cap B_1}(g_{1,12}, z_{1\tau_1(1)}, z_{1\tau_1(2)}).$$

(2) *If $P_{\mathfrak{d}\#} = P_2 \times B_2$,*

$$\int_{U_{\mathfrak{d}\#\mathbb{A}}/U_{\mathfrak{d}\#k}} E_B(g_{\mathfrak{d}} n(u_{\mathfrak{d}}), z) du_{\mathfrak{d}} = \sum_{\substack{\tau_1 \in \mathfrak{W}_2 \\ \tau_2 = 1,(1,2)}} M_\tau(z) \lambda'^{\tau z + \rho}_{\mathfrak{d}} E_{M_{\mathfrak{d}12} \cap B_1}(g_{1,23}, z_{1\tau_1(2)}, z_{1\tau_1(3)}).$$

The following lemma is a consequence of the Mellin inversion formula.

Lemma (11.1.5) *Suppose that $f(s)$ is a holomorphic function which is rapidly decreasing on any vertical strip contained in the domain $\{s \mid \mathrm{Re}(s) > 0\}$. Let $E(g^0, s)$ be the Eisenstein series on $\mathrm{GL}(2)^0_{\mathbb{A}}$, and ω a character of $\mathbb{A}^\times/k^\times$. Then*

$$\frac{1}{2\pi\sqrt{-1}} \int_{\mathrm{GL}(2)^0_{\mathbb{A}}/\mathrm{GL}(2)_k} \int_{\mathrm{Re}(s)=r>0} \omega(\det g^0) f(s) E(g^0, s) dg^0 ds = \delta(\omega) f(1).$$

Let $W = \mathrm{Sym}^2 k^2$, i.e. the space of binary quadratic forms. For $\Psi \in \mathscr{S}(W_{\mathbb{A}})$, $\omega = (\omega_1, \omega_2)$ a character of $(\mathbb{A}^\times/k^\times)^2$, and $s \in \mathbb{C}$, let

$$\Sigma_W(\Psi, \omega, s), \Sigma_{W+}(\Psi, \omega, s), \Sigma_{W,\mathrm{ad}}(\Psi, \omega, s)$$

be the functions which correspond to $Z_{V_2}(\Psi, \omega, s), Z_{V_2}(\Psi, \omega, s), Z_{V_2,\mathrm{ad}}(\Psi, \omega, s)$ in §4.2. Let $Z_{W,0}, Z'_{W,0} \subset W$ be the subspaces which correspond to Z_0, Z'_0 in §4.2 respectively. We define $Z^{\mathrm{ss}}_{W,0k}$ etc. in the same manner as in §4.2. Let $\widetilde{R}_{W,0}, R'_{W,0}$ be the operators which correspond to $\widetilde{R}_0, R'_0$ in §4.2. Let H_W be the group which corresponds to H in §4.2. Let $X_W \subset \mathbb{A}^1 \times \mathrm{GL}(2)^0_{\mathbb{A}}$ be the subset which corresponds to X_V in §4.2. Let $T_W(\widetilde{R}_{W,0}\Psi, \omega, s, s_1)$ etc. be the adjusting term etc. We use the same function $\alpha(u)$ as in §2.2.

Let $\widetilde{g} \in \mathbb{A} \times \mathrm{GL}(2)^0_{\mathbb{A}} \cong \mathbb{R}_+ \times \mathbb{A}^1 \times \mathrm{GL}(2)^0_{\mathbb{A}}$. For $\Psi \in \mathscr{S}(W_{\mathbb{A}})$, we define

$$\text{(11.1.6)} \qquad \Theta_{Z'_{W,0}}(\Psi, \widetilde{g}) = \sum_{x \in Z'^{\mathrm{ss}}_{W,0k}} \Psi(\widetilde{g}x),$$

$$\Theta_{W,\mathrm{st}}(\Psi, \widetilde{g}) = \sum_{x \in Z^{\mathrm{ss}}_{W,\mathrm{st}k}} \Psi(\widetilde{g}x).$$

Then

$$\text{(11.1.7)} \qquad \Theta_{W,\mathrm{st}}(\Psi, \widetilde{g}) = \sum_{\gamma \in G_k/H_{W,k}} \Theta_{Z'_{W,0}}(\Psi, \widetilde{g}\gamma).$$

Next, we consider distributions associated with unstable strata.

(a) $\mathfrak{d} = (\beta_1)$.

We can identify $Z_\mathfrak{d}$ with $\mathrm{Sym}^2 k^3$, i.e. the space of ternary quadratic forms. For $\Psi \in \mathscr{S}(Z_{\mathfrak{d}\mathbb{A}})$, $\omega = (\omega_1, \omega_2)$ a character of $(\mathbb{A}^\times/k^\times)^2$, and $s \in \mathbb{C}$, let

$$\Sigma_\mathfrak{d}(\Psi, \omega, s), \Sigma_{\mathfrak{d}+}(\Psi, \omega, s)$$

be the functions which correspond to $Z_{V_3}(\Psi, \omega, s), Z_{V_3+}(\Psi, \omega, s)$ in Chapter 4.

(b) $\mathfrak{d} = (\beta_4)$.

We can identify $Z_\mathfrak{d}$ with $M(2,2)$, i.e. the space of 2×2 matrices. For $\Psi \in \mathscr{S}(Z_{\mathfrak{d}\mathbb{A}})$, $\omega = (\omega_1, \omega_2)$ a character of $(\mathbb{A}^\times/k^\times)^2$, and $s \in \mathbb{C}$, let $\Sigma_\mathfrak{d}(\Psi, \omega, s)$ be the integral (3.1.8) for $L = Z^{\mathrm{ss}}_{\mathfrak{d}k}$. We discussed this case in §3.8.

(c) $\mathfrak{d} = (\beta_6), (\mathfrak{d}_8)$.

We can identify $Z_\mathfrak{d}$ with $\mathrm{Sym}^2 k^2 \oplus k$ or $\mathrm{Sym}^2 k^2 \oplus k^2$. We considered these representations in Chapters 6 and 7. For $\Psi \in \mathscr{S}(Z_{\mathfrak{d}\mathbb{A}})$, $\omega = (\omega_1, \omega_2, \omega_3)$ a character of $(\mathbb{A}^\times/k^\times)^3$, χ a principal quasi-character of $G^1_\mathbb{A}/G_k$, and $s \in \mathbb{C}$, let

$$\Sigma_\mathfrak{d}(\Psi, \omega, \chi, s), \Sigma_{\mathfrak{d}+}(\Psi, \omega, \chi, s)$$

be the functions which correspond to $Z_V(\Phi, \omega, \chi, s), Z_{V+}(\Phi, \omega, \chi, s)$ for the case $V = \mathrm{Sym}^2 k^2 \oplus k$ or $\mathrm{Sym}^2 k^2 \oplus k^2$ in Chapters 5 and 6. Let $t_2(\lambda_2)$ be as in §6.2. Let $\chi_\mathfrak{d}$ be the principal quasi-character of $G^1_\mathbb{A}/G_k$ such that $\chi_\mathfrak{d}(t_2(\lambda_2)) = \lambda_2^{-1}$ for (β_6), and $\chi_\mathfrak{d}(t_2(\lambda_2)) = \lambda_2^2$ for (β_8). We use the notation $\kappa_{V_1}, \kappa_{V_2}, \mathscr{F}_{V_1}, \mathscr{F}_{V_2}$ instead of $\kappa_{V_{\mathfrak{d}1}}, \kappa_{V_{\mathfrak{d}2}}, \mathscr{F}_{V_{\mathfrak{d}1}}, \mathscr{F}_{V_{\mathfrak{d}2}}$ as in Chapter 9.

We identify $a_{\beta_6}(\underline{\lambda}_2), a_{\beta_8}(\underline{\lambda}_2)$ in §8.2 with $t_2(\lambda_2)$. Then $\kappa_{\mathfrak{d}1}(t_2(\lambda_2)) = \lambda_2^{-4}$ for (β_6), and $\kappa_{\mathfrak{d}1}(t_2(\lambda_2)) = \lambda_2^{-6}$ for (β_8). We consider $\kappa_{\mathfrak{d}1}$ as a principal quasi-character of the group in §6.2.

For $\mathfrak{d} = (\beta_6)$, let $\Sigma_{\mathfrak{d},\mathrm{ad}}(\Psi, \omega, \chi, s)$ be the adjusted zeta function. We denote by $R_{\mathfrak{d},0}, R'_{\mathfrak{d},0}$ the operators which correspond to $R_{V,0}, R'_{V,0}$ in §6.2. We define $T_\mathfrak{d}(R_{\mathfrak{d},0}\Psi, \omega, s, s_1)$ etc. similarly. Let $H_\mathfrak{d} \subset M_\mathfrak{d}$ be the subgroup generated by T and the element $((1,2),1)$. Let $t^0 = a(\underline{\lambda}_{11}, \underline{\lambda}_{12}, \underline{\lambda}_{13}; \underline{\lambda}_{21}, \underline{\lambda}_{22})\widehat{t}^0 \in T^0_\mathbb{A}$ where $\lambda_{ij} \in \mathbb{R}_+$ for all i, j. We define

$$X_\mathfrak{d} = \{kn'(u_0)t^0 \mid k \in K \cap M_{\mathfrak{d}\mathbb{A}}, u_0 \in \mathbb{A}, |\underline{\lambda}_{11}\underline{\lambda}_{12}^{-1}| \geq \alpha(u_0)\}. \tag{11.1.8}$$

Then $G^0_\mathbb{A} \cap M_{\mathfrak{d}\mathbb{A}}/H_{\mathfrak{d}k} \cong X_\mathfrak{d}/T_k$.

(d) $\mathfrak{d} = (\beta_{10})$.

We can identify $Z_\mathfrak{d}$ with $\mathrm{Sym}^2 k^2 \otimes k^2$. For $\Psi \in \mathscr{S}(Z_{\mathfrak{d}\mathbb{A}})$, $\omega = (\omega_1, \omega_2)$ a character of $(\mathbb{A}^\times/k^\times)^2$, and $s \in \mathbb{C}$, let

$$\Sigma_\mathfrak{d}(\Psi, \omega, \chi, s), \Sigma_{\mathfrak{d}+}(\Psi, \omega, \chi, s)$$

be the functions which correspond to $Z_V(\Phi, \omega, s), Z_{V+}(\Phi, \omega, s)$ for the case $V = \mathrm{Sym}^2 k^2 \otimes k^2$ in Chapter 5. Let $Y_{\mathfrak{d},0}, Z_{\mathfrak{d},0}, Z'_{\mathfrak{d},0} \subset Z_\mathfrak{d}$ be the subspaces which correspond to $Y_{V,0}, Z_{V,0}, Z'_{V,0}$ in Chapter 5 respectively. We define $R_{\mathfrak{d},0}$ etc. similarly. Let $\Sigma_{\mathfrak{d},\mathrm{ad}}(\Psi, \omega, s)$ be the adjusted zeta function and $T_\mathfrak{d}(R_{\mathfrak{d},0}\Psi, \omega, s, s_1)$ etc. the adjusting term. Let $H_\mathfrak{d} \subset M_\mathfrak{d}$ be the subgroup generated T and the element $((2,3),(1,2))$. Let $t^0 = a(\underline{\lambda}_{11}, \underline{\lambda}_{12}, \underline{\lambda}_{13}; \underline{\lambda}_{21}, \underline{\lambda}_{22})\widehat{t}^0 \in T^0_\mathbb{A}$. We define

$$X_\mathfrak{d} = \{kn''(u_0)t^0 n_{B_2}(u_1) \mid k \in K \cap M_{\mathfrak{d}\mathbb{A}}, u_0 \in \mathbb{A}, |\underline{\lambda}_{12}\underline{\lambda}_{13}^{-1}| \geq \alpha(u_0)\}. \tag{11.1.9}$$

Then $G^0_{\mathbb{A}} \cap M_{\mathfrak{d}\mathbb{A}}/H_{\mathfrak{d}k} \cong X_{\mathfrak{d}}/T_k$. Let $h = (\mu_1, \mu_2, q_1, q_2, u_0)$ and $h^1 = (1, \mu_2, q_1, q_2, u_0)$ where $\mu_1, \mu_2 \in \mathbb{R}_+$, $q_1, q_2 \in \mathbb{A}^1$, $u_0 \in \mathbb{A}$. We define dh in the obvious manner. For $\Psi \in \mathscr{S}(Z_{\mathfrak{d},0\mathbb{A}})$, we define

(11.1.10) $$\widetilde{\Theta}_{Z_{\mathfrak{d},0}}(\Psi, h) = \sum_{x_{1,33}, x_{2,22} \in k^\times} \Psi(\underline{\mu}_1 \underline{\mu}_2^{-1} q_1 x_{1,33}, \underline{\mu}_1 \underline{\mu}_2 q_2 x_{2,22}, 2\underline{\mu}_1 \underline{\mu}_2 q_2 x_{2,22} u_0).$$

(e) As before, we use the notation $\Sigma_{W,\text{ad},(i)}(\Psi, \omega, s_0), \Sigma_{\mathfrak{d},(i)}(\Psi, \omega, s_0)$ etc. for the i-th coefficient of the Laurent expansion of $\Sigma_{W,\text{ad}}(\Psi, \omega, s), \Sigma_{\mathfrak{d}}(\Psi, \omega, s)$ etc. In the next three chapters, $g_{\mathfrak{d}}$ denotes an element of $G^0_{\mathbb{A}} \cap M_{\mathfrak{d}\mathbb{A}}$, and $g^0_{\mathfrak{d}}$ an element of $M^0_{\mathfrak{d}\mathbb{A}}$. Let $dg_{\mathfrak{d}}, dg^0_{\mathfrak{d}}$ be the measures which we defined in §3.3. Let $d^\times t^0$ be the measure on $T^0_{\mathbb{A}}$ such that $dg^0 = dk d^\times t^0 du$ where dk, du are the canonical measures on $K, N_{\mathbb{A}}$.

(f) Next, we define some distributions associated with the space W of binary quadratic forms, and distributions which are variations of the standard L-functions.

Let $\tau = (\tau_1, \tau_2)$ be a Weyl group element. Let

$$z = (z_1, z_2) = (z_{11}, z_{12}, z_{13}, z_{21}, z_{22}) \in \mathfrak{t}^*_{\mathbb{C}}.$$

Consider the following substitution

(11.1.11) $$s_{\tau 11} = z_{1\tau_1(3)} - z_{1\tau_1(2)}, \; s_{\tau 12} = z_{1\tau_1(2)} - z_{1\tau_1(1)}, \; s_{\tau 2} = z_{2\tau_2(2)} - z_{2\tau_2(1)}.$$

Let $s_\tau = (s_{\tau 11}, s_{\tau 12}, s_{\tau 2})$, and $s_{\tau 1} = (s_{\tau 11}, s_{\tau 12})$. We define

$$ds_{\tau 1} = ds_{\tau 11} ds_{\tau 12}, \; ds_\tau = ds_{\tau 1} ds_{\tau 2}.$$

Then $dz = ds_\tau$.

We define functions $\widetilde{L}_1(s_{\tau 1}), \widetilde{M}_{\tau_1}(s_{\tau 1})$ of $s_{\tau_1} \in \mathbb{C}^2$ by the following table.

Table (11.1.12)

τ_1	$\widetilde{L}_1(s_{\tau 1})$	$\widetilde{M}_{\tau_1}(s_{\tau 1})$
1	$-2(s_{\tau 11} + s_{\tau 12})$	1
(1,2)	$-2s_{\tau 11}$	$\phi(s_{\tau 12})$
(2,3)	$-2s_{\tau 12}$	$\phi(s_{\tau 11})$
(1,2,3)	$2s_{\tau 11}$	$\phi(s_{\tau 11})\phi(s_{\tau 11} + s_{\tau 12})$
(1,3,2)	$2s_{\tau 12}$	$\phi(s_{\tau 12})\phi(s_{\tau 11} + s_{\tau 12})$
(1,3)	$2(s_{\tau 11} + s_{\tau 12})$	$\phi(s_{\tau 11})\phi(s_{\tau 12})\phi(s_{\tau 11} + s_{\tau 12})$

We also define

(11.1.13) $$\widetilde{L}_2(s_{\tau 2}) = \begin{cases} -s_{\tau 2} & \tau_2 = 1, \\ s_{\tau 2} & \tau_2 = (1,2), \end{cases}$$

$$\widetilde{M}_{\tau_2}(s_{\tau 2}) = \begin{cases} 1 & \tau_2 = 1, \\ \phi(s_{\tau 2}) & \tau_2 = (1,2), \end{cases}$$

$$\widetilde{L}(s_\tau) = \widetilde{L}_1(s_{\tau 1}) + C\widetilde{L}_2(s_{\tau 2}), \; \widetilde{M}_\tau(s_\tau) = \widetilde{M}_{\tau_1}(s_{\tau 1})\widetilde{M}_{\tau_2}(s_{\tau 2}),$$

$$\widetilde{\Lambda}(w; s_\tau) = \frac{\psi(z)}{w - \widetilde{L}(s_\tau)}, \; \widetilde{\Lambda}_\tau(w; s_\tau) = \widetilde{M}_\tau(s_\tau)\widetilde{\Lambda}(w; s_\tau),$$

$$\widetilde{\Lambda}_1(w; s_{\tau 1}) = \frac{\psi(z_1, \frac{1}{2}, -\frac{1}{2})}{w - \widetilde{L}_{1\tau_1}(s_{\tau 1}) - C}, \; \widetilde{\Lambda}_{1\tau_1}(w; s_{\tau 1}) = \widetilde{\Lambda}_1(w; s_{\tau 1})\widetilde{M}_{\tau_1}(s_{\tau 1}),$$

where z and s_τ are related as in (11.1.11). It is easy to see that $L(z) = \widetilde{L}(s_\tau)$. We remind the reader that $w_0 = 4 + C$. We also remind the reader that we are still assuming that

$$\psi(-\tau_G z) = \psi(z),\ \psi(\rho) \neq 0.$$

We consider the domain which corresponds to D_τ for $\mathrm{Re}(s_\tau)$ and use the same notation D_τ. For example, if $\tau = ((1,2,3),(1,2))$, a point $r = (r_1, r_2, r_3) \in \mathbb{R}^3$ belongs to D_τ if and only if $r_1 > 1, r_1 + r_2 > 1, r_3 > 1$.

Let $\rho_\tau \in \mathbb{R}^3$ be the element which corresponds to ρ by the substitution (11.1.11). Similarly, let $\rho_{1\tau_1} \in \mathbb{R}^2, \rho_{2\tau_2} \in \mathbb{R}$ be the elements which correspond to ρ_1, ρ_2 by the substitution (11.1.11). It is easy to see that $\rho_{2\tau_2} = -1$ if $\tau_2 = 1$, and $\rho_{2\tau_2} = 1$ if $\tau_2 = (1,2)$. Also

$$\rho_{1\tau_1} = (-1,-1), (-2,1), (1,-2), (2,-1), (-1,2), (1,1)$$

for $\tau_1 = 1, (1,2), (2,3), (1,2,3), (1,3,2), (1,3)$, respectively.

Definition (11.1.14) *Let i be a positive integer and $\Psi \in \mathscr{S}(\mathbb{A}^i)$. Let $l_1(s_\tau), \cdots, l_i(s_\tau)$ be linear functions of s_τ, and $l_{i+1}(s_\tau)$ a polynomial of s_τ.*

We define

$$\Sigma_{i,\mathrm{sub}}(\Psi, l, s_\tau) = l_{i+1}(s_\tau)^{-1} \Sigma_i(\Psi, l_1(s_\tau), \cdots, l_i(s_\tau)).$$

('sub' stands for 'substitution.')

Let $G = \mathrm{GL}(2)$, and $G^0_\mathbb{A} = \mathrm{GL}(2)^0_\mathbb{A}$. Let $l_1(s_\tau)$ be a linear function, and $l_2(s_\tau)$ a polynomial. We assume that $l_i(-s_{\tau 11}, s_{\tau 11} + s_{\tau 12}, s_{\tau 2}) = l_i(s_{\tau 11}, s_{\tau 12}, s_{\tau 2})$ for $i = 1,2$. Let $g^0 \in \mathrm{GL}(2)^0_\mathbb{A}, t \in \mathbb{A}^\times$.

Definition (11.1.15) *Let $\tau = (\tau_1, \tau_2)$ be a Weyl group element such that $\tau_1 \in \mathfrak{W}_2$. For $\Psi \in \mathscr{S}(W_\mathbb{A})$, we define*

$$(1)\qquad E_{\mathrm{sub}}(l, t, g^0, s_\tau) = \frac{|t|^{l_1(s_\tau)}}{l_2(s_\tau)} E_B(t, g^0, -s_{\tau 11}),$$

$$(2)\qquad \mathscr{E}_{\mathrm{sub},\tau}(l, t, g^0, w) = \left(\frac{1}{2\pi\sqrt{-1}}\right)^3 \int_{\mathrm{Re}(s_\tau)=r} E_{\mathrm{sub}}(l, t, g^0, s_\tau) \widetilde{\Lambda}_\tau(w; s_\tau) ds_\tau,$$

$$(3)\qquad \Omega_{W,\tau}(\Psi, l, w) = \int_{\substack{\mathbb{A}^\times/k^\times \times G^0_\mathbb{A}/G_k \\ |t|\geq 1}} \Theta_{W_k^{ss}}(\Psi, tg^0) \mathscr{E}_{\mathrm{sub},\tau}(l, t, g^0, w) d^\times t dg^0,$$

$$(4)\qquad \Omega^s_{W,\tau}(\Psi, l, w) = \int_{\substack{\mathbb{A}^\times/k^\times \times G^0_\mathbb{A}/G_k \\ |t|\geq 1}} \Theta_{W^s}(\Psi, tg^0) \mathscr{E}_{\mathrm{sub},\tau}(l, t, g^0, w) d^\times t dg^0,$$

$$(5)\qquad \Omega_{W,\mathrm{st},\tau}(\Psi, l, w) = \int_{\substack{\mathbb{A}^\times/k^\times \times G^0_\mathbb{A}/G_k \\ |t|\geq 1}} \Theta_{W,\mathrm{st}}(\Psi, tg^0) \mathscr{E}_{\mathrm{sub},\tau}(l, t, g^0, w) d^\times t dg^0,$$

$$(6)\qquad \Sigma^s_{W,\mathrm{sub}+}(\Psi, l, s_\tau) = \int_{\substack{\mathbb{A}^\times/k^\times \times G^0_\mathbb{A}/G_k \\ |t|\geq 1}} \Theta_{W^s}(\Psi, tg^0) E_{\mathrm{sub},\tau}(l, t, g^0, s_\tau) d^\times t dg^0,$$

where in (2), *we choose $r = (r_1, r_2, r_3) \in \mathbb{R}^3$ so that $r \in D_\tau, l_2(r) > 0$, and $r_1 < -1$.*

For the rest of this book, when we consider contour integrals with respect to s_τ, r is of the form $r = (r_1, r_2, r_3) \in \mathbb{R}^3$.

The distribution $\Sigma^s_{W,\mathrm{sub}+}(\Psi, s_\tau)$ is well defined as long as $\mathrm{Re}(s_{\tau 11}) < -1$, and

$$\Omega^s_{W,\tau}(\Psi, l, w) = \left(\frac{1}{2\pi\sqrt{-1}}\right)^3 \int_{\substack{\mathrm{Re}(s_\tau)=r \\ r_1<-1,\ l_2(r)>0}} \Sigma^s_{W,\mathrm{sub}+}(\Psi, l, s_\tau)\widetilde{\Lambda}_\tau(w; s_\tau)ds_\tau.$$

Clearly,

$$\Omega_{W,\tau}(\Psi, l, w) = \Omega^s_{W,\tau}(\Psi, l, w) + \Omega_{W,\mathrm{st},\tau}(\Psi, l, w).$$

Since we have to deal with integrals in (11.1.15) many times in Chapter 12, we consider them simultaneously in this section and the next section.

Consider (11.1.15)(5). Let $E'_{\mathrm{sub}}(l, t, g^0, s_\tau)$ (resp. $E''_{\mathrm{sub}}(l, t, g^0, s_\tau)$) be the constant (resp. non-constant) term of $E_{\mathrm{sub}}(l, t, g^0, s_\tau)$. It is easy to see that the non-constant term $E''_{\mathrm{sub}}(l, t, g^0, s_\tau)$ is holomorphic as long as $\mathrm{Re}(s_{\tau 11}) < 0$.

Definition (11.1.16)

(1) $$\Sigma''_{W,\mathrm{st},\mathrm{sub}+}(\Psi, l, s_\tau) = \int_{\substack{\mathbb{A}^\times/k^\times \times G^0_{\mathbb{A}}/G_k \\ |t|\geq 1}} \Theta_{W,\mathrm{st}}(\Psi, tg^0)E''_{\mathrm{sub}}(l, t, g^0, s_\tau)d^\times t dg^0.$$

(2) $$\Omega''_{W,\mathrm{st},\tau}(\Psi, l, w) = \int_{\substack{\mathrm{Re}(s_\tau)=r \\ r_1<0,\ l_2(r)>0}} \Sigma''_{W,\mathrm{st},\mathrm{sub}+}(\Psi, l, s_\tau)\widetilde{\Lambda}_\tau(w; s_\tau)ds_\tau.$$

The distribution $\Sigma''_{W,\mathrm{st},\mathrm{sub}+}(\Psi, l, s_\tau)$ is well defined as long as $\mathrm{Re}(s_{\tau 11}) < 0$.

We consider the constant term. Let $t^0 = a_2(\underline{\mu} t_1, \underline{\mu}^{-1} t_2)$ where $\mu \in \mathbb{R}_+, t_1, t_2 \in \mathbb{A}^1$. It is well known that

$$E'_{\mathrm{sub}}(l, t, t^0, s_\tau) = \frac{|t|^{l_1(s_\tau)}}{l_2(s_\tau)}(\mu^{1-s_{\tau 11}} + \phi(-s_{\tau 11})\mu^{1+s_{\tau 11}}).$$

Let $\tau' = (\tau'_1, \tau_2)$ be a Weyl group element such that $\tau'_1 = 1$ or $(2,3)\tau_1$. We consider $s_{\tau'}$ as in (11.1.11) for τ' also. If $\tau'_1 = (2,3)\tau_1$, then $s_{\tau' 11} = -s_{\tau 11}, s_{\tau' 12} = s_{\tau 11} + s_{\tau 12}$, and $\phi(-s_{\tau 11})\widetilde{\Lambda}_\tau(w; s_\tau) = \widetilde{\Lambda}_{\tau'}(w; s_{\tau'})$. Therefore, by the assumption on l,

$$\begin{aligned} &\int_{\mathrm{Re}(s_\tau)=r} E'_{\mathrm{sub}}(l, t, t^0, s_\tau)\widetilde{\Lambda}_\tau(w; s_\tau)ds_\tau \\ &= \sum_{\tau'=\tau,((2,3),1)\tau} \int_{\mathrm{Re}(s_{\tau'})=r'} \frac{|t|^{l_1(s_{\tau'})}\mu^{1-s_{\tau' 11}}}{l_2(s_{\tau'})}\widetilde{\Lambda}_{\tau'}(w; s_{\tau'})ds_{\tau'}. \end{aligned}$$

Let B be the Borel subgroup of $\mathrm{GL}(1) \times \mathrm{GL}(2)$, $b = (t, n_2(u_0)a_2(\underline{\mu} t_1, \underline{\mu}^{-1} t_2))$, and $db = d^\times \mu d^\times t d^\times t_1 d^\times t_2 du_0$. We identify the maximal torus of $\mathrm{GL}(1) \times \mathrm{GL}(2)$ with $\mathrm{GL}(1)^3$. By a similar argument as in §4.2,

$$\int_{\substack{X_W \cap B_{\mathbb{A}}/(k^\times)^3 \\ |t|\geq 1}} \frac{|t|^{l_1(s_{\tau'})}\mu^{1-s_{\tau' 11}}}{l_2(s_{\tau'})}\Theta_{Z'_{W,0}}(\Psi, b)db = \frac{T_{W+}(\widetilde{R}_{W,0}\Psi, l_1(s_{\tau'}), \frac{1}{2}(1 - s_{\tau' 11}))}{(s_{\tau' 11} - 1)l_2(s_{\tau'})}.$$

Let

$$(11.1.17)\quad \Sigma_{W,\mathrm{st,sub}+}(\Psi,l,s_{\tau'}) = \frac{T_{W+}(\widetilde{R}_{W,0}\Psi, l_1(s_{\tau'}), \frac{1}{2}(1-s_{\tau'11}))}{(s_{\tau'11}-1)l_2(s_{\tau'})},$$

$$\Omega'_{W,\mathrm{st},\tau'}(\Psi,w) = \left(\frac{1}{2\pi\sqrt{-1}}\right)^3 \int_{\mathrm{Re}(s_{\tau'})=r} \Sigma_{W,\mathrm{st,sub}+}(\Psi,l,s_{\tau'})\widetilde{\Lambda}_{\tau'}(w;s_{\tau'})ds_{\tau'}.$$

Then

$$(11.1.18)\qquad \Omega_{W,\mathrm{st},\tau}(\Psi,l,w) = \sum_{\tau'=\tau,((2,3),1)\tau} \Omega'_{W,\mathrm{st},\tau'}(\Psi,l,w) + \Omega''_{W,\mathrm{st},\tau}(\Psi,l,w).$$

§11.2 Technical lemmas

Consider a Weyl group element $\tau = (\tau_1,\tau_2)$ such that $\tau_1 \in \mathfrak{W}_2$. First we consider the distributions $\Omega^s_{W,\tau}(\Psi,l,w)$ etc. in §11.1. We consider the following three possibilities for l_2:

(1) $l_2(s_\tau) = -s_{\tau 11} - 2s_{\tau 12} + 2s_{\tau 2} - 1$,
(2) $l_2(s_\tau) = s_{\tau 11} + 2s_{\tau 12} + 2s_{\tau 2} - 7$,
(3) $l_2(s_\tau) = s_{\tau 11} + 2s_{\tau 12} + 2s_{\tau 2} - 9$.

If $\tau_2 = 1$, we fix $r_1 = -2, r_2 = 4$, and choose $r_3 \gg 0$. Then $\widetilde{L}(r) \le 8 - Cr_3 \ll 0$. Therefore, $\Omega^s_{W,\tau}(\Psi,l,w), \Omega''_{W,\mathrm{st},\tau}(\Psi,l,w) \sim 0$ if $\tau_2 = 1$. So we assume that $\tau_2 = (1,2)$.

Lemma (11.2.1) *Suppose that $\tau_1 \in \mathfrak{W}_2$ and $\tau_2 = (1,2)$. Then by changing ψ if necessary,*
(1) $\Omega^s_{W,\tau}(\Psi,l,w), \Omega''_{W,\mathrm{st},\tau}(\Psi,l,w) \sim 0$ unless $\tau_1 = (1,3,2)$,
(2) If $\tau_1 = (1,3,2)$, $\Omega''_{W,\mathrm{st},\tau}(\Psi,l,w) \sim 0$, and

$$\Omega^s_{W,\tau}(\Psi,l,w) \sim C_G\Lambda(w;\rho)\frac{\Sigma_{W+}(\Psi,l_1(-1,2,1))}{l_1(-1,2,1)}.$$

Proof. If $\tau_2 = (1,2)$, $l_2(\rho_\tau) \neq 0$ for all the cases. Therefore, by the passing principle (3.6.1), we can ignore the set $l_2(s_\tau) = 0$ when we move the contour. So for the case (1), we choose r so that $l_2(r) < 0$. We move the contour so that $r_3 < 1$ as follows.

$$\begin{aligned}
&\Omega^s_{W,\tau}(\Psi,l,w) \\
&= \left(\frac{1}{2\pi\sqrt{-1}}\right)^3 \int_{\mathrm{Re}(s_\tau)=(-2,6,\frac{1}{2})} \Sigma^s_{W,\mathrm{sub}+}(\Psi,l,s_\tau)\widetilde{\Lambda}_\tau(w;s_\tau)ds_\tau \\
&\quad + \varrho\left(\frac{1}{2\pi\sqrt{-1}}\right)^2 \int_{\substack{\mathrm{Re}(s_{\tau 1})=(r_1,r_2) \\ r_1+2r_2>7}} \Sigma^s_{W,\mathrm{sub}+}(\Psi,l,s_{\tau 1},1)\widetilde{\Lambda}_{1\tau_1}(w;s_{\tau 1})ds_{\tau 1}.
\end{aligned}$$

Note that $l_2(-2,6,\frac{1}{2}) \neq 0$ for all the cases.

Since $C > 100$, $\widetilde{L}(-2,6,\frac{1}{2}) \le 12 + \frac{1}{2}C < 4 + C$. So, we can ignore the first term. The same argument works for $\Omega''_{W,\mathrm{st},\tau}(\Psi,l,w)$ also.

Consider the second term. If $\tau_1 = 1$ or $(1,2)$, we fix $r_1 = -\frac{3}{2}$, and choose $r_2 \gg 0$. Then $\widetilde{L}(r) \ll 0$ or $\widetilde{L}(r) = 3 + C$. Therefore, we can ignore these cases. This proves (1).

Consider $\tau_1 = (1,3,2)$. We move the contour to $\mathrm{Re}(s_{\tau 1}) = (-1-\delta, 2+\delta)$ where $\delta > 0$ is a small number. For (2), (3), we can use (3.6.1), and ignore the line $l_2(s_{\tau 1}, 1) = 0$ Therefore, in all the cases,

$$
\begin{aligned}
&\varrho \left(\frac{1}{2\pi\sqrt{-1}}\right)^2 \int_{\substack{\mathrm{Re}(s_{\tau 1})=(r_1,r_2) \\ r_1+2r_2>7}} \Sigma^s_{W,\mathrm{sub}+}(\Psi, l, s_{\tau 1}, 1)\widetilde{\Lambda}_{1\tau_1}(w; s_{\tau 1}) ds_{\tau 1} \\
&= \varrho \left(\frac{1}{2\pi\sqrt{-1}}\right)^2 \int_{\mathrm{Re}(s_{\tau 1})=(-1-\delta, 2+\delta)} \Sigma^s_{W,\mathrm{sub}+}(\Psi, l, s_{\tau 1}, 1)\widetilde{\Lambda}_{1\tau_1}(w; s_{\tau 1}) ds_{\tau 1} \\
&= \varrho \left(\frac{1}{2\pi\sqrt{-1}}\right)^2 \int_{\mathrm{Re}(s_{\tau 1})=(-\frac{1}{2}, \frac{3}{2})} \Sigma^s_{W,\mathrm{sub}+}(\Psi, l, s_{\tau 1}, 1)\widetilde{\Lambda}_{1\tau_1}(w; s_{\tau 1}) ds_{\tau 1} \\
&\quad - \frac{\varrho}{2\pi\sqrt{-1}} \int_{\mathrm{Re}(s_{\tau 12})=2+\delta} \operatorname*{Res}_{s_{\tau 11}=-1} \Sigma^s_{W,\mathrm{sub}+}(\Psi, l, s_{\tau 1}, 1)\widetilde{\Lambda}_{1\tau_1}(w; -1, s_{\tau 12}) ds_{\tau 12}.
\end{aligned}
$$

Note that since we move the contour from the left to the right, the sign of the second term is negative.

Since $\widetilde{L}_1(-\frac{1}{2}, \frac{3}{2}) = 3$, we can ignore the first term. Since $\widetilde{M}_{\tau_1}(-1, s_{\tau 12}) = \phi_2(s_{\tau 12} - 1)$, and

$$
\operatorname*{Res}_{s_{\tau 11}=-1} E_{\mathrm{sub}}(l, t, g^0, s_{\tau 1}, 1) = -\frac{\varrho |t|^{l_1(-1, s_{\tau 12}, 1)}}{l_2(-1, s_{\tau 12}, 1)},
$$

we only have to move the contour to $1 < \mathrm{Re}(s_{\tau 12}) < 2$.

Q.E.D.

Lemma (11.2.2) *Let $\tau' = (\tau'_1, \tau'_2)$ be a Weyl group element and $\delta > 0$ a small number. Then by changing ψ if necessary,*
(1) $\Omega'_{W,\mathrm{st},\tau'}(\Psi, l, w) \sim 0$ unless $\tau' = \tau_G$,
(2) If $\tau' = \tau_G$,

$$
\begin{aligned}
&\Omega'_{W,\mathrm{st},\tau'}(\Psi, l, w) \\
&\sim \frac{\varrho^2}{2\pi\sqrt{-1}} \int_{\mathrm{Re}(s_{\tau' 11})=1+\delta} \Sigma_{W,\mathrm{st},\mathrm{sub}+}(\Psi, l, s_{\tau' 11}, 1, 1)\phi_2(s_{\tau' 11})\widetilde{\Lambda}_1(w; s_{\tau' 11}, 1) ds_{\tau' 11}.
\end{aligned}
$$

Proof. Similarly as in (11.2.1), we can ignore the case $\tau'_2 = 1$, and assume that $\tau'_2 = (1,2)$. Then

$$
\begin{aligned}
&\Omega'_{W,\mathrm{st},\tau'}(\Psi, l, w) \\
&\sim \varrho \left(\frac{1}{2\pi\sqrt{-1}}\right)^2 \int_{\substack{\mathrm{Re}(s_{\tau' 1})=(r_1,r_2) \\ r_1+2r_2>7}} \Sigma_{W,\mathrm{st},\mathrm{sub}+}(\Psi, l, s_{\tau' 1}, 1)\widetilde{\Lambda}_{1\tau'_1}(w; s_{\tau' 1}) ds_{\tau' 1}.
\end{aligned}
$$

(We have to use the passing principle (3.6.1) for the case (1).)

If $\tau_1' = 1, (1,2)$ or $(1,3)$, we choose $r_1, r_2 \gg 0$. Then $\widetilde{L}(r_1, r_2, 1) \ll 0$. If $\tau_1' = (1,2,3)$, we fix $r_1 = \frac{3}{2}$, and choose $r_2 \gg 0$, and if $\tau_1' = (1,3,2)$, we fix $r_2 = \frac{3}{2}$, and choose $r_1 \gg 0$. In both cases, $\widetilde{L}(r_1, r_2, 1) = 3 + C < 4 + C$. This proves (1).

Suppose $\tau_1 = (1,3)$. Then $\rho_\tau = (1,1,1)$. Clearly, $l_2(1,1,1) \neq 0$ for all the cases. Therefore, we can move the contour crossing this line. So

$$\begin{aligned}
&\left(\frac{1}{2\pi\sqrt{-1}}\right)^2 \int_{\substack{\mathrm{Re}(s_{\tau'1}) = (r_1, r_2) \\ r_1 + 2r_2 > 7}} \Sigma_{W,\mathrm{st},\mathrm{sub}+}(\Psi, l, s_{\tau'1}, 1) \widetilde{\Lambda}_{1\tau'1}(w; s_{\tau'1}) ds_{\tau'1} \\
&= \left(\frac{1}{2\pi\sqrt{-1}}\right)^2 \int_{\mathrm{Re}(s_{\tau'1}) = (\frac{5}{4}, \frac{1}{2})} \Sigma_{W,\mathrm{st},\mathrm{sub}+}(\Psi, l, s_{\tau'1}, 1) \widetilde{\Lambda}_{1\tau_1'}(w; s_{\tau'1}) ds_{\tau'1} \\
&\quad + \frac{\varrho}{2\pi\sqrt{-1}} \int_{\mathrm{Re}(s_{\tau'11}) = 1+\delta} \Sigma_{W,\mathrm{st},\mathrm{sub}+}(\Psi, l, s_{\tau'11}, 1, 1) \\
&\qquad\qquad \times \phi_2(s_{\tau'11}) \widetilde{\Lambda}_1(w; s_{\tau'11}, 1) ds_{\tau'11},
\end{aligned}$$

where $\delta > 0$ is a small number.

We can ignore the first term by the usual argument. This proves the lemma.

Q.E.D.

The following lemma follows from these considerations.

Lemma (11.2.3) *Suppose that $\tau' = \tau_G$ and $\delta > 0$ is a small number. Then by changing ψ if necessary*

$$\begin{aligned}
&\sum_{\substack{\tau_1 \in \mathfrak{W}_2 \\ \tau_2 = 1, (1\,2)}} \Omega_{W,\mathrm{st},\tau}(\Psi, l, w) \\
&\sim C_G \Lambda(w; \rho) \frac{\Sigma_{W+}(\Psi, l_1(-1,2,1))}{l_2(-1,2,1)} \\
&\quad + \frac{\varrho^2}{2\pi\sqrt{-1}} \int_{\mathrm{Re}(s_{\tau'11}) = 1+\delta} \Sigma_{W,\mathrm{st},\mathrm{sub}+}(\Psi, l, s_{\tau'11}, 1, 1) \phi_2(s_{\tau'11}) \widetilde{\Lambda}_1(w; s_{\tau'11}, 1) ds_{\tau'11}
\end{aligned}$$

Now we consider slightly different contour integrals. Let $l = (\frac{1}{2}(s_{\tau 11} + 1), l_2)$ where l_2 is one of the following:

(1) $l_2(s_\tau) = (-s_{\tau 11} - 2s_{\tau 12} + 2s_{\tau 2} - 1)(s_{\tau 12} + 1)$,
(2) $l_2(s_\tau) = (-s_{\tau 11} - 2s_{\tau 12} + 2s_{\tau 2} - 1)(s_{\tau 11} + s_{\tau 12} - 1)$,
(3) $l_2(s_\tau) = (s_{\tau 11} + 2s_{\tau 12} - 2s_{\tau 2} + 1)(-s_{\tau 11} + 2s_{\tau 2} + 1)$,
(4) $l_2(s_\tau) = (s_{\tau 11} + 2s_{\tau 12} - 2s_{\tau 2} + 1)(s_{\tau 11} + 2s_{\tau 2} - 3)$,
(5) $l_2(s_\tau) = (s_{\tau 11} + 2s_{\tau 12} + 2s_{\tau 2} - 7)(s_{\tau 12} + 1)$,
(6) $l_2(s_\tau) = (s_{\tau 11} + 2s_{\tau 12} + 2s_{\tau 2} - 7)(s_{\tau 11} + s_{\tau 12} - 1)$,
(7) $l_2(s_\tau) = (s_{\tau 11} + 2s_{\tau 12} + 2s_{\tau 2} - 7)(5s_{\tau 11} + 4s_{\tau 12} - 2s_{\tau 2} + 1)$,
(8) $l_2(s_\tau) = (s_{\tau 11} + 2s_{\tau 12} + 2s_{\tau 2} - 7)(s_{\tau 11} - 4s_{\tau 12} + 2s_{\tau 2} - 13)$,
(9) $l_2(s_\tau) = (s_{\tau 11} + 2s_{\tau 12} + 2s_{\tau 2} - 7)(s_{\tau 11} - 2s_{\tau 12} + 2s_{\tau 2} + 1)$,
(10) $l_2(s_\tau) = (s_{\tau 11} + 2s_{\tau 12} + 2s_{\tau 2} - 7)(3s_{\tau 11} + 2s_{\tau 12} - 2s_{\tau 2} - 9)$,
(11) $l_2(s_\tau) = (s_{\tau 11} + 2s_{\tau 12} + 2s_{\tau 2} - 7)(-s_{\tau 11} + 2s_{\tau 2} + 1)$,
(12) $l_2(s_\tau) = (s_{\tau 11} + 2s_{\tau 12} + 2s_{\tau 2} - 7)(s_{\tau 11} + 2s_{\tau 2} - 3)$,
(13) $l_2(s_\tau) = (s_{\tau 11} + 2s_{\tau 12} + 2s_{\tau 2} - 9)(5s_{\tau 11} + 4s_{\tau 12} - 2s_{\tau 2} - 9)$,
(14) $l_2(s_\tau) = (s_{\tau 11} + 2s_{\tau 12} + 2s_{\tau 2} - 9)(s_{\tau 11} - 4s_{\tau 12} + 2s_{\tau 2} - 3)$,

(15) $l_2(s_\tau) = (s_{\tau 11} + 2s_{\tau 12} + 2s_{\tau 2} - 9)(s_{\tau 11} - 2s_{\tau 12} + 2s_{\tau 2} - 5)$,
(16) $l_2(s_\tau) = (s_{\tau 11} + 2s_{\tau 12} + 2s_{\tau 2} - 9)(3s_{\tau 11} + 2s_{\tau 12} - 2s_{\tau 2} - 3)$.

For these cases, let l_{21} (resp. l_{22}) be the first factor (resp. second factor) of l_2. Let $\Psi \in \mathscr{S}(\mathbb{A})$. We consider

$$\left(\frac{1}{2\pi\sqrt{-1}}\right)^3 \int_{\mathrm{Re}(s_\tau)=r} \Sigma_{1,\mathrm{sub}}(\Psi, l, s_\tau)\widetilde{\Lambda}_\tau(w; s_\tau)ds_\tau,$$

where we choose the contour so that

$$r = (r_1, r_2, r_3) \in D_\tau, r_2 > 1, \text{and } l_{21}(r), l_{22}(r) > 0.$$

Lemma (11.2.4) *Suppose that* $\tau' = \tau_G$ *and* $\delta > 0$ *is a small number. Then by changing* ψ *if necessary,*

$$\sum_\tau \left(\frac{1}{2\pi\sqrt{-1}}\right)^3 \int_{\mathrm{Re}(s_\tau)=r} \Sigma_{1,\mathrm{sub}}(\Psi, l, s_\tau)\widetilde{\Lambda}_\tau(w; s_\tau)ds_\tau$$
$$\sim \frac{\varrho^2}{2\pi\sqrt{-1}} \int_{\mathrm{Re}(s_{\tau'11})=1+\delta} \Sigma_{1,\mathrm{sub}}(\Psi, l, s_{\tau'11}, 1, 1)\phi_2(s_{\tau'11})\widetilde{\Lambda}_1(w; s_{\tau'11}, 1)ds_{\tau'11}.$$

We devote the rest of this section to the proof of this lemma.

Lemma (11.2.5) *Suppose that* $l(s_\tau) = (\frac{1}{2}(s_{\tau 11}+1), l_2(s_\tau))$ *where* $l_2(s_\tau)$ *is one of* (1)–(16). *Then by changing* ψ *if necessary,*
(1) *if* $\tau_2 = 1$,

$$\left(\frac{1}{2\pi\sqrt{-1}}\right)^3 \int_{\mathrm{Re}(s_\tau)=r} \Sigma_{1,\mathrm{sub}}(\Psi, l, s_\tau)\widetilde{\Lambda}_\tau(w; s_\tau)ds_\tau \sim 0, \textit{ and}$$

(2) *if* $\tau_2 = (1,2)$,

$$\left(\frac{1}{2\pi\sqrt{-1}}\right)^3 \int_{\mathrm{Re}(s_\tau)=r} \Sigma_{1,\mathrm{sub}}(\Psi, l, s_\tau)\widetilde{\Lambda}_\tau(w; s_\tau)ds_\tau$$
$$\sim \varrho \left(\frac{1}{2\pi\sqrt{-1}}\right)^2 \int_{\mathrm{Re}(s_{\tau 1})=(r_1,r_2)} \Sigma_{1,\mathrm{sub}}(\Psi, l, s_{\tau 1}, 1)\widetilde{\Lambda}_{1\tau_1}(w; s_{\tau 1}, 1)ds_{\tau 1},$$

where we choose the contour so that

$$(r_1, r_2) \in D_{\tau_1}, r_1 > 1, l_{21}(r_1, r_2, 1) < 0, l_{22}(r_1, r_2, 1) > 0$$

for cases (1), (2), *and*

$$(r_1, r_2) \in D_{\tau_1}, r_1 > 1, l_{2i}(r_1, r_2, 1) > 0 \textit{ for } i = 1, 2$$

for cases (3)–(16).

Proof. If $\tau_2 = 1$, we can choose $r_3 \gg 0$, and $\widetilde{L}(r) \ll 0$. Therefore, (1) is clear.

We assume that $\tau_2 = (1,2)$. Then $\rho_{2\tau_2} = 1$. For cases (1) and (2), it is easy to check that the point $\rho_{1\tau_1}$ is not on the lines $l_{21}(s_{\tau 1},1) = 0, l_{22}(s_{\tau 1},1) = 0$. Therefore, by the passing principle, we can move the contour crossing these lines. So we can assume that $l_{21}(r_1,r_2,1) < 0, l_{22}(r_1,r_2,1) > 0$ for these cases.

Consider the following values for (r_1,r_2).

$$(r_1,r_2) = \begin{cases} (3,5) & (1),(2),(4)-(7),(10),(12),(13),(16), \\ (\frac{3}{2},20) & (3),(11), \\ (15,2) & (8),(9),(14),(15). \end{cases}$$

Then for cases (1) and (2), $l_{21}(r_1,r_2,r_3) < 0, l_{22}(r_1,r_2,r_3) > 0$ for $r_3 = \frac{1}{2},1,2$, and for cases (3)–(16), $l_{2i}(r_1,r_2,r_3) > 0$ for $i = 1,2$ and $r_3 = \frac{1}{2},1,2$.

Therefore,

$$\begin{aligned} &\left(\frac{1}{2\pi\sqrt{-1}}\right)^3 \int_{\mathrm{Re}(s_\tau)=(r_1,r_2,2)} \Sigma_{1,\mathrm{sub}}(\Psi,l,s_\tau)\widetilde{\Lambda}_\tau(w;s_\tau)ds_\tau \\ &= \left(\frac{1}{2\pi\sqrt{-1}}\right)^3 \int_{\mathrm{Re}(s_\tau)=(r_1,r_2,\frac{1}{2})} \Sigma_{1,\mathrm{sub}}(\Psi,l,s_\tau)\widetilde{\Lambda}_\tau(w;s_\tau)ds_\tau \\ &\quad + \varrho\left(\frac{1}{2\pi\sqrt{-1}}\right)^2 \int_{\mathrm{Re}(s_{\tau 1})=(r_1,r_2)} \Sigma_{1,\mathrm{sub}}(\Psi,l,s_{\tau 1},1)\widetilde{\Lambda}_{1\tau_1}(w;s_{\tau 1})ds_{\tau 1}. \end{aligned}$$

In all the cases, $\widetilde{L}(r_1,r_2,\frac{1}{2}) \le 43 + \frac{1}{2}C < 4 + C$ since $C > 100$. Therefore, we can ignore the first term. This proves statement (2).

Q.E.D.

If $\tau_1 = 1,(1,2),(2,3)$, $\widetilde{L}_1(r_1,r_2) \le 0$ for the above choices of r_1,r_2. Therefore,

$$\left(\frac{1}{2\pi\sqrt{-1}}\right)^2 \int_{\mathrm{Re}(s_{\tau 1})=(r_1,r_2)} \Sigma_{1,\mathrm{sub}}(\Psi,l,s_{\tau 1},1)\widetilde{\Lambda}_{1\tau_1}(w;s_{\tau 1},1)ds_{\tau 1} \sim 0$$

for these cases.

Suppose that $\tau_1 = (1,2,3),(1,3,2)$ or $(1,3)$. Let $\delta,\delta_1 > 0$ be small numbers such that $\delta_1\delta^{-1} \gg 0$.

Lemma (11.2.6) *By changing ψ if necessary,*

$$\begin{aligned} &\left(\frac{1}{2\pi\sqrt{-1}}\right)^2 \int_{\mathrm{Re}(s_{\tau 1})=(r_1,r_2)} \Sigma_{1,\mathrm{sub}}(\Psi,l,s_{\tau 1},1)\widetilde{\Lambda}_{1\tau_1}(w;s_{\tau 1},1)ds_{\tau 1} \\ &\sim \left(\frac{1}{2\pi\sqrt{-1}}\right)^2 \int_{\mathrm{Re}(s_{\tau 1})=(1+\delta,1+\delta)} \Sigma_{1,\mathrm{sub}}(\Psi,l,s_{\tau 1},1)\widetilde{\Lambda}_{1\tau_1}(w;s_{\tau 1},1)ds_{\tau 1}. \end{aligned}$$

Proof. For cases (1) and (2), the domain $\{(r_1,r_2) \mid l_{21}(r_1,r_2,1) < 0, l_{22}(r_1,r_2,1) > 0\}$ contains $(1+\delta,1+\delta)$. For cases (3) and (4), the domain $\{(r_1,r_2) \mid l_{21}(r_1,r_2,1) > 0, l_{22}(r_1,r_2,1) > 0\}$ contains $(1+\delta,1+\delta)$.

For $\tau_1 = (1,2,3),(1,3,2),(1,3)$, $\rho_{1\tau_1} = (2,-1),(-1,2),(1,1)$. For cases (6)–(11), (13)–(15), $l_{2i}(\rho_{1\tau_1},1) \ne 0$ for $i = 1,2$. For cases (5), (12), (16), $l_{21}(\rho_{1\tau_1},1) \ne 0$, and the line segment joining (r_1,r_2) and $(1+\delta,1+\delta)$ does not meet the line

$l_{22}(s_{\tau 11}, s_{\tau 12}, 1) = 0$. Therefore, the lemma follows from the passing principle (3.6.1) for these cases.

Q.E.D.

If $\tau_1 \neq (1,3)$, $\widetilde{L}_1(1+\delta, 1+\delta) < 4$ as long as $\delta > 0$ is small. So

$$\left(\frac{1}{2\pi\sqrt{-1}}\right)^2 \int_{\mathrm{Re}(s_{\tau 1})=(1+\delta,1+\delta)} \Sigma_{1,\mathrm{sub}}(\Psi, l, s_{\tau 1}, 1)\widetilde{\Lambda}_{1\tau_1}(w; s_{\tau 1})ds_{\tau 1} \sim 0.$$

Suppose $\tau_1 = (1,3)$. Then

$$\begin{aligned}
&\left(\frac{1}{2\pi\sqrt{-1}}\right)^2 \int_{\mathrm{Re}(s_{\tau 1})=(1+\delta,1+\delta)} \Sigma_{1,\mathrm{sub}}(\Psi, l, s_{\tau 1}, 1)\widetilde{\Lambda}_{1\tau_1}(w; s_{\tau 1})ds_{\tau 1} \\
&= \left(\frac{1}{2\pi\sqrt{-1}}\right)^2 \int_{\mathrm{Re}(s_{\tau 1})=(1+\delta,1-\delta_1)} \Sigma_{1,\mathrm{sub}}(\Psi, l, s_{\tau 1}, 1)\widetilde{\Lambda}_{1\tau_1}(w; s_{\tau 1})ds_{\tau 1} \\
&\quad + \frac{\varrho}{2\pi\sqrt{-1}} \int_{\mathrm{Re}(s_{\tau 11})=1+\delta} \Sigma_{1,\mathrm{sub}}(\Psi, l, s_{\tau 11}, 1, 1)\phi_2(s_{\tau 11})\widetilde{\Lambda}_1(w; s_{\tau 11}, 1)ds_{\tau 11} \\
&\sim \frac{\varrho}{2\pi\sqrt{-1}} \int_{\mathrm{Re}(s_{\tau 11})=1+\delta} \Sigma_{1,\mathrm{sub}}(\Psi, l, s_{\tau 11}, 1, 1)\phi_2(s_{\tau 11})\widetilde{\Lambda}_1(w; s_{\tau 11}, 1)ds_{\tau 11},
\end{aligned}$$

because the first term can be ignored by the assumption on δ, δ_1. This proves (11.2.4).

Q.E.D.

Chapter 12 Contributions from unstable strata

§12.1 The case $\mathfrak{d} = (\beta_1)$

We now start the analysis of the constant terms $\Xi_{\mathfrak{p}}(\Phi, \omega, w)$. We remind the reader that we are still assuming that $\Phi = M_{V,\omega}\Phi$, so if $\delta_{\#}(\omega) = 1$, Φ is K-invariant.

In this section, we consider paths $\mathfrak{p} = (\mathfrak{d}, \mathfrak{s})$ whose β-sequences start with β_1. Since $\mathscr{E}((g^0)^{\iota}, w) = \mathscr{E}(g^0, w)$ and the characters $e_{\mathfrak{p}i}$ do not depend on $\mathfrak{s}(1)$, we only consider $\mathfrak{p} = (\mathfrak{d}, \mathfrak{s})$ such that $\mathfrak{s}(1) = 0$. For paths with $\mathfrak{s}(1) = 1$, all the results are valid replacing Φ by $\widehat{\Phi}$ and ω by ω^{-1}, and changing the sign. All the β-sequences which start with β_1 satisfy Condition (3.4.16)(1). Let $\mathfrak{p}_0 = (\mathfrak{d}_0, \mathfrak{s}_0)$ where $\mathfrak{d}_0 = (\beta_1)$. Let $\mathfrak{p}_{11} = (\mathfrak{d}_1, \mathfrak{s}_{11}), \mathfrak{p}_{12} = (\mathfrak{d}_1, \mathfrak{s}_{12})$ where $\mathfrak{d}_1 = (\beta_1, \beta_{1,1})$, and $\mathfrak{s}_{11}(2) = 0, \mathfrak{s}_{12}(2) = 1$. Let $\mathfrak{p}_{21} = (\mathfrak{d}_2, \mathfrak{s}_{21}), \mathfrak{p}_{22} = (\mathfrak{d}_2, \mathfrak{s}_{22})$ where $\mathfrak{d}_2 = (\beta_1, \beta_{1,2})$, and $\mathfrak{s}_{21}(2) = 0, \mathfrak{s}_{22}(2) = 1$. Let $\mathfrak{d}_3 = (\beta_1, \beta_{11}, \beta_{1,1,1})$, and $\mathfrak{p}_{3i} = (\mathfrak{d}_3, \mathfrak{s}_{3i})$ for $i = 1, 2, 3, 4$ where $(\mathfrak{s}_{3i}(2), \mathfrak{s}_{3i}(3)) = (0,0), (0,1), (1,0), (1,1)$ for $i = 1, 2, 3, 4$ in that order. We consider contributions from these β-sequences. Throughout this section, we assume that $\Psi = R_{\mathfrak{d}_0}\Phi_{\mathfrak{p}_0}$.

Let

$$
\begin{aligned}
d_0(\lambda_1) &= a(1,1,1;\lambda_1^{-1},\underline{\lambda}_1),\\
d_1(\lambda_1,\lambda_2) &= a(\underline{\lambda}_2^{-2},\underline{\lambda}_2,\underline{\lambda}_2;\lambda_1^{-1},\underline{\lambda}_1),\\
d_2(\lambda_1,\lambda_2) &= a(\underline{\lambda}_2^{-1},\underline{\lambda}_2^{-1},\underline{\lambda}_2^{2};\underline{\lambda}_1^{-1},\underline{\lambda}_1),\\
d_3(\lambda_1,\lambda_2,\lambda_3) &= a(\underline{\lambda}_2^{-2},\underline{\lambda}_2\underline{\lambda}_3^{-1},\underline{\lambda}_2\underline{\lambda}_3;\underline{\lambda}_1^{-1},\underline{\lambda}_1),
\end{aligned}
\tag{12.1.1}
$$

for $\lambda_1, \lambda_2, \lambda_3 \in \mathbb{R}_+$.

Easy computations show the following two lemmas.

Lemma (12.1.2)
(1) $e_{\mathfrak{p}_{11}1}(d_1(\lambda_1,\lambda_2)) = \lambda_1$, $e_{\mathfrak{p}_{11}2}(d_1(\lambda_1,\lambda_2)) = \lambda_1\lambda_2^2$.
(2) $e_{\mathfrak{p}_{12}1}(d_1(\lambda_1,\lambda_2)) = \lambda_1^{-1}$, $e_{\mathfrak{p}_{12}2}(d_1(\lambda_1,\lambda_2)) = \lambda_1\lambda_2^2$.
(3) $e_{\mathfrak{p}_{21}1}(d_2(\lambda_1,\lambda_2)) = \lambda_1$, $e_{\mathfrak{p}_{22}1}(d_2(\lambda_1,\lambda_2)) = \lambda_1^{-1}$.
(4) $e_{\mathfrak{p}_{31}1}(d_3(\lambda_1,\lambda_2,\lambda_3)) = \lambda_1$, $e_{\mathfrak{p}_{31}2}(d_3(\lambda_1,\lambda_2,\lambda_3)) = \lambda_1\lambda_2^2$.
(5) $e_{\mathfrak{p}_{32}1}(d_3(\lambda_1,\lambda_2,\lambda_3)) = \lambda_1^{-1}$, $e_{\mathfrak{p}_{32}2}(d_3(\lambda_1,\lambda_2,\lambda_3)) = (\lambda_1\lambda_2^2)^{-1}$.
(6) $e_{\mathfrak{p}_{33}1}(d_3(\lambda_1,\lambda_2,\lambda_3)) = \lambda_1^{-1}$, $e_{\mathfrak{p}_{33}2}(d_3(\lambda_1,\lambda_2,\lambda_3)) = \lambda_1\lambda_2^2$.
(7) $e_{\mathfrak{p}_{34}1}(d_3(\lambda_1,\lambda_2,\lambda_3)) = \lambda_1$, $e_{\mathfrak{p}_{34}2}(d_3(\lambda_1,\lambda_2,\lambda_3)) = (\lambda_1\lambda_2^2)^{-1}$.

Lemma (12.1.3)
(1) $\sigma_{\mathfrak{p}_{11}}(d_1(\lambda_1,\lambda_2)) = \lambda_1^2\lambda_2^6$, $\sigma_{\mathfrak{p}_{12}}(d_1(\lambda_1,\lambda_2)) = \lambda_1^4\lambda_2^6$, $\kappa_{\mathfrak{d}_1 1}(d_1(\lambda_1,\lambda_2)) = \lambda_1^{-3}\lambda_2^{-6}$.
(2) $\sigma_{\mathfrak{p}_{21}}(d_2(\lambda_1,\lambda_2)) = \lambda_1^2\lambda_2^6$, $\sigma_{\mathfrak{p}_{22}}(d_2(\lambda_1,\lambda_2)) = \lambda_1^4\lambda_2^6$.
(3) $\sigma_{\mathfrak{p}_{31}}(d_3(\lambda_1,\lambda_2,\lambda_3)) = \lambda_1^2\lambda_2^6\lambda_3^2$, $\sigma_{\mathfrak{p}_{32}}(d_3(\lambda_1,\lambda_2,\lambda_3)) = \lambda_1\lambda_3^2$,
$\sigma_{\mathfrak{p}_{33}}(d_3(\lambda_1,\lambda_2,\lambda_3)) = \lambda_1^4\lambda_1^6\lambda_3^2$, $\sigma_{\mathfrak{p}_{34}}(d_3(\lambda_1,\lambda_2,\lambda_3)) = \lambda_1^{-1}\lambda_3^2$.

Let $\mathfrak{P}_1 = \{\mathfrak{p}_0, \mathfrak{p}_{11}, \mathfrak{p}_{12}\}$, $\mathfrak{P}_2 = \{\mathfrak{p}_{21}, \mathfrak{p}_{22}, \mathfrak{p}_{31}, \mathfrak{p}_{32}, \mathfrak{p}_{33}, \mathfrak{p}_{34}\}$. By (3.5.9),

$$
\begin{aligned}
\epsilon_{\mathfrak{p}_0}\Xi_{\mathfrak{p}_0}(\Phi,\omega,w) &= \sum_{\mathfrak{p}\in\mathfrak{P}_1} \epsilon_{\mathfrak{p}} \left(\Xi_{\mathfrak{p}+}(\Phi,\omega,w) + \widehat{\Xi}_{\mathfrak{p}+}(\Phi,\omega,w)\right)\\
&\quad + \sum_{\mathfrak{p}\in\mathfrak{P}_1} \epsilon_{\mathfrak{p}} \left(\widehat{\Xi}_{\mathfrak{p}\#}(\Phi,\omega,w) - \Xi_{\mathfrak{p}\#}(\Phi,\omega,w)\right)\\
&\quad + \sum_{\mathfrak{p}\in\mathfrak{P}_2} \epsilon_{\mathfrak{p}}\Xi_{\mathfrak{p}}(\Phi,\omega,w).
\end{aligned}
\tag{12.1.4}
$$

(a) $\Xi_{\mathfrak{p}_0+}(\Phi,\omega,w)$, $\widehat{\Xi}_{\mathfrak{p}_0+}(\Phi,\omega,w)$.

The element $d_0(\lambda_1)$ acts on $Z_{\mathfrak{d}_0\mathrm{A}}$ by multiplication by $\underline{\lambda}_1$. It is easy to see that $\sigma_{\mathfrak{p}_0}(d_0(\lambda_1))=\lambda_1^2$, $\kappa_{\mathfrak{d}_0 1}(d_0(\lambda_1))=\lambda_1^{-6}$. Therefore, the following proposition follows from (3.5.13).

Proposition (12.1.5)

(1) $$\Xi_{\mathfrak{p}_0+}(\Phi,\omega,w)\sim C_G\Lambda(w;\rho)\delta_{\mathfrak{d}_0}(\omega)\Sigma_{\mathfrak{d}_0+}(\Psi,\omega_{\mathfrak{d}_0},2).$$

(2) $$\widehat{\Xi}_{\mathfrak{p}_0+}(\Phi,\omega,w)\sim C_G\Lambda(w;\rho)\delta_{\mathfrak{d}_0}(\omega)\Sigma_{\mathfrak{d}_0+}(\mathscr{F}_{\mathfrak{d}_0}\Psi,\omega_{\mathfrak{d}_0}^{-1},4).$$

(b) $\Xi_{\mathfrak{p}_0\#}(\Phi,\omega,w)$, $\widehat{\Xi}_{\mathfrak{p}_0\#}(\Phi,\omega,w)$.

Proposition (12.1.6)

(1) $$\Xi_{\mathfrak{p}_0\#}(\Phi,\omega,w)\sim C_G\Lambda(w;\rho)\frac{\mathfrak{V}_3\delta_{\#}(\omega)\Psi(0)}{2}.$$

(2) $$\widehat{\Xi}_{\mathfrak{p}_0\#}(\Phi,\omega,w)\sim -C_G\Lambda(w;\rho)\frac{\mathfrak{V}_3\delta_{\#}(\omega)\mathscr{F}_{\mathfrak{d}_0}\Psi(0)}{4}.$$

Proof. Let $\mathfrak{d}=\mathfrak{d}_0,\mathfrak{p}=\mathfrak{p}_0$. Let $g_{\mathfrak{d}}^0\in M_{\mathfrak{d}}^0$, $g_{\mathfrak{d}}=d_0(\lambda_1)g_{\mathfrak{d}}^0$. Then $dg_{\mathfrak{d}}=d^\times\lambda_1 dg_{\mathfrak{d}}^0$. In this case, $M_{\mathfrak{d}\mathrm{A}}^1=M_{\mathfrak{d}\mathrm{A}}^0$. It is easy to see that

$$\Xi_{\mathfrak{p}\#}(\Phi,\omega,w)=\Psi(0)\int_{\substack{\mathbb{R}_+\times M_{\mathfrak{d}\mathrm{A}}^0/M_{\mathfrak{d}k}\\ \lambda_1\le 1}}\omega(g_{\mathfrak{d}})\lambda_1^2\mathscr{E}_{\mathfrak{p}}(g_{\mathfrak{d}},w)dg_{\mathfrak{d}},$$

$$\widehat{\Xi}_{\mathfrak{p}\#}(\Phi,\omega,w)=\mathscr{F}_{\mathfrak{d}}\Psi(0)\int_{\substack{\mathbb{R}_+\times M_{\mathfrak{d}\mathrm{A}}^0/M_{\mathfrak{d}k}\\ \lambda_1\le 1}}\omega(g_{\mathfrak{d}})\lambda_1^{-4}\mathscr{E}_{\mathfrak{p}}(g_{\mathfrak{d}},w)dg_{\mathfrak{d}}.$$

Let $\tau=(1,\tau_2)$ and $\tau_2=1$ or $(1,2)$. Let $s_{\tau 2}$ be as in (11.1.11). By a similar consideration as in (3.5.20),

$$\Xi_{\mathfrak{p}\#}(\Phi,\omega,w)=\sum_{\tau_2}\frac{\delta_{\#}(\omega)}{2\pi\sqrt{-1}}\int_{\mathrm{Re}(s_{\tau 2})=r_3>1}\widetilde{M}_{\tau_2}(s_{\tau 2})\frac{\Lambda(w;\rho_1,z_2)}{s_{\tau 2}+1}ds_{\tau 2},$$

where z_2 and $s_{\tau 2}$ are related as in (11.1.11). If $\tau_2=1$, $L_2(\mathrm{Re}(z_2))=-r_3<-1$ on the above contour. Therefore, we only have to consider the case $\tau_2=(1,2)$. We then move the contour to $0<r_3<1$, and (1) follows. The proof of (2) is similar. Q.E.D.

(c) $\Xi_{\mathfrak{p}_{11}+}(\Phi,\omega,w)$, $\widehat{\Xi}_{\mathfrak{p}_{11}+}(\Phi,\omega,w)$.

In (c), (d), let $\mathfrak{d}=\mathfrak{d}_1,\mathfrak{p}=\mathfrak{p}_{11}$, and $\Psi_1=R_{\mathfrak{d}_1}\Phi_{\mathfrak{p}_{11}}$. Let $g_{\mathfrak{d}}=d_1(\lambda_1,\lambda_2)g_{\mathfrak{d}}^0$ where $g_{\mathfrak{d}}^0\in M_{\mathfrak{d}\mathrm{A}}^0$ is as in the second element of (11.1.3). Also in this case, $M_{\mathfrak{d}\mathrm{A}}^1=M_{\mathfrak{d}\mathrm{A}}^0$. It is easy to see that $dg_{\mathfrak{d}}=2d^\times\lambda_1 d^\times\lambda_2 dg_{\mathfrak{d}}^0$. Let $\mu=\lambda_1\lambda_2^2$. Then $2d^\times\lambda_1 d^\times\lambda_2=d^\times\lambda_1 d^\times\mu$.

Let $\tau=(\tau_1,\tau_2)$ be a Weyl group element, and s_τ as in (11.1.11).

Definition (12.1.7)
(1) $l_{\mathfrak{p}_{11}}=(l_{\mathfrak{p}_{11},1},l_{\mathfrak{p}_{11},2})$ *where* $l_{\mathfrak{p}_{11},1}(s_\tau)=\frac{1}{2}(s_{\tau 11}+2s_{\tau 12}+3)$, $l_{\mathfrak{p}_{11},2}(s)=-s_{\tau 11}-2s_{\tau 12}+2s_{\tau 2}-1$,

(2) $\widehat{l}_{\mathfrak{p}_{11}} = (\widehat{l}_{\mathfrak{p}_{11},1}, \widehat{l}_{\mathfrak{p}_{11},2})$ where $\widehat{l}_{\mathfrak{p}_{11},1}(s_\tau) = 3 - l_{\mathfrak{p}_{11},1}(s_\tau)$, $\widehat{l}_{\mathfrak{p}_{11},2}(s_\tau) = l_{\mathfrak{p}_{11},2}(s_\tau)$.

Note that $l_{\mathfrak{p}_{11}}(-s_{\tau 11}, s_{\tau 11}+s_{\tau 12}, s_{\tau 2}) = l_{\mathfrak{p}_{11}}(s_{\tau 11}, s_{\tau 12}, s_{\tau 2})$ etc.

By (12.1.3)(1),

(12.1.8) $$\sigma_{\mathfrak{p}}(d_1(\lambda_1,\lambda_2))d_1(\lambda_1,\lambda_2)^{\tau z+\rho} = \lambda_1^{s_{\tau 2}-1}\lambda_2^{s_{\tau 11}+2s_{\tau 12}-3} = \lambda_1^{\frac{1}{2}l_{\mathfrak{p},2}(s_\tau)}\mu^{l_{\mathfrak{p},1}(s_\tau)},$$
$$\sigma_{\mathfrak{p}}(d_1(\lambda_1,\lambda_2))d_1(\lambda_1,\lambda_2)^{\tau z+\rho}\kappa_{\mathfrak{d}1}(d_1(\lambda_1,\lambda_2)) = \lambda_1^{\frac{1}{2}l_{\mathfrak{p},2}(s_\tau)}\mu^{l_{\mathfrak{p},1}(s_\tau)-3}.$$

The elements $d_1(\lambda_1,\lambda_2), \theta_{\mathfrak{d}}(d_1(\lambda_1,\lambda_2))$ act on $Z_{\mathfrak{d}\mathrm{A}}$ by multiplication by $\underline{\mu}, \underline{\mu}^{-1}$ respectively. Also the integrals defining $\Xi_{\mathfrak{p}_{11}+}(\Phi,\omega,w), \widehat{\Xi}_{\mathfrak{p}_{11}+}(\Phi,\omega,w)$ do not depend on t_{11}, t_{21}. Therefore, by (12.1.8),

$$\Xi_{\mathfrak{p}_{11}+}(\Phi,\omega,w) = 2\delta_\#(\omega)\sum_{\substack{\tau_1\in\mathfrak{W}_2\\ \tau_2=1,(1,2)}}\Omega_{W,\tau}(\Psi_1, l_{\mathfrak{p}_{11}}, w),$$
$$\widehat{\Xi}_{\mathfrak{p}_{11}+}(\Phi,\omega,w) = 2\delta_\#(\omega)\sum_{\substack{\tau_1\in\mathfrak{W}_2\\ \tau_2=1,(1,2)}}\Omega_{W,\tau}(\mathscr{F}_W\Psi_1, \widehat{l}_{\mathfrak{p}_{11}}, w).$$

The following proposition follows from (11.2.3).

Proposition (12.1.9) *Suppose that $\tau=\tau_G$* and *$\delta>0$ is a small number. Then by changing ψ if necessary,*

$$\begin{aligned}\Xi_{\mathfrak{p}_{11}+}(\Phi,\omega,w) \sim &- C_G\Lambda(w;\rho)\delta_\#(\omega)\Sigma_{W+}(\Psi_1,3)\\ &+\frac{2\varrho^2\delta_\#(\omega)}{2\pi\sqrt{-1}}\int_{\mathrm{Re}(s_{\tau 11})=1+\delta}\Sigma_{W,\mathrm{st,sub}+}(\Psi_1, l_{\mathfrak{p}_{11}}, s_{\tau 11},1,1)\\ &\qquad\times\phi_2(s_{\tau 11})\widetilde{\Lambda}_1(w;s_{\tau 11},1)ds_{\tau 11},\\ \widehat{\Xi}_{\mathfrak{p}_{11}+}(\Phi,\omega,w) \sim &- C_G\Lambda(w;\rho)\delta_\#(\omega)\Sigma_{W+}(\mathscr{F}_W\Psi_1,0)\\ &+\frac{2\varrho^2\delta_\#(\omega)}{2\pi\sqrt{-1}}\int_{\mathrm{Re}(s_{\tau 11})=1+\delta}\Sigma_{W,\mathrm{st,sub}+}(\mathscr{F}_W\Psi_1, \widehat{l}_{\mathfrak{p}_{11}}, s_{\tau 11},1,1)\\ &\qquad\times\phi_2(s_{\tau 11})\widetilde{\Lambda}_1(w;s_{\tau 11},1)ds_{\tau 11}.\end{aligned}$$

(d) $\Xi_{\mathfrak{p}_{11}\#}(\Phi,\omega,w), \widehat{\Xi}_{\mathfrak{p}_{11}\#}(\Phi,\omega,w)$.

We define

$$\Sigma_{\mathfrak{p}_{11}\#,\tau}(\Phi,\omega,w,s_{\tau 12},s_{\tau 2}) = \delta_\#(\omega)\Psi_1(0)\frac{\widetilde{\Lambda}_\tau(w;-1,s_{\tau 12},s_{\tau 2})}{(s_{\tau 12}+1)(-s_{\tau 12}+s_{\tau 2})},$$
$$\widehat{\Sigma}_{\mathfrak{p}_{11}\#,\tau}(\Phi,\omega,w,s_{\tau 12},s_{\tau 2}) = \delta_\#(\omega)\mathscr{F}_W\Psi_1(0)\frac{\widetilde{\Lambda}_\tau(w;-1,s_{\tau 12},s_{\tau 2})}{(s_{\tau 12}-2)(-s_{\tau 12}+s_{\tau 2})}.$$

Then by (11.1.4), (11.1.5), (12.1.8), and the Mellin inversion formula, the distributions $\Xi_{\mathfrak{p}_{11}\#}(\Phi,\omega,w), \widehat{\Xi}_{\mathfrak{p}_{11}\#}(\Phi,\omega,w)$ are equal to

(12.1.10) $$\sum_{\substack{\tau_1\in\mathfrak{W}_2\\ \tau_2=1,(1,2)}}\left(\frac{1}{2\pi\sqrt{-1}}\right)^2\int_{\substack{\mathrm{Re}(s_{\tau 12},s_{\tau 2})=(r_2,r_3)\\ r_2>2,\ r_3>r_2}}\Sigma_{\mathfrak{p}_{11}\#,\tau}(\Phi,\omega,w,s_{\tau 12},s_{\tau 2})ds_{\tau 12},s_{\tau 2},$$
$$\sum_{\substack{\tau_1\in\mathfrak{W}_2\\ \tau_2=1,(1,2)}}\left(\frac{1}{2\pi\sqrt{-1}}\right)^2\int_{\substack{\mathrm{Re}(s_{\tau 12},s_{\tau 2})=(r_2,r_3)\\ r_2>2,\ r_3>r_2}}\widehat{\Sigma}_{\mathfrak{p}_{11}\#,\tau}(\Phi,\omega,w,s_{\tau 12},s_{\tau 2})ds_{\tau 12},s_{\tau 2}.$$

Proposition (12.1.11) *Suppose that $\tau = \tau_G$ and $\delta > 0$ is a small number. Then by changing ψ if necessary,*

$$(1) \quad \Xi_{\mathfrak{p}_{11}\#}(\Phi, \omega, w) \sim -C_G \Lambda(w; \rho) \frac{\mathfrak{V}_2 \delta_\#(\omega) \Psi_1(0)}{3},$$

$$(2) \quad \widehat{\Xi}_{\mathfrak{p}_{11}\#}(\Phi, \omega, w) \sim \frac{\varrho \delta_\#(\omega) \mathscr{F}_W \Psi_1(0)}{2\pi\sqrt{-1}} \int_{\mathrm{Re}(s_{\tau 11})=1+\delta} \frac{\phi_2(s_{\tau 11}) \widetilde{\Lambda}_1(w; s_{\tau 11}, 1)}{s_{\tau 11}(1 - s_{\tau 11})} ds_{\tau 11}.$$

Proof. Consider the term in (12.1.10) which corresponds to τ. If $\tau_2 = 1$, we can ignore such a term by the usual argument. So, we assume that $\tau_2 = 1$. If $\tau_1 = 1, (1,2)$ $\rho_\tau = (-1,-1,1), (-2,1,1)$. Therefore, by (3.6.1), we can move the contour crossing the line $s_{\tau 12} = 2$. So in both cases, we can move the contour so that $r_2 = 1 + \delta, r_3 = 1 + 2\delta$ where $\delta > 0$ is a small number. Then $\widetilde{L}(-1, r_2, r_3) \leq 2 + (1+2\delta)C < 4 + C$ if δ is sufficiently small. Therefore, we can ignore these cases.

Suppose $\tau_1 = (1,3,2)$. The point $\rho_\tau = (-1,2,1)$ is not on the lines $s_{\tau 12} + 1 = 0, -s_{\tau 12} + s_{\tau 2} = 0$. So we can move the contour crossing these lines by (3.6.1) so that $r_2 < r_3$. Then for (1), the term in (12.1.10) which corresponds to $\tau = ((1,3,2),(1,2))$ is equal to

$$\begin{aligned}
&\left(\frac{1}{2\pi\sqrt{-1}}\right)^2 \int_{\substack{\mathrm{Re}(s_{\tau 12}, s_{\tau 2}) = (r_2, r_3) \\ r_3 > r_2 > 2}} \Sigma_{\mathfrak{p}_{11}\#,\tau}(\Phi, \omega, w, s_{\tau 12}, s_{\tau 2}) ds_{\tau 12} ds_{\tau 2} \\
&= \left(\frac{1}{2\pi\sqrt{-1}}\right)^2 \int_{\mathrm{Re}(s_{\tau 12}, s_{\tau 2}) = (3, \frac{1}{2})} \Sigma_{\mathfrak{p}_{11}\#,\tau}(\Phi, \omega, w, s_{\tau 12}, s_{\tau 2}) ds_{\tau 12} ds_{\tau 2} \\
&\quad + \frac{1}{2\pi\sqrt{-1}} \int_{\mathrm{Re}(s_{\tau 12}) = r_2 > 2} \operatorname*{Res}_{s_{\tau 2}=1} \Sigma_{\mathfrak{p}_{11}\#,\tau}(\Phi, \omega, w, s_{\tau 12}, s_{\tau 2}) ds_{\tau 12}.
\end{aligned}$$

We can ignore the first term as usual. If $\tau_1 = 1$ or $(1,2)$, $\widetilde{L}(-1,3,1) \leq 2 + C < 4 + C$. Therefore, we can ignore both these terms. Suppose $\tau_1 = (1,3,2)$. Since $\widetilde{M}_{\tau_1}(-1, s_{\tau 12}) = \phi_2(s_{\tau 12} - 1)$ and this function has a simple pole at $s_{\tau 12} = 2$, we only have to move the contour to $1 < \mathrm{Re}(s_{\tau 11}) < 2$. This proves (1). The proof of (2) is similar except that the integrand of the second term has a pole of order 2 at $s_{\tau 12} = 2$, and we leave as it is. If $\tau' = \tau_G$, $s_{\tau' 12} = s_{\tau 12} - 1$. We consider the substitution $z \to -\tau_G z$. Then $s_{\tau' 11}, s_{\tau' 12}$ are exchanged, and $\widetilde{\Lambda}_1(w; s_{\tau' 11}, 1) = \widetilde{\Lambda}_1(w; 1, s_{\tau' 11})$ by the assumption $\psi(-\tau_G z) = \psi(z)$. Thus, after these substitutions, we get the expression of (2) in terms of $s_{\tau' 11}$.

Q.E.D.

(e) $\Xi_{\mathfrak{p}_{12}+}(\Phi, \omega, w)$, $\widehat{\Xi}_{\mathfrak{p}_{12}+}(\Phi, \omega, w)$.

In (e)–(g), $\mathfrak{d} = \mathfrak{d}_1, \mathfrak{p} = \mathfrak{p}_{12}$, and $\Psi_2 = R_{\mathfrak{d}_1} \Phi_{\mathfrak{p}_{12}}$. Let $g_\mathfrak{d} = d_1(\lambda_1, \lambda_2) g_\mathfrak{d}^0$ where $g_\mathfrak{d}^0 \in M_{\mathfrak{d}\mathbb{A}}^0$ is as in the second element of (11.1.3).

By (12.1.2)(2), (12.1.3)(1),

$$\begin{aligned}
&\Xi_{\mathfrak{p}_{12}+}(\Phi, \omega, w) \\
&= 2 \int_{\substack{\mathbb{R}_+^2 \times M_{\mathfrak{d}\mathbb{A}}^0 / M_{\mathfrak{d}k} \\ \lambda_1 \geq 1, \ \lambda_1 \lambda_2^2 \geq 1}} \omega(g_\mathfrak{d})^{-1} \lambda_1^4 \lambda_2^6 \Theta_{Z_\mathfrak{d}}(\Psi_2, g_\mathfrak{d}) \mathscr{E}_{\mathfrak{p}_{12}}(g_\mathfrak{d}, w) d^\times \lambda_1 d^\times \lambda_2 dg_\mathfrak{d}^0,
\end{aligned}$$

$$\widehat{\Xi}_{\mathfrak{p}_{12}+}(\Phi,\omega,w)$$
$$= 2\int_{\substack{\mathbb{R}_+^2\times M_{\mathfrak{d}\mathbb{A}}^0/M_{\mathfrak{d}k}\\ \lambda_1\geq 1,\ \lambda_1\lambda_2^2\leq 1}} \omega(g_{\mathfrak{d}})^{-1}\lambda_1\Theta_{Z_{\mathfrak{d}}}(\mathscr{F}_W\Psi_2,\theta_{\mathfrak{d}}(g_{\mathfrak{d}}))\mathscr{E}_{\mathfrak{p}_{12}}(g_{\mathfrak{d}},w)d^\times\lambda_1 d^\times\lambda_2 dg_{\mathfrak{d}}^0.$$

Let $\mu=\lambda_1\lambda_2^2$. Then $\lambda_1^4\lambda_2^6=\lambda_1\mu^3$. The functions $\Theta_{Z_{\mathfrak{d}}}(\Psi,g_{\mathfrak{d}}),\Theta_{Z_{\mathfrak{d}}}(\mathscr{F}_{\mathfrak{d}}\Psi,\theta_{\mathfrak{d}}(g_{\mathfrak{d}}))$, and $\mathscr{E}_{\mathfrak{p}_{12}}(g_{\mathfrak{d}},w)$ do not depend on t_{11},t_{21}. Therefore, these distributions are 0 unless ω is trivial.

Let τ,s_τ be as before. Easy computations show the following lemma.

Lemma (12.1.12) $\theta_{\mathfrak{p}_{12}}(d_2(\lambda_1,\lambda_2))^{\tau z+\rho}=\lambda_1^{\frac{1}{2}(-s_{\tau 11}-2s_{\tau 12}-2s_{\tau 2}+5)}\mu^{\frac{1}{2}(s_{\tau 11}+2s_{\tau 12}-3)}$.

Definition (12.1.13)
(1) $l_{\mathfrak{p}_{12}}=(l_{\mathfrak{p}_{12},1},l_{\mathfrak{p}_{12},2})$ *where* $l_{\mathfrak{p}_{12},1}(s_\tau)=\frac{1}{2}(s_{\tau 11}+2s_{\tau 12}+3)$, $l_{\mathfrak{p}_{12},2}(s_\tau)=s_{\tau 11}+2s_{\tau 12}+2s_{\tau 2}-7$,
(2) $\widehat{l}_{\mathfrak{p}_{12}}=(\widehat{l}_{\mathfrak{p}_{12},1},\widehat{l}_{\mathfrak{p}_{12},2})$ *where* $\widehat{l}_{\mathfrak{p}_{12},1}(s_\tau)=3-l_{\mathfrak{p}_{12},1}(s_\tau)$, $\widehat{l}_{\mathfrak{p}_{12},2}(s_\tau)=l_{\mathfrak{p}_{12},2}(s_\tau)$.

Note that $l_{\mathfrak{p}_{12}}(-s_{\tau 11},s_{\tau 11}+s_{\tau 12},s_{\tau 2})=l_{\mathfrak{p}_{12}}(s_{\tau 11},s_{\tau 12},s_{\tau 2})$ etc.

By (11.1.4), (11.1.5), (12.1.12),

$$\Xi_{\mathfrak{p}_{12}+}(\Phi,\omega,w)=2\delta_\#(\omega)\sum_{\substack{\tau_1\in\mathfrak{W}_2\\ \tau_2=1,(1,\ 2)}}\Omega_{W,\tau}(\Psi_2,l_{\mathfrak{p}_{12}},w),$$
$$\widehat{\Xi}_{\mathfrak{p}_{12}+}(\Phi,\omega,w)=2\delta_\#(\omega)\sum_{\substack{\tau_1\in\mathfrak{W}_2\\ \tau_2=1,(1,\ 2)}}\Omega_{W,\tau}(\mathscr{F}_W\Psi_2,\widehat{l}_{\mathfrak{p}_{12}},w).$$

The following proposition follows from (11.2.3).

Proposition (12.1.14) *Suppose that* $\tau=\tau_G$ *and* $\delta>0$ *is a small number. Then by changing* ψ *if necessary,*

$$\begin{aligned}\Xi_{\mathfrak{p}_{12}+}(\Phi,\omega,w)&\sim -C_G\Lambda(w;\rho)\delta_\#(\omega)\Sigma_{W+}(\Psi_2,3)\\&\quad+\frac{2\varrho^2}{2\pi\sqrt{-1}}\int_{\mathrm{Re}(s_{\tau 11})=1+\delta}\Sigma_{W,\mathrm{st,sub}+}(\Psi_2,l_{\mathfrak{p}_{12}},s_{\tau 11},1,1)\\&\qquad\times\phi_2(s_{\tau 11})\widetilde{\Lambda}_1(w;1,s_{\tau 12})ds_{\tau 12},\\ \widehat{\Xi}_{\mathfrak{p}_{12}+}(\Phi,\omega,w)&\sim -C_G\Lambda(w;\rho)\delta_\#(\omega)\Sigma_{W+}(\mathscr{F}_W\Psi_2,0)\\&\quad+\frac{2\varrho^2}{2\pi\sqrt{-1}}\int_{\mathrm{Re}(s_{\tau 11})=1+\delta}\Sigma_{W,\mathrm{st,sub}+}(\mathscr{F}_W\Psi_2,\widehat{l}_{\mathfrak{p}_{12}},s_{\tau 11},1,1)\\&\qquad\times\phi_2(s_{\tau 11})\widetilde{\Lambda}_1(w;1,s_{\tau 12})ds_{\tau 12}.\end{aligned}$$

(f) $\Xi_{\mathfrak{p}_{12}\#}(\Phi,\omega,w)$, $\widehat{\Xi}_{\mathfrak{p}_{12}\#}(\Phi,\omega,w)$

These distributions are 0 unless ω is trivial for a similar reason. We define

$$(12.1.15)\quad \Sigma_{\mathfrak{p}_{12}\#,\tau}(\Phi,\omega,w,s_{\tau 12},s_{\tau 2})=\delta_\#(\omega)\Psi_2(0)\frac{\widetilde{\Lambda}_\tau(w;-1,s_{\tau 12},s_{\tau 2})}{(s_{\tau 12}+s_{\tau 2}-4)(s_{\tau 12}+1)},$$
$$\widehat{\Sigma}_{\mathfrak{p}_{12}\#,\tau}(\Phi,\omega,w,s_{\tau 12},s_{\tau 2})=\delta_\#(\omega)\mathscr{F}_W\Psi_2(0)\frac{\widetilde{\Lambda}_\tau(w;-1,s_{\tau 12},s_{\tau 2})}{(s_{\tau 12}+s_{\tau 2}-4)(s_{\tau 12}-2)}.$$

Then by (11.1.4), (11.1.5), (12.1.12), and the Mellin inversion formula, the distributions $\Xi_{\mathfrak{p}_{12}\#}(\Phi,\omega,w), \widehat{\Xi}_{\mathfrak{p}_{12}\#}(\Phi,\omega,w)$ are equal to

$$\sum_{\substack{\tau_1\in\mathfrak{W}_2 \\ \tau_2=1,(1,2)}} \left(\frac{1}{2\pi\sqrt{-1}}\right)^2 \int_{\substack{\mathrm{Re}(s_{\tau 12},s_{\tau 2})=(r_2,r_3) \\ r_2>2,\ r_2+r_3>4}} \Sigma_{\mathfrak{p}_{12}\#,\tau}(\Psi,\omega,w,s_{\tau 12},s_{\tau 2})ds_{\tau 12}ds_{\tau 2},$$

$$\sum_{\substack{\tau_1\in\mathfrak{W}_2 \\ \tau_2=1,(1,2)}} \left(\frac{1}{2\pi\sqrt{-1}}\right)^2 \int_{\substack{\mathrm{Re}(s_{\tau 12},s_{\tau 2})=(r_2,r_3) \\ r_2>2,\ r_2+r_3>4}} \widehat{\Sigma}_{\mathfrak{p}_{12}\#,\tau}(\Psi,\omega,w,s_{\tau 12},s_{\tau 2})ds_{\tau 12}ds_{\tau 2}.$$

Proposition (12.1.16) *Suppose that $\tau=\tau_G$ and $\delta>0$ is a small number. Then by changing ψ if necessary,*

$$(1)\quad \Xi_{\mathfrak{p}_{12}\#}(\Phi,\omega,w) \sim -C_G\Lambda(w;\rho)\frac{\mathfrak{V}_2\delta_\#(\omega)\Psi_2(0)}{3},$$

$$(2)\quad \widehat{\Xi}_{\mathfrak{p}_{12}\#}(\Phi,\omega,w) \sim \frac{\varrho\delta_\#(\omega)\mathscr{F}_W\Psi_2(0)}{2\pi\sqrt{-1}} \int_{\mathrm{Re}(s_{\tau 11})=1+\delta} \frac{\phi_2(s_{\tau 11})\widetilde{\Lambda}_1(w;s_{\tau 11},1)}{(s_{\tau 11}-1)(s_{\tau 11}-2)}ds_{\tau 11}.$$

Proof. As usual we can ignore the case $\tau_2=1$, and assume that $\tau_2=(1,2)$. We can ignore the cases $\tau_1=1,(1,2)$ by the same argument as in (12.1.11). Suppose $\tau_1=(1,3,2)$. Consider (1).

We move the contour in the following manner:

$$\begin{aligned}
&\left(\frac{1}{2\pi\sqrt{-1}}\right)^2 \int_{\substack{\mathrm{Re}(s_{\tau 12},s_{\tau 2})=(r_2,r_3) \\ r_2>2,\ r_2+r_3>4}} \Sigma_{\mathfrak{p}_{12}\#,\tau}(\Psi_1,\omega,w,s_{\tau 12},s_{\tau 2})ds_{\tau 12}ds_{\tau 2} \\
&= \left(\frac{1}{2\pi\sqrt{-1}}\right)^2 \int_{\mathrm{Re}(s_{\tau 12},s_{\tau 2})=(4,\frac12)} \Sigma_{\mathfrak{p}_{12}\#,\tau}(\Psi_1,\omega,w,s_{\tau 12},s_{\tau 2})ds_{\tau 12}ds_{\tau 2} \\
&\quad + \frac{1}{2\pi\sqrt{-1}} \int_{\mathrm{Re}(s_{\tau 12})=r_2>3} \operatorname*{Res}_{s_{\tau 2}=1} \Sigma_{\mathfrak{p}_{12}\#,\tau}(\Psi_1,\omega,w,s_{\tau 12},s_{\tau 2})ds_{\tau 12}.
\end{aligned}$$

We can ignore the first term because $\widetilde{L}(r)\le 8+\frac12 C<4+C$. The point $\rho_\tau=(-1,2,1)$ does not satisfy the condition $s_{\tau 12}=3$. Therefore, we can ignore the pole of $\mathrm{Res}_{s_{\tau 2}=1}\Sigma_{\mathfrak{p}_{12}\#,\tau}(\Psi_1,\omega,w,s_{\tau 12},s_{\tau 2})$ at $s_{\tau 12}=3$ by (3.6.1). Hence,

$$\begin{aligned}
&\frac{1}{2\pi\sqrt{-1}} \int_{\mathrm{Re}(s_{\tau 12})=r_2>3} \operatorname*{Res}_{s_{\tau 2}=1} \Sigma_{\mathfrak{p}_{12}\#,\tau}(\Psi_1,\omega,w,s_{\tau 12},s_{\tau 2})ds_{\tau 12} \\
&\sim \frac{1}{2\pi\sqrt{-1}} \int_{\substack{\mathrm{Re}(s_{\tau 12})=r_2 \\ 1<r_2<2}} \operatorname*{Res}_{s_{\tau 2}=1} \Sigma_{\mathfrak{p}_{12}\#,\tau}(\Psi_1,\omega,w,s_{\tau 12},s_{\tau 2})ds_{\tau 12} \\
&\quad - C_G\Lambda(w;\rho)\frac{\mathfrak{V}_2\delta_\#(\omega)\Psi_1(0)}{3}.
\end{aligned}$$

We can ignore the first term in the usual manner. This proves (1), and the proof of (2) is similar to (12.1.11).

Q.E.D.

(g) $\Xi_{\mathfrak{p}_{21}}(\Phi,\omega,w)$, $\Xi_{\mathfrak{p}_{22}}(\Phi,\omega,w)$.

Let $\mathfrak{d}=\mathfrak{d}_2$. Let $g_{\mathfrak{d}}=d_2(\lambda_1,\lambda_2)g_{\mathfrak{d}}^0$ where $g_{\mathfrak{d}}^0$ is as in the first element of (11.1.3). It is easy to see that $dg_{\mathfrak{d}}=2d^\times\lambda_1 d^\times\lambda_2 dg_{\mathfrak{d}}^0$. Let $\Psi_i=R_{\mathfrak{d}_2}R_{\mathfrak{p}_{2i}}\Phi$ for $i=1,2$. In this case, $P_{\mathfrak{d}\#}=P_1\times B_2$.

By (12.1.2), (12.1.3),

$$\Xi_{\mathfrak{p}_{21}}(\Phi,\omega,w)=2\int_{\substack{\mathbb{R}_+^2\times M_{\mathfrak{d}\mathbb{A}}^0/M_{\mathfrak{d}k}\\ \lambda_1\le 1}}\omega(g_{\mathfrak{d}})\lambda_1^2\lambda_2^6\Theta_{Z_{\mathfrak{d}}}(\Psi_1,g_{\mathfrak{d}})\mathscr{E}_{\mathfrak{p}_{21}}(g_{\mathfrak{d}},w)d^\times\lambda_1 d^\times\lambda_2 dg_{\mathfrak{d}}^0,$$

$$\Xi_{\mathfrak{p}_{22}}(\Phi,\omega,w)=2\int_{\substack{\mathbb{R}_+^2\times M_{\mathfrak{d}\mathbb{A}}^0/M_{\mathfrak{d}k}\\ \lambda_1\ge 1}}\omega(g_{\mathfrak{d}})^{-1}\lambda_1^4\lambda_2^6\Theta_{Z_{\mathfrak{d}}}(\Psi_2,g_{\mathfrak{d}})\mathscr{E}_{\mathfrak{p}_{22}}(g_{\mathfrak{d}},w)d^\times\lambda_1 d^\times\lambda_2 dg_{\mathfrak{d}}^0.$$

Let $\mu=\lambda_1\lambda_2^4$. Then $\lambda_1^2\lambda_2^6=\lambda_1^{\frac12}\mu^{\frac32}$, $\lambda_1^4\lambda_2^6=\lambda_1^{\frac52}\mu^{\frac32}$, and $d^\times\lambda_1 d^\times\lambda_2=\frac14 d^\times\lambda_1 d^\times\mu$.

Let $\tau=(\tau_1,\tau_2)$ be a Weyl group element such that $\tau_1\in\mathfrak{W}_1$. Easy computations show the following lemma.

Lemma (12.1.17) *Suppose* $s_{\tau 12}=-1$. *Then*
(1) $\theta_{\mathfrak{p}_{21}}(d_2(\lambda_1,\lambda_2))^{\tau z+\rho}=\lambda_1^{\frac12(-s_{\tau 11}+2s_{\tau 2})}\mu^{\frac12(s_{\tau 11}-2)}$,
(2) $\theta_{\mathfrak{p}_{22}}(d_2(\lambda_1,\lambda_2))^{\tau z+\rho}=\lambda_1^{\frac12(-s_{\tau 11}-2s_{\tau 2}+4)}\mu^{\frac12(s_{\tau 11}-2)}$.

Definition (12.1.18) *Let* $l_{\mathfrak{p}_{2i}}=(\frac12(s_{\tau 11}+1),l_{\mathfrak{p}_{2i},2})$ *for* $i=1,2$ *where*
(1) $l_{\mathfrak{p}_{21},2}(s_\tau)=-s_{\tau 11}+2s_{\tau 2}+1$,
(2) $l_{\mathfrak{p}_{22},2}(s_\tau)=s_{\tau 11}+2s_{\tau 2}-9$.

Note that the functions $\Theta_{Z_{\mathfrak{d}}}(\Psi_i,g_{\mathfrak{d}}),\mathscr{E}_{\mathfrak{p}_{2i}}(g_{\mathfrak{d}},w)$ do not depend on $g_{1,12},t_{21}$ for $i=1,2$. Therefore, the distributions $\Xi_{\mathfrak{p}_{21}}(\Phi,\omega,w),\Xi_{\mathfrak{p}_{22}}(\Phi,\omega,w)$ are 0 unless ω is trivial. We define

$$\Sigma_{\mathfrak{p}_{2i}}(\Phi,w,s_{\tau 11},s_{\tau 2})=[\Sigma_{1,\mathrm{sub}}(\Psi_i,l_{\mathfrak{p}_{2i}},s_\tau)\widetilde{\Lambda}_\tau(w;s_\tau)]_{s_{\tau 12}=-1}$$

for $i=1,2$. Then $\Xi_{\mathfrak{p}_{2i}}(\Phi,\omega,w)$ is equal to

$$\delta_\#(\omega)\sum_{\substack{\tau_1\in\mathfrak{W}_1\\ \tau_2=1,(1,2)}}\left(\frac{1}{2\pi\sqrt{-1}}\right)^2\int_{\mathrm{Re}(s_{\tau 11},s_{\tau 2})=(r_1,r_3)}\Sigma_{\mathfrak{p}_{2i}}(\Phi,w,s_\tau)ds_{\tau 11}ds_{\tau 2}$$

for $i=1,2$, where we choose the contour so that $r_1>2,2r_3>r_1-1$ for $i=1$ and $r_1>2,2r_3>9-r_1$ for $i=2$.

Proposition (12.1.19)

(1) $$\Xi_{\mathfrak{p}_{21}}(\Phi,\omega,w)\sim C_G\Lambda(w;\rho)\mathfrak{V}_2\delta_\#(\omega)\Sigma_1(R_{\mathfrak{d}_2}\Phi_{\mathfrak{p}_{21}},\frac32).$$

(2) $$\Xi_{\mathfrak{p}_{22}}(\Phi,\omega,w)\sim -C_G\Lambda(w;\rho)\frac{\mathfrak{V}_2\delta_\#(\omega)}{5}\Sigma_1(R_{\mathfrak{d}_2}\Phi_{\mathfrak{p}_{22}},\frac32).$$

Proof. Let $r_1'=\frac94,r_3'=\frac34$ for $i=1$ and $r_1'=9,r_3'=\frac12$ for $i=2$. Then

$$\begin{aligned}&\left(\frac{1}{2\pi\sqrt{-1}}\right)^2\int_{\mathrm{Re}(s_{\tau 11},s_{\tau 2})=(r_1,r_3)}\Sigma_{\mathfrak{p}_{2i}}(\Phi,w,s_\tau)ds_{\tau 11}ds_{\tau 2}\\ &=\left(\frac{1}{2\pi\sqrt{-1}}\right)^2\int_{\mathrm{Re}(s_{\tau 11},s_{\tau 2})=(r_1',r_3')}\Sigma_{\mathfrak{p}_{2i}}(\Phi,w,s_\tau)ds_{\tau 11}ds_{\tau 2}\\ &\quad+\frac{1}{2\pi\sqrt{-1}}\int_{\mathrm{Re}(s_{\tau 11})=r_1}\operatorname*{Res}_{s_{\tau 2}=1}\Sigma_{\mathfrak{p}_{2i}}(\Phi,w,s_\tau)ds_{\tau 11}.\end{aligned}$$

Since $\widetilde{L}(r_1', -1, r_3') \leq 18 + \frac{3}{4}C < w_0$ for all the cases, we can ignore the first term. The rest of the argument is similar as before.

Q.E.D.

(h) $\Xi_{\mathfrak{p}_{3i}}(\Phi, \omega, w)$ for $i = 1, \cdots, 4$.

Let $\mathfrak{d} = \mathfrak{d}_3$. Let $t^0 = d_3(\lambda_1, \lambda_2, \lambda_3)\widehat{t}^0$. Then $d^\times t^0 = 2d^\times\lambda_1 d^\times\lambda_2 d^\times\lambda_3 d^\times\widehat{t}^0$. By (12.1.2)(4)–(7) and (12.1.3)(3),

$$\Xi_{\mathfrak{p}_{31}}(\Phi, \omega, w) = \int_{\substack{T^0_{\mathbb{A}}/T_k \\ \lambda_1 \leq 1,\ \lambda_1\lambda_2^2 \leq 1}} \omega(t^0)\lambda_1^2\lambda_2^6\lambda_3^2\Theta_{Z_{\mathfrak{d}}}(R_{\mathfrak{d}}\Phi_{\mathfrak{p}_{33}}, t^0)\mathscr{E}_{\mathfrak{p}_{31}}(t^0, w)d^\times t^0,$$

$$\Xi_{\mathfrak{p}_{32}}(\Phi, \omega, w) = \int_{\substack{T^0_{\mathbb{A}}/T_k \\ \lambda_1 \geq 1,\ \lambda_1\lambda_2^2 \geq 1}} \omega(t^0)^{-1}\lambda_1\lambda_3^2\Theta_{Z_{\mathfrak{d}}}(R_{\mathfrak{d}}\Phi_{\mathfrak{p}_{33}}, t^0)\mathscr{E}_{\mathfrak{p}_{32}}(t^0, w)d^\times t^0,$$

$$\Xi_{\mathfrak{p}_{33}}(\Phi, \omega, w) = \int_{\substack{T^0_{\mathbb{A}}/T_k \\ \lambda_1 \geq 1,\ \lambda_1\lambda_2^2 \leq 1}} \omega(t^0)^{-1}\lambda_1^4\lambda_2^6\lambda_3^2\Theta_{Z_{\mathfrak{d}}}(R_{\mathfrak{d}}\Phi_{\mathfrak{p}_{33}}, t^0)\mathscr{E}_{\mathfrak{p}_{33}}(t^0, w)d^\times t^0,$$

$$\Xi_{\mathfrak{p}_{34}}(\Phi, \omega, w) = \int_{\substack{T^0_{\mathbb{A}}/T_k \\ \lambda_1 \leq 1,\ \lambda_1\lambda_2^2 \geq 1}} \omega(t^0)\lambda_1^{-1}\lambda_3^2\Theta_{Z_{\mathfrak{d}}}(R_{\mathfrak{d}}\Phi_{\mathfrak{p}_{34}}, t^0)\mathscr{E}_{\mathfrak{p}_{34}}(t^0, w)d^\times t^0.$$

For a similar reason as before, these distributions are 0 unless ω is trivial Let $\mu_1 = \lambda_1\lambda_2^2$, and $\mu_2 = \lambda_1\lambda_2^2\lambda_3^2$. Then

$$\lambda_1^2\lambda_2^6\lambda_3^2 = \lambda_1^{-1}\mu_1^2\mu_2,\ \lambda_1\lambda_3^2 = \lambda_1\mu_1^{-1}\mu_2,$$
$$\lambda_1^4\lambda_2^6\lambda_3^2 = \lambda_1\mu_1^2\mu_2, \text{ and } \lambda_1^{-1}\lambda_3^2 = \lambda_1^{-1}\mu_1^{-1}\mu_2.$$

It is easy to see that $d^\times t^0 = \frac{1}{2}d^\times\lambda_1 d^\times\mu_1 d^\times\mu_2 d^\times\widehat{t}^0$.

Easy computations show the following lemma.

Lemma (12.1.20)

(1) $\theta_{\mathfrak{p}_{31}}(d_3(\lambda_1, \lambda_2, \lambda_3))^{\tau z+\rho} = \lambda_1^{\frac{1}{2}(-s_{\tau 11}-2s_{\tau 12}-2s_{\tau 2}+1)}\mu_1^{s_{\tau 12}-1}\mu_2^{\frac{1}{2}(s_{\tau 11}-1)}$.

(2) $\theta_{\mathfrak{p}_{32}}(d_3(\lambda_1, \lambda_2, \lambda_3))^{\tau z+\rho} = \lambda_1^{\frac{1}{2}(s_{\tau 11}+2s_{\tau 12}+2s_{\tau 2}-1)}\mu_1^{-s_{\tau 11}-s_{\tau 12}+2}\mu_2^{\frac{1}{2}(s_{\tau 11}-1)}$.

(3) $\theta_{\mathfrak{p}_{33}}(d_3(\lambda_1, \lambda_2, \lambda_3))^{\tau z+\rho} = \lambda_1^{\frac{1}{2}(-s_{\tau 11}-2s_{\tau 12}-2s_{\tau 2}+5)}\mu_1^{s_{\tau 12}-1}\mu_2^{\frac{1}{2}(s_{\tau 11}-1)}$.

(4) $\theta_{\mathfrak{p}_{34}}(d_3(\lambda_1, \lambda_2, \lambda_3))^{\tau z+\rho} = \lambda_1^{\frac{1}{2}(s_{\tau 11}+2s_{\tau 12}+2s_{\tau 2}-5)}\mu_1^{-s_{\tau 11}-s_{\tau 12}+2}\mu_2^{\frac{1}{2}(s_{\tau 11}-1)}$.

Definition (12.1.21) *Let* $l_{\mathfrak{p}_{4i}} = (\frac{1}{2}(s_{\tau 11}+1), l_{\mathfrak{p}_{4i},2})$ *where*

(1) $l_{\mathfrak{p}_{31},2}(s_\tau) = (-s_{\tau 11} - 2s_{\tau 12} + 2s_{\tau 2} - 1)(s_{\tau 12} + 1)$,

(2) $l_{\mathfrak{p}_{32},2}(s_\tau) = (-s_{\tau 11} - 2s_{\tau 12} + 2s_{\tau 2} - 1)(s_{\tau 11} + s_{\tau 12} - 1)$,

(3) $l_{\mathfrak{p}_{33},2}(s_\tau) = (s_{\tau 11} + 2s_{\tau 12} + 2s_{\tau 2} - 7)(s_{\tau 12} + 1)$,

(4) $l_{\mathfrak{p}_{34},2}(s_\tau) = (s_{\tau 11} + 2s_{\tau 12} + 2s_{\tau 2} - 7)(s_{\tau 11} + s_{\tau 12} - 1)$.

By (12.1.20),

$$\Xi_{\mathfrak{p}_{3i}}(\Phi, \omega, w) = \delta_{\#}(\omega)\sum_{\tau}\left(\frac{1}{2\pi\sqrt{-1}}\right)^3 \int_{\mathrm{Re}(s_\tau)=r} \Sigma_{1,\mathrm{sub}}(\Phi, l_{\mathfrak{p}_{3i}}, s_\tau)\widetilde{\Lambda}_\tau(w; s_\tau)ds_\tau$$

for $i = 1, \cdots, 4$ where we choose the contour so that $r_1, r_2, r_3 > 1$ and both factors of $l_{\mathfrak{p}_{3i}}(r)$ are positive.

Let

$$\Sigma_{\mathfrak{p}_{31}}(\Phi, s_{\tau 11}) = -\frac{\Sigma_1(R_{\mathfrak{d}_3}\Phi_{\mathfrak{p}_{31}}, \frac{1}{2}(s_{\tau 11}+1))}{2(s_{\tau 11}+1)},$$
$$\Sigma_{\mathfrak{p}_{32}}(\Phi, s_{\tau 11}) = -\frac{\Sigma_1(R_{\mathfrak{d}_3}\Phi_{\mathfrak{p}_{32}}, \frac{1}{2}(s_{\tau 11}+1))}{s_{\tau 11}(s_{\tau 11}+1)},$$
$$\Sigma_{\mathfrak{p}_{33}}(\Phi, s_{\tau 11}) = \frac{\Sigma_1(R_{\mathfrak{d}_3}\Phi_{\mathfrak{p}_{33}}, \frac{1}{2}(s_{\tau 11}+1))}{2(s_{\tau 11}-3)},$$
$$\Sigma_{\mathfrak{p}_{34}}(\Phi, s_{\tau 11}) = \frac{\Sigma_1(R_{\mathfrak{d}_3}\Phi_{\mathfrak{p}_{34}}, \frac{1}{2}(s_{\tau 11}+1))}{s_{\tau 11}(s_{\tau 11}-3)}.$$

The following proposition follows from cases (1), (2), (5), (6) of (11.2.4).

Proposition (12.1.22) *Suppose that $\tau = \tau_G$ and $\delta > 0$ is a small number. Then by changing ψ if necessary,*

$$\Xi_{\mathfrak{p}_{3i}}(\Phi, \omega, w) \sim \frac{\varrho^2 \delta_\#(\omega)}{2\pi\sqrt{-1}} \int_{\mathrm{Re}(s_{\tau 11})=1+\delta} \Sigma_{\mathfrak{p}_{3i}}(\Phi, s_{\tau 11})\phi_2(s_{\tau 11})\widetilde{\Lambda}_1(w; s_{\tau 11}, 1)ds_{\tau 11}$$

for $i = 1, 2, 3, 4$.

(i) Now we combine the computations in this section. Let $\mathfrak{d} = \mathfrak{d}_0, \mathfrak{p} = \mathfrak{p}_0$. We define

$$\begin{aligned}
J_1(\Phi, \omega) = {} & \delta_{\mathfrak{d}_0}(\omega)\left(\Sigma_{\mathfrak{d}_0+}(R_{\mathfrak{d}_0}\Phi_{\mathfrak{p}_0}, \omega_{\mathfrak{d}_0}, 2) + \Sigma_{\mathfrak{d}_0+}(\mathscr{F}_{\mathfrak{d}_0}R_{\mathfrak{d}_0}\Phi_{\mathfrak{p}_0}, \omega_{\mathfrak{d}_0}^{-1}, 4)\right) \\
& - \mathfrak{V}_3\delta_\#(\omega)\left(\frac{R_{\mathfrak{d}_0}\Phi_{\mathfrak{p}_0}(0)}{2} + \frac{\mathscr{F}_{\mathfrak{d}_0}R_{\mathfrak{d}_0}\Phi_{\mathfrak{p}_0}(0)}{4}\right) \\
& - \mathfrak{V}_2\delta_\#(\omega)\left(\Sigma_1(R_{\mathfrak{d}_2}\Phi_{\mathfrak{p}_{21}}, \frac{3}{2}) + \frac{1}{5}\Sigma_1(R_{\mathfrak{d}_2}\Phi_{\mathfrak{p}_{22}}, \frac{3}{2})\right), \\
J_2(\Phi, \omega) = {} & (\Sigma_{W+}(R_{\mathfrak{d}_1}\Phi_{\mathfrak{p}_{11}}, 3) + \Sigma_{W+}(\mathscr{F}_W R_{\mathfrak{d}_1}\Phi_{\mathfrak{p}_{11}}, 0)) \\
& - (\Sigma_{W+}(R_{\mathfrak{d}_1}\Phi_{\mathfrak{p}_{12}}, 3) + \Sigma_{W+}(\mathscr{F}_W R_{\mathfrak{d}_1}\Phi_{\mathfrak{p}_{12}}, 0)) \\
& + \frac{\mathfrak{V}_2}{3}(-R_{\mathfrak{d}_1}\Phi_{\mathfrak{p}_{11}}(0) + R_{\mathfrak{d}_1}\Phi_{\mathfrak{p}_{12}}(0)), \\
\Upsilon_{\mathfrak{d},1}(\Phi, \omega) = {} & J_1(\Phi, \omega) + \delta_\#(\omega)J_2(\Phi), \\
J_3(\Phi, s_{\tau 11}) = {} & \frac{2T_{W+}(\widetilde{R}_{W,0}R_{\mathfrak{d}_1}\Phi_{\mathfrak{p}_{11}}, \frac{1}{2}(s_{\tau 11}+5), \frac{1}{2}(1-s_{\tau 11}))}{(s_{\tau 11}-1)(s_{\tau 11}+1)} \\
& + \frac{2T_{W+}(\widetilde{R}_{W,0}\mathscr{F}_W R_{\mathfrak{d}_1}\Phi_{\mathfrak{p}_{11}}, \frac{1}{2}(1-s_{\tau 11}), \frac{1}{2}(1-s_{\tau 11}))}{(s_{\tau 11}-1)(s_{\tau 11}+1)}, \\
J_4(\Phi, s_{\tau 11}) = {} & \frac{2T_{W+}(\widetilde{R}_{W,0}R_{\mathfrak{d}_1}\Phi_{\mathfrak{p}_{12}}, \frac{1}{2}(s_{\tau 11}+5), \frac{1}{2}(1-s_{\tau 11}))}{(s_{\tau 11}-1)(s_{\tau 11}-3)} \\
& + \frac{2T_{W+}(\widetilde{R}_{W,0}\mathscr{F}_W R_{\mathfrak{d}_1}\Phi_{\mathfrak{p}_{12}}, \frac{1}{2}(1-s_{\tau 11}), \frac{1}{2}(1-s_{\tau 11}))}{(s_{\tau 11}-1)(s_{\tau 11}-3)},
\end{aligned}$$

$$\begin{aligned}\Upsilon_{\mathfrak{d},2}(\Phi,\omega,s_{\tau 11}) &= \delta_{\#}(\omega)(J_3(\Phi,s_{\tau 11})+J_4(\Phi,s_{\tau 11}))\\ &\quad+\mathfrak{V}_2\delta_{\#}(\omega)\left(\frac{\mathscr{F}_W R_{\mathfrak{d}_1}\Phi_{\mathfrak{p}_{11}}(0)}{s_{\tau 11}(s_{\tau 11}-1)}+\frac{\mathscr{F}_W R_{\mathfrak{d}_1}\Phi_{\mathfrak{p}_{12}}(0)}{(s_{\tau 11}-1)(s_{\tau 11}-2)}\right)\\ &\quad+\delta_{\#}(\omega)(\Sigma_{\mathfrak{p}_{31}}(\Phi,s_{\tau 11})-\Sigma_{\mathfrak{p}_{32}}(\Phi,s_{\tau 11})\\ &\quad+\delta_{\#}(\omega)(-\Sigma_{\mathfrak{p}_{33}}(\Phi,s_{\tau 11})+\Sigma_{\mathfrak{p}_{34}}(\Phi,s_{\tau 11})).\end{aligned}$$

Then by (12.1.4)–(12.1.6), (12.1.9), (12.1.11), (12.1.14), (12.1.16), (12.1.19), (12.1.22), and by changing ψ if necessary,

$$\begin{aligned}\Xi_{\mathfrak{p}_0}(\Phi,\omega,w) &\sim C_G\Lambda(w;\rho)\Upsilon_{\mathfrak{d},1}(\Phi,\omega)\\ &\quad+\frac{\varrho^2}{2\pi\sqrt{-1}}\int_{\mathrm{Re}(s_{\tau 11})=1+\delta}\Upsilon_{\mathfrak{d},2}(\Phi,\omega,s_{\tau 11})\phi_2(s_{\tau 11})\widetilde{\Lambda}_1(w;s_{\tau 11},1)ds_{\tau 11}.\end{aligned}$$

Let $\Upsilon_{\mathfrak{d},2,(i)}(\Phi,\omega,1)$, $J_{3,(i)}(\Phi,1)$ etc. be the i-th coefficient of the Laurent expansion.

We define

$$\begin{aligned}J_5(\Phi) &= -\frac{1}{2}(T_{W+}(\widetilde{R}_{W,0}R_{\mathfrak{d}_1}\Phi_{\mathfrak{p}_{11}},3)+T_{W+}(\widetilde{R}_{W,0}\mathscr{F}_W R_{\mathfrak{d}_1}\Phi_{\mathfrak{p}_{11}},0)),\\ J_6(\Phi) &= \frac{1}{2}(T_{W+}(\widetilde{R}_{W,0}\mathscr{F}_W R_{\mathfrak{d}_1}\Phi_{\mathfrak{p}_{11}},3)+T_{W+}(\widetilde{R}_{W,0}\mathscr{F}_W R_{\mathfrak{d}_1}\Phi_{\mathfrak{p}_{12}},0)).\end{aligned}$$

Then

$$\begin{aligned}J_{3,(0)}(\Phi,1) &= J_5(\Phi)+2\frac{d}{ds_{\tau 11}}\bigg|_{s_{\tau 11}=1}\frac{T_{W+}(\widetilde{R}_{W,0}R_{\mathfrak{d}_1}\Phi_{\mathfrak{p}_{11}},\frac{1}{2}(s_{\tau 11}+5),0)}{s_{\tau 11}+1}\\ &\quad+2\frac{d}{ds_{\tau 11}}\bigg|_{s_{\tau 11}=1}\frac{T_{W+}(\widetilde{R}_{W,0}\mathscr{F}_W R_{\mathfrak{d}_1}\Phi_{\mathfrak{p}_{11}},\frac{1}{2}(1-s_{\tau 11}),0)}{s_{\tau 11}+1},\\ J_{4,(0)}(\Phi,1) &= J_6(\Phi)+2\frac{d}{ds_{\tau 11}}\bigg|_{s_{\tau 11}=1}\frac{T_{W+}(\widetilde{R}_{W,0}R_{\mathfrak{d}_1}\Phi_{\mathfrak{p}_{12}},\frac{1}{2}(s_{\tau 11}+5),0)}{s_{\tau 11}-3}\\ &\quad+2\frac{d}{ds_{\tau 11}}\bigg|_{s_{\tau 11}=1}\frac{T_{W+}(\widetilde{R}_{W,0}\mathscr{F}_W R_{\mathfrak{d}_1}\Phi_{\mathfrak{p}_{12}},\frac{1}{2}(1-s_{\tau 11}),0)}{s_{\tau 11}-3}.\end{aligned}$$

By the principal part formula for the standard L-function in one variable, if Φ is K-invariant,

$$\begin{aligned}&T_{W+}(\widetilde{R}_{W,0}R_{\mathfrak{d}_1}\Phi_{\mathfrak{p}_{11}},\frac{1}{2}(s_{\tau 11}+5),0)+T_{W+}(\widetilde{R}_{W,0}\mathscr{F}_W R_{\mathfrak{d}_1}\Phi_{\mathfrak{p}_{11}},\frac{1}{2}(1-s_{\tau 11}),0)\\ &=\Sigma_1(R'_{W,0}\Phi_{\mathfrak{p}_{11}},\frac{1}{2}(s_{\tau 11}+3))+\frac{2\mathscr{F}_{\mathfrak{d}_3}R_{\mathfrak{d}_3}\Phi_{\mathfrak{p}_{31}}(0)}{s_{\tau 11}+3}-\frac{2\mathscr{F}_{\mathfrak{d}_3}R_{\mathfrak{d}_3}\Phi_{\mathfrak{p}_{32}}(0)}{s_{\tau 11}+1},\\ &T_{W+}(\widetilde{R}_{W,0}R_{\mathfrak{d}_1}\Phi_{\mathfrak{p}_{12}},\frac{1}{2}(s_{\tau 11}+5),0)+T_{W+}(\widetilde{R}_{W,0}\mathscr{F}_W R_{\mathfrak{d}_1}\Phi_{\mathfrak{p}_{12}},\frac{1}{2}(1-s_{\tau 11}),0)\\ &=\Sigma_1(R'_{W,0}\Phi_{\mathfrak{p}_{12}},\frac{1}{2}(s_{\tau 11}+3))+\frac{2\mathscr{F}_{\mathfrak{d}_3}R_{\mathfrak{d}_3}\Phi_{\mathfrak{p}_{33}}(0)}{s_{\tau 11}+3}-\frac{2\mathscr{F}_{\mathfrak{d}_3}R_{\mathfrak{d}_3}\Phi_{\mathfrak{p}_{34}}(0)}{s_{\tau 11}+1}.\end{aligned}$$

Therefore,

$$J_{3,(0)}(\Phi,1) = J_5(\Phi) + \frac{1}{2}(-\Sigma_1(R'_{W,0}\Phi_{\mathfrak{p}_{11}},2) + \Sigma_{1,(1)}(R'_{W,0}\Phi_{\mathfrak{p}_{11}},2))$$
$$- \frac{3}{8}\mathscr{F}_{\mathfrak{d}_3} R_{\mathfrak{d}_3}\Phi_{\mathfrak{p}_{31}}(0) + \mathscr{F}_{\mathfrak{d}_3} R_{\mathfrak{d}_3}\Phi_{\mathfrak{p}_{32}}(0),$$
$$J_{4,(0)}(\Phi,1) = J_6(\Phi) - \frac{1}{2}(\Sigma_1(R'_{W,0}\Phi_{\mathfrak{p}_{12}},2) + \Sigma_{1,(1)}(R'_{W,0}\Phi_{\mathfrak{p}_{12}},2))$$
$$- \frac{1}{8}\mathscr{F}_{\mathfrak{d}_3} R_{\mathfrak{d}_3}\Phi_{\mathfrak{p}_{33}}(0).$$

It is easy to see that

$$\frac{1}{s_{\tau 11}(s_{\tau 11}-1)} = \frac{1}{s_{\tau 11}-1} - 1 + O(s_{\tau 11}-1),$$
$$\frac{1}{(s_{\tau 11}-1)(s_{\tau 11}-2)} = -\frac{1}{s_{\tau 11}-1} - 1 + O(s_{\tau 11}-1).$$

Also

$$\Sigma_{\mathfrak{p}_{31},(0)}(\Phi,1) = \frac{1}{4}\Sigma_{1,(-1)}(R_{\mathfrak{d}_3}\Phi_{\mathfrak{p}_{31}},1) - \frac{1}{4}\Sigma_{1,(0)}(R_{\mathfrak{d}_3}\Phi_{\mathfrak{p}_{31}},1),$$
$$\Sigma_{\mathfrak{p}_{32},(0)}(\Phi,1) = \frac{3}{2}\Sigma_{1,(-1)}(R_{\mathfrak{d}_3}\Phi_{\mathfrak{p}_{32}},1) - \frac{1}{2}\Sigma_{1,(0)}(R_{\mathfrak{d}_3}\Phi_{\mathfrak{p}_{32}},1),$$
$$\Sigma_{\mathfrak{p}_{33},(0)}(\Phi,1) = -\frac{1}{4}\Sigma_{1,(-1)}(R_{\mathfrak{d}_3}\Phi_{\mathfrak{p}_{33}},1) - \frac{1}{4}\Sigma_{1,(0)}(R_{\mathfrak{d}_3}\Phi_{\mathfrak{p}_{33}},1),$$
$$\Sigma_{\mathfrak{p}_{34},(0)}(\Phi,1) = \frac{1}{2}\Sigma_{1,(-1)}(R_{\mathfrak{d}_3}\Phi_{\mathfrak{p}_{34}},1) - \frac{1}{2}\Sigma_{1,(0)}(R_{\mathfrak{d}_3}\Phi_{\mathfrak{p}_{34}},1).$$

We define

$$J_7(\Phi) = -\frac{1}{8}\Sigma_{1,(-1)}(R_{\mathfrak{d}_3}\Phi_{\mathfrak{p}_{31}},1) - \frac{1}{4}\Sigma_{1,(0)}(R_{\mathfrak{d}_3}\Phi_{\mathfrak{p}_{31}},1)$$
$$- \frac{1}{2}\Sigma_{1,(-1)}(R_{\mathfrak{d}_3}\Phi_{\mathfrak{p}_{32}},1) + \frac{1}{2}\Sigma_{1,(0)}(R_{\mathfrak{d}_3}\Phi_{\mathfrak{p}_{32}},1)$$
$$+ \frac{1}{8}\Sigma_{1,(-1)}(R_{\mathfrak{d}_3}\Phi_{\mathfrak{p}_{33}},1) + \frac{1}{4}\Sigma_{1,(0)}(R_{\mathfrak{d}_3}\Phi_{\mathfrak{p}_{33}},1)$$
$$+ \frac{1}{2}\Sigma_{1,(-1)}(R_{\mathfrak{d}_3}\Phi_{\mathfrak{p}_{34}},1) - \frac{1}{2}\Sigma_{1,(0)}(R_{\mathfrak{d}_3}\Phi_{\mathfrak{p}_{34}},1),$$
$$J_8(\Phi) = \frac{1}{2}\left(-3\Sigma_{1,(0)}(R'_{W,0}\Phi_{\mathfrak{p}_{11}},2) + \Sigma_{1,(1)}(R'_{W,0}\Phi_{\mathfrak{p}_{11}},2)\right)$$
$$- \frac{1}{2}\left(3\Sigma_{1,(0)}(R'_{W,0}\Phi_{\mathfrak{p}_{12}},2) + \Sigma_{1,(1)}(R'_{W,0}\Phi_{\mathfrak{p}_{12}},2)\right).$$

Clearly,

$$\mathscr{F}_{\mathfrak{d}_3} R_{\mathfrak{d}_3}\Phi_{\mathfrak{p}_{3i}}(0) = \Sigma_{1,(-1)}(R_{\mathfrak{d}_3}\Phi_{\mathfrak{p}_{3i}},1)$$

for all i. Also if Φ is K-invariant, by (4.4.11),

$$\mathfrak{V}_2\mathscr{F}_W R_{\mathfrak{d}_1}\Phi_{\mathfrak{p}_{11}}(0) = \Sigma_1(R'_{W,0}\Phi_{\mathfrak{p}_{12}},2),\ \mathfrak{V}_2\mathscr{F}_W R_{\mathfrak{d}_1}\Phi_{\mathfrak{p}_{12}}(0) = \Sigma_1(R'_{W,0}\Phi_{\mathfrak{p}_{11}},2).$$

Therefore,

$$\Upsilon_{\mathfrak{d},2,(0)}(\Phi,\omega,1)=\delta_{\#}(\omega)\sum_{i=5}^{8}J_i(\Phi).$$

By the principal part formula (4.2.15),

$$J_2(\Phi)+J_5(\Phi)+J_6(\Phi)+J_7(\Phi)=\Sigma_{W,\mathrm{ad},(0)}(R_{\mathfrak{d}_1}\Phi_{\mathfrak{p}_{11}},3)-\Sigma_{W,\mathrm{ad},(0)}(R_{\mathfrak{d}_1}\Phi_{\mathfrak{p}_{12}},3).$$

Hence,

$$\begin{aligned}&\Upsilon_{\mathfrak{d},1}(\Phi,\omega)+\Upsilon_{\mathfrak{d},2,(0)}(\Phi,\omega,1)\\&=J_1(\Phi,\omega)+\delta_{\#}(\omega)J_8(\Phi)\\&\quad+\delta_{\#}(\omega)\left(\Sigma_{W,\mathrm{ad},(0)}(R_{\mathfrak{d}_1}\Phi_{\mathfrak{p}_{11}},3)-\Sigma_{W,\mathrm{ad},(0)}(R_{\mathfrak{d}_1}\Phi_{\mathfrak{p}_{12}},3)\right).\end{aligned}$$

By Theorem (4.0.1), we get the following proposition.

Proposition (12.1.23)

$$\Upsilon_{\mathfrak{d},1}(\Phi,\omega)+\Upsilon_{\mathfrak{d},2,(0)}(\Phi,\omega,1)=\delta_{\mathfrak{d}}(\omega)\Sigma_{\mathfrak{d}}(R_{\mathfrak{d}}\Phi_{\mathfrak{p}},\omega_{\mathfrak{d}},2).$$

§12.2 The case $\mathfrak{d}=(\beta_2)$

We prove the following proposition in this section.

Proposition (12.2.1) *Let $\mathfrak{p}=(\mathfrak{d},\mathfrak{s})$ be a path such that $\mathfrak{d}=(\beta_2)$. Then by changing ψ if necessary, $\Xi_{\mathfrak{p}}(\Phi,\omega,w)\sim 0$.*

Proof. Since $\mathscr{E}((g^0)^{\iota},w)=\mathscr{E}(g^0,w)$, we only consider $\mathfrak{p}$ such that $\mathfrak{s}(1)=0$. Let $g^0_{\mathfrak{d}}$ be as in the second element of (1.1.3). Let $\tau=(\tau_1,\tau_2)$ be a Weyl group element such that $\tau_1\in\mathfrak{W}_2$.

Consider the situation in (3.5.17). In this case, we choose

$$\begin{aligned}\lambda_{\mathfrak{d}}&=a(\underline{\lambda_1}^{-4},\underline{\lambda}_1^2,\underline{\lambda}_1^2;\underline{\lambda_1}^{-3},\underline{\lambda}_1^3),\\\lambda_{\mathfrak{d}}^{(2)}&=a(\underline{\lambda_2}^{-2},\underline{\lambda}_2,\underline{\lambda}_2;\underline{\lambda}_2^2,\underline{\lambda_2}^{-2}),\end{aligned}$$

where $\lambda_1,\lambda_2\in\mathbb{R}_+$ ($\lambda_{\mathfrak{d}}^{(3)}=1$). Then

$$\begin{aligned}\sigma_{\mathfrak{p}}(\lambda_{\mathfrak{d}})\lambda_{\mathfrak{d}}^{\tau z+\rho}&=\lambda_1^{2s_{\tau 11}+4s_{\tau 12}+3s_{\tau 2}+9},\\\sigma_{\mathfrak{p}}(\lambda_{\mathfrak{d}}^{(2)})(\lambda_{\mathfrak{d}}^{(2)})^{\tau z+\rho}&=\lambda_2^{s_{\tau 11}+2s_{\tau 12}-2s_{\tau 2}+1}.\end{aligned}$$

Therefore, $LS_{\mathfrak{p},\tau}+h$ in (3.5.17) is $\{s_\tau\mid 1+s_{\tau 11}+2s_{\tau 12}-2s_{\tau 2}=0\}$. Let $g_{1,23}=ka_2(\underline{\mu}t_1,\underline{\mu}^{-1}t_2)n_2(u)$ be the Iwasawa decomposition of $g_{1,23}$ where $\mu\in\mathbb{R}_+,t_1,t_2\in\mathbb{A}^1$. Then $t(g^0_{\mathfrak{d}})^{\sigma\tau z+\rho}=\mu^{-s_{\tau 11}+1}$ if $\sigma=1$, and $t(g^0_{\mathfrak{d}})^{\sigma\tau z+\rho}=\mu^{s_{\tau 11}+1}$ if $\sigma=(2,3)$.

We choose $r_1=3,r_2=2,r_3=4$ if $\sigma=1$, and $r_1=-3,r_2=5,r_3=4$ if $\sigma=(2,3)$. Then

$$|\sigma_{\mathfrak{p}}(\lambda_{\mathfrak{d}})\lambda_{\mathfrak{d}}^{\tau z+\rho}t(g^0_{\mathfrak{d}})^{\sigma\tau z+\rho}|=\lambda_1^{35}\mu^{-2}.$$

If $q=\mathrm{Re}(z)$, $q\in D_{\sigma\tau}$ for the above choice of r. Moreover, if $\tau_2=1$, $\widetilde{L}(r)\le 22-4C<4+C$ since $C>100$.

It is easy to see that $\lambda_{\mathfrak{d}}$ acts on $Z_{\mathfrak{d}\mathrm{A}}$ by multiplication by $\underline{\lambda}_1^{11}$. Hence, by (4.1.3)(2), the condition (3.5.16) is satisfied for all σ, τ, and $\Xi_{\mathfrak{p},\tau}(\Phi,\omega,w) \sim 0$ unless $\tau_2 = (1,2)$. If $\tau_2 = (1,2)$, $\rho_\tau = (-1,-1,1),(-2,1,1)$ or $(-1,2,1)$. These points do not belong to the set $\{s_\tau \mid 1+s_{\tau 11}+2s_{\tau 12}-2s_{\tau 2}=0\}$. Therefore, by (3.5.19), we can replace ψ if necessary and assume that $\Xi_{\mathfrak{p},\tau}(\Phi,\omega,w)=0$. This proves the proposition.

Q.E.D.

§12.3 The case $\mathfrak{d} = (\beta_3)$

We prove the following proposition in this section.

Proposition (12.3.1) *Let $\mathfrak{p} = (\mathfrak{d},\mathfrak{s})$ be a path such that $\mathfrak{d} = (\beta_3)$. Then by changing ψ if necessary, $\Xi_{\mathfrak{p}}(\Phi,\omega,w) \sim 0$.*

Proof. As in §2.2, we only consider $\mathfrak{p}$ such that $\mathfrak{s}(1)=0$. For $\lambda = (\lambda_1,\lambda_2) \in \mathbb{R}_+$, let $d(\lambda) = a(\underline{\lambda}_1^{-1},\underline{\lambda}_1^{-1},\underline{\lambda}_1^{2};\underline{\lambda}_2^{-1},\underline{\lambda}_2)$. Let $g_{\mathfrak{d}} = d(\lambda)g_{\mathfrak{d}}^0$ where $g_{\mathfrak{d}}^0$ is as in the first element of (1.1.3). Then $dg_{\mathfrak{d}} = 2d^\times\lambda_1 d^\times\lambda_2 dg_{\mathfrak{d}}^0$. Let $\tau = (\tau_1,\tau_2)$ be a Weyl group element such that $\tau_1 \in \mathfrak{W}_1$. Let s_τ be as before.

Consider the substitution $\mu = \lambda_1^4\lambda_2$. Then

$$d(\lambda)^{\tau z-\rho} = \lambda_1^{2s_{\tau 11}+s_{\tau 12}+3}\lambda_2^{s_{\tau 2}+1} = \lambda_1^{2s_{\tau 11}+s_{\tau 12}-4s_{\tau 2}-1}\mu^{s_{\tau 2}+1}. \tag{12.3.2}$$

Also $d^\times\lambda_1 d^\times\lambda_2 = d^\times\lambda_1 d^\times\mu$.

Let $\Psi = R_{\mathfrak{d}}\Phi_{\mathfrak{p}}$. The functions $\Theta_{Z_{\mathfrak{d}}}(\Psi,g_{\mathfrak{d}}), \mathscr{E}_{\mathfrak{p}}(g_{\mathfrak{d}},w)$ do not depend on t_{13},t_{21}. So $\Xi_{\mathfrak{p}}(\Phi,\omega,w)=0$ unless ω_2 is trivial.

By (11.1.4), (11.1.5),

$$\begin{aligned}&\Xi_{\mathfrak{p}}(\Phi,\omega,w)\\ &= \sum_{\substack{\tau_1\in\mathfrak{W}_1\\ \tau_2=1,(1,2)}} \frac{\delta_\#(\omega)}{2\pi\sqrt{-1}}\int_{\mathrm{Re}(s_{\tau 2})=r_3>1} \Sigma_1(\Psi,s_{\tau 2}+1)\widetilde{\Lambda}_\tau(w;2s_{\tau 2}+1,-1,s_{\tau 2})ds_{\tau 2}.\end{aligned}$$

Let

$$\Xi_{\mathfrak{p},\tau}(\Phi,\omega,w) = \frac{\delta_\#(\omega)}{2\pi\sqrt{-1}}\int_{\mathrm{Re}(s_{\tau 2})=r_3>1} \Sigma_1(\Psi,s_{\tau 2}+1)\widetilde{\Lambda}_\tau(w;2s_{\tau 2}+1,-1,s_{\tau 2})ds_{\tau 2}.$$

As usual, we can ignore the case $\tau_2 = 1$. Suppose $\tau_2 = (1,2)$. If $\tau_1 = 1,(2,3),(1,2,3)$, $\rho_\tau = (-1,-1,1),(1,-2,1),(2,-1,1)$. So in all the cases, ρ_τ does not belong to the set $\{s_\tau \mid s_{\tau 11} = 2s_{\tau 2}+1\}$. Therefore, by (3.5.19), $\Xi_{\mathfrak{p},\tau}(\Phi,\omega,w) \sim 0$

Q.E.D.

§12.4 The case $\mathfrak{d} = (\beta_4)$

Let $d_1(\lambda_1) = d_{P_1}(\lambda_1)$ and $d_2(\lambda_1,\lambda_2,\lambda_3) = a(\underline{\lambda}_1^{-1}\underline{\lambda}_2^{-1},\underline{\lambda}_1^{-1}\underline{\lambda}_2,\underline{\lambda}_1^{2};\underline{\lambda}_3^{-1},\underline{\lambda}_3)$ for $\lambda_1,\lambda_2,\lambda_3 \in \mathbb{R}_+$. We consider paths $\mathfrak{p}_0 = (\mathfrak{d}_0,\mathfrak{s}_0), \mathfrak{p}_{11} = (\mathfrak{d}_1,\mathfrak{s}_{21}), \mathfrak{p}_{12} = (\mathfrak{d}_1,\mathfrak{s}_{22})$ such that $\mathfrak{d}_0 = (\beta_4), \mathfrak{d}_1 = (\beta_4,\beta_{4,1})$, and $\mathfrak{s}_{21}(2)=0, \mathfrak{s}_{22}(2)=1$. As in previous sections, we only consider the case $\mathfrak{s}_0(1)=\mathfrak{s}_{21}(1)=\mathfrak{s}_{22}(1)=0$.

Since $Y_{\mathfrak{d}} = Z_{\mathfrak{d}}$ for $\mathfrak{d} = \mathfrak{d}_1$, $\Xi_{\mathfrak{p}_{1i}}(\Phi, \omega, w)$ is well defined for $\mathrm{Re}(w) \gg 0$ for $i = 1, 2$. Therefore, by (3.5.9),

$$(12.4.1) \qquad \begin{aligned} \Xi_{\mathfrak{p}_0}(\Phi, \omega, w) &= \Xi_{\mathfrak{p}_0+}(\Phi, \omega, w) + \widehat{\Xi}_{\mathfrak{p}_0+}(\Phi, \omega, w) \\ &\quad + \widehat{\Xi}_{\mathfrak{p}_0\#}(\Phi, \omega, w) - \Xi_{\mathfrak{p}_0\#}(\Phi, \omega, w) \\ &\quad + \Xi_{\mathfrak{p}_{12}}(\Phi, \omega, w) - \Xi_{\mathfrak{p}_{11}}(\Phi, \omega, w). \end{aligned}$$

Easy computations show the following two lemmas.

Lemma (12.4.2) $e_{\mathfrak{p}_{11}1}(d_2(\lambda_1, \lambda_2, \lambda_3)) = \lambda_1$, $e_{\mathfrak{p}_{12}1}(d_2(\lambda_1, \lambda_2, \lambda_3)) = \lambda_1^{-1}$.

Lemma (12.4.3)
(1) $\sigma_{\mathfrak{p}_0}(d_1(\lambda_1)) = \lambda_1^{-2}$, $\kappa_{\mathfrak{p}_0 1}(d_1(\lambda_1)) = \lambda_1^{-4}$.
(2) $\sigma_{\mathfrak{p}_{11}}(d_2(\lambda_1, \lambda_2, \lambda_3)) = \lambda_1^{-2}\lambda_2^2\lambda_3^2$, $\sigma_{\mathfrak{p}_{12}}(d_2(\lambda_1, \lambda_2, \lambda_3)) = \lambda_1^6\lambda_2^2\lambda_3^2$.

(a) $\Xi_{\mathfrak{p}_0+}(\Phi, \omega, w)$, $\widehat{\Xi}_{\mathfrak{p}_0+}(\Phi, \omega, w)$.

In (a), (b), $\mathfrak{d} = \mathfrak{d}_0, \mathfrak{p} = \mathfrak{p}_0$, and $\Psi = R_{\mathfrak{d}}\Phi_{\mathfrak{p}}$. Let $g_{\mathfrak{d}} = d_1(\lambda_1)g_{\mathfrak{d}}^0$ where $g_{\mathfrak{d}}$ is as in the first element of (11.1.3). Then $dg_{\mathfrak{d}} = 2d^\times\lambda_1 dg_{\mathfrak{d}}^0$.

The element $d_1(\lambda_1)$ acts on $Z_{\mathfrak{d}\mathbb{A}}$ by multiplication by $\underline{\lambda}_1$. Therefore, the following proposition follows from (3.4.14) and (12.4.3)(1).

Proposition (12.4.4)

$$(1) \qquad \Xi_{\mathfrak{p}_0+}(\Phi, \omega, w) \sim 2C_G\Lambda(w;\rho)\delta_{\mathfrak{d}_0}(\omega)\Sigma_{\mathfrak{d}_0+}(R_{\mathfrak{d}_0}\Phi_{\mathfrak{p}_0}, \omega_{\mathfrak{d}_0}, -2).$$

$$(2) \qquad \widehat{\Xi}_{\mathfrak{p}_0+}(\Phi, \omega, w) \sim 2C_G\Lambda(w;\rho)\delta_{\mathfrak{d}_0}(\omega)\Sigma_{\mathfrak{d}_0+}(\mathscr{F}_{\mathfrak{d}_0}R_{\mathfrak{d}_0}\Phi_{\mathfrak{p}_0}, \omega_{\mathfrak{d}_0}^{-1}, 6).$$

(b) $\Xi_{\mathfrak{p}_0\#}(\Phi, \omega, w)$, $\widehat{\Xi}_{\mathfrak{p}_0\#}(\Phi, \omega, w)$.

Consider $\tau = (\tau_1, 1)$ such that $\tau_1 \in \mathfrak{W}_1$. It is easy to see that $d_1(\lambda_1)^{\tau z+\rho} = \lambda_1^{2s_{\tau 11}+s_{\tau 12}-3}$.

By (11.1.3), (11.1.4), (12.4.3)(1),

$$(12.4.5) \quad \Xi_{\mathfrak{p}_0\#}(\Phi, \omega, w) = \sum_{\tau_1 \in \mathfrak{W}_1} \frac{\delta_\#(\omega)\Psi(0)}{2\pi\sqrt{-1}} \int_{\mathrm{Re}(s_{\tau 11})=r_1>3} \frac{\widetilde{\Lambda}_{1\tau_1}(w; s_{\tau 11}, -1)}{s_{\tau 11}-3} ds_{\tau 11},$$

$$\widehat{\Xi}_{\mathfrak{p}_0\#}(\Phi, \omega, w) = \sum_{\tau_1 \in \mathfrak{W}_1} \frac{\delta_\#(\omega)\mathscr{F}_{\mathfrak{d}}\Psi(0)}{2\pi\sqrt{-1}} \int_{\mathrm{Re}(s_{\tau 11})=r_1>2} \frac{\widetilde{\Lambda}_{1\tau_1}(w; s_{\tau 11}, -1)}{s_{\tau 11}-5} ds_{\tau 11}.$$

Note that $\widetilde{\Lambda}_{1\tau_1}(w; s_{\tau 11}, -1) = \widetilde{\Lambda}_{(\tau_1,1)}(w; s_{\tau 11}, -1, -1)$.

Proposition (12.4.6) *By changing* ψ *if necessary,*

$$(1) \qquad \Xi_{\mathfrak{p}_0\#}(\Phi, \omega, w) \sim -C_G\Lambda(w;\rho)\mathfrak{V}_2^2\delta_\#(\omega)R_{\mathfrak{d}_0}\Phi_{\mathfrak{p}_0}(0),$$

$$(2) \qquad \widehat{\Xi}_{\mathfrak{p}_0\#}(\Phi, \omega, w) \sim -C_G\Lambda(w;\rho)\frac{\mathfrak{V}_2^2\delta_\#(\omega)\mathscr{F}_{\mathfrak{d}_0}R_{\mathfrak{d}_0}\Phi_{\mathfrak{p}_0}(0)}{3}.$$

Proof. Since the proof is similar, we only prove (1). If $\tau_1 = 1$ or $(2,3)$, we choose $r_1 \gg 0$ in (12.4.5). Then, $\widetilde{L}_1(r_1, -1) \ll 0$ if $\tau_1 = 1$, and $\widetilde{L}_1(r_1, -1) = 2 < 4$ if $\tau_1 = (2,3)$. Therefore, we can ignore these cases. If $\tau_1 = (1,2,3)$, $\widetilde{M}_{\tau_1}(s_{\tau 11}, -1) =$

$\phi_2(s_{\tau 11}-1)$. The point $\rho_\tau = (2,-1,-1)$ does not satisfy the condition $s_{\tau 11} = 3$. Therefore, we can move the contour crossing this point. So,

$$\int_{\substack{\mathrm{Re}(s_{\tau 11})=r_1 \\ r_1>3}} \frac{\widetilde{\Lambda}_{1\tau_1}(w; s_{\tau 11}, -1)}{s_{\tau 11}-3} ds_{\tau 11} = \int_{\substack{\mathrm{Re}(s_{\tau 11})=r_1 \\ 1<r_1<2}} \frac{\widetilde{\Lambda}_{1\tau_1}(w; s_{\tau 11}, -1)}{s_{\tau 11}-3} ds_{\tau 11} - C_G \Lambda(w;\rho)\mathfrak{V}_2^2.$$

If $r_1 < 2$, $\widetilde{L}_1(r_1,-1) < 4$. So we can ignore the first term. Therefore, this proves the proposition.

Q.E.D.

(c) $\Xi_{\mathfrak{p}_{11}}(\Phi,\omega,w)$, $\Xi_{\mathfrak{p}_{12}}(\Phi,\omega,w)$.

Let $\mathfrak{d} = \mathfrak{d}_1$, $\Psi_i = R_{\mathfrak{d}_1}\Phi_{\mathfrak{p}_{1i}}$ for $i=1,2$. Let $t^0 = d_2(\lambda_1,\lambda_2,\lambda_3)\widehat{t}^0$. Then $d^\times t^0 = 2d^\times\lambda_1 d^\times\lambda_2 d^\times\lambda_3 d^\times\widehat{t}^0$. By (12.4.2), (12.4.3)(2),

$$\text{(12.4.7)}\quad \Xi_{\mathfrak{p}_{11}}(\Phi,\omega,w) = \int_{\substack{T_{\mathbb{A}}^0/T_k \\ \lambda_1\le 1}} \omega(t^0)\lambda_1^{-2}\lambda_2^2\lambda_3^2\Theta_{Z_\mathfrak{d}}(\Psi_1,t^0)\mathscr{E}_{\mathfrak{p}_{11}}(t^0,w)d^\times t^0,$$

$$\Xi_{\mathfrak{p}_{12}}(\Phi,\omega,w) = \int_{\substack{T_{\mathbb{A}}^0/T_k \\ \lambda_1\ge 1}} \omega(t^0)^{-1}\lambda_1^6\lambda_2^2\lambda_3^2\Theta_{Z_\mathfrak{d}}(\Psi_2,t^0)\mathscr{E}_{\mathfrak{p}_{12}}(t^0,w)d^\times t^0.$$

Let $\mu = \lambda_1\lambda_2\lambda_3$. Then $\lambda_1^{-2}\lambda_2^2\lambda_3^2 = \lambda_1^{-4}\mu^2$, and $\lambda_1^6\lambda_2^2\lambda_3^2 = \lambda_1^4\mu^2$. Also

$$d^\times\lambda_1 d^\times\lambda_2 d^\times\lambda_3 = d^\times\lambda_1 d^\times\lambda_2 d^\times\mu.$$

The functions $\Theta_{Z_\mathfrak{d}}(R_\mathfrak{d}\Phi_{\mathfrak{p}_{11}},t^0)$, $\mathscr{E}_{\mathfrak{p}_{11}}(t^0,w)$ etc. do not depend on t_{11}, t_{21}. Therefore, these distributions are 0 unless ω is trivial.

Easy computations show the following lemma.

Lemma (12.4.8)
(1) $\theta_{\mathfrak{p}_{11}}(d_2(\lambda_1,\lambda_2,\lambda_3))^{\tau z+\rho} = \lambda_1^{2s_{\tau 11}+s_{\tau 12}-s_{\tau 2}-2}\lambda_2^{s_{\tau 12}-s_{\tau 2}}\mu^{s_{\tau 2}-1}$.
(2) $\theta_{\mathfrak{p}_{12}}(d_2(\lambda_1,\lambda_2,\lambda_3))^{\tau z+\rho} = \lambda_1^{-2s_{\tau 11}-s_{\tau 12}-s_{\tau 2}+4}\lambda_2^{s_{\tau 12}-s_{\tau 2}}\mu^{s_{\tau 2}-1}$.

Definition (12.4.9) *Let* $l_{\mathfrak{p}_{2i}} = (l_{\mathfrak{p}_{2i},1}, l_{\mathfrak{p}_{2i},2})$ *where*
(1) $l_{\mathfrak{p}_{11},1}(s_\tau) = s_{\tau 2}+1$, $l_{\mathfrak{p}_{11},2}(s_\tau) = s_{\tau 11}-3$,
(2) $l_{\mathfrak{p}_{12},1}(s_\tau) = s_{\tau 2}+1$, $l_{\mathfrak{p}_{12},2}(s_\tau) = s_{\tau 11}+s_{\tau 12}-4$.

The function $\Theta_{Z_\mathfrak{d}}(\Psi_i,t^0)$ does not depend on λ_1,λ_2. So by the Mellin inversion formula, $\Xi_{\mathfrak{p}_{2i}}(\Phi,\omega,w)$ is equal to

$$\delta_\#(\omega)\sum_\tau\left(\frac{1}{2\pi\sqrt{-1}}\right)^2\int_{\substack{\mathrm{Re}(s_{\tau 1})=(r_1,r_2) \\ r_1,r_2\gg 0}} \Sigma_{1,\mathrm{sub}}(\Psi_i, l_{\mathfrak{p}_{2i}}, s_{\tau 1}, s_{\tau 12})\widetilde{\Lambda}_\tau(w; s_{\tau 1}, s_{\tau 12})ds_{\tau 1}$$

for $i=1,2$.

Proposition (12.4.10) *By changing* ψ *if necessary,*

$$\Xi_{\mathfrak{p}_{12}}(\Phi,\omega,w) - \Xi_{\mathfrak{p}_{11}}(\Phi,\omega,w)$$
$$\sim C_G\Lambda(w;\rho)\delta_\#(\omega)\left(-\frac{\Sigma_{1,(0)}(\Psi_2,2)}{4} + \frac{\Sigma_{1,(1)}(\Psi_1,2)}{2} - \frac{\Sigma_{1,(1)}(\Psi_2,2)}{2}\right).$$

Proof. We define

$$\Sigma^{(1)}(\Phi, s_{\tau 1}) = \Sigma_{1,\mathrm{sub}}(\Psi_i, l_{\mathfrak{p}_{22}}, s_{\tau 1}, s_{\tau 12}) - \Sigma_{1,\mathrm{sub}}(\Psi_i, l_{\mathfrak{p}_{21}}, s_{\tau 1}, s_{\tau 12}),$$
$$\Sigma^{(2)}(\Phi, s_{\tau 12}) = \Sigma^{(1)}(\Phi, 1, s_{\tau 12}).$$

Then

$$\Sigma^{(2)}(\Phi, s_{\tau 12}) = \frac{\Sigma_1(\Psi_1, s_{\tau 12}+1)}{2} + \frac{\Sigma_1(\Psi_2, s_{\tau 12}+1)}{s_{\tau 12}-3},$$

and

$$\Xi_{\mathfrak{p}_{12}}(\Phi, \omega, w) - \Xi_{\mathfrak{p}_{11}}(\Phi, \omega, w)$$
$$= \delta_{\#}(\omega) \sum_{\tau} \left(\frac{1}{2\pi\sqrt{-1}}\right)^2 \int_{\substack{\mathrm{Re}(s_{\tau 1})=(r_1, r_2) \\ r_1, r_2 \gg 0}} \Sigma^{(1)}(\Phi, s_{\tau 1}) \widetilde{\Lambda}_{\tau}(w; s_{\tau 1}, s_{\tau 12}) ds_{\tau 1}.$$

We can ignore the case $\tau_2 = 1$ as usual, and we assume that $\tau_2 = (1,2)$. If $\tau_1 = 1, (1,2), (2,3)$, Then $\widetilde{L}(r) \ll 0$. Therefore, we can ignore these cases, and we assume that $\tau_1 = (1,2,3), (1,3,2)$ or $(1,3)$. The point ρ_τ does not satisfy the conditions $s_{\tau 11} = 3, s_{\tau 11} + s_{\tau 12} = 4$. Therefore, by (3.6.1), we can move the contour crossing these lines by replacing ψ if necessary so that $(r_1, r_2) = (1+\delta, 1+\delta)$ where $\delta > 0$ is a small number. If $\tau_1 = (1,2,3), (1,3,2)$, $\widetilde{L}(r_1, r_2, r_2) = (2+\delta)(1+C) < w_0 = 4+C$ if δ is sufficiently small. Therefore, we can ignore these cases also.

Suppose that $\tau_1 = (1,3)$. Then

$$\left(\frac{1}{2\pi\sqrt{-1}}\right)^2 \int_{\substack{\mathrm{Re}(s_{\tau 1})=(1+\delta, 1+\delta) \\ r_1, r_2 \gg 0}} \Sigma^{(1)}(\Phi, s_{\tau 1}) \widetilde{\Lambda}_{\tau}(w; s_{\tau 1}, s_{\tau 12}) ds_{\tau 1}$$
$$= \left(\frac{1}{2\pi\sqrt{-1}}\right)^2 \int_{\mathrm{Re}(s_{\tau 1})=(1-\delta_1, 1+\delta)} \Sigma^{(1)}(\Phi, s_{\tau 1}) \widetilde{\Lambda}_{\tau}(w; s_{\tau 1}, s_{\tau 12}) ds_{\tau 1}$$
$$+ \frac{\varrho}{2\pi\sqrt{-1}} \int_{\mathrm{Re}(s_{\tau 12})=1+\delta} \Sigma^{(2)}(\Phi, s_{\tau 12}) \phi(s_{\tau 12}) \phi_2(s_{\tau 12}) \widetilde{\Lambda}(w; 1, s_{\tau 12}, s_{\tau 12}) ds_{\tau 12}$$
$$\sim \frac{\varrho}{2\pi\sqrt{-1}} \int_{\mathrm{Re}(s_{\tau 12})=1+\delta} \Sigma^{(2)}(\Phi, s_{\tau 12}) \phi(s_{\tau 12}) \phi_2(s_{\tau 12}) \widetilde{\Lambda}(w; 1, s_{\tau 12}, s_{\tau 12}) ds_{\tau 12},$$

where we choose $\delta, \delta_1 > 0$ small and $\delta_1 \delta^{-1} \gg 0$.

We proved that $\Sigma_1(\Psi_1, 2) = \Sigma_1(\Psi_2, 2)$ in §3.8. But if ω is trivial, Φ is K-invariant by assumption. Therefore, this is a consequence of (4.4.7) also. This implies that $\Sigma^{(1)}(\Phi, 1) = 0$.

Since

$$\frac{1}{s_{\tau 12}-3} = -\frac{1}{2} - \frac{1}{4}(s_{12}-1) + O((s_{\tau 12}-1)^2),$$
$$\Sigma^{(2)}(\Phi, s_{\tau 12}) = \Sigma^{(3)}(\Phi)(s_{\tau 12}-1) + O((s_{\tau 12}-1)^2),$$

where

$$\Sigma^{(3)}(\Phi) = -\frac{\Sigma_{1,(0)}(\Psi_2, 2)}{4} - \frac{\Sigma_{1,(1)}(\Psi_2, 2)}{2} + \frac{\Sigma_{1,(1)}(\Psi_1, 2)}{2}.$$

Therefore,

$$\Xi_{\mathfrak{p}_{12}}(\Phi, \omega, w) - \Xi_{\mathfrak{p}_{11}}(\Phi, \omega, w) \sim C_G \Lambda(w; \rho) \delta_{\#}(\omega) \Sigma^{(3)}(\Phi). \qquad \text{Q.E.D.}$$

The following proposition follows from (12.4.4), (12.4.6), (12.4.10), (3.8.10).

Proposition (12.4.11) *By changing ψ if necessary,*

$$\Xi_{\mathfrak{p}_0}(\Phi,\omega,w) \sim 2C_G\Lambda(w;\rho)\delta_{\mathfrak{d}_0}(\omega)\Sigma_{\mathfrak{d}_0}(R_{\mathfrak{d}_0}\Phi_{\mathfrak{p}_0},\omega_{\mathfrak{d}_0},-2).$$

§12.5 The case $\mathfrak{d} = (\beta_5)$

We prove the following proposition in this section.

Proposition (12.5.1) *Let $\mathfrak{p} = (\mathfrak{d},\mathfrak{s})$ be a path such that $\mathfrak{d} = (\beta_5)$. Then by changing ψ if necessary, $\Xi_{\mathfrak{p}}(\Phi,\omega,w) \sim 0$.*

Proof. As in previous sections, we only consider a path such that $\mathfrak{s}(1) = 0$. Let $\Psi = R_{\mathfrak{d}}\Phi_{\mathfrak{p}}$.

Suppose that $f(q)$ is a function of $q = (q_1,q_2) \in (\mathbb{A}^1/k^\times)^2$. Let $\widehat{t}^0$ be as before. Let $d^\times q, d^\times\widehat{t}^0$ be the usual measures. An easy computation shows that

$$\int_{(\mathbb{A}^1/k^\times)^5} \omega(\widehat{t}^0)f(\gamma_{1,33}(\widehat{t}^0),\gamma_{2,23}(\widehat{t}^0))d^\times\widehat{t}^0 = \delta_\#(\omega)\int_{(\mathbb{A}^1/k^\times)^2}\omega_{\mathfrak{d}}(q)f(q)d^\times q.$$

For $\mu = (\mu_1,\mu_2,\mu_3) \in \mathbb{R}_+^3$, we define

$$d(\mu) = d(\mu_1,\mu_2,\mu_3) = a(\underline{\mu}_2^{-1}\underline{\mu}_3, \underline{\mu}_1^{-\frac{1}{2}}\underline{\mu}_2\underline{\mu}_3^{-\frac{3}{2}}, \underline{\mu}_1^{\frac{1}{2}}\underline{\mu}_3^{\frac{1}{2}}; \underline{\mu}_3^{-1},\underline{\mu}_3).$$

Let $t^0 = d(\mu)\widehat{t}^0$. Then $d^\times t^0 = d^\times\mu_1 d^\times\mu_2 d^\times\mu_3 d^\times\widehat{t}^0$, and $\mu_1 = |\gamma_{1,33}(t^0)|$, $\mu_2 = |\gamma_{2,23}(t^0)|$. Let $\tau, s_\tau = (s_{\tau 12}, s_{\tau 11}, s_{\tau 2})$ be as before. It is easy to see that $(t^0)^{-\rho} = \mu_1^{\frac{1}{2}}\mu_2\mu_3^{\frac{1}{2}}$, $\kappa_{\mathfrak{d}2}(t^0) = \mu_1^{-1}\mu_3^{-2}$, and

$$(t^0)^{\tau z-\rho} = \mu_1^{\frac{1}{2}(s_{\tau 11}+1)}\mu_2^{s_{\tau 12}+1}\mu_3^{\frac{1}{2}(s_{\tau 11}-2s_{\tau 12}+2s_{\tau 2}+1)}. \tag{12.5.2}$$

Since $\Theta_{Z_{\mathfrak{d}}}(\Psi,t^0)$ does not depend on μ_3, by (12.5.2),

$$\begin{aligned}&\int_{\mathbb{R}_+^2\times(\mathbb{A}^1/k^\times)^5}\omega(t^0)\kappa_{\mathfrak{d}2}(t^0)(t^0)^{\tau z-\rho}\Theta_{Z_{\mathfrak{d}}}(\Psi,t^0)d\mu_1 d\mu_2 d^\times\widehat{t}^0\\ &= \delta_\#(\omega)\mu_3^{\frac{s_{\tau 11}}{2}-s_{\tau 12}+s_{\tau 2}-\frac{3}{2}}\Sigma_2(\Psi,\frac{1}{2}(s_{\tau 11}-1),s_{\tau 12}+1).\end{aligned}$$

By the Mellin inversion formula,

$$\begin{aligned}&\frac{1}{2\pi\sqrt{-1}}\int_{\mathrm{Re}(s_{\tau 2})=r_3\gg 0}\mu_3^{\frac{s_{\tau 11}}{2}-s_{\tau 12}+s_{\tau 2}-\frac{3}{2}}\widetilde{\Lambda}_\tau(w;s_\tau)ds_{\tau 2}\\ &= \widetilde{\Lambda}_\tau(w;s_{\tau 12},s_{\tau 11},-\frac{s_{\tau 11}}{2}+s_{\tau 12}+\frac{3}{2}).\end{aligned}$$

Let

$$\Sigma_{\mathfrak{d}}(\Psi,w,s_{\tau 1}) = \Sigma_2(\Psi,\frac{1}{2}(s_{\tau 11}-1),s_{\tau 12}+1)\widetilde{\Lambda}_\tau(w;s_{\tau 12},s_{\tau 11},-\frac{s_{\tau 11}}{2}+s_{\tau 12}+\frac{3}{2}).$$

Then

$$\Xi_{\mathfrak{p}}(\Phi,\omega,w)=\delta_{\#}(\omega)\sum_{\tau}\left(\frac{1}{2\pi\sqrt{-1}}\right)^2\int_{\substack{\mathrm{Re}(s_{\tau 11})=(r_1,r_2)\\ r_1>3,\ r_2>1,\ 2r_2+1>r_1}}\Sigma_{\mathfrak{d}}(\Psi,w,s_{\tau 1})ds_{\tau 1}.$$

We can ignore the case $\tau_2=1$ as usual, and assume that $\tau_2=(1,2)$.

If $\tau_1=1$, we choose $2r_1=r_2\gg 0$. Then $\widetilde{L}(r_1,r_2,-\frac{r_1}{2}+r_2+\frac{3}{2})=-2(r_1+r_2)+\frac{3}{2}C\ll 0$, so we can ignore this case. If $\tau_1\neq 1$, ρ_τ is not on the hyperplane $s_{\tau 2}=-\frac{s_{\tau 11}}{2}+s_{\tau 12}+\frac{3}{2}$. Therefore, we can ignore the rest of the cases by (3.5.19).

Q.E.D.

§12.6 The case $\mathfrak{d}=(\beta_6)$

Let $\mathfrak{d}_0=(\beta_6)$, $\mathfrak{d}_i=(\beta_6,\beta_{6,i})$ for $i=1,\cdots,4$. Let $\mathfrak{p}_0=(\mathfrak{d}_0,\mathfrak{s}_0),\mathfrak{p}_{11}=(\mathfrak{d}_1,\mathfrak{s}_{11}),\mathfrak{p}_{12}=(\mathfrak{d}_1,\mathfrak{s}_{12})$ be paths such that $\mathfrak{s}_{11}(2)=0$, $\mathfrak{s}_{12}(2)=1$. As in previous sections, we only consider such paths satisfying $\mathfrak{s}_0(1)=\mathfrak{s}_{11}(1)=\mathfrak{s}_{12}(1)=0$. Throughout this section, $\Psi=R_{\mathfrak{d}_0}\Phi_{\mathfrak{p}_0}$. Let

$$d_1(\lambda_1,\lambda_2)=a(\underline{\lambda}_1^{-1}\underline{\lambda}_2,\underline{\lambda}_1^{-1}\underline{\lambda}_2,\underline{\lambda}_1^2\underline{\lambda}_2^{-2};\underline{\lambda}_1^{-3}\underline{\lambda}_2^{-1},\underline{\lambda}_1^3\underline{\lambda}_2),$$
$$d_2(\lambda_1,\lambda_2,\lambda_3)=a(\underline{\lambda}_1^{-1}\underline{\lambda}_2\underline{\lambda}_3^{-1},\underline{\lambda}_1^{-1}\underline{\lambda}_2\underline{\lambda}_3,\underline{\lambda}_1^2\underline{\lambda}_2^{-2};\underline{\lambda}_1^{-3}\underline{\lambda}_2^{-1},\underline{\lambda}_1^3\underline{\lambda}_2)$$

for $\lambda_1,\lambda_2,\lambda_3\in\mathbb{R}_+$.

Let $Z_{\mathfrak{d}_0}=V_1\oplus V_2$ where $V_{1k}\cong\mathrm{Sym}^2k^2$, $V_{2k}\cong k$. Then $d_1(\lambda_1,\lambda_2)$ acts on $V_{1\mathbb{A}},V_{2\mathbb{A}}$ by multiplication by $\underline{\lambda}_1\underline{\lambda}_2^3,\underline{\lambda}_1\underline{\lambda}_2^{-5}$ respectively. Let $\chi_{\mathfrak{d}_0},\kappa_{V_1}$ etc. be as in §11.1.

(a) $\Xi_{\mathfrak{p}_0+}(\Phi,\omega,w)$ etc.

Let $\mathfrak{p}=\mathfrak{p}_0,\mathfrak{d}=\mathfrak{d}_0$. Let $g_{\mathfrak{d}}=d_1(\lambda_1,\lambda_2)g_{\mathfrak{d}}^0$ where $g_{\mathfrak{d}}^0$ is as in the first element of (11.1.3). Then $dg_{\mathfrak{d}}=8d^\times\lambda_1 d^\times\lambda_2 dg_{\mathfrak{d}}^0$, and $d_1(\lambda_1,\lambda_2)^{-2\rho}=\lambda_1^{12}\lambda_2^{-4}$.

Easy computations show the following lemma.

Lemma (12.6.1)
(1) $\sigma_{\mathfrak{p}_0}(d_1(\lambda_1,\lambda_2))=\lambda_1^{-3}\lambda_2^{-1}$.
(2) $\kappa_{\mathfrak{d}_0 1}(d_1(\lambda_1,\lambda_2))=\lambda_1^{-4}\lambda_2^{-4}$.
(3) $\kappa_{\mathfrak{d}_0 2}(d_1(\lambda_1,\lambda_2))=\lambda_1^{-15}\lambda_2^3$.

The following proposition follows from (3.5.13).

Proposition (12.6.2)

(1) $$\Xi^s_{\mathfrak{p}_0+}(\Phi,\omega,w)\sim 8C_G\Lambda(w;\rho)\delta_{\mathfrak{d}_0}(\omega)\Sigma_{\mathfrak{d}_0+}(\Psi,\omega_{\mathfrak{d}_0},\chi_{\mathfrak{d}_0},-3).$$

(2) $$\widehat{\Xi}^s_{\mathfrak{p}_0+}(\Phi,\omega,w)\sim 8C_G\Lambda(w;\rho)\delta_{\mathfrak{d}_0}(\omega)\Sigma_{\mathfrak{d}_0+}(\mathscr{F}_{\mathfrak{d}_0}\Psi,\omega_{\mathfrak{d}_0}^{-1},\kappa_{\mathfrak{d}_0 1}^{-1}\chi_{\mathfrak{d}_0}^{-1},7).$$

Next, we consider $\Xi_{\mathfrak{p},\mathrm{st}+}(\Phi,\omega,w)$.

Let $t^0=d_2(\lambda_1,\lambda_2,\lambda_3)\widehat{t}^0$. Then $d^\times t^0=8d^\times\lambda_1 d^\times\lambda_2 d^\times\lambda_3 d^\times\widehat{t}^0$. Let $H_{\mathfrak{d}},X_{\mathfrak{d}}$ be as in §11.1. Let $\widetilde{g}_{\mathfrak{d}}=n'(u_0)t^0$. Then $\widetilde{g}_{\mathfrak{d}}\in X_{\mathfrak{d}}$ if and only if $\lambda_3\le\alpha(u_0)^{-\frac{1}{2}}$. Let $d\widetilde{g}_{\mathfrak{d}}=d^\times t^0 du_0$.

By (12.6.1),

$$\Xi_{\mathfrak{p},\mathrm{st}+}(\Phi,\omega,w)=\int_{X_{\mathfrak{d}}\cap B_{\mathbb{A}}/T_k}\omega(\widetilde{g}_{\mathfrak{d}})\lambda_1^{-3}\lambda_2^{-1}\Theta_{Z'_{\mathfrak{d},0}}(\Psi,\widetilde{g}_{\mathfrak{d}})\mathscr{E}_{\mathfrak{p}}(\widetilde{g}_{\mathfrak{d}},w)d\widetilde{g}_{\mathfrak{d}},$$

$$\widehat{\Xi}_{\mathfrak{p},\mathrm{st}+}(\Phi,\omega,w)=\int_{X_{\mathfrak{d}}\cap B_{\mathbb{A}}/T_k}\omega(\widetilde{g}_{\mathfrak{d}})^{-1}\lambda_1^{-7}\lambda_2^{-5}\Theta_{Z'_{\mathfrak{d},0}}(\mathscr{F}_{\mathfrak{d}}\Psi,\theta_{\mathfrak{d}}(\widetilde{g}_{\mathfrak{d}}))\mathscr{E}_{\mathfrak{p}}(\widetilde{g}_{\mathfrak{d}},w)d\widetilde{g}_{\mathfrak{d}}.$$

If $f(q)$ is a function of $q=(q_1,q_2)\in(\mathbb{A}^1/k^\times)^2$,

$$\int_{(\mathbb{A}^1/k^\times)^5}\omega(\widehat{t}^0)f(\gamma_{2,12}(\widehat{t}^0),\gamma_{1,33}(\widehat{t}^0))d^\times\widehat{t}^0=\delta_\#(\omega)\int_{(\mathbb{A}^1/k^\times)^2}f(q)d^\times q.$$

Therefore, these distributions are 0 unless ω is trivial.

We will show that we can replace $\mathscr{E}_{\mathfrak{p}}(g_{\mathfrak{d}},w)$ by $\mathscr{E}_N(t^0,w)$ in the above integrals. It is easy to see that $\mathscr{E}_{\mathfrak{p}}(g_{\mathfrak{d}},w)$ is the constant term with respect to $P_1\times B_2$.

Lemma (12.6.3)

(1) $\Xi_{\mathfrak{p},\mathrm{st}+}(\Phi,\omega,w)\sim\delta_\#(\omega)\int_{X_{\mathfrak{d}}\cap B_{\mathbb{A}}/T_k}\lambda_1^{-3}\lambda_2^{-1}\Theta_{Z'_{\mathfrak{d},0}}(\Psi,g_{\mathfrak{d}})\mathscr{E}_N(t^0,w)d^\times t^0du_0.$

(2) $\widehat{\Xi}_{\mathfrak{p},\mathrm{st}+}(\Phi,\omega,w)\sim\delta_\#(\omega)\int_{X_{\mathfrak{d}}\cap B_{\mathbb{A}}/T_k}\lambda_1^{-7}\lambda_2^{-5}\Theta_{Z'_{\mathfrak{d},0}}(\mathscr{F}_{\mathfrak{d}}\Psi,\theta_{\mathfrak{d}}(g_{\mathfrak{d}}))\mathscr{E}_N(t^0,w)d^\times t^0du_0.$

Proof. Fix a compact set $C\subset\mathbb{A}^1$ so that C surjects to $\mathbb{A}^1/k^\times$. Clearly, $|\gamma_{2,12}(t^0)|=\lambda_1\lambda_2^3$, $|\gamma_{1,33}(t^0)|=\lambda_1\lambda_2^{-5}$. Therefore, there exist Schwartz–Bruhat functions $0\le\Psi_1\in\mathscr{S}(\mathbb{A}^2)$, $0\le\Psi_2\in\mathscr{S}(\mathbb{A})$ such that

$$\begin{aligned}\Psi(g_{\mathfrak{d}}x)&\ll\Psi_1(\underline{\lambda}_1\underline{\lambda}_2^3x_{2,12},\underline{\lambda}_1\underline{\lambda}_2^3x_{2,12}u_0)\Psi_2(\underline{\lambda}_1\underline{\lambda}_2^{-5}x_{1,33}),\\ \Psi(\theta_{\mathfrak{d}}(g_{\mathfrak{d}})x)&\ll\Psi_1((\underline{\lambda}_1\underline{\lambda}_2^3)^{-1}x_{2,12},(\underline{\lambda}_1\underline{\lambda}_2^3)^{-1}x_{2,12}u_0)\Psi_2((\underline{\lambda}_1\underline{\lambda}_2^{-5})^{-1}x_{1,33})\end{aligned}$$

for $x\in Z'_{\mathfrak{d},0k}$, $t_{ij}\in C$. Therefore, for any $N\ge1$,

$$\begin{aligned}\Theta_{Z'_{\mathfrak{d},0}}(\Psi,g_{\mathfrak{d}})&\ll(\lambda_1\lambda_2^{-5})^{-N}\sum_{x\in k^\times}\Psi_1(\underline{\lambda}_1\underline{\lambda}_2^3x,\underline{\lambda}_1\underline{\lambda}_2^3xu_0),\\ \Theta_{Z'_{\mathfrak{d},0}}(\mathscr{F}_{\mathfrak{d}}\Psi,\theta_{\mathfrak{d}}(g_{\mathfrak{d}}))&\ll(\lambda_1\lambda_2^{-5})^{N}\sum_{x\in k^\times}\Psi_1((\underline{\lambda}_1\underline{\lambda}_2^3)^{-1}x,(\underline{\lambda}_1\underline{\lambda}_2^3)^{-1}xu_0).\end{aligned}$$

By the consideration as in §2.1,

$$|\mathscr{E}_{\mathfrak{p}}(g_{\mathfrak{d}},w)-\mathscr{E}_N(t^0,w)|\ll\sum\widetilde{\mathscr{E}}_{I,\tau,1}(g_{\mathfrak{d}},w),$$

where the summation is over $I=(I_1,I_2)$ such that $I_1=\{1\}$. So by (3.4.30), there exists a slowly increasing function $h(t^0)$ and a constant $\delta>0$ such that if $M>w_0$ and $l\gg0$,

$$|\mathscr{E}_{\mathfrak{p}}(g_{\mathfrak{d}},w)-\mathscr{E}_N(t^0,w)|\ll h(t^0)\lambda_3^l$$

for $\le w_0-\delta\le\mathrm{Re}(w)\le M$.

Let

$$X_+ = \{(\lambda_1, \lambda_2, \lambda_3, u_0) \in \mathbb{R}_+^3 \times \mathbb{A} \mid \lambda_3 \le \alpha(u_0)^{-\frac{1}{2}}\}.$$

We substitute $\mu = \lambda_1\lambda_2^3$. Then the differences of the left hand side and the right hand side of (1), (2) are bounded by constant multiples of the following integrals

$$\int_{\substack{X_+ \\ \lambda_1 \ge 1}} h(t^0)\lambda_1^{\frac{8}{3}(-N-1)}\mu^{\frac{1}{3}(5N-1)}\lambda_3^l \sum_{x\in k^\times} \Psi_1(\underline{\mu}x, \underline{\mu}xu_0)d^\times\lambda_1 d^\times\mu du_0,$$

$$\int_{\substack{X_+ \\ \lambda_1 \le 1}} h(t^0)\lambda_1^{\frac{8}{3}(N-2)}\mu^{\frac{5}{3}(-N-1)}\lambda_3^l \sum_{x\in k^\times} \Psi_1(\underline{\mu}^{-1}x, \underline{\mu}^{-1}xu_0)d^\times\lambda_1 d^\times\mu du_0.$$

These integrals converge absolutely if $N \gg l$. This proves the lemma.

Q.E.D.

Definition (12.6.4)
(1) $l_{\mathfrak{p}_0,\mathrm{st}}(s_\tau) = (s_{\tau 11} + 2s_{\tau 12} + 3s_{\tau 2} - 9, -s_{\tau 11} - 2s_{\tau 12} + s_{\tau 2} + 1, \frac{1}{2}(1 - s_{\tau 11}))$,
(2) $\widehat{l}_{\mathfrak{p}_0,\mathrm{st}}(s_\tau) = (-s_{\tau 11} - 2s_{\tau 12} - 3s_{\tau 2} + 13, s_{\tau 11} + 2s_{\tau 12} - s_{\tau 2} + 3, \frac{1}{2}(1 - s_{\tau 11}))$.

We define

$$\Sigma_{\mathfrak{p},\mathrm{st}+}(\Phi, s_\tau) = \frac{T_{\mathfrak{d}+}(R_{\mathfrak{d},0}\Psi, l_{\mathfrak{p}_0,\mathrm{st}}(s_\tau))}{s_{\tau 11} - 1},$$

$$\widehat{\Sigma}_{\mathfrak{p},\mathrm{st}+}(\Phi, s_\tau) = \frac{T_{\mathfrak{d}+}(R_{\mathfrak{d},0}\mathscr{F}_{\mathfrak{d}}\Psi, \widehat{l}_{\mathfrak{p}_0,\mathrm{st}}(s_\tau))}{s_{\tau 11} - 1}.$$

It is easy to see that

$$(t^0)^{\tau z+\rho} = \lambda_1^{2s_{\tau 11}+s_{\tau 12}+3s_{\tau 2}-6}\lambda_2^{-2s_{\tau 11}-s_{\tau 12}+s_{\tau 2}+2}\lambda_3^{s_{\tau 12}-1}. \tag{12.6.5}$$

If $\tau' = ((1,3),1)\tau((1,3),1)$,

$$\widetilde{\Lambda}_\tau(w; s_{\tau 12}, s_{\tau 11}, s_{\tau 2}) = \widetilde{\Lambda}_{\tau'}(w; s_\tau),$$

because $\psi(-\tau_G z) = \psi(z)$. Therefore, by (12.6.5), and exchanging $s_{\tau 11}, s_{\tau 12}$,

$$\begin{aligned}
&\Xi_{\mathfrak{p},\mathrm{st}+}(\Phi, \omega, w) \\
&= 8\delta_\#(\omega)\sum_\tau \left(\frac{1}{2\pi\sqrt{-1}}\right)^3 \int_{\substack{\mathrm{Re}(s_\tau)=r \\ r\in D_\tau,\ r_1>1}} \Sigma_{\mathfrak{p},\mathrm{st}+}(\Phi, s_{\tau 12}, s_{\tau 11}, s_{\tau 12})\widetilde{\Lambda}_\tau(w; s_\tau)ds_\tau \\
&= 8\delta_\#(\omega)\sum_\tau \left(\frac{1}{2\pi\sqrt{-1}}\right)^3 \int_{\substack{\mathrm{Re}(s_\tau)=r \\ r\in D_\tau,\ r_1>1}} \Sigma_{\mathfrak{p},\mathrm{st}+}(\Phi, s_\tau)\widetilde{\Lambda}_\tau(w; s_\tau)ds_\tau.
\end{aligned}$$

Similarly,

$$\widehat{\Xi}_{\mathfrak{p},\mathrm{st}+}(\Phi, \omega, w) \sim 8\delta_\#(\omega)\sum_\tau \left(\frac{1}{2\pi\sqrt{-1}}\right)^3 \int_{\substack{\mathrm{Re}(s_\tau)=r \\ r\in D_\tau,\ r_1>1}} \widehat{\Sigma}_{\mathfrak{p},\mathrm{st}+}(\Phi, s_\tau)\widetilde{\Lambda}_\tau(w; s_\tau)ds_\tau.$$

Since $\Sigma_{\mathfrak{p},\mathrm{st}+}(\Phi, s_\tau), \widehat{\Sigma}_{\mathfrak{p},\mathrm{st}+}(\Phi, s_\tau)$ are holomorphic for $\mathrm{Re}(s_{\tau 11}) > 1$, the following proposition follows by the usual argument.

Proposition (12.6.6) *Suppose that $\tau = \tau_G$ and $\delta > 0$ is a small number. Then*

(1) $\Xi_{\mathfrak{p}_0,\mathrm{st}+}(\Phi, \omega, w)$

$$\sim \frac{8\varrho^2\delta_{\#}(\omega)}{2\pi\sqrt{-1}} \int_{\mathrm{Re}(s_{\tau 11})=1+\delta} \Sigma_{\mathfrak{p}_0,\mathrm{st}+}(\Phi, s_{\tau 11}, 1, 1)\phi_2(s_{\tau 11})\widetilde{\Lambda}_1(w; s_{\tau 11}, 1) ds_{\tau 11},$$

(2) $\widehat{\Xi}_{\mathfrak{p}_0,\mathrm{st}+}(\Phi, \omega, w)$

$$\sim \frac{8\varrho^2\delta_{\#}(\omega)}{2\pi\sqrt{-1}} \int_{\mathrm{Re}(s_{\tau 11})=1+\delta} \widehat{\Sigma}_{\mathfrak{p}_0,\mathrm{st}+}(\Phi, s_{\tau 11}, 1, 1)\phi_2(s_{\tau 11})\widetilde{\Lambda}_1(w; s_{\tau 11}, 1) ds_{\tau 11}.$$

(b) $\Xi_{\mathfrak{p}_{1i}}(\Phi, \omega, w)$

Let $\mathfrak{d} = \mathfrak{d}_1$, and $\Psi_i = R_{\mathfrak{d}_1}\Phi_{\mathfrak{p}_{1i}}$ for $i = 1, 2$. Let

$$\mu_1 = |\gamma_{2,22}(d_2(\lambda_1, \lambda_2, \lambda_3)| = \lambda_1\lambda_2^3\lambda_3^2,\ \mu_2 = |\gamma_{1,33}(d_2(\lambda_1, \lambda_2, \lambda_3)| = \lambda_1\lambda_2^{-5}.$$

Then $\lambda_2 = \lambda_1^{\frac{1}{5}}\mu_2^{-\frac{1}{5}}$, $\lambda_3 = \lambda_1^{-\frac{4}{5}}\mu_1^{\frac{1}{2}}\mu_2^{\frac{3}{10}}$ and $d^\times\lambda_1 d^\times\lambda_2 d^\times\lambda_3 = \frac{1}{10}d^\times\lambda_1 d^\times\mu_1 d^\times\mu_2$.

Easy computations show the following lemma.

Lemma (12.6.7)

(1) $\sigma_{\mathfrak{p}_{11}}(d_2(\lambda_1, \lambda_2, \lambda_3)) = \lambda_1^{-3}\lambda_2^{-1}\lambda_3^2 = \lambda_1^{-\frac{24}{5}}\mu_1\mu_2^{\frac{4}{5}}$.

(2) $\sigma_{\mathfrak{p}_{12}}(d_2(\lambda_1, \lambda_2, \lambda_3)) = \lambda_1^7\lambda_2^5\lambda_3^2 = \lambda_1^{\frac{32}{5}}\mu_1\mu_2^{-\frac{2}{5}}$.

(3) $\theta_{\mathfrak{p}_{11}}(d_2(\lambda_1, \lambda_2, \lambda_3))^{\tau z+\rho} = \lambda_1^{\frac{8}{5}(s_{\tau 11}+2s_{\tau 2}-3)}\mu_1^{\frac{1}{2}(s_{\tau 12}-1)}\mu_2^{\frac{1}{10}(4s_{\tau 11}+5s_{\tau 12}-2s_{\tau 2}-7)}$.

(4) $\theta_{\mathfrak{p}_{12}}(d_2(\lambda_1, \lambda_2, \lambda_3))^{\tau z+\rho} = \lambda_1^{\frac{8}{5}(-s_{\tau 11}-s_{\tau 12}-2s_{\tau 2}+4)}\mu_1^{\frac{1}{2}(s_{\tau 12}-1)} \times \mu_1^{\frac{1}{10}(-4s_{\tau 11}+s_{\tau 12}+2s_{\tau 2}+1)}$.

Definition (12.6.8)

(1) $l_{\mathfrak{p}_{11},1}(s_\tau) = (\frac{1}{2}(s_{\tau 11}+1), \frac{1}{10}(5s_{\tau 11}+4s_{\tau 12}-2s_{\tau 2}+1), s_{\tau 12}+2s_{\tau 2}-6)$.

(2) $l_{\mathfrak{p}_{12},1}(s_\tau) = (\frac{1}{2}(s_{\tau 11}+1), \frac{1}{10}(s_{\tau 11}-4s_{\tau 12}+2s_{\tau 2}-3), s_{\tau 11}+s_{\tau 12}+2s_{\tau 2}-8)$.

By (12.6.7), and exchanging $s_{\tau 11}, s_{\tau 12}$,

$$\Xi_{\mathfrak{p}_{1i}}(\Phi, \omega, w) = \frac{\delta_{\#}(\omega)}{2}\sum_\tau \left(\frac{1}{2\pi\sqrt{-1}}\right)^3 \int_{\substack{\mathrm{Re}(s_\tau)=r\\ r_1 \gg r_2,\ r_3>1}} \Sigma_{2,\mathrm{sub}}(\Psi_i, l_{\mathfrak{p}_{1i}}, s_\tau)\widetilde{\Lambda}_\tau(w; s_\tau) ds_\tau$$

for $i = 1, 2$ where the first (resp. second) coordinate of Ψ_i corresponds to $x_{2,22}$ (resp. $x_{1,33}$). We define

$$\Sigma_{\mathfrak{p}_{11}}(\Phi, s_{\tau 11}) = -\frac{1}{6}\Sigma_2(R_{\mathfrak{d}_1}\Phi_{\mathfrak{p}_{11}}, \frac{1}{2}(s_{\tau 11}+1), \frac{1}{10}(5s_{\tau 11}+3)),$$

$$\Sigma_{\mathfrak{p}_{12}}(\Phi, s_{\tau 11}) = \frac{\Sigma_2(R_{\mathfrak{d}_1}\Phi_{\mathfrak{p}_{12}}, \frac{1}{2}(s_{\tau 11}+1), \frac{1}{10}(s_{\tau 11}-5))}{2(s_{11}-5)}.$$

Proposition (12.6.9) *Suppose that $\tau = \tau_G$ and $\delta > 0$ is a small number. Then by changing ψ if necessary,*

$$\Xi_{\mathfrak{p}_{1i}}(\Phi, \omega, w) \sim \frac{\varrho^2\delta_{\#}(\omega)}{2\pi\sqrt{-1}} \int_{\mathrm{Re}(s_{\tau 11})=1+\delta} \Sigma_{\mathfrak{p}_{1i}}(\Phi, s_{\tau 11})\phi_2(s_{\tau 11})\widetilde{\Lambda}_1(w; s_{\tau 11}, 1) ds_{\tau 11}$$

for $i = 1, 2$.

Proof. We can ignore the case $\tau_2 = 1$ as usual, and assume that $\tau_2 = (1,2)$. Let $r = (r_1, r_2, r_3) = (20, 2, 2)$. Then by the usual argument,

$$\begin{aligned}&\left(\frac{1}{2\pi\sqrt{-1}}\right)^3 \int_{\mathrm{Re}(s_\tau)=r} \Sigma_{2,\mathrm{sub}}(\Psi_i, l_{\mathfrak{p}1i}, s_\tau)\widetilde{\Lambda}(w; s_\tau)ds_\tau \\ &\sim \left(\frac{1}{2\pi\sqrt{-1}}\right)^2 \int_{\mathrm{Re}(s_{\tau 1})=(r_1,r_2)} \Sigma_{2,\mathrm{sub}}(\Psi_i, l_{\mathfrak{p}1i}, s_{\tau 1}, 1)\widetilde{\Lambda}_{1\tau_1}(w; s_{\tau 1})ds_{\tau 1}.\end{aligned}$$

If $\tau_1 = 1, (1,2), (2,3)$, we choose $r_1 \gg r_2 \gg 0$. Then $\widetilde{L}(r_1, r_2, 1) \ll 0$. Therefore, we can ignore these cases.

The point $\rho_{1\tau_1}$ is $(2,-1), (-1,2), (1,1)$ for $\tau_1 = (1,2,3), (1,3,2), (1,3)$. So $\rho_{1\tau_1}$ is not on the lines $s_{\tau 12} = 4$, $5s_{\tau 11} + 4s_{\tau 12} = 11$ for these cases. Therefore, we can move the contour crossing these lines by (3.6.1) so that $(r_1, r_2) = (1+\delta, 1+\delta)$. Then we only have to move the contour so that $(r_1, r_2) = (1-\delta_1, 1+\delta)$ where $\delta, \delta_1 > 0$ are small numbers and $\delta_1\delta^{-1} \gg 0$.

Q.E.D.

(c) $\Xi_{\mathfrak{p}_0,\mathfrak{d}_2,i}(\Phi, \omega, w)$

Let $\mathfrak{d} = \mathfrak{d}_0, \mathfrak{p} = \mathfrak{p}_0$, and $\mathfrak{d}' = \mathfrak{d}_2$. Let $\mu_1 = \lambda_1\lambda_2^3$. Then $d^\times\lambda_1 d^\times\lambda_2 = \frac{1}{3}d^\times\lambda_1 d^\times\mu_1$. Easy computations show the following lemma.

Lemma (12.6.10)

(1) $d_1(\lambda_1,\lambda_2)^{\tau z+\rho} = \lambda_1^{\frac{4}{3}(2s_{\tau 11}+s_{\tau 12}+2s_{\tau 2}-5)}\mu_1^{\frac{1}{3}(-2s_{\tau 11}-s_{\tau 12}+s_{\tau 2}+2)}$.

(2) $\sigma_{\mathfrak{p}}(d_1(\lambda_1,\lambda_2)) = \lambda_1^{-\frac{8}{3}}\mu_1^{-\frac{1}{3}}$.

(3) $\kappa_{\mathfrak{d}1}(d_1(\lambda_1,\lambda_2)) = \lambda_1^{-\frac{8}{3}}\mu_1^{-\frac{4}{3}}$.

(4) $\kappa_{V_1}(d_1(\lambda_1,\lambda_2))) = (\lambda_1\lambda_2^3)^{-3} = \mu_1^{-3}$.

(5) $\kappa_{V_2}(d_1(\lambda_1,\lambda_2)) = (\lambda_1\lambda_2^{-5})^{-1} = \lambda_1^{-\frac{8}{3}}\mu_1^{\frac{5}{3}}$.

Definition (12.6.11) *Let* $l_{2i} = (l_{2i,1}, l_{2i,2})$ *for* $i = 1,2,3,4$ *where*

(1) $l_{21,1}(s_\tau) = \frac{1}{3}(-s_{\tau 11} - 2s_{\tau 12} + s_{\tau 2} + 6)$, $l_{21,2}(s_\tau) = s_{\tau 11} + 2s_{\tau 12} + 2s_{\tau 2} - 9$,

(2) $l_{22,1}(s_\tau) = \frac{1}{3}(-s_{\tau 11} - 2s_{\tau 12} + s_{\tau 2} + 1)$, $l_{22,2}(s_\tau) = s_{\tau 11} + 2s_{\tau 12} + 2s_{\tau 2} - 7$,

(3) $l_{23,1}(s_\tau) = 3 - l_{21,1}(s_\tau)$, $l_{23,2}(s_\tau) = l_{21,2}(s_\tau)$,

(4) $l_{24,1}(s_\tau) = 3 - l_{22,1}(s_\tau)$, $l_{24,2}(s_\tau) = l_{22,2}(s_\tau)$.

Note that $l_{2i}(-s_{\tau 11}, s_{\tau 11} + s_{\tau 12}, s_{\tau 2}) = l_{2i}(s_{\tau 11}, s_{\tau 12}, s_{\tau 2})$ for $i = 1,2,3,4$.

Let

$$\Psi_1 = R_{\mathfrak{d}'}\mathscr{F}_{V_1}\mathscr{F}_{\mathfrak{d}}\Psi,\ \Psi_2 = R_{\mathfrak{d}'}\Psi,\ \Psi_3 = R_{\mathfrak{d}'}\mathscr{F}_{\mathfrak{d}}\Psi,\ \Psi_4 = R_{\mathfrak{d}'}\mathscr{F}_{V_1}\Psi.$$

By (12.6.10), and exchanging $s_{\tau 11}, s_{\tau 12}$,

$$\Xi_{\mathfrak{p},\mathfrak{d}',i}(\Phi, \omega, w) = 2\delta_{\#}(\omega) \sum_{\substack{\tau_1 \in \mathfrak{W}_2 \\ \tau_2 = 1,(1,2)}} \Omega_{W,l_{2i},\tau}(\Psi_i, w)$$

for $i = 1,2,3,4$.

The following proposition follows from (11.2.3).

Proposition (12.6.12) *Suppose that $\tau=\tau_G$ and $\delta>0$ is a small number. Then by changing ψ if necessary,*

$$\begin{aligned}\Xi_{\mathfrak{p},\mathfrak{d}',i}(\Phi,\omega,w) &\sim 2C_G\Lambda(w;\rho)\frac{\delta_{\#}(\omega)\Sigma_{W+}(\Psi_i,l_{2i,1}(-1,2,1))}{l_{2i,2}(-1,2,1)}\\ &\quad+\frac{2\rho^2\delta_{\#}(\omega)}{2\pi\sqrt{-1}}\int_{\mathrm{Re}(s_{\tau 11})=1+\delta}\Sigma_{W,\mathrm{st},\mathrm{sub}+}(\Psi_i,l_{2i},s_{\tau 11},1,1)\\ &\qquad\qquad\times\phi_2(s_{\tau 11})\widetilde{\Lambda}_1(w;s_{\tau 11},1)ds_{\tau 11}\end{aligned}$$

for $i=1,2,3,4$.

(d) $\Xi_{\mathfrak{p}_0,\mathfrak{d}_3,i}(\Phi,\omega,w)$

Let $\mathfrak{d}'=\mathfrak{d}_3$. For the cases $i=1,2$, we make the change of variable $\mu_1=\lambda_1\lambda_2^3$, $\mu_2=\lambda_1\lambda_2^3\lambda_3^2=\mu_1\lambda_3^2$. Then $\lambda_3=\mu_1^{-\frac{1}{2}}\mu_2^{\frac{1}{2}}$, and

$$\begin{aligned}(12.6.13)\quad & d_2(\lambda_1,\lambda_2,\lambda_3)^{\tau z+\rho}\\ &=\lambda_1^{\frac{4}{3}(2s_{\tau 11}+s_{\tau 12}+2s_{\tau 2}-5)}\mu_1^{\frac{1}{6}(-4s_{\tau 11}-5s_{\tau 12}+2s_{\tau 2}+7)}\mu_2^{\frac{1}{2}(s_{\tau 12}-1)}.\end{aligned}$$

For the cases $i=3,4$, we make the change of variable $\mu_1=\lambda_1\lambda_2^3$, $\mu_2=(\lambda_1\lambda_2^3)^{-1}\lambda_3^2=\mu_1^{-1}\lambda_3^2$. Then $\lambda_3=\mu_1^{\frac{1}{2}}\mu_2^{\frac{1}{2}}$, and

$$\begin{aligned}(12.6.14)\quad & d_2(\lambda_1,\lambda_2,\lambda_3)^{\tau z+\rho}\\ &=\lambda_1^{\frac{4}{3}(2s_{\tau 11}+s_{\tau 12}+2s_{\tau 2}-5)}\mu_1^{\frac{1}{6}(-4s_{\tau 11}+s_{\tau 12}+2s_{\tau 2}+1)}\mu_2^{\frac{1}{2}(s_{\tau 12}-1)}.\end{aligned}$$

In both cases, $d^\times\lambda=\frac{1}{6}d^\times\lambda_1 d^\times\mu_1 d^\times\mu_2$.

It is easy to see from (12.6.1) that $\sigma_{\mathfrak{p}}(d_1(\lambda_1,\lambda_2))\lambda_3^2=\lambda_1^{-\frac{8}{3}}\mu_1^{-\frac{4}{3}}\mu_2$ for the cases $i=1,2$, and $\sigma_{\mathfrak{p}}(d_1(\lambda_1,\lambda_2))\lambda_3^2=\lambda_1^{-\frac{8}{3}}\mu_1^{\frac{2}{3}}\mu_2$ for the cases $i=3,4$.

Definition (12.6.15) *Let $l_{3i}=(\frac{1}{2}(s_{\tau 11}+1),l_{3i,2})$ where*
(1) $l_{31,2}(s_\tau)=(s_{\tau 11}+2s_{\tau 12}+2s_{\tau 2}-9)(5s_{\tau 11}+4s_{\tau 12}-2s_{\tau 2}-9)$,
(2) $l_{32,2}(s_\tau)=(s_{\tau 11}+2s_{\tau 12}+2s_{\tau 2}-7)(5s_{\tau 11}+4s_{\tau 12}-2s_{\tau 2}+1)$,
(3) $l_{33,2}(s_\tau)=(s_{\tau 11}+2s_{\tau 12}+2s_{\tau 2}-9)(s_{\tau 11}-4s_{\tau 12}+2s_{\tau 2}-3)$,
(4) $l_{34,2}(s_\tau)=(s_{\tau 11}+2s_{\tau 12}+2s_{\tau 2}-7)(s_{\tau 11}-4s_{\tau 12}+2s_{\tau 2}-13)$.

Let

$$\Psi_1=R_{\mathfrak{d}'}\mathscr{F}_{V_1}\mathscr{F}_{\mathfrak{d}}\Psi,\ \Psi_2=R_{\mathfrak{d}'}\Psi,\ \Psi_3=R_{\mathfrak{d}'}\mathscr{F}_{\mathfrak{d}}\Psi,\ \Psi_4=R_{\mathfrak{d}'}\mathscr{F}_{V_1}\Psi.$$

By (12.6.1), (12.6.12), (12.6.13), and exchanging $s_{\tau 11},s_{\tau 12}$,

$$\begin{aligned}&\Xi_{\mathfrak{p},\mathfrak{d}',i}(\Phi,\omega,w)\\ &=6\delta_{\#}(\omega)\sum_\tau\left(\frac{1}{2\pi\sqrt{-1}}\right)^3\int_{\substack{\mathrm{Re}(s_\tau)=r\\ r_1\gg r_2,\ r_3>1}}\Sigma_{1,\mathrm{sub}}(\Psi_i,l_{3i},s_\tau)\widetilde{\Lambda}_\tau(w;s_\tau)ds_\tau\end{aligned}$$

for $i=1,2,3,4$.

We define

$$\Sigma_{\mathfrak{p}_{31}}(\Phi, s_{\tau 11}) = \frac{6\Sigma_1(R_{\mathfrak{d}_3}\mathscr{F}_{V_1}\mathscr{F}_{\mathfrak{d}}\Psi, \frac{1}{2}(s_{\tau 11}+1))}{(s_{\tau 11}-5)(5s_{\tau 11}-7)},$$

$$\Sigma_{\mathfrak{p}_{32}}(\Phi, s_{\tau 11}) = \frac{6\Sigma_1(R_{\mathfrak{d}_3}\Psi, \frac{1}{2}(s_{\tau 11}+1))}{(s_{\tau 11}-3)(5s_{\tau 11}+3)},$$

$$\Sigma_{\mathfrak{p}_{33}}(\Phi, s_{\tau 11}) = \frac{6\Sigma_1(R_{\mathfrak{d}_3}\mathscr{F}_{\mathfrak{d}}\Psi, \frac{1}{2}(s_{\tau 11}+1))}{(s_{\tau 11}-5)^2},$$

$$\Sigma_{\mathfrak{p}_{34}}(\Phi, s_{\tau 11}) = \frac{6\Sigma_1(R_{\mathfrak{d}_3}\mathscr{F}_{V_1}\Psi, \frac{1}{2}(s_{\tau 11}+1))}{(s_{\tau 11}-3)(s_{\tau 11}-15)}.$$

The following proposition follows from cases (7), (8), (13), (14) of (11.2.4).

Proposition (12.6.16) *Suppose that $\tau = \tau_G$ and $\delta > 0$ is a small number. Then by changing ψ if necessary,*

$$\Xi_{\mathfrak{p},\mathfrak{d}',i}(\Phi, \omega, w) \sim \frac{\varrho^2\delta_{\#}(\omega)}{2\pi\sqrt{-1}}\int_{\mathrm{Re}(s_{\tau 11})=1+\delta} \Sigma_{\mathfrak{p}_{3i}}(\Phi, s_{\tau 11})\phi_2(s_{\tau 11})\widetilde{\Lambda}_1(w, s_{\tau 11}, 1)ds_{\tau 11}$$

for $i = 1, 2, 3, 4$.

(e) $\Xi_{\mathfrak{p}_0,\mathfrak{d}_4,i}(\Phi, \omega, w)$

Let $\mathfrak{d}' = \mathfrak{d}_4$. Let $\mu_2 = \lambda_1^{\frac{8}{3}}$, $\mu_3 = (\lambda_1\lambda_2^3)^{-\frac{5}{3}}$. Then $d^\times\lambda_1 d^\times\lambda_2 = \frac{3}{40}d^\times\mu_2 d^\times\mu_3$. Let $\mu = (\mu_2, \mu_3)$, and $d^\times\mu = d^\times\mu_2 d^\times\mu_3$.

Let $\tau = (\tau_1, \tau_2)$ be a Weyl group element such that $\tau_1 \in \mathfrak{W}_1$. The following lemma is easy to prove and the proof is left to the reader.

Lemma (12.6.17)

(1) $\sigma_{\mathfrak{p}}(d_1(\lambda_1, \lambda_2)) = \mu_2^{-1}\mu_3^{\frac{1}{5}}$.

(2) $\kappa_{\mathfrak{d}1}(d_1(\lambda_1, \lambda_2)) = \mu_2^{-1}\mu_3^{\frac{4}{5}}$.

(3) $\kappa_{V_1}(d_1(\lambda_1, \lambda_2)) = \mu_3^{\frac{9}{5}}$.

(4) $\kappa_{V_2}(d_1(\lambda_1, \lambda_2)) = \mu_2^{-1}\mu_3^{-1}$.

It is easy to see that

$$[d_1(\lambda_1, \lambda_2)^{\tau z+\rho}]_{s_{\tau 12}=-1} = \mu_2^{s_{\tau 11}+s_{\tau 2}-3}\mu_3^{\frac{1}{5}(2s_{\tau 12}-s_{\tau 2}-3)}. \tag{12.6.18}$$

Let $\Psi_1 = R_{\mathfrak{d}'}\mathscr{F}_{\mathfrak{d}}\Psi, \Psi_2 = R_{\mathfrak{d}'}\mathscr{F}_{V_2}\Psi, \Psi_3 = R_{\mathfrak{d}'}\mathscr{F}_{V_2}\mathscr{F}_{\mathfrak{d}}\Psi, \Psi_4 = R_{\mathfrak{d}'}\Psi$. We define distributions $\Sigma_{\mathfrak{p},\mathfrak{d}',i}(\Phi, \omega, s_{\tau 11}, s_{\tau 2})$'s for $i = 1, 2, 3, 4$ by the following integrals in that order.

$$\frac{3\delta_{\#}(\omega)}{5}\int_{\substack{\mathbb{R}_+^2\times\mathbb{A}^1/k^\times \\ \mu_2\le 1,\ \mu_3\le 1}} \mu_2^{s_{\tau 11}+s_{\tau 2}-5}\mu_3^{\frac{1}{5}(2s_{\tau 11}-s_{\tau 2}+2)}\Theta_1(\Psi_1, (\underline{\mu}_2\underline{\mu}_3)^{-1}t)d^\times\mu d^\times t,$$

$$\frac{3\delta_{\#}(\omega)}{5}\int_{\substack{\mathbb{R}_+^2\times\mathbb{A}^1/k^\times \\ \mu_2\le 1,\ \mu_3\le 1}} \mu_2^{s_{\tau 11}+s_{\tau 2}-5}\mu_3^{\frac{1}{5}(2s_{\tau 11}-s_{\tau 2}-7)}\Theta_1(\Psi_2, (\underline{\mu}_2\underline{\mu}_3)^{-1}t)d^\times\mu d^\times t,$$

$$\frac{3\delta_{\#}(\omega)}{5}\int_{\substack{\mathbb{R}_+^2\times\mathbb{A}^1/k^\times \\ \mu_2\le 1,\ \mu_3\ge 1}} \mu_2^{s_{\tau 11}+s_{\tau 2}-4}\mu_3^{\frac{1}{5}(2s_{\tau 11}-s_{\tau 2}+7)}\Theta_1(\Psi_3, \underline{\mu}_2\underline{\mu}_3 t)d^\times\mu d^\times t,$$

$$\frac{3\delta_{\#}(\omega)}{5}\int_{\substack{\mathbb{R}_+^2\times\mathbb{A}^1/k^\times \\ \mu_2\le 1,\ \mu_3\ge 1}} \mu_2^{s_{\tau 11}+s_{\tau 2}-4}\mu_3^{\frac{1}{5}(2s_{\tau 11}-s_{\tau 2}-2)}\Theta_1(\Psi_4, \underline{\mu}_2\underline{\mu}_3 t)d^\times\mu d^\times t.$$

We define

$$\widetilde{\Sigma}_{\mathfrak{p},\mathfrak{d}',i}(\Phi,\omega,w,s_{\tau 11},s_{\tau 2}) = \Sigma_{\mathfrak{p},\mathfrak{d}',i}(\Phi,\omega,s_{\tau 11},s_{\tau 2})\widetilde{\Lambda}_\tau(w;s_{\tau 11},-1,s_{\tau 2}).$$

Let $s'_\tau = (s_{\tau 11}, s_{\tau 2})$ and $ds'_\tau = ds_{\tau 11}ds_{\tau 2}$. By (12.6.17), the Mellin inversion formula,

$$\Xi_{\mathfrak{p},\mathfrak{d}',i}(\Phi,\omega,w) = \sum_{\substack{\tau_1\in\mathfrak{W}_1\\ \tau_2=1,(1,2)}} \left(\frac{1}{2\pi\sqrt{-1}}\right)^2 \int_{\mathrm{Re}(s'_\tau)=(r_1,r_3)} \widetilde{\Sigma}_{\mathfrak{p},\mathfrak{d}',i}(\Phi,\omega,w,s'_\tau)ds'_\tau,$$

where we choose the contour so that $(r_1,-1,r_3)\in D_\tau$, and $r_1 \gg r_3 \gg 0$.

For $\lambda\in\mathbb{R}_+$ and $i=1,2,3,4$, we define $\Psi_{i\lambda}(x)=\Psi_i(\underline{\lambda}x)$. Clearly, Ψ_3 (resp. Ψ_4) is the Fourier transform of Ψ_1 (resp. Ψ_2) with respect to the character $< >$. Therefore, for any $c_1,c_2\in\mathbb{C}$,

$$\begin{aligned}
&\int_0^1 \lambda^{c_1}\Sigma_{1+}(\Psi_{i+2\lambda},c_2)d^\times\lambda + \int_0^1 \lambda^{c_1-1}\Sigma_{1+}(\Psi_{i\lambda^{-1}},1-c_2)d^\times\lambda \\
&= \frac{1}{c_1-c_2}\Sigma_1(\Psi_{i+2\lambda},c_2) + \frac{\Psi_{i+2}(0)}{c_1c_2} - \frac{\Psi_i(0)}{(c_1-1)(c_2-1)}.
\end{aligned}$$

Hence,

$$\begin{aligned}
&\Sigma_{\mathfrak{p},\mathfrak{d}',1}(\Phi,\omega,s_{\tau 11},s_{\tau 2}) + \Sigma_{\mathfrak{p},\mathfrak{d}',3}(\Phi,\omega,s_{\tau 11},s_{\tau 2}) \\
&= \delta_\#(\omega)\frac{\Sigma_1(\Psi_3,\frac{1}{5}(2s_{\tau 11}-s_{\tau 2}+7))}{s_{\tau 11}+2s_{\tau 2}-9} \\
&\quad + 3\delta_\#(\omega)\frac{\Psi_3(0)}{(s_{\tau 11}+s_{\tau 2}-4)(2s_{\tau 11}-s_{\tau 2}+7)}, \\
&\quad - 3\delta_\#(\omega)\frac{\Psi_1(0)}{(s_{\tau 11}+s_{\tau 2}-5)(2s_{\tau 11}-s_{\tau 2}+2)}, \\
&\Sigma_{\mathfrak{p},\mathfrak{d}',2}(\Phi,\omega,s_{\tau 11},s_{\tau 2}) + \Sigma_{\mathfrak{p},\mathfrak{d}',4}(\Phi,\omega,s_{\tau 11},s_{\tau 2}) \\
&= \delta_\#(\omega)\frac{\Sigma_1(\Psi_4,\frac{1}{5}(2s_{\tau 11}-s_{\tau 2}-2))}{s_{\tau 11}+2s_{\tau 2}-6} \\
&\quad + 3\delta_\#(\omega)\frac{\Psi_4(0)}{(s_{\tau 11}+s_{\tau 2}-4)(2s_{\tau 11}-s_{\tau 2}-2)}, \\
&\quad - 3\delta_\#(\omega)\frac{\Psi_2(0)}{(s_{\tau 11}+s_{\tau 2}-5)(2s_{\tau 11}-s_{\tau 2}-7)}.
\end{aligned}$$

Proposition (12.6.19) *By changing ψ if necessary,*

(1) $$\begin{aligned}\Xi_{\mathfrak{p},\mathfrak{d}',1}(\Phi,\omega,w) + \Xi_{\mathfrak{p},\mathfrak{d}',3}(\Phi,\omega,w) &\sim -C_G\Lambda(w;\rho)\frac{\mathfrak{V}_2\delta_\#(\omega)\Sigma_1(R_{\mathfrak{d}'}\Psi_1,-1)}{5} \\ &\quad + 3C_G\Lambda(w;\rho)\delta_\#(\omega)\left(-\frac{\Psi_3(0)}{10}+\frac{\Psi_1(0)}{10}\right),\end{aligned}$$

(2) $$\begin{aligned}\Xi_{\mathfrak{p},\mathfrak{d}',2}(\Phi,\omega,w) + \Xi_{\mathfrak{p},\mathfrak{d}',4}(\Phi,\omega,w) &\sim -C_G\Lambda(w;\rho)\frac{\mathfrak{V}_2\delta_\#(\omega)\Sigma_1(R_{\mathfrak{d}'}\Psi_4,-\frac{1}{5})}{2} \\ &\quad - 3C_G\Lambda(w;\rho)\delta_\#(\omega)\left(\Psi_4(0)-\frac{\Psi_2(0)}{8}\right).\end{aligned}$$

Proof. As usual, we can ignore the case $\tau_2 = 1$, and assume that $\tau_2 = (1,2)$.

By the usual argument,

$$\left(\frac{1}{2\pi\sqrt{-1}}\right)^2 \int_{\mathrm{Re}(s'_\tau)=(r_1,r_3)} \left(\widetilde{\Sigma}_{\mathfrak{p},\mathfrak{d}',i}(\Phi,\omega,w,s'_\tau) + \widetilde{\Sigma}_{\mathfrak{p},\mathfrak{d}',i+2}(\Phi,\omega,w,s'_\tau)\right) ds'_\tau$$
$$\sim \frac{\varrho}{2\pi\sqrt{-1}} \int_{\mathrm{Re}(s_{\tau 11})=r_1} \operatorname*{Res}_{s_{\tau 2}=1} \left(\widetilde{\Sigma}_{\mathfrak{p},\mathfrak{d}',i}(\Phi,\omega,w,s'_\tau) + \widetilde{\Sigma}_{\mathfrak{p},\mathfrak{d}',i+2}(\Phi,\omega,w,s'_\tau)\right) ds_{\tau 11}$$

for $i = 1,2$.

The poles of the function

$$[\Sigma_{\mathfrak{p},\mathfrak{d}',i}(\Phi,\omega,s_{\tau 11},s_{\tau 2}) + \Sigma_{\mathfrak{p},\mathfrak{d}',i+2}(\Phi,\omega,s_{\tau 11},s_{\tau 2})]_{s_{\tau 2}=1}$$

for $i = 1,2$ are $s_{\tau 11} = \frac{1}{2}, \frac{3}{2}, 3, 4, 6, 7$. But $\rho_{1\tau_1}$ does not satisfy these conditions. Therefore, by (3.6.1), we can move the contour so that $r_1 = 1+\delta$ where $\delta > 0$ is a small number. Then $\widetilde{L}(1+\delta,1,1) < w_0$ if $\tau_1 = 1$ or $(2,3)$. So we can ignore these cases. Suppose $\tau_1 = (1,2,3)$. Then by the usual argument,

$$\frac{\varrho}{2\pi\sqrt{-1}} \int_{\mathrm{Re}(s_{\tau 11})=1+\delta} \operatorname*{Res}_{s_{\tau 2}=1} \left(\widetilde{\Sigma}_{\mathfrak{p},\mathfrak{d}',i}(\Phi,\omega,w,s'_\tau) + \widetilde{\Sigma}_{\mathfrak{p},\mathfrak{d}',i+2}(\Phi,\omega,w,s'_\tau)\right) ds_{\tau 11}$$
$$\sim C_G \Lambda(w;\rho)\mathfrak{V}_2 \left[\Sigma_{\mathfrak{p},\mathfrak{d}',i}(\Phi,\omega,s'_\tau) + \Sigma_{\mathfrak{p},\mathfrak{d}',i+2}(\Phi,\omega,s'_\tau)\right]_{s'_\tau=(2,1)}.$$

Q.E.D.

(f) Now we combine the computations in this section. Let $\mathfrak{d} = \mathfrak{d}_0, \mathfrak{p} = \mathfrak{p}_0$.

We define

$$\begin{aligned}
J_1(\Phi,\omega) &= 8\delta_{\mathfrak{d}}(\omega)\left(\Sigma_{\mathfrak{d}+}(\Psi,\omega_{\mathfrak{d}},\chi,-3) + \Sigma_{\mathfrak{d}+}(\mathscr{F}_{\mathfrak{d}}\Psi,\omega_{\mathfrak{d}}^{-1},\kappa_{\mathfrak{d}1}^{-1}\chi_{\mathfrak{d}}^{-1},7)\right) \\
&\quad + \mathfrak{V}_2\delta_{\#}(\omega)\left(-\frac{1}{5}\Sigma_1(R_{\mathfrak{d}_4}\mathscr{F}_{\mathfrak{d}}\Psi,-1) + \frac{1}{2}\Sigma_1(R_{\mathfrak{d}_4}\Psi,-\frac{1}{5})\right), \\
J_2(\Phi,\omega) &= -\frac{\delta_{\#}(\omega)}{2}\left(\Sigma_{W+}(R_{\mathfrak{d}_2}\mathscr{F}_{\mathfrak{d}}\Psi,\frac{5}{3}) + \Sigma_{W+}(\mathscr{F}_W R_{\mathfrak{d}_2}\mathscr{F}_{\mathfrak{d}}\Psi,\frac{4}{3})\right) \\
&\quad + \delta_{\#}(\omega)\left(\Sigma_{W+}(R_{\mathfrak{d}_2}\Psi,-\frac{1}{3}) + \Sigma_{W+}(\mathscr{F}_W R_{\mathfrak{d}_2}\Psi,\frac{10}{3})\right) \\
&\quad + 3\mathfrak{V}_2\delta_{\#}(\omega)\left(\frac{\mathscr{F}_{\mathfrak{d}_4}R_{\mathfrak{d}_4}\Psi(0)}{8} + \frac{R_{\mathfrak{d}_4}\mathscr{F}_{\mathfrak{d}}\Psi(0)}{10}\right) \\
&\quad + 3\mathfrak{V}_2\delta_{\#}(\omega)\left(-\frac{\mathscr{F}_{\mathfrak{d}_4}R_{\mathfrak{d}_4}\mathscr{F}_{\mathfrak{d}}\Psi(0)}{10} + R_{\mathfrak{d}_4}\Psi(0)\right), \\
\Upsilon_{\mathfrak{d},1}(\Phi,\omega) &= J_1(\Phi,\omega) + J_2(\Phi,\omega), \\
J_3(\Phi,s_{\tau 11}) &= \frac{8T_{\mathfrak{d}+}(R_{\mathfrak{d},0}\Psi,s_{\tau 11}-4,-s_{\tau 11},\frac{1}{2}(1-s_{\tau 11}))}{s_{\tau 11}-1} \\
&\quad + \frac{8T_{\mathfrak{d}+}(R_{\mathfrak{d},0}\mathscr{F}_{\mathfrak{d}}\Psi,8-s_{\tau 11},s_{\tau 11}+4,\frac{1}{2}(1-s_{\tau 11}))}{s_{\tau 11}-1}, \\
J_4(\Phi,s_{\tau 11}) &= \frac{2T_{W+}(\widetilde{R}_{W,0}R_{\mathfrak{d}_2}\mathscr{F}_{\mathfrak{d}}\Psi,\frac{1}{3}(s_{\tau 11}+4),\frac{1}{2}(1-s_{\tau 11}))}{(s_{\tau 11}-1)(s_{\tau 11}-5)} \\
&\quad + \frac{2T_{W+}(\widetilde{R}_{W,0}\mathscr{F}_W R_{\mathfrak{d}_2}\mathscr{F}_{\mathfrak{d}}\Psi,\frac{1}{3}(5-s_{\tau 11}),\frac{1}{2}(1-s_{\tau 11}))}{(s_{\tau 11}-1)(s_{\tau 11}-5)},
\end{aligned}$$

$$
\begin{aligned}
J_5(\Phi) = &- \frac{2T_{W+}(\widetilde{R}_{W,0}R_{\mathfrak{d}_2}\Psi, -\frac{s_{\tau 11}}{3}, \frac{1}{2}(1-s_{\tau 11}))}{(s_{\tau 11}-1)(s_{\tau 11}-3)} \\
&- \frac{2T_{W+}(\widetilde{R}_{W,0}\mathscr{F}_W R_{\mathfrak{d}_2}\Psi, \frac{1}{3}(s_{\tau 11}+9), \frac{1}{2}(1-s_{\tau 11}))}{(s_{\tau 11}-1)(s_{\tau 11}-3)},
\end{aligned}
$$

$$
\begin{aligned}
\Upsilon_{\mathfrak{d},2}(\Phi,\omega,s_{\tau 11}) = {}& \delta_{\#}(\omega)(\sum_{i=3}^{5} J_i(\Phi,s_{\tau 11}) + \Sigma_{\mathfrak{p}_{12}}(\Phi,s_{\tau 11}) - \Sigma_{\mathfrak{p}_{11}}(\Phi,s_{\tau 11})) \\
&+ \delta_{\#}(\omega)\sum_{i=1}^{4}(-1)^{i+1}\Sigma_{\mathfrak{p}_{3i}}(\Phi,s_{\tau 11}).
\end{aligned}
$$

The following proposition follows from (9.1.10), (12.6.2), (12.6.6), (12.6.9), (12.6.12), (12.6.16), (12.6.19).

Proposition (12.6.20) *Suppose that $\tau = \tau_G$ and $\delta > 0$ is a small number. Then by changing ψ if necessary,*

$$
\begin{aligned}
\Xi_{\mathfrak{p}_0}(\Phi,\omega,w) \sim{}& C_G\Lambda(w;\rho)\Upsilon_{\mathfrak{d},1}(\Phi,\omega) \\
&+ \frac{\varrho^2}{2\pi\sqrt{-1}}\int_{\mathrm{Re}(s_{\tau 11})=1+\delta} \Upsilon_{\mathfrak{d},2}(\Phi,\omega,s_{\tau 11})\phi_2(s_{\tau 11})\widetilde{\Lambda}_1(w;s_{\tau 11},1)ds_{\tau 11}.
\end{aligned}
$$

Let $\Upsilon_{\mathfrak{d},2,(i)}(\Phi,\omega,1)$ etc. be the i-th coefficient of the Laurent expansion. We define

$$
\begin{aligned}
J_6(\Phi) &= -4\left(T_{\mathfrak{d}+}(R_{\mathfrak{d},0}\Psi,-3,-1) + T_{\mathfrak{d}+}(R_{\mathfrak{d},0}\mathscr{F}_{\mathfrak{d}}\Psi,7,5)\right), \\
J_7(\Phi) &= \frac{1}{4}\left(T_{W+}(\widetilde{R}_{W,0}\mathscr{F}_W R_{\mathfrak{d}_2}\mathscr{F}_{\mathfrak{d}}\Psi,\frac{4}{3}) + T_{W+}(\widetilde{R}_{W,0}\mathscr{F}_{\mathfrak{d}} R_{\mathfrak{d}_2}\Psi,\frac{5}{3})\right), \\
J_8(\Phi) &= -\frac{1}{2}\left(T_{W+}(\widetilde{R}_{W,0}R_{\mathfrak{d}_2}\Psi,-\frac{1}{3}) + T_{W+}(\widetilde{R}_{W,0}\mathscr{F}_W R_{\mathfrak{d}_2}\Psi,\frac{10}{3})\right).
\end{aligned}
$$

It is easy to see that

$$
\begin{aligned}
J_{3,(0)}(\Phi,1) = {}& J_6(\Phi) + 8\frac{d}{ds_{\tau 11}}\bigg|_{s_{\tau 11}=1} T_{\mathfrak{d}+}(R_{\mathfrak{d},0}\Psi, s_{\tau 11}-4, -s_{\tau 11}, 0) \\
&+ 8\frac{d}{ds_{\tau 11}}\bigg|_{s_{\tau 11}=1} T_{\mathfrak{d}+}(R_{\mathfrak{d},0}\mathscr{F}_{\mathfrak{d}}\Psi, 8-s_{\tau 11}, s_{\tau 11}+4, 0)), \\
J_{4,(0)}(\Phi,1) = {}& J_7(\Phi) + +2\frac{d}{ds_{\tau 11}}\bigg|_{s_{\tau 11}=1} \frac{T_{W+}(\widetilde{R}_{W,0}\mathscr{F}_W R_{\mathfrak{d}_2}\mathscr{F}_{\mathfrak{d}}\Psi, \frac{1}{3}(5-s_{\tau 11}), 0)}{s_{\tau 11}-5} \\
&+ 2\frac{d}{ds_{\tau 11}}\bigg|_{s_{\tau 11}=1} \frac{T_{W+}(\widetilde{R}_{W,0}R_{\mathfrak{d}_2}\mathscr{F}_{\mathfrak{d}}\Psi, \frac{1}{3}(s_{\tau 11}+4), 0)}{s_{\tau 11}-5}, \\
J_{5,(0)}(\Phi,1) = {}& J_8(\Phi) - 2\frac{d}{ds_{\tau 11}}\bigg|_{s_{\tau 11}=1} \frac{T_{W+}(\widetilde{R}_{W,0}R_{\mathfrak{d}_2}\Psi, -\frac{s_{\tau 11}}{3}, 0)}{s_{\tau 11}-3} \\
&- 2\frac{d}{ds_{\tau 11}}\bigg|_{s_{\tau 11}=1} \frac{T_{W+}(\widetilde{R}_{W,0}\mathscr{F}_W R_{\mathfrak{d}_2}\Psi, \frac{1}{3}(s_{\tau 11}+9), 0)}{s_{\tau 11}-3}.
\end{aligned}
$$

The distributions

$$T_{\mathfrak{d}+}(R_{\mathfrak{d},0}\Psi, s_{\tau 11}-4, -s_{\tau 11}, 0), T_{\mathfrak{d}+}(R_{\mathfrak{d},0}\mathscr{F}_{\mathfrak{d}}\Psi, 8-s_{\tau 11}, s_{\tau 11}+4, 0)$$

are equal to the following integrals

$$\int_{\substack{\mathbb{R}_+^2\times(\mathbb{A}^1/k^\times)^2\\ \lambda_1\geq 1}} \lambda_1^{s_{\tau 11}-5}\lambda_2^{-s_{\tau 11}-3}\Theta_2(R'_{\mathfrak{d},0}\Psi, \underline{\lambda}_1\underline{\lambda}_2^3 q_1, \underline{\lambda}_1\underline{\lambda}_2^{-5}q_2)d^\times\lambda_1 d^\times\lambda_2 d^\times q_1 d^\times q_2,$$

$$\int_{\substack{\mathbb{R}_+^2\times(\mathbb{A}^1/k^\times)^2\\ \lambda_1\geq 1}} \lambda_1^{7-s_{\tau 11}}\lambda_2^{s_{\tau 11}+1}\Theta_2(R'_{\mathfrak{d},0}\mathscr{F}_{\mathfrak{d}}\Psi, \underline{\lambda}_1\underline{\lambda}_2^3 q_1, \underline{\lambda}_1\underline{\lambda}_2^{-5}q_2)d^\times\lambda_1 d^\times\lambda_2 d^\times q_1 d^\times q_2,$$

where the first (resp. second) coordinate of $R'_{\mathfrak{d},0}\Psi, R'_{\mathfrak{d},0}\mathscr{F}_{\mathfrak{d}}\Psi$ corresponds to $x_{2,12}$ (resp. $x_{1,33}$).

For $\widetilde{\Psi}\in\mathscr{S}(Z'_{\mathfrak{d},0\mathbb{A}})$, we define

$$\mathscr{F}'_{\mathfrak{d},0}\widetilde{\Psi}(x_{2,12},x_{1,33}) = \int_{\mathbb{A}^2}\widetilde{\Psi}(y_{2,12},y_{1,33}) < x_{2,12}y_{2,12}+x_{1,33}y_{1,33} > dy_{2,12}dy_{1,33}.$$

Then if Φ is K-invariant, $R'_{\mathfrak{d},0}\mathscr{F}_{\mathfrak{d}}\Psi = \mathscr{F}'_{\mathfrak{d},0}R'_{\mathfrak{d},0}\Psi$. For $\widetilde{\Psi}\in\mathscr{S}(Z'_{\mathfrak{d},0\mathbb{A}})$, let R_1, R_2 be the restrictions to the first coordinate and the second coordinate respectively. Also let

$$\text{(12.6.21)}\qquad \begin{aligned}\mathscr{F}_1\widetilde{\Psi}(x_{2,12}) &= \int_{\mathbb{A}}\widetilde{\Psi}(y_{2,12},0) < x_{2,12}y_{2,12} > dy_{2,12},\\ \mathscr{F}_2\widetilde{\Psi}(x_{1,33}) &= \int_{\mathbb{A}}\widetilde{\Psi}(0,y_{1,33}) < x_{1,33}y_{1,33} > dy_{1,33}.\end{aligned}$$

Lemma (12.6.22) *Suppose that Φ is K-invariant. Then*

$$\begin{aligned}&8T_{\mathfrak{d}+}(R_{\mathfrak{d},0}\Psi, s_{\tau 11}-4, -s_{\tau 11}, 0) + 8T_{\mathfrak{d}+}(R_{\mathfrak{d},0}\mathscr{F}_{\mathfrak{d}}\Psi, 8-s_{\tau 11}, s_{\tau 11}+4, 0)\\ &= \Sigma_2(R'_{\mathfrak{d},0}\Psi, \frac{1}{2}(s_{\tau 11}-7), \frac{1}{2}(s_{\tau 11}-3))\\ &\quad -\frac{2\Sigma_1(R_1R'_{\mathfrak{d},0}\mathscr{F}_{\mathfrak{d}}\Psi, \frac{1}{3}(s_{\tau 11}+1))}{s_{\tau 11}-5} - \frac{2\Sigma_1(\mathscr{F}_2R'_{\mathfrak{d},0}\mathscr{F}_{\mathfrak{d}}\Psi, \frac{1}{5}(s_{\tau 11}+6))}{s_{\tau 11}-9}\\ &\quad +\frac{2\Sigma_1(\mathscr{F}_1R'_{\mathfrak{d},0}\Psi, \frac{1}{3}(s_{\tau 11}+6))}{s_{\tau 11}-3} + \frac{2\Sigma_1(R_2R'_{\mathfrak{d},0}\Psi, \frac{1}{5}(s_{\tau 11}+3))}{s_{\tau 11}-7}.\end{aligned}$$

Proof. By (6.1.2),

$$\begin{aligned}&\Theta_2(R'_{\mathfrak{d},0}\Psi, \underline{\lambda}_1\underline{\lambda}_2^3 q_1, \underline{\lambda}_1\underline{\lambda}_2^{-5}q_2) - \lambda_1^{-2}\lambda_2^2\Theta_2(\mathscr{F}'_{\mathfrak{d},0}R'_{\mathfrak{d},0}\Psi, \underline{\lambda}_1^{-1}\underline{\lambda}_2^{-3}q_1^{-1}, \underline{\lambda}_1^{-1}\underline{\lambda}_2^5 q_2^{-1})\\ &= \lambda_1^{-1}\lambda_2^5\Theta_1(\mathscr{F}_1R'_{\mathfrak{d},0}\mathscr{F}_{\mathfrak{d}}\Psi, \underline{\lambda}_1\underline{\lambda}_2^3 q_1) + \lambda_1^{-2}\lambda_2^2\Theta_1(R_2R'_{\mathfrak{d},0}\mathscr{F}_{\mathfrak{d}}\Psi, \underline{\lambda}_1^{-1}\underline{\lambda}_2^5 q_2^{-1})\\ &\quad -\Theta_1(R_1R'_{\mathfrak{d},0}\Psi, \underline{\lambda}_1\underline{\lambda}_2^3 q_1) - \lambda_1^{-1}\lambda_2^5\Theta_1(\mathscr{F}_2R'_{\mathfrak{d},0}\Psi, \underline{\lambda}_1^{-1}\underline{\lambda}_2^5 q_2^{-1})\\ &= \lambda_1^{-2}\lambda_2^2\Theta_1(R_1R'_{\mathfrak{d},0}\mathscr{F}_{\mathfrak{d}}\Psi, \underline{\lambda}_1^{-1}\underline{\lambda}_2^{-3}q_1^{-1}) + \lambda_1^{-1}\lambda_2^{-3}\Theta_1(\mathscr{F}_2R'_{\mathfrak{d},0}\mathscr{F}_{\mathfrak{d}}\Psi, \underline{\lambda}_1\underline{\lambda}_2^{-5}q_2)\\ &\quad -\lambda_1^{-1}\lambda_2^{-3}\Theta_1(\mathscr{F}_1R'_{\mathfrak{d},0}\Psi, \underline{\lambda}_1^{-1}\underline{\lambda}_2^{-3}q_1^{-1}) - \Theta_1(R_2R'_{\mathfrak{d},0}\Psi, \underline{\lambda}_1\underline{\lambda}_2^{-5}q_2).\end{aligned}$$

We divide the integral according as $\lambda_2 \geq 1$ or $\lambda_2 \leq 1$. If $\lambda_2 \geq 1$, we use the first equation, and if $\lambda_2 \leq 1$, we use the second equation. Then (12.6.22) is equal to the summation of

$$\Sigma_2(R'_{\mathfrak{d},0}\Psi, \frac{1}{2}(s_{\tau 11}-7), \frac{1}{2}(s_{\tau 11}-3))$$

and the integral of the following function

$$\begin{aligned}
&-\frac{\lambda_1^{s_{\tau 11}-6}}{3}\Sigma_{1+}\left((\mathscr{F}_1 R'_{\mathfrak{d},0}\mathscr{F}_{\mathfrak{d}}\Psi)_{\lambda_1}, \frac{-s_{\tau 11}+2}{3}\right)\\
&-\frac{\lambda_1^{s_{\tau 11}-7}}{5}\Sigma_{1+}\left((R_2 R'_{\mathfrak{d},0}\mathscr{F}_{\mathfrak{d}}\Psi)_{\lambda_1^{-1}}, \frac{-s_{\tau 11}-1}{5}\right)\\
&+\frac{\lambda_1^{s_{\tau 11}-5}}{3}\Sigma_{1+}\left((R_1 R'_{\mathfrak{d},0}\Psi)_{\lambda_1}, \frac{-s_{\tau 11}-3}{3}\right)\\
&+\frac{\lambda_1^{s_{\tau 11}-6}}{5}\Sigma_{1+}\left((\mathscr{F}_2 R'_{\mathfrak{d},0}\Psi)_{\lambda_1^{-1}}, \frac{-s_{\tau 11}+2}{5}\right)\\
&-\frac{\lambda_1^{s_{\tau 11}-7}}{3}\Sigma_{1+}\left((R_1 R'_{\mathfrak{d},0}\mathscr{F}_{\mathfrak{d}}\Psi)_{\lambda_1^{-1}}, \frac{s_{\tau 11}+1}{3}\right)\\
&-\frac{\lambda_1^{s_{\tau 11}-6}}{5}\Sigma_{1+}\left((\mathscr{F}_2 R'_{\mathfrak{d},0}\mathscr{F}_{\mathfrak{d}}\Psi)_{\lambda_1}, \frac{s_{\tau 11}+6}{5}\right)\\
&+\frac{\lambda_1^{s_{\tau 11}-6}}{3}\Sigma_{1+}\left((\mathscr{F}_1 R'_{\mathfrak{d},0}\Psi)_{\lambda_1^{-1}}, \frac{s_{\tau 11}+6}{3}\right)\\
&+\frac{\lambda_1^{s_{\tau 11}-5}}{5}\Sigma_{1+}\left((R_2 R'_{\mathfrak{d},0}\Psi)_{\lambda_1}, \frac{s_{\tau 11}+3}{5}\right)
\end{aligned}$$

over $\lambda_1 \leq 1$.

By the principal part formula for the standard L-function in one variable,

$$\begin{aligned}
&\frac{\lambda_1^{s_{\tau 11}-6}}{3}\Sigma_{1+}\left((\mathscr{F}_1 R'_{\mathfrak{d},0}\mathscr{F}_{\mathfrak{d}}\Psi)_{\lambda_1}, \frac{-s_{\tau 11}+2}{3}\right)\\
&\quad+\frac{\lambda_1^{s_{\tau 11}-7}}{3}\Sigma_{1+}\left((R_1 R'_{\mathfrak{d},0}\mathscr{F}_{\mathfrak{d}}\Psi)_{\lambda_1^{-1}}, \frac{s_{\tau 11}+1}{3}\right)\\
&=\frac{\lambda_1^{\frac{4s_{\tau 11}-20}{3}}}{3}\Sigma_1\left(R_1 R'_{\mathfrak{d},0}\mathscr{F}_{\mathfrak{d}}\Psi, \frac{s_{\tau 11}+1}{3}\right)\\
&\quad+\frac{\lambda_1^{s_{\tau 11}-6}\mathscr{F}_1 R'_{\mathfrak{d},0}\mathscr{F}_{\mathfrak{d}}\Psi(0)}{-s_{\tau 11}+2}+\frac{\lambda_1^{s_{\tau 11}-7}R_1 R'_{\mathfrak{d},0}\mathscr{F}_{\mathfrak{d}}\Psi(0)}{s_{\tau 11}+1},\\
&\frac{\lambda_1^{s_{\tau 11}-7}}{5}\Sigma_{1+}\left((R_2 R'_{\mathfrak{d},0}\mathscr{F}_{\mathfrak{d}}\Psi)_{\lambda_1^{-1}}, \frac{-s_{\tau 11}-1}{5}\right)\\
&\quad+\frac{\lambda_1^{s_{\tau 11}-6}}{5}\Sigma_{1+}\left((\mathscr{F}_2 R'_{\mathfrak{d},0}\mathscr{F}_{\mathfrak{d}}\Psi)_{\lambda_1}, \frac{s_{\tau 11}+6}{5}\right)\\
&=\frac{\lambda_1^{\frac{4s_{\tau 11}-36}{5}}}{5}\Sigma_1\left(\mathscr{F}_2 R'_{\mathfrak{d},0}\mathscr{F}_{\mathfrak{d}}\Psi, \frac{s_{\tau 11}+6}{5}\right)\\
&\quad-\frac{\lambda_1^{s_{\tau 11}-7}R_2 R'_{\mathfrak{d},0}\mathscr{F}_{\mathfrak{d}}\Psi(0)}{s_{\tau 11}+1}+\frac{\lambda_1^{s_{\tau 11}-6}\mathscr{F}_2 R'_{\mathfrak{d},0}\mathscr{F}_{\mathfrak{d}}\Psi(0)}{s_{\tau 11}+6},
\end{aligned}$$

$$\frac{\lambda_1^{s_{\tau 11}-5}}{3}\Sigma_{1+}\left((R_1 R'_{\mathfrak{d},0}\Psi)_{\lambda_1}, \frac{-s_{\tau 11}-3}{3}\right)$$
$$+\frac{\lambda_1^{s_{\tau 11}-6}}{3}\Sigma_{1+}\left((\mathscr{F}_1 R'_{\mathfrak{d},0}\Psi)_{\lambda_1^{-1}}, \frac{s_{\tau 11}+6}{3}\right)$$
$$=\frac{\lambda_1^{4s_{\tau 11}-12}}{3}\Sigma_1\left(\mathscr{F}_1 R'_{\mathfrak{d},0}\Psi, \frac{s_{\tau 11}+6}{3}\right)$$
$$-\frac{\lambda_1^{s_{\tau 11}-5}R_1 R'_{\mathfrak{d},0}\Psi(0)}{s_{\tau 11}+3}+\frac{\lambda_1^{s_{\tau 11}-6}\mathscr{F}_1 R'_{\mathfrak{d},0}\Psi(0)}{s_{\tau 11}+6},$$
$$\frac{\lambda_1^{s_{\tau 11}-6}}{5}\Sigma_{1+}\left((\mathscr{F}_2 R'_{\mathfrak{d},0}\Psi)_{\lambda_1^{-1}}, \frac{-s_{\tau 11}+2}{5}\right)$$
$$+\frac{\lambda_1^{s_{\tau 11}-5}}{5}\Sigma_{1+}\left((R_2 R'_{\mathfrak{d},0}\Psi)_{\lambda_1}, \frac{s_{\tau 11}+3}{5}\right)$$
$$=\frac{\lambda_1^{4s_{\tau 11}-28}}{5}\Sigma_1\left(R_2 R'_{\mathfrak{d},0}\Psi, \frac{s_{\tau 11}+3}{5}\right)$$
$$+\frac{\lambda_1^{s_{\tau 11}-6}\mathscr{F}_2 R'_{\mathfrak{d},0}\Psi(0)}{-s_{\tau 11}+2}+\frac{\lambda_1^{s_{\tau 11}-5}R_2 R'_{\mathfrak{d},0}\Psi(0)}{s_{\tau 11}+3}.$$

Note that $\mathscr{F}_1 R'_{\mathfrak{d},0}\mathscr{F}_{\mathfrak{d}}\Psi$ is the Fourier transform of $R_1 R'_{\mathfrak{d},0}\mathscr{F}_{\mathfrak{d}}\Psi$ with respect to the standard bilinear form on $\mathbb{A}$ etc.

If Φ is K-invariant,

$$\begin{aligned}
\mathscr{F}_1 R'_{\mathfrak{d},0}\mathscr{F}_{\mathfrak{d}}\Psi(0) &= \mathscr{F}_2 R'_{\mathfrak{d},0}\Psi(0),\\
R_2 R'_{\mathfrak{d},0}\mathscr{F}_{\mathfrak{d}}\Psi(0) &= R_1 R'_{\mathfrak{d},0}\mathscr{F}_{\mathfrak{d}}\Psi(0),\\
R_1 R'_{\mathfrak{d},0}\Psi(0) &= R_2 R'_{\mathfrak{d},0}\Psi(0),\\
\mathscr{F}_2 R'_{\mathfrak{d},0}\mathscr{F}_{\mathfrak{d}}\Psi(0) &= \mathscr{F}_1 R'_{\mathfrak{d},0}\Psi(0).
\end{aligned}$$

Now (12.6.22) easily follows from these considerations.

Q.E.D.

If Φ is K-invariant, by the principal part formula for the standard L-function in one variable,

$$T_{W+}(\widetilde{R}_{W,0}\mathscr{F}_W R_{\mathfrak{d}_2}\mathscr{F}_{\mathfrak{d}}\Psi, \frac{1}{3}(5-s_{\tau 11}), 0)+T_{W+}(\widetilde{R}_{W,0}R_{\mathfrak{d}_2}\mathscr{F}_{\mathfrak{d}}\Psi, \frac{1}{3}(s_{\tau 11}+4), 0)$$
$$=\Sigma_1(R_1 R'_{\mathfrak{d},0}\mathscr{F}_{\mathfrak{d}}\Psi, \frac{1}{3}(s_{\tau 11}+1))+\frac{3\Sigma_{1,(-1)}(R_{\mathfrak{d}_3}\mathscr{F}_W R_{\mathfrak{d}_2}\mathscr{F}_{\mathfrak{d}}\Psi, 1)}{2-s_{\tau 11}}$$
$$+\frac{3\Sigma_{1,(-1)}(R_{\mathfrak{d}_3}\mathscr{F}_{\mathfrak{d}}\Psi, 1)}{s_{\tau 11}+1},$$
$$T_{W+}(\widetilde{R}_{W,0}R_{\mathfrak{d}_2}\Psi, -\frac{s_{\tau 11}}{3}, 0)+T_{W+}(\widetilde{R}_{W,0}\mathscr{F}_W R_{\mathfrak{d}_2}\Psi, \frac{1}{3}(s_{\tau 11}+9), 0)$$
$$=\Sigma_1(\mathscr{F}_1 R'_{\mathfrak{d},0}\Psi, \frac{1}{3}(s_{\tau 11}+6))+\frac{3\Sigma_{1,(-1)}(R_{\mathfrak{d}_3}\mathscr{F}_W R_{\mathfrak{d}_2}\Psi, 1)}{s_{\tau 11}+6}$$
$$-\frac{3\Sigma_{1,(-1)}(R_{\mathfrak{d}_3}\Psi, 1)}{s_{\tau 11}+3}.$$

Note that if Φ is K-invariant,

$$\Sigma_1(\mathscr{F}_2 R'_{\mathfrak{d},0}\mathscr{F}_{\mathfrak{d}}\Psi, \frac{1}{5}(s_{\tau 11}+6)) = \operatorname*{Res}_{s'=1}\Sigma_2(R_{\mathfrak{d}_1}\Phi_{\mathfrak{p}_{12}}, s', -\frac{1}{5}(s_{\tau 11}+1)),$$
$$\Sigma_1(R_2 R'_{\mathfrak{d},0}\Psi, \frac{1}{5}(s_{\tau 11}+3)) = \operatorname*{Res}_{s'=1}\Sigma_2(R_{\mathfrak{d}_1}\Phi_{\mathfrak{p}_{11}}, s', \frac{1}{5}(s_{\tau 11}+3)).$$

Therefore,

$$\begin{aligned}\sum_{i=3}^{5} J_{i,(0)}(\Phi) = & \sum_{i=6}^{9} J_i(\Phi) \\ & - \frac{1}{15}\Sigma_{2,(-1,1)}(R_{\mathfrak{d}_1}\Phi_{\mathfrak{p}_{11}}, 1, \frac{4}{5}) - \frac{1}{18}\Sigma_{2,(-1,0)}(R_{\mathfrak{d}_1}\Phi_{\mathfrak{p}_{11}}, 1, \frac{4}{5}) \\ & - \frac{1}{20}\Sigma_{2,(-1,1)}(R_{\mathfrak{d}_1}\Phi_{\mathfrak{p}_{12}}, 1, -\frac{2}{5}) + \frac{1}{32}\Sigma_{2,(-1,0)}(R_{\mathfrak{d}_1}\Phi_{\mathfrak{p}_{12}}, 1, -\frac{2}{5}) \\ & - \frac{3}{16}\Sigma_{1,(-1)}(R_{\mathfrak{d}_3}\Psi, 1) + \frac{15}{98}\Sigma_{1,(-1)}(R_{\mathfrak{d}_3}\mathscr{F}_W R_{\mathfrak{d}_2}\Psi, 1) \\ & + \frac{3}{16}\Sigma_{1,(-1)}(R_{\mathfrak{d}_3}\mathscr{F}_{\mathfrak{d}}\Psi, 1) - \frac{15}{8}\Sigma_{1,(-1)}(R_{\mathfrak{d}_3}\mathscr{F}_W R_{\mathfrak{d}_2}\mathscr{F}_{\mathfrak{d}}\Psi, 1),\end{aligned}$$

where

$$J_9(\Phi) = \frac{1}{2}\Sigma_{2,(1,0)}(R'_{\mathfrak{d},0}\Psi, -3, -1) + \frac{1}{2}\Sigma_{2,(0,1)}(R'_{\mathfrak{d},0}\Psi, -3, -1).$$

Also,

$$\begin{aligned}\Sigma_{\mathfrak{p}_{11},(0)}(\Phi, 1) = & -\frac{1}{6}\Sigma_{2,(-1,1)}(R_{\mathfrak{d}_1}\Phi_{\mathfrak{p}_{11}}, 1, \frac{4}{5}) - \frac{1}{6}\Sigma_{2,(0,0)}(R_{\mathfrak{d}_1}\Phi_{\mathfrak{p}_{11}}, 1, \frac{4}{5}), \\ \Sigma_{\mathfrak{p}_{12},(0)}(\Phi, 1) = & -\frac{1}{40}\Sigma_{2,(-1,1)}(R_{\mathfrak{d}_1}\Phi_{\mathfrak{p}_{12}}, 1, -\frac{2}{5}) - \frac{1}{16}\Sigma_{2,(-1,0)}(R_{\mathfrak{d}_1}\Phi_{\mathfrak{p}_{12}}, 1, -\frac{2}{5}) \\ & - \frac{1}{8}\Sigma_{2,(0,0)}(R_{\mathfrak{d}_1}\Phi_{\mathfrak{p}_{12}}, 1, -\frac{2}{5}), \\ \Sigma_{\mathfrak{p}_{31},(0)}(\Phi, 1) = & \frac{33}{8}\Sigma_{1,(-1)}(R_{\mathfrak{d}_3}\mathscr{F}_W R_{\mathfrak{d}_2}\mathscr{F}_{\mathfrak{d}}\Psi, 1) + \frac{3}{4}\Sigma_{1,(0)}(R_{\mathfrak{d}_3}\mathscr{F}_W R_{\mathfrak{d}_2}\mathscr{F}_{\mathfrak{d}}\Psi, 1), \\ \Sigma_{\mathfrak{p}_{32},(0)}(\Phi, 1) = & \frac{3}{32}\Sigma_{1,(-1)}(R_{\mathfrak{d}_3}\Psi, 1) - \frac{3}{8}\Sigma_{1,(0)}(R_{\mathfrak{d}_3}\Psi, 1), \\ \Sigma_{\mathfrak{p}_{33},(0)}(\Phi, 1) = & \frac{3}{8}\Sigma_{1,(-1)}(R_{\mathfrak{d}_3}\mathscr{F}_{\mathfrak{d}}\Psi, 1) + \frac{3}{8}\Sigma_{1,(0)}(R_{\mathfrak{d}_3}\mathscr{F}_{\mathfrak{d}}\Psi, 1), \\ \Sigma_{\mathfrak{p}_{33},(0)}(\Phi, 1) = & \frac{12}{49}\Sigma_{1,(-1)}(R_{\mathfrak{d}_3}\mathscr{F}_W R_{\mathfrak{d}_2}\Psi, 1) + \frac{3}{14}\Sigma_{1,(0)}(R_{\mathfrak{d}_3}\mathscr{F}_W R_{\mathfrak{d}_2}\Psi, 1).\end{aligned}$$

We define

$$\begin{aligned}J_{10}(\Phi) = & \frac{1}{6}\Sigma_{2,(0,0)}(R_{\mathfrak{d}_1}\Phi_{\mathfrak{p}_{11}}, 1, \frac{4}{5}) - \frac{1}{8}\Sigma_{2,(0,0)}(R_{\mathfrak{d}_1}\Phi_{\mathfrak{p}_{12}}, 1, -\frac{2}{5}) \\ & + \frac{1}{10}\Sigma_{2,(-1,1)}(R_{\mathfrak{d}_1}\Phi_{\mathfrak{p}_{11}}, 1, \frac{4}{5}) - \frac{3}{40}\Sigma_{2,(-1,1)}(R_{\mathfrak{d}_1}\Phi_{\mathfrak{p}_{12}}, 1, -\frac{2}{5}) \\ & - \frac{1}{18}\Sigma_{2,(-1,0)}(R_{\mathfrak{d}_1}\Phi_{\mathfrak{p}_{11}}, 1, \frac{4}{5}) + \frac{1}{32}\Sigma_{2,(-1,0)}(R_{\mathfrak{d}_1}\Phi_{\mathfrak{p}_{12}}, 1, -\frac{2}{5}),\end{aligned}$$

$$\begin{aligned}J_{11}(\Phi) =& \frac{9}{16}\Sigma_{1,(-1)}(R_{\mathfrak{d}_3}\mathscr{F}_{\mathfrak{d}}\Psi,1) + \frac{3}{8}\Sigma_{1,(0)}(R_{\mathfrak{d}_3}\mathscr{F}_{\mathfrak{d}}\Psi,1) \\ &+ \frac{9}{4}\Sigma_{1,(-1)}(R_{\mathfrak{d}_3}\mathscr{F}_W R_{\mathfrak{d}_2}\mathscr{F}_{\mathfrak{d}}\Psi,1) + \frac{3}{4}\Sigma_{1,(0)}(R_{\mathfrak{d}_3}\mathscr{F}_W R_{\mathfrak{d}_2}\mathscr{F}_{\mathfrak{d}}\Psi,1) \\ &- \frac{9}{32}\Sigma_{1,(-1)}(R_{\mathfrak{d}_3}\Psi,1) + \frac{3}{8}\Sigma_{1,(0)}(R_{\mathfrak{d}_3}\Psi,1) \\ &- \frac{9}{98}\Sigma_{1,(-1)}(R_{\mathfrak{d}_3}\mathscr{F}_W R_{\mathfrak{d}_2}\Psi,1) - \frac{3}{14}\Sigma_{1,(0)}(R_{\mathfrak{d}_3}\mathscr{F}_W R_{\mathfrak{d}_2}\Psi,1).\end{aligned}$$

By the above considerations, we get

$$\Upsilon_{\mathfrak{d},2,(0)}(\Phi,\omega,1) = \delta_{\#}(\omega)\sum_{i=6}^{11} J_i(\Phi).$$

By the principal part formula (4.2.15),

$$J_2(\Phi) + J_7(\Phi) + J_8(\Phi) + J_{11}(\Phi) = \Sigma_{W,\mathrm{ad}}(R_{\mathfrak{d}_2}\Psi, -\frac{1}{3}) - \frac{1}{2}\Sigma_{W,\mathrm{ad}}(R_{\mathfrak{d}_2}\mathscr{F}_{\mathfrak{d}}\Psi, \frac{5}{3}).$$

Therefore,

$$\begin{aligned}&\Upsilon_{\mathfrak{d},1}(\Phi,\omega) + \Upsilon_{\mathfrak{d},2,(0)}(\Phi,\omega,1) \\ &= J_1(\Phi,\omega) + \delta_{\#}(\omega)(J_6(\Phi) + J_9(\Phi) + J_{10}(\Phi)) \\ &\quad + \delta_{\#}(\omega)(\Sigma_{W,\mathrm{ad}}(R_{\mathfrak{d}_2}\Psi, -\frac{1}{3}) - \frac{1}{2}\Sigma_{W,\mathrm{ad}}(R_{\mathfrak{d}_2}\mathscr{F}_{\mathfrak{d}}\Psi, \frac{5}{3})).\end{aligned}$$

By (6.3.11), we get the following proposition.

Proposition (12.6.23)

$$\Upsilon_{\mathfrak{d},1}(\Phi,\omega) + \Upsilon_{\mathfrak{d},2,(0)}(\Phi,\omega) = 8\delta_{\mathfrak{d}}(\omega)\Sigma_{\mathfrak{d},\mathrm{ad}}(R_{\mathfrak{d}}\Phi_{\mathfrak{p}},\omega_{\mathfrak{d}},\chi_{\mathfrak{d}},-3) + \delta_{\#}(\omega)J_9(\Phi).$$

§12.7 The case $\mathfrak{d} = (\beta_7)$

Let $\mathfrak{p} = (\mathfrak{d}, \mathfrak{s})$ be a path where $\mathfrak{d} = (\beta_7)$. As in previous sections, we only consider such path such that $\mathfrak{s}(1) = 0$. Let $\Psi = R_{\mathfrak{d}}\Phi_{\mathfrak{p}}$ throughout this section. Suppose that $f(q)$ is a function of $q = (q_1, q_2, q_3) \in (\mathbb{A}^1/k^\times)^3$. An easy consideration shows that

$$\begin{aligned}&\int_{(\mathbb{A}^1/k^\times)^5} \omega(\widehat{t}^0) f(\gamma_{1,33}(\widehat{t}^0), \gamma_{2,13}(\widehat{t}^0), \gamma_{2,22}(\widehat{t}^0)) d^\times \widehat{t}^0 \\ &= \delta_{\mathfrak{d}}(\omega)\int_{(\mathbb{A}^1/k^\times)^3} \omega_{\mathfrak{d}}(q) f(q) d^\times q.\end{aligned}$$

Let

$$d(\mu) = d(\mu_1,\mu_2,\mu_3) = a(\mu_1^{-\frac{1}{2}}\mu_3^{-\frac{1}{2}}, \mu_2^{-\frac{1}{3}}\mu_3^{\frac{1}{3}}, \mu_1^{\frac{1}{2}}\mu_2^{\frac{1}{3}}\mu_3^{\frac{1}{6}}; \mu_2^{-\frac{2}{3}}\mu_3^{-\frac{1}{3}}, \mu_2^{\frac{2}{3}}\mu_3^{\frac{1}{3}}).$$

Let $t^0 = d(\mu)\widehat{t}^0$. Then $d^\times t^0 = \frac{1}{6} d^\times \mu d^\times \widehat{t}^0$, and

$$\mu_1 = |\gamma_{1,33}(t^0)|,\ \mu_2 = |\gamma_{2,13}(t^0)|,\ \mu_3 = |\gamma_{2,22}(t^0)|.$$

Let τ, s_τ be as before.

Easy computations show the following lemma.

Lemma (12.7.1)
(1) $\kappa_{\mathfrak{d}2}(t^0) = \mu_1^{-\frac{3}{2}}\mu_2^{-2}\mu_3^{-\frac{3}{2}}$.
(2) $(t^0)^{\tau z-\rho} = \mu_1^{\frac{1}{2}(s_{\tau 11}+s_{\tau 12}+2)}\mu_2^{\frac{1}{3}(s_{\tau 11}+2s_{\tau 2}+3)}\mu_3^{\frac{1}{6}(s_{\tau 11}+3s_{\tau 12}+2s_{\tau 2}+6)}$.

Definition (12.7.2) *Let* $l_{\mathfrak{p}} = (l_{\mathfrak{p},1}, l_{\mathfrak{p},2}, l_{\mathfrak{p},3}, 1)$ *where*
$l_{\mathfrak{p},1}(s_\tau) = \frac{1}{2}(s_{\tau 11} + s_{\tau 12} - 1)$, $l_{\mathfrak{p},2}(s_\tau) = \frac{1}{3}(s_{\tau 11} + 2s_{\tau 2} - 3)$, $l_{\mathfrak{p},3}(s_\tau) = \frac{1}{6}(s_{\tau 11} + 3s_{\tau 12} + 2s_{\tau 2} - 3)$.

By (12.7.1),

$$(12.7.3)\quad \int_{T^0_{\mathbb{A}}/T_k} \omega(t^0)\kappa_{\mathfrak{d}2}(t^0)(t^0)^{\tau z-\rho}\Theta_{Z_{\mathfrak{d}}}(\Psi, t^0)d^\times t^0 = \frac{\delta_{\mathfrak{d}}(\omega)}{6}\Sigma_{3,\mathrm{sub}}(\Psi, l_{\mathfrak{p}}, \omega_{\mathfrak{d}}, s_\tau).$$

Let $D = \{r \mid r_1 + r_2 > 3, r_1 + 2r_3 > 4, \frac{r_1}{6} + \frac{r_2}{2} + \frac{r_3}{3} > \frac{3}{2}\}$. By (12.7.3),

$$\Xi_{\mathfrak{p}}(\Psi, \omega, w) = \delta_{\mathfrak{d}}(\omega)\sum_\tau \left(\frac{1}{2\pi\sqrt{-1}}\right)^3 \int_{\mathrm{Re}(s_\tau)=r} \Sigma_{3,\mathrm{sub}}(\Psi, l_{\mathfrak{p}}, \omega_{\mathfrak{d}}, s_\tau)\widetilde{\Lambda}_\tau(w; s_\tau)ds_\tau,$$

where we choose the contour so that $r \in D \cap D_\tau$.

We define

$$\Upsilon_{\mathfrak{d},2}(\Phi, \omega, s_{\tau 11}) = \frac{\delta_{\mathfrak{d}}(\omega)}{6}\Sigma_3(\Psi, \omega_{\mathfrak{d}}, \frac{s_{\tau 11}}{2}, \frac{1}{3}(s_{\tau 11} - 1), \frac{1}{6}(s_{\tau 11} + 2)).$$

Proposition (12.7.4) *Suppose that* $\tau = \tau_G$ *and* $\delta > 0$ *is a small number. Then by changing* ψ *if necessary,*

$$\Xi_{\mathfrak{p}}(\Phi, \omega, w) \sim \frac{\varrho^2\delta_{\mathfrak{d}}(\omega)}{2\pi\sqrt{-1}} \int_{\mathrm{Re}(s_{\tau 11})=1+\delta} \Upsilon_{\mathfrak{d},2}(\Phi, \omega, s_{\tau 11})\phi_2(s_{\tau 11})\widetilde{\Lambda}_1(w; s_{\tau 11}, 1)ds_{\tau 11}.$$

Proof. We can ignore the case $\tau_2 = 1$ as usual, and assume that $\tau_1 = (1, 2)$.

The pole of $\Sigma_{3,\mathrm{sub}}(\Psi, l_{\mathfrak{p}}, \omega_{\mathfrak{d}}, s_\tau)$ in the set $\{s_\tau \mid \mathrm{Re}(s_{\tau 1}) = (6, 3), \mathrm{Re}(s_{\tau 2}) > 0\}$ is $s_{\tau 2} = 1$. So

$$\begin{aligned}
&\left(\frac{1}{2\pi\sqrt{-1}}\right)^3 \int_{\mathrm{Re}(s_\tau)=(6,3,2)} \Sigma_{3,\mathrm{sub}}(\Psi, l_{\mathfrak{p}}, \omega_{\mathfrak{d}}, s_\tau)\widetilde{\Lambda}_\tau(w; s_\tau)ds_\tau \\
&= \left(\frac{1}{2\pi\sqrt{-1}}\right)^3 \int_{\mathrm{Re}(s_\tau)=(6,3,\frac{1}{2})} \Sigma_{3,\mathrm{sub}}(\Psi, l_{\mathfrak{p}}, \omega_{\mathfrak{d}}, s_\tau)\widetilde{\Lambda}_\tau(w; s_\tau)ds_\tau \\
&\quad + \varrho\left(\frac{1}{2\pi\sqrt{-1}}\right)^2 \int_{\mathrm{Re}(s_{\tau 1})=(6,3)} \Sigma_{3,\mathrm{sub}}(\Psi, l_{\mathfrak{p}}, \omega_{\mathfrak{d}}, s_{\tau 1}, 1)\widetilde{\Lambda}_{1\tau_1}(w; s_{\tau 1})ds_{\tau 1}.
\end{aligned}$$

Since $C > 100$, $\widetilde{L}(6, 3, \frac{1}{2}) \leq 18 + \frac{1}{2}C < 4 + C$. Therefore, we can ignore the first term.

If $\tau_1 = 1, (1, 2)$, or $(1, 3)$, we choose $r_1, r_2 \gg 0$. If $\tau_1 = (1, 2, 3)$, we fix $r_1 = \frac{3}{2}$ and choose $r_2 \gg 0$. If $\tau_1 = (1, 3, 2)$, we fix $r_2 = \frac{3}{2}$ and choose $r_1 \gg 0$. For these cases, $\widetilde{L}(r_1, r_2, 1) < w_0$. Therefore, we only have to consider the case $\tau = \tau_G$.

The pole structure of $\Sigma_{3,\mathrm{sub}}(\Psi, l_{\mathfrak{p}}, \omega_{\mathfrak{d}}, s_{\tau 1}, 1)$ is as follows.

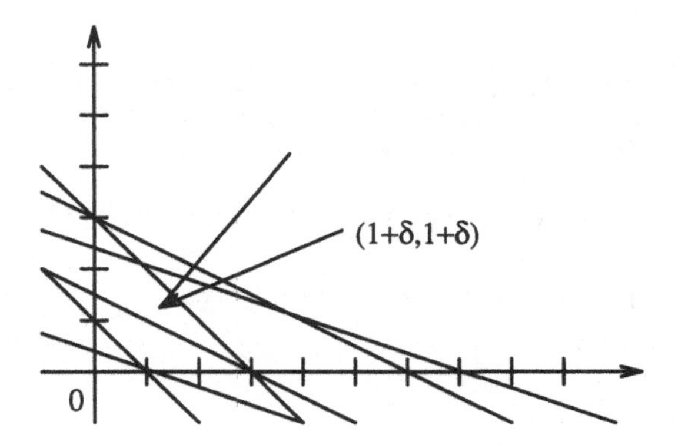

The point $\rho_{1\tau_1}$ is not on the lines $s_{\tau 11} = 4,\ 3s_{\tau 12} + s_{\tau 11} = 1, 7,\ s_{\tau 12} + s_{\tau 11} = 3$. Therefore, by (1.2.1), we can move the contour crossing these lines so that $(r_1, r_2) = (1+\delta, 1+\delta)$. Then we only have to move the contour so that $\mathrm{Re}(s_{\tau 1}) = (1+\delta, 1-\delta_1)$ where $\delta, \delta_1 > 0$ are small and $\delta_1 \delta^{-1} \gg 0$. Since

$$\Sigma_{3,\mathrm{sub}}(\Psi, l_{\mathfrak{p}}, \omega_{\mathfrak{d}}, s_{\tau 11}, 1, 1) = \Upsilon_{\mathfrak{d},2}(\Phi, \omega, s_{\tau 11}),$$

the proposition follows.

Q.E.D.

Let $\Upsilon_{\mathfrak{d},2,(i)}(\Phi, \omega, 1)$ be the i-th coefficient of the Laurent expansion. The following proposition is clear.

Proposition (12.7.5)

$$\begin{aligned}\Upsilon_{\mathfrak{d},2,(0)}(\Phi, \omega) =& \frac{\delta_{\mathfrak{d}}(\omega)}{6} \Sigma_{3,(0,0,0)}(\Psi, \omega_{\mathfrak{d}}, \frac{1}{2}, 0, \frac{1}{2}) \\ &+ \frac{\delta(\omega_1)\delta(\omega_2^2)}{4} \Sigma_{3,(1,-1,0)}(\Psi, (\omega_2, 1, \omega_2), \frac{1}{2}, 0, \frac{1}{2}) \\ &+ \frac{\delta(\omega_1)\delta(\omega_2^2)}{12} \Sigma_{3,(0,-1,1)}(\Psi, (\omega_2, 1, \omega_2), \frac{1}{2}, 0, \frac{1}{2}).\end{aligned}$$

§12.8 The case $\mathfrak{d} = (\beta_8)$

Let $\mathfrak{d}_0 = (\beta_8)$, and $\mathfrak{d}_i = (\beta_8, \beta_{8,i})$ for $i = 1, \cdots, 6$. Let $\mathfrak{p}_0 = (\mathfrak{d}_0, \mathfrak{s}_0)$. Let $\mathfrak{p}_{i1} = (\mathfrak{d}_i, \mathfrak{s}_{i1}), \mathfrak{p}_{i2} = (\mathfrak{d}_i, \mathfrak{s}_{i2})$ where $\mathfrak{s}_{i1}(2) = 0, \mathfrak{s}_{i2}(2) = 1$ for $i = 1, 2, 3$. As in previous sections, we only consider such paths such that $\mathfrak{s}_0(1) = \mathfrak{s}_{i1}(1) = \mathfrak{s}_{i2}(1) = 0$. Let $\Psi = R_{\mathfrak{d}_0} \Phi_{\mathfrak{p}_0}$ throughout this section.

Let

$$\begin{aligned} d_1(\lambda_1, \lambda_2) &= a(\underline{\lambda}_1^{-4} \underline{\lambda}_2^{-2}, \underline{\lambda}_1^2 \underline{\lambda}_2, \underline{\lambda}_1^2 \underline{\lambda}_2; \underline{\lambda}_1^{-3} \underline{\lambda}_2^2, \underline{\lambda}_1^3 \underline{\lambda}_2^{-2}), \\ d_2(\lambda_1, \lambda_2, \lambda_3) &= a(\underline{\lambda}_1^{-4} \underline{\lambda}_2^{-2}, \underline{\lambda}_1^2 \underline{\lambda}_2 \underline{\lambda}_3^{-1}, \underline{\lambda}_1^2 \underline{\lambda}_2 \underline{\lambda}_3; \underline{\lambda}_1^{-3} \underline{\lambda}_2^2, \underline{\lambda}_1^3 \underline{\lambda}_2^{-2}). \end{aligned}$$

Let $Z_{\mathfrak{d}} = V_1 \oplus V_2$ where $V_{1k} \cong \mathrm{Sym}^2 k^2$, $V_{2k} \cong k^2$. Then $d_1(\lambda_1, \lambda_2)$ acts on $V_{1\mathbb{A}}, V_{2\mathbb{A}}$ by multiplication by $\underline{\lambda}_1\underline{\lambda}_2^4, \underline{\lambda}_1\underline{\lambda}_2^{-3}$ respectively. Let $\chi_{\mathfrak{d}}, \kappa_{V_1}$ etc. be as in §11.1.

(a) $\Xi_{\mathfrak{p}_0}(\Phi, \omega, w)$

Let $g_{\mathfrak{d}} = d_1(\lambda_1, \lambda_2)g_{\mathfrak{d}}^0$ where $g_{\mathfrak{d}}^0$ is as in the second element of (11.1.3). It is easy to see that $dg_{\mathfrak{d}} = 14 d^\times \lambda_1 d^\times \lambda_2 dg_{\mathfrak{d}}^0$, and $d_1(\lambda_1, \lambda_2, \lambda_3)^{-2\rho} = \lambda_1^{18}\lambda_2^2\lambda_3^2$.

Let τ, s_τ be as before. An easy computation shows that

$$d_2(\lambda_1, \lambda_2, \lambda_3)^{\tau z + \rho} = \lambda_1^{2s_{\tau 11}+4s_{\tau 12}+3s_{\tau 2}-9}\lambda_2^{s_{\tau 11}+2s_{\tau 12}-2s_{\tau 2}-1}\lambda_3^{s_{\tau 11}-1}. \tag{12.8.1}$$

Easy computations show the following lemma.

Lemma (12.8.2)

(1) $\sigma_{\mathfrak{p}_0}(d_1(\lambda_1, \lambda_2)) = \lambda_1^{-3}\lambda_2^2$.

(2) $\kappa_{\mathfrak{d}_0 1}(d_1(\lambda_1, \lambda_2)) = \lambda_1^{-5}\lambda_2^{-6}$.

(3) $\kappa_{\mathfrak{d}_0 2}(d_1(\lambda_1, \lambda_2)) = \lambda_1^{-21}$.

Since $Z_{\mathfrak{d}k}^{\mathrm{ss}} = Z_{\mathfrak{d}k}^{s}$, the following proposition follows from (3.5.13) and (12.8.1).

Proposition (12.8.3)

(1) $\qquad \Xi_{\mathfrak{p}_0+}(\Phi, \omega, w) \sim 14 C_G \Lambda(w; \rho)\delta_{\mathfrak{d}_0}(\omega)\Sigma_{\mathfrak{d}_0+}(\Psi, \omega_{\mathfrak{d}_0}, \chi_{\mathfrak{d}_0}, -3),$

(2) $\qquad \widehat{\Xi}_{\mathfrak{p}_0+}(\Phi, \omega, w) \sim 14 C_G \Lambda(w; \rho)\delta_{\mathfrak{d}_0}(\omega)\Sigma_{\mathfrak{d}_0+}(\mathscr{F}_{\mathfrak{d}_0}\Psi, \omega_{\mathfrak{d}_0}^{-1}, \kappa_{\mathfrak{d}_0 1}^{-1}\chi_{\mathfrak{d}_0}^{-1}, 8).$

(b) $\Xi_{\mathfrak{p}_{1i}}(\Phi, \omega, w)$

Let $\mathfrak{d} = \mathfrak{d}_1$. In (b)–(d), $t^0 = d_2(\lambda_1, \lambda_2, \lambda_3)\widehat{t}^0$. Let

$$\mu_1 = |\gamma_{1,33}(t^0)| = \lambda_1\lambda_2^4\lambda_3^2, \ \mu_2 = |\gamma_{2,12}(t^0)| = \lambda_1\lambda_2^{-3}\lambda_3^{-1}.$$

Then $\lambda_2 = \lambda_1^{\frac{3}{2}}\mu_1^{-\frac{1}{2}}\mu_2^{-1}$, $\lambda_3 = \lambda_1^{-\frac{7}{2}}\mu_1^{\frac{3}{2}}\mu_2^2$. It is easy to see that

$$d^\times t^0 = 7 d^\times \lambda_1 d^\times \mu_1 d^\times \mu_2 d^\times \widehat{t}^0.$$

Easy computations show the following two lemmas.

Lemma (12.8.4)

(1) $e_{\mathfrak{p}_{i1}}(d_2(\lambda_1, \lambda_2, \lambda_3)) = \lambda_1$, $e_{\mathfrak{p}_{i2}}(d_2(\lambda_1, \lambda_2, \lambda_3)) = \lambda_1^{-1}$ for $i = 1, 2, 3$.

Lemma (12.8.5)

(1) $\sigma_{\mathfrak{p}_{11}}(d_2(\lambda_1, \lambda_2, \lambda_3)) = \lambda_1^{-4}\lambda_2^5\lambda_3 = \mu_1^{-1}\mu_2^{-3}$.

(2) $\sigma_{\mathfrak{p}_{12}}(d_2(\lambda_1, \lambda_2, \lambda_3)) = \lambda_1^7\lambda_2^7\lambda_3 = \lambda_1^{14}\mu_1^{-2}\mu_2^{-5}$.

(3) $\theta_{\mathfrak{p}_{11}}(d_2(\lambda_1, \lambda_2, \lambda_3))^{\tau z + \rho} = \lambda_1^{7s_{\tau 12}-7}\mu_1^{s_{\tau 11}-s_{\tau 12}+s_{\tau 2}-1}\mu_2^{s_{\tau 11}-2s_{\tau 12}+2s_{\tau 2}-1}$.

(4) $\theta_{\mathfrak{p}_{12}}(d_2(\lambda_1, \lambda_2, \lambda_3))^{\tau z + \rho} = \lambda_1^{-7s_{\tau 11}-7s_{\tau 12}+14}\mu_1^{2s_{\tau 11}+s_{\tau 12}-s_{\tau 2}-2}\mu_2^{3s_{\tau 11}+2s_{\tau 12}-2s_{\tau 2}-3}$.

Definition (12.8.6) *Let* $l_{\mathfrak{p}_{1i}} = (l_{\mathfrak{p}_{1i},1}, l_{\mathfrak{p}_{1i},2}, l_{\mathfrak{p}_{1i},3})$ *where*

(1) $l_{\mathfrak{p}_{11},1}(s_\tau) = -s_{\tau 11} + s_{\tau 12} + s_{\tau 2} - 2$, $l_{\mathfrak{p}_{11},2}(s_\tau) = -2s_{\tau 11} + s_{\tau 12} + 2s_{\tau 2} - 4$, $l_{\mathfrak{p}_{11},3}(s_\tau) = s_{\tau 11} - 1$,

(2) $l_{\mathfrak{p}_{12},1}(s_\tau) = s_{\tau 11} + 2s_{\tau 12} - s_{\tau 2} - 4$, $l_{\mathfrak{p}_{12},2}(s_\tau) = 2s_{\tau 11} + 3s_{\tau 12} - 2s_{\tau 2} - 8$, $l_{\mathfrak{p}_{12},3}(s_\tau) = s_{\tau 11} + s_{\tau 12} - 4$.

By (12.8.1), (12.8.4), (12.8.5), and exchanging $s_{\tau 11}, s_{\tau 12}$,

$$\Xi_{\mathfrak{p}_1 i}(\Phi,\omega,w) = \delta_{\#}(\omega)\sum_{\tau}\left(\frac{1}{2\pi\sqrt{-1}}\right)^3 \int_{\substack{\mathrm{Re}(s_\tau)=r \\ r_1\gg r_2,\, r_3>1}} \Sigma_{2,\mathrm{sub}}(R_{\mathfrak{d}_1}\Phi_{\mathfrak{p}_{1i}}, l_{\mathfrak{p}_{1i}}, s_\tau)\widetilde{\Lambda}_\tau(w;s_\tau)ds_\tau,$$

where the first (resp. second) coordinate of $R_{\mathfrak{d}_1}\Phi_{\mathfrak{p}_{1i}}$ corresponds to $x_{1,33}$ (resp. $x_{2,12}$).

We define

$$\Sigma_{\mathfrak{p}_{11}}(\Phi, s_{\tau 11}) = \frac{\Sigma_2(R_{\mathfrak{d}_1}\Phi_{\mathfrak{p}_{11}}, -s_{\tau 11}, -2s_{\tau 11}-1)}{s_{\tau 11}-1},$$

where the first (resp. second) coordinate of $R_{\mathfrak{d}_1}\Phi_{\mathfrak{p}_{11}}$ corresponds to $x_{1,33}$ (resp. $x_{2,12}$).

Proposition (12.8.7) *Suppose that $\tau=\tau_G$ and $\delta>0$ is a small number. Then by changing ψ if necessary,*

(1) $\Xi_{\mathfrak{p}_{11}}(\Phi,\omega,w) \sim \dfrac{\varrho^2\delta_{\#}(\omega)}{2\pi\sqrt{-1}}\displaystyle\int_{\mathrm{Re}(s_{\tau 11})=1+\delta} \Sigma_{\mathfrak{p}_{11}}(\Phi, s_{\tau 11})\phi_2(s_{\tau 11})\widetilde{\Lambda}_1(w; s_{\tau 11}, 1)ds_{\tau 11},$

(2) $\Xi_{\mathfrak{p}_{12}}(\Phi,\omega,w) \sim -C_G\Lambda(w;\rho)\dfrac{\delta_{\#}(\omega)\Sigma_2(R_{\mathfrak{d}_1}\Phi_{\mathfrak{p}_{12}}, -2, -5)}{2},$

where in (2) the first (resp. second) coordinate of $R_{\mathfrak{d}_1}\Phi_{\mathfrak{p}_{11}}$ corresponds to $x_{1,33}$ (resp. $x_{2,12}$).

Proof. The proof of this proposition is similar to (12.6.9). The only difference is that we have to check that the points $(2,-1), (-1,2), (1,1)$ are not on the lines

$$s_{\tau 11}-s_{\tau 12}=1,2,\ s_{\tau 11}-2s_{\tau 12}=2,3,\ 2s_{\tau 11}+s_{\tau 12}=6,\ 3s_{\tau 11}+2s_{\tau 12}=10,11,\ s_{\tau 11}=3$$

in order to use (3.6.1).

Q.E.D.

(c) $\Xi_{\mathfrak{p}_{2i}}(\Phi,\omega,w)$

Let $\mathfrak{d}=\mathfrak{d}_2$. Let

$$\mu_1 = |\gamma_{1,23}(t^0)| = \lambda_1\lambda_2^4,\ \mu_2 = |\gamma_{2,13}(t^0)| = \lambda_1\lambda_2^{-3}\lambda_3.$$

Then $\lambda_2 = \lambda_1^{-\frac14}\mu_1^{\frac14}$, $\lambda_3 = \lambda_1^{-\frac74}\mu_1^{\frac34}\mu_2$. It is easy to see that

$$d^\times t^0 = \frac{7}{2}d^\times\lambda_1 d^\times\mu_1 d^\times\mu_2 d^\times\widehat{t}^0.$$

Easy computations show the following lemma.

Lemma (12.8.8)

(1) $\sigma_{\mathfrak{p}_{21}}(d_2(\lambda_1,\lambda_2,\lambda_3)) = \lambda_1^{-4}\lambda_2^{-2} = \lambda_1^{-\frac72}\mu_1^{-\frac12}$.

(2) $\sigma_{\mathfrak{p}_{22}}(d_2(\lambda_1,\lambda_2,\lambda_3)) = \lambda_1^{7}$.

(3) $\theta_{\mathfrak{p}_{21}}(d_2(\lambda_1,\lambda_2,\lambda_3))^{\tau z+\rho} = \lambda_1^{\frac72(s_{\tau 12}+s_{\tau 2}-2)}\mu_1^{\frac12(2s_{\tau 11}+s_{\tau 12}-s_{\tau 2}-2)}\mu_2^{s_{\tau 11}-1}$.

(4) $\theta_{\mathfrak{p}_{22}}(d_2(\lambda_1,\lambda_2,\lambda_3))^{\tau z+\rho}=\lambda_1^{\frac{7}{2}(-s_{\tau 11}-s_{\tau 12}-s_{\tau 2}+3)}\mu_1^{\frac{1}{2}(s_{\tau 11}-s_{\tau 12}+s_{\tau 2}-1)}\mu_2^{s_{\tau 11}-1}$.

Definition (12.8.9) *Let* $l_{\mathfrak{p}_{2i}}=(l_{\mathfrak{p}_{2i},1},l_{\mathfrak{p}_{2i},2},l_{\mathfrak{p}_{2i},3})$ *where*
(1) $l_{\mathfrak{p}_{21},1}(s_\tau)=\frac{1}{2}(2s_{\tau 11}+s_{\tau 12}-s_{\tau 2}-3)$, $l_{\mathfrak{p}_{21},2}(s_\tau)=s_{\tau 11}-1$, $l_{\mathfrak{p}_{21},3}(s_\tau)=s_{\tau 12}+s_{\tau 2}-3$,
(2) $l_{\mathfrak{p}_{22},1}(s_\tau)=\frac{1}{2}(s_{\tau 11}-s_{\tau 12}+s_{\tau 2}-1)$, $l_{\mathfrak{p}_{22},2}(s_\tau)=s_{\tau 11}-1$, $l_{\mathfrak{p}_{22},3}(s_\tau)=s_{\tau 11}+s_{\tau 12}+s_{\tau 2}-5$.

By (12.8.8), and a similar argument as in previous sections,

$$\begin{aligned}&\Xi_{\mathfrak{p}_{2i}}(\Phi,\omega,w)\\&=\delta_\#(\omega)\sum_\tau\left(\frac{1}{2\pi\sqrt{-1}}\right)^3\int_{\substack{\mathrm{Re}(s_\tau)=r\\ r_1\gg r_2,\ r_3>1}}\Sigma_{2,\mathrm{sub}}(R_{\mathfrak{d}_2}\Phi_{\mathfrak{p}_{2i}},l_{\mathfrak{p}_{2i}},s_\tau)\widetilde{\Lambda}_\tau(w;s_\tau)ds_\tau,\end{aligned}$$

where the first (resp. second) coordinate of $R_{\mathfrak{d}_2}\Phi_{\mathfrak{p}_{2i}}$ corresponds to $x_{1,23}$ (resp. $x_{2,13}$) for $i=1,2$

Let $\Phi^{(1)}_{\mathfrak{p}_{21}}\in\mathscr{S}(\mathbb{A})$ be a function defined by the formula

$$\Phi^{(1)}_{\mathfrak{p}_{21}}(x_{1,23})=\int_{\mathbb{A}^2}\Psi(0,x_{1,23},x_{1,33},0,x_{2,13})dx_{1,33}dx_{2,13}.$$

We define $\Phi^{(1)}_{\mathfrak{p}_{22}}$ similarly for $R_{\mathfrak{d}_2}\Phi_{\mathfrak{p}_{22}}$. If Φ is K-invariant, $\Phi^{(1)}_{\mathfrak{p}_{21}}$ is the Fourier transform of $\Phi^{(1)}_{\mathfrak{p}_{22}}$ with respect to the character $<\ >$.

We define

$$\begin{aligned}\Sigma^{(1)}_{\mathfrak{p}_{21}}(\Phi,s_{\tau 11})&=\Sigma_2(R_{\mathfrak{d}_2}\Phi_{\mathfrak{p}_{21}},\frac{1}{2}(2s_{\tau 11}-3),s_{\tau 11}-1),\\ \Sigma^{(2)}_{\mathfrak{p}_{21}}(\Phi,s_{\tau 11})&=\Sigma_2(R_{\mathfrak{d}_2}\Phi_{\mathfrak{p}_{21}},s_{\tau 11}-1,s_{\tau 11}-1),\\ \Sigma_{\mathfrak{p}_{22}}(\Phi,s_{\tau 11})&=\frac{\Sigma_2(R_{\mathfrak{d}_2}\Phi_{\mathfrak{p}_{22}},\frac{1}{2}(s_{\tau 11}-1),s_{\tau 11}-1)}{s_{\tau 11}-3}.\end{aligned}$$

Let $\mathfrak{W}'=\{(1,2,3),(1,3,2),(1,3)\}$.

Proposition (12.8.10) *Suppose that* $\tau=\tau_G$ *and* $\delta>0$ *is a small number. Then by changing* ψ *if necessary,*

$$\begin{aligned}(1)\quad&\Xi_{\mathfrak{p}_{21}}(\Phi,\omega,w)\\&\sim-\frac{\varrho^2\delta_\#(\omega)}{2\pi\sqrt{-1}}\int_{\mathrm{Re}(s_{\tau 11})=1+\delta}\Sigma^{(1)}_{\mathfrak{p}_{21}}(\Phi,s_{\tau 11})\phi_2(s_{\tau 11})\widetilde{\Lambda}_1(w;s_{\tau 11},1)ds_{\tau 11}\\&+\sum_{\substack{\tau'=(\tau_1',(1,2))\\ \tau_1'\in\mathfrak{W}'}}\frac{\varrho\delta_\#(\omega)}{2\pi\sqrt{-1}}\int_{\mathrm{Re}(s_{\tau' 11})=r_1}\Sigma^{(2)}_{\mathfrak{p}_{21}}(\Phi,s_{\tau' 11})\widetilde{\Lambda}_{1\tau_1'}(w;s_{\tau' 11},2)ds_{\tau' 11}\\&+\sum_{\substack{\tau'=(\tau_1',(1,2))\\ \tau_1'\in\mathfrak{W}'}}\frac{\varrho\delta_\#(\omega)}{2\pi\sqrt{-1}}\int_{\mathrm{Re}(s_{\tau' 12})=r_2}\frac{\Sigma_1(\Phi^{(1)}_{\mathfrak{p}_{21}},\frac{s_{\tau' 12}}{2})}{s_{\tau' 12}-2}\widetilde{\Lambda}_{1\tau_1'}(w;2,s_{\tau' 12})ds_{\tau' 12},\end{aligned}$$

(2) $\Xi_{\mathfrak{p}_{22}}(\Phi,\omega,w)$

$$\sim \frac{\varrho^2\delta_{\#}(\omega)}{2\pi\sqrt{-1}}\int_{\mathrm{Re}(s_{\tau 11})=1+\delta}\Sigma_{\mathfrak{p}_{22}}(\Phi,s_{\tau 11})\phi_2(s_{\tau 11})\widetilde{\Lambda}_1(w;s_{\tau 11},1)ds_{\tau 11}$$
$$+\sum_{\substack{\tau'=(\tau_1',(1,2))\\ \tau_1'\in\mathfrak{W}'}}\frac{\varrho\delta_{\#}(\omega)}{2\pi\sqrt{-1}}\int_{\mathrm{Re}(s_{\tau'12})=r_2}\frac{\Sigma_1(\Phi^{(1)}_{\mathfrak{p}_{22}},1-\frac{s_{\tau'12}}{2})}{s_{\tau'12}-2}\widetilde{\Lambda}_{1\tau_1'}(w;2,s_{\tau'12})ds_{\tau'12}.$$

Proof. We can ignore the case $\tau_2=1$ as usual, and assume that $\tau_2=(1,2)$. Let $r=(r_1,r_2,r_3)=(10,2,2)$. By the usual argument,

$$\left(\frac{1}{2\pi\sqrt{-1}}\right)^3\int_{\mathrm{Re}(s_\tau)=r}\Sigma_{2,\mathrm{sub}}(R_{\mathfrak{d}_2}\Phi_{\mathfrak{p}_{2i}},l_{\mathfrak{p}_{2i}},s_\tau)\widetilde{\Lambda}_\tau(w;s_\tau)ds_\tau$$
$$\sim\varrho\left(\frac{1}{2\pi\sqrt{-1}}\right)^2\int_{\mathrm{Re}(s_{\tau 1})=(r_1,r_2)}\Sigma_{2,\mathrm{sub}}(R_{\mathfrak{d}_2}\Phi_{\mathfrak{p}_{2i}},l_{\mathfrak{p}_{2i}},s_{\tau 1},1)\widetilde{\Lambda}_{1\tau_1}(w;s_{\tau 1})ds_{\tau 1}$$

for $i=1,2$.

The pole structure of $\Sigma_{2,\mathrm{sub}}(R_{\mathfrak{d}_2}\Phi_{\mathfrak{p}_{2i}},l_{\mathfrak{p}_{2i}},s_{\tau 1},1)$ for $i=1,2$ is as follows.

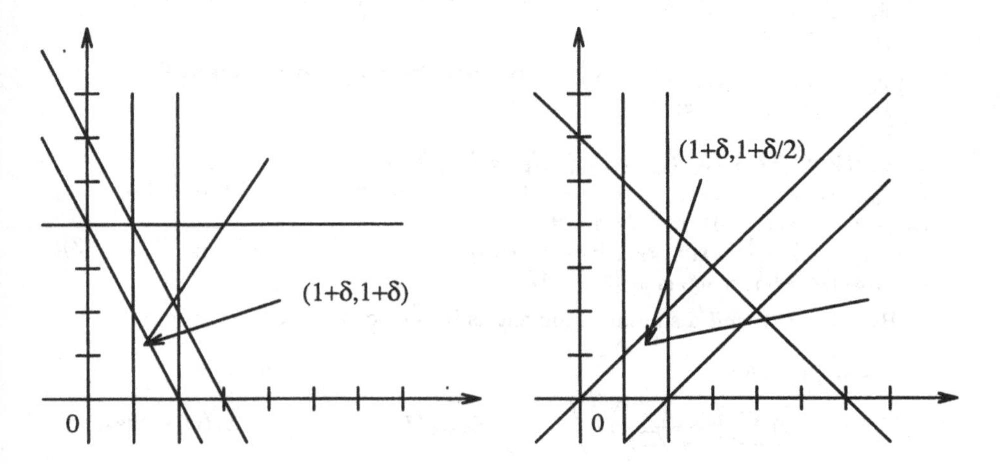

If $\tau_1=1,(1,2),(2,3)$, we choose $r_1\gg r_2\gg 0$. Then $\widetilde{L}(r_1,r_2,1)\ll 0$. So we can ignore these cases. The point $\rho_{1\tau_1}$ is $(2,-1),(-1,2),(1,1)$ for $\tau_1=(1,2,3),(1,3,2),(1,3)$. The point $\rho_{1\tau_1}$ is not on the lines $2s_{\tau 11}+s_{\tau 12}=4,6$, $s_{\tau 11}=s_{\tau 12}+2$, $s_{\tau 11}+s_{\tau 12}=4$ for these cases. So we can move the contour crossing these lines by (3.6.1). We move the contour to $\mathrm{Re}(s_{\tau 1})=(1+\delta,1+\delta)$ for $i=1$, and $\mathrm{Re}(s_{\tau 1})=(1+\delta,1+\frac{\delta}{2})$ for $i=2$. In this process, we have to cross the lines $s_{\tau 11}=2, s_{\tau 12}=2$, and the condition of (3.6.1) is not satisfied for these lines. When we cross these lines, we pick up the third term of (1), and the second term of (2).

If $\tau_1=(1,2,3),(1,3,2)$,

$$\left(\frac{1}{2\pi\sqrt{-1}}\right)^2\int_{\mathrm{Re}(s_{\tau 1})=(1+\delta,1+\delta)}\Sigma_{2,\mathrm{sub}}(R_{\mathfrak{d}_2}\Phi_{\mathfrak{p}_{21}},l_{\mathfrak{p}_{21}},s_{\tau 1},1)\widetilde{\Lambda}_{1\tau_1}(w;s_{\tau 1})ds_{\tau 1}\sim 0,$$

because $\widetilde{L}(1+\delta,1+\delta,1)<w_0$ for these cases. Similarly, we can ignore the cases $\tau_1=(1,2,3),(1,3,2)$ for $i=2$. Then we only have to move the contour so that $(r_1,r_2)=(1+\delta,1-\delta_1)$ where $\delta_1>0$ is a small number such that $\delta_1\delta^{-1}\gg 0$. This proves the proposition.

Q.E.D.

(d) $\Xi_{\mathfrak{p}_{3i}}(\Phi,\omega,w)$

Let $\mathfrak{d}=\mathfrak{d}_3$. Let

$$\mu_1=|\gamma_{1,33}(t^0)|=\lambda_1\lambda_2^4\lambda_3^2,\ \mu_2=|\gamma_{2,13}(t^0)|=\lambda_1\lambda_2^{-3}\lambda_3.$$

Then $\lambda_2=\lambda_1^{\frac{1}{10}}\mu_1^{\frac{1}{10}}\mu_2^{-\frac{1}{5}}$, $\lambda_3=\lambda_1^{-\frac{7}{10}}\mu_1^{\frac{3}{10}}\mu_2^{\frac{2}{5}}$. It is easy to see that

$$d^\times t^0=\frac{7}{5}d^\times\lambda_1 d^\times\mu_1 d^\times\mu_2 d^\times\widehat{t}^0.$$

Easy computations show the following lemma.

Lemma (12.8.11)

(1) $\sigma_{\mathfrak{p}_{31}}(d_2(\lambda_1,\lambda_2,\lambda_3))=\lambda_1^{-3}\lambda_2^2\lambda_3^2=\lambda_1^{-\frac{21}{5}}\mu_1^{\frac{4}{5}}\mu_2^{\frac{2}{5}}$.

(2) $\sigma_{\mathfrak{p}_{32}}(d_2(\lambda_1,\lambda_2,\lambda_3))=\lambda_1^8\lambda_2^4\lambda_3^2=\lambda_1^7\mu_1$.

(3) $\theta_{\mathfrak{p}_{31}}(d_2(\lambda_1,\lambda_2,\lambda_3))^{\tau z+\rho}=\lambda_1^{\frac{7}{5}(s_{\tau 11}+3s_{\tau 12}+2s_{\tau 2}-6)}\mu_1^{\frac{1}{5}(2s_{\tau 11}+s_{\tau 12}-s_{\tau 2}-2)}$
$\times\mu_2^{\frac{1}{5}(s_{\tau 11}-2s_{\tau 12}+2s_{\tau 2}-1)}$.

(4) $\theta_{\mathfrak{p}_{32}}(d_2(\lambda_1,\lambda_2,\lambda_3))^{\tau z+\rho}=\lambda_1^{\frac{7}{5}(-2s_{\tau 11}-3s_{\tau 12}-2s_{\tau 2}+7)}\mu_1^{\frac{1}{5}(s_{\tau 11}-s_{\tau 12}+s_{\tau 2}-1)}$
$\times\mu_2^{\frac{1}{5}(3s_{\tau 11}+2s_{\tau 12}-2s_{\tau 2}-3)}$.

Definition (12.8.12) $l_{\mathfrak{p}_{3i}}=(l_{\mathfrak{p}_{3i},1},l_{\mathfrak{p}_{3i},2},l_{\mathfrak{p}_{3i},3})$ *where*

(1) $l_{\mathfrak{p}_{31},1}(s_\tau)=\frac{1}{5}(2s_{\tau 11}+s_{\tau 12}-s_{\tau 2}+2)$, $l_{\mathfrak{p}_{31},2}(s_\tau)=\frac{1}{5}(s_{\tau 11}-2s_{\tau 12}+2s_{\tau 2}+1)$, $l_{\mathfrak{p}_{31},3}(s_\tau)=s_{\tau 11}+3s_{\tau 12}+2s_{\tau 2}-9$,

(2) $l_{\mathfrak{p}_{32},1}(s_\tau)=\frac{1}{5}(s_{\tau 11}-s_{\tau 12}+s_{\tau 2}+4)$, $l_{\mathfrak{p}_{32},2}(s_\tau)=\frac{1}{5}(3s_{\tau 11}+2s_{\tau 12}-2s_{\tau 2}-3)$, $l_{\mathfrak{p}_{32},3}(s_\tau)=2s_{\tau 11}+3s_{\tau 12}+2s_{\tau 2}-12$.

By (12.8.11), and a similar argument as in previous sections,

$$\begin{aligned}&\Xi_{\mathfrak{p}_{3i}}(\Phi,\omega,w)\\&=\delta_\#(\omega)\sum_\tau\left(\frac{1}{2\pi\sqrt{-1}}\right)^3\int_{\mathrm{Re}(s_\tau)=r}\Sigma_{2,\mathrm{sub}}(R_{\mathfrak{d}_3}\Phi_{\mathfrak{p}_{3i}},l_{\mathfrak{p}_{3i}},s_\tau)\widetilde{\Lambda}_\tau(w;s_\tau)ds_\tau,\end{aligned}$$

where the first (resp. second) coordinate of $R_{\mathfrak{d}}\Phi_{\mathfrak{p}_{3i}}$ corresponds to $x_{1,33}$ (resp. $x_{2,23}$) for $i=1,2$.

Let

$$\Sigma_{\mathfrak{p}_{32}}(\Phi,s_{\tau 11})=\frac{\Sigma_2(R_{\mathfrak{d}_3}\Phi_{\mathfrak{p}_{32}},\frac{1}{5}(s_{\tau 11}+4),\frac{1}{5}(3s_{\tau 11}-3))}{2s_{\tau 11}-7}.$$

Proposition (12.8.13) *Suppose that $\tau=\tau_G$ and $\delta>0$ is a small number. Then by changing ψ if necessary,*

$$\Xi_{\mathfrak{p}_{31}}(\Phi,\omega,w)\sim -C_G\Lambda(w;\rho)\frac{\delta_\#(\omega)\Sigma_2(R_{\mathfrak{d}_3}\Phi_{\mathfrak{p}_{31}},\frac{4}{5},\frac{2}{5})}{3},$$

$$\Xi_{\mathfrak{p}_{32}}(\Phi,\omega,w)\sim\frac{\varrho^2\delta_\#(\omega)}{2\pi\sqrt{-1}}\int_{\mathrm{Re}(s_{\tau 11})=1+\delta}\Sigma_{\mathfrak{p}_{32}}(\Phi,s_{\tau 11})\phi_2(s_{\tau 11})\widetilde{\Lambda}_1(w;s_{\tau 11},1)ds_{\tau 11}.$$

Proof. The proof of this proposition is similar to (12.6.9). The only difference is that we have to check that the points $(2,-1),(-1,2),(1,1)$ are not on lines $s_{\tau 11} = 2s_{\tau 12}+2,\ s_{\tau 11}+3s_{\tau 12}=7,\ s_{\tau 11}+2s_{\tau 12}=4,\ 2s_{\tau 11}+3s_{\tau 12}=10,\ 3s_{\tau 11}+2s_{\tau 12}=10$ in order to use (3.6.1).

Q.E.D.

(e) $\Xi_{\mathfrak{p},\mathfrak{d}_4,i}(\Phi,\omega,w)$

Let $\mathfrak{d}=\mathfrak{d}_0, \mathfrak{d}'=\mathfrak{d}_4$, and $\mathfrak{p}=\mathfrak{p}_0$.

Let $g_{\mathfrak{d}} = d_1(\lambda_1,\lambda_2)dg^0_{\mathfrak{d}}$, and $\mu_1 = \lambda_1\lambda_2^4$. Then $dg_{\mathfrak{d}} = 2d^\times\lambda_1 d^\times\lambda_2 dg^0_{\mathfrak{d}}$, $\lambda_2 = \lambda_1^{-\frac{1}{4}}\mu_1^{\frac{1}{4}}$, and

$$d_1(\lambda_1,\lambda_2)^{\tau z+\rho} = \lambda_1^{\frac{7}{4}(s_{\tau 11}+2s_{\tau 12}+2s_{\tau 2}-5)}\mu_1^{\frac{1}{4}(s_{\tau 11}+2s_{\tau 12}-2s_{\tau 2}-1)}. \tag{12.8.14}$$

Easy computations show the following lemma.

Lemma (12.8.15)

(1) $\sigma_{\mathfrak{p}_1}(d_1(\lambda_1,\lambda_2)) = \lambda_1^{-3}\lambda_2^2 = \lambda_1^{-\frac{7}{2}}\mu_1^{\frac{1}{2}}$.

(2) $\kappa_{\mathfrak{d}1}(d_1(\lambda_1,\lambda_2)) = \lambda_1^{-5}\lambda_2^{-6} = \lambda_1^{-\frac{7}{2}}\mu_1^{-\frac{3}{2}}$.

(3) $\kappa_{V_1}(d_1(\lambda_1,\lambda_2)) = (\lambda_1\lambda_2^4)^{-3} = \mu_1^{-3}$.

(4) $\kappa_{V_2}(d_1(\lambda_1,\lambda_2)) = (\lambda_1\lambda_2^{-3})^{-2} = \lambda_1^{-\frac{7}{2}}\mu_1^{\frac{3}{2}}$.

Definition (12.8.16) *Let* $l_{4i} = (l_{4i,1}, l_{4i,2})$ *where*

(1) $l_{41,1}(s_\tau) = \frac{1}{4}(s_{\tau 11}+2s_{\tau 12}-2s_{\tau 2}+7)$, $l_{41,2}(s_\tau) = s_{\tau 11}+2s_{\tau 12}+2s_{\tau 2}-9$,

(2) $l_{42,1}(s_\tau) = \frac{1}{4}(s_{\tau 11}+2s_{\tau 12}-2s_{\tau 2}+1)$, $l_{42,2}(s_\tau) = s_{\tau 11}+2s_{\tau 12}+2s_{\tau 2}-7$,

(3) $l_{43,1}(s_\tau) = 3 - l_{41,1}(s_\tau)$, $l_{43,2}(s_\tau) = l_{41,2}(s_\tau)$,

(4) $l_{44,1}(s_\tau) = 3 - l_{42,1}(s_\tau)$, $l_{44,2}(s_\tau) = l_{42,2}(s_\tau)$.

Note that $l_{4i}(-s_{\tau 11}, s_{\tau 11}+s_{\tau 12}, s_{\tau 2}) = l_{4i}(s_{\tau 11}, s_{\tau 12}, s_{\tau 2})$ for $i=1,2,3,4$.

Let

$$\Psi_1 = R_{\mathfrak{d}_4}\mathscr{F}_{V_1}\mathscr{F}_{\mathfrak{d}}\Psi,\ \Psi_2 = R_{\mathfrak{d}_4}\Psi,\ \Psi_3 = R_{\mathfrak{d}_4}\mathscr{F}_{\mathfrak{d}}\Psi,\ \Psi_4 = R_{\mathfrak{d}_4}\mathscr{F}_{V_1}\Psi.$$

By (12.8.14), (12.8.15), and a similar argument as before,

$$\Xi_{\mathfrak{p},\mathfrak{d}',i}(\Phi,\omega,w) = 2\delta_\#(\omega)\sum_\tau \Omega_{W,l_{41},\tau}(\Psi_i,w)$$

for $i=1,2,3,4$.

The following proposition follows from (11.2.3).

Proposition (12.8.17) *Suppose that* $\tau=\tau_G$ *and* $\delta>0$ *is a small number. Then by changing* ψ *if necessary,*

$$\begin{aligned}\Xi_{\mathfrak{p},\mathfrak{d}',i}(\Phi,\omega,w) \sim{}& \frac{2\Sigma_{W+}(\Psi_i, l_{4i,1}(-1,2,1))}{l_{4i,2}(-1,2,1)}C_G\Lambda(w;\rho)\\ &+\frac{2\varrho^2\delta_\#(\omega)}{2\pi\sqrt{-1}}\int_{\mathrm{Re}(s_{\tau 11})=1+\delta}\Sigma_{W,\mathrm{st,sub}+}(\Psi_i, l_{4i}, s_{\tau 11},1,1)\\ &\qquad\times\phi_2(s_{\tau 11})\widetilde{\Lambda}_1(w;s_{\tau 11},1)ds_{\tau 11}\end{aligned}$$

for $i=1,2,3,4$.

(f) $\Xi_{\mathfrak{p},\mathfrak{d}_5,i}(\Phi,\omega,w)$

Let $\mathfrak{d}'=\mathfrak{d}_5$. Let t^0 be as in (b). When we consider the cases $i=1,2$, we make the change of variable $\mu_1=\lambda_1\lambda_2^4$, $\mu_2=\lambda_1\lambda_2^4\lambda_3^2=\mu_1\lambda_3^2$. Then

$$(12.8.18)\qquad d_3(\lambda)^{\tau z+\rho}=\lambda_1^{\frac{7}{4}(s_{\tau 11}+2s_{\tau 12}+2s_{\tau 2}-5)}\mu_1^{\frac{1}{4}(-s_{\tau 11}+2s_{\tau 12}-2s_{\tau 2}+1)}\mu_2^{\frac{1}{2}(s_{\tau 11}-1)}.$$

$\kappa_{V_1}(d_2(\lambda_1,\lambda_2,\lambda_3))$ etc. are the same as in (12.8.15).

When we consider the cases $i=3,4$, we make the change of variable $\mu_1=\lambda_1\lambda_2^4$, $\mu_2=(\lambda_1\lambda_2^4)^{-1}\lambda_3^2=\mu_1^{-1}\lambda_3^2$. Then

$$(12.8.19)\qquad d_3(\lambda)^{\tau z+\rho}=\lambda_1^{\frac{7}{4}(s_{\tau 11}+2s_{\tau 12}+2s_{\tau 2}-5)}\mu_1^{\frac{1}{4}(3s_{\tau 11}+2s_{\tau 12}-2s_{\tau 2}-3)}\mu_2^{\frac{1}{2}(s_{\tau 11}-1)}.$$

It is easy to see that

$$(12.8.20)\qquad \sigma_{\mathfrak{p}}(d_2(\lambda_1,\lambda_2,\lambda_3))\lambda_3^2=\begin{cases}\lambda_1^{-\frac{7}{2}}\mu_1^{-\frac{1}{2}}\mu_2 & i=1,2,\\ \lambda_1^{-\frac{7}{2}}\mu_1^{\frac{3}{2}}\mu_2 & i=3,4.\end{cases}$$

In both cases, $d^\times t^0=\frac{7}{4}d^\times\lambda_1 d^\times\mu_1 d^\times\mu_2 d^\times\widehat{t}^0$.

Definition (12.8.21) *Let* $l_{5i}=(\frac{1}{2}(s_{\tau 11}+1),l_{5i,2})$ *where*

(1) $l_{51,1}(s_\tau)=(s_{\tau 11}+2s_{\tau 12}+2s_{\tau 2}-9)(s_{\tau 11}-2s_{\tau 12}+2s_{\tau 2}-5)$,
(2) $l_{52,1}(s_\tau)=(s_{\tau 11}+2s_{\tau 12}+2s_{\tau 2}-7)(s_{\tau 11}-2s_{\tau 12}+2s_{\tau 2}+1)$,
(3) $l_{53,1}(s_\tau)=(s_{\tau 11}+2s_{\tau 12}+2s_{\tau 2}-9)(3s_{\tau 11}+2s_{\tau 12}-2s_{\tau 2}-3)$,
(4) $l_{54,1}(s_\tau)=(s_{\tau 11}+2s_{\tau 12}+2s_{\tau 2}-7)(3s_{\tau 11}+2s_{\tau 12}-2s_{\tau 2}-9)$.

Let Ψ_i be as in (e). Then by (12.8.18)–(12.8.20), and a similar argument as before,

$$\begin{aligned}&\Xi_{\mathfrak{p},\mathfrak{d}',i}(\Phi,\omega,w)\\ &=4\delta_\#(\omega)\sum_\tau\left(\frac{1}{2\pi\sqrt{-1}}\right)^3\int_{\substack{\mathrm{Re}(s_\tau)=r\\ r_3>1,\ r_1\gg r_2}}\Sigma_{1,\mathrm{sub}}(\Psi_i,l_{5i},s_\tau)\widetilde{\Lambda}_\tau(w;s_\tau)ds_\tau\end{aligned}$$

for $i=1,2,3,4$.

We define

$$\begin{aligned}\Sigma_{\mathfrak{p},\mathfrak{d}_5,1}(\Phi,s_{\tau 11})&=\frac{4\Sigma_1(R_{\mathfrak{d}_5}\mathscr{F}_{V_1}\mathscr{F}_{\mathfrak{d}}\Psi,\frac{1}{2}(s_{\tau 11}+1))}{(s_{\tau 11}-5)^2},\\ \Sigma_{\mathfrak{p},\mathfrak{d}_5,2}(\Phi,s_{\tau 11})&=\frac{4\Sigma_1(R_{\mathfrak{d}_5}\Psi,\frac{1}{2}(s_{\tau 11}+1))}{(s_{\tau 11}-3)(s_{\tau 11}+1)},\\ \Sigma_{\mathfrak{p},\mathfrak{d}_5,3}(\Phi,s_{\tau 11})&=\frac{4\Sigma_1(R_{\mathfrak{d}_5}\mathscr{F}_{\mathfrak{d}}\Psi,\frac{1}{2}(s_{\tau 11}+1))}{3(s_{\tau 11}-5)(s_{\tau 11}-1)},\\ \Sigma_{\mathfrak{p},\mathfrak{d}_5,4}(\Phi,s_{\tau 11})&=\frac{4\Sigma_1(R_{\mathfrak{d}_5}\mathscr{F}_{V_1}\Psi,\frac{1}{2}(s_{\tau 11}+1))}{3(s_{\tau 11}-3)^2}.\end{aligned}$$

The following proposition follows from cases (9), (10), (15), (16) of (11.2.4).

Proposition (12.8.22) *Suppose that $\tau = \tau_G$ and $\delta > 0$ is a small number. Then by changing ψ if necessary,*

$$\Xi_{\mathfrak{p},\mathfrak{d}',i}(\Phi,\omega,w) \sim \frac{\varrho^2\delta_{\#}(\omega)}{2\pi\sqrt{-1}}\int_{\mathrm{Re}(s_{\tau 11})=1+\delta} \Sigma_{\mathfrak{p},\mathfrak{d}_5,i}(\Phi,s_{\tau 11})\phi_2(s_{\tau 11})\widetilde{\Lambda}_1(w;s_{\tau 11},1)ds_{\tau 11}$$

for $i = 1,2,3,4$.

(g) $\Xi_{\mathfrak{p},\mathfrak{d}_6,i}(\Phi,\omega,w)$

Let t^0 be as in (b). We make the change of variable

$$\mu_1 = \lambda_1\lambda_2^4,\ \mu_2 = (\lambda_1\lambda_2^{-3})^{-1}\lambda_3 = \lambda_1^{-1}\lambda_2^3\lambda_3$$

for the cases $i = 1,2$. Then $\lambda_3 = \lambda_1^{\frac{7}{4}}\mu_1^{-\frac{3}{4}}\mu_2$, and

$$d_3(\lambda)^{\tau z+\rho} = \lambda_1^{\frac{7}{2}(s_{\tau 11}+s_{\tau 12}+s_{\tau 2}-3)}\mu_1^{\frac{1}{2}(-s_{\tau 11}+s_{\tau 12}-s_{\tau 2}+1)}\mu_2^{s_{\tau 11}-1}. \tag{12.8.23}$$

We make the change of variable

$$\mu_1 = \lambda_1\lambda_2^4,\ \mu_2 = \lambda_1\lambda_2^{-3}\lambda_3 = \lambda_1\lambda_2^{-3}\lambda_3$$

for the cases $i = 3,4$. Then $\lambda_3 = \lambda_1^{-\frac{7}{4}}\mu_1^{\frac{3}{4}}\mu_2$, and

$$d_3(\lambda)^{\tau z+\rho} = \lambda_1^{\frac{7}{2}(s_{\tau 12}+s_{\tau 2}-2)}\mu_1^{\frac{1}{2}(2s_{\tau 11}+s_{\tau 12}-s_{\tau 2}-2)}\mu_2^{s_{\tau 11}-1}. \tag{12.8.24}$$

It is easy to see that

$$\sigma_{\mathfrak{p}}(d_2(\lambda_1,\lambda_2,\lambda_3))\lambda_3^2 = \begin{cases} \mu_1^{-1}\mu_2^2 & i=1,2, \\ \lambda_1^{-7}\mu_1^2\mu_2^2 & i=3,4. \end{cases} \tag{12.8.25}$$

In both cases $d^\times t^0 = 7d^\times\lambda_1 d^\times\mu_1 d^\times\mu_2 d^\times\widehat{t}^0$.

Definition (12.8.26) *Let $l_{6i} = (s_{\tau 11}+1, l_{6i,2})$ where*
(1) $l_{61,1}(s_\tau) = (s_{\tau 11}+s_{\tau 12}+s_{\tau 2}-4)(s_{\tau 11}-s_{\tau 12}+s_{\tau 2}+4)$,
(2) $l_{62,1}(s_\tau) = (s_{\tau 11}+s_{\tau 12}+s_{\tau 2}-4)(s_{\tau 11}-s_{\tau 12}+s_{\tau 2}-2)$,
(3) $l_{63,1}(s_\tau) = (s_{\tau 12}+s_{\tau 2}-4)(2s_{\tau 11}+s_{\tau 12}-s_{\tau 2}-4)$,
(4) $l_{64,1}(s_\tau) = (s_{\tau 12}+s_{\tau 2}-4)(2s_{\tau 11}+s_{\tau 12}-s_{\tau 2}+2)$.

Let

$$\Psi_1 = R_{\mathfrak{d}_6}\mathscr{F}_{\mathfrak{d}}\Psi,\ \Psi_2 = R_{\mathfrak{d}_6}\mathscr{F}_{V_2}\Psi,\ \Psi_3 = R_{\mathfrak{d}_6}\mathscr{F}_{V_2}\mathscr{F}_{\mathfrak{d}}\Psi,\ \Psi_4 = R_{\mathfrak{d}_6}\Psi.$$

Then by (12.8.23)–(12.8.25), and a similar argument as before,

$$\begin{aligned}&\Xi_{\mathfrak{p},\mathfrak{d}_6,i}(\Phi,\omega,w)\\ &= 2\delta_{\#}(\omega)\sum_\tau\left(\frac{1}{2\pi\sqrt{-1}}\right)^3\int_{\substack{\mathrm{Re}(s_\tau)=r\\ r_1\gg r_2,\ r_3>2}} \Sigma_{1,\mathrm{sub}}(\Psi_i,l_{6i},s_\tau)\widetilde{\Lambda}_\tau(w;s_\tau)ds_\tau\end{aligned}$$

for $i = 1,2,3,4$.

Proposition (12.8.27) *By changing ψ if necessary,*

$$\Xi_{\mathfrak{p},\mathfrak{d}_6,i}(\Phi,\omega,w) \sim 2C_G\Lambda(w;\rho)\delta_{\#}(\omega)\Sigma_{1,\mathrm{sub}}(\Psi_i,l_{6i},1,1,1)$$

for $i=1,2,3,4$.

Proof. We can ignore the case $\tau_2=1$ as usual, and assume that $\tau_2=1$. Let $r=(r_1,r_2,r_3)=(10,5,3)$. By the usual argument,

$$\begin{aligned}
&\left(\frac{1}{2\pi\sqrt{-1}}\right)^3\int_{\mathrm{Re}(s_\tau)=r}\Sigma_{1,\mathrm{sub}}(\Psi_i,l_{6i},s_\tau)\widetilde{\Lambda}_\tau(w;s_\tau)ds_\tau \\
&\sim \varrho\left(\frac{1}{2\pi\sqrt{-1}}\right)^2\int_{\mathrm{Re}(s_{\tau 1})=(r_1,r_2)}\Sigma_{1,\mathrm{sub}}(\Psi_i,l_{6i},s_{\tau 1},1)\widetilde{\Lambda}_{1\tau_1}(w;s_{\tau 1})ds_{\tau 1}.
\end{aligned}$$

The points $(2,-1),(-1,2),(1,1)$ are not on the lines $s_{\tau 11}+s_{\tau 12}=3$, $s_{\tau 12}=3$, $s_{\tau 11}=s_{\tau 12}+1$, $2s_{\tau 11}+s_{\tau 12}=5$. Therefore, by (3.6.1), we can move the contour crossing these lines so that $(r_1,r_2)=(1+\delta,1+\delta)$. Since $\Sigma_{1,\mathrm{sub}}(\Psi_i,l_{6i},s_\tau)$ is holomorphic at $(1,1,1)$, the proposition follows by a similar argument as before. Q.E.D.

(h) Now we combine the computations in this section. Let $\mathfrak{d}=\mathfrak{d}_0,\mathfrak{p}=\mathfrak{p}_0$.

We define

$$\begin{aligned}
J_1(\Phi,\omega) &= 14\delta_{\mathfrak{d}}(\omega)\left(\Sigma_{\mathfrak{d}+}(\Psi,\omega_{\mathfrak{d}},\chi_{\mathfrak{d}},-3)+\Sigma_{\mathfrak{d}+}(\mathscr{F}_{\mathfrak{d}}\Psi,\omega_{\mathfrak{d}}^{-1},\kappa_{\mathfrak{d}1}^{-1}\chi_{\mathfrak{d}}^{-1},8)\right) \\
&\quad -\frac{\delta_{\#}(\omega)}{2}\Sigma_2(R_{\mathfrak{d}_1}\Phi_{\mathfrak{p}_{12}},-2,-5)+\frac{\delta_{\#}(\omega)}{3}\Sigma_2(R_{\mathfrak{d}}\Phi_{\mathfrak{p}_{31}},\frac{4}{5},\frac{2}{5}), \\
J_2(\Phi) &= -\frac{1}{2}\Sigma_{W+}(R_{\mathfrak{d}_4}\mathscr{F}_{\mathfrak{d}}\Psi,1)-\frac{1}{2}\Sigma_{W+}(\mathscr{F}_W R_{\mathfrak{d}_4}\mathscr{F}_{\mathfrak{d}}\Psi,2) \\
&\quad +\Sigma_{W+}(R_{\mathfrak{d}_4}\Psi,\frac{1}{2})+\Sigma_{W+}(\mathscr{F}_W R_{\mathfrak{d}_4}\Psi,\frac{5}{2}) \\
&\quad +\frac{\Sigma_1(\mathscr{F}_{\mathfrak{d}_6}R_{\mathfrak{d}_6}\mathscr{F}_{\mathfrak{d}}\Psi,2)}{2}+\frac{\Sigma_1(R_{\mathfrak{d}_6}\Psi,2)}{4} \\
&\quad -\frac{2}{5}\Sigma_1(R_{\mathfrak{d}_6}\mathscr{F}_{\mathfrak{d}}\Psi,2)-2\Sigma_1(\mathscr{F}_{\mathfrak{d}_6}R_{\mathfrak{d}_6}\Psi,2), \\
\Upsilon_{\mathfrak{d},1}(\Phi,\omega) &= J_1(\Phi,\omega)+\delta_{\#}(\omega)J_2(\Phi), \\
J_3(\Phi,s_{\tau 11}) &= \frac{2T_{W+}(\widetilde{R}_{W,0}R_{\mathfrak{d}_4}\mathscr{F}_{\mathfrak{d}}\Psi,\frac{1}{4}(5-s_{\tau 11}),\frac{1}{2}(1-s_{\tau 11}))}{(s_{\tau 11}-1)(s_{\tau 11}-5)} \\
&\quad +\frac{2T_{W+}(\widetilde{R}_{W,0}\mathscr{F}_W R_{\mathfrak{d}_4}\mathscr{F}_{\mathfrak{d}}\Psi,\frac{1}{4}(s_{\tau 11}+7),\frac{1}{2}(1-s_{\tau 11}))}{(s_{\tau 11}-1)(s_{\tau 11}-5)}, \\
J_4(\Phi,s_{\tau 11}) &= -\frac{2T_{W+}(\widetilde{R}_{W,0}R_{\mathfrak{d}_4}\Psi,\frac{1}{4}(s_{\tau 11}+1),\frac{1}{2}(1-s_{\tau 11}))}{(s_{\tau 11}-1)(s_{\tau 11}-3)} \\
&\quad -\frac{2T_{W+}(\widetilde{R}_{W,0}\mathscr{F}_W R_{\mathfrak{d}_4}\Psi,\frac{1}{4}(11-s_{\tau 11}),\frac{1}{2}(1-s_{\tau 11}))}{(s_{\tau 11}-1)(s_{\tau 11}-3)}, \\
\Upsilon_{\mathfrak{d},2}(\Phi,\omega,s_{\tau 11}) &= \delta_{\#}(\omega)(-\Sigma_{\mathfrak{p}_{11}}(\Phi,s_{\tau 11})+\Sigma_{\mathfrak{p}_{21}}^{(1)}(\Phi,s_{\tau 11})) \\
&\quad +\delta_{\#}(\omega)(\Sigma_{\mathfrak{p}_{22}}(\Phi,s_{\tau 11})+\Sigma_{\mathfrak{p}_{32}}(\Phi,s_{\tau 11})) \\
&\quad +\delta_{\#}(\omega)(J_3(\Phi,s_{\tau 11})+J_4(\Phi,s_{\tau 11})) \\
&\quad +\delta_{\#}(\omega)\sum_{i=1}^{4}(-1)^{i+1}\Sigma_{\mathfrak{p},\mathfrak{d}_5,i}(\Phi,s_{\tau 11}).
\end{aligned}$$

Since the third term of (12.8.10)(1) and the second term of (12.8.10)(2) cancel out, we get the following proposition by (9.1.10), (12.8.3), (12.8.7), (12.8.10), (12.8.13), (12.8.17), (12.8.22), (12.8.27).

Proposition (12.8.28) *Suppose that $\tau = \tau_G$ and $\delta > 0$ is a small number. Then by changing ψ if necessary,*

$$\begin{aligned}
&\Xi_{\mathfrak{p}_0}(\Phi,\omega,w)\\
&\sim \Upsilon_{\mathfrak{d},1}(\Phi,\omega)C_G\Lambda(w;\rho)\\
&\quad + \frac{\varrho^2}{2\pi\sqrt{-1}}\int_{\mathrm{Re}(s_{\tau 11})=1+\delta}\Upsilon_{\mathfrak{d},2}(\Phi,\omega,s_{\tau 11})\widetilde{\Lambda}_1(w;s_{\tau 11},1)ds_{\tau 11}\\
&\quad + \sum_{\substack{\tau'=(\tau'_1,(1,2))\\ \tau'_1\in\mathfrak{W}'}}\frac{\varrho\delta_{\#}(\omega)}{2\pi\sqrt{-1}}\int_{\mathrm{Re}(s_{\tau' 11})=r_2}\Sigma^{(2)}_{\mathfrak{p}_{21}}(\Phi,s_{\tau' 11})\widetilde{\Lambda}_{1\tau'_1}(w;s_{\tau' 11},2)ds_{\tau' 11}\\
&\quad - \sum_{i=1,2}\widetilde{\Xi}_{\mathfrak{p}_{i1}}(\Phi,\omega,w).
\end{aligned}$$

Note that we have proved in Chapter 10 that $\widetilde{\Xi}_{\mathfrak{p}_{ij}}(\Phi,\omega,w)$ is well defined for $i,j=1,2$ and $\widetilde{\Xi}_{\mathfrak{p}_{i2}}(\Phi,\omega,w)\sim 0$ for $i=1,2$.

Let $\Upsilon_{\mathfrak{d},2,(i)}(\Phi,\omega,1)$ etc. be the i-th coefficient of the Laurent expansion as before. It is easy to see that

$$\begin{aligned}
\Sigma_{\mathfrak{p}_{11},(0)}(\Phi,1) &= -\Sigma_{2,(1,0)}(R_{\mathfrak{d}_1}\Phi_{\mathfrak{p}_{11}},-1,-3) - 2\Sigma_{2,(0,1)}(R_{\mathfrak{d}_1}\Phi_{\mathfrak{p}_{11}},-1,-3),\\
\Sigma^{(1)}_{\mathfrak{p}_{21},(0)}(\Phi,1) &= \Sigma_{2,(1,-1)}(R_{\mathfrak{d}_2}\Phi_{\mathfrak{p}_{21}},-\tfrac{1}{2},0) + \Sigma_{2,(0,0)}(R_{\mathfrak{d}_2}\Phi_{\mathfrak{p}_{21}},-\tfrac{1}{2},0),\\
\Sigma_{\mathfrak{p}_{22},(0)}(\Phi,1) &= -\frac{1}{2}\Sigma_{2,(0,0)}(R_{\mathfrak{d}_2}\Phi_{\mathfrak{p}_{22}},0,0) - \frac{1}{2}\Sigma_{2,(-1,0)}(R_{\mathfrak{d}_2}\Phi_{\mathfrak{p}_{22}},0,0)\\
&\quad - \frac{1}{4}\Sigma_{2,(0,-1)}(R_{\mathfrak{d}_2}\Phi_{\mathfrak{p}_{22}},0,0) - \frac{1}{4}\Sigma_{2,(-1,-1)}(R_{\mathfrak{d}_2}\Phi_{\mathfrak{p}_{22}},0,0)\\
&\quad - \frac{1}{4}\Sigma_{2,(1,-1)}(R_{\mathfrak{d}_2}\Phi_{\mathfrak{p}_{22}},0,0) - \Sigma_{2,(-1,1)}(R_{\mathfrak{d}_2}\Phi_{\mathfrak{p}_{22}},0,0),\\
\Sigma_{\mathfrak{p}_{32},(0)}(\Phi,1) &= -\frac{1}{5}\Sigma_{2,(0,0)}(R_{\mathfrak{d}_3}\Phi_{\mathfrak{p}_{32}},1,0) - \frac{2}{5}\Sigma_{2,(-1,0)}(R_{\mathfrak{d}_3}\Phi_{\mathfrak{p}_{32}},1,0)\\
&\quad - \frac{2}{15}\Sigma_{2,(0,-1)}(R_{\mathfrak{d}_3}\Phi_{\mathfrak{p}_{32}},1,0) - \frac{4}{15}\Sigma_{2,(-1,-1)}(R_{\mathfrak{d}_3}\Phi_{\mathfrak{p}_{32}},1,0)\\
&\quad - \frac{1}{15}\Sigma_{2,(1,-1)}(R_{\mathfrak{d}_3}\Phi_{\mathfrak{p}_{32}},1,0) - \frac{3}{5}\Sigma_{2,(-1,1)}(R_{\mathfrak{d}_3}\Phi_{\mathfrak{p}_{32}},1,0).
\end{aligned}$$

We define

$$\begin{aligned}
J_5(\Phi) &= \frac{1}{4}(T_{W+}(\widetilde{R}_{W,0}\mathscr{F}_W R_{\mathfrak{d}_4}\mathscr{F}_{\mathfrak{d}}\Psi,2) + T_{W+}(\widetilde{R}_{W,0}R_{\mathfrak{d}_4}\mathscr{F}_{\mathfrak{d}}\Psi,1)),\\
J_6(\Phi) &= -\frac{1}{2}(T_{W+}(\widetilde{R}_{W,0}R_{\mathfrak{d}_4}\Psi,\tfrac{1}{2}) + T_{W+}(\widetilde{R}_{W,0}\mathscr{F}_W R_{\mathfrak{d}_4}\Psi,\tfrac{5}{2})).
\end{aligned}$$

Then

$$J_{3,(0)}(\Phi,1) = J_5(\Phi) + 2\frac{d}{ds_{\tau 11}}\bigg|_{s_{\tau 11}=1} \frac{T_{W+}(\widetilde{R}_{W,0}\mathscr{F}_W R_{\mathfrak{d}_4}\mathscr{F}_{\mathfrak{d}}\Psi, \frac{1}{4}(s_{\tau 11}+7),0)}{s_{\tau 11}-5}$$
$$+ 2\frac{d}{ds_{\tau 11}}\bigg|_{s_{\tau 11}=1} \frac{T_{W+}(\widetilde{R}_{W,0}R_{\mathfrak{d}_4}\mathscr{F}_{\mathfrak{d}}\Psi, \frac{1}{4}(5-s_{\tau 11}),0)}{s_{\tau 11}-5},$$
$$J_{4,(0)}(\Phi,1) = J_6(\Phi) - 2\frac{d}{ds_{\tau 11}}\bigg|_{s_{\tau 11}=1} \frac{T_{W+}(\widetilde{R}_{W,0}R_{\mathfrak{d}_4}\Psi, \frac{1}{4}(s_{\tau 11}+1),0)}{s_{\tau 11}-3}$$
$$- 2\frac{d}{ds_{\tau 11}}\bigg|_{s_{\tau 11}=1} \frac{T_{W+}(\widetilde{R}_{W,0}\mathscr{F}_W R_{\mathfrak{d}_4}\Psi, \frac{1}{4}(11-s_{\tau 11}),0)}{s_{\tau 11}-3}.$$

Easy computations show that

$$T_{W+}(\widetilde{R}_{W,0}\mathscr{F}_W R_{\mathfrak{d}_4}\mathscr{F}_{\mathfrak{d}}\Psi, \frac{1}{4}(s_{\tau 11}+7),0) + T_{W+}(\widetilde{R}_{W,0}R_{\mathfrak{d}_4}\mathscr{F}_{\mathfrak{d}}\Psi, \frac{1}{4}(5-s_{\tau 11}),0)$$
$$= \frac{4\Sigma_{1,(-1)}(R_{\mathfrak{d}_5}\mathscr{F}_{\mathfrak{d}}\Psi,1)}{1-s_{\tau 11}} + \frac{4\Sigma_{1,(-1)}(R_{\mathfrak{d}_5}\mathscr{F}_W R_{\mathfrak{d}_4}\mathscr{F}_{\mathfrak{d}}\Psi,1)}{s_{\tau 11}+3}$$
$$- \operatorname*{Res}_{s'=0} \Sigma_2(R_{\mathfrak{d}_2}\Phi_{\mathfrak{p}_{22}}, \frac{1}{4}(1-s_{\tau 11}), s'),$$
$$T_{W+}(\widetilde{R}_{W,0}R_{\mathfrak{d}_4}\Psi, \frac{1}{4}(s_{\tau 11}+1),0) + T_{W+}(\widetilde{R}_{W,0}\mathscr{F}_W R_{\mathfrak{d}_4}\Psi, \frac{1}{4}(11-s_{\tau 11}),0)$$
$$= \frac{4\Sigma_{1,(-1)}(R_{\mathfrak{d}_5}\Psi,1)}{s_{\tau 11}-3} + \frac{4\Sigma_{1,(-1)}(R_{\mathfrak{d}_5}\mathscr{F}_W R_{\mathfrak{d}_4}\Psi,1)}{7-s_{\tau 11}}$$
$$- \operatorname*{Res}_{s'=0} \Sigma_2(R_{\mathfrak{d}_2}\Phi_{\mathfrak{p}_{21}}, \frac{1}{4}(s_{\tau 11}-3), s').$$

It is easy to see that

$$- \operatorname*{Res}_{s'=0} \Sigma_2(R_{\mathfrak{d}_2}\Phi_{\mathfrak{p}_{22}}, \frac{1}{4}(1-s_{\tau 11}), s') + \frac{4\Sigma_{1,(-1)}(R_{\mathfrak{d}_5}\mathscr{F}_{\mathfrak{d}}\Psi,1)}{1-s_{\tau 11}}$$
$$= -\Sigma_{2,(0,-1)}(R_{\mathfrak{d}_2}\Phi_{\mathfrak{p}_{22}},0,0) + \frac{1}{4}\Sigma_{2,(0,-1)}(R_{\mathfrak{d}_2}\Phi_{\mathfrak{p}_{22}}0,0)(s_{\tau 11}-1)$$
$$+ O((s_{\tau 11}-1))^2).$$

Therefore,

$$J_{3,(0)}(\Phi,1) = J_5(\Phi) + \frac{1}{8}\Sigma_{2,(0,-1)}(R_{\mathfrak{d}_2}\Phi_{\mathfrak{p}_{22}},0,0) - \frac{1}{8}\Sigma_{2,(1,-1)}(R_{\mathfrak{d}_2}\Phi_{\mathfrak{p}_{22}},0,0),$$
$$J_{4,(0)}(\Phi,1) = J_6(\Phi) - \frac{1}{2}\Sigma_{2,(0,-1)}(R_{\mathfrak{d}_2}\Phi_{\mathfrak{p}_{21}},-\frac{1}{2},0) - \frac{1}{4}\Sigma_{2,(1,-1)}(R_{\mathfrak{d}_2}\Phi_{\mathfrak{p}_{21}},-\frac{1}{2},0)$$
$$- 2\Sigma_{1,(-1)}(R_{\mathfrak{d}_5}\Psi,1) + \frac{4}{9}\Sigma_{1,(-1)}(R_{\mathfrak{d}_5}\mathscr{F}_W R_{\mathfrak{d}_4}\Psi,1).$$

Also it is easy to see that

$$\begin{aligned}
\Sigma_{\mathfrak{p},\mathfrak{d}_5,1,(0)}(\Phi,1) &= \frac{1}{4}\Sigma_{1,(-1)}(R_{\mathfrak{d}_5}\mathscr{F}_W R_{\mathfrak{d}_4}\mathscr{F}_{\mathfrak{d}}\Psi,1) + \frac{1}{4}\Sigma_{1,(0)}(R_{\mathfrak{d}_5}\mathscr{F}_W R_{\mathfrak{d}_4}\mathscr{F}_{\mathfrak{d}}\Psi,1),\\
\Sigma_{\mathfrak{p},\mathfrak{d}_5,2,(0)}(\Phi,1) &= \Sigma_{1,(0)}(R_{\mathfrak{d}_5}\Psi,1),\\
\Sigma_{\mathfrak{p},\mathfrak{d}_5,3,(0)}(\Phi,1) &= -\frac{1}{24}\Sigma_{1,(-1)}(R_{\mathfrak{d}_5}\mathscr{F}_{\mathfrak{d}}\Psi,1) - \frac{1}{12}\Sigma_{1,(0)}(R_{\mathfrak{d}_5}\mathscr{F}_{\mathfrak{d}}\Psi,1)\\
&\quad - \frac{1}{6}\Sigma_{1,(1)}(R_{\mathfrak{d}_5}\mathscr{F}_{\mathfrak{d}}\Psi,1),\\
\Sigma_{\mathfrak{p},\mathfrak{d}_5,4,(0)}(\Phi,1) &= \frac{2}{3}\Sigma_{1,(-1)}(R_{\mathfrak{d}_5}\mathscr{F}_W R_{\mathfrak{d}_4}\Psi,1) + \frac{1}{3}\Sigma_{1,(0)}(R_{\mathfrak{d}_5}\mathscr{F}_W R_{\mathfrak{d}_4}\Psi,1).
\end{aligned}$$

We define

$$\begin{aligned}
J_7(\Phi) =&\Sigma_{2,(0,0)}(R_{\mathfrak{d}_2}\Phi_{\mathfrak{p}_{21}},-\frac{1}{2},0) + \frac{3}{4}\Sigma_{2,(1,-1)}(R_{\mathfrak{d}_2}\Phi_{\mathfrak{p}_{21}},-\frac{1}{2},0)\\
&- \frac{1}{2}\Sigma_{2,(0,-1)}(R_{\mathfrak{d}_2}\Phi_{\mathfrak{p}_{21}},-\frac{1}{2},0)\\
&- \frac{1}{2}\Sigma_{2,(0,0)}(R_{\mathfrak{d}_2}\Phi_{\mathfrak{p}_{22}},0,0) - \frac{3}{8}\Sigma_{2,(1,-1)}(R_{\mathfrak{d}_2}\Phi_{\mathfrak{p}_{22}},0,0)\\
&- \frac{2}{5}\Sigma_{2,(-1,1)}(R_{\mathfrak{d}_2}\Phi_{\mathfrak{p}_{22}},0,0) - \frac{1}{8}\Sigma_{2,(0,-1)}(R_{\mathfrak{d}_2}\Phi_{\mathfrak{p}_{22}},0,0)\\
&- \frac{1}{10}\Sigma_{2,(-1,0)}(R_{\mathfrak{d}_2}\Phi_{\mathfrak{p}_{22}},0,0) - \frac{1}{40}\Sigma_{2,(-1,-1)}(R_{\mathfrak{d}_2}\Phi_{\mathfrak{p}_{22}},0,0)\\
&- \frac{1}{5}\Sigma_{2,(0,0)}(R_{\mathfrak{d}_3}\Phi_{\mathfrak{p}_{32}},1,0) - \frac{3}{20}\Sigma_{2,(1,-1)}(R_{\mathfrak{d}_3}\Phi_{\mathfrak{p}_{32}},1,0)\\
&- \frac{1}{20}\Sigma_{2,(0,-1)}(R_{\mathfrak{d}_3}\Phi_{\mathfrak{p}_{32}},1,0),\\
J_8(\Phi) =&\frac{1}{4}\Sigma_{1,(0)}(R_{\mathfrak{d}_5}\mathscr{F}_W R_{\mathfrak{d}_4}\mathscr{F}_{\mathfrak{d}}\Psi,1) + \frac{1}{4}\Sigma_{1,(-1)}(R_{\mathfrak{d}_5}\mathscr{F}_W R_{\mathfrak{d}_4}\mathscr{F}_{\mathfrak{d}}\Psi,1)\\
&+ \Sigma_{1,(0)}(R_{\mathfrak{d}_5}\Psi,1) - 2\Sigma_{1,(-1)}(R_{\mathfrak{d}_5}\Psi,1)\\
&- \frac{1}{3}\Sigma_{1,(0)}(R_{\mathfrak{d}_5}\mathscr{F}_W R_{\mathfrak{d}_4}\Psi,1) - \frac{2}{9}\Sigma_{1,(-1)}(R_{\mathfrak{d}_5}\mathscr{F}_W R_{\mathfrak{d}_4}\Psi,1).
\end{aligned}$$

If Φ is K-invariant,

$$\begin{aligned}
\Sigma_{2,(-1,-1)}(R_{\mathfrak{d}_2}\Phi_{\mathfrak{p}_{22}},0,0) &= \Sigma_{1,(-1)}(R_{\mathfrak{d}_5}\mathscr{F}_{\mathfrak{d}}\Psi,1),\\
\Sigma_{2,(i,-1)}(R_{\mathfrak{d}_3}\Phi_{\mathfrak{p}_{32}},1,0) &= -\Sigma_{1,(i)}(R_{\mathfrak{d}_5}\mathscr{F}_{\mathfrak{d}}\Psi,1),\\
\Sigma_{2,(-1,i)}(R_{\mathfrak{d}_3}\Phi_{\mathfrak{p}_{32}},1,0) &= -\Sigma_{2,(-1,i)}(R_{\mathfrak{d}_2}\Phi_{\mathfrak{p}_{22}},0,0)
\end{aligned}$$

for all i.

Therefore, these considerations show that

$$\begin{aligned}
\Upsilon_{\mathfrak{d},2,(0)}(\Phi,\omega,1) =&\delta_{\#}(\omega)(J_7(\Phi) + J_8(\Phi) - \frac{1}{4}\Sigma_{1,(1)}(R_{\mathfrak{d}_5}\mathscr{F}_{\mathfrak{d}}\Psi,1))\\
&+ \delta_{\#}(\omega)\Sigma_{2,(1,0)}(R_{\mathfrak{d}_1}\Phi_{\mathfrak{p}_{11}},-1,-3)\\
&+ 2\delta_{\#}(\omega)\Sigma_{2,(0,1)}(R_{\mathfrak{d}_1}\Phi_{\mathfrak{p}_{11}},-1,-3).
\end{aligned}$$

By the principal part formula (4.2.15) for $\Sigma_{W,\mathrm{ad},(0)}(R_{\mathfrak{d}_4}\mathscr{F}_{\mathfrak{d}}\Psi,1)$ etc.,

$$J_2(\Phi)+J_8(\Phi)=-\frac{1}{2}\Sigma_{W,\mathrm{ad},(0)}(R_{\mathfrak{d}_4}\mathscr{F}_{\mathfrak{d}}\Psi,1)+\Sigma_{W,\mathrm{ad}}(R_{\mathfrak{d}_4}\Psi,\frac{1}{2}).$$

Note that by (4.4.7), if Φ is K-invariant,

$$\begin{aligned}
&\Sigma_1(\mathscr{F}_{\mathfrak{d}_6}R_{\mathfrak{d}_6}\mathscr{F}_{\mathfrak{d}}\Psi,2)=\mathfrak{V}_2\mathscr{F}_{\mathfrak{d}}\Psi(0),\\
&\Sigma_1(R_{\mathfrak{d}_6}\Psi,2)=\mathfrak{V}_2\mathscr{F}_W R_{\mathfrak{d}_4}\mathscr{F}_{\mathfrak{d}}\Psi(0),\\
&\Sigma_1(R_{\mathfrak{d}_6}\mathscr{F}_{\mathfrak{d}}\Psi,2)=\mathfrak{V}_2\mathscr{F}_W R_{\mathfrak{d}_4}\Psi(0),\\
&\Sigma_1(\mathscr{F}_{\mathfrak{d}_6}R_{\mathfrak{d}_6}\Psi,2)=\mathfrak{V}_2\Psi(0).
\end{aligned}$$

So

$$\begin{aligned}
\Upsilon_{\mathfrak{d},1}(\Phi,w)+\Upsilon_{\mathfrak{d},2,(0)}(\Phi,w)=&J_1(\Phi,\omega)+\delta_{\#}(\omega)(J_7(\Phi)-\frac{1}{4}\Sigma_{1,(1)}(R_{\mathfrak{d}_5}\mathscr{F}_{\mathfrak{d}}\Psi,1))\\
&-\frac{\delta_{\#}(\omega)}{2}\Sigma_{W,\mathrm{ad},(0)}(R_{\mathfrak{d}_4}\mathscr{F}_{\mathfrak{d}}\Psi,1)\\
&+\delta_{\#}(\omega)\Sigma_{W,\mathrm{ad}}(R_{\mathfrak{d}_4}\Psi,\frac{1}{2})\\
&+\delta_{\#}(\omega)\Sigma_{2,(1,0)}(R_{\mathfrak{d}_1}\Phi_{\mathfrak{p}_{11}},-1,-3)\\
&+2\delta_{\#}(\omega)\Sigma_{2,(0,1)}(R_{\mathfrak{d}_1}\Phi_{\mathfrak{p}_{11}},-1,-3)).
\end{aligned}$$

By (7.3.7), we get the following proposition.

Proposition (12.8.29)

$$\begin{aligned}
\Upsilon_{\mathfrak{d},1}(\Phi,w)+\Upsilon_{\mathfrak{d},2,(0)}(\Phi,w)=&14\delta_{\mathfrak{d}}(\omega)\Sigma_{\mathfrak{d},(0)}(R_{\mathfrak{d}}\Phi_{\mathfrak{p}}\Phi,\omega_{\mathfrak{d}},\chi_{\mathfrak{d}},-3)\\
&+\delta_{\#}(\omega)\Sigma_{2,(1,0)}(R_{\mathfrak{d}_1}\Phi_{\mathfrak{p}_{11}},-1,-3)\\
&+2\delta_{\#}(\omega)\Sigma_{2,(0,1)}(R_{\mathfrak{d}_1}\Phi_{\mathfrak{p}_{11}},-1,-3).
\end{aligned}$$

§12.9 The case $\mathfrak{d}=(\beta_9)$

Let $\mathfrak{p}=(\mathfrak{d},\mathfrak{s})$ be a path such that $\mathfrak{d}=(\beta_9)$. As in previous sections, we only consider such path such that $\mathfrak{s}(1)=0$. Let $\Psi=R_{\mathfrak{d}}\Phi_{\mathfrak{p}}$ throughout this section. Let $\mathfrak{W}'$ be as in §12.8.

Suppose that $f(q)$ is a function of $q=(q_1,q_2,q_3)\in(\mathbb{A}^1/k^\times)^3$. An easy consideration shows that

$$\begin{aligned}
&\int_{(\mathbb{A}^1/k^\times)^5}\omega(\widehat{t}^0)f(\gamma_{1,23}(\widehat{t}^0),\gamma_{2,13}(\widehat{t}^0),\gamma_{2,22}(\widehat{t}^0))d^\times\widehat{t}^0\\
&=\delta_{\mathfrak{d}}(\omega)\int_{(\mathbb{A}^1/k^\times)^3}\omega_{\mathfrak{d}}(\widehat{t}^0)f(q)d^\times q.
\end{aligned}$$

Let

$$d(\mu)=d(\mu_1,\mu_2,\mu_3)=a(\underline{\mu}_1^{-1}\underline{\mu}_2^{-\frac{2}{3}}\underline{\mu}_3^{-\frac{1}{3}},\underline{\mu}_2^{-\frac{1}{3}}\underline{\mu}_3^{\frac{1}{3}},\underline{\mu}_1\underline{\mu}_2;\underline{\mu}_2^{-\frac{2}{3}}\underline{\mu}_3^{-\frac{1}{3}},\underline{\mu}_2^{\frac{2}{3}}\underline{\mu}_3^{\frac{1}{3}}).$$

Let $t^0 = d(\mu)\widehat{t}^0$. Then $d^\times t^0 = \frac{1}{3} d^\times \mu d^\times \widehat{t}^0$. It is easy to see that

$$\mu_1 = |\gamma_{1,23}(t^0)|,\ \mu_2 = |\gamma_{2,13}(t^0)|,\ \mu_3 = |\gamma_{2,22}(t^0)|.$$

Let τ, s_τ be as before. Then $\kappa_{\mathfrak{d}2}(t^0) = \mu_1^{-5} \mu_2^{-\frac{16}{3}} \mu_3^{-\frac{2}{3}}$, and

$$d(\mu)^{\tau z - \rho} = \mu_1^{s_{\tau 11}+s_{\tau 12}+2} \mu_2^{\frac{1}{3}(3s_{\tau 11}+2s_{\tau 12}+2s_{\tau 2}+7)} \mu_3^{\frac{1}{3}(s_{\tau 12}+s_{\tau 2}+2)}. \tag{12.9.1}$$

Definition (12.9.2) *Let $l_\mathfrak{p} = (l_{\mathfrak{p},1}, l_{\mathfrak{p},2}, l_{\mathfrak{p},3}, 1)$ where*

$$l_{\mathfrak{p},1}(s_\tau) = s_{\tau 11} + s_{\tau 12} - 3,\ l_{\mathfrak{p},2}(s_\tau) = \frac{1}{3}(3s_{\tau 11} + 2s_{\tau 12} + 2s_{\tau 2} - 9),$$
$$l_{\mathfrak{p},3}(s_\tau) = \frac{1}{3}(s_{\tau 12} + s_{\tau 2}).$$

Let $D = \{r \mid r_1 + r_2 > 4, \frac{r_1+2(r_2+r_3)}{3} - 3, \frac{r_2+r_3}{3} > 1\}$. Then a similar consideration as in §12.7 shows that

$$\Xi_\mathfrak{p}(\Phi, \omega, w) = \frac{\delta_\mathfrak{d}(\omega)}{3} \sum_\tau \left(\frac{1}{2\pi\sqrt{-1}}\right)^3 \int_{\mathrm{Re}(s_\tau)=r} \Sigma_{3,\mathrm{sub}}(\Psi, l_\mathfrak{p}, \omega_\mathfrak{d}, s_\tau) \widetilde{\Lambda}_\tau(w; s_\tau) ds_\tau,$$

where we choose r from $D \cap D_\tau$.

As usual, we can ignore the case $\tau_2 = 1$, and assume that $\tau_2 = (1,2)$. The point $(3,3,2)$ belongs to the domain $D \cap D_\tau$. By moving the contour to $\mathrm{Re}(s_\tau) = (3,3,\frac{1}{2})$,

$$\begin{aligned}
&\left(\frac{1}{2\pi\sqrt{-1}}\right)^3 \int_{\mathrm{Re}(s_\tau)=(3,3,2)} \Sigma_{3,\mathrm{sub}}(\Psi, l_\mathfrak{p}, \omega_\mathfrak{d}, s_\tau) \widetilde{\Lambda}_\tau(w; s_\tau) ds_\tau \\
&= \left(\frac{1}{2\pi\sqrt{-1}}\right)^3 \int_{\mathrm{Re}(s_\tau)=(3,3,\frac{1}{2})} \Sigma_{3,\mathrm{sub}}(\Psi, l_\mathfrak{p}, \omega_\mathfrak{d}, s_\tau) \widetilde{\Lambda}_\tau(w; s_\tau) ds_\tau \\
&\quad + \varrho \left(\frac{1}{2\pi\sqrt{-1}}\right)^2 \int_{\mathrm{Re}(s_{\tau 1})=(3,3)} \Sigma_{3,\mathrm{sub}}(\Psi, l_\mathfrak{p}, \omega_\mathfrak{d}, s_{\tau 1}, 1) \widetilde{\Lambda}_{1\tau_1}(w; s_{\tau 1}) ds_{\tau 1}.
\end{aligned}$$

Since $C > 100$, $L_\tau(3,3,\frac{1}{2}) \le 12 + \frac{1}{2}C < 4 + C$. Therefore, we can ignore the first term.

If $\tau_1 = 1, (1,2)$ or $(2,3)$, we choose $r_1, r_2 \gg 0$. Then $\widetilde{L}_1(r_1, r_2) < 0$ for these cases.

The function $\Sigma_{3,\mathrm{sub}}(\Psi, l_\mathfrak{p}, \omega_\mathfrak{d}, s_{\tau 1}, 1)$ has the following pole structure.

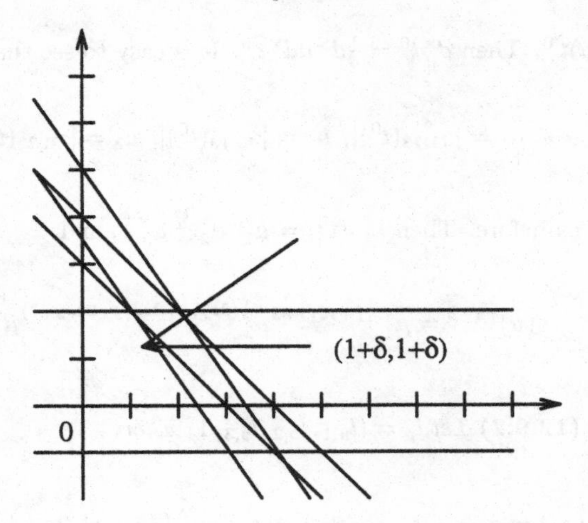

The points $(-1,2),(1,1)$ are not on any of the lines except for the line $s_{\tau 12}=2$. Therefore, by the passing principle (3.6.1), we can move the contour from the point $(r_1,r_2)=(3,3)$ to $(1+\delta,1+\delta)$ where $\delta>0$ is a small number crossing other lines.

We define $\Phi_{\mathfrak{p}}^{(1)} \in \mathscr{S}(\mathbb{A})$ by the integral

$$\Phi_{\mathfrak{p}}^{(1)}(x_{1,23},x_{2,13}) = \int_{\mathbb{A}} \Psi(x_{1,23},x_{2,13},x_{2,22})dx_{2,22}.$$

Let

$$f(\Phi,w,s_{\tau'11}) = \Sigma_2(\Phi_{\mathfrak{p}}^{(1)}, s_{\tau'11}-1, s_{\tau'11}-1)\Lambda_{1\tau'_1}(w;s_{\tau'11},2),$$

and $r=(1+\delta,1+\delta)$. Then by the above consideration,

$$\begin{aligned}
&\Xi_{\mathfrak{p}}(\Phi,\omega,w)\\
&\sim \frac{\varrho\delta_{\mathfrak{d}}(\omega)}{3}\sum_{\tau}\left(\frac{1}{2\pi\sqrt{-1}}\right)^2\int_{\mathrm{Re}(s_{\tau 1})=r}\Sigma_{3,\mathrm{sub}}(\Psi,l_{\mathfrak{p}},\omega_{\mathfrak{d}},s_{\tau 1},1)\widetilde{\Lambda}_{1\tau_1}(w;s_{\tau 1})ds_{\tau 1}\\
&\quad + \sum_{\substack{\tau'=(\tau'_1,(1,2))\\ \tau'_1\in\mathfrak{W}'}}\frac{\varrho\delta_{\#}(\omega)}{2\pi\sqrt{-1}}\int_{\mathrm{Re}(s_{\tau'11})=r_1>1} f(\Phi,w,s_{\tau'11})ds_{\tau'11}.
\end{aligned}$$

If $\tau_1=(1,2,3),(1,3,2)$, $\widetilde{L}_1(1+\delta,1+\delta)<4$. Therefore, we can ignore these cases in the first term. Since the function $\Sigma_{3,\mathrm{sub}}(\Psi,l_{\mathfrak{p}},\omega_{\mathfrak{d}},s_{\tau 1},1)$ is holomorphic at $s_{\tau 1}=(1,1)$, we get the following proposition.

Proposition (12.9.3) *By changing ψ if necessary,*

$$\begin{aligned}
\Xi_{\mathfrak{p}}(\Phi,\omega,w) &\sim C_G\Lambda(w;\rho)\frac{\delta_{\mathfrak{d}}(\omega)}{3}\Sigma_3(R_{\mathfrak{d}}\Phi_{\mathfrak{p}},\omega_{\mathfrak{d}},-1,-\frac{2}{3},\frac{2}{3})\\
&\quad + \sum_{\substack{\tau'=(\tau'_1,(1,2))\\ \tau'_1\in\mathfrak{W}'}}\frac{\varrho\delta_{\#}(\omega)}{2\pi\sqrt{-1}}\int_{\mathrm{Re}(s_{\tau'11})=r_1>1} f(\Phi,w,s_{\tau'11})ds_{\tau'11}.
\end{aligned}$$

§12.10 The case $\mathfrak{d} = (\beta_{10})$

In this section, $\mathfrak{d}_0 = (\beta_{10}), \mathfrak{d}_1 = (\beta_{10}, \beta_{10,1}), \mathfrak{d}_2 = (\beta_{10}, \beta_{10,2}), \mathfrak{d}_3 = (\beta_{10}, \beta_{10,3})$, and $\mathfrak{d}_4 = (\beta_{10}, \beta_{10,2}, \beta_{10,2,1})$. Let $\mathfrak{p}_0 = (\mathfrak{d}_0, \mathfrak{s}_0), \mathfrak{p}_{11} = (\mathfrak{d}_1, \mathfrak{s}_{11}), \mathfrak{p}_{12} = (\mathfrak{d}_1, \mathfrak{s}_{12}), \mathfrak{p}_{21} = (\mathfrak{d}_2, \mathfrak{s}_{21}), \mathfrak{p}_{22} = (\mathfrak{d}_2, \mathfrak{s}_{22}), \mathfrak{p}_{31} = (\mathfrak{d}_3, \mathfrak{s}_{31}), \mathfrak{p}_{32} = (\mathfrak{d}_3, \mathfrak{s}_{32})$ where $(\mathfrak{s}_{i1}(2), \mathfrak{s}_{i2}(2)) = (0,1)$ for $i = 1, \cdots, 4$. Let $\mathfrak{p}_{4i} = (\mathfrak{d}_4, \mathfrak{s}_{4i})$ for $i = 1, \cdots, 4$ where $(\mathfrak{s}_{4i}(2), \mathfrak{s}_{4i}(3)) = (0,0), (0,1), (1,0), (1,1)$ for $i = 1,2,3,4$ in that order. As in previous sections, we only consider such paths such that $\mathfrak{p}_0(1) = 0$ etc. Let $\Psi = R_{\mathfrak{d}_0}\Phi_{\mathfrak{p}_0}$ throughout this section.

Let

$$d_1(\lambda_1) = d_{P_2}(\lambda_1),$$

$$d_2(\lambda_1, \lambda_2) = a(\underline{\lambda}_1^{-2}, \underline{\lambda}_1, \underline{\lambda}_1; \underline{\lambda}_2^{-1}, \underline{\lambda}_2),$$

$$d_3(\lambda_1, \lambda_2, \lambda_3) = a(\underline{\lambda}_1^{-2}, \underline{\lambda}_1\underline{\lambda}_3^{-1}, \underline{\lambda}_1\underline{\lambda}_3; \underline{\lambda}_2^{-1}, \underline{\lambda}_2).$$

Easy computations show the following two lemmas.

Lemma (12.10.1)

(1) $e_{\mathfrak{p}_{i1}1}(d_3(\lambda_1,\lambda_2,\lambda_3)) = \lambda_1,\ e_{\mathfrak{p}_{i2}1}(d_3(\lambda_1,\lambda_2,\lambda_3)) = \lambda_1^{-1}$ *for* $i = 1, 3$.
(2) $e_{\mathfrak{p}_{21}1}(d_2(\lambda_1,\lambda_2)) = \lambda_1,\ e_{\mathfrak{p}_{22}1}(d_2(\lambda_1,\lambda_2)) = \lambda_1^{-1}$.
(3) $e_{\mathfrak{p}_{41}1}(d_3(\lambda_1,\lambda_2,\lambda_3)) = \lambda_1,\ e_{\mathfrak{p}_{41}2}(d_3(\lambda_1,\lambda_2,\lambda_3)) = \lambda_1^2\lambda_2$.
(4) $e_{\mathfrak{p}_{42}1}(d_3(\lambda_1,\lambda_2,\lambda_3)) = \lambda_1^{-1},\ e_{\mathfrak{p}_{41}2}(d_3(\lambda_1,\lambda_2,\lambda_3)) = (\lambda_1^2\lambda_2)^{-1}$.
(5) $e_{\mathfrak{p}_{43}1}(d_3(\lambda_1,\lambda_2,\lambda_3)) = \lambda_1^{-1},\ e_{\mathfrak{p}_{41}2}(d_3(\lambda_1,\lambda_2,\lambda_3)) = \lambda_1^2\lambda_2$.
(6) $e_{\mathfrak{p}_{44}1}(d_3(\lambda_1,\lambda_2,\lambda_3) = \lambda_1,\ e_{\mathfrak{p}_{41}2}(d_3(\lambda_1,\lambda_2,\lambda_3)) = (\lambda_1^2\lambda_2)^{-1}$.

Lemma (12.10.2)

(1) $\sigma_{\mathfrak{p}_0}(d_1(\lambda_1)) = \lambda_1^6$.
(2) $\sigma_{\mathfrak{p}_{11}}(d_3(\lambda_1,\lambda_2,\lambda_3)) = \sigma_{\mathfrak{p}_{12}}(d_3(\lambda_1,\lambda_2,\lambda_3)) = \lambda_1^4\lambda_2$.
(3) $\sigma_{\mathfrak{p}_{21}}(d_2(\lambda_1,\lambda_2)) = \sigma_{\mathfrak{p}_{22}}(d_2(\lambda_1,\lambda_2)) = \lambda_1^6\lambda_2^2$.
(4) $\sigma_{\mathfrak{p}_{31}}(d_3(\lambda_1,\lambda_2,\lambda_3)) = \sigma_{\mathfrak{p}_{32}}(d_3(\lambda_1,\lambda_2,\lambda_3)) = \lambda_1^4\lambda_2$.
(5) $\sigma_{\mathfrak{p}_{41}}(d_3(\lambda_1,\lambda_2,\lambda_3)) = \sigma_{\mathfrak{p}_{43}}(d_3(\lambda_1,\lambda_2,\lambda_3)) = \lambda_1^6\lambda_2^2\lambda_3^2$.
(6) $\sigma_{\mathfrak{p}_{42}}(d_3(\lambda_1,\lambda_2,\lambda_3)) = \sigma_{\mathfrak{p}_{44}}(d_3(\lambda_1,\lambda_2,\lambda_3)) = \lambda_2\lambda_3^2$.

Let $\mathfrak{P}_1 = \{\mathfrak{p}_0, \mathfrak{p}_{21}, \mathfrak{p}_{22}\}$, $\mathfrak{P}_2 = \{\mathfrak{p}_{11}, \mathfrak{p}_{12}\}$, $\mathfrak{P}_3 = \{\mathfrak{p}_{31}, \mathfrak{p}_{32}, \mathfrak{p}_{41}, \mathfrak{p}_{42}, \mathfrak{p}_{43}, \mathfrak{p}_{44}\}$. The paths $\mathfrak{p}_0, \mathfrak{p}_{21}, \mathfrak{p}_{22}, \mathfrak{p}_{4i}$ for $i = 1,2,3,4$ satisfy Condition (3.4.16)(1). The paths $\mathfrak{p}_{11}, \mathfrak{p}_{12}$ satisfy Condition (3.4.16)(2). For the paths $\mathfrak{p} = \mathfrak{p}_{31}, \mathfrak{p}_{32}$, $\widetilde{\Xi}_{\mathfrak{p}}(\Phi, \omega, w) = 0$. Therefore, by (3.5.9),

$$\begin{aligned}(12.10.3) \qquad \epsilon_{\mathfrak{p}_0}\Xi_{\mathfrak{p}_0}(\Phi,\omega,w) = &\sum_{\mathfrak{p}\in\mathfrak{P}_1} \epsilon_{\mathfrak{p}}(\Xi_{\mathfrak{p}+}(\Phi,\omega,w) + \widehat{\Xi}_{\mathfrak{p}+}(\Phi,\omega,w)) \\ &+ \sum_{\mathfrak{p}\in\mathfrak{P}_1} \epsilon_{\mathfrak{p}}(\widehat{\Xi}_{\mathfrak{p}\#}(\Phi,\omega,w) - \Xi_{\mathfrak{p}\#}(\Phi,\omega,w)) \\ &+ \sum_{\mathfrak{p}\in\mathfrak{P}_2} \epsilon_{\mathfrak{p}}(\Xi_{\mathfrak{p}}(\Phi,\omega,w) + \widetilde{\Xi}_{\mathfrak{p}}(\Phi,\omega,w)) \\ &+ \sum_{\mathfrak{p}\in\mathfrak{P}_3} \epsilon_{\mathfrak{p}}\Xi_{\mathfrak{p}}(\Phi,\omega,w),\end{aligned}$$

and all the distributions in the above formula are well defined for $\mathrm{Re}(w) \gg 0$.

(a) $\Xi_{\mathfrak{p}_0}(\Phi,\omega,w)$

Let $\mathfrak{d}=\mathfrak{d}_0, \mathfrak{p}=\mathfrak{p}_0$. Let $\mathfrak{d}_\mathfrak{d}=d_1(\lambda_1)g_\mathfrak{d}^0$ where $g_\mathfrak{d}^0 \in M_{\mathfrak{d}\mathbb{A}}^0$. Then $dg_\mathfrak{d}=2d^\times\lambda_1 dg_\mathfrak{d}^0$. It is easy to see that $\kappa_{\mathfrak{d}1}(d_1(\lambda_1))=\lambda_1^{-12}$. Since $d_1(\lambda_1)$ acts on $Z_{\mathfrak{d}\mathbb{A}}$ by multiplication by λ_1^2, the following proposition follows from (3.5.13) and (12.10.2)(1).

Proposition (12.10.4)

(1) $$\Xi^s_{\mathfrak{p}_0+}(\Phi,\omega,w)\sim C_G\Lambda(w;\rho)\delta_{\mathfrak{d}_0}(\omega)\Sigma_{\mathfrak{d}_0+}(\Psi,\omega_{\mathfrak{d}_0},3).$$

(2) $$\widehat{\Xi}^s_{\mathfrak{p}_0+}(\Phi,\omega,w)\sim C_G\Lambda(w;\rho)\delta_{\mathfrak{d}_0}(\omega)\Sigma_{\mathfrak{d}_0+}(\mathscr{F}_{\mathfrak{d}_0}\Psi,\omega_{\mathfrak{d}_0}^{-1},3).$$

Let $t^0=d_3(\lambda_1,\lambda_2,\lambda_3)\widehat{t}^0$, and $\widetilde{g}_\mathfrak{d}=n''(u_0)t^0 n_{B_2}(u_4)$ for $u_0,u_4\in\mathbb{A}$. It is easy to see that $d^\times t^0=2d^\times\lambda_1 d^\times\lambda_2 d^\times\lambda_3 d^\times\widehat{t}^0$. Let $d\widetilde{g}_\mathfrak{d}=\lambda_3^{-2}d^\times t^0 du_0 du_4$. Clearly, $\mathscr{E}_\mathfrak{p}(g_\mathfrak{d},w)$ is the constant term with respect to P_2. Let $\mathscr{E}_N(t^0,w)$ be the constant term with respect to the Borel subgroup as before. We show that we can replace $\mathscr{E}_\mathfrak{p}(g_\mathfrak{d},w)$ by $\mathscr{E}_N(t^0,w)$.

Lemma (12.10.5)

(1) $$\Xi_{\mathfrak{p},\mathrm{st}+}(\Phi,\omega,w)\sim\int_{\substack{X_\mathfrak{d}\cap B_\mathbb{A}/T_k\\ \lambda_1\geq 1}}\omega(t^0)\sigma_\mathfrak{p}(t^0)\Theta_{Z'_{\mathfrak{d},0}}(\Psi,\widetilde{g}_\mathfrak{d})\mathscr{E}_N(t^0,w)d\widetilde{g}_\mathfrak{d}.$$

(2) $$\widehat{\Xi}_{\mathfrak{p},\mathrm{st}+}(\Phi,\omega,w)\sim\int_{\substack{X_\mathfrak{d}\cap B_\mathbb{A}/T_k\\ \lambda_1\leq 1}}\omega(t^0)\sigma_\mathfrak{p}(t^0)\kappa_{\mathfrak{d}1}(t^0)\Theta_{Z'_{\mathfrak{d},0}}(\mathscr{F}_\mathfrak{d}\Psi,\theta_\mathfrak{d}(\widetilde{g}_\mathfrak{d}))\mathscr{E}_N(t^0,w)d\widetilde{g}_\mathfrak{d}.$$

Proof. Consider (1). By the consideration as in §2.1,

$$|\mathscr{E}_\mathfrak{p}(g_\mathfrak{d},w)-\mathscr{E}_N(t^0,w)|\leq\sum\widetilde{\mathscr{E}}_{I,\tau,1}(t^0,w),$$

where the summation is over $I=(I_1,I_2)$ such that $I_1=\{2\}$ or $I_2=\{1\}$.

By (3.4.30), there exist a slowly increasing function $h(t^0)$ and a constant $\delta>0$ such that if $M>w_0, l_1, l_2\gg 0$,

$$\sum\widetilde{\mathscr{E}}_{I,\tau,1}(t^0,w)\ll h(t^0)(\lambda_2^{l_1}+\lambda_3^{l_2})$$

for $w_0-\delta\leq\mathrm{Re}(w)\leq M$.

We choose Schwartz–Bruhat functions $0\leq\Psi_1\in\mathscr{S}(\mathbb{A}), 0\leq\Psi_2\in\mathscr{S}(\mathbb{A}^2)$ so that

$$\begin{aligned}&\int_\mathbb{A}|\Theta_{Z'_{\mathfrak{d},0}}(\Psi,\widetilde{g}_\mathfrak{d})|du_4\\ &\leq|\gamma_{2,33}(t^0)|^{-1}\Theta_1(\Psi_1,\gamma_{1,33}(t^0))\sum_{x\in k^\times}\Psi_2(\gamma_{2,22}(t^0)x,\gamma_{2,22}(t^0)xu_0).\end{aligned}$$

Then by (1.2.6), for any $N\geq 1, l_1, l_2\gg 0$, the difference of the left hand side and the right hand side of (1) is bounded by a constant multiple of the following integral

$$\begin{aligned}\int_{\substack{T^0_\mathbb{A}/T_k\times\mathbb{A}\\ \lambda_1\geq 1,\ \lambda_3\leq\sqrt{\alpha(u_0)}^{-1}}}&h'(t^0)(\lambda_1^2\lambda_2^{-1}\lambda_3^2)^{-N}(\lambda_2^{l_1}+\lambda_3^{l_2})\\ &\times\sum_{x\in k^\times}\Psi_2(\gamma_{2,22}(t^0)x,\gamma_{2,22}(t^0)xu_0)d^\times t^0 du_0,\end{aligned}$$

where $h'(t^0)$ is a slowly increasing function independent of N, l_1, l_2.

Let $\mu = \lambda_1^2\lambda_2\lambda_3^{-2}$. Then

$$(\lambda_1^2\lambda_2^{-1}\lambda_3^2)^{-N}(\lambda_2^{l_1}+\lambda_3^{l_2}) = \mu^N\lambda_1^{-4N}((\mu\lambda_1^{-2}\lambda_3^2)^{l_1}+\lambda_3^{l_2}).$$

If $N \gg l_1, l_2 \gg 0$, the above integral converges absolutely. This proves (1). Since $\theta_{\mathfrak{d}}(d_3(\lambda_1,\lambda_2,\lambda_3)) = d_3(\lambda_1^{-1},\lambda_2,\lambda_3)$, and the integral is for $\lambda_1 \leq 1$, the proof of (2) is similar.

Q.E.D.

Let τ, s_τ be as before. Let $\mu = \lambda_2\lambda_3^{-2}$. Then

(12.10.6)

$$(t^0)^{\tau z-\rho} = \lambda_1^{s_{\tau 11}+2s_{\tau 12}+3}\lambda_2^{s_{\tau 2}+1}\lambda_3^{s_{\tau 11}+1} = \lambda_1^{s_{\tau 11}+2s_{\tau 12}+3}\mu^{s_{\tau 2}+1}\lambda_3^{s_{\tau 11}+2s_{\tau 2}+3}.$$

Definition (12.10.7)

(1) $l_{\mathrm{st},1}(s_\tau) = (\frac{1}{2}(s_{\tau 11}+2s_{\tau 12}+1), s_{\tau 2}, \frac{1}{2}(-s_{\tau 11}-2s_{\tau 2}+3)$.

(2) $\widehat{l}_{\mathrm{st},1}(s_\tau) = \frac{1}{2}(7-s_{\tau 11}-2s_{\tau 12}), s_{\tau 2}, \frac{1}{2}(-s_{\tau 11}-2s_{\tau 2}+3))$.

(3) $l_{\mathrm{st},2}(s_\tau) = \widehat{l}_{\mathrm{st},2}(s_\tau) = s_{\tau 11}+2s_{\tau 2}-3$.

We define

$$\Sigma_{\mathfrak{p},\mathrm{st}+}(\Phi,\omega,s_\tau) = \delta_{\mathfrak{d},\mathrm{st}}(\omega)\frac{T_{\mathfrak{d}+}(R_{\mathfrak{d},0}\Psi,\omega_{\mathfrak{d},\mathrm{st}},l_{\mathrm{st},1}(s_\tau))}{l_{\mathrm{st},2}(s_\tau)},$$

$$\widehat{\Sigma}_{\mathfrak{p},\mathrm{st}+}(\Phi,\omega,s_\tau) = \delta_{\mathfrak{d},\mathrm{st}}(\omega)\frac{T_{\mathfrak{d}+}(R_{\mathfrak{d},0}\mathscr{F}_{\mathfrak{d}}\Psi,\omega_{\mathfrak{d},\mathrm{st}}^{-1},\widehat{l}_{\mathrm{st},2}(s_\tau))}{\widehat{l}_{\mathrm{st},2}(s_\tau)}.$$

Since $|\gamma_{2,33}(t^0)| = \lambda_1^2\lambda_2\lambda_3^2 = \lambda_1^2\mu\lambda_3^4$, by (12.10.6) and a similar consideration as before,

$$\Xi_{\mathfrak{p},\mathrm{st}+}(\Phi,\omega,w) \sim \sum_\tau\left(\frac{1}{2\pi\sqrt{-1}}\right)^3\int_{\mathrm{Re}(s_\tau)=r}\Sigma_{\mathfrak{p},\mathrm{st}+}(\Phi,\omega,s_\tau)\widetilde{\Lambda}_\tau(w;s_\tau)ds_\tau,$$

$$\widehat{\Xi}_{\mathfrak{p},\mathrm{st}+}(\Phi,\omega,w) \sim \sum_\tau\left(\frac{1}{2\pi\sqrt{-1}}\right)^3\int_{\mathrm{Re}(s_\tau)=r}\widehat{\Sigma}_{\mathfrak{p},\mathrm{st}+}(\Phi,\omega,s_\tau)\widetilde{\Lambda}_\tau(w;s_\tau)ds_\tau,$$

where we choose the contour so that $r \in D_\tau, r_1+r_2 > 3$.

Since $\Sigma_{\mathfrak{p},\mathrm{st}+}(\Phi,\omega,s_{\tau 1},1), \widehat{\Sigma}_{\mathfrak{p},\mathrm{st}+}(\Phi,\omega,s_{\tau 1},1)$ are holomorphic for $\mathrm{Re}(s_{\tau 11}) > 1$, we get the following proposition by the usual argument.

Proposition (12.10.8) *Suppose that $\tau = \tau_G$ and $\delta > 0$ is a small number. Then*

(1) $\Xi_{\mathfrak{p},\mathrm{st}+}(\Phi,\omega,w)$

$$\sim \frac{\varrho^2\delta_{\mathfrak{d},\mathrm{st}}(\omega)}{2\pi\sqrt{-1}}\int_{\mathrm{Re}(s_{\tau 11})=1+\delta}\Sigma_{\mathfrak{p},\mathrm{st}+}(\Phi,\omega,s_{\tau 11},1,1)\phi_2(s_{\tau 11})\widetilde{\Lambda}_1(w;s_{\tau 11},1)d_{\tau 11},$$

(2) $\widehat{\Xi}_{\mathfrak{p},\mathrm{st}+}(\Phi,\omega,w)$

$$\sim \frac{\varrho^2\delta_{\mathfrak{d},\mathrm{st}}(\omega)}{2\pi\sqrt{-1}}\int_{\mathrm{Re}(s_{\tau 11})=1+\delta}\widehat{\Sigma}_{\mathfrak{p},\mathrm{st}+}(\Phi,\omega,s_{\tau 11},1,1)\phi_2(s_{\tau 11})\widetilde{\Lambda}_1(w;s_{\tau 11},1)ds_{\tau 11}.$$

(b) $\Xi_{\mathfrak{p}_0\#}(\Phi,\omega,w)$, $\widehat{\Xi}_{\mathfrak{p}_0\#}(\Phi,\omega,w)$

Let $\tau=(\tau_1,(1,2))$. Let s_τ be as before. It is easy to see that

$$[d_1(\lambda_1)^{\tau z+\rho}]_{s_{\tau 11}=-1,s_{\tau 2}=1}=\lambda_1^{2s_{\tau 12}-4}.$$

$\kappa_{\mathfrak{d}_0}(d_1(\lambda_1))$ is as in (a). By (11.1.4), (11.1.5), (12.10.2), and the Mellin inversion formula,

$$\Xi_{\mathfrak{p}_0\#}(\Phi,\omega,w)=\sum_{\substack{\tau=(\tau_1,(1,2))\\ \tau_1\in\mathfrak{W}_2}}\frac{\delta_\#(\omega)\Psi(0)}{2\pi\sqrt{-1}}\int_{\mathrm{Re}(s_{\tau 12})=r_2>2}\frac{\widetilde{\Lambda}_{1\tau_1}(w;-1,s_{\tau 12})}{s_{\tau 12}+1}ds_{\tau 2},$$

$$\widehat{\Xi}_{\mathfrak{p}_0\#}(\Phi,\omega,w)=\sum_{\substack{\tau=(\tau_1,(1,2))\\ \tau_1\in\mathfrak{W}_2}}\frac{\delta_\#(\omega)\mathscr{F}_{\mathfrak{d}_0}\Psi(0)}{2\pi\sqrt{-1}}\int_{\mathrm{Re}(s_{\tau 12})=r_2>2}\frac{\widetilde{\Lambda}_{1\tau_1}(w;-1,s_{\tau 12})}{s_{\tau 12}-5}ds_{\tau 2}.$$

Proposition (12.10.9) *By changing ψ if necessary,*

$$\Xi_{\mathfrak{p}_0\#}(\Phi,\omega,w)\sim C_G\Lambda(w;\rho)\frac{\mathfrak{V}_2^2\delta_\#(\omega)\Psi(0)}{3},$$

$$\widehat{\Xi}_{\mathfrak{p}_0\#}(\Phi,\omega,w)\sim -C_G\Lambda(w;\rho)\frac{\mathfrak{V}_2^2\delta_\#(\omega)\mathscr{F}_{\mathfrak{d}_0}\Psi(0)}{3}.$$

Proof. If $\tau_1=1,(1,2)$, we choose $r_2\gg 0$, and we can ignore these cases. Suppose $\tau_1=(1,3,2)$. Then $\frac{1}{s_{\tau 2}+1},\frac{1}{s_{\tau 2}-4}$ are holomorphic at $\rho_{1\tau 1}=(-1,2)$. Therefore, the proposition follows from the passing principle (3.6.1) and the usual argument.

Q.E.D.

(c) $\Xi_{\mathfrak{p}_{11}}(\Phi,\omega,w)$, $\Xi_{\mathfrak{p}_{12}}(\Phi,\omega,w)$.

Let $\mathfrak{d}=\mathfrak{d}_1$. Let $t^0=d_3(\lambda_1,\lambda_2,\lambda_3)\widehat{t}^0$. We make the change of variables

$$\mu_1=|\gamma_{1,33}(t^0)|=\lambda_1^2\lambda_2^{-1}\lambda_3^2,\ \mu_2=|\gamma_{2,23}(t^0)|=\lambda_1^2\lambda_2.$$

Then it is easy to see that $d^\times t^0=d^\times\lambda_1 d^\times\mu_1 d^\times\mu_2 d^\times\widehat{t}^0$. Also $\lambda_1^4\lambda_2=\lambda_1^2\mu_2$.

Easy computations show the following lemma.

Lemma (12.10.10)

(1) $\theta_{\mathfrak{p}_{11}}(d_3(\lambda_1,\lambda_2,\lambda_3))^{\tau z+\rho}=\lambda_1^{-s_{\tau 11}+2s_{\tau 12}-2s_{\tau 2}+1}\mu_1^{\frac{1}{2}(s_{\tau 11}-1)}\mu_2^{\frac{1}{2}(s_{\tau 11}+2s_{\tau 2}-3)}$.

(2) $\theta_{\mathfrak{p}_{12}}(d_3(\lambda_1,\lambda_2,\lambda_3))^{\tau z+\rho}=\lambda_1^{-3s_{\tau 11}-2s_{\tau 12}-2s_{\tau 2}+7}\mu_1^{\frac{1}{2}(s_{\tau 11}-1)}\mu_2^{\frac{1}{2}(s_{\tau 11}+2s_{\tau 2}-3)}$.

Definition (12.10.11) *Let* $l_{\mathfrak{p}_{2i}}=(\frac{1}{2}(s_{\tau 11}-1),\frac{1}{2}(s_{\tau 11}+2s_{\tau 2}-1),l_{\mathfrak{p}_{2i},3})$ *where*

(1) $l_{\mathfrak{p}_{11},3}(s_\tau)=-s_{\tau 11}+2s_{\tau 12}-2s_{\tau 2}+3$.

(2) $l_{\mathfrak{p}_{12},3}(s_\tau)=3s_{\tau 11}+2s_{\tau 12}+2s_{\tau 2}-9$.

By (12.10.1), (12.10.2), (12.10.10) and a similar consideration as before,

$$\Xi_{\mathfrak{p}_{1i}}(\Phi,\omega,w)$$

$$=\delta_\#(\omega)\sum_\tau\left(\frac{1}{2\pi\sqrt{-1}}\right)^3\int_{\mathrm{Re}(s_\tau)=r}\Sigma_{2,\mathrm{sub}}(R_{\mathfrak{d}_1}\Phi_{\mathfrak{p}_{1i}},l_{\mathfrak{p}_{1i}},s_\tau)\widetilde{\Lambda}_\tau(w;s_\tau)ds_\tau,$$

where we choose the contour so that $r \in D_\tau, r_1 > 3, r_1 + 2r_3 > 3$, and $l_{\mathfrak{p}_{1i}}(r) > 0$ for $i = 1, 2$. Also the first (resp. second) coordinate of $R_{\mathfrak{d}_1}\Phi_{\mathfrak{p}_{1i}}$ corresponds to $x_{1,33}$ (resp. $x_{2,23}$).

We define

$$\Sigma_{\mathfrak{p}_{11}}(\Phi, s_{\tau 11}) = \frac{\Sigma_2(R_{\mathfrak{d}}\Phi_{\mathfrak{p}_{11}}, \frac{1}{2}(s_{\tau 11} - 1), \frac{1}{2}(s_{\tau 11} + 1))}{3 - s_{\tau 11}},$$
$$\Sigma_{\mathfrak{p}_{12}}(\Phi, s_{\tau 11}) = \frac{\Sigma_2(R_{\mathfrak{d}}\Phi_{\mathfrak{p}_{12}}, \frac{1}{2}(s_{\tau 11} - 1), \frac{1}{2}(s_{\tau 11} + 1))}{3s_{\tau 11} - 5}.$$

Proposition (12.10.12) *Suppose that $\tau = \tau_G$ and $\delta > 0$ is a small number. Then by changing ψ if necessary,*

$$(1)\quad \Xi_{\mathfrak{p}_{11}}(\Phi, \omega, w) \sim \frac{\varrho^2 \delta_\#(\omega)}{2\pi\sqrt{-1}} \int_{\mathrm{Re}(s_{\tau 11}) = 1+\delta} \Sigma_{\mathfrak{p}_{11}}(\Phi, s_{\tau 11}) \phi_2(s_{\tau 11}) \widetilde{\Lambda}_1(w; s_{\tau 11}, 1) ds_{\tau 11},$$
$$(2)\quad \Xi_{\mathfrak{p}_{12}}(\Phi, \omega, w) \sim \frac{\varrho^2 \delta_\#(\omega)}{2\pi\sqrt{-1}} \int_{\mathrm{Re}(s_{\tau 11}) = 1+\delta} \Sigma_{\mathfrak{p}_{12}}(\Phi, s_{\tau 11}) \phi_2(s_{\tau 11}) \widetilde{\Lambda}_1(w; s_{\tau 11}, 1) ds_{\tau 11}.$$

Proof. The proof of this proposition is similar to (12.6.9). The only difference is that we have to check that the points $(2, -1), (-1, 2), (1, 1)$ are not on lines $s_{\tau 11} = 2s_{\tau 12} + 1$, $2s_{\tau 12} + 3s_{\tau 11} = 7$ in order to use (3.6.1).

Q.E.D.

(d) $\Xi_{\mathfrak{p}_{21}+}(\Phi, \omega, w)$, $\widehat{\Xi}_{\mathfrak{p}_{21}+}(\Phi, \omega, w)$ etc.

In (d), (e), $\mathfrak{d} = \mathfrak{d}_2$. Let $g_{\mathfrak{d}} = d_2(\lambda_1, \lambda_2) g_{\mathfrak{d}}^0$. Then $dg_{\mathfrak{d}} = 2d^\times\lambda_1 d^\times\lambda_2 dg_{\mathfrak{d}}^0$. Let $\lambda = d_2(\lambda_1, \lambda_2)$, and $d^\times\lambda = d^\times\lambda_1 d^\times\lambda_2$. Let $\mu = \lambda_1^2\lambda_2$. Then $\lambda_1^6\lambda_2^2 = \lambda_1^2\mu^2$, $\lambda_2^{-1} = \lambda_1^2\mu^{-1}$, and $d^\times\lambda_1 d^\times\lambda_2 = d^\times\lambda_1 d^\times\mu$.

By (12.10.1), (12.10.2),

$$\Xi_{\mathfrak{p}_{21}+}(\Phi, \omega, w),\ \widehat{\Xi}_{\mathfrak{p}_{21}+}(\Phi, \omega, w),\ \Xi_{\mathfrak{p}_{22}+}(\Phi, \omega, w),\ \widehat{\Xi}_{\mathfrak{p}_{22}+}(\Phi, \omega, w)$$

are equal to

$$\int_{\substack{\mathbb{R}_+^2 \times M_{\mathfrak{d}\mathbb{A}}^0 / M_{\mathfrak{d}k} \\ \lambda_1 \leq 1,\ \lambda_1^2\lambda_2 \geq 1}} \omega(g_{\mathfrak{d}}) \lambda_1^6 \lambda_2^2 \Theta_{Z_{\mathfrak{d}}}(R_{\mathfrak{d}}\Phi_{\mathfrak{p}_{21}}, g_{\mathfrak{d}}) \mathscr{E}_{\mathfrak{p}_{21}}(g_{\mathfrak{d}}, w) d^\times\lambda dg_{\mathfrak{d}}^0,$$
$$\int_{\substack{\mathbb{R}_+^2 \times M_{\mathfrak{d}\mathbb{A}}^0 / M_{\mathfrak{d}k} \\ \lambda_1 \leq 1,\ \lambda_1^2\lambda_2 \leq 1}} \omega(g_{\mathfrak{d}}) \lambda_2^{-1} \Theta_{Z_{\mathfrak{d}}}(\mathscr{F}_{\mathfrak{d}} R_{\mathfrak{d}}\Phi_{\mathfrak{p}_{22}}, \theta_{\mathfrak{d}}(g_{\mathfrak{d}})) \mathscr{E}_{\mathfrak{p}_{21}}(g_{\mathfrak{d}}, w) d^\times\lambda dg_{\mathfrak{d}}^0,$$
$$\int_{\substack{\mathbb{R}_+^2 \times M_{\mathfrak{d}\mathbb{A}}^0 / M_{\mathfrak{d}k} \\ \lambda_1 \geq 1,\ \lambda_1^2\lambda_2 \geq 1}} \omega(g_{\mathfrak{d}})^{-1} \lambda_1^6 \lambda_2^2 \Theta_{Z_{\mathfrak{d}}}(R_{\mathfrak{d}}\Phi_{\mathfrak{p}_{21}}, g_{\mathfrak{d}}) \mathscr{E}_{\mathfrak{p}_{22}}(g_{\mathfrak{d}}, w) d^\times\lambda dg_{\mathfrak{d}}^0,$$
$$\int_{\substack{\mathbb{R}_+^2 \times M_{\mathfrak{d}\mathbb{A}}^0 / M_{\mathfrak{d}k} \\ \lambda_1 \geq 1,\ \lambda_1^2\lambda_2 \leq 1}} \omega(g_{\mathfrak{d}})^{-1} \lambda_2^{-1} \Theta_{Z_{\mathfrak{d}}}(\mathscr{F}_{\mathfrak{d}} R_{\mathfrak{d}}\Phi_{\mathfrak{p}_{22}}, \theta_{\mathfrak{d}}(g_{\mathfrak{d}})) \mathscr{E}_{\mathfrak{p}_{22}}(g_{\mathfrak{d}}, w) d^\times\lambda dg_{\mathfrak{d}}^0,$$

respectively.

Easy computations show the following lemma.

Lemma (12.10.13)
(1) $\theta_{\mathfrak{p}_{21}}(d_2(\lambda_1,\lambda_2))^{\tau z+\rho} = \lambda_1^{s_{\tau 11}+2s_{\tau 12}-2s_{\tau 2}-1}\mu^{s_{\tau 2}-1}$.
(2) $\theta_{\mathfrak{p}_{22}}(d_2(\lambda_1,\lambda_2))^{\tau z+\rho} = \lambda_1^{-s_{\tau 11}-2s_{\tau 12}-2s_{\tau 2}+5}\mu^{s_{\tau 2}-1}$.

Let

Definition (12.10.14) *Let* $l_{\mathfrak{p}_{2i}} = (l_{\mathfrak{p}_{2i},1}, l_{\mathfrak{p}_{2i},2})$, $\widehat{l}_{\mathfrak{p}_{2i}} = (\widehat{l}_{\mathfrak{p}_{2i},1}, \widehat{l}_{\mathfrak{p}_{2i},2})$ *be linear functions for* $i=1,2$ *where*
(1) $l_{\mathfrak{p}_{21},1}(s_\tau) = s_{\tau 2}+1$, $l_{\mathfrak{p}_{21},2}(s_\tau) = s_{\tau 11}+2s_{\tau 12}-2s_{\tau 2}+1$,
(2) $\widehat{l}_{\mathfrak{p}_{21},1}(s_\tau) = 3-l_{\mathfrak{p}_{21},1}(s_\tau)$, $\widehat{l}_{\mathfrak{p}_{21},2}(s_\tau) = l_{\mathfrak{p}_{21},2}(s_\tau)$,
(3) $l_{\mathfrak{p}_{22},1}(s_\tau) = s_{\tau 2}+1$, $l_{\mathfrak{p}_{22},2}(s_\tau) = s_{\tau 11}+2s_{\tau 12}+2s_{\tau 2}-7$,
(4) $\widehat{l}_{\mathfrak{p}_{22},1}(s_\tau) = 3-l_{\mathfrak{p}_{22},1}(s_\tau)$, $\widehat{l}_{\mathfrak{p}_{22},2}(s_\tau) = l_{\mathfrak{p}_{22},2}(s_\tau)$.

Note that $l_{\mathfrak{p}_{21}}(-s_{\tau 11}, s_{\tau 11}+s_{\tau 12}, s_{\tau 2}) = l_{\mathfrak{p}_{21}}(s_{\tau 11}, s_{\tau 12}, s_{\tau 2})$ etc.

By (12.10.1), (12.10.2) and a similar consideration as before,

$$\Xi_{\mathfrak{p}_{2i}+}(\Phi,\omega,w) = \delta_{\#}(\omega)\sum_{\substack{\tau_1\in\mathfrak{W}_2\\ \tau_2=1,(1,2)}} \Omega_{W,l_{\mathfrak{p}_{2i}},\tau}(R_{\mathfrak{d}}\Phi_{\mathfrak{p}_{2i}}, w),$$

$$\widehat{\Xi}_{\mathfrak{p}_{2i}+}(\Phi,\omega,w) = \delta_{\#}(\omega)\sum_{\substack{\tau_1\in\mathfrak{W}_2\\ \tau_2=1,(1,2)}} \Omega_{W,\widehat{l}_{\mathfrak{p}_{2i}},\tau}(\mathscr{F}_{\mathfrak{d}}R_{\mathfrak{d}}\Phi_{\mathfrak{p}_{2i}}, w),$$

for $i=1,2$.

We define

$$\Sigma_{\mathfrak{p}_{2i}+}(\Phi, s_{\tau 11}) = 2\Sigma_{W,\text{st},\text{sub}+}(R_{\mathfrak{d}_2}\Phi_{\mathfrak{p}_{2i}}, l_{\mathfrak{p}_{2i}}, s_{\tau 11}, 1, 1),$$
$$\widehat{\Sigma}_{\mathfrak{p}_{2i}+}(\Phi, s_{\tau 11}) = 2\Sigma_{W,\text{st},\text{sub}+}(\mathscr{F}_{\mathfrak{d}_2}R_{\mathfrak{d}_2}\Phi_{\mathfrak{p}_{2i}}, \widehat{l}_{\mathfrak{p}_{2i}}, s_{\tau 11}, 1, 1)$$

for $i=1,2$.

The following proposition follows from (11.2.3).

Proposition (12.10.15) *Suppose that* $\tau=\tau_G$ *and* $\delta>0$ *is a small number. Then by changing* ψ *if necessary,*

(1) $$\begin{aligned}&\Xi_{\mathfrak{p}_{2i}+}(\Phi,\omega,w)\\ &\sim 2C_G\Lambda(w;\rho)\delta_{\#}(\omega)\frac{\Sigma_{W+}(R_{\mathfrak{d}}\Phi_{\mathfrak{p}_{21}}, l_{\mathfrak{p}_{2i},1}(-1,2,1))}{l_{\mathfrak{p}_{2i},2}(-1,2,1)}\\ &\quad+\frac{\varrho^2\delta_{\#}(\omega)}{2\pi\sqrt{-1}}\int_{\mathrm{Re}(s_{\tau 11})=1+\delta}\Sigma_{\mathfrak{p}_{2i}+}(\Phi,s_{\tau 11})\phi_2(s_{\tau 11})\widetilde{\Lambda}_1(w;s_{\tau 11},1)ds_{\tau 11},\end{aligned}$$

(2) $$\begin{aligned}&\widehat{\Xi}_{\mathfrak{p}_{21}+}(\Phi,\omega,w)\\ &\sim 2C_G\Lambda(w;\rho)\delta_{\#}(\omega)\frac{\Sigma_{W+}(\mathscr{F}_{\mathfrak{d}_2}R_{\mathfrak{d}_2}\Phi_{\mathfrak{p}_{21}}, \widehat{l}_{\mathfrak{p}_{2i},1}(-1,2,1))}{\widehat{l}_{\mathfrak{p}_{2i},2}(-1,2,1)}\\ &\quad+\frac{\varrho^2\delta_{\#}(\omega)}{2\pi\sqrt{-1}}\int_{\mathrm{Re}(s_{\tau 11})=1+\delta}\widehat{\Sigma}_{\mathfrak{p}_{2i}+}(\Phi,s_{\tau 11})\phi_2(s_{\tau 11})\widetilde{\Lambda}_1(w;s_{\tau 11},1)ds_{\tau 11}\end{aligned}$$

for $i=1,2$.

(e) $\Xi_{\mathfrak{p}_{21}\#}(\Phi,\omega,w)$, $\Xi_{\mathfrak{p}_{22}\#}(\Phi,\omega,w)$ etc.

Let $\mu=\lambda_1^2\lambda_2$. Then $\lambda_1^6\lambda_2^2=\lambda_1^2\mu^2$, $\lambda_2^{-1}=\lambda_1^2\mu^{-1}$.

By (12.10.1), (12.10.2),

$$\Xi_{\mathfrak{p}_{21}\#}(\Phi,\omega,w) = R_{\mathfrak{d}_2}\Phi_{\mathfrak{p}_{21}}(0)\int_{\substack{\mathbb{R}^2_+\times M^0_{\mathfrak{d}\mathbb{A}}/M_{\mathfrak{d}k}\\ \lambda_1\leq 1,\ \mu\leq 1}} \omega(g^1_{\mathfrak{d}})\lambda_1^6\lambda_2^2\mathscr{E}_{\mathfrak{p}_{21}}(g_{\mathfrak{d}},w)d^\times\lambda_1 d^\times\mu dg^0_{\mathfrak{d}},$$

$$\widehat{\Xi}_{\mathfrak{p}_{21}\#}(\Phi,\omega,w) = \mathscr{F}_{\mathfrak{d}}R_{\mathfrak{d}_2}\Phi_{\mathfrak{p}_{21}}(0)\int_{\substack{\mathbb{R}^2_+\times M^0_{\mathfrak{d}\mathbb{A}}/M_{\mathfrak{d}k}\\ \lambda_1\leq 1,\ \mu\leq 1}} \omega(g^1_{\mathfrak{d}})\lambda_2^{-1}\mathscr{E}_{\mathfrak{p}_{21}}(g_{\mathfrak{d}},w)d^\times\lambda_1 d^\times\mu dg^0_{\mathfrak{d}},$$

$$\Xi_{\mathfrak{p}_{22}\#}(\Phi,\omega,w) = R_{\mathfrak{d}_2}\Phi_{\mathfrak{p}_{22}}(0)\int_{\substack{\mathbb{R}^2_+\times M^0_{\mathfrak{d}\mathbb{A}}/M_{\mathfrak{d}k}\\ \lambda_1\geq 1,\ \mu\leq 1}} \omega(g^1_{\mathfrak{d}})^{-1}\lambda_1^6\lambda_2^2\mathscr{E}_{\mathfrak{p}_{22}}(g_{\mathfrak{d}},w)d^\times\lambda_1 d^\times\mu dg^0_{\mathfrak{d}},$$

$$\widehat{\Xi}_{\mathfrak{p}_{22}\#}(\Phi,\omega,w) = \mathscr{F}_{\mathfrak{d}}R_{\mathfrak{d}_2}\Phi_{\mathfrak{p}_{22}}(0)\int_{\substack{\mathbb{R}^2_+\times M^0_{\mathfrak{d}\mathbb{A}}/M_{\mathfrak{d}k}\\ \lambda_1\geq 1,\ \mu\leq 1}} \omega(g^1_{\mathfrak{d}})^{-1}\lambda_2^{-1}\mathscr{E}_{\mathfrak{p}_{22}}(g_{\mathfrak{d}},w)d^\times\lambda_1 d^\times\mu dg^0_{\mathfrak{d}}.$$

Easy computations show the following lemma.

Lemma (12.10.16)
(1) $\theta_{\mathfrak{p}_{21}}(d_2(\lambda_1,\lambda_2))^{\tau z+\rho} = \lambda_1^{s_{\tau 11}+2s_{\tau 12}-3}\lambda_2^{s_{\tau 2}-1} = \lambda_1^{s_{\tau 11}+2s_{\tau 12}-2s_{\tau 2}-1}\mu^{s_{\tau 2}-1}$.
(2) $\theta_{\mathfrak{p}_{22}}(d_2(\lambda_1,\lambda_2))^{\tau z+\rho} = \lambda_1^{-s_{\tau 11}-2s_{\tau 12}+3}\lambda_2^{s_{\tau 2}-1} = \lambda_1^{-s_{\tau 11}-2s_{\tau 12}-2s_{\tau 2}+5}\mu^{s_{\tau 2}-1}$.

Let $s'_\tau = (s_{\tau 12}, s_{\tau 2})$ and $ds'_\tau = ds_{\tau 12}ds_{\tau 2}$. We define

$$\Sigma_{\mathfrak{p}_{21}\#,\tau}(\Phi,\omega,w,s'_\tau) = 2\delta_\#(\omega)R_{\mathfrak{d}_2}\Phi_{\mathfrak{p}_{21}}(0)\frac{\widetilde{\Lambda}_\tau(w;s_{\tau 12},-1,s_{\tau 2})}{(s_{\tau 12}-s_{\tau 2})(s_{\tau 2}+1)},$$

$$\widehat{\Sigma}_{\mathfrak{p}_{21}\#,\tau}(\Phi,\omega,w,s'_\tau) = 2\delta_\#(\omega)\mathscr{F}_{\mathfrak{d}}R_{\mathfrak{d}_2}\Phi_{\mathfrak{p}_{21}}(0)\frac{\widetilde{\Lambda}_\tau(w;s_{\tau 12},-1,s_{\tau 2})}{(s_{\tau 12}-s_{\tau 2})(s_{\tau 2}-2)},$$

$$\Sigma_{\mathfrak{p}_{22}\#,\tau}(\Phi,\omega,w,s'_\tau) = 2\delta_\#(\omega)R_{\mathfrak{d}_2}\Phi_{\mathfrak{p}_{22}}(0)\frac{\widetilde{\Lambda}_\tau(w;s_{\tau 12},-1,s_{\tau 2})}{(s_{\tau 12}+s_{\tau 2}-4)(s_{\tau 2}+1)},$$

$$\widehat{\Sigma}_{\mathfrak{p}_{22}\#,\tau}(\Phi,\omega,w,s'_\tau) = 2\delta_\#(\omega)\mathscr{F}_{\mathfrak{d}}R_{\mathfrak{d}_2}\Phi_{\mathfrak{p}_{22}}(0)\frac{\widetilde{\Lambda}_\tau(w;s_{\tau 12},-1,s_{\tau 2})}{(s_{\tau 12}+s_{\tau 2}-4)(s_{\tau 2}-2)}.$$

Then by (12.10.15) and a similar consideration as before,

$$\Xi_{\mathfrak{p}_{2i}\#}(\Phi,\omega,w) = \sum_{\substack{\tau_1\in\mathfrak{W}_2\\ \tau_2=1,(1,2)}}\left(\frac{1}{2\pi\sqrt{-1}}\right)^2\int_{\mathrm{Re}(s'_\tau)=(r_2,r_3)}\Sigma_{\mathfrak{p}_{2i}\#,\tau}(\Phi,\omega,w,s'_\tau)ds'_\tau,$$

$$\widehat{\Xi}_{\mathfrak{p}_{2i}\#}(\Phi,\omega,w) = \sum_{\substack{\tau_1\in\mathfrak{W}_2\\ \tau_2=1,(1,2)}}\left(\frac{1}{2\pi\sqrt{-1}}\right)^2\int_{\mathrm{Re}(s'_\tau)=(r_2,r_3)}\Sigma_{\mathfrak{p}_{2i}\#,\tau}(\Phi,\omega,w,s'_\tau)ds'_\tau,$$

where we choose the contour so that $(-1,r_2,r_3)\in D_\tau$ and $r_2>r_3$ etc.

Proposition (12.10.17) *By changing ψ if necessary,*

(1) $$\Xi_{\mathfrak{p}_{21}\#}(\Phi,\omega,w) \sim C_G\Lambda(w;\rho)\mathfrak{V}_2\delta_\#(\omega)R_{\mathfrak{d}_2}\Phi_{\mathfrak{p}_{21}}(0),$$

(2) $$\widehat{\Xi}_{\mathfrak{p}_{21}\#}(\Phi,\omega,w) \sim -2C_G\Lambda(w;\rho)\mathfrak{V}_2\delta_\#(\omega)\mathscr{F}_{\mathfrak{d}}R_{\mathfrak{d}_2}\Phi_{\mathfrak{p}_{21}}(0),$$

(3) $$\Xi_{\mathfrak{p}_{22}\#}(\Phi,\omega,w) \sim -C_G\Lambda(w;\rho)\mathfrak{V}_2\delta_\#(\omega)R_{\mathfrak{d}_2}\Phi_{\mathfrak{p}_{22}}(0),$$

(4) $$\widehat{\Xi}_{\mathfrak{p}_{22}\#}(\Phi,\omega,w) \sim 2C_G\Lambda(w;\rho)\mathfrak{V}_2\delta_\#(\omega)\mathscr{F}_{\mathfrak{d}}R_{\mathfrak{d}_2}\Phi_{\mathfrak{p}_{22}}(0).$$

Proof. We can ignore the case $\tau_2 = 1$ as usual, and assume that $\tau_2 = (1,2)$.

If $\tau_1 = 1, (1,2)$, we choose $r_2 \gg 0, r_3 = 1+\delta$ where $\delta > 0$ is a small number. Then $\widetilde{L}(-1, r_2, r_3) \le 2 + (1+\delta)C < w_0$ if δ is sufficiently small. Therefore, we can ignore these cases.

Suppose $\tau_1 = (1,3,2)$. Then the point $\rho_\tau = (-1,2,1)$ does not satisfy the conditions $s_{\tau 12} = s_{\tau 2}, s_{\tau 12} + s_{\tau 2} = 4, s_{\tau 2} = 2, -1$. Therefore, we can move the contour crossing these lines by the passing principle (3.6.1). Since $\Sigma_{\mathfrak{p}_{21}\#,\tau}(\Phi, \omega, w, s'_\tau)$ etc. are holomorphic at $s'_\tau = (2,1)$, the proposition follows.

Q.E.D.

(f) $\Xi_{\mathfrak{p}_{31}}(\Phi,\omega,w)$, $\Xi_{\mathfrak{p}_{32}}(\Phi,\omega,w)$.

Let $\mathfrak{d} = \mathfrak{d}_3$. Let t^0 be as in (c). Let $\mu = |\gamma_{2,33}(t^0)| = \lambda_1^2\lambda_2\lambda_3^2$. Then $\lambda_1^6\lambda_2^2\lambda_3^2 = \lambda_1^4\lambda_2\mu$, and $d^\times t^0 = d^\times\lambda_1 d^\times\lambda_2 d^\times\mu d^\times\widehat{t}^0$.

By (12.10.1), (12.10.2),

$$\Xi_{\mathfrak{p}_{31}}(\Phi,\omega,w) = \int_{\substack{T^0_{\mathbb{A}}/T_k \\ \lambda_1 \le 1}} \omega(t^0)^{-1}\lambda_1^6\lambda_2^2\lambda_3^2\Theta_{Z_\mathfrak{d}}(R_\mathfrak{d}\Phi_{\mathfrak{p}_{31}}, t^0)\mathscr{E}_{\mathfrak{p}_{31}}(t^0,w)d^\times t^0,$$

$$\Xi_{\mathfrak{p}_{32}}(\Phi,\omega,w) = \int_{\substack{T^0_{\mathbb{A}}/T_k \\ \lambda_1 \ge 1}} \omega(t^0)\lambda_1^6\lambda_2^2\lambda_3^2\Theta_{Z_\mathfrak{d}}(R_\mathfrak{d}\Phi_{\mathfrak{p}_{32}}, t^0)\mathscr{E}_{\mathfrak{p}_{32}}(t^0,w)d^\times t^0.$$

Easy computations show the following lemma.

Lemma (12.10.18)

(1) $\theta_{\mathfrak{p}_{31}}(d_3(\lambda_1,\lambda_2,\lambda_3))^{\tau z+\rho} = \lambda_1^{s_{\tau 12}-1}\lambda_2^{\frac{1}{2}(-s_{\tau 11}+2s_{\tau 2}+1)}\mu^{\frac{1}{2}(s_{\tau 11}-1)}$.

(2) $\theta_{\mathfrak{p}_{32}}(d_3(\lambda_1,\lambda_2,\lambda_3))^{\tau z+\rho} = \lambda_1^{-1s_{\tau 11}-s_{\tau 12}+4}\lambda_2^{\frac{1}{2}(-s_{\tau 11}+2s_{\tau 2}+1)}\mu^{\frac{1}{2}(s_{\tau 11}-1)}$.

Let

$$\Sigma_{\mathfrak{p}_{31},\tau}(\Phi,\omega,w,s_{\tau 12},s_{\tau 2}) = 2\delta_\#(\omega)\Sigma_1(R_\mathfrak{d}\Phi_{\mathfrak{p}_{31}}, s_{\tau 2}+1)\frac{\widetilde{\Lambda}_\tau(w; 2s_{\tau 2}+1, s_{\tau 12}, s_{\tau 2})}{s_{\tau 12}+1},$$

$$\Sigma_{\mathfrak{p}_{32},\tau}(\Phi,\omega,w,s_{\tau 12},s_{\tau 2}) = 2\delta_\#(\omega)\Sigma_1(R_\mathfrak{d}\Phi_{\mathfrak{p}_{32}}, s_{\tau 2}+1)\frac{\widetilde{\Lambda}_\tau(w; 2s_{\tau 2}+1, s_{\tau 12}, s_{\tau 2})}{s_{\tau 12}+2s_{\tau 2}-3}.$$

Then by (12.10.7) and a similar consideration as before,

$$\Xi_{\mathfrak{p}_{3i}}(\Phi,\omega,w) = \sum_\tau\left(\frac{1}{2\pi\sqrt{-1}}\right)^2\int_{\mathrm{Re}(s_{\tau 12},s_{\tau 2})=(r_1,r_3)}\Sigma_{\mathfrak{p}_{3i},\tau}(\Phi,\omega,w,s_{\tau 12},s_{\tau 2})ds_{\tau 12}ds_{\tau 2},$$

where we choose the contour so that $(2r_2+1, r_2, r_3) \in D_\tau, r_2 > 0$, and $r_2 > -1, r_2+2r_3 > 3$ for $i = 1,2$ respectively.

Proposition (12.10.19) *By changing ψ if necessary, $\Xi_{\mathfrak{p}_{3i}}(\Phi,\omega,w) \sim 0$ for $i = 1,2$.*

Proof. We can ignore the case $\tau_2 = 1$ as usual and assume that $\tau_2 = (1,2)$. Then the point ρ_τ is not on the line $s_{\tau 11} = 2s_{\tau 2}+1$ for all the cases. Therefore, the proposition follows from (3.5.19).

Q.E.D.

(g) $\Xi_{\mathfrak{p}_{4i}}(\Phi,\omega,w)$ for $i=1,\cdots,4$.

Let $\mathfrak{d}=\mathfrak{d}_4$. Let t^0 be as in (c). Let $\mu_1=\lambda_1^2\lambda_2$, and $\mu_2=\lambda_1^2\lambda_2\lambda_3^2$. Then $\lambda_1^6\lambda_2^2\lambda_3^2=\lambda_1^2\mu_1\mu_2$, $\lambda_2\lambda_3^2=\lambda_1^{-2}\mu_2$. It is easy to see that $d^\times t^0=d^\times\lambda_1 d^\times\mu_1 d^\times\mu_2 d^\times\widehat{t}^0$.

By (12.10.1), (12.10.2),

$$\Xi_{\mathfrak{p}_{41}}(\Phi,\omega,w)=\int_{\substack{T^0_{\mathbb{A}}/T_k\\ \lambda_1\le 1,\ \lambda_1^2\lambda_2\le 1}}\omega(t^0)\lambda_1^6\lambda_2^2\lambda_3^2\Theta_{Z_\mathfrak{d}}(R_\mathfrak{d}\Phi_{\mathfrak{p}_{41}},t^0)\mathscr{E}_{\mathfrak{p}_{41}}(t^0,w)d^\times t^0,$$

$$\Xi_{\mathfrak{p}_{42}}(\Phi,\omega,w)=\int_{\substack{T^0_{\mathbb{A}}/T_k\\ \lambda_1\ge 1,\ \lambda_1^2\lambda_2\ge 1}}\omega(t^0)^{-1}\lambda_2\lambda_3^2\Theta_{Z_\mathfrak{d}}(R_\mathfrak{d}\Phi_{\mathfrak{p}_{42}},t^0)\mathscr{E}_{\mathfrak{p}_{42}}(t^0,w)d^\times\lambda_1 d^\times t^0,$$

$$\Xi_{\mathfrak{p}_{43}}(\Phi,\omega,w)=\int_{\substack{T^0_{\mathbb{A}}/T_k\\ \lambda_1\ge 1,\ \lambda_1^2\lambda_2\le 1}}\omega(t^0)^{-1}\lambda_1^6\lambda_2^2\lambda_3^2\Theta_{Z_\mathfrak{d}}(R_\mathfrak{d}\Phi_{\mathfrak{p}_{43}},t^0)\mathscr{E}_{\mathfrak{p}_{43}}(t^0,w)d^\times t^0,$$

$$\Xi_{\mathfrak{p}_{44}}(\Phi,\omega,w)=\int_{\substack{T^0_{\mathbb{A}}/T_k\\ \lambda_1\le 1,\ \lambda_1^2\lambda_2\ge 1}}\omega(t^0)\lambda_2\lambda_3^2\Theta_{Z_\mathfrak{d}}(R_\mathfrak{d}\Phi_{\mathfrak{p}_{44}},t^0)\mathscr{E}_{\mathfrak{p}_{44}}(t^0,w)d^\times\lambda_1 d^\times t^0.$$

Easy computations show the following lemma.

Lemma (12.10.20)

(1) $\theta_{\mathfrak{p}_{41}}(t^0)^{\tau z+\rho}=\lambda_1^{s_{\tau 11}+2s_{\tau 12}-2s_{\tau 2}-1}\mu_1^{\frac12(-s_{\tau 11}+2s_{\tau 2}-1)}\mu_2^{\frac12(s_{\tau 11}-1)}$.
(2) $\theta_{\mathfrak{p}_{42}}(t^0)^{\tau z+\rho}=\lambda_1^{-s_{\tau 11}-2s_{\tau 12}+2s_{\tau 2}+1}\mu_1^{\frac12(-s_{\tau 11}-2s_{\tau 2}+3)}\mu_2^{\frac12(s_{\tau 11}-1)}$.
(3) $\theta_{\mathfrak{p}_{43}}(t^0)^{\tau z+\rho}=\lambda_1^{-s_{\tau 11}-2s_{\tau 12}-2s_{\tau 2}+5}\mu_1^{\frac12(-s_{\tau 11}+2s_{\tau 2}-1)}\mu_2^{\frac12(s_{\tau 11}-1)}$.
(4) $\theta_{\mathfrak{p}_{44}}(t^0)^{\tau z+\rho}=\lambda_1^{s_{\tau 11}+2s_{\tau 12}+2s_{\tau 2}-5}\mu_1^{\frac12(-s_{\tau 11}-2s_{\tau 2}+3)}\mu_2^{\frac12(s_{\tau 11}-1)}$.

Definition (12.10.21) *Let* $l_{\mathfrak{p}_{4i}}=(\frac12(s_{\tau 11}+1),l_{\mathfrak{p}_{4i},2})$ *where*
(1) $l_{\mathfrak{p}_{41},2}(s_\tau)=(s_{\tau 11}+2s_{\tau 12}-2s_{\tau 2}+1)(-s_{\tau 11}+2s_{\tau 2}+1)$,
(2) $l_{\mathfrak{p}_{42},2}(s_\tau)=(s_{\tau 11}+2s_{\tau 12}-2s_{\tau 2}+1)(s_{\tau 11}+2s_{\tau 2}-3)$,
(3) $l_{\mathfrak{p}_{43},2}(s_\tau)=(s_{\tau 11}+2s_{\tau 12}+2s_{\tau 2}-7)(-s_{\tau 11}+2s_{\tau 2}+1)$,
(4) $l_{\mathfrak{p}_{44},2}(s_\tau)=(s_{\tau 11}+2s_{\tau 12}+2s_{\tau 2}-7)(s_{\tau 11}+2s_{\tau 2}-3)$.

By (12.10.20) and a similar consideration as before,

$$\begin{aligned}&\Xi_{\mathfrak{p}_{4i}}(\Phi,\omega,w)\\&=2\delta_\#(\omega)\sum_\tau\left(\frac{1}{2\pi\sqrt{-1}}\right)^3\int_{\mathrm{Re}(s_\tau)=r}\Sigma_{1,\mathrm{sub}}(R_\mathfrak{d}\Phi_{\mathfrak{p}_{4i}},l_{\mathfrak{p}_{4i}},s_\tau)\widetilde{\Lambda}_\tau(w;s_\tau)ds_\tau,\end{aligned}$$

where we choose the contour so that $r\in D_\tau, r_1>1$ and both factors of $l_{\mathfrak{p}_{4i},2}$ are positive for $i=1,2,3,4$.

We define

$$\Sigma_{\mathfrak{p}_{41}}(\Phi,s_{\tau 11})=\frac{2\Sigma_1(R_{\mathfrak{d}_4}\Phi_{\mathfrak{p}_{41}},\frac12(s_{\tau 11}+1))}{(s_{\tau 11}+1)(3-s_{\tau 11})},$$

$$\Sigma_{\mathfrak{p}_{42}}(\Phi,s_{\tau 11})=\frac{2\Sigma_1(R_{\mathfrak{d}_4}\Phi_{\mathfrak{p}_{42}},\frac12(s_{\tau 11}+1))}{(s_{\tau 11}+1)(s_{\tau 11}-1)},$$

$$\Sigma_{\mathfrak{p}_{43}}(\Phi,s_{\tau 11})=\frac{2\Sigma_1(R_{\mathfrak{d}_4}\Phi_{\mathfrak{p}_{43}},\frac12(s_{\tau 11}+1))}{(s_{\tau 11}-3)(3-s_{\tau 11})},$$

$$\Sigma_{\mathfrak{p}_{44}}(\Phi,s_{\tau 11})=\frac{2\Sigma_1(R_{\mathfrak{d}_4}\Phi_{\mathfrak{p}_{44}},\frac12(s_{\tau 11}+1))}{(s_{\tau 11}-3)(s_{\tau 11}-1)}.$$

The following proposition follows from cases (3), (4), (11), (12) of (11.2.4).

Proposition (12.10.22) *By changing ψ if necessary,*

$$\Xi_{\mathfrak{p}_{4i}}(\Phi,\omega,w) \sim \frac{2\varrho^2\delta_{\#}(\omega)}{2\pi\sqrt{-1}} \int_{\mathrm{Re}(s_{\tau 11})=1+\delta} \Sigma_{\mathfrak{p}_{4i}}(\Phi, s_{\tau 11})\widetilde{\Lambda}_{1\tau_1}(w; s_{\tau 11}, 1)ds_{\tau 11}$$

for $i = 1, 2, 3, 4$.

(h) Now we combine the computations in this section. Let $\mathfrak{d} = \mathfrak{d}_0, \mathfrak{p} = \mathfrak{p}_0$. We define

$$\begin{aligned}
J_1(\Phi,\omega) =& \delta_{\mathfrak{d}}(\omega)\left(\Sigma_{\mathfrak{d}+}(\Psi,\omega_{\mathfrak{d}},3) + \Sigma_{\mathfrak{d}+}(\mathscr{F}_{\mathfrak{d}}\Psi,\omega_{\mathfrak{d}}^{-1},3)\right) \\
& - \frac{\mathfrak{V}_2^2\delta_{\#}(\omega)}{3}(R_{\mathfrak{d}}\Phi_{\mathfrak{p}}(0) + \mathscr{F}_{\mathfrak{d}}R_{\mathfrak{d}}\Phi_{\mathfrak{p}}(0)), \\
J_2(\Phi) =& -\left(\Sigma_{W+}(R_{\mathfrak{d}_2}\Phi_{\mathfrak{p}_{21}},2) + \Sigma_{W+}(\mathscr{F}_W R_{\mathfrak{d}_2}\Phi_{\mathfrak{p}_{21}},1)\right) \\
& - \left(\Sigma_{W+}(R_{\mathfrak{d}_2}\Phi_{\mathfrak{p}_{22}},2) + \Sigma_{W+}(\mathscr{F}_W R_{\mathfrak{d}_2}\Phi_{\mathfrak{p}_{22}},1)\right) \\
& + \mathfrak{V}_2\left(R_{\mathfrak{d}_2}\Phi_{\mathfrak{p}_{21}}(0) + 2\mathscr{F}_W R_{\mathfrak{d}_2}\Phi_{\mathfrak{p}_{21}}(0)\right) \\
& + \mathfrak{V}_2\left(R_{\mathfrak{d}_2}\Phi_{\mathfrak{p}_{22}}(0) + 2\mathscr{F}_W R_{\mathfrak{d}_2}\Phi_{\mathfrak{p}_{22}}(0)\right), \\
\Upsilon_{\mathfrak{d},1}(\Phi,\omega) =& J_1(\Phi,\omega) + \delta_{\#}(\omega)J_2(\Phi), \\
J_3(\Phi,\omega,s_{\tau 11}) =& \frac{T_{\mathfrak{d}+}(R_{\mathfrak{d},0}\Psi,\omega_{\mathfrak{d},\mathrm{st}},\frac{1}{2}(s_{\tau 11}+3),1,\frac{1}{2}(1-s_{\tau 11}))}{s_{\tau 11}-1} \\
& + \frac{T_{\mathfrak{d}+}(R_{\mathfrak{d},0}\mathscr{F}_{\mathfrak{d}}\Psi,\omega_{\mathfrak{d},\mathrm{st}}^{-1},\frac{1}{2}(5-s_{\tau 11}),1,\frac{1}{2}(1-s_{\tau 11}))}{s_{\tau 11}-1}, \\
J_4(\Phi,s_{\tau 11}) =& -\frac{2T_{W+}(\widetilde{R}_{W,0}R_{\mathfrak{d}_2}\Phi_{\mathfrak{p}_{21}},2,\frac{1}{2}(1-s_{\tau 11}))}{(s_{\tau 11}+1)(s_{\tau 11}-1)} \\
& - \frac{2T_{W+}(\widetilde{R}_{W,0}\mathscr{F}_W R_{\mathfrak{d}_2}\Phi_{\mathfrak{p}_{21}},1,\frac{1}{2}(1-s_{\tau 11}))}{(s_{\tau 11}+1)(s_{\tau 11}-1)}, \\
J_5(\Phi,s_{\tau 11}) =& \frac{2T_{W+}(\widetilde{R}_{W,0}R_{\mathfrak{d}_2}\Phi_{\mathfrak{p}_{22}},2,\frac{1}{2}(1-s_{\tau 11}))}{(s_{\tau 11}-3)(s_{\tau 11}-1)} \\
& + \frac{2T_{W+}(\widetilde{R}_{W,0}\mathscr{F}_W R_{\mathfrak{d}_2}\Phi_{\mathfrak{p}_{22}},1,\frac{1}{2}(1-s_{\tau 11}))}{(s_{\tau 11}-3)(s_{\tau 11}-1)}, \\
\Upsilon_{\mathfrak{d},2}(\Phi,\omega,s_{\tau 11}) =& \delta_{\mathfrak{d},\mathrm{st}}(\omega)J_3(\Phi,\omega,s_{\tau 11}) + \delta_{\#}(\omega)(J_4(\Phi,s_{\tau 11}) + J_5(\Phi,s_{\tau 11})) \\
& + \delta_{\#}(\omega)\sum_{i=1,2}(-1)^i\Sigma_{\mathfrak{p}_{1i}}(\Phi,s_{\tau 11}) \\
& + \delta_{\#}(\omega)(\Sigma_{\mathfrak{p}_{41}}(\Phi,s_{\tau 11}) - \Sigma_{\mathfrak{p}_{42}}(\Phi,s_{\tau 11})) \\
& + \delta_{\#}(\omega)(-\Sigma_{\mathfrak{p}_{43}}(\Phi,s_{\tau 11}) + \Sigma_{\mathfrak{p}_{44}}(\Phi,s_{\tau 11})).
\end{aligned}$$

By Theorem (10.0.1) and (12.10.3), (12.10.4), (12.10.8), (12.10.9), (12.10.12), (12.10.15), (12.10.17), (12.10.19), (12.10.22), we get the following proposition.

Proposition (12.10.23) *Suppose that $\tau = \tau_G$ and $\delta > 0$ is a small number. Then by changing ψ if necessary,*

$$\begin{aligned}
\Xi_{\mathfrak{p}_0}(\Phi,\omega,w) \sim & C_G\Lambda(w;\rho)\Upsilon_{\mathfrak{d},1}(\Phi,\omega,w) \\
& + \frac{\varrho^2}{2\pi\sqrt{-1}}\int_{\mathrm{Re}(s_{\tau 11})=1+\delta}\Upsilon_{\mathfrak{d},2}(\Phi,\omega,s_{\tau 11})\phi_2(s_{\tau 11})\widetilde{\Lambda}_1(w;s_{\tau 11},1)ds_{\tau 11}.
\end{aligned}$$

Let $\Upsilon_{\mathfrak{d},2,(0)}(\Phi,\omega,1)$ etc. be the i-th coefficient of the Laurent expansion as before. We define

$$J_6(\Phi,\omega) = -\frac{1}{2}(T_{\mathfrak{d}+}(R_{\mathfrak{d},0}\Psi,\omega_{\mathfrak{d},\mathrm{st}},2,1) + \frac{1}{2}T_{\mathfrak{d}+}(R_{\mathfrak{d},0}\mathscr{F}_{\mathfrak{d}}\Psi,\omega_{\mathfrak{d},\mathrm{st}}^{-1},4,1)).$$

Then it is easy to see that

$$\begin{aligned} J_{3,(0)}(\Phi,\omega,1) =& J_6(\Phi,\omega) + \frac{d}{ds_{\tau 11}}\bigg|_{s_{\tau 11}=1} T_{\mathfrak{d}+}(R_{\mathfrak{d},0}\Psi,\omega_{\mathfrak{d},\mathrm{st}},\frac{1}{2}(s_{\tau 11}+3),1,0) \\ &+ \frac{d}{ds_{\tau 11}}\bigg|_{s_{\tau 11}=1} T_{\mathfrak{d}+}(R_{\mathfrak{d},0}\mathscr{F}_{\mathfrak{d}}\Psi,\omega_{\mathfrak{d},\mathrm{st}}^{-1},\frac{1}{2}(5-s_{\tau 11}),1,0). \end{aligned}$$

Also,

$$\begin{aligned} &T_{\mathfrak{d}+}(R_{\mathfrak{d},0}\Psi,\omega_{\mathfrak{d},\mathrm{st}},\frac{1}{2}(s_{\tau 11}+3),1,0), \\ &= \int_{\substack{\mathbb{R}_+^2\times(\mathbb{A}^1/k^\times)^2 \\ \mu_1\geq 1}} \mu_1^{\frac{s_{\tau 11}+1}{2}} \Theta_2(R'_{\mathfrak{d},0}\Psi,\underline{\mu}_1\underline{\mu}_2^{-1}q_1,\underline{\mu}_1\underline{\mu}_2 q_2)d^\times\mu_1 d^\times\mu_2 d^\times q_1 d^\times q_2, \\ &T_{\mathfrak{d}+}(R_{\mathfrak{d},0}\mathscr{F}_{\mathfrak{d}}\Psi,\omega_{\mathfrak{d},\mathrm{st}}^{-1},\frac{1}{2}(5-s_{\tau 11}),1,0) \\ &= \int_{\substack{\mathbb{R}_+^2\times(\mathbb{A}^1/k^\times)^2 \\ \mu_1\geq 1}} \mu_1^{\frac{3-s_{\tau 11}}{2}} \Theta_2(R'_{\mathfrak{d},0}\mathscr{F}_{\mathfrak{d}}\Psi,\underline{\mu}_1\underline{\mu}_2^{-1}q_1,\underline{\mu}_1\underline{\mu}_2 q_2)d^\times\mu_1 d^\times\mu_2 d^\times q_1 d^\times q_2. \end{aligned}$$

For $\widetilde{\Psi}\in\mathscr{S}(Z'_{\mathfrak{d},0\mathbb{A}})$, We define

$$\mathscr{F}'_{\mathfrak{d},0}\widetilde{\Psi}(x_{1,33},x_{2,22}) = \int_{\mathbb{A}^2}\widetilde{\Psi}(y_{1,33},y_{2,12}) < x_{2,12}y_{2,12}+x_{1,33}y_{1,33} > dy_{2,12}dy_{1,33}.$$

Then if Φ is K-invariant,

$$R'_{\mathfrak{d},0}\mathscr{F}_{\mathfrak{d}}R_{\mathfrak{d}}\Phi(x_{1,33},x_{2,22}) = \mathscr{F}'_{\mathfrak{d},0}R'_{\mathfrak{d},0}R_{\mathfrak{d}}\Phi(x_{2,22},x_{1,33}).$$

For $\widetilde{\Psi}\in\mathscr{S}(Z'_{\mathfrak{d},0\mathbb{A}})$, let R_1,R_2 be the restrictions to the first coordinate and the second coordinate respectively. Also let

$$\begin{aligned} &\mathscr{F}_1\widetilde{\Psi}(x_{1,33}) = \int_{\mathbb{A}}\widetilde{\Psi}(y_{1,33},0) < x_{1,33}y_{1,33} > dy_{1,33}, \\ &\mathscr{F}_2\widetilde{\Psi}(x_{2,22}) = \int_{\mathbb{A}}\widetilde{\Psi}(0,y_{2,22}) < x_{2,22}y_{2,22} > dy_{2,22}. \end{aligned} \tag{12.10.24}$$

The proof of the following lemma is easy, and is left to the reader.

Lemma (12.10.25) *Suppose that Φ is K-invariant. Then*

(1) $\mathscr{F}_1R'_{\mathfrak{d},0}\mathscr{F}_{\mathfrak{d}}\Psi(0) = \mathscr{F}_1R'_{\mathfrak{d},0}\Psi(0)$, $\mathscr{F}_2R'_{\mathfrak{d},0}\mathscr{F}_{\mathfrak{d}}\Psi(0) = \mathscr{F}_2R'_{\mathfrak{d},0}\Psi(0)$,

(2) $\Sigma_{2,(-1,-1)}(R_{\mathfrak{d}_1}\Phi_{\mathfrak{p}_{11}},0,1) = -\Sigma_{1,(-1)}(R_{\mathfrak{d}_4}\Phi_{\mathfrak{p}_{42}},1)$,

(3) $\Sigma_{2,(-1,-1)}(R_{\mathfrak{d}_1}\Phi_{\mathfrak{p}_{12}},0,1) = -\Sigma_{1,(-1)}(R_{\mathfrak{d}_4}\Phi_{\mathfrak{p}_{44}},1)$,

(4) $\Sigma_{1,(i)}(\mathscr{F}_1R'_{\mathfrak{d},0}\Psi,1) = (-1)^i\Sigma_{2,(i,-1)}(R_{\mathfrak{d}_1}\Phi_{\mathfrak{p}_{11}},0,1)$ for all i,

(5) $\Sigma_{1,(i)}(\mathscr{F}_1R'_{\mathfrak{d},0}\mathscr{F}_{\mathfrak{d}}\Psi,1) = (-1)^i\Sigma_{2,(i,-1)}(R_{\mathfrak{d}_1}\Phi_{\mathfrak{p}_{12}},0,1)$ for all i.

Lemma (12.10.26)

$$\begin{aligned}
&\delta_{\mathfrak{d},\mathrm{st}}(\omega)T_{\mathfrak{d}+}(R_{\mathfrak{d},0}\Psi,\omega_{\mathfrak{d},\mathrm{st}},\frac{1}{2}(s_{\tau 11}+3),1,0)\\
&\quad+\delta_{\mathfrak{d},\mathrm{st}}(\omega)T_{\mathfrak{d}+}(R_{\mathfrak{d},0}\mathscr{F}_{\mathfrak{d}}\Psi,\omega_{\mathfrak{d},\mathrm{st}}^{-1},\frac{1}{2}(5-s_{\tau 11}),1,0)\\
&=\frac{\delta_{\mathfrak{d},\mathrm{st}}(\omega)}{2}\Sigma_2(R'_{\mathfrak{d},0}\Psi,\omega_{\mathfrak{d},\mathrm{st}},\frac{s_{\tau 11}+1}{4},\frac{s_{\tau 11}+1}{4})\\
&\quad+\delta_{\#}(\omega)\left(\frac{4\Sigma_{2,(-1,-1)}(R_{\mathfrak{d}_1}\Phi_{\mathfrak{p}_{11}},0,1)}{(s_{\tau 11}+1)^2}+\frac{2\Sigma_{2,(0,-1)}(R_{\mathfrak{d}_1}\Phi_{\mathfrak{p}_{11}},0,1)}{s_{\tau 11}+1}\right)\\
&\quad+\delta_{\#}(\omega)\left(\frac{4\Sigma_{2,(-1,-1)}(R_{\mathfrak{d}_1}\Phi_{\mathfrak{p}_{12}},0,1)}{(s_{\tau 11}-3)^2}-\frac{2\Sigma_{2,(0,-1)}(R_{\mathfrak{d}_1}\Phi_{\mathfrak{p}_{12}},0,1)}{s_{\tau 11}-3}\right)\\
&\quad+\delta_{\#}(\omega)\left(-\frac{4\Sigma_{1,(-1)}(R_{\mathfrak{d}_4}\Phi_{\mathfrak{p}_{42}},1)}{(s_{\tau 11}+1)^2}+\frac{2\Sigma_{1,(0)}(R_{\mathfrak{d}_4}\Phi_{\mathfrak{p}_{42}},1)}{s_{\tau 11}+1}\right)\\
&\quad-\delta_{\#}(\omega)\left(\frac{4\Sigma_{1,(-1)}(R_{\mathfrak{d}_4}\Phi_{\mathfrak{p}_{44}},1)}{(s_{\tau 11}-3)^2}+\frac{2\Sigma_{1,(0)}(R_{\mathfrak{d}_4}\Phi_{\mathfrak{p}_{44}},1)}{s_{\tau 11}-3}\right).
\end{aligned}$$

Proof. The functions

$$T_{\mathfrak{d}+}(R_{\mathfrak{d},0}\Psi,\omega_{\mathfrak{d},\mathrm{st}},\frac{1}{2}(s_{\tau 11}+3),1,0),\ T_{\mathfrak{d}+}(R_{\mathfrak{d},0}\mathscr{F}_{\mathfrak{d}}\Psi,\omega_{\mathfrak{d},\mathrm{st}}^{-1},\frac{1}{2}(5-s_{\tau 11}),1,0))$$

are equal to the following integrals

$$\int_{\substack{\mathbb{R}_+^2\times(\mathbb{A}^1/k^\times)^2\\ \lambda_1\geq 1}}\omega_{\mathfrak{d},\mathrm{st}}(q_1,q_2)\lambda_1^{\frac{s_{\tau 11}+1}{2}}\Theta_2(R'_{\mathfrak{d},0}\Psi,\underline{\lambda}_1\underline{\lambda}_2^{-1},\underline{\lambda}_1\underline{\lambda}_2)d^\times\lambda_1 d^\times\lambda_2 d^\times q_1 d^\times q_2,$$

$$\int_{\substack{\mathbb{R}_+^2\times(\mathbb{A}^1/k^\times)^2\\ \lambda_1\geq 1}}\omega_{\mathfrak{d},\mathrm{st}}^{-1}(q_1,q_2)\lambda_1^{\frac{3-s_{\tau 11}}{2}}\Theta_2(R'_{\mathfrak{d},0}\mathscr{F}_{\mathfrak{d}}\Psi,\underline{\lambda}_1\underline{\lambda}_2^{-1},\underline{\lambda}_1\underline{\lambda}_2)d^\times\lambda_1 d^\times\lambda_2 d^\times q_1 d^\times q_2.$$

Note that $\omega_{\mathfrak{d},\mathrm{st}}(q_1,q_2)=\omega_{\mathfrak{d},\mathrm{st}}(q_2,q_1)$.

By (6.1.2),

$$\begin{aligned}
&\lambda_1^{\frac{s_{\tau 11}+1}{2}}\Theta_2(R'_{\mathfrak{d},0}\Psi,\underline{\lambda}_1\underline{\lambda}_2^{-1}q_1,\underline{\lambda}_1\underline{\lambda}_2q_2)-\lambda_1^{\frac{s_{\tau 11}-3}{2}}\Theta_2(\mathscr{F}'_{\mathfrak{d},0}R'_{\mathfrak{d},0}\Psi,\underline{\lambda}_1^{-1}\underline{\lambda}_2q_1^{-1},\underline{\lambda}_1\underline{\lambda}_2q_2^{-1})\\
&=\lambda_1^{\frac{s_{\tau 11}-1}{2}}\lambda_2^{-1}\Theta_1(\mathscr{F}_2R'_{\mathfrak{d},0}\mathscr{F}_{\mathfrak{d}}\Psi,\underline{\lambda}_1\underline{\lambda}_2^{-1}q_1)+\lambda_1^{\frac{s_{\tau 11}-3}{2}}\Theta_1(R_1R'_{\mathfrak{d},0}\mathscr{F}_{\mathfrak{d}}\Psi,\underline{\lambda}_1^{-1}\underline{\lambda}_2^{-1}q_2^{-1})\\
&\quad-\lambda_1^{\frac{s_{\tau 11}+1}{2}}\lambda_2^{-1}\Theta_1(R_1R'_{\mathfrak{d},0}\Psi,\underline{\lambda}_1\underline{\lambda}_2^{-1}q_1)-\lambda_1^{\frac{s_{\tau 11}-1}{2}}\lambda_2^{-1}\Theta_1(\mathscr{F}_2R'_{\mathfrak{d},0}\Psi,\underline{\lambda}_1^{-1}\underline{\lambda}_2^{-1}q_2^{-1})\\
&=\lambda_1^{\frac{s_{\tau 11}-3}{2}}\lambda_2^{-1}\Theta_1(R_2R'_{\mathfrak{d},0}\mathscr{F}_{\mathfrak{d}}\Psi,\underline{\lambda}_1^{-1}\underline{\lambda}_2q_1^{-1})+\lambda_1^{\frac{s_{\tau 11}-1}{2}}\lambda_2\Theta_1(\mathscr{F}_1R'_{\mathfrak{d},0}\mathscr{F}_{\mathfrak{d}}\Psi,\underline{\lambda}_1\underline{\lambda}_2q_2)\\
&\quad-\lambda_1^{\frac{s_{\tau 11}-1}{2}}\lambda_2\Theta_1(\mathscr{F}_1R'_{\mathfrak{d},0}\Psi,\underline{\lambda}_1^{-1}\underline{\lambda}_2q_1^{-1})-\lambda_1^{\frac{s_{\tau 11}+1}{2}}\lambda_2\Theta_1(R_2R'_{\mathfrak{d},0}\Psi,\underline{\lambda}_1\underline{\lambda}_2q_2).
\end{aligned}$$

Note that each term is independent of either q_1 or q_2.

We divide the integral according to $\lambda_2\leq 1$ or $\lambda_2\geq 1$. If $\lambda_2\leq 1$, we use the first equation, and if $\lambda_2\geq 1$, we use the second equation. Then (12.10.26) is equal to the sum of the function

$$\frac{\delta_{\mathfrak{d},\mathrm{st}}(\omega)}{2}\Sigma_2(R'_{\mathfrak{d},0}\Psi,\omega_{\mathfrak{d},\mathrm{st}},\frac{s_{\tau 11}+1}{4},\frac{s_{\tau 11}+1}{4}),$$

and the integral of the function

$$
\begin{aligned}
&-\delta_{\#}(\omega)(\lambda_1^{\frac{s_{\tau 11}-1}{2}}\Sigma_{1+}((\mathscr{F}_2R'_{\mathfrak{d},0}\mathscr{F}_{\mathfrak{d}}\Psi)_{\lambda_1},1)+\lambda_1^{\frac{s_{\tau 11}-3}{2}}\Sigma_{1+}((R_1R'_{\mathfrak{d},0}\mathscr{F}_{\mathfrak{d}}\Psi)_{\lambda_1^{-1}},0))\\
&+\delta_{\#}(\omega)(\lambda_1^{\frac{s_{\tau 11}+1}{2}}\Sigma_{1+}((R_1R'_{\mathfrak{d},0}\Psi)_{\lambda_1},0)+\lambda_1^{\frac{s_{\tau 11}-1}{2}}\Sigma_{1+}((\mathscr{F}_2R'_{\mathfrak{d},0}\Psi)_{\lambda_1^{-1}},1))\\
&-\delta_{\#}(\omega)(\lambda_1^{\frac{s_{\tau 11}-3}{2}}\Sigma_{1+}((R_2R'_{\mathfrak{d},0}\mathscr{F}_{\mathfrak{d}}\Psi)_{\lambda_1^{-1}},0)+\lambda_1^{\frac{s_{\tau 11}-1}{2}}\Sigma_{1+}((\mathscr{F}_1R'_{\mathfrak{d},0}\mathscr{F}_{\mathfrak{d}}\Psi)_{\lambda_1},1))\\
&+\delta_{\#}(\omega)(\lambda_1^{\frac{s_{\tau 11}-1}{2}}\Sigma_{1+}((\mathscr{F}_1R'_{\mathfrak{d},0}\Psi)_{\lambda_1^{-1}},1)+\lambda_1^{\frac{s_{\tau 11}+1}{2}}\Sigma_{1+}((R_2R'_{\mathfrak{d},0}\Psi)_{\lambda_1},0))
\end{aligned}
$$

over $0\le\lambda_1\le 1$.

By the principal part formula for the standard L-function in one variable,

$$
\begin{aligned}
&\lambda_1^{\frac{s_{\tau 11}-1}{2}}\Sigma_{1+}((\mathscr{F}_2R'_{\mathfrak{d},0}\mathscr{F}_{\mathfrak{d}}\Psi)_{\lambda_1},1)+\lambda_1^{\frac{s_{\tau 11}-3}{2}}\Sigma_{1+}((R_2R'_{\mathfrak{d},0}\mathscr{F}_{\mathfrak{d}}\Psi)_{\lambda_1^{-1}},0)\\
&=\lambda_1^{\frac{s_{\tau 11}-1}{2}}\Sigma_{1,(0)}((\mathscr{F}_2R'_{\mathfrak{d},0}\mathscr{F}_{\mathfrak{d}}\Psi)_{\lambda_1},1)+\lambda_1^{\frac{s_{\tau 11}-1}{2}}\mathscr{F}_2R'_{\mathfrak{d},0}\mathscr{F}_{\mathfrak{d}}\Psi(0),\\
&\lambda_1^{\frac{s_{\tau 11}-3}{2}}\Sigma_{1+}((R_1R'_{\mathfrak{d},0}\mathscr{F}_{\mathfrak{d}}\Psi)_{\lambda_1^{-1}},0)+\lambda_1^{\frac{s_{\tau 11}-1}{2}}\Sigma_{1+}((\mathscr{F}_1R'_{\mathfrak{d},0}\mathscr{F}_{\mathfrak{d}}\Psi)_{\lambda_1},1)\\
&=\lambda_1^{\frac{s_{\tau 11}-1}{2}}\Sigma_{1,(0)}((\mathscr{F}_1R'_{\mathfrak{d},0}\mathscr{F}_{\mathfrak{d}}\Psi)_{\lambda_1},1)+\lambda_1^{\frac{s_{\tau 11}-1}{2}}\mathscr{F}_1R'_{\mathfrak{d},0}\mathscr{F}_{\mathfrak{d}}\Psi(0),\\
&\lambda_1^{\frac{s_{\tau 11}+1}{2}}\Sigma_{1+}((R_1R'_{\mathfrak{d},0}\Psi)_{\lambda_1},0)+\lambda_1^{\frac{s_{\tau 11}-1}{2}}\Sigma_{1+}((\mathscr{F}_1R'_{\mathfrak{d},0}\Psi)_{\lambda_1^{-1}},1)\\
&=\lambda_1^{\frac{s_{\tau 11}-1}{2}}\Sigma_{1,(0)}((\mathscr{F}_1R'_{\mathfrak{d},0}\Psi)_{\lambda_1^{-1}},1)+\lambda_1^{\frac{s_{\tau 11}-1}{2}}\mathscr{F}_1R'_{\mathfrak{d},0}\Psi(0),\\
&\lambda_1^{\frac{s_{\tau 11}-1}{2}}\Sigma_{1+}((\mathscr{F}_2R'_{\mathfrak{d},0}\Psi)_{\lambda_1^{-1}},1)+\lambda_1^{\frac{s_{\tau 11}+1}{2}}\Sigma_{1+}((R_2R'_{\mathfrak{d},0}\Psi)_{\lambda_1},0)\\
&=\lambda_1^{\frac{s_{\tau 11}-1}{2}}\Sigma_{1,(0)}((\mathscr{F}_2R'_{\mathfrak{d},0}\Psi)_{\lambda_1^{-1}},1)+\lambda_1^{\frac{s_{\tau 11}-1}{2}}\mathscr{F}_2R'_{\mathfrak{d},0}\Psi(0).
\end{aligned}
$$

Note that $\mathscr{F}_2R'_{\mathfrak{d},0}\mathscr{F}_{\mathfrak{d}}\Psi$ is the Fourier transform of $R_2R'_{\mathfrak{d},0}\mathscr{F}_{\mathfrak{d}}\Psi$ with respect to the standard bilinear form on $\mathbb{A}$ etc.

By (12.10.25), terms like $\lambda_1^{\frac{s_{\tau 11}-1}{2}}\mathscr{F}_2\mathscr{F}_{\mathfrak{d}}\Psi(0)$ cancel out. It is easy to see that

$$
\begin{aligned}
\Sigma_{1,(0)}((\mathscr{F}_2R'_{\mathfrak{d},0}\mathscr{F}_{\mathfrak{d}}\Psi)_{\lambda_1},1)=&-\lambda_1^{-1}(\log\lambda_1)\Sigma_{1,(-1)}(\mathscr{F}_2R'_{\mathfrak{d},0}\mathscr{F}_{\mathfrak{d}}\Psi,1)\\
&+\lambda_1^{-1}\Sigma_{1,(0)}(\mathscr{F}_2R'_{\mathfrak{d},0}\mathscr{F}_{\mathfrak{d}}\Psi,1),\\
\Sigma_{1,(0)}((\mathscr{F}_1R'_{\mathfrak{d},0}\mathscr{F}_{\mathfrak{d}}\Psi)_{\lambda_1},1)=&-\lambda_1^{-1}(\log\lambda_1)\Sigma_{1,(-1)}(\mathscr{F}_1R'_{\mathfrak{d},0}\mathscr{F}_{\mathfrak{d}}\Psi,1)\\
&+\lambda_1^{-1}\Sigma_{1,(0)}(\mathscr{F}_1R'_{\mathfrak{d},0}\mathscr{F}_{\mathfrak{d}}\Psi,1),\\
\Sigma_{1,(0)}((\mathscr{F}_1R'_{\mathfrak{d},0}\Psi)_{\lambda_1^{-1}},1)=&\lambda_1(\log\lambda_1)\Sigma_{1,(-1)}(\mathscr{F}_1R'_{\mathfrak{d},0}\Psi,1)\\
&+\lambda_1\Sigma_{1,(0)}(\mathscr{F}_1R'_{\mathfrak{d},0}\Psi,1),\\
\Sigma_{1,(0)}((\mathscr{F}_2R'_{\mathfrak{d},0}\Psi)_{\lambda_1^{-1}},1)=&\lambda_1(\log\lambda_1)\Sigma_{1,(0)}(\mathscr{F}_2R'_{\mathfrak{d},0}\Psi,1)\\
&+\lambda_1\Sigma_{1,(0)}(\mathscr{F}_2R'_{\mathfrak{d},0}\Psi,1).
\end{aligned}
$$

If Φ is K-invariant,

$$
\mathscr{F}_2R'_{\mathfrak{d},0}\mathscr{F}_{\mathfrak{d}}\Psi=R_{\mathfrak{d}_4}\Phi_{\mathfrak{p}_{44}},\quad \mathscr{F}_2R'_{\mathfrak{d},0}\Psi=R_{\mathfrak{d}_4}\Phi_{\mathfrak{p}_{42}}.
$$

Therefore, integrating over $\lambda_1 \le 1$, and using (12.10.25), we get the proposition. Q.E.D.

We define

$$\begin{aligned} J_7(\Phi,\omega) =& \frac{1}{8}(\Sigma_{2,(1,0)}(R'_{\mathfrak{d},0}\Psi,\omega_{\mathfrak{d},\mathrm{st}},\frac{1}{2},\frac{1}{2}) + \Sigma_{2,(0,1)}(R'_{\mathfrak{d},0}\Psi,\omega_{\mathfrak{d},\mathrm{st}},\frac{1}{2},\frac{1}{2})), \\ J_8(\Phi) =& -\Sigma_{2,(-1,-1)}(R_{\mathfrak{d}_1}\Phi_{\mathfrak{p}_{11}},0,1) - \frac{1}{2}\Sigma_{2,(0,-1)}(R_{\mathfrak{d}_1}\Phi_{\mathfrak{p}_{11}},0,1) \\ &+ \Sigma_{2,(-1,-1)}(R_{\mathfrak{d}_1}\Phi_{\mathfrak{p}_{12}},0,1) + \frac{1}{2}\Sigma_{2,(0,-1)}(R_{\mathfrak{d}_1}\Phi_{\mathfrak{p}_{12}},0,1) \\ &+ \Sigma_{1,(-1)}(R_{\mathfrak{d}_4}\Phi_{\mathfrak{p}_{42}},1) - \frac{1}{2}\Sigma_{1,(0)}(R_{\mathfrak{d}_4}\Phi_{\mathfrak{p}_{42}},1) \\ &- \Sigma_{1,(-1)}(R_{\mathfrak{d}_4}\Phi_{\mathfrak{p}_{44}},1) + \frac{1}{2}\Sigma_{1,(0)}(R_{\mathfrak{d}_4}\Phi_{\mathfrak{p}_{44}},1), \\ J_9(\Phi) =& \frac{1}{2}(T_{W+}(\widetilde{R}_{W,0}R_{\mathfrak{d}_2}\Phi_{\mathfrak{p}_{21}},2) + T_{W+}(\widetilde{R}_{W,0}\mathscr{F}_W R_{\mathfrak{d}_2}\Phi_{\mathfrak{p}_{21}},1)), \\ J_{10}(\Phi) =& \frac{1}{2}(T_{W+}(\widetilde{R}_{W,0}R_{\mathfrak{d}_2}\Phi_{\mathfrak{p}_{22}},2) + T_{W+}(\widetilde{R}_{W,0}\mathscr{F}_W R_{\mathfrak{d}_2}\Phi_{\mathfrak{p}_{22}},1)). \end{aligned}$$

Then by (12.10.26),

$$\delta_{\mathfrak{d},\mathrm{st}}(\omega)J_{3,(0)}(\Phi,\omega,1) = \delta_{\mathfrak{d},\mathrm{st}}(\omega)(J_6(\Phi,\omega) + J_7(\Phi,\omega)) + \delta_{\#}(\omega)J_8(\Phi).$$

Also,

$$\begin{aligned} &J_{4,(0)}(\Phi,1) \\ &= J_9(\Phi) + \frac{1}{2}(T_{W+}(\widetilde{R}_{W,0}R_{\mathfrak{d}_2}\Phi_{\mathfrak{p}_{21}},2,0) + T_{W+}(\widetilde{R}_{W,0}\mathscr{F}_W R_{\mathfrak{d}_2}\Phi_{\mathfrak{p}_{21}},1,0)), \\ &J_{5,(0)}(\Phi,1) \\ &= J_{10}(\Phi) - \frac{1}{2}(T_{W+}(\widetilde{R}_{W,0}R_{\mathfrak{d}_2}\Phi_{\mathfrak{p}_{22}},2,0) + T_{W+}(\widetilde{R}_{W,0}\mathscr{F}_W R_{\mathfrak{d}_2}\Phi_{\mathfrak{p}_{22}},1,0)). \end{aligned}$$

By the principal part formula for the standard L-function in one variable,

$$\begin{aligned} &T_{W+}(\widetilde{R}_{W,0}R_{\mathfrak{d}_2}\Phi_{\mathfrak{p}_{21}},2,0) + T_{W+}(\widetilde{R}_{W,0}\mathscr{F}_W R_{\mathfrak{d}_2}\Phi_{\mathfrak{p}_{21}},1,0) \\ &= -\Sigma_{2,(-1,0)}(R_{\mathfrak{d}_1}\Phi_{\mathfrak{p}_{11}},0,1) + \Sigma_{1,(-1)}(R_{\mathfrak{d}_4}\Phi_{\mathfrak{p}_{41}},1), \\ &T_{W+}(\widetilde{R}_{W,0}R_{\mathfrak{d}_2}\Phi_{\mathfrak{p}_{22}},2,0) + T_{W+}(\widetilde{R}_{W,0}\mathscr{F}_W R_{\mathfrak{d}_2}\Phi_{\mathfrak{p}_{22}},1,0) \\ &= -\Sigma_{2,(-1,0)}(R_{\mathfrak{d}_1}\Phi_{\mathfrak{p}_{12}},0,1) + \Sigma_{1,(-1)}(R_{\mathfrak{d}_4}\Phi_{\mathfrak{p}_{43}},1). \end{aligned}$$

Therefore,

$$\begin{aligned} J_{4,(0)}(\Phi,1) &= \delta_{\#}(\omega)(J_9(\Phi) - \frac{1}{2}\Sigma_{2,(-1,0)}(R_{\mathfrak{d}_1}\Phi_{\mathfrak{p}_{11}},0,1) + \frac{1}{2}\Sigma_{1,(-1)}(R_{\mathfrak{d}_4}\Phi_{\mathfrak{p}_{41}},1)), \\ J_{5,(0)}(\Phi,1) &= \delta_{\#}(\omega)(J_{10}(\Phi) + \frac{1}{2}\Sigma_{2,(-1,0)}(R_{\mathfrak{d}_1}\Phi_{\mathfrak{p}_{12}},0,1) - \frac{1}{2}\Sigma_{1,(-1)}(R_{\mathfrak{d}_4}\Phi_{\mathfrak{p}_{43}},1)). \end{aligned}$$

It is easy to see that

$$\begin{aligned}\Sigma_{\mathfrak{p}_{11},(0)}(\Phi,1) =& \frac{1}{2}\Sigma_{2,(0,0)}(R_{\mathfrak{d}_1}\Phi_{\mathfrak{p}_{11}},0,1) + \frac{1}{2}\Sigma_{2,(1,-1)}(R_{\mathfrak{d}_1}\Phi_{\mathfrak{p}_{11}},0,1)\\ &+ \frac{1}{2}\Sigma_{2,(-1,1)}(R_{\mathfrak{d}_1}\Phi_{\mathfrak{p}_{11}},0,1) + \frac{1}{2}\Sigma_{2,(0,-1)}(R_{\mathfrak{d}_1}\Phi_{\mathfrak{p}_{11}},0,1)\\ &+ \frac{1}{2}\Sigma_{2,(-1,0)}(R_{\mathfrak{d}_1}\Phi_{\mathfrak{p}_{11}},0,1) + \frac{1}{2}\Sigma_{2,(-1,-1)}(R_{\mathfrak{d}_1}\Phi_{\mathfrak{p}_{11}},0,1),\\ \Sigma_{\mathfrak{p}_{12},(0)}(\Phi,1) =& -\frac{1}{2}\Sigma_{2,(0,0)}(R_{\mathfrak{d}_1}\Phi_{\mathfrak{p}_{12}},0,1) - \frac{1}{2}\Sigma_{2,(1,-1)}(R_{\mathfrak{d}_1}\Phi_{\mathfrak{p}_{12}},0,1)\\ &- \frac{1}{2}\Sigma_{2,(-1,1)}(R_{\mathfrak{d}_1}\Phi_{\mathfrak{p}_{12}},0,1) - \frac{3}{2}\Sigma_{2,(0,-1)}(R_{\mathfrak{d}_1}\Phi_{\mathfrak{p}_{12}},0,1)\\ &- \frac{3}{2}\Sigma_{2,(-1,0)}(R_{\mathfrak{d}_1}\Phi_{\mathfrak{p}_{12}},0,1) - \frac{9}{2}\Sigma_{2,(-1,-1)}(R_{\mathfrak{d}_1}\Phi_{\mathfrak{p}_{12}},0,1).\end{aligned}$$

Also

$$\begin{aligned}\Sigma_{\mathfrak{p}_{41},(0)}(\Phi,1) =& \frac{1}{2}\Sigma_{1,(0)}(R_{\mathfrak{d}_4}\Phi_{\mathfrak{p}_{41}},1),\\ \Sigma_{\mathfrak{p}_{42},(0)}(\Phi,1) =& \frac{1}{2}\Sigma_{1,(1)}(R_{\mathfrak{d}_4}\Phi_{\mathfrak{p}_{42}},1) - \frac{1}{2}\Sigma_{1,(0)}(R_{\mathfrak{d}_4}\Phi_{\mathfrak{p}_{42}},1)\\ &+ \frac{1}{2}\Sigma_{1,(-1)}(R_{\mathfrak{d}_4}\Phi_{\mathfrak{p}_{42}},1),\\ \Sigma_{\mathfrak{p}_{43},(0)}(\Phi,1) =& -\frac{1}{2}\Sigma_{1,(0)}(R_{\mathfrak{d}_4}\Phi_{\mathfrak{p}_{43}},1) - \Sigma_{1,(-1)}(R_{\mathfrak{d}_4}\Phi_{\mathfrak{p}_{43}},1),\\ \Sigma_{\mathfrak{p}_{44},(0)}(\Phi,1) =& -\frac{1}{2}\Sigma_{1,(1)}(R_{\mathfrak{d}_4}\Phi_{\mathfrak{p}_{44}},1) - \frac{1}{2}\Sigma_{1,(0)}(R_{\mathfrak{d}_4}\Phi_{\mathfrak{p}_{44}},1)\\ &- \frac{1}{2}\Sigma_{1,(-1)}(R_{\mathfrak{d}_4}\Phi_{\mathfrak{p}_{44}},1).\end{aligned}$$

We define

$$\begin{aligned}J_{11}(\Phi) =& -\frac{1}{2}\Sigma_{2,(0,0)}(R_{\mathfrak{d}_1}\Phi_{\mathfrak{p}_{11}},0,1) - \frac{1}{2}\Sigma_{2,(1,-1)}(R_{\mathfrak{d}_1}\Phi_{\mathfrak{p}_{11}},0,1)\\ &- \frac{1}{2}\Sigma_{2,(-1,1)}(R_{\mathfrak{d}_1}\Phi_{\mathfrak{p}_{11}},0,1) - \Sigma_{2,(0,-1)}(R_{\mathfrak{d}_1}\Phi_{\mathfrak{p}_{11}},0,1)\\ &- \Sigma_{2,(-1,0)}(R_{\mathfrak{d}_1}\Phi_{\mathfrak{p}_{11}},0,1) - 2\Sigma_{2,(-1,-1)}(R_{\mathfrak{d}_1}\Phi_{\mathfrak{p}_{11}},0,1)\\ &- \frac{1}{2}\Sigma_{2,(0,0)}(R_{\mathfrak{d}_1}\Phi_{\mathfrak{p}_{12}},0,1) - \frac{1}{2}\Sigma_{2,(1,-1)}(R_{\mathfrak{d}_1}\Phi_{\mathfrak{p}_{12}},0,1)\\ &- \frac{1}{2}\Sigma_{2,(-1,1)}(R_{\mathfrak{d}_1}\Phi_{\mathfrak{p}_{12}},0,1) - \Sigma_{2,(0,-1)}(R_{\mathfrak{d}_1}\Phi_{\mathfrak{p}_{12}},0,1)\\ &- \Sigma_{2,(-1,0)}(R_{\mathfrak{d}_1}\Phi_{\mathfrak{p}_{12}},0,1) - 2\Sigma_{2,(-1,-1)}(R_{\mathfrak{d}_1}\Phi_{\mathfrak{p}_{12}},0,1)\\ &- \frac{1}{2}\Sigma_{1,(1)}(R_{\mathfrak{d}_4}\Phi_{\mathfrak{p}_{42}},1) - \frac{1}{2}\Sigma_{1,(1)}(R_{\mathfrak{d}_4}\Phi_{\mathfrak{p}_{44}},1),\\ J_{12}(\Phi) =& \frac{1}{2}\Sigma_{1,(0)}(R_{\mathfrak{d}_4}\Phi_{\mathfrak{p}_{41}},1) + \frac{1}{2}\Sigma_{1,(-1)}(R_{\mathfrak{d}_4}\Phi_{\mathfrak{p}_{41}},1)\\ &+ \frac{1}{2}\Sigma_{1,(0)}(R_{\mathfrak{d}_4}\Phi_{\mathfrak{p}_{43}},1) + \frac{1}{2}\Sigma_{1,(-1)}(R_{\mathfrak{d}_4}\Phi_{\mathfrak{p}_{43}},1).\end{aligned}$$

Then by the above considerations,

$$\Upsilon_{\mathfrak{d},2,0}(\Phi,\omega) = \delta_{\mathfrak{d},\mathrm{st}}(\omega)\sum_{i=6,7} J_i(\Phi,\omega) + \delta_{\#}(\omega)\sum_{i=9}^{12} J_i(\Phi).$$

By the principal part formula (4.2.15),

$$\sum_{i=2,9,10,12} J_i(\Phi) = -\Sigma_{W,\mathrm{ad},(0)}(R_{\mathfrak{d}_2}\Phi_{\mathfrak{p}_{21}}, 2) - \Sigma_{W,\mathrm{ad},(0)}(R_{\mathfrak{d}_2}\Phi_{\mathfrak{p}_{22}}, 2).$$

Therefore,

$$\begin{aligned}&\Upsilon_{\mathfrak{d},1}(\Phi,\omega) + \Upsilon_{\mathfrak{d},2,0}(\Phi,\omega)\\ &= \delta_{\mathfrak{d},\mathrm{st}}(\omega)\sum_{i=6,7} J_i(\Phi,\omega) + \delta_{\#}(\omega)J_{11}(\Phi)\\ &\quad - \delta_{\#}(\omega)(\Sigma_{W,\mathrm{ad},(0)}(R_{\mathfrak{d}_2}\Phi_{\mathfrak{p}_{21}}, 2) + \Sigma_{W,\mathrm{ad},(0)}(R_{\mathfrak{d}_2}\Phi_{\mathfrak{p}_{22}}, 2)).\end{aligned}$$

By (5.6.4), we get the following proposition.

Proposition (12.10.27)

$$\begin{aligned}&\Upsilon_{\mathfrak{d},1}(\Phi,\omega) + \Upsilon_{\mathfrak{d},2,(0)}(\Phi,\omega)\\ &= \delta_{\mathfrak{d}}(\omega)\Sigma_{\mathfrak{d},\mathrm{ad}}(R_{\mathfrak{d}}\Phi_{\mathfrak{p}}, \omega_{\mathfrak{d}}, 3)\\ &\quad + \frac{\delta_{\mathfrak{d},\mathrm{st}}(\omega)}{8}\left(\Sigma_{2,(1,0)}(R'_{\mathfrak{d},0}\Psi, \omega_{\mathfrak{d},\mathrm{st}}, \frac{1}{2}, \frac{1}{2}) + \Sigma_{2,(0,1)}(R'_{\mathfrak{d},0}\Psi, \omega_{\mathfrak{d},\mathrm{st}}, \frac{1}{2}, \frac{1}{2})\right).\end{aligned}$$

Chapter 13 The main theorem

§13.1 The cancellations of distributions

In this chapter, $\mathfrak{d} = (\beta_i)$ for $i = 1, \cdots, 10$, $\mathfrak{d}_{8,1} = (\beta_8, \beta_{8,1})$, and $\mathfrak{d}_{8,2} = (\beta_8, \beta_{8,2})$. Let $\mathfrak{p}_{8,2} = (\mathfrak{d}_{8,2}, \mathfrak{s}_{8,2})$, $\mathfrak{p}_9 = (\mathfrak{d}_9, \mathfrak{s}_9)$ where $\mathfrak{d}_{8,2} = (\beta_8, \beta_{8,2})$, $\mathfrak{d}_9 = (\beta_9)$ and $\mathfrak{s}_{8,2}(1) = \mathfrak{s}_9(1)$, $\mathfrak{s}_{8,2}(2) = 0$. Then $\epsilon_{\mathfrak{p}_{8,2}} = -\epsilon_{\mathfrak{p}_9}$.

An easy consideration shows that $R_{\mathfrak{d}_1}\Phi_{\mathfrak{p}_1} = \Phi^{(1)}_{\mathfrak{p}_9}$ Therefore, the third term of (12.8.28) and the second term of (12.9.2) cancel out.

We define

$$\begin{aligned}\Upsilon_1(\Phi,\omega) = & \sum_{i=1,6,8,10} (\Upsilon_{\mathfrak{d}_i,1}(\widehat{\Phi},\omega^{-1}) - \Upsilon_{\mathfrak{d}_i,1}(\Phi,\omega)) \\ & + 2\delta_{\mathfrak{d}_4}(\omega)(\Sigma_{\mathfrak{d}_4}(R_{\mathfrak{d}_4}\widehat{\Phi},\omega_{\mathfrak{d}}^{-1},-2) - \Sigma_{\mathfrak{d}_4}(R_{\mathfrak{d}_4}\Phi,\omega_{\mathfrak{d}},-2)) \\ & + \frac{\delta_{\mathfrak{d}_9}(\omega)}{3}\left(\Sigma_3(R_{\mathfrak{d}_9}\widehat{\Phi},\omega_{\mathfrak{d}_9}^{-1},-1,-\frac{2}{3},\frac{2}{3}) - \Sigma_3(R_{\mathfrak{d}_9}\Phi,\omega_{\mathfrak{d}_9},-1,-\frac{2}{3},\frac{2}{3})\right) \\ & + \mathfrak{V}_2\mathfrak{V}_3\delta_{\#}(\omega)(\widehat{\Phi}(0) - \Phi(0)),\end{aligned}$$

and

$$\Upsilon_2(\Phi,\omega,s_{\tau 11}) = \sum_{i=1,6,7,8,10} (\Upsilon_{\mathfrak{d}_i,2}(\widehat{\Phi},\omega^{-1},s_{\tau 11}) - \Upsilon_{\mathfrak{d}_i,2}(\Phi,\omega,s_{\tau 11})).$$

Suppose that $\tau = \tau_G$ and $\delta > 0$ is a small number. Then by (10.1.1), (10.1.2), and the results in Chapter 12, by changing ψ if necessary,

$$\begin{aligned} I(\Phi,\omega,w) \sim & C_G\Lambda(w;\rho)\Upsilon_1(\Phi,\omega) \\ & + \frac{\varrho^2}{2\pi\sqrt{-1}}\int_{\mathrm{Re}(s_{\tau 11})=1+\delta} \Upsilon_2(\Phi,\omega,s_{\tau 11})\phi_2(s_{\tau 11})\widetilde{\Lambda}_1(w;s_{\tau 11},1)ds_{\tau 11}.\end{aligned}$$

We get the following lemma by Wright's principle.

Lemma (13.1.1) *The distribution $\Upsilon_2(\Phi,\omega,s_{\tau 11})$ is holomorphic at $s_{\tau 11} = 1$.*

Therefore, the cancellations of higher order terms of $\Upsilon_2(\Phi,\omega,s_{\tau 11})$ are guaranteed, and we are not obliged to check them. Hence, we get the following proposition.

Proposition (13.1.2) $I^0(\Phi,\omega) = \Upsilon_1(\Phi,\omega) + \Upsilon_2(\Phi,\omega,1)$.

Clearly,

$$\Upsilon_2(\Phi,\omega,1) = \sum_{i=1,6,7,8,10} (\Upsilon_{\mathfrak{d}_i,2,(0)}(\widehat{\Phi},\omega^{-1},1) - \Upsilon_{\mathfrak{d}_i,2,(0)}(\Phi,\omega,1)).$$

Let $\mathfrak{p}_{8,1} = (\mathfrak{d}_{8,1}, \mathfrak{s}_{8,1})$, $\mathfrak{p}_6 = (\mathfrak{d}_6, \mathfrak{s}_6)$ be paths such that $\mathfrak{s}_{8,1}(1) = \mathfrak{s}_6(1)$, $\mathfrak{s}_{8,1}(2) = 0$. Then it is easy to see that

$$\Sigma_{2,(i,j)}(R_{\mathfrak{d}_{8,1}}\Phi_{\mathfrak{p}_{8,1}},-1,-3) = \Sigma_{2,(j,i)}(R'_{\mathfrak{d}_{6,0}}\Phi_{\mathfrak{p}_6},-3,-1)$$

for all i, j. Also if $\mathfrak{p}_7 = (\mathfrak{d}_7, \mathfrak{s}_7), \mathfrak{p}_{10} = (\mathfrak{d}_{10}, \mathfrak{s}_{10})$ are paths such that $\mathfrak{s}_7(1) = \mathfrak{s}_{10}(1)$,

$$\Sigma_{3,(1,-1,0)}(R_{\mathfrak{d}_7}\Phi_{\mathfrak{p}_7}, \omega_{\mathfrak{d}_7}, (\omega_2, 1, \omega_2), \frac{1}{2}, 0, \frac{1}{2}) = -\Sigma_{2,(1,0)}(R'_{\mathfrak{d}_{10},0}\Phi_{\mathfrak{p}_{10}}, \omega_{\mathfrak{d},\mathrm{st}}, \frac{1}{2}, \frac{1}{2}),$$
$$\Sigma_{3,(0,-1,1)}(R_{\mathfrak{d}_7}\Phi_{\mathfrak{p}_7}, \omega_{\mathfrak{d}_7}, (\omega_2, 1, \omega_2), \frac{1}{2}, 0, \frac{1}{2}) = -\Sigma_{2,(0,1)}(R'_{\mathfrak{d}_{10},0}\Phi_{\mathfrak{p}_{10}}, \omega_{\mathfrak{d},\mathrm{st}}, \frac{1}{2}, \frac{1}{2}).$$

We define

$$
\begin{aligned}
(13.1.3)\qquad f(\Phi,\omega) =\,& \delta_{\mathfrak{d}_1}(\omega)\Sigma_{\mathfrak{d}_1}(R_{\mathfrak{d}_1}\Phi, \omega_{\mathfrak{d}_1}, 2) \\
&+ 2\delta_{\mathfrak{d}_4}(\omega)\Sigma_{\mathfrak{d}_4}(R_{\mathfrak{d}_4}\Phi, \omega_{\mathfrak{d}_4}, -2) \\
&+ 8\delta_{\mathfrak{d}_6}(\omega)\Sigma_{\mathfrak{d}_6,\mathrm{ad}}(R_{\mathfrak{d}_6}\Phi, \omega_{\mathfrak{d}_6}, \chi_{\mathfrak{d}_6}, -3) \\
&+ \frac{5}{2}\delta_{\#}(\omega)\Sigma_{2,(1,0)}(R'_{\mathfrak{d}_6,0}\Phi, -3, -1) \\
&+ \frac{3}{2}\delta_{\#}(\omega)\Sigma_{2,(0,1)}(R'_{\mathfrak{d}_6,0}\Phi, -3, -1) \\
&+ \frac{\delta_{\mathfrak{d}_7}(\omega)}{6}\Sigma_{3,(0,0,0)}(R_{\mathfrak{d}_7}\Phi, \omega_{\mathfrak{d}_7}, \frac{1}{2}, 0, \frac{1}{2}) \\
&+ 14\delta_{\mathfrak{d}_8}(\omega)\Sigma_{\mathfrak{d}_8,(0)}(R_{\mathfrak{d}_8}\Phi, \omega_{\mathfrak{d}_8}, \chi_{\mathfrak{d}_8}, -3) \\
&+ \frac{\delta_{\mathfrak{d}_9}(\omega)}{3}\Sigma_3(R_{\mathfrak{d}_9}\Phi, \omega_{\mathfrak{d}_9}, -1, -\frac{2}{3}, \frac{2}{3}) \\
&+ \delta_{\mathfrak{d}_{10}}(\omega)\Sigma_{\mathfrak{d}_{10},\mathrm{ad}}(R_{\mathfrak{d}_{10}}\Phi, \omega_{\mathfrak{d}_{10}}, 3) \\
&- \frac{\delta_{\mathfrak{d}_{10},\mathrm{st}}(\omega)}{8}\Sigma_{2,(1,0)}(R'_{\mathfrak{d}_{10},0}\Phi, \omega_{\mathfrak{d}_{10},\mathrm{st}}, \frac{1}{2}, \frac{1}{2}) \\
&+ \frac{\delta_{\mathfrak{d}_{10},\mathrm{st}}(\omega)}{24}\Sigma_{2,(0,1)}(R'_{\mathfrak{d}_{10},0}\Phi, \omega_{\mathfrak{d}_{10},\mathrm{st}}, \frac{1}{2}, \frac{1}{2}).
\end{aligned}
$$

By (13.1.2), we get the following proposition.

Proposition (13.1.4) $I^0(\Phi,\omega) = f(\widehat{\Phi}, \omega^{-1}) - f(\Phi,\omega)$.

§13.2 The principal part formula

We are finally ready to prove the principal part formula for the zeta function for the space of pairs of ternary quadratic forms. For $\Phi \in \mathscr{S}(V_{\mathbb{A}})$, $\lambda \in \mathbb{R}_+$, let $\Phi_\lambda(x) = \Phi(\underline{\lambda}x)$ as before. The following relations are easy to verify, and the proof is left to the reader.

Lemma (13.2.1)

(1) $$\Phi_\lambda(0) = \Phi(0).$$

(2) $$\Sigma_{\mathfrak{d}_1}(R_{\mathfrak{d}_1}\Phi_\lambda, \omega_{\mathfrak{d}_1}, 2) = \lambda^{-2}\Sigma_{\mathfrak{d}_1}(R_{\mathfrak{d}_1}\Phi, \omega_{\mathfrak{d}_1}, 2).$$

(3) $$\Sigma_{\mathfrak{d}_4}(R_{\mathfrak{d}_4}\Phi_\lambda, \omega_{\mathfrak{d}_4}, -2) = \Sigma_{\mathfrak{d}_4}(R_{\mathfrak{d}_4}\Phi, \omega_{\mathfrak{d}_4}, -2).$$

(4) $$\Sigma_{\mathfrak{d}_6,\mathrm{ad}}(R_{\mathfrak{d}_6}\Phi_\lambda, \omega_{\mathfrak{d}_6}, \chi_{\mathfrak{d}_6}, -3) = \Sigma_{\mathfrak{d}_6,\mathrm{ad}}(R_{\mathfrak{d}_6}\Phi, \omega_{\mathfrak{d}_6}, \chi_{\mathfrak{d}_6}, -3).$$

(5) $$\Sigma_{2,(1,0)}(R'_{\mathfrak{d}_6,0}\Phi_\lambda, -3, -1) = \Sigma_{2,(1,0)}(R'_{\mathfrak{d}_6,0}\Phi, -3, -1) - (\log\lambda)\Sigma_2(R'_{\mathfrak{d}_6,0}\Phi, -3, -1).$$

(6) $$\Sigma_{2,(0,1)}(R'_{\mathfrak{d}_6,0}\Phi_\lambda, -3, -1) = \Sigma_{2,(0,1)}(R'_{\mathfrak{d}_6,0}\Phi, -3, -1) - (\log\lambda)\Sigma_2(R'_{\mathfrak{d}_6,0}\Phi, -3, -1).$$

$$(7)\ \delta_{\mathfrak{d}_7}(\omega)\Sigma_{3,(0,0,0)}(R_{\mathfrak{d}_7}\Phi_\lambda,\omega_{\mathfrak{d}_7},\frac{1}{2},0,\frac{1}{2}) = \lambda^{-3}\delta_{\mathfrak{d}_7}(\omega)\Sigma_{3,(0,0,0)}(R_{\mathfrak{d}_7}\Phi,\omega_{\mathfrak{d}_7},\frac{1}{2},0,\frac{1}{2}) + \lambda^{-3}(\log\lambda)\delta_{\mathfrak{d}_{10},\mathrm{st}}(\omega)\Sigma_2(R'_{\mathfrak{d}_{10},0}\Phi,\omega_{\mathfrak{d}_{10},\mathrm{st}},\frac{1}{2},\frac{1}{2}).$$

$$(8)\ \delta_{\mathfrak{d}_8}(\omega)\Sigma_{\mathfrak{d}_8,(0)}(R_{\mathfrak{d}_8}\Phi_\lambda,\omega_{\mathfrak{d}_8},\chi_{\mathfrak{d}_8},-3) = \delta_{\mathfrak{d}_8}(\omega)\Sigma_{\mathfrak{d}_8,(0)}(R_{\mathfrak{d}_8}\Phi,\omega_{\mathfrak{d}_8},\chi_{\mathfrak{d}_8},-3) + \frac{1}{2}\lambda(\log\lambda)\delta_\#(\omega)\Sigma_2(R'_{\mathfrak{d}_6,0}R_{\mathfrak{d}_6}\Phi,-3,-1).$$

$$(9)\qquad \Sigma_3(R_{\mathfrak{d}_9}\Phi_\lambda,\omega_{\mathfrak{d}_9},-1,-\frac{2}{3},\frac{2}{3}) = \lambda^{-2}\Sigma_3(R_{\mathfrak{d}_9}\Phi,\omega_{\mathfrak{d}_9},-1,-\frac{2}{3},\frac{2}{3}).$$

$$(10)\qquad \Sigma_{\mathfrak{d}_{10},\mathrm{ad}}(R_{\mathfrak{d}_{10}}\Phi_\lambda,\omega_{\mathfrak{d}_{10}},3) = \lambda^{-3}\Sigma_{\mathfrak{d}_{10},\mathrm{ad}}(R_{\mathfrak{d}_{10}}\Phi,\omega_{\mathfrak{d}_{10}},3).$$

$$(11)\qquad \Sigma_{2,(1,0)}(R'_{\mathfrak{d}_{10},0}\Phi_\lambda,\omega_{\mathfrak{d}_{10},\mathrm{st}},\frac{1}{2},\frac{1}{2}) = \lambda^{-3}\Sigma_{2,(1,0)}(R'_{\mathfrak{d}_{10},0}\Phi,\omega_{\mathfrak{d}_{10},\mathrm{st}},\frac{1}{2},\frac{1}{2}) - \lambda^{-3}(\log\lambda)\Sigma_2(R'_{\mathfrak{d}_{10},0}\Phi,\omega_{\mathfrak{d}_{10},\mathrm{st}},\frac{1}{2},\frac{1}{2}).$$

Note that (8) of the above lemma follows from (7.3.7). Similar relations hold for $\widehat{\Phi}$ also.

We define

$$\begin{aligned}
F_{(-1)}(\Phi,\omega,0) =& \mathfrak{V}_2\mathfrak{V}_3\delta_\#(\omega)\Phi(0) + 2\delta_{\mathfrak{d}_4}(\omega)\Sigma_{\mathfrak{d}_4}(R_{\mathfrak{d}_4}\Phi,\omega_{\mathfrak{d}_4},-2))\\
&+ 8\delta_{\mathfrak{d}_6}(\omega)\Sigma_{\mathfrak{d}_6,\mathrm{ad}}(R_{\mathfrak{d}_6}\Phi,\omega_{\mathfrak{d}_6},\chi_{\mathfrak{d}_6},-3)\\
&+ 14\delta_{\mathfrak{d}_8}(\omega)\Sigma_{\mathfrak{d}_8,(0)}(R_{\mathfrak{d}_8}\Phi,\omega_{\mathfrak{d}_8},\chi_{\mathfrak{d}_8},-3)\\
&+ \frac{\delta_\#(\omega)}{2}(5\Sigma_{2,(1,0)}(R'_{\mathfrak{d}_6,0}\Phi,-3,-1) + 3\Sigma_{2,(0,1)}(R'_{\mathfrak{d}_6,0}\Phi,-3,-1)),\\
F_{(-2)}(\Phi,\omega,0) =& -3\delta_\#(\omega)\Sigma_2(R'_{\mathfrak{d}_6,0}\Phi,-3,-1),\\
F_{(-1)}(\Phi,\omega,2) =& \delta_{\mathfrak{d}_1}(\omega)\Sigma_{\mathfrak{d}_1}(R_{\mathfrak{d}_1}\Phi,\omega_{\mathfrak{d}_1},2) + \frac{\delta_{\mathfrak{d}_9}(\omega)}{3}\Sigma_3(R_{\mathfrak{d}_9}\Phi,\omega_{\mathfrak{d}_9},-1,-\frac{2}{3},\frac{2}{3}),\\
F_{(-1)}(\Phi,\omega,3) =& \frac{\delta_{\mathfrak{d}_7}(\omega)}{6}\Sigma_{3,(0,0,0)}(R_{\mathfrak{d}_7}\Phi,\omega_{\mathfrak{d}_7},\frac{1}{2},0,\frac{1}{2})\\
&+ \delta_{\mathfrak{d}_{10}}(\omega)\Sigma_{\mathfrak{d}_{10},\mathrm{ad}}(R_{\mathfrak{d}_{10}}\Phi,\omega_{\mathfrak{d}_{10}},3))\\
&- \frac{\delta_{\mathfrak{d}_{10},\mathrm{st}}(\omega)}{8}\Sigma_{2,(1,0)}(R'_{\mathfrak{d}_{10},0}\Phi,\omega_{\mathfrak{d}_{10},\mathrm{st}},\frac{1}{2},\frac{1}{2})\\
&+ \frac{\delta_{\mathfrak{d}_{10},\mathrm{st}}(\omega)}{24}\Sigma_{2,(0,1)}(R'_{\mathfrak{d}_{10},0}\Phi,\omega_{\mathfrak{d}_{10},\mathrm{st}},\frac{1}{2},\frac{1}{2}),\\
F_{(-2)}(\Phi,\omega,3) =& -\frac{\delta_{\mathfrak{d}_{10},\mathrm{st}}(\omega)}{4}\Sigma_2(R'_{\mathfrak{d}_{10},0}\Phi,\omega_{\mathfrak{d}_{10},\mathrm{st}},\frac{1}{2},\frac{1}{2}).
\end{aligned}$$

We define

$$F(\Phi,\omega,s) = \frac{F_{(-1)}(\Phi,\omega,2)}{s-2} + \sum_{\substack{i=0,3\\ j=1,2}} \frac{F_{(-j)}(\Phi,\omega,i)}{(s-i)^j}.$$

By replacing Φ in (3.1.4) by Φ_λ and integrating over $0 \le \lambda \le 1$, we get the following principal part formula for our zeta function.

Theorem (13.2.2)

$$Z_V(\Phi,\omega,s) = Z_{V+}(\Phi,\omega,s) + Z_{V+}(\widehat{\Phi},\omega^{-1},12-s) - F(\widehat{\Phi},\omega^{-1},12-s) - F(\Phi,\omega,s).$$

§13.3 Concluding remarks

We used the passing principle (3.6.1) in various places to move contours. Instead of establishing cancellations of various distributions, it enabled us to ignore them as long as we can check some easy conditions. We also used Wright's principle to ignore higher order terms of certain distributions. However, it is possible to study the poles without these two methods, and in fact, the author's original approach was to establish cancellations explicitly. It is an interesting exercise to establish cancellations of various contour integrals without using the passing principle (3.6.1) or Wright's principle, because one can see that the cancellations between different paths are happening. They correspond to cancellations between divergent integrals, and our two methods and the choice of the constants in the definition of the smoothed Eisenstein series enabled us to handle most of them without establishing the cancellations.

In Chapter 12, we computed contributions from unstable strata, and obtained more or less special values of the zeta functions for the corresponding representations. Those zeta functions have higher order poles, so we had to recover the higher order terms. When we analyzed the constant terms in Chapter 12, they came not only from the unstable strata, but from the adjusting terms also. The author does not know if this is a special case of something more general. But for this reason, we had to compute the contributions from unstable strata explicitly, and compare with principal part formulas of the corresponding zeta functions. This was rather annoying, and the author hopes to find a better interpretation of this phenomenon in the future.

If the center of the stabilizer of a point in $Z_{\mathfrak{d}k}^{ss}$ contains a split torus, we could ignore $\Xi_{\mathfrak{p}}(\Phi, \omega, w)$ in most cases. However, this is not always the case, for example, in the case $G = \mathrm{GL}(2) \times \mathrm{GL}(2)$, $V = k^2 \otimes k^2$. This is another reason why we still cannot handle all the cases simultaneously.

We restricted ourselves to groups which are products of $\mathrm{GL}(n)$'s, mainly because we do not have the uniform estimates of Whittaker functions. As far as the formulation of equivariant Morse theory is concerned, there does not seem to be any major problem in generalizing our formulation. However, for other groups, we still do not know how we can describe the Eisenstein series in terms of generalized Whittaker functions, and until we find out the answer to this problem, we cannot handle other groups. On the other hand, there is no doubt that the cases where the group is a product of $\mathrm{GL}(n)$'s contain the most interesting cases (see [84]), and we hope our approach enables us to handle more cases.

It is clear from the computations in this book that the complexity of our computations increases exponentially with the rank of the group. Therefore, the number of the distributions we have to handle can be quite large, and may require a computer calculation. However, we still do not know when the significant cancellations as in §10.3, §10.6 and §13.1 will occur. Also, in order to do such a task, our theory has to be generalized so that we can handle reducible representations. However, these are questions to be answered in the future.

Bibliography

[1] Borel, A. Some finiteness properties of adele groups over number fields. *Publ. Math. IHES*, n° 16:101–126, 1963.

[2] Borel, A. *Linear algebraic groups, 2nd enlarged edition.* Springer, Berlin, Heidelberg, New York, 1991.

[3] Borevich, Z. I, and I. R. Shafarevich. *Number theory.* Academic Press, 1966.

[4] Brioschi, F. Sulla risoluzione delle equazioni del quinto grado. In *Opere mathematiche*, volume I, pages 335–341. Ulrico Hoepi, Milano, 1904.

[5] Bump, D. *Automorphic forms on* $\mathrm{GL}(3,\mathbb{R})$, volume 1083. Springer Lecture Notes in Mathematics, Berlin, Heidelberg, New York, 1984.

[6] Bump, D. The Rankin-Selberg method: a survey. In *Number theory, trace formula and discrete groups (Oslo 1987)*, pages 49–109. Academic Press, Boston, 1989.

[7] Casselman, W., and J. Shalika. The unramified principal series of p-adic groups II: The Whittaker function. *Compositio Math.*, 41:207–231, 1980.

[8] Cogdell, J. W. Congruence zeta functions for $\mathrm{M}_2(\mathbb{Q})$ and their associated modular forms. *Math. Annalen*, 266:141–198, 1983.

[9] Datskovsky, B. A mean value theorem for class numbers of quadratic extensions. To appear in the memorial volume for Emil Grosswald, 1992.

[10] Datskovsky, B., and D. J. Wright. The adelic zeta function associated with the space of binary cubic forms II: Local theory. *J. Reine Angew. Math.*, 367:27–75, 1986.

[11] Datskovsky, B., and D. J. Wright. Density of discriminants of cubic extensions. *J. Reine Angew. Math.*, 386:116–138, 1988.

[12] Davenport, H. On the class-number of binary cubic forms I and II. *London Math. Soc.*, 26:183–198, 1951. Corrigendum: ibid., 27:512, 1952.

[13] Davenport, H., and H. Heilbronn. On the density of discriminants of cubic fields I. *Bull. London Math. Soc.*, 1:345–348, 1961.

[14] Davenport, H. and H. Heilbronn. On the density of discriminants of cubic fields. II. *Proc. Royal Soc.*, A322,:405–420, 1971.

[15] Denef, J. On the evaluation of certain p-adic integrals. In *Progress in Math. No. 59*, pages 25–47. Birkhäuser, 1985.

[16] Gauss, C. F. *Disquisitiones arithmeticae.* Yale University Press, New Haven, London, 1966.

[17] Goldfeld, D., and J. Hoffstein. Eisenstein series of $\frac{1}{2}$-integral weight and the mean value of real Dirichlet series. *Invent. Math.*, 80:185–208, 1985.

[18] Gray, A., and G. B. Mathews. *A treatise on Bessel functions and their applications to physics.* Macmillan, London, 1952.

[19] Gyoja, A. Gauss sums of prehomogeneous vector spaces. *Proc. Japan Acad.*, Math. Sci. 61 Ser. A no. 1:19–22, 1985.

[20] Hermite, C. Sur la résolution de l'équation du cinquième degré. In *Oevres*, volume II, pages 5–12. Gauthier-Villars, Paris, 1908.

[21] Hironaka, Y., and F. Sato. Spherical functions and local densities of alternating forms. *Amer. J. Math.*, 110:473–512, 1988.

[22] Igusa, J. Some results on p-adic complex powers. *Amer. J. Math.*, 106:1013–1032, 1984.

[23] Igusa, J. On functional equations of complex powers. *Invent. Math.*, 85:1–29, 1986.

[24] Igusa, J. On a certain class of prehomogeneous vector spaces. *J. of Pure and Applied Algebra*, 47:265–282, 1987.

[25] Igusa, J. Zeta distributions associated with some invarinats. *Amer. J. Math.*, 109:1–34, 1987.

[26] Igusa, J. *b*-functions and *p*-adic integrals. In *Algebraic analysis*, pages 231–241. Academic Press, 1988.

[27] Igusa, J. On the arithmetic of a singular invariant. *Amer. J. Math.*, 110:197–233, 1988.

[28] Kempf, G. Instability in invariant theory. *Ann. of Math.*, 108:299–316, 1978.

[29] Kempf, G., and L. Ness. The length of vectors in representation spaces. In *Algebraic Geometry, Proceedings, Copenhagen*, volume 732, pages 233–242. Springer Lecture Notes in Mathematics, Berlin, Heidelberg, New York, 1978.

[30] Kimura, T. *Study of irreducible prehomogeneous vector spaces.* Master thesis in Japanese, Univ. Tokyo, 1974.

[31] Kimura, T. The *b*-functions and holonomy diagrams of irreducible regular prehomogeneous vector spaces. *Nagoya Math. J.*, 85:1–80, 1982.

[32] Kimura, T., and I. Ozeki. On the micro-local structure of the regular prehomogeneous vector space associated with Spin(10) × GL(3). *Proc. Japan Acad.*, Math. Sci. 58 Ser. A:18–21, 1980.

[33] Kimura, T., and M. Muro. On some series of regular irreducible prehomogeneous vector spaces. *Proc. Japan Acad.*, Math. Sci. 55 Ser. A:384–389, 1979.

[34] Kimura, T., F. Sato, and X.-W. Zhu. On the poles of *p*-adic complex powers and the *b*-functions of prehomogeneous vector spaces. *Amer. J. Math.*, 112:423–437, 1990.

[35] Kirwan, F. C. *Cohomology of quotients in symplectic and algebraic geometry.* Mathematical Notes, Princeton University Press, 1984.

[36] Klein, F. Weitere Untersuchungen über das Ikosaeder. In *Abhandlungen*, volume II, pages 321–384. Springer, Berlin, 1922.

[37] Klein, F. *Lectures on the icosahedron and the solution of equations of the fifth degree.* Dover, New York, 1956.

[38] Kneser, M. *Lectures on Galois cohomology of classical groups.* Tata Lecture Note, Bombay, 1969.

[39] Kronecker, L. Sur la résolution de l'équation du cinquième degré (extrait d'une lettre adressée à M. Hermite). In *Werke*, volume IV, pages 43–47. Teubner, Leipzig, Berlin, 1929.

[40] Langlands, R. The volume of the fundamental domain for some arithmetical subgroups of Chevalley groups. In *Proceedings of symposium in pure mathematics*, volume 9. 1966.

[41] Langlands, R. *On the functional equations satisfied by Eisenstein series*, volume 544. Springer Lecture Notes in Mathematics, Berlin, Heidelberg, New York, 1976.

[42] Lipschitz, R. In *Sitzungsber*, pages 174–185. Akad. Berlin, 1865.

[43] Loeser, F. Fonctions d'Igusa *p*-adiques et polynômes de Bernstein. *Amer. J. Math.*, 110:1–22, 1988.

[44] Milnor, J. *Morse theory.* Annals of Mathematical Studies, Princeton University Press, 1969.

[45] Muller, I. Théorie des groupes—décomposition orbitale des espaces préhomogènes réguliers de type parabolique commutatif et application. *C. R. Acad. Sci. Paris*, t. 303 Sériè I, n° 11, 1986.

[46] Mumford, D., and J. Fogarty. *Geometric invariant theory, 2nd edition.* Springer, Berlin, Heidelberg, New York, 1982.

[47] Narkiewicz, W. *Elementary and analytic theory of algebraic numbers.* PWN, Warszawa, 1974.

[48] Ness, L. A stratification of the null cone via the moment map. *Amer. J. Math.*, 106:1281–1325, 1984.

[49] Ozeki, I. On the micro-local structure of a regular prehomogeneous vector space associated with SL(5) × GL(4) I. *Proc. Japan Acad.*, Math. Sci. 55 Ser. A:37–40, 1979.

[50] Ozeki, I. On the micro-local structure of a regular prehomogeneous vector space associated with GL(8). *Proc. Japan Acad.*, Math. Sci. 56 Ser. A:18–21, 1980.

[51] Piatetski-Shapiro, I. I. Euler subgroups. In *Lie Groups and their Representations*, pages 597–620. John Wiley and Sons, New York, Toronto, 1975.

[52] Rubenthaler, H. *Espaces préhomogènes de type parabolique.* Thèse, Université Strasbourg, 1982.

[53] Rubenthaler, H. Espaces préhomogènes de type parabolique. *Lect. Math. Kyoto Univ.*, 14:189–221, 1982.

[54] Rubenthaler, H., and G. Schiffmann. Opérateurs différentiels de Shimura et espaces préhomogènes. *Invent. Math.*, 90:409–442, 1987.

[55] Sato, F. On zeta functions of ternary zero forms. *J. Fac. Sci. Univ. Tokyo, Sect IA*, 28:585–604, 1981.

[56] Sato, F. Zeta functions in several variables associated with prehomogeneous vector spaces I: Functional equations. *Tôhoku Math. J.*, (2) 34:no. 3 437–483, 1982.

[57] Sato, F. Zeta functions in several variables associated with prehomogeneous vector spaces III: Eisenstein series for indefinite quadratic forms. *Ann. of Math.*, 116:177–212, 1982.

[58] Sato, F. Zeta functions in several variables associated with prehomogeneous vector spaces II: A convergence criterion. *Tôhoku Math. J.*, (2) 35 no. 1:77–99, 1983.

[59] Sato, M., and T. Kimura. A classification of irreducible prehomogeneous vector spaces and their relative invariants. *Nagoya Math. J.*, 65:1–155, 1977.

[60] Sato, M., and T. Shintani. On zeta functions associated with prehomogeneous vector spaces. *Ann. of Math.*, 100:131–170, 1974.

[61] Sato, M., M. Kashiwara, T. Kimura, and T. Oshima. Micro-local analysis of prehomogeneous vector spaces. *Invent. Math.*, 62:117–179, 1980.

[62] Serre, J. P. *Cohomologie galoisienne*, volume 5. Springer Lecture Notes in Mathematics, Berlin, Heiderberg, New York, 1965.

[63] Serre, J. P. Extensions icosaédrique. In *Oeuvres*, volume III, pages 550–554. Springer, Berlin, Heiderberg, New York, 1986.

[64] Shintani, T. On Dirichlet series whose coefficients are class-numbers of integral binary cubic forms. *J. Math. Soc. Japan*, 24:132–188, 1972.

[65] Shintani, T. On zeta-functions associated with vector spaces of quadratic forms. *J. Fac. Sci. Univ. Tokyo, Sect IA*, 22:25–66, 1975.

[66] Shintani, T. On an explicit formula for class 1 Whittaker functions on GL_n over p-adic fields. *Proc. Japan Acad.*, 52:180–182, 1976.

[67] Siegel, C. L. Über die analytische Theorie der quadratischer Formen. *Ann. of Math.*, 36:527–606, 1935.

[68] Siegel, C. L. Über die Zetafunctionen indefiniter quadratischer Formen. *Math. Z.*, 43:682–708, 1938.

[69] Siegel, C. L. The average measure of quadratic forms with given discriminant and signature. *Ann. of Math.*, 45:667–685, 1944.

[70] Siegel, C. L. On the theory of indefinite quadratic forms. *Ann. of Math.*, 45:577–622, 1944.

[71] Stade, E. On explicit integral formulas for $GL(n, \mathbb{R})$-Whittaker functions. *Duke Math. J.*, 60:313–362, 1990.

[72] Suzuki, M. *Group theory*, volume I. Springer, Berlin, Heiderberg, New York, 1982.

[73] Tate, J. *Fourier analysis in number fields and Hecke's zeta function.* Ph.D. thesis, Princeton University, 1950.

[74] Van der Waerden, B. *History of algebra.* Springer, Berlin, Heidelberg, New York, 1985.

[75] Vinberg, E. B. On the classification of the nilpotent elements of graded Lie algebras. *Soviet Math. Dokl.*, 16:1517–1520, 1975.

[76] Vinogradov, I. M. *Selected works.* Springer, Berlin, Heidelberg, New York, 1985.

[77] Weil, A. Sur certains groupes d'opérateurs unitaires. *Acta Math.*, 111:143–211, 1964.

[78] Weil, A. Sur la formule de Siegel dans la théorie des groupes classiques. *Acta Math.*, 113:1–87, 1965.

[79] Weil, A. *Basic number theory.* Springer, Berlin, Heidelberg, New York, 1974.

[80] Weil, A. *Adeles and algebraic groups.* Birkhäuser, Boston, Stuttgart, 1982.

[81] Weyl, H. *The classical groups.* Princeton Univ. Press, 1946.

[82] Whittaker, E. T., and G. N. Watson. *A course of modern analysis.* Cambridge Univ. Press, 1927.

[83] Wright, D. J. The adelic zeta function associated to the space of binary cubic forms part I: Global theory. *Math. Ann.*, 270:503–534, 1985.

[84] Wright, D. J., and A. Yukie. Prehomogeneous vector spaces and field extensions. *Invent. Math.*, 110:283–314, 1992.

[85] Yukie, A. Prehomogeneous vector spaces Eisenstein series and invariant theory. preprint 1992.

[86] Yukie, A. On the Shintani zeta function for the space of binary quadratic forms. *Math. Ann.*, 292:355–374, 1992.

List of symbols

$f(x) \ll g(x)$ §0.1
V_R^{ss} §0.2
$o(x), o_\infty(x), \mu(x), \mu_\infty(x), b_{x,v}$ §0.3
$Z_{x,v}(\Phi, s), X_{x,v}(\Phi, s), \xi(\Phi, s)$ §0.3 $b(s), b^*(s)$ (0.3.4)
$a_n(t_1, \cdots, t_n), n_n(u)$ (1.1.1)
$K = \otimes_v K_v$ §1.1
ρ (half the sum of positive weights) §1.1
$\phi(s), \phi_n(s)$ (1.1.2)
ϱ after (1.1.2)
$\mathfrak{V}_n$ (1.1.3)
N_τ, N_τ^- (2.1.4)
I_τ after (2.1.5)
$c_v(u), s_v(u)$ (2.2.1)
$\alpha(u), \alpha_v(u_v)$ (2.2.2)
$E_B(g^0, z)$ (2.2.3)
$M_v(z), M(z)$ (2.2.8), (2.2.10)
τ_G before (2.3.1), after (3.1.8)
$W_n(\alpha, z)$ etc. (2.3.1)
$\lambda_{n,v}(t, z)$ (2.3.7)
$\sigma_{n,v}(\alpha, z)$ (2.3.12)
$D(n, r)$ (2.3.16)
$K_\nu(x)$ (2.3.18)
$\delta(\alpha, t)$ (2.3.25)
$W_\tau(\alpha, z)$ (2.4.2)
$E_{I,\tau,\nu}(g^0, z)$ (2.4.6)
$D_\tau, D_{I,\tau}$ etc. (2.4.7)
T_+^1 before (3.1.1)
$\Theta_L(\Phi, g)$ (3.1.1)
$Z_L(\Phi, \omega, \chi, s), Z_{L+}(\Phi, \omega, \chi, s)$ (3.1.2)
V_k^{s} (3.1.3)
$\mathrm{rd}_{i,M}(\mu)$ (3.1.5)
$\omega(g^0)$ (3.1.7)
$[x, y]_V$ (3.1.9),
$\mathscr{F}_V$ (3.1.10)
$\widetilde{g}^\iota$ after (3.1.10)
$M_{V,\omega}$ (3.1.11)
$\mathfrak{t}, (\ ,\ ), \|\ \|$ §3.2
$\mathfrak{B}$ (3.2.3)
$\mathfrak{d} = (\beta_1, \cdots, \beta_a)$ (3.2.6)
$P_\mathfrak{d}, M_\mathfrak{d}$ after (3.2.7)
$M_\mathfrak{d}''$ the end of §3.2
$\mathscr{F}_\mathfrak{d}$ after (3.3.1)
$R_\mathfrak{d}\Psi, \widetilde{R}_\mathfrak{d}\Psi$ (3.3.2)
$\Phi_\mathfrak{p}, \theta_\mathfrak{p}$ (3.3.3), (3.3.4), (3.3.5)
$P_{\mathfrak{d}\#}, U_{\mathfrak{d}\#}, \rho_\mathfrak{d}$ (3.3.6)

$P^1_{\mathfrak{d}\mathrm{A}}, M^1_{\mathfrak{d}\mathrm{A}}$ etc. (3.3.7)
$\widehat{\Omega}_{\mathfrak{d}}, T^0_{\mathfrak{d}\eta}$ after (3.3.7)
$e_{\mathfrak{p}i}$ (3.3.9), (3.3.10)
$A_{\mathfrak{p}0}$ etc. (3.3.11)
$g_{\mathfrak{d}}, dg_{\mathfrak{d}}$ after (3.3.11), before (3.3.13)
$\epsilon_{\mathfrak{p}}$ (3.3.12)
$\omega_{\mathfrak{p}}$ after (3.3.12)
$\sigma_{\mathfrak{p}}$ (3.3.13)
t^0 the beginning of §3.4
$(\ ,\ )_0, \|\ \|_0$ after (3.4.1)
$L(z), \Lambda(w;z), \Lambda_\tau(w;z)$ (3.4.2), (3.4.3), (3.4.6)
$M_\tau(z)$ (3.4.5)
$\mathscr{E}(g^0,w,\psi)$ (3.4.7)
$\mathscr{E}_{\mathfrak{p}}(g_{\mathfrak{d}},w), \widetilde{\mathscr{E}}_{\mathfrak{p}'}(g_{\mathfrak{d}'}n(u_{\mathfrak{d}'}),w)$ (3.4.8), (3.4.9), (3.4.10)
$\mathscr{E}_U(g^0,w), \mathscr{E}_{U,\tau}(g^0,w)$ (3.4.12), (3.4.14)
$\tau_{\mathfrak{p}}$ etc. after (3.4.13)
$(\lambda'_{\mathfrak{d}} g^0_{\mathfrak{d}})^r$ (3.4.17)
$C(A'_{\mathfrak{p}0} M^0_{\mathfrak{d}\mathrm{A}}/M_{\mathfrak{d}k}, r)$ (3.4.18)
$\mathscr{E}_{I,\tau,\nu}(g^0,w)$ etc. (3.4.27)
C_G (3.4.31)
$f_1(w) \sim f_2(w)$ (3.4.32)
$\Theta_{S_{\mathfrak{d}}}(\Psi_1, g_{\mathfrak{d}})$ etc. (3.5.1)
$\Xi_{\mathfrak{p}}(\Phi,\omega,w)$ etc. (3.5.2)
$J_{\mathfrak{d}}(\Phi,g_{\mathfrak{d}}), J'_{\mathfrak{d}}(\Phi,g_{\mathfrak{d}})$ (3.5.6)
$\widetilde{\Xi}_{\mathfrak{p}}(\Phi,\omega,w)$ (3.5.8)
Φ_λ after (3.5.9)
$s_\tau, \widetilde{\Lambda}(w;s_\tau)$ §3.6
ρ_τ §3.6
$\Sigma_i(\Psi_i,s), \Sigma_{i,(j_1,\cdots,j_i)}(\Psi_i,s_0)$ (3.8.1), (3.8.2)
$S_{n,i}$ etc. §4.0.
W (the space of binary quadratic forms) the beginning of Chapter 5.
X_V (4.2.2), §5.5
$T_{V_2}(\Psi,\omega,s,s_1)$ etc. (4.2.3)
$R'_0, \widetilde{R}_0, R_{V,0}, \widetilde{R}_{\mathfrak{d}_2,0}$ etc. (4.2.10), (5.3.2)
$Z_{V,\mathrm{ad}}(\Phi,\omega,s)$ (4.2.14), (5.6.3)
$\mathfrak{F}_{\mathfrak{d}1}$ etc. (4.3.1)
$f_{\mathfrak{d}}, h_{\mathfrak{d}}$ (4.3.2)
$c_{\mathfrak{p}}$ (4.3.3), (4.3.4)
$\mathfrak{P}_1$ etc. after (4.3.4)
$Z_{V_2,\mathrm{ad},(i)}(\Psi,\omega,s_0)$ etc. (4.4.1)
$R'_{\mathfrak{d},0}, \widetilde{R}_{\mathfrak{d},0}$ (4.4.2)
$I_{\mathfrak{p}}(\Phi,\omega)$ etc. (4.4.5)
$\Theta_{Z'_{\mathfrak{d},0}}(R_{\mathfrak{d}}\Phi_{\mathfrak{p}}, g_{\mathfrak{d}})$ etc. (4.6.1)
$\widetilde{X}_{\mathfrak{d}}$ etc. after (4.6.2)
$\widehat{t}^0, t^0$ (5.1.2), (9.2.4)
$T_V(\Phi,\omega,s,s_1,s_2)$ etc. (5.2.1), (6.3.1)
$\widetilde{\Theta}_{Z_{V,0}}(\Psi_1,t',u_0)$ etc. (5.2.4)

Index

Printed in the United States
By Bookmasters